Cell Biology

Cell Biology

Gerald Karp

Professor of Microbiology and Cell Science
I.F.A.S.
University of Florida

McGraw-Hill Book Company

New York   St. Louis   San Francisco   Auckland   Bogotá   Düsseldorf
Johannesburg   London   Madrid   Mexico   Montreal   New Delhi
Panama   Paris   São Paulo   Singapore   Sydney   Tokyo   Toronto

Cell Biology

1 2 3 4 5 6 7 8 9 0 VHVH 7 8 3 2 1 0 9

This book was set in Caledonia by Ruttle, Shaw & Wetherill, Inc.
The editors were James E. Vastyan and James W. Bradley;
the designer was Merrill Haber;
the production supervisor was Charles Hess.
The drawings were done by J & R Services, Inc.
Von Hoffmann Press, Inc., was printer and binder.

Cover: Ben Shahn, "The Scientist" 1959, © Estate of Ben Shahn 1979. (Visual Artists and Galleries Association, Inc.). The photographer was Geoffrey Clements, courtesy of Kenneth W. Prescott.

Library of Congress Cataloging in Publication Data

Karp, Gerald.
    Cell biology.

    Includes bibliographies and index.
    1. Cytology. I. Title.
QH581.2.K37   574.8'7   78-26822
ISBN 0-07-033341-6

To Patsy

O sweet spontaneous
earth how often have
the
doting

     fingers of
prurient philosophers pinched
and
poked

thee
, has the naughty thumb
of science prodded
thy

    beauty . how
often have religions taken
thee upon their scraggy knees
squeezing and

buffeting thee that thou mightest conceive
gods
      (but
true

to the incomparable
couch of death thy
rhythmic
lover
     thou answerest

them only with

       spring)

CONTENTS

and Protein Synthesis / DNA Cloning and Sequencing / Appendix on
Properties of Messenger RNA as Determined by Molecular Hybridization /
References

PREFACE

When I first began work on this text, I envisioned the preface as a section of the book in which many pearls of wisdom should reside. However, with the completion of the book, I find myself rather empty of profound thoughts and have decided simply to point out a few of the book's salient aspects. Since the primary function of this book is to serve as a text in an undergraduate course in cell biology, it is toward this purpose that the preface is directed. To a large degree, the arrangement of topics in a cell biology course varies with the perspective of the instructor. This book was written essentially in order from Chapter 1 to Chapter 19, but I believe it can be read following alternative sequences. There are basically five sections to the text: Chapters 1 to 4 cover subjects that are basic to the understanding of the remainder of the book; Chapters 5 to 9 cover membranes and membrane-mediated activities (including mitochondrial and chloroplast metabolism); Chapters 10 to 14 cover genetic functions; Chapters 15 and 16 cover events involving motility; and Chapters 17 to 19 cover such miscellaneous subjects as organelle genetics, the immune system, and cancer and aging. Although I can envision numerous problems if one attempts to rearrange the order of chapters *within* each section, one could cover the middle three sections in any order desired. When related topics are discussed in more than one chapter, the pages are generally cross-referenced.

Based on my experience in teaching cell biology, it seems reasonable to assume a great deal of heterogeneity in the backgrounds of the readers. Much of this heterogeneity reflects the fact that similar subjects are dealt with in several different course areas: microbiology, biochemistry, physiology, and developmental biology, as well as plant and animal cell biology per se. In writing this book, I have assumed relatively little prior scientific knowledge. The reader should have had courses in introductory biology, including a first contact with the major cell organelles, and general chemistry, including a bit of knowledge of organic molecules.

Just as there is a variety of backgrounds among students in a given class, so there is a variety of cell biology courses being taught. To the author of a cell biology text, this variety is reflected in the extent of coverage. Cell biology courses range from a single quarter to a full year, and it is difficult for a single text to satisfy these different demands. As I went along, I realized that I was writing what would likely be a fairly lengthy text. The length seemed attributable to several factors: the number of topics covered, the depth of coverage of certain subjects (particularly membrane structure and eukaryotic molecular genetics), and my tendency to write lengthy explanations. Since it is impossible to tailor a manuscript to a particular audience and is equally unthinkable to leave out information that I know to be of vital importance (regardless of whether others find it trivial), I have divided the text within each chapter into two parts. The bulk of each chapter (approximately 75%) is presented as essential reading. The remaining text, consisting of peripheral subjects, additional examples, in-depth experimental analyses, or background information in one case (that of Mendelian genetics in Chapter 10), is differentiated by a smaller typeface but is retained in place within each chapter. This separation of material is designed to allow the instructor more flexibility in making assignments, in keeping with the particular length or emphases of a given course. Most importantly, material that is not assigned remains available for the student to further pursue a topic of interest.

A list of references is included at the end of each chapter. In each case, the list has been drawn with one primary purpose in mind, that of presenting readers with recent review articles from which specific research papers can be tracked down. In addition, it is hoped that students will take advantage of the several fine scientific indexes available to locate further reading material. A brief visit with the Science Citation Index, Index Medicus, or Biological Abstracts will open up numerous pathways to the literature which can be followed.

Overall, this book is written with an experimental approach in mind. Students will, it is hoped, obtain a meaningful understanding of cellular events, rather than simply a mass of information. It is also hoped that the interested student will become familiar with the types of questions that experimenters have asked, how some of the answers have been obtained, and the current state of the field. Most of all, I have tried to instill an interest in and appreciation for cellular complexity and the means by which it is studied.

I am grateful to the following persons for their careful reading of a large part of or all of the manuscript and for their helpful comments: Professors E. R. Vyse of Montana State University, Curtis Williams of the State University of New York at Purchase, J. R. C. Brown of the University of

Maryland, Clemer K. Bartell of the University of New Orleans, Joseph Tupper of Syracuse University, Ernest R. Stout of the Virginia Polytechnic Institute and State University, James J. Mrotek of North Texas State University, and Stanley J. Roux of the University of Texas at Austin.

I am also grateful to those persons all over the world who have contributed photographs from their work for use as illustrations in this text and those authors of other books whose diagrams I have used in this text. In most cases, if I found a diagram in another book which I felt expressed, particularly well, a point I was making, I used that diagram (with permission) rather than attempting to construct a similar one or an inferior one simply to have a piece of original artwork. In this regard, I would like to express particular thanks to the authors and publishers of two books from which several figures were taken: "Biochemistry" by A. L. Lehninger, published by Worth, and "The Molecular Biology of the Gene" by J. D. Watson, published by W. A. Benjamin. In those cases in which the findings of a particular study is illustrated, I have tried to reproduce the graph or other figure as it appeared in the original research report.

I owe a particular debt of gratitude to the people at McGraw-Hill, particularly Jim Bradley, Merrill Haber, and Jim Vastyan, who worked so hard on the preparation of this book. A special thanks is due John Cordes of J & R Services, who converted a slowly arriving pile of scribbles into a superb art program.

Finally, I would like to apologize, in advance, for any errors that may occur in the text, and express my heartfelt embarassment. Any comments or criticisms from the readers would be greatly appreciated.

Gerald Karp

TO THE STUDENT

I am taking the liberty here to offer the reader some advice before beginning the text: Do not accept everything you read as being true. There are
several reasons for urging such skepticism. Undoubtedly, there are errors
in this text which reflect the author's ignorance of or misinterpretation of
some aspect of the scientific literature. But, more importantly, there is
reason to briefly consider the nature of biological research. We all like to
think of scientists as seekers of the "Truth." In reality, there is little truth
to be obtained. Rather, there is only a greater or lesser amount of evidence
in support of a particular belief. Nothing in biology is ever *proved*. We
merely compile data concerning a particular cell organelle, metabolic reaction, intracellular movement, etc., and draw some type of conclusion.
Generally, the more data, the better the conclusion; but in many cases
there is considerable disagreement about the facts themselves. Observations made by different laboratories, often working with different types of
cells and procedures, will inevitably be in some disagreement.

Since the study of cell function generally requires the use of considerable instrumentation, the investigator is quite far removed from the subject
being studied. To a large degree, cells are like tiny black boxes, of which
nothing was known a brief 200 years ago. Since then we have developed
many ways to probe these small dark boxes, but invariably we are groping
in an area which cannot be fully illuminated. A discovery is made or a new
technique developed and a new thin beam of light penetrates into the box.
With further work, our understanding of the nature of the structure or event

is broadened, but we are always left with questions which relate to remaining uncertainties. We generate more complete and sophisticated constructions, but we can never be sure how closely our views approach the reality we assume is there. In this regard, the study of cell biology can be compared to the study of an elephant as conducted by six blind men in an old Indian fable. The six travel to a nearby palace to learn about the nature of elephants. When they arrive, each approaches the elephant and begins to touch it. The first blind man touches the side of the elephant and concludes that an elephant is smooth like a wall. The second touches the trunk and decides that an elephant is round like a snake. The other members of the group touch the tusk, leg, ear, and tail of the elephant, and each forms his impression of the animal based on his own limited experiences.

Cell biologists are limited in a similar manner to what they can learn by using a particular technique or experimental approach. They hope that each piece of new information adds to the preexisting body of knowledge to provide a better concept of the nature of the activity being studied. But, regardless of the effort, the total picture remains uncertain. As a consequence, throughout this book the reader will find attempts to qualify statements. One reads that a finding or a theory is "generally believed" or "widely accepted." Even if there is a consensus of agreement concerning the "facts" regarding a particular phenomenon, there are often several possible interpretations of the data. Hypotheses are put forth and generally stimulate further research, thereby leading to a reevaluation of the original proposal. Most hypotheses that remain valid undergo a sort of evolution and, when presented in the text, should not be considered wholly correct or incorrect. A theory is constructed in terms of the concepts and prevailing perspectives of the time. As new techniques and information become available, new insights are made.

At present, information in nearly every area of cell biology is accumulating at a rapid pace; yet so much remains to be discovered that, with each passing year, entirely new types of cell structures and cell activities are described. A few brief examples will illustrate this point. In the last couple of years, the use of special enzymes and DNA cloning techniques have revealed hitherto unsuspected aspects of eukaryotic gene structure (page 555). Similarly, the development of the high-voltage electron microscope has led to the portrayal of the cytoplasm as a much more organized and structured entity than previously believed (page 756). In one area, that of brain cell function, the surface has only been scratched. There is a great feeling of excitement among cell biologists at present, for the prospects of future discovery remain very high.

Gerald Karp

Cell Biology

CHAPTER ONE

Introduction to the Study of Cell Biology

Human evolution, both biological and social, has coincided with a gradual development of the means by which to manipulate the environment. In order to accomplish this task, humans have acquired the skills to make increasingly more complex tools. In many ways, our technological accomplishments have had a very mixed impact on the quality of our lives in our increasingly "unnatural" environment. It is not the intention of this introduction to decide whether we have created a "wasteland," as some would advocate, or a "bold new frontier." Rather, the intention is to emphasize that technological developments manifest themselves along many diverse paths, one of which is cell biology, the subject of this text. Virtually the entire scope of cell biology is concerned with subjects that we cannot directly see, touch, or hear. Yet in spite of this tremendous handicap, cells are the subject of thousands of articles a year with virtually every aspect of their minuscule structure coming under scrutiny. In many ways the study of cell biology stands as a tribute to human curiosity in seeking to discover, and to human creative intelligence in devising the means by which these discoveries can be made. This is not to imply that cell biologists have a monopoly on these noble traits. At one end of the scientific spectrum, we are learning of objects at the outer fringe of the universe, ones with properties very different from those on earth. At the other end of the spectrum, investigators are focusing their attention on particles of subatomic dimensions, ones with equally inconceivable properties. Clearly, our universe consists of worlds within worlds, all aspects of which make for fascinating study. In this vein, the overriding goal of this text is to generate within students an appreciation for both cells and their study.

As will be apparent throughout this book, cellular and molecular biology is highly reductionist in its view that knowledge of the parts of the whole can explain the entire character of that whole. One of the inevitable consequences of this viewpoint is the tendency to replace one's feeling for the wonder and mysticism of life with the need to explain everything in terms of the workings of the "machinery of the living system." To the degree to which this occurs, it is hoped that this loss can be replaced by an equally strong appreciation for the beauty and complexity of the mechanisms underlying cellular activity.

THE DISCOVERY OF CELLS

As a result of the very small dimensions of cells, the study of their properties awaited the development of the microscope and thus began much later than that of other aspects of the natural sciences. Robert Hooke, the person credited with the discovery of cells, made his discovery by examining a very thin slice of cork cut with a pen knife (Fig. 1-1). As shown in one of his drawings (Fig. 1-1*b*), Hooke observed the honeycomblike nature of cork. He used the term *cell* to describe the compartments he saw and concluded that each was a completely enclosed space without interconnecting passageways. What Hooke had observed were the empty cell walls of dead plant tissue, walls that had originally been produced by the living cells they surrounded.

Although several contemporaries of Hooke were soon able to observe a variety of different types of plant cells as well as living, single-celled organisms, the realization that all plants and animals, regardless of their diverse outward appearance, were composed of component cells was not actually made until the 1830s. In 1838, Matthias Schleiden published the results of his research on plants. He concluded that, regardless of a particular tissue's appearance, plants were made up of cells and that the plant embryo arose from a single cell. In 1839, a collaborator of Schleiden's, Theodor Schwann, published a more comprehensive report on the cellular basis of animal life. Schwann also proposed that all tissues, whether they be muscle or nerve, elastic or horny, were composed of cells. Schwann furthermore concluded that the cells of plants and animals were entirely analogous structures and that, as such, cells were the functional *units* of all living organisms. Schwann realized that each cell was an "Individual, an independent Whole," which somehow operated together with other cells in a harmonious manner.

Besides formulating the "cell theory," the reports of Schleiden and Schwann focused attention on the cell nucleus, which had been discovered a few years earlier by Robert Brown, as a structure of particular importance in cell function. The formulation of the cell theory in the first half of the

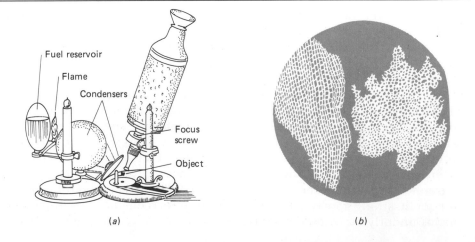

Figure 1-1. *(a)* Microscope used by Robert Hooke, with lamp and condensor for illumination of the object *(From T. I. Storer et al., "General Zoology," 5th ed., McGraw-Hill, 1972.)* *(b)* Hooke's drawing of a thin slice of cork.

Fuel reservoir

Flame

Condensers

Focus screw

Object

(a)

(b)

nineteenth century set the stage for the real beginnings of the analysis of cell structure and function in the latter half of the century. The first important point to be established concerned the origin of cells. In this regard, Schleiden and Schwann promoted an entirely erroneous concept, namely that cells could arise from noncellular materials. Given the prominence that these two German scientists held in the scientific world, it took a number of years before observations by several other biologists were accepted as demonstrating that cells did not arise in this manner any more than organisms arose by spontaneous generation. By 1855, it had been clearly stated that cells arose in only one manner, by division from a preexisting cell. This concept of the origin of cells set the stage for subsequent investigations into the nature of cell heredity and the cellular link between generations of organisms. These investigations, which will be discussed in some detail in Chapter 10, form the basis upon which much of the information in this book rests.

CELLULAR PROPERTIES AND ORGANIZATION

As first described by Schleiden and Schwann, cells are typically referred to as "units of life." What properties would one expect from such a unit? First of all, we would expect that all life forms, regardless of their nature, would have a cellular basis. This supposition has been amply confirmed. Examination of a great variety of organisms indicates that all of them are cellular in nature, and, furthermore, that the cells are remarkably similar. As is described in a following section, there appear to be two main types of cells among living organisms, prokaryotic cells and eukaryotic cells, each representing a different level of structural and functional complexity. We might also hope that cells could maintain a semiautonomous existence, one which would allow them to be removed from a multicellular organism and be kept alive and healthy outside of that organism. It has been shown that cells from essentially any organism, including humans, can be cultured outside the body, i.e., in vitro, under conditions in which they will remain alive and prosperous, long after the organism from which they were removed has died. Certain human cells, for example, have been kept in culture for several decades and can be made available to an investigator simply by removing them from the freezer.

A brief examination of a typical cell's activities suggests that, to a large degree, the properties of an entire multicellular organism are a reflection of the properties of its microscopic component cells. Organisms take up food, digest and assimilate it, and excrete waste products; they take up oxygen and release carbon dioxide; they maintain a particular water content; they are capable of growth, reproduction, and movement; they respond to external stimulation; they expend energy to carry out their activities; they inherit a genetic program from their parents and pass it on to their offspring; finally they die. These are the activities that we associate with the life of an organism, and all of them are activities that are carried out by its individual cells. In an important sense an organism is the sum of its parts, and its activities the sum of the activities of its component cells. However, in another sense the organism is much more than simpy a collection of individual cells. New properties emerge with each increasing level of complexity. Just as molecules have different properties from the atoms of which they are built, tissues have different properties from their component cells, organs different properties from their component tissues, and an organism different properties from its component organ systems. Even though a single nerve cell may have the capacity to conduct impulses, presumably it cannot think. Similarly, a single kidney cell cannot filter blood nor can a single retinal cell construct an image. It is apparent, therefore, that information gained from the study of individual cells and their organelles—the level of organization at which this book is focused—is inherently limited in terms of the phenomena it can explain. Nevertheless, the capabilities of a tissue, or organ, or organ system are ultimately dependent upon its cells. An understanding of the basic structural and functional properties of cells provides an essential foundation for the analysis of questions in organ physiology, developmental biology, histology, or any related area. The study of cell biology is, therefore, not only important in its own right, but is invaluable in studying any aspect of the biological sciences.

In the previous paragraph a number of essential properties of life were listed. There is one overriding property of life, one which forms the foundation for all of life's characteristics. This remaining property is that of complexity or organization. A structure's ability to perform a given task of a certain difficulty is in some ways a measure of its complexity. As the demands upon the equipment increase, so too does its complexity and organization. A brief glance at the capability of the human condition makes it abundantly clear that the human machinery must be vastly more complex than any spaceship or computer. Our ability to think, create, decide, and act upon our decisions in an infinite variety of ways has no true parallel in the manmade world.

Complexity is one of those properties that is always evident but difficult to delineate. In biological systems, complexity is probably best defined in terms of the information content of the genetic material, but this type of definition doesn't help much in trying to visualize the nature of cellular organization for the first time. For our purposes, complexity is better thought of in terms of order or predictability. The more complex a structure, the greater the number of parts that must be in their proper place, the less tolerance of errors in the nature and interactions of the parts, and the more regulation or control which must be exerted to maintain the system. During the course of this book we will have occasion to consider the complexity of life at several different levels. We will discuss the organization of atoms into small molecular weight compounds, the organization of these molecules into larger molecular weight polymers, the organization of these macromolecules into supramolecular complexes, which are in turn organized into subcellular organelles and finally into cells. As will be apparent, there is a great deal of predictability present at every level. Each type of cell has a predictable appearance in the electron microscope, i.e., its organelles have a particular shape and location. Similarly, the biochemist finds that each type of cell has a predictable variety of proteins, and each protein a predictable sequence of amino acids, and so on.

Regulation

One of the most important features of living systems, whether an entire plant or animal or an individual cell, is their ability to self-regulate. The term that is usually used to denote this capacity within the whole organism is *homeostasis*. If one measures the amount of a particular substance in the blood, saliva, or cerebrospinal fluid, or its pH or osmolarity, one finds that these properties are maintained within relatively strict limits. Each of these variables must be monitored by the appropriate sensory system within the body and adjustments continually made to maintain the necessary value. Since the organism cannot depend on receiving a tune-up by some outside agent, it must possess all the necessary information to keep itself healthy. The following example from the endocrine system will serve to illustrate this point.

Among the best-studied types of homeostatic control mechanisms are the feedback controls by which the levels of various hormones are regulated (Fig. 1-2). For example, the adrenal gland produces the hormone cortisol in response to the presence of adrenocorticotropic hormone (ACTH) produced by the pituitary gland at the base of the brain. In turn, secretion of ACTH by the pituitary occurs in response to the production and secretion of yet another hormone, the corticotropic releasing factor (CRF). CRF is secreted by the tips of nerve cells that project from the hypothalamus, located within the brain. If cortisol production were to become excessive, blood levels of this hormone would rise, an event which is monitored by the hypothalamus. The increased blood levels of cortisol would lead to the decreased secretion of CRF by the hypothalamus, then the decreased secretion of ACTH by the pituitary, and finally the decreased secretion of cortisol by the adrenal gland. The feedback interactions among the various components can also be demonstrated in another way. If one of the adrenal glands is removed, the drop in cortisol (and other hormones) leads to an increased hypothalamic and pituitary response which ultimately results in the enlargement of the remaining adrenal gland causing an increase in its secretory output.

As in the body at large, there are many different control mechanisms operating within each cell. A considerable fraction of this book will concern itself with these various regulatory mechanisms. For example, we will discuss the manner in which ATP concentrations regulate the level of ATP production (Chapter 4), the mechanism by which the nucleotide sequence of DNA is proofread for errors during its synthesis or repaired upon damage (Chapter 14), and the manner in which calcium levels are able to cause contraction or relaxation of a muscle fiber (Chapter 16). We are gradually learning more and more about how the cell controls its activities, but there is much more left to discover. The magnitude of the problem can be illustrated by briefly describing an experiment performed in the nineteenth century by Hans Driesch (Fig. 1-3). Driesch, who

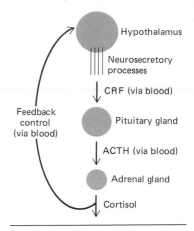

Figure 1-2. Schematic illustration of the levels within the endocrine system involved in the control of cortisol secretion.

Figure 1-3. The isolation of the first four cells of a sea urchin embryo. The larva on the left represents a normal-sized control while the four smaller larvae are the products of isolated cells.

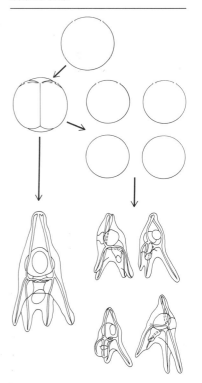

5

worked with sea urchin embryos, found that he could completely separate the first two or four cells of the embryo and each would proceed to develop into a normal, intact embryo. In other words, even though each of these cells normally would have developed into a certain portion of the embryo had the embryo been left undisturbed, once the cells were separated from one another, each recognized the fact that its neighbors were no longer present and it went on to form the entire organism. Even though nearly one hundred years have passed since this very important experiment was performed, we are not in a much better position to explain it today.

Throughout this book we will be discussing processes which require a series of ordered steps, much like the assembly line construction of an automobile in which workers add, remove, or make specific adjustments as it moves along. In the cell the blueprints for the products are contained in the nucleic acids, and the workers to construct them are primarily the proteins. It is the presence of these two types of macromolecules which, more than any other factor, sets the chemistry of the cell apart from that of the nonliving world. In the cell, the workers must act without the benefit of a consciousness to direct them. Each step of the process must occur spontaneously and in such a way that the next step is automatically triggered. In many ways the functions of a cell operate in a manner analogous to the orange-squeezing machine discovered by Professor Butts, shown in Fig. 1-4. In other words, all the information to direct a particular activity (whether it be the synthesis of a protein, the secretion of a hormone, or the contraction of a muscle fiber) must be present within the system itself.

THE FOUNDATIONS OF CELL CHEMISTRY

During the course of this book we will have many occasions to discuss the structure of a large number of different molecules and their interactions. In order to better understand the manner in which atoms and molecules interact, it is worthwhile to briefly consider the types of bonds which atoms can form with one another. It will be assumed during the course of this discussion that the reader is already familiar with the electronic structure of the atom, the concept of electron shells surrounding the atomic nucleus, and the basis for an atom's valence.

Covalent Bonds

Atoms react with one another to achieve a more stable configuration, one in which the outer shell of electrons is filled. With the exception of the inert noble gases (helium, neon, argon, etc.), the atoms which make up the elements are characterized by having outer shells which are not filled with electrons (Fig. 1-5a). In contrast, examination of the electronic configuration of atoms contained *within molecules* reveals that sufficient electrons are now present in each atom to completely fill the outer shell. In the case of a hydrogen atom, its outer (and only) shell is filled when two electrons are present; in the case of other atoms, the outer shell is filled when it contains eight electrons. Covalent bonds are those in which pairs of electrons are shared between two different atoms. Consider the reac-

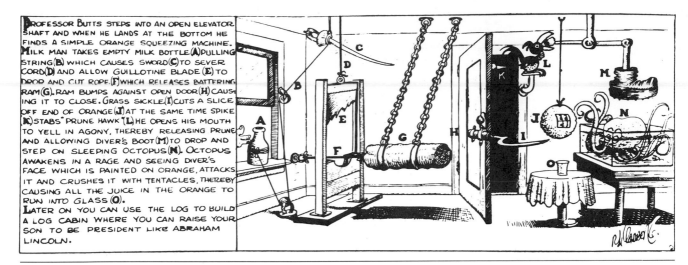

PROFESSOR BUTTS STEPS INTO AN OPEN ELEVATOR SHAFT AND WHEN HE LANDS AT THE BOTTOM HE FINDS A SIMPLE ORANGE SQUEEZING MACHINE. MILK MAN TAKES EMPTY MILK BOTTLE(A)PULLING STRING(B) WHICH CAUSES SWORD(C)TO SEVER CORD(D) AND ALLOW GUILLOTINE BLADE (E) TO DROP AND CUT ROPE(F)WHICH RELEASES BATTERING RAM(G). RAM BUMPS AGAINST OPEN DOOR(H) CAUSING IT TO CLOSE. GRASS SICKLE(I)CUTS A SLICE OFF END OF ORANGE(J) AT THE SAME TIME SPIKE (K)STABS PRUNE HAWK(L) HE OPENS HIS MOUTH TO YELL IN AGONY, THEREBY RELEASING PRUNE AND ALLOWING DIVER'S BOOT(M)TO DROP AND STEP ON SLEEPING OCTOPUS(N). OCTOPUS AWAKENS IN A RAGE AND SEEING DIVER'S FACE WHICH IS PAINTED ON ORANGE, ATTACKS IT AND CRUSHES IT WITH TENTACLES, THEREBY CAUSING ALL THE JUICE IN THE ORANGE TO RUN INTO GLASS (O).
LATER ON YOU CAN USE THE LOG TO BUILD A LOG CABIN WHERE YOU CAN RAISE YOUR SON TO BE PRESIDENT LIKE ABRAHAM LINCOLN.

Figure 1-4. A cartoon drawn by Rube Goldberg depicting a particular chain reaction of events. *(Reprinted by permission of King Features Syndicate.)*

tion between two atoms of hydrogen to form molecular hydrogen gas (H_2). Each hydrogen atom contains a single electron. When two such atoms are combined, each receives the benefit of having two orbital electrons rather than just one. Both atoms become stabilized by having the outer shell filled, even though the electron pair is actually being shared by both constituents. The stabilization process is accompanied by the release of energy. The amount of energy released by the reactants as they combine varies with the particular reaction but is constant for that reaction. The same amount of energy must be absorbed by the atoms at some later time if the bond between them is to be broken.

The number of electrons in the outer shell of those atoms (other than hydrogen) which are commonly involved in covalent bond formation within living systems are as follows: carbon atoms, four outer electrons; nitrogen atoms, five outer electrons; oxygen atoms, six outer electrons; phosphorus atoms, five outer electrons; and sulfur atoms, six outer electrons (Fig. 1-5a). Since each of these atoms requires eight outer-shell electrons to achieve a stable configuration, it can be seen that carbon will tend to share four pairs of electrons, nitrogen and phosphorus three pairs of electrons, and oxygen and sulfur two pairs of electrons. Figure 1.5b shows the structure of three molecules and the electrons present in each atom's outer shell.

In many cases, two atoms can be covalently bound to each other in such a way that more than one pair of electrons is shared. If two electron pairs are involved in a given bond it is termed a *double bond,* and if three pairs of electrons are shared it is a *triple bond.* The presence of a single vs. a double or triple bond between atoms has important consequences for both its structure and reactivity. One of the differences, for example, is the ability of the atoms of a single bond to rotate relative to one another; the atoms of double and triple bonds lack this ability.

The chemistry of life very clearly centers around the chemistry of the carbon atom. The essential quality of carbon which has allowed it to play this role is the incredible number of molecules (several hundred thousand

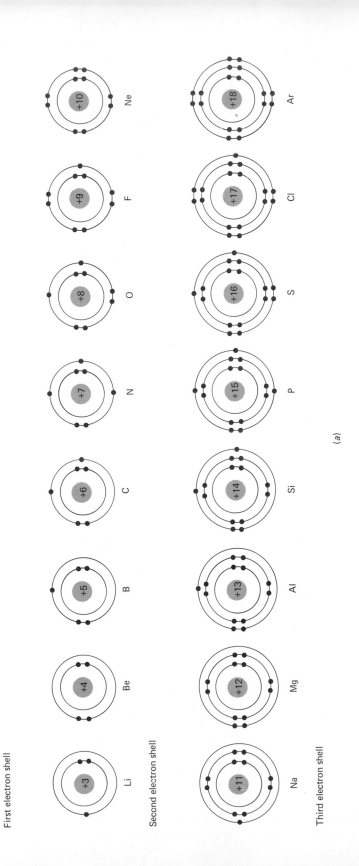

Figure 1-5. Electronic structure of selected atoms (a) and molecules (b). (From J. B. Hendrickson, D. J. Cram, and G. S. Hammond, "Organic Chemistry," 3d ed., McGraw-Hill, 1970.)

8

of which are known) it can form. Having four outer electrons, a carbon atom can bond with four other atoms in order to achieve its stable electronic configuration. This feature, by itself, provides carbon atoms with the opportunity to participate in the formation of a large variety of molecules. More important, however, is its ability to form bonds with other carbon atoms so as to construct molecules with backbones containing long chains of carbon atoms. Both the size and electronic structure of carbon make it uniquely suited for generating large numbers of molecules. Silicon, for example, which is just below carbon in the periodic table, is apparently too large for its positively charged nucleus to attract the outer shell electrons of neighboring atoms with sufficient force to hold such large molecules together.

The ability of carbon atoms to form four single bonds with other atoms allows the same collection of atoms to be organized in more than one way. Consider a molecule with the formula C_2H_6O, i.e., one containing two carbon atoms, six hydrogen atoms, and one oxygen atom. There are two different structures that can be drawn with this collection of atoms.

$$\begin{array}{cccc} & H & H & \\ & | & | & \\ H-&C-&C-&OH \\ & | & | & \\ & H & H & \end{array} \qquad \begin{array}{ccccc} & H & & H & \\ & | & & | & \\ H-&C-&O-&C-&H \\ & | & & | & \\ & H & & H & \end{array}$$

One of these molecules (on the left) is ordinary ethyl alcohol, the other is methyl ether, normally a gas. These two molecules have very different properties as a result of the manner in which the various atoms are bonded to one another. Since these two molecules have the same formula (C_2H_6O), they are said to be *structural isomers* to one another. Molecules made of larger numbers of atoms have increasingly greater numbers of structural isomers.

Carbon atoms form covalent bonds with a characteristic structural relationship to one another. The four bonds are placed so as to form a tetrahedron with the carbon atom located at its center (Fig. 3-2). The tetrahedral structure of a bonded carbon atom results in a different type of isomerism, termed *stereoisomerism*, between molecules having the same structural formula. The nature of stereoisomerism will be discussed on page 74, in connection with the structure of amino acids.

Ionization

When two of the same types of atoms (such as H_2) unite with one another, the electron pairs of the outer shell will be equally shared between the two partner atoms. However, when two unlike atoms are combined with one another, it is inevitable that the positively charged nucleus of one atom will exert a greater attractive force on the outer electrons than will its bonded atom. Consequently, the shared electrons will tend to be located more closely to the atom with the greater attractive force. Since the atom with the greater share of the electrons will be more negatively charged, it is said to be more *electronegative.* The electronegativity of an atom is dependent upon two factors, the number of positive charges in

TABLE 1-1
Electronegativities of Atoms

Kernel Charge				
+1	+4	+5	+6	+7
H	C	N	O	F
2.2	2.5	3.0	3.5	4.0
	Si	P	S	Cl
	1.9	2.2	2.5	3.0
				Br
				2.8
				I
				2.5

SOURCE: J. B. Hendrickson et al., "Organic Chemistry," 3d ed., McGraw-Hill, 1970.

its nucleus (the more protons the more electronegative) and the distance of the outer electrons from the nucleus (the greater the distance the less electronegative) (Table 1-1). When a reaction occurs between a weakly electronegative atom (such as those in the first column of the periodic table) and a strongly electronegative atom (such as those in the next to the last column, the halogens), the electronegative atom is able to completely "capture" the shared electrons and the molecule falls apart, i.e., *ionizes,* into its two components. The formation of NaCl from sodium metal and chlorine gas is a typical reaction of this sort. Once the reaction has taken place, the two atoms of the molecule dissociate from one another when present in an aqueous solution. Since the Cl atom has an extra electron (relative to the number of protons in its nucleus) it has a negative charge and is termed an *anion.* The Na atom which has lost an electron has an extra positive charge and is termed a *cation.* Reactions in which electrons are transferred from one atom to another are termed *oxidation-reduction reactions* and are considered in further detail on Page 180.

Noncovalent (Weak) Bonds

In most cases the differences in the electronegativity between two bonded atoms is not large enough to cause the complete transfer of an electron from one to the other. Instead, the electrons are simply held more closely to one of the atoms. A bond of this type is said to be *polarized* such that one of the atoms has a partial negative charge and the other a partial positive charge. This is generally denoted in the following manner.

$$\overset{\displaystyle|}{\underset{\displaystyle|}{C}}\overset{\delta^+\ \delta^-\quad\ \delta^+}{}\!-\!O\!-\!H$$

A partial separation of the charge across a bond can be measured as a *dipole moment,* i.e., a specific orientation of the atoms when placed in an electric field. If the individual bond moments of a molecule are situated so that they do not cancel one another, then the entire molecule has a dipole moment and will orient in an electric field. Molecules of this type, i.e., ones with an asymmetric distribution of charge, are said to be *polar* molecules. Molecules which lack polarized bonds (or ones in which the polarized bonds are symmetrically placed and cancel each other out) are said to be *nonpolar.*

The presence of polarized bonds is of the utmost importance in determining the reactivity of molecules. Molecules which *lack* electronegative atoms, i.e., molecules which lack N, S, O, P, etc., tend to be relatively inert. The hydrocarbons, molecules made only of carbon and hydrogen, for example, do not take part in many types of reactions.

Up to this point we have been discussing chemical bonds within molecules, i.e., covalent bonds. Interactions *between* molecules are based on a variety of different types of bonds, ones having considerably less energy and greater bond length. Bonds of this type are termed *weak bonds* (or secondary bonds) to denote the ease with which they can be broken, a feature which allows them to mediate the dynamic interactions which occur among molecules in the cell. Even though an individual bond may be readily broken, when numerous weak bonds are present together between molecules (or between different parts of the same large molecule), their attractive forces are additive, and, taken as a whole, they provide the structure with considerable stability. These aspects will be considered in further detail when the structure of proteins and nucleic acids is discussed in Chapters 3 and 10.

Interactions between molecules are based on the same overriding principle which governs interactions between atoms, namely, that unlike charges attract one another. In this case, the attraction is between a negative center in one molecule and a positive center in another molecule. For example, the ability of a compound to dissolve in water can be understood by considering the electronic configuration of both the solute and the aqueous solvent. Water is composed of two weakly electronegative hydrogen atoms bonded to a strongly electronegative oxygen atom (Fig. 1-6a). Water molecules, consequently, have an internal separation of charge, a property which causes them to associate with other polar molecules. The positive regions of the polar solute molecules associate with the negative oxygen atom of water, while the negative regions of the solute associate with the hydrogen atoms of water. The electrostatic interactions between polar solutes and water are responsible for separating the solute molecules, which greatly enhances their solubility. In contrast, if a nonpolar hydrocarbon, such as paraffin oil, is mixed with water, it will not be dissolved.

In addition to its ability to dissolve polar molecules, water is also an excellent solvent for ionic compounds such as NaCl. The basis for this important aspect of an aqueous solvent is the very high *dielectric constant* of water, 78.6 at 25°C. Considering the fact that opposite charges attract one another, one would expect that an ionic compound such as NaCl would remain as such rather than falling apart into oppositely charged ions. The attractive forces between two opposite charges is defined by Coulomb's law,

$$F = \frac{q_1 q_2}{D r^2}$$

where q_1 and q_2 are the magnitude of the charges, r the distance between the charges, and D the dielectric constant of the solvent. The higher the dielectric constant, the less the attractive force between two oppositely charged ions at a given distance apart. In addition, the polar nature of water molecules cause them to form clusters (*hydration shells*) around ions (Fig. 1-6 b) in a similar manner to their association with polar solutes.

The solvent properties of water are essential determinants of the reactions occurring within the cell. In most cases, over 80% of the weight of the cell is water, and essentially all of the chemical reactants are dissolved in it. Although it is less obvious, the insolubility of nonpolar molecules in water also plays a vital role within cells. For example, the insolubility of lipids in water has allowed for the evolution of lipid-

Figure 1-6. (*a*) Diagram showing the spatial arrangement of the atoms within a water molecule. (*b*) Diagrammatic representation of the hydration of solute particles, in this case positively and negatively charged ions.

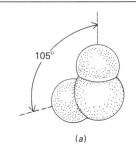

(a)

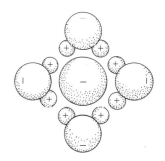

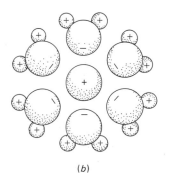

(b)

containing membranes which can remain together to form insoluble walls which act to divide the cell into separate aqueous compartments. Similarly, the presence of nonpolar amino acids in protein molecules allows them to form domains within the protein which exclude water and take on special properties.

The same types of principles that govern the interaction between water molecules and their solutes are also involved in the interactions between the solutes themselves. Weak bonds between molecules (or between different parts of very large molecules) are dependent upon the attraction between oppositely charged centers and are generally referred to as an *electrostatic* bond. In comparison to covalent bonds, electrostatic bonds are very weak. For example, it generally requires approximately 83 kcals of energy to disrupt a covalent bond between two carbon atoms in one mole of molecules, whereas it requires only a few kcals to disrupt an equivalent number of electrostatic bonds. This type of weak attractive force can occur between fully charged groups on the interacting molecules or between partially charged regions as was discussed above for polar solutes in polar solvents. The term *ionic bond* is commonly used to refer to an interaction between fully charged groups, such as a positively charged amino group in one molecule and a negatively charged carboxyl group in another molecule. In the presence of water, ionic bonds are quite weak.

One type of electrostatic bond of particular importance in living cells is the *hydrogen bond.* Hydrogen bonds are formed when a hydrogen atom becomes associated with two electronegative atoms as in Fig. 10-21. In this type of weak bond, the hydrogen atom is covalently bound to one of the electronegative atoms and weakly bound (the hydrogen bond) to the other one. In the cell, hydrogen bonds usually occur between hydrogen atoms and oxygen or nitrogen atoms. Hydrogen atoms are able to form these types of bonds because of their lack of orbital electrons. The pair of electrons shared between the electronegative atom and the hydrogen atom in the covalent bond is greatly displaced toward the nucleus of the electronegative atom. This displacement leaves the bare, positively charged nucleus of the hydrogen atom in a position to form an attractive interaction with an unshared pair of outer electrons of a second electronegative atom.

The tendency for polar molecules to form weak bonds with one another as the result of electrostatic attractions between the molecules has important consequences for nonpolar molecules (or nonpolar parts of large molecules) that happen to be present in a polar solvent. Since the tendency for polar molecules to form ionic and hydrogen bonds inevitably leads to the exclusion of nonpolar groups, the latter tend to become associated with one another by what is termed a *hydrophobic interaction.* Hydrophobic, which means "water-hating," refers to the inability of nonpolar groups to associate with water. Hydrophobic interactions are best illustrated by considering the properties of molecules which are composed of both polar (hydrophilic or "water-loving") and nonpolar (hydrophobic) regions. In the present discussion we will use soaps as an example. Soaps are molecules having a long nonpolar hydrocarbon chain and a polar salt group at one end (Fig. 1-7). Molecules of this sort are able to remove grease or dirt from a surface because the hydrocarbon chain of the soap associates with the nonpolar oily material, while the polar salt group at the other end associates with the aqueous solvents. The ionic bonds between the sodium salt of the soap and the water molecules spontaneously force the nonpolar sections of the soap molecules toward the center of the aggregate. The forced interactions among hydrophobic parts of molecules are extremely important in molecular interactions within living cells.

Hydrophobic interactions of the type just described cannot be classified as true bonds since the energy required to disrupt them results from the need to disperse the surrounding water molecules rather than the internal nonpolar groups. In addition to this type of interaction, hydrophobic groups are capable of forming weak bonds with one another. As in the case of the ionic and hydrogen bonds of polar molecules, the weak bonds between nonpolar groups also result from electrostatic attraction. Polar molecules associate because they contain permanent asymmetric charge distributions within their structure. A closer examination of the covalent bonds in a nonpolar molecule (such as in H_2 gas or methane) reveals that electron distributions are not constantly symmetrical. The location of an electron around an atom at any given instant is a statistical matter and, therefore, will vary from one instant to the next. In other words, at one time the electron density may happen to be greater on one side of an atom, even though the atom shares the electrons equally with some other atom. These transient asymmetries in electron distribution result in momentary separations of charge, i.e., dipoles, within the molecule. If two molecules with transitory dipoles are very close to one another, and oriented in the appropriate manner, they will experience a weak attractive force (termed a *van der Waals force*) which can serve to bond them together. In addition, the formation of a temporary separation of charge in one molecule can serve to *induce* a similar separation in an adjacent molecule. In this way, additional attractive forces can be generated among nonpolar molecules.

In summary, we have seen that several different types of interactions among atoms have been found to exist, all of which are based upon the attraction of unlike charges. Covalent bonds, which are the strongest type of interatomic associations, require the greatest input of energy to break. Weaker bonds, i.e., ionic bonds, hydrogen bonds, van der Waals attractive forces, and hydrophobic interactions, are important in the cell even though they may require very little energy to bring about their disruption.

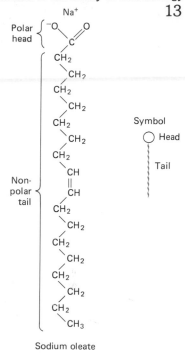

Sodium oleate

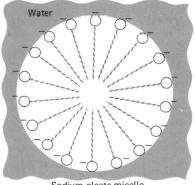

Sodium oleate micelle

Figure 1-7. Formation of a soap micelle in water. The nonpolar tails of the sodium oleate are hidden from the water, whereas the negatively charged carboxyl groups are exposed. (*From A. L. Lehninger, "Biochemistry," 2d ed., Worth, 1975.*)

ACIDS, BASES, AND BUFFERS

In a previous section we discussed a type of bond in which an electron pair was totally "captured" by one of the atoms, resulting in the dissociation of the members of the bond to form ions. In this section we will consider briefly one type of ionizeable bond, that involving a hydrogen ion, i.e., a proton. Since a hydrogen atom has a minimal electronegativity, it readily gives up its electron to a strongly electronegative atom and dissociates to form a proton. Any molecule which is capable of releasing (donating) a hydrogen ion is termed an *acid*. Conversely, any molecule which is capable of accepting a proton is termed a *base*. Hydrogen chloride, for example, is an acid since it dissociates *when dissolved in water* into a proton and a chloride ion. The proton released in water associates with a water molecule to form a hydronium ion, H_3O^+. In this case, the water is acting as a base in accepting the hydrogen ion. The hydronium ion, which can serve as a proton donor, is defined as an acid. It is apparent that acids and bases exist in pairs or *couples*. When the acid loses a proton it forms a base (termed the *conjugate base* of that acid). Similarly, when the base accepts a proton it forms an acid (termed the *conjugate acid* of that base). The acid always contains one more positive charge than its conjugate base. Water is an example of an *amphoteric* molecule, i.e., one that can serve both as an acid or base.

$$H_3O^+ \rightleftharpoons H^+ + H_2O \rightleftharpoons OH^- + H^+$$

Acid $\qquad\qquad$ Amphoteric $\qquad$ Base

A given water molecule acts as an acid when it donates a proton, thereby becoming a hydroxyl, or as a base when it accepts a proton. We will discuss another important type of amphoteric molecule, the amino acids, in Chapter 3.

TABLE 1-2
Strengths of Acids and Bases

	Acids		Bases	
Strong	HCl	Cl^-	Very weak
	H_2SO_4	SO_4^{2-}	
	H_3O^+	H_2O	
Weak	$HC_2H_3O_2$	$C_2H_3O_2^- \leftrightarrow$ Weak	
	H_2CO_3	CO_3^{2-}	
	H_2S	S^{2-}	
	NH_4^+	NH_3	
Very weak	H_2O	OH^-	Strong

SOURCE: K. Krauskopf, "Fundamentals of Physical Science," 3d ed., McGraw-Hill, 1953.

If one compares various acids, i.e., proton donors, one finds considerable variation with respect to the ease with which the molecule will give up its proton. The more readily the proton is lost, i.e., the less strong the attraction of the conjugate base for its proton, the stronger the acid. Hydrogen chloride is a very strong acid, one which will readily transfer its proton to water molecules when dissolved. The conjugate base of a strong acid, such as HCl, is a weak base (Table 1-2). Acetic acid, in contrast, is a relatively weak acid since, for the most part, it remains undissociated when dissolved in water. In a sense, one can consider the degree of dissociation of an acid as a measure of the competition for protons among the components of a solution. Water is a better competitor, i.e., a stronger base, than chloride ion, so HCl completely dissociates.

In contrast, acetate ion is a stronger base than water, so it remains largely undissociated.

The acidity of a solution is measured by the concentration of hydronium ions. Since the concentration of hydronium ions in an aqueous solution can vary over a tremendous range, acidity is more readily expressed in terms of the log of the hydronium ion concentration using the term pH.

$$pH = -\log [H_3O^+]$$

For example, a solution having a pH of 5 contains a hydronium ion concentration of 10^{-5} M.

In the absence of any solute, a water molecule can ionize into a hydroxyl ion and a proton which becomes associated with another water molecule.

$$H_2O \rightleftharpoons H^+ + OH^-$$

The equilibrium constant for this reaction (a dissociation reaction) is

$$K_{eq} = \frac{[H^+][OH^-]}{[H_2O]}$$

Since the water concentration is constant (55.51 M), we can generate a new constant, K_w, the *ion product constant* for water.

$$K_w = [H_3O^+][OH^-]$$

which is equal to 10^{-14}. In other words, in an aqueous solution the product of these two ions will always equal 10^{-14}. In the absence of solute, the concentration of each of these species is approximately 10^{-7} M. The extremely low level of dissociation of water indicates that it is a very weak acid. In the presence of an acid, the concentration of hydronium ion rises and the concentration of hydroxyl ions drops (as a result of a combination with protons to form water) so that the ion product remains at 10^{-14}.

The change in pH which occurs upon the addition of a certain amount of acid (or base) depends upon the nature of the solution to which the acid (or base) is added. If a solution contains a *buffer*, it tends to resist a change in pH. Buffered solutions are usually ones which contain either a weak acid together with a salt of that acid, or a weak base together with the salt of that base. Consider, for example, a solution containing acetic acid together with sodium acetate. If one were to add a limited amount of strong acid, such as HCl, to this solution, the added protons would be neutralized by the presence of the acetate ions acting as a base. If a limited amount of NaOH were added to the solution instead, the excess hydroxyl ions would be neutralized by the protons donated by acetic acid molecules. An equally effective buffer could be made of a weak base, such as ammonium hydroxide, and the salt of that base, such as ammonium chloride. In all of these cases there is the equilibrium

$$Acid + H_2O \rightleftharpoons base + H_3O^+$$

The dissociation constant for this reaction (corrected for the concentration of water) is

$$K_a = \frac{[\text{Base}][\text{H}_3\text{O}^+]}{[\text{Acid}]}$$

which can be rearranged to give

$$[\text{H}_3\text{O}^+] = K_a \frac{[\text{Acid}]}{[\text{Base}]}$$

If we take the log of both sides, and substitute pH for the negative log of the hydronium ion concentration and pK_a for the negative log of the dissociation constant, we arrive at

$$\text{pH} = \text{p}K_a + \log \frac{[\text{Base}]}{[\text{Acid}]}$$

an expression termed the *Henderson-Hasselbach equation*, from which one can calculate the pH of a solution in the presence of a particular buffer system. Living systems are particularly sensitive to changes in pH, so it is important that physiological fluids are well buffered. The predominate buffer system in the blood is the carbonic acid–bicarbonate (H_2CO_3–HCO_3^-) couple. Within the cell, certain of the amino acid residues of the proteins serve as weak acids and bases (page 77) and are, therefore, capable of providing buffer action.

THE REACTIVITY OF BIOLOGICAL MOLECULES

During the course of this book we will consider a large variety of molecules and their interactions. Nearly all of these molecules will be composed of a carbon-containing skeleton to which various "functional groups" are covalently attached. The functional groups are composed of specific combinations of atoms which make them reactive to a varying degree depending upon the situation. The basis for the reactivity of a particular group rests on the properties of atoms we have discussed in this chapter. The functional groups of greatest importance among biological molecules are illustrated in Table 1-3.

TABLE 1-3
Important Functional Groups in Biological Molecules

Group	Molecular Example	Biological Significance
—C—H Methyl	Alanine	Highly insoluble in water; does not form hydrogen bonds
—OH Hydroxyl	Ethanol (ethyl alcohol)	Water soluble; forms hydrogen bonds

TABLE 1-3 Continued

Group	Molecular Example	Biological Significance
Carboxyl	Acetic acid	Usually charged: $$-C\overset{O}{\underset{O^-}{\diagdown}} \rightleftharpoons -C\overset{O}{\underset{O^-}{\diagdown}} + H^+$$ good acceptor of hydrogen bonds
Amino	Glycine	Usually charged: $-NH_2 + H^+ \rightleftharpoons -NH_3^+$ forms hydrogen bonds
Phosphate	Glyceraldehyde-3-phosphate	Always negatively charged; forms hydrogen bonds with water, thus very water soluble
Carbonyl	Acetaldehyde	Forms hydrogen bonds; usually exists in keto form: $$R-\underset{\underset{O}{\|}}{C}-CH_3$$
	Pyruvate	as opposed to the enol form: $$R-\underset{\underset{}{}}{\overset{OH}{\overset{\|}{C}}}=CH_2$$
Amide	Asparagine	Usually charged. Forms hydrogen bonds.
Sulfhydryl	Cysteine	Two SH groups are easily oxidized to form S—S (disulfide bonds); SH groups form very weak hydrogen bonds

SOURCE: J. D. Watson, "Molecular Biology of the Gene," 3d ed., W. A. Benjamin, 1976.

REFERENCES

Bronowski, J. J., *The Commonsense of Science.* Howard University Press, 1953.

———, *Science and Human Values.* Harper & Row, Torchbooks, 1959.

Chargaff, E., Persp. Biol. Med. **16**, 486–502, 1973. "Bitter Fruits from the Tree of Knowledge: Remarks on the Current Revulsion of Science."

Dubos, René, *Dreams of Reason.* Columbia University Press, 1961.

Eisley, L., *The Immense Journey.* Random House, Vintage Books, 1957.

Gabriel, M. L., and Fogel, S., eds., *Great Experiments in Biology.* Prentice-Hall, 1955.

Greene, M., and Mendelsohn, E., *Topics in the Philosophy of Biology.* Reidel, 1976.

Guttman, B. S., Bioscience **26**, 112–113, 1976. "Is Levels of Organization A Useful Biological Concept?"

Monod, J., *Chance and Necessity.* Knopf, 1971.

Pauling, L., *The Nature of the Chemical Bond.* 3d ed., Cornell University Press, 1960.

Rosenberg, E., *Cell and Molecular Biology: An Appreciation.* Holt, Rinehart and Winston, 1971.

Roszak, T., *Where the Wasteland Ends.* Doubleday, 1972.

Russell, B., *The Future of Science.* Wisdom Library, Philosophical Library, Inc., 1959.

Sayers, G., chairman, Federation Proc. **36**, 2087–2109, 1977. Symposium on "The Nature of Corticotropin Releasing Factor."

Schally, A. V., Arimura, A., and Kastion, A. J., Science **179**, 341–350, 1973. "Hypothalamic Regulatory Hormones."

Stent, G. S., Sci. Am. **227**, 84–93, Dec. 1972. "Prematurity and Uniqueness in Scientific Discovery."

———, Science **187**, 1052–1057, 1975. "Limits to the Scientific Understanding of Man."

Thomas, L., *Lives of a Cell: Notes of a Biology Watcher.* Viking, 1974.

Young, L., ed., *The Mystery of Matter.* Oxford University Press, 1965.

CHAPTER TWO

The Organization of Living Systems

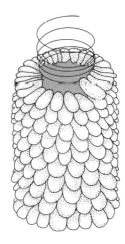

The emphasis in the introductory chapter centered around the complexity and organization of living systems. A closer examination indicates that there are basically three levels of complexity among such systems: the viral, prokaryotic, and eukaryotic levels. The differences among these three groups are very large and there is no ambiguity as to which group a particular "organism" belongs. Even though the groups are nonoverlapping, there is very little doubt that all life forms on earth are evolutionarily related, since they all share precisely the same genetic system. All of them transmit their genetic inheritance through a DNA or an RNA coding system which contains precisely the same code words to specify one or another of the twenty amino acids that are incorporated into proteins.

VIRUSES

Viruses, which are noncellular in nature, are vastly less complex than prokaryotic or eukaryotic cellular systems. Although as a group they are extremely heterogeneous, all viruses share certain basic properties. All of them are obligatory parasites, i.e., they cannot reproduce unless present within some host cell. Depending upon the virus, the host can be a bacterial, plant, or animal cell. In addition, viruses exist in two very different states, one within a host cell and the other outside the confines of a cell. Outside the cell, the virus exists as a particle, or *virion*, in which its genetic material (often together with a few proteins) is enclosed within some non-

19

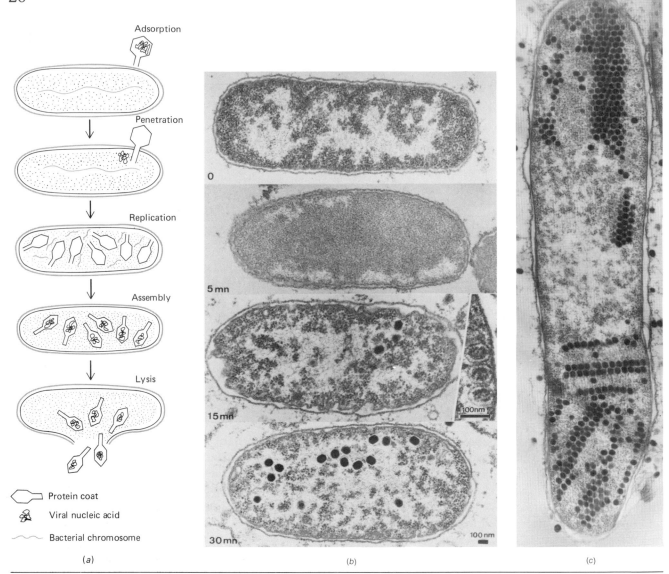

Adsorption

Penetration

Replication

Assembly

Lysis

Protein coat

Viral nucleic acid

Bacterial chromosome

(a)

0

5 mn

15 mn

30 mn

100nm

100 nm

(b)

(c)

Figure 2-1. *(a)* Steps which occur during the infection of a bacterium by a bacteriophage. *(From M. J. Pelczar, R. D. Reid, and E. C. S. Chan, "Microbiology," 4th ed., McGraw-Hill, 1977.) (b)* The course of a bacteriophage infection as seen in a series of electron micrographs taken at successive times (indicated at lower left) during the process. The top picture shows an uninfected cell; the clear areas contain the bacterial DNA. At 5 minutes postinfection, nuclear disruption is seen, accompanied by the hydrolysis of bacterial DNA and synthesis of phage DNA. By about 8 minutes, phage structural proteins appear, and a few minutes later the first infective phage are found within the cell. The photograph taken at 15 minutes shows the presence of a few mature particles. The insert shows the presence of "preheads," a precursor stage (devoid of DNA) along the assembly path leading to head formation. The photograph at 30 minutes shows a later stage containing an increasing number of finished heads. *(Courtesy of B. Menge, J. v.d. Broek, H. Wunderli, M. Wurtz, and E. Kellenberger.) (c)* A late stage in the infection process showing the ordered accumulation of a large number of particles. *(Courtesy of J. Pisani and G. Chapman.)*

genetic container. During the infective stage, the genetic material is released from its container into the cell where it can have various properties depending upon the type of virus and the conditions at the time. In some cases, when inside a host cell, the genetic material of the virus directs the formation of new virus particles leading to the death of the cell. The course of events that occurs during infection of a bacterial cell by a bacterial virus (a bacteriophage) is shown in Fig. 2-1. The final outcome of the infection is the release of several hundred new virus particles. Unlike cellular reproduction, the formation of new virus particles is always indirect via some type of infective intermediate. In addition, there is never any type of growth stage—viruses are packaged (or assembled) from the components directly into the mature-sized virion.

In other cases, a viral infection may not lead to the death of the host cell, but rather the virus may remain in the cell in a much less conspicuous manner. For example, there are viruses (termed *lysogenic phage*) that exist within a host bacterial cell yet have essentially no demonstrable effect on the cell's activities (Fig. 2-2). Even in these cases, the viral presence can be readily revealed; under the appropriate experimental conditions (such as illuminating the cell with ultraviolet radiation) the viruses escape from their repressed state and kill the cell. In other cases, a virus can remain within a cell and drastically affect the behavior of that cell without killing it. This is what occurs when a *tumor virus* infects a suitable host cell and converts it from a normal cell into a malignant one. In these cases there is generally no evidence of viral particle production within the cell but, rather, the presence of the virus is manifested in the altered growth properties of the cell which lead to the formation of tumors.

When present in the extracellular virion stage, viruses are little more than macromolecular aggregates. They cannot metabolize, reproduce, and so forth, but are only capable of attaching to a suitable host cell to begin an infection. Virus particles come in a great variety of shapes, sizes, and constructions. All of them contain a limited amount of genetic information. The genetic program of the particle may consist of DNA (either double- or single-stranded) or RNA (either double- or single-stranded). The extent to which the genetic material is packed into the very small particle is dramatically shown in electron micrographs of ruptured particles (Fig. 2-3). Remarkably, some viruses seem to have as few as 3 different genetic functions, while some of the more complex ones may have as many as 500 different genes. The fewer the genes, the more the virus relies on materials produced by the host's genes in order to carry out the host's own destruction. Regardless of the size of the *genome* (one set of genetic material), the virus relies upon the host to provide the material for the construction of new viruses. The role of the viral genome in a general sense is to supervise the activities so that the efforts of the cell are directed toward meeting the needs of the virus rather than the host.

The capsule (or *capsid*) in which the genetic material of the virus is contained is generally made up of a specific number of subunits. There are numerous advantages to construction by subunit, one of the most

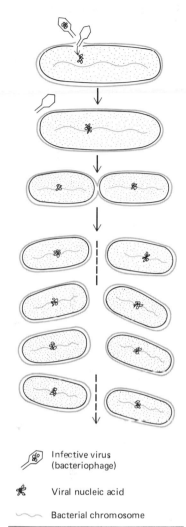

Infective virus (bacteriophage)

Viral nucleic acid

Bacterial chromosome

Figure 2-2. In the lysogenic state, the viral DNA is integrated into the host chromosome where it is replicated along with the host DNA. In this state, activity of the viral DNA (termed a *prophage*) is suppressed and no virus particles are formed. *(From M. J. Pelczar, R. D. Reid, and E. C. S. Chan, "Microbiology," 4th ed., McGraw-Hill, 1977.)*

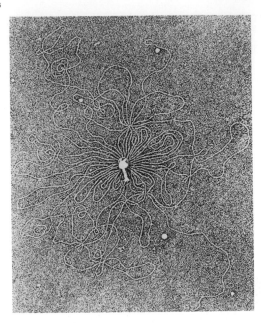

Figure 2-3. DNA being released from the ruptured head of a T$_2$ bacteriophage. This linear DNA molecule (note the two free ends) has a molecular weight of approximately 1.2×10^8 daltons (180,000 base pairs) and measures 68 μm in length. [*From A. K. Kleinschmidt et al.*, Bioc. Biop. Acta, **61:861** *(1962).*]

apparent being the economy it affords the virus in terms of information storage. If a viral coat is made of only one protein (present in many copies), the virus needs only one gene to code for this protein. This one gene can be used many times during the production of the specific protein molecules. In contrast, if the coat were made of numerous different proteins, or one tremendously large protein, an equivalent amount of information would have to be included in the viral DNA (or RNA).

Since the capsule is made of subunits, the subunits must be packed together into some type of precise arrangement. As will be discussed at greater length in Chapter 3, the assembly of many virus particles occurs by having the various pieces simply fit together on their own by virtue of their shape, a process termed *self-assembly*. In order to self-assemble, viral particles must be organized in some geometrical pattern so that the pieces fit together in an obvious ordered manner. In certain forms, the capsid has an elongated helical organization (Fig. 2-4b), one in which the subunits attach to one another to form a type of spiraling enclosure around the genetic material. In other cases, the capsid is organized into a *polyhedron*, i.e., a structure having plane faces. A particularly common polyhedral shape found in viruses is the icosahedron (Fig. 2-4a) as exemplified by the adenovirus (Fig. 2-5). Viruses having larger icosahedral particles contain capsids with a greater number of subunits but the same overall geometrical lattice. The unvarying geometrical organization of viruses is revealed most dramatically by their ability to be crystallized as was first accomplished by Wendell Stanley for the tobacco mosaic virus in 1935. The ability to crystallize certain viruses serves to call further attention to their molecular level of organization. These crystals can be dissolved and the individual particles are fully able to infect new cells.

Many viruses have components in addition to the polyhedral cover

over their genetic program. For example, the familiar T-bacterial viruses (Fig. 2-4c) have a stalk, base plate, and tail fibers, which together cause the particle to appear not unlike a landing module for the moon. Many animal viruses are covered by an outer wrapper which is derived largely from the host cell's own outer membrane.

The study of viruses has had a long and interesting history. It was a viral disease, namely smallpox, for which Edward Jenner was able to develop the first vaccine in 1798. It would be approximately 150 years before anyone would have an actual glimpse at what these agents actually looked like. The research by Jenner, Louis Pasteur, Peyton Rous, and many other early investigators of viruses serves as a fine example of the ability of cell biologists to determine some of the basic characteristics of objects too small to be observed directly. The basic question to be solved was the nature of the agent responsible for various types of diseases ranging from influenza to polio or measles (and more recently, in some cases at least, cancer). For a long period it was assumed that the agents responsible for all types of diseases were bacterial. However, near the end of the nineteenth century, it became apparent that there were some basic differences among disease-causing "microorganisms." It was found that the agents responsible for certain diseases, such as foot-and-mouth disease in cattle, were so small that they could pass through the pores in filters that had been designed to hold back known bacteria. As it turns out, size alone is a poor criterion to distinguish viruses from true cellular microorganisms since some of the very large viruses (such as vaccinia) are equivalent in size to the very small bacteria (such as *Mycoplasma*). However, except for a few cases, viruses are much smaller than cellular organisms.

What is the evolutionary relationship between viruses and cellular life forms? At first glance one might conclude that viruses represent a primitive life form, one possibly similar to those that existed on earth before the evolution of the much more complex prokaryotic cells. However, if one considers that the "lives" of viruses are totally dependent upon the presence of a cell to invade, it becomes obvious that, at least in their present state, viruses could

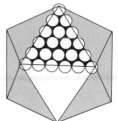

Icosahedron, with one possible arrangement of capsomeres shown top centre. Other arrangements also known

(a) Cubic symmetry; e.g., adenovirus

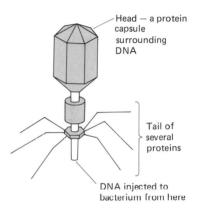

Head — a protein capsule surrounding DNA

Tail of several proteins

DNA injected to bacterium from here

(b) Complex symmetry; T-2 bacteriophage

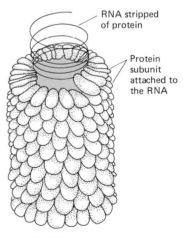

RNA stripped of protein

Protein subunit attached to the RNA

(c) Helical symmetry, e.g., tobacco mosaic virus

Figure 2-4. The structure of three major types of viral particles. *(From N. A. Edwards and K. S. Hassal, "Cellular Biochemistry and Physiology," McGraw-Hill, 1971.)*

Figure 2-5. Electron micrograph of a negatively stained preparation of simian adenovirus particles. Insert shows a model of an icosahedral particle. *(Courtesy of K. O. Smith and M. D. Trousdale.)*

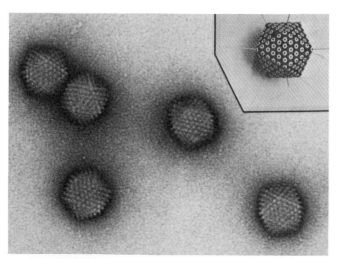

not have arrived on the scene before their hosts. Since viruses share the same genetic basis with one another as well as with prokaryotic and eukaryotic cells, they could not have arisen independently as a primitive form *after* other cells had evolved. It seems most reasonable to assume that viruses represent a degenerate form, i.e., one derived from a more complex organism. Since viruses are so simple relative to the simplest cellular organisms, it is unlikely that they evolved from cells by some gradual process involving the loss of characteristics in the way that multicellular parasites (such as flukes and tapeworms) tend to evolve from closely related free-living forms (such as turbellarian flatworms). It would appear more likely that the viral genome was derived originally from some small fragment of a cellular genome, a fragment which could maintain some type of autonomous existence within the cell. Natural selection could have acted then on this part of a cellular genome to cause it to become modified into a totally different type of genetic element. Considering the tremendous diversity among viruses, it is not unlikely that different groups evolved independently from one another, i.e., were originally derived from different cellular organisms.

PROKARYOTES VERSUS EUKARYOTES

The gap between the viruses and the cellular prokaryotes is a large one. Although there are a number of prokaryotes which are very small and carry out their reproductive processes within other cells (such as the *Rickettsia*, the *Chlamydias*, and the *Mycoplasma*), all of them represent a self-contained unit of life surrounded by a membrane and complete with all of the intracellular machinery necessary to carry out their own metabolic reactions, their own self-duplication, their own energy capture and utilization, and so forth.

The division of the world into a prokaryotic and a eukaryotic branch was largely the result of the examination of cells with the electron microscope. The electron microscope revealed for the first time the structural nature of the internal contents of tiny bacterial cells. Careful scrutiny of many different bacterial, plant, and animal cells suggested that a fundamental dichotomy existed among cellular organisms. The more carefully and completely cells were examined, the more support was gained for the concept of two distinct types of cells. On the prokaryotic side, there are diverse forms of bacteria and a group generally termed the *blue-green algae*. The term *algae* was applied to these organisms on the basis of their photosynthetic activities before their structural relationship to bacteria was uncovered with the electron microscope. They would be more aptly referred to as blue-green bacteria. Although blue-green algae are clearly prokaryotic in their organization, they do have a well-developed membranous photosynthetic apparatus which bears a great similarity to the structural components of the chloroplasts of higher cells. Like plant cells, and unlike bacteria, photosynthesis leads to the production of molecular oxygen.

There are many basic differences between prokaryotic (Fig. 2-6) and eukaryotic cells (Fig. 2-7), as well as many basic similarities. In both types of cells the living protoplasm is surrounded by a membrane which

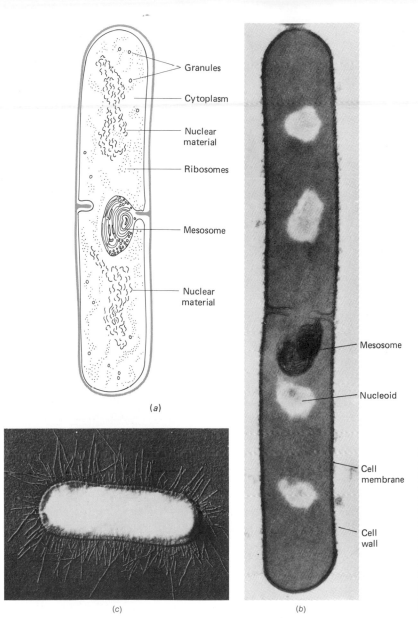

(a)

(c)   (b)

Figure 2-8. The prokaryotic cell. *(a)* Diagrammatic representation of the structure of a bacterium. *M. J. Pelczar, R. D. Reid, and E. C. S. Chan, "Microbiology," 4th ed., McGraw-Hill, 1977.)* *(b)* Electron micrograph of a thin section of the bacterium, *Bacillus subtilis*, showing nuclear material (light-appearing areas) in addition to cell wall, cytoplasmic membrane, mesosome, and initial stage of cross-wall formation. *(Courtesy of S. F. Zane and G. B. Chapman.)* *(c)* The bacterium, *Salmonella anatum*. Note the numerous fimbriae projecting from the surface. [*From J. P. Duguid, E. S. Anderson, and J. Campbell,* J. Pathol. Bacteriol., **92:**107 (1966).]

serves as a selectively permeable barrier between the living and nonliving worlds. The structure of this lipid-protein membrane is very similar in all types of cells (Fig. 5-14), although there is considerable variation in the particular types of lipids and proteins of the organelle. Functionally, the two cell membranes can be quite different. In prokaryotic cells, the cell membrane is the site of energy-yielding reactions, which in eukaryotic cells are restricted to a complex cytoplasmic organelle, the *mitochondrion* (Fig. 8-1). The nature of the region just outside the cell membrane is

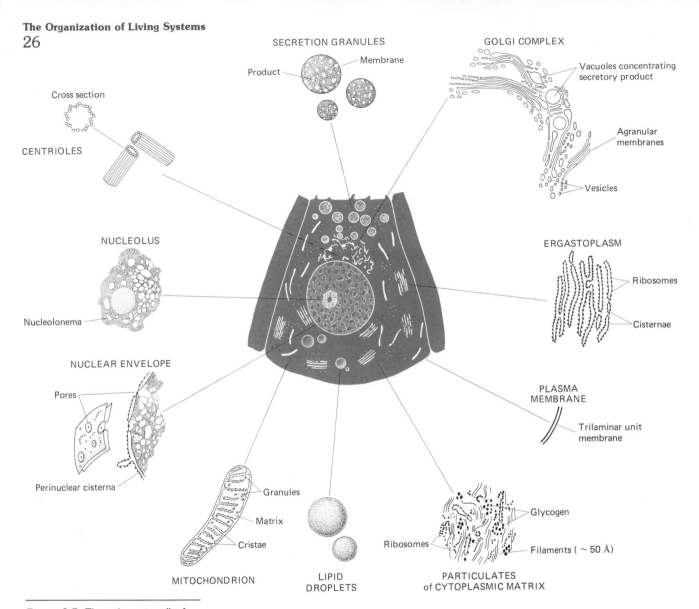

SECRETION GRANULES
Product
Membrane

GOLGI COMPLEX
Vacuoles concentrating secretory product
Agranular membranes
Vesicles

CENTRIOLES
Cross section

NUCLEOLUS
Nucleolonema

ERGASTOPLASM
Ribosomes
Cisternae

NUCLEAR ENVELOPE
Pores
Perinuclear cisterna

PLASMA MEMBRANE
Trilaminar unit membrane

MITOCHONDRION
Granules
Matrix
Cristae

LIPID DROPLETS

PARTICULATES of CYTOPLASMIC MATRIX
Glycogen
Ribosomes
Filaments (∼ 50 Å)

Figure 2-7. The eukaryotic cell. In the center of this figure is a diagram of the cell illustrating the form of its organelles and inclusions as they appear by light microscopy. Around the periphery are representations of the finer structure of these same components as seen in electron micrographs. The ergastoplasm is now also called the *rough endoplasmic reticulum. (From W. Bloom and D. W. Fawcett, "A Textbook of Histology," 10th ed., Saunders, 1975.)*

very different in the two types of cells. With the exception of one primitive group of prokaryotes, the *Mycoplasma* (or PPLO as they were formerly called), all prokaryotic cells possess a *cell wall* (Fig. 6-6) which performs numerous invaluable services to the cell it surrounds. Although the cells of many eukaryotes (the algae, fungi, and plants) are also bounded by a cell wall, the chemical composition of the walls of the two types of cells is very different.

Both prokaryotic and eukaryotic cells contain a cytoplasmic and a nuclear region. The nuclear regions contain the genetic material of the cell from which information is transmitted to direct the synthetic activities of the entire cell. The nature of the process of information storage and utilization is very similar in both types of cells. However, there are numerous basic differences between nuclei of the two types. Prokaryotic cells contain much smaller amounts of DNA (from about 0.25 mm in length in the *Mycoplasma* to about 3 mm in length in the blue-green algae). Although there are some eukaryotic cells, such as yeast, which have only slightly more DNA (4.6 mm in yeast) than that of the blue-greens, most eukaryotic cells (even those of eukaryotic microorganisms) contain several orders of magnitude more genetic information. The DNA of eukaryotic cells is tightly associated with various types of proteins to form a structurally complex *chromosome* (Fig. 15-12). In contrast, the DNA of prokaryotic cells is present in what is essentially a naked chromosome made up only of the DNA itself. Another basic feature of eukaryotic nuclei is the presence of a distinct boundary membrane, termed the *nuclear envelope* (Fig. 13-35), a structure which is totally lacking in the prokaryotic cell.

The cytoplasm of the two types of cells is also very different. The cytoplasm of eukaryotic cells contains a great diversity of structures as is readily apparent from the most superficial examination of an electron micrograph of nearly any cell of the body. There are several distinct types of organelles, some of which like the chloroplast, mitochondria, lysosome, Golgi complex, endoplasmic reticulum, or peroxisome are either a membranous structure or at least membrane-bound. In addition to these distinct types of organelles there is a variety of membrane-bound vesicles and vacuoles of varying dimension. In the case of plants, a single vacuole can occupy most of the cell (Fig. 9-23). In addition, eukaryotic cells contain organelles which lack a membraneous component. Included in this latter group are the elongated microtubules (Fig. 15 1) and microfilaments (Fig. 16-14) which are involved in cell contractility, movement, and support. The other prominent nonmembranous component of the cytoplasm is the ribosome (Fig. 3-20), a particle of about 250 Å found either free or attached to membranes. Of the eukaryotic structures just mentioned, the only one also found in prokaryotic cells is the ribosome. Even though ribosomes of prokaryotic and eukaryotic cells have considerably different dimensions (those of prokaryotes are smaller), these organelles perform the same role by essentially the same mechanism in both types of cells. In other words, the process of protein synthesis, like metabolism in general, is very similar in all cells of all organisms.

Whereas the cytoplasm of eukaryotic cells is filled with a diverse array of membranous organelles, the "cytoplasm" of prokaryotic cells is essentially devoid of these types of structures. In most prokaryotes, the only cytoplasmic membranes present [termed *mesosomes* (Fig. 2-6b)] are derived simply from infoldings of the plasma membrane. As mentioned above, the energy functions of the mitochondria of eukaryotic cells are taken over by the bacterial cell membrane, while the functions of the other eukaryotic cell membranes seem not to be required in the prokaryote. For the most part, the cytoplasmic membranes of eukaryotic cells form a system of interconnecting channels and vesicles which are involved in carrying substances from one part of a cell to another, as well as between the inside of the cell and its environment. These types of functions appear to be lacking in bacteria and blue-green algae. Prokaryotic cells are much

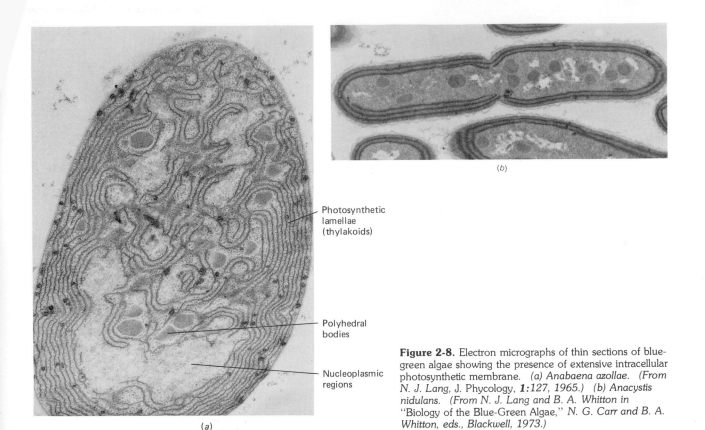

Photosynthetic
lamellae
(thylakoids)

Polyhedral
bodies

Nucleoplasmic
regions

(a)

(b)

Figure 2-8. Electron micrographs of thin sections of blue-green algae showing the presence of extensive intracellular photosynthetic membrane. *(a) Anabaena azollae.* *(From N. J. Lang,* J. Phycology, **1:**127, 1965.) *(b) Anacystis nidulans. (From N. J. Lang and B. A. Whitton in* "Biology of the Blue-Green Algae," *N. G. Carr and B. A. Whitton, eds., Blackwell, 1973.)*

smaller than cells of eukaryotes and directed intracytoplasmic communication is less important. In addition, much of the flow of materials through a eukaryotic cell reflects one or another specialized function of that cell type in a multicellular organism. For example, secretory cells, such as those of the pancreas or salivary gland, are responsible for pouring their products out into adjoining ducts. The opposite type of process is seen upon examination of the activities of white blood cells or macrophages which are responsible for picking up various types of debris within the animal and taking it into the cell enclosed in a membranous vacuole. Bacterial calls do not take up particulate material and their limited secretory activities do not require specialized internal membrane. Substances entering and leaving the bacterial cell do so by passing directly through the membrane itself. The most prominent exception to the statement that prokaryotic cells lack internal membranous organelles (other than that of the membrane derived from the surface) is found among the blue-green algae (Fig. 2-8) which contain a true membranous photosynthetic apparatus whose components, termed *thylakoids*, are quite similar to the elements of the plant cell chloroplast.

Other major differences between eukaryotic and prokaryotic cells can be found upon examination of their reproductive behavior, their means of cell division, and their locomotory apparatus. For the most part, prokaryotes are nonsexual organisms. They contain only one copy of their single chromosome, which is duplicated before division and then separated into two daughter cells. There is no process comparable to meiosis, or gamete formation, or true fertilization. Even though true sexual reproduction is lacking among prokaryotes, there does exist a phenomenon, termed *conjugation* (Fig. 11-5), whereby certain bacteria can pass a piece of DNA from one cell to another. However, the recipient almost never receives a whole chromosome from the donor, and the condition in which the recipient cell contains both its own and its partner's DNA is a transient one. The cell soon reverts back to having a single chromosome, i.e., the haploid state. Not only do prokaryotes lack meiosis, they lack mitosis as well. The term *mitosis* refers to a particular type of process by which duplicated chromosomes of eukaryotic cells are separated into daughter cells. In eukaryotic cells the chromosomes are duplicated in the period between mitoses. The duplicated chromosomes undergo a type of condensation process prior to separation and the duplicates are then pulled apart during mitosis by a microtubule-containing *mitotic spindle* (Fig. 15-18). In prokaryotes there is no condensation of the chromosome and no spindle. The DNA is duplicated and the two copies (which are attached to the cell membrane) are separated by the growth of the cell membrane between the two points of attachment (Fig. 14-30).

The manner in which movement occurs in the two types of cells is also very different. Whereas eukaryotic cells have evolved highly complex and variable types of locomotor functions, those of prokaryotes have remained very primitive. The most distinctive type of movement among prokaryotes is accomplished by a very thin protein filament, termed a *flagellum* (Fig. 16-26), which protrudes from the cell and rotates. The rotations of the flagellum exert pressure against the surrounding fluid medium causing the cell to be propelled forward. Certain eukaryotic cells (such as spermatozoa) also possess flagella, but in this latter case the organelles are much more complex (Fig. 16-29). In addition to flagella, many bacteria contain another type of protruding rod-shaped structure, termed *pili* (Fig. 2-6c), which are not involved in locomotion but rather appear to mediate the adhesion of bacteria to one another.

In the preceding paragraphs, the most important differences between the prokaryotic and eukaryotic condition were pointed out. Each of these points will be elaborated upon in the respective chapter in which that type of organelle or cell function is discussed. The differences between prokaryotic and eukaryotic cells extend beyond the presence or absence of organelles. In one way, that of metabolism, the prokaryotes are very sophisticated organisms. A bacterium, such as *Escherichia coli,* has the ability to live and prosper on a culture medium containing a simple carbon and nitrogen source and a few inorganic ions. The bacterial cell contains all of the enzymes necessary to convert one or two small molecular weight organic compounds into hundreds of substances which the cell must have. This capability allows one or another type of bacteria to survive in virtually any environment on earth. In contrast, eukaryotic cells typically require a much greater variety of organic compounds, including a number of "vitamins" and other essential substances which it cannot make on its own. In all other ways, the prokaryotic cell is primitive by eukaryotic standards and its level of organization has placed very stringent limits on the potential of these cells. For the most part, prokaryotic cells are limited to activities such as metabolism, growth, and reproduction. At best, some of them can move either toward or away from a few types of stimuli. In contrast, eukaryotic cells have evolved in a myriad different ways. Even in those eukaryotic organisms composed of a single cell, the *protists* (includes algae, protozoa, fungi, etc.), one finds the cells capable of much more complex functions and a much broader interaction with their environment.

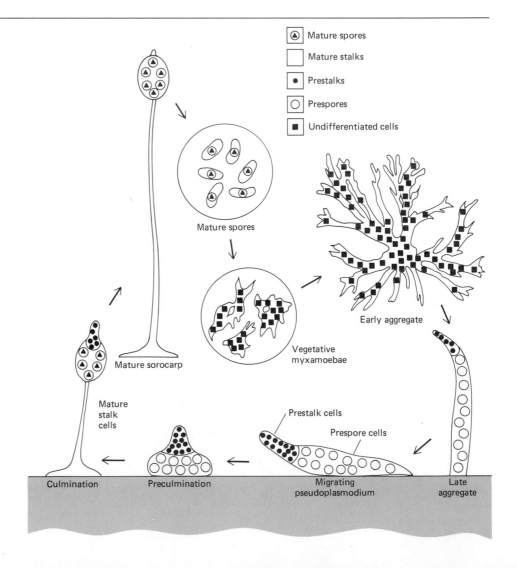

Figure 2-9. Life cycle of the slime mold, *Dictyostelium discoideum.* *(Courtesy of J. H. Gregg.)*

CELLULAR SPECIALIZATION

With the evolution of multicellular plants and animals, a new level of organization could be achieved. Whereas the single-celled organism must be composed of a cell in which all types of activities are housed, the multicellular organism possesses the potential for cell specialization in which particular groups of cells are suited for one or another type of activity. If one considers the requirements for cellular contraction, versus those for secretion, versus those for conduction of nerve impulses, and so forth, it becomes apparent that one cell cannot carry out all of these different processes. Consequently, the evolution of cell specialization gave living organisms a whole new potential for the exploitation of their environment, a level totally beyond the reach of prokaryotes. The examination of one of the most primitive multicellular eukaryotes, the cellular slime mold *Dictyostelium,* will illustrate the basic advantage of the division of labor among cells.

At most times during their life cycle, (Fig. 2-9), cells of *Dictyostelium* exist as solitary, independent amoebae wandering about over their substrate. Each cell represents a complete, self-sufficient organism. However, when the food supplies become scarce, an entirely new type of activity is triggered among the cells, and they stream toward each other to form an aggregate of a highly characteristic appearance (Fig. 2-10*a*).

Figure 2-10. Photographs of various stages in the life cycle of the slime mold. *(a)* Amoebae undergoing aggregation by migrating toward a common center. *(Courtesy of J. T. Bonner.)* *(b)* Side view of slug (pseudoplasmodium) near the close of the migrating period, showing the tip lifting from the substratum and a trail of mucus behind. *(Courtesy of D. Francis.)* *(c)* An early culmination stage. The mass of cells has rounded up and is beginning to lift off the substratum. Cells that will become stalk are seen at the anterior tip. *(Courtesy of K. B. Raper.)* *(d)* The mature fruiting body *(sorocarp)* with the elongate stalk holding up the mass of spores before its release. *(Courtesy of W. F. Loomis.)*

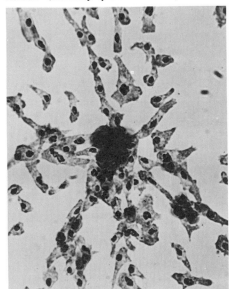

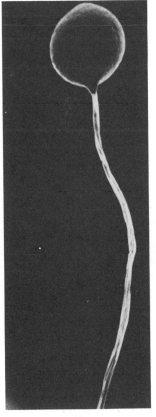

(a)

(b)

(c)

(d)

Each single-celled organism has now become a small part of a multi-cellular individual. This new stage in the life cycle is termed a *pseudoplasmodium,* or simply a slug (Fig. 2-10*b*), which migrates for a period over the surface of its container. Closer examination of the cells of the slug reveals that the cells are no longer a homogeneous population. Rather, the cells of the anterior third of the slug (termed *prestalk* cells) can be distinguished from those of the posterior section (termed *prespore* cells) by a number of criteria, such as those shown in Fig. 2-11. If one waits for a while longer a dramatic series of events occurs: the slug stops its migration, rounds up on the substrate (Fig. 2-10*c*), and then begins to rise into the air as an elongated fruiting body (Fig. 2-10*d*). The fruiting body is composed of a slender stalk supporting a rounded mass of cells toward its tip (Fig. 2-10*d*). Closer examination indicates that the cells of the stalk are very different from the remainder of the cells which are present as *spores,* i.e., inactive cells surrounded by a thick wall or capsule. The stalk cells of the fruiting body are derived from the anterior cells of the slug while the spore cells are derived from its posterior end. These two types of cells not only appear different under the microscope, but have a very different function to perform. The stalk cells support the spore mass above the substrate, while the spore cells are destined to carry on the life of the population into the next generation. The process whereby a cell, in this case a slime mold amoeba, becomes a more specialized cell, such as a stalk cell or a spore cell, is termed *differentiation.*

In the case of the slime mold, there are two alternate paths of differentiation open to a particular amoeba as it enters into its aggregation state.

Figure 2-11. Autoradiographs showing the progress of cellular differentiation in *Dictyostelium.* The incorporation of the monosaccharide, ³H-fucose, serves as a very early marker for the differentiation of prespore cells. The black regions (due to the presence of silver grains) reveal the sites of ³H-sugar incorporation. *(a)* A section through a late aggregation stage in which a very few cells (arrows) in the aggregate are beginning to show the prespore characteristic. *(b)* A later stage in the life cycle in which the prespore region is clearly demarcated by the cells' incorporation activity. [*From J. H. Gregg and G. C. Karp,* Exp. Cell Res., **112:31** *(1978).*]

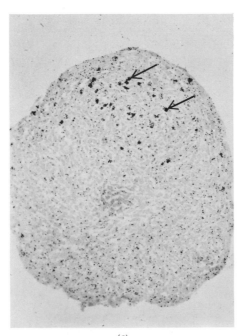

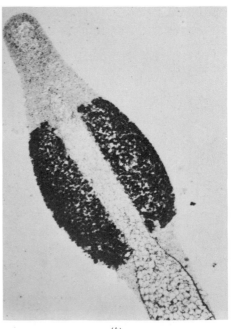

(a)   (b)

In contrast, when an egg of a higher animal becomes fertilized and develops gradually into an adult organism there are numerous different pathways for differentiation (Fig. 2-12). Some cells become part of a digestive gland, others part of a muscular system, and so on. Each of these specialized cells can be traced back to the relatively unspecialized fertilized egg, a product of the combined efforts of two different organisms from the previous generation. If one studies the various types of cells of a multicellular organism with the electron microscope, one sees in most cases the same types of organelles. Mitochondria, for example, are found in essentially all types of cells. In one type of cell, however, they may be rounded while in another they may be elongated. Similarly, mitochondria of one cell may be scattered throughout the cytoplasm, whereas in another cell they may be concentrated near a particular surface. In each case, the number, appearance, and location are dictated by the needs of the particular cell type.

In addition to the presence of the ubiquitous organelles, cells also contain structures and substances characteristic of that particular cell type, ones which may not be found in other cells. For example, cartilage cells contain and secrete materials for their extracellular environment, materials which give cartilage its tough, yet flexible mechanical properties. Red blood cells contain high concentrations of hemoglobin, the oxygen-binding protein involved in respiration. At the same time, all cells share those basic activities which are required to maintain life itself. In other words, we can conceive of each specialized cell as carrying out, in a sense,

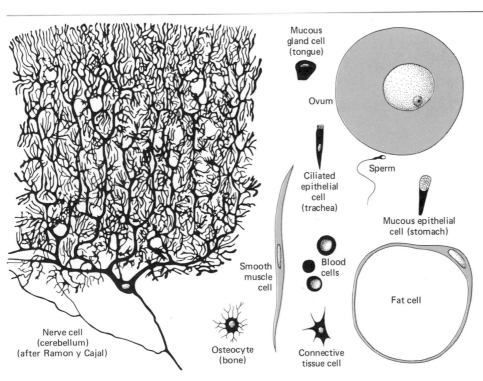

Figure 2-12. Diagrammatic representation of a variety of cell types. All of the cells are drawn to scale. (*From W. M. Copenhaver, R. P. Bunge, and M. B. Bunge, "Bailey's Textbook of Histology,"* 16th ed., *The William & Wilkins Company, 1971.*)

Mucous gland cell (tongue)

Ovum

Nerve cell (cerebellum) (after Ramon y Cajal)

Smooth muscle cell

Ciliated epithelial cell (trachea)

Sperm

Mucous epithelial cell (stomach)

Blood cells

Fat cell

Osteocyte (bone)

Connective tissue cell

two sets of activities, one directed toward serving its own needs in remaining alive and the other directed toward serving the needs of the organism as a whole.

UNITS OF MEASUREMENT

Since cells are generally of microscopic dimensions, the units most commonly used in this book will be correspondingly small. There are two units of linear measure which are most commonly used in connection with structures at the cellular level, the micrometer, μm (often referred to as the micron), and the angstrom, Å. One angstrom is equivalent to 0.1 nanometer (nm), 10^{-4} μm, or 10^{-8} cm, or approximately the diameter of a hydrogen atom. Angstroms and nanometers are units of measurement used when discussing molecular or supramolecular levels of organization. A typical globular protein molecule (such as myoglobin) is approximately 45 Å $\times$ 35 Å $\times$ 25 Å, highly elongated proteins (such as collagen or myosin) are over 1000 Å in length, and DNA is approximately 20 Å in width. The angstrom is a convenient unit of linear measure when describing small subcellular organelles seen in the electron microscope. Ribosomes, microtubules, and microfilaments, for example, are all between 50 and 250 Å in diameter. Larger organelles, such as nuclei or mitochondria, are more easily defined in terms of micrometers. A typical liver cell may be approximately 20 μm in diameter, a nucleus approximately 10 μm in diameter, and a mitochondrion approximately 2 μm in length.

THE SIZE OF CELLS

Figure 2-13 shows the relative size of objects of interest in cellular biology. Nearly all cells are of microscopic dimensions and the presence of a few particularly large ones such as the eggs of vertebrates is somewhat misleading. These giant cells actually contain a very small amount of what one would consider "living protoplasm." The remainder of the egg is filled with inert materials which are used as nutrition for the developing embryo. In nearly every case, the relative size of different multicellular organisms can be explained on the basis of their cell number, rather than the size of their component cells. There are numerous reasons for the maintenance of cells at microscopic dimensions. In the following discussion we will briefly consider two of these reasons.

To a considerable degree, the activities that occur within a cell are under the control of the genetic material housed in the nucleus of that cell. As will be discussed at some length in later chapters, there is a very limited number of copies of each gene present in a given nucleus, regardless of how large that nucleus might be. The presence of a limited number of genes places a strict limit on the amount of information-carrying messages which can be sent by the nucleus out into the cytoplasm to direct its activity. The greater the bulk of cytoplasm in a cell, the more limiting is its genetic content.

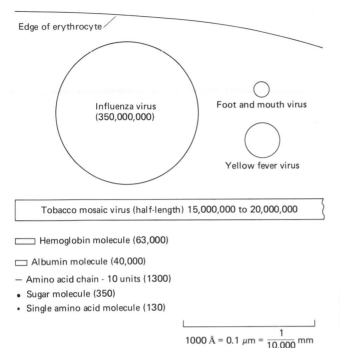

Edge of erythrocyte

Influenza virus
(350,000,000)

Foot and mouth virus

Yellow fever virus

Tobacco mosaic virus (half-length) 15,000,000 to 20,000,000

☐ Hemoglobin molecule (63,000)

☐ Albumin molecule (40,000)

— Amino acid chain - 10 units (1300)

• Sugar molecule (350)
• Single amino acid molecule (130)

$1000 \text{ Å} = 0.1 \ \mu m = \frac{1}{10,000} \text{ mm}$

Figure 2-13. Relative sizes of molecules, viruses, and the erythrocyte of humans. Line at lower right shows scale and is 0.1 μm long. Molecular weights of viruses and molecules are given in parentheses. *(From J. A. V. Butler, "Inside the Living Cell," reproduced by permission of George Allen & Unwin Ltd., 1959.)*

Another very important basis for the limitation of cell's size becomes apparent when we consider the consequences of an increase in cell size on the surface area/volume ratio. The outer surface of a cell is covered by a thin plasma membrane which serves to separate the contents of the cell from its surroundings. All nutrients, oxygen, regulatory substances, and so on, which enter a cell must do so by passing across this outer boundary. Since the rate of passage of materials across the cell membrane is limited, the surface area of a cell places a limit on the exchange of substances between the cell and its environment. A quick calculation of volumes and surface areas of objects will serve to illustrate how the surface area becomes increasingly limiting as linear dimensions increase. Consider the case of two cubes having linear dimensions of 1 and 10 cm. Whereas one of the edges of the larger cube is 10 times that of the smaller cube, the surface area of the larger cube is 100 times as great (600 cm^2 vs. 6 cm^2), and its volume is 1000 times as large (1000 cm^3 vs. 1 cm^3). It is apparent from this type of calculation that as a body increases in size, its surface area (which increases with the square of the linear dimensions), cannot keep pace with its volume (which increases with the cube of the linear dimensions). In the case of a cell, the mass of its protoplasm is proportional to its volume, while its exchange surface, i.e., its plasma membrane, is proportional to its surface area. Since the needs of a cell are determined by its volume and the ability of substances to enter the cell by its surface area, the size of a cell becomes limited.

If one examines the shape of cells in a multicellular organism, one finds there are certain ways in which cells have at least partially overcome the difficulties posed by the surface area/volume effect. In many cases, the plasma membrane of the cell is thrown into rather deep folds or protrusions, a feature which greatly increases the surface area of the membrane. The cells of the intestine, for example (Fig. 6-1), contain finger-like microvilli at the surface across which nutrients are absorbed. In other cases, cells are seen to be highly flattened or elongated (as in the case of a nerve cell). When a cell assumes one of these unorthodox shapes, all of its protoplasm remains in close association with its surface membrane and, therefore, in a position to receive sufficient oxygen, nutrients, and so forth. A nerve cell, for example, which may stretch several feet in length, generally remains microscopic in its diameter.

THE MICROSCOPE

Since cells are almost exclusively of dimensions beneath those visible with the naked eye, it was essential, if a knowledge of their nature were to be gained, that an instrument be designed to increase their visibility. By the end of the seventeenth century a few persons, namely Robert Hooke, Nehemiah Grew, and Anton van Leeuwenhoek, had constructed microscopes capable of observing small biological specimens; but the lenses they had ground had serious problems and were very limited in the degree to which they could be used to observe subcellular structure. During the first half of the nineteenth century a large effort was made to improve the quality of the lenses and construction of the microscope, and by the end of the nineteenth century microscope manufacture and tissue preparation had reached a level comparable to that of present-day efforts. By the late 1930s, the theoretical observations concerning the nature of the electron had already been put to practical application, and the development of the electron microscope in Europe and the United States was underway.

The development of the electron microscope has been a remarkable technological accomplishment that has provided biologists with a view of the cellular world totally beyond the reach of the light microscope (Fig. 2-14). Before the development of the electron microscope, information on cell structure was lacking relative to information on cell function. With the widespread use of this instrument, this situation has been reversed and we are presently in possession of a detailed picture of subcellular anatomy and a more limited understanding of cell physiology. A knowledge of cell structure is important in its own right, but the interrelationship between cell structure and cell function is so important that it is essential if we are to understand how the parts work.

We are all accustomed to the sight of the double-layered nuclear membrane with its periodically spaced pores, or the sausage-shaped mitochondrion with its internal cristae, or the cytoplasm packed with bipartite ribosomes, or the many other features common to various cell types. In the world of the cytologist of only a few decades ago, the structures within a cell were either unknown or recognized at a poorly defined level (Fig.

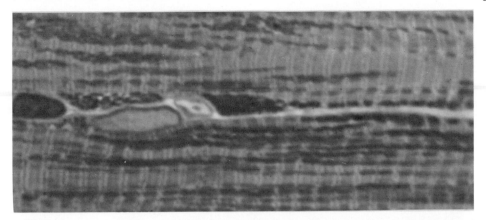

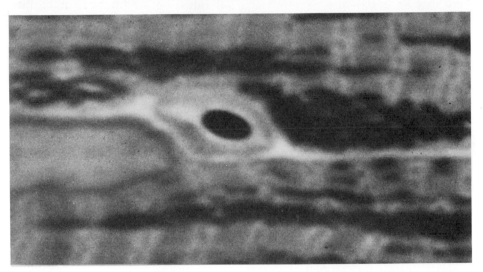

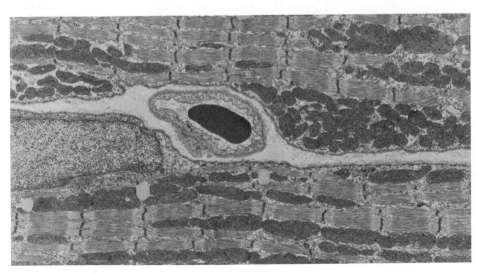

Figure 2-14. Photographs of three adjacent sections of skeletal muscle embedded in plastic and stained for light microscopy (top two figures) or transmission electron microscopy. In the top figure, a 2-μm-thick section has been photographed by light microscopy using a high dry objective lens and ideal illumination. Details of two muscle fibers and interposed capillary are visible. Muscle fibrillar striations are just discernible ($\times$2100). In the middle figure, the next adjacent section has been cut at 1 μm and photographed under oil immersion and higher magnification in an effort to achieve better resolution. In fact, while magnification and contrast are greater, there has been no gain (and perhaps some loss) in resolution of fine detail. This is an example of "empty magnification" ($\times$6500). In the bottom figure, the next adjacent section, cut at about 1/40 μm in thickness, has been prepared for electron microscopy. The resulting image reflects a one- to two-hundred-fold increase in resolving power at the same magnification. Note particularly details of muscle myofibrils, mitochondria, and the red blood cell contained by the capillary ($\times$6500). The topics of magnification and resolution are discussed at length on pages 42–43. *(Courtesy of D. E. Kelly and M. A. Cahill.)*

2-14). These statements are not meant to imply that the electron microscope has totally replaced the light microscope, making it obsolete. Each has its advantages. The light microscope has obvious practical advantages: it is inexpensive by comparison, it takes much less maintenance, it takes much less experience to operate and time to set up. In addition, there are many more staining procedures available for the light microscope which means that there is a greater opportunity to cause a particular type of material to stand out and be localized. There are more quantitative procedures with the light microscope that allow the amount of a particular substance to be measured. More importantly, there is a level of organization that is better suited for light microscopy than electron microscopy. Sections prepared for the light microscope are typically several cell layers thick and include a much larger piece of tissue, large enough, for example, to show an entire cross section of a rat intestine, or trachea, or kidney. Consequently, the observer, at a single glance, can see the interrelationships of the many types of cells of the various types of tissues that make up an entire organ. In the electron microscope, this overall picture must be pieced together from a great number of smaller images, making it much more difficult to obtain. Equally as important, the light microscope can be used to watch the dynamic state of the living cell, and time-lapse motion pictures can be taken to better appreciate cellular activities. The environment that exists within the column of the electron microscope is not suitable for living specimens due to the vacuum required for electron beam propagation.

The resolution obtainable with the electron microscope places it in a very important position among available techniques. Information about molecular activities is primarily received from biochemical studies, which are conducted on materials below the resolution of the electron microscope. When large macromolecules begin to interact with one another, complexes are formed which begin to be visible in the electron microscope. From this level up to that where many cells are interacting to form tissues, the electron microscope, with its tremendous range in magnification, can provide invaluable information.

With the light microscope many of the cell's organelles were unresolvable or poorly revealed, whereas with the electron microscope the detailed structure, i.e., the *ultrastructure* or *fine structure,* of the cell can be seen (Fig. 2-14). Consequently studies have revealed the fine structure of a tremendous variety of cells from a large number of plants and animals. But beyond this, the electron microscope has been one of the most powerful tools for understanding how cells function. For example, during secretion, membranous structures fuse with one another; during contraction, protein filaments slide across one another; during mitosis, microtubules appear by polymerization. These are structures clearly seen in electron micrographs. The problem in the analysis of their function is that any given thin section provides a view of a cell that has been frozen in time. There is no way to watch the process unfold, so at best we must find examples that show the process at its various stages and fit the pieces together. There are a number of techniques that increase the analytic powers of the microscope. These include the use of radioactive isotopes and giant molecules, both of which can be made visible in the electron microscope (see pages 224 and 315 for examples).

In another capacity the electron microscope is often used as a tool to monitor the progress of a biochemical analysis. For example, if the proteins of the plasma membrane are being studied, one of the first steps is to obtain a preparation of purified plasma membranes. An examination of the preparation in the electron microscope will reveal how pure it is, i.e., to what degree it is contaminated by structures other than plasma membranes. This is very important in studies coupled to the biochemical analysis of organelle function. Taking advantage of its maximum resolving power, the electron microscope has been very important in examining preparations of isolated macromolecules. Questions, for example, that concern the shape of various RNA molecules, or the lengths of DNA molecules present in chromosomes, or the structure of multienzyme complexes have been tackled with the electron microscope. A detailed discussion of the theory and techniques in light and electron microscopy can be found in the appendix to this chapter.

APPENDIX ON MICROSCOPY

Light

By the end of the nineteenth century the concept of the nature of light as a wave phenomenon was firmly established. Wave phenomena are those which oscillate in a repeating manner with respect to distance and time, whether they are waves formed on the surface of a lake, waves moving along a plucked guitar string, sound waves, or electromagnetic waves traveling at approximately 186,000 miles/second through empty space. In the case of electromagnetic radiation, a plot of the intensity of the associated electromagentic fields versus distance generates a sine wave as shown in Fig. 2-15. The distance between corresponding points, such as the crests, on two successive waves is termed the *wavelength*. The number of waves that pass by per second is termed the *frequency*. The greater the wavelength, the less will be the frequency. The wavelength of electromagnetic radiation can theoretically vary from zero to infinity. Wavelengths between 4000 Å and 7000 Å represent the visible light spectrum, i.e., that range to which our photoreceptors have evolved a sensitivity. This is clearly a subjective definition; without our presence, there would be no light, only electromagnetic radiation. Similarly, without our presence, there would be no color, only electromagnetic radiation of specified wavelengths.

At the beginning of the century, observations were made that could not be explained on the basis of classical principles of physics. Proposals were made by Planck and Einstein that altered the basic concept of matter and energy. One of these proposals, the "wave-particle duality," suggests that phenomena classically described as waves, such as light, can be shown to possess particulate properties and that particles, such as electrons, can be shown to behave as waves under certain conditions of measurement. The particulate, or quantal, nature of electromagnetic radiation is best illustrated by considering the energy content of light radiation and the manner in which light interacts with matter. This will be described when we consider the absorption of light energy by chloroplasts during photosynthesis in Chapter 9.

Figure 2-15. The wave form of electromagnetic radiation.

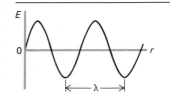

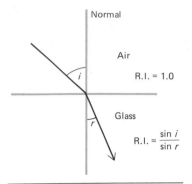

Figure 2-16. The refraction of a ray of light as it passes into a medium of greater refractive index.

The wave nature of electrons will be considered later in this section when the operation of the electron microscope is discussed. For the present we are concerned only with the wave properties of light, which are sufficient to explain the principles behind the operation of the light microscope.

The Light Microscope

It was mentioned in the preceding paragraphs that electromagnetic radiation travels through empty space with a velocity of approximately 186,000 miles/second. If, however, such radiation enters a material substance, its velocity is decreased, the degree of decrement being related to the nature of the substance and the frequency of the radiation. The ratio of the velocity of light in a vacuum to that in a given substance is the refractive index (n) of that substance. The velocity of light in low-density gases, such as air, is very little changed from that in a vacuum, but the velocity in materials such as glass can be greatly reduced. This property is the basis of the optical function of the light microscope. When a ray of light obliquely strikes a boundary surface between two transparent media of different refractive indexes, the direction taken by the ray through the second medium is altered (Fig. 2-16); i.e., it is refracted. The degree of refraction depends upon the angle formed by the ray and the boundary, known as the angle of incidence. According to Snell's law:

$$\frac{\text{Sine of angle of incidence}}{\text{Sine of angle of refraction}} = \frac{\text{velocity of light in first medium}}{\text{velocity of light in second medium}}$$

The angle of refraction for a ray passing into a medium of greater refractive index (lesser velocity), e.g., air to glass, is always less than the angle of incidence. In other words the ray is bent toward the normal (Fig. 2-16).

The light microscope utilizes the refraction of light rays by a system of lenses to accomplish the magnification of the specimen (object) by forming an image of the object that is larger than the object itself. A simple convex lens having two curved surfaces where refraction can occur is shown in Fig. 2-17. If the surfaces of this lens are ground so as to give the desired curva-

Figure 2-17. The formation of an image by a simple lens. Information concerning these diagrams is presented in the text.

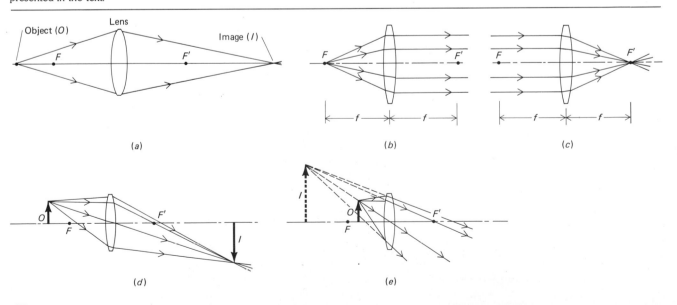

ture, the light rays impinging on the lens from one side can be caused to converge on the opposite side and, thereby, brought to focus (Fig. 2-17a). The points F and F' are the two *focal points* of the lens which serve as important landmarks along the lens axis. If an object point is placed at F of Fig. 2-17b, no image is formed by the lens, but rather the light rays leaving the lens remain parallel to one another. Conversely, if parallel rays of light impinge on this lens from the left side, they will be brought to focus at the other focal point, F' (Fig. 2-17c). The other cases shown in Fig. 2-17 illustrate that the position and size of the image formed is dependent upon the position of the object (an arrow) relative to the first focal point of the lens. In Fig. 2-17d the object is placed just beyond F and an enlarged real image (though inverted) is produced. If the object is placed within the focal point of the lens, an entirely different result is obtained: an enlarged, virtual image (not inverted) is produced on the same side of the lens as the object (Fig. 2-17e). The difference between a real and virtual image is that the former represents an actual location in space of the object's plane of focus and this can be visualized if an appropriate screen is placed in that plane. The image seen on a movie screen is a real image produced by the lenses of the projector. An image in a mirror is a virtual image. It appears to exist at a distance behind the mirror, though no such plane of focus is present at that position. Both types of images can be used as objects for a second lens to form a second image as will be seen shortly when the compound microscope is considered.

The diagram of the light microscope shown in Fig. 2-18 indicates the most important components. A light source, which may be external to the micro-

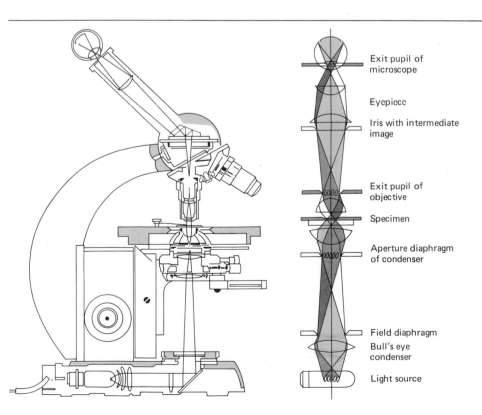

Figure 2-18. Sectional diagram through a compound microscope. Path of the light rays shown is that obtained by Köhler illumination. *(After R. Schenk and G. Kistler,* "Mikrophotographie," *Karger, Basel, 1960.)*

Exit pupil of microscope

Eyepiece

Iris with intermediate image

Exit pupil of objective

Specimen

Aperture diaphragm of condenser

Field diaphragm
Bull's eye condenser

Light source

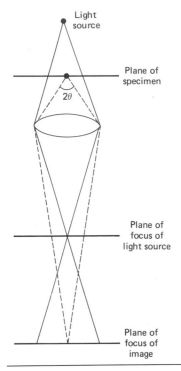

Plane of
focus of
image

Figure 2-19. Illustration of the different paths taken by light rays forming the image of the specimen (dotted line) and the background light of the field (solid line).

Figure 2-20. Diagram illustrating image formation in a compound microscope. The path of two light rays (labeled 1 and 2) from a point Q on the object is followed to the eye. *(From R. Barer, "Lecture Notes on the Use of the Microscope," 3d ed., F. A. Davis, 1968.)*

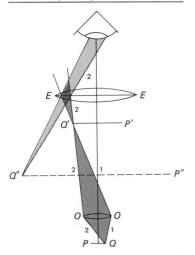

scope or built into its base, is needed to provide illumination for the specimen. The substage condenser is required to gather in the diffuse light rays from the source and illuminate the specimen with a small cone of light with sufficient intensity to allow very small parts of the specimen to be seen after magnification. The manner in which the specimen is illuminated can be of critical importance in determining the quality of the image obtained. The objective lens is responsible for collecting light rays.

From this point we need to consider two sets of light rays that enter the objective lens, those which the specimen has altered and those which it has not (Fig. 2-19). The latter group constitutes a cone of light from the condenser lens that passes directly into the objective lens and forms the background light of the visual field. The former group of light rays are ones that emanate, in a sense, from the many points of which the object is composed. Light rays from the object are brought to focus by the objective lens to form an image of the object within the column of the microscope. The construction of the microscope is such that the object is located just beyond the first focal point of the objective lens (Fig. 2-20), so that the image formed is real and enlarged. In actuality an objective lens consists of a series of lenses constructed to produce a single lens system. The image formed by the objective lens is made to fall just inside the focal point of a second lens (Fig. 2-20), the ocular, which then uses this image as an object for the formation of an enlarged virtual image. A third lens system, that of the front part of the eye, then uses the virtual image produced by the ocular as an object and a real image is produced upon the retina. Focusing of the light microscope changes the relative distances between the specimen and the objective lens so that the final image comes to focus on the retina of the eye. The overall magnification attained is the product of the magnification produced by the objective lens and that produced by the ocular lens. As a consequence of using a light microscope, the area of the retina that is covered by an image of the specimen is greatly enlarged over that obtained by examining the specimen with the naked eye.

Resolution

Up to this point we have considered only magnification of an object using the refractive properties of lenses; we have paid no attention to the quality of the image produced, i.e., the extent to which the detail of the specimen is retained. Suppose you are looking at a structure in the microscope using a relatively high-power objective (for example, 63×) and an ocular which magnifies the image of the objective lens another five-fold (a 5× ocular). Suppose the field is composed of chromosomes and it is important to determine the number present but some of them are very close together and cannot be distinguished as separate structures (Fig. 2-21a). The apparent problem seems like one of magnification and one solution might be to change oculars to increase the size of the object being viewed. If you were to switch to a 10× ocular in this case, you would most likely increase your ability to determine the number of chromosomes present (Fig. 2-21b), the reason being that you have now spread the image of the field produced by the objective lens over a greater part of your retina. The picture our eyes provide us is one made up of the information sent to the brain from a finite number (thousands) of tiny photoreceptors. The more receptors that are included to provide information concerning the image, the more detail present in the image will be seen (Fig. 2-22). If, however, you switch to a 20× ocular, you are not likely to see any more detail, though what you had previously observed will be larger (Fig. 2-21c), i.e., will occupy more retinal surface. The reason for the lack of additional discrimination after this second switch is that the image produced by the objective lens upon which the ocular is acting does not possess any further detail to be seen. Increasing the image size beyond the point at which the retina is given more information is simply *empty magnification* (illustrated in Fig. 2-14). This problem is not restricted to microscope use but is present in essentially all visual processes. A television screen (or more elegantly a Seurat painting) illustrates a similar principle. Both contain images that are composed of a finite number of dots. If you start far enough away and begin to walk toward the picture, more and more detail is seen until a point is soon reached where a deterioration in the quality of the overall picture results as our eyes are able to discern the composite nature of the picture by seeing empty space around each structural element. The point at which the individual components are seen as such depends upon the size and density of the photoreceptors of the eye. If the observer had been a falcon, for example, whose visual acuity is much greater than our own due to its having a much greater receptor density in its retina, this point would have been reached at a greater distance from the picture.

The optical quality of a given objective lens is reflected in the extent to which the fine detail present in the specimen can be discriminated, or resolved. The term *resolution* is stated most simply as a measure of the ability to see two neighboring points as distinct entities. If two distinct parts of an object are not separated by a sufficient distance (as in Fig. 2-22), they will be seen as one structure, i.e., they will not be resolved. The factors that determine what a sufficient distance might be are numerous; some limiting factors are inherent in all lens systems whereas others can be corrected for in better-quality microscopes. The question to be considered is why there should be any loss of information in the image as compared to the object.

Loss of resolution by an optical system is a theoretical subject having several different aspects to consider. Ultimately we need to explain why, when we look at a specimen through even the finest microscope, there is a limit to the detail that can be seen. The easiest way to visualize the problem is to consider an object under observation as being made up of an infinite number of points and to consider what happens when one point of light is examined. It can be shown that the light emitted from a point source on a slide can never be brought back to focus as a point in the image, but rather at best as a small disk. This lack of fidelity in the formation of the image is not a defect of the lens but rather a result of diffraction, a wave property of light. Without considering the details (available in any physics book), diffraction can be thought of as an edge effect and is best illustrated (as was first done by Young in 1800) by a beam of light passing through a fine slit onto a screen behind (Fig. 2-23). The result that you might expect from this experiment would be a bright bar of light (Fig. 2-23a) appearing on the screen covering an area equal to that of the slit through which the light has passed. However, the result is the appearance on the screen of alternating light and dark fringes

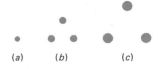

Figure 2-21. Magnification versus resolution. The transition from *(a)* to *(b)* provides the viewer with increased magnification and resolution, while the transition from *(b)* to *(c)* provides only increased magnification (empty magnification).

Figure 2-22. A highly schematic illustration of the relationship between the stimulation of individual photoreceptors and the resulting scene one would perceive. The diagram illustrates the value of having the image fall over a sufficient area of the retina.

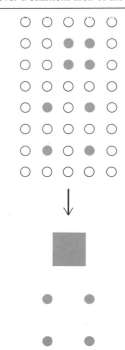

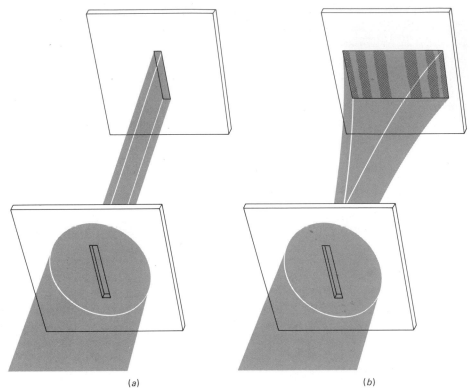

Figure 2-23. *(a)* Geometrical "shadow" of a slit. *(b)* Diffraction pattern of a slit. The slit width has been greatly exaggerated. *(From F. W. Sears and M. W. Zemansky, "College Physics," 3d ed., © 1960, Addison-Wesley, Reading, Massachusetts. P. 918. Reprinted with permission.)*

(a)   (b)

(Fig. 2-23*b*) extending into what would be expected to be the dark shadow behind the rim of the slit. It is as if light were being "bent" outward as it passes through the opening. If the opening is a hole rather than a slit, the screen is illuminated by a bright disk surrounded by concentric light and dark rings which diminish in intensity as they extend from the center. The larger the hole, the closer the rings are to the center, i.e., the smaller the diffraction pattern that is formed.

The image of a point light source seen through the microscope is the very same type of fringed diffraction disk (termed an *Airy disk*) (Fig. 2-24) and can be considered as a consequence of light passing through the aperture (opening) of the lens, rather than a slit or hole. As in the above case, *the larger the aperture of the lens, the smaller are the diffraction disks*. The size of the disks formed from a point source of light can be considered a direct measure of the resolving power of the lens. The smaller the disk, the more the image approaches a true point, as was originally the case for the object. Similarly, the smaller the diffraction disks, the closer the points in the object can be together and still be seen as two points in the image (Fig. 2-25). If the diffraction rings of two adjacent points overlap to a sufficient degree, these points will not be resolved.

The diffraction of light by the lens, which limits resolution, is one aspect of this topic, the diffraction of light by the object is another. Consider illuminating a tiny spherical object, such as a diatom, present on a slide. Light passing through this pointlike object will be diffracted forming rings of light

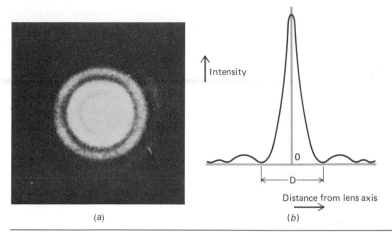

Intensity

0

D

Distance from lens axis

(a)

(b)

Figure 2-24. *(a)* Photograph of the diffraction pattern (termed an *Airy disk*) of a point source of light. *(From A. C. Hardy and F. H. Perrin, "Principles of Optics," McGraw-Hill, 1932.)* *(b)* The intensity of light in the Airy disk at increasing distance from the lens axis. The areas under the curves indicate the relative amounts of light in the central disk and the surrounding rings.

which travel away from the object at increasing angles (Fig. 2-26). All of the light emerging from the object is important because all of it contains information about the nature of the object. The extent to which such rays are not intercepted by the objective lens is a measure of the information that is lost to the lens in forming the image. Part of the problem of loss of resolution is that inevitably only a fraction of the light that leaves an object can be taken

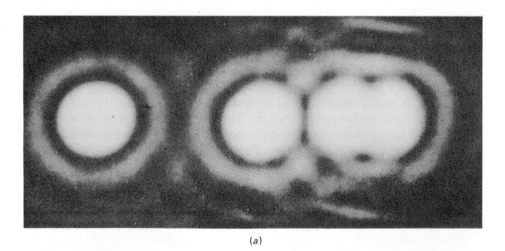

(a)

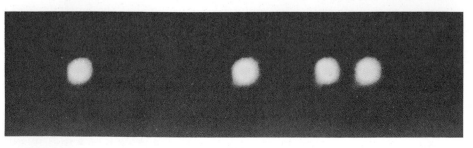

(b)

Figure 2-25. Diffraction patterns of four "point" sources, with a circular opening in front of the lens. In *(a)* the opening is so small that the patterns at the right are just resolved. Increasing the aperture *(b)* decreases the size of the diffraction patterns and increases the resolution. *(From F. W. Sears and M. W. Zemansky, "College Physics," 3d ed., © 1960, Addison-Wesley, Reading, Massachusetts. P. 930. Reprinted with permission.)*

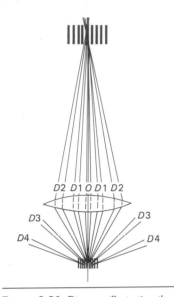

Figure 2-26. Diagram illustrating the loss of diffracted light by an objective lens. The larger the lens aperture, the more light collected. *(From W. G. Hartley, "How to Use the Microscope," Natural History Press, 1964.)*

Figure 2-27. The role of oil immersion in increasing resolution. *(From R. Barer, "Lecture Notes on the Use of the Microscope," 3d ed., F. A. Davis, 1968.)*

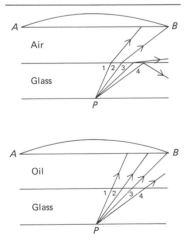

in by the objective lens. However, the greater the aperture of the objective lens, i.e., the greater its diameter, the greater will be the percentage of the light from the object that is collected.

There are two other factors that determine the percentage of the light from the object that is gathered in by the objective lens, and thus the resolution of the lens. These are the refractive index of the medium between the object and the lens, and the wavelength of the light. The relationship between refractive index and resolution is shown in Fig. 2-27. The refractive index determines the direction that light rays coming from the object will take. If there is air between the object and the lens, then rays passing from the glass coverslip into the air will be entering a medium of lesser refractive index (as compared to glass) and will be bent outward, away from being captured by the lens. If immersion oil is present between the coverslip and the objective lens, the path of light rays leaving the coverslip is unchanged because the refractive index of immersion oil is equal to that of glass (1.5). As a result, more information from the object can be utilized in the formation of the resulting image.

The various optical properties of light are manifestations of its wave phenomena and are, therefore, affected by its wavelength. In the case of diffraction, the longer the wavelength the more the deflection and the larger the diffraction disks. Under normal circumstances objects to be viewed under the microscope are illuminated with white light, a mixture of all wavelengths. By the use of special illuminating sources, or more easily by placing filters in the light path, shorter wavelength light (toward the violet end of the spectrum) can be employed and resolution improved. Most importantly, in this respect, it can be seen that *the resolution attainable by a microscope is limited by the wavelength of the illumination.*

These various factors are combined in the following relationship:

$$D = \frac{0.61\,\lambda}{n\,\sin\theta}$$

In this equation D is the diameter of the diffraction disk (measured from the center of the disk to the first dark ring) formed as an image of a point source. The smaller the value for D, the greater the resolution. The wavelength of the light is λ (527 nm is used for white light), and n is the refractive index of the medium. Theta is the angle shown in Fig. 2-19, which is a measure of the light-gathering ability of the lens and is directly related to its aperture. The denominator is usually termed the *numerical aperture* or *N.A.* The numerical aperture is a constant for each lens, a measure of its light-gathering qualities. For an objective that is designed for use in air, the maximum possible N.A. is 1, since the sine of the maximum angle of theta possible, 90°, is 1 and the refractive index of air is 1. For an objective designed to be immersed in oil, the maximum N.A. is approximately 1.5. A common rule of thumb is that a useful magnification for a microscope is approximately 1000 times the numerical aperture of the objective lens being used. An oil immersion lens with a numerical aperture of 1.4 can be usefully employed to produce images of overall magnification of about 1400. Beyond this point empty magnification occurs and image quality deteriorates. High numerical aperture is achieved by using lenses with short focal lengths and placing the lens and the specimen very close to one another. One drawback of lenses of high numerical aperture is that only a very thin section can be in focus at one time.

If we substitute the minimum possible wavelength and the greatest possible numerical aperture in the above equation, we can determine the *limit of resolution* of the light microscope. The value of slightly less than 0.2 μm is obtained. Another way to express this is in terms of theoretical resolving power. Approximately 125,000 geometrical lines (infinitely thin) per inch could theoretically be distinguished. The limit of resolution of the naked eye, which has a numerical aperture of about 0.004, is approximately 0.0036 inch, or 0.1 mm. The use of a light microscope can increase resolution at least 500 times.

Up to this point in the discussion, factors limiting resolution in the light microscope have been theoretical ones common to all lenses. In addition, a number of aberrations can exist in a lens which can markedly affect its resolving power. There are seven important aberrations and they serve as the handicaps for lensmakers to overcome in producing objective lenses whose actual resolving power approaches that of the theoretical limits. The reason that objective lenses are made of a complex series of lenses rather than a single converging type is to eliminate these aberrations. Typically, one lens unit affords the magnification while the others compensate for errors in the first lens to provide an overall corrected image.

We will only briefly consider two types, spherical aberration and chromatic aberration, the others being less important. In both cases, the image deteriorates because not all light rays are coming to focus in one image plane. Consider once again the formation of an image of a point light source. With a lens having spherical aberration, those rays that pass through the central part are brought to focus at a different distance from the lens than rays passing through the lens periphery. In other words, the lens has more than one focal point and the result is a blurred image. With a lens having chromatic aberration, light of different wavelengths is brought to focus at different distances, the violet end of the spectrum focusing closest to the lens. The resulting image shows shifting color patterns as the microscope focus is adjusted. The basis for this aberration is that the shorter the wavelength, the greater the deviation of its path by refraction. This is the basis for the dispersion of white light into the color spectrum by a prism or the formation of a rainbow. Since image formation by converging lenses is a result of refraction, use of an uncorrected lens will cause a dispersion of the various wavelengths. Identification of lenses as achromatic or, even beter, apochromatic indicates they are corrected to a large degree for chromatic aberration.

Visibility

On the more practical side of microscopy from that of resolution is the topic of visibility, which is concerned with factors that allow an object actually to be observed. This may seem like a trivial matter; if an object is there it should be capable of being seen. Consider the case of a glass bead. Under most conditions, i.e., against most backgrounds, it is clearly visible. If, however, such a bead is dropped into a beaker of immersion oil having the same refractive index, it disappears from view; it no longer affects the light in any obvious manner that is different from the background fluid. Anyone who has spent any time searching for an amoeba can appreciate the problem of visibility when using the light microscope. What we see, through a window or through a microscope, are those objects which affect the light differently from their background. Another term for visibility in this sense of the word is *contrast;* the contrast between adjacent parts of an object or an object and its background. To gain visibility in the macroscopic world we examine objects by having the light fall on them, i.e., incident light, and we observe the light that is reflected back to our eyes. In microscopy we place the object between the light and our eyes and view the light that is transmitted through the object. If you take an object and go into a room with one light source in it and hold the object between the light source and your eye, you can appreciate part of the difficulty in such illumination; it requires that the object being examined be nearly transparent, i.e., translucent. Therein lies another aspect of the problem since objects that are "nearly transparent" can be difficult to see. A microscope, however, is well suited for observing alterations of light caused by very thin objects and various special techniques of illumination have been devised to make such objects even more obvious. Some of these are discussed below.

An object becomes visible by affecting the light in several ways, most importantly by diffraction and absorption. We have already considered the scattering of light by diffraction, the most important property for seeing an unstained object. There are two ways in which the abosorption of light enhances contrast. Consider the wing of a fly as a specimen. Those parts of the wing that are thicker will absorb more light than the thinner parts, which will appear brighter. As a result of the difference in intensity of the transmitted light from the parts of the wing, its image is seen. In a different way, contrast is achieved because different objects *selectively* absorb light of different wavelengths, i.e., they appear colored. In a previous example, a glass bead was dropped into immersion oil thereby becoming "invisible." If that bead had been colored, it would have remained in view. A microscopic object that appears colored is transmitting those wavelengths that are not being absorbed. If white light is being used for illumination, what we are seeing represents light of all wavelengths minus those being removed by the object. If an object is blue-green in appearance, for example, it indicates that light at the red end of the spectrum is being absorbed selectively.

Generally, biological materials are relatively lacking in color, especially when visibility at the subcellular level is being considered. To make up for this, a great variety of dyes have become available which are suited for staining cells and tissues. Different dyes bind to different types of molecules and, therefore, not only do these procedures increase contrast between the parts of the specimen, they can also indicate where in the cells or tissues different types of substance are found. A good example of this latter case is the Feulgen stain, which is specific for DNA, causing it to appear pink under the microscope. Since the cell nucleus is the repository for DNA, it is that structure which becomes visible using the Fuelgen stain. If the DNA is removed (digested away with an enzyme) prior to the staining procedure, no such contrast is achieved.

One of the main problems with stains is that they generally cannot be used with living cells; they are toxic, or the staining conditions are toxic, or they do not penetrate. The Feulgen stain, for example, requires that the tissue be hydrolyzed in acid before the stain is applied. There are a few stains that are suitable for use with living cells; these are called *vital dyes*. A commonly performed experiment in introductory biology labs involves feeding *Paramecia* yeast particles that are stained with Congo red. The yeast particles are taken up by these ciliated protozoans and become packaged in food vacuoles. The stain becomes a marker for these membrane-bound vacuoles. If the fate of these stained particles is followed, they eventually change color from red to blue in response to the changing pH of the food vacuole that accompanies digestion. In this case the dye is a pH indicator and can be used to follow a physiological process.

The use of a microscope in which light from the illuminating source is caused to converge on the specimen by the substage condenser thereby forming a cone of bright light that can enter the objective lens is termed *bright-field microscopy*. This main beam of light is seen as a bright background against which the image of the specimen must be contrasted. It is ideally suited for specimens of high contrast such as stained sections of tissues, but it may not provide optimal visibility for other specimens. One means of increasing contrast with bright-field illumination is to close the diaphragm of the substage condenser which, in effect, greatly decreases the background light relative to the diffracted light which contains information about the object. As a result, an image of the object stands out against a darkened background. However, a marked loss of resolution occurs when this is done and a distorted image, as if viewing a ghost, results if the aperture is excessively closed. In the following section various alternate means of microscopy are considered that have become employed commonly for special purposes.

Phase-Contrast Microscopy

As described above, bright-field microscopy relies heavily upon the absorption of light by the object. Small, unstained material, such as a living cell, absorbs very little light and can be difficult to see (Fig. 2-28*a*). The development of the phase-contrast microscope has provided an ingenious technique to get around certain of the difficulties and make highly transparent objects more visible (Fig. 2-28*b*). The ability to see different parts of an object depends upon their capacity to affect light in different manners. One basis upon which intracellular organelles differ is their refractive index. Cell organelles are made up of different proportions of various molecules: DNA, RNA, protein, lipid, carbohydrate, salts, and water. Regions of different composition are likely to have different refractive indexes. Normally, however, such differences cannot be detected by our eyes. The differences that can be seen when viewing a specimen through a microscope are differences in intensity, i.e., relative brightness and darkness. The phase-contrast microscope converts refractive index differences into intensity differences, which are then visible to the eye. The basis for this "conversion" centers on the ability of light waves to interact with one another, a property termed *interference*.

The outcome of the interaction of two waves of light depends upon the phase of the cycle of the waves when they meet. There are two extremes that can occur. If the two waves meet in exactly the same phase they will

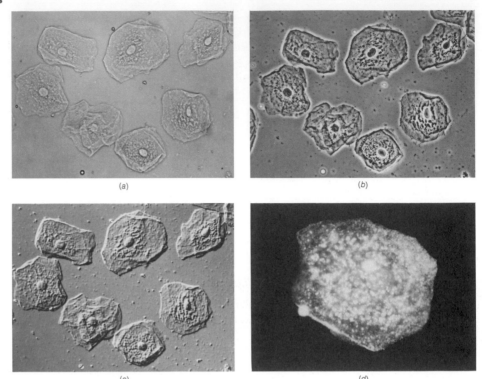

(a)

(b)

(c)

(d)

Figure 2-28. Light photomicrographs of oral epithelial cells (cells scraped from the inner cheek surface) as observed under different types of illumination: *(a)* Brightfield; *(b)* phase-contrast; *(c)* Nomarski interference; *(d)* dark-field. *(a–c, courtesy of Carl Zeiss, Inc.)*

interfere in a completely constructive manner, i.e., they will be additive (Fig. 2-29a) and the wave that is formed will have an amplitude (displacement) that is greater than that of either of the two component waves. If the two waves reach the point such that they are one-half wavelength out of phase, the interference will be totally destructive; i.e., the two will cancel each other out (Fig. 2-29b) and no waves will be propagated from the common point. Any other combination of waves will produce intermediate results. The interaction of light waves results in interference in an analogous manner. The diffraction pattern shown in Fig. 2-24 is in fact the result of interference. The light fringes are locations where a maximum addition has occurred and the dark bands where a complete canceling out has occurred.

The construction of the phase-contrast microscope is such that it accomplishes two tasks that the bright-field microscope does not: it separates the direct light (the background light of the field) from the light diffracted by the object, and it causes these two types of waves to be approximately one-half

Figure 2-29. *(a)* Constructive interference. The two waves are in phase and they combine to produce a wave having twice the amplitude. *(b)* Destructive interference. The two waves are one-half wavelength out of phase and they combine to cancel each other out.

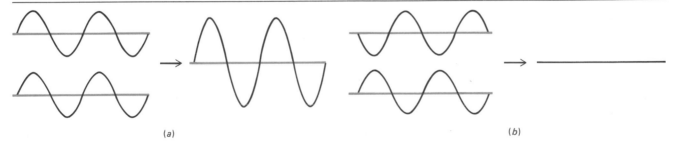

(a)

(b)

wavelength out of phase with one another so that they can destructively interact and cause changes in intensity. The separation of light is accomplished in the following way. A disk is placed in the substage condenser that is solid except for the presence of a thin ring capable of allowing light to pass through (Fig. 2-30*a*). As a result of this annulus, the condenser produces a *hollow* cone of light (a ring of light) to illuminate the field and enter the objective lens.

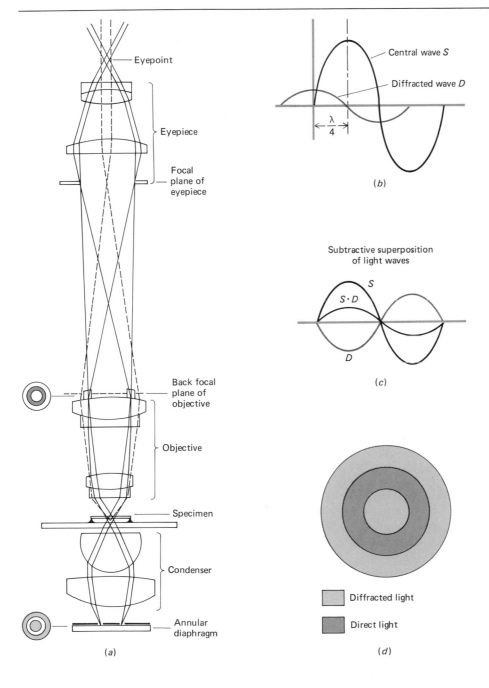

Eyepoint

Eyepiece

Focal plane of eyepiece

Back focal plane of objective

Objective

Specimen

Condenser

Annular diaphragm

(*a*)

Central wave *S*

Diffracted wave *D*

$\frac{\lambda}{4}$

(*b*)

Subtractive superposition of light waves

S

S·D

D

(*c*)

Diffracted light

Direct light

(*d*)

Figure 2-30. (*a*) The path of light through a phase-contrast microscope. The dotted lines represent light diffracted by the specimen, the solid lines represent the direct light. (*b*) The one-quarter wavelength difference in phase that normally occurs between diffracted light and direct light. (*c*) Destructive interference between the two waves when the direct light is retarded an additional one-quarter wavelength. (*d*) The distribution of the direct and diffracted light as it passes through the objective lens. [(*a*)–(*c*) *Courtesy of the American Optical Company.*]

Since the illuminating light is in the shape of a ring, it is a ring of direct light that enters the objective after passing through the slide. This can be seen clearly by removing the ocular and looking down the viewing tube at the back focal plane of the objective. Phase-contrast objective lenses are constructed such that they have a ring, termed a *phase plate*, at their back focal plane which corresponds geometrically with the position of the ring of direct light from the condenser (Fig. 2-30*a*). Since all of the direct light is in this ring, all of it must pass through the phase plate. In contrast, the light which is diffracted by the parts of the object is scattered from the points of diffraction across the entire surface of the lens (Fig. 2-30*d*). In consequence, these two types of light have been separated to a large extent. The only area of overlap is the diffracted light which fortuitously goes through the ring and becomes lost for use in interference.

The interference between the diffracted light and the direct light is accomplished in the following way. When light is diffracted by a transparent object it is *retarded* one-fourth wavelength relative to the undiffracted, direct background light (Fig. 2-30*b*). What needs to be done to obtain destructive interference is to retard it an additional one-quarter wavelength relative to the direct light so that it can destructively interfere with it (Fig. 2-30*c*). This is accomplished by the phase plate present in the objective. This plate is actually a place in the lens where the glass is thinner and the *direct light* in passing through this decreased path of glass is *advanced* an additional one-quarter wavelength which places it one-half wavelength out of phase with the diffracted light.

Assume for a minute that all parts of the object being examined have exactly the same refractive index as the medium in which they are suspended. If this were the case, all of the diffracted light and direct light would be one-half wavelength out of phase with one another and interference would result in a uniform decrease in intensity. Now consider an object whose various parts have differing refractive indexes, all of which are greater than the surrounding medium. When a light ray crosses a surface between two transparent media of different refractive indexes, a change in velocity occurs. The effect on the wave profile of a light ray passing into and out of a medium of higher refractive index is shown in Fig. 2-31. A decrease in velocity results in an elongation of each cycle with respect to time, i.e., in a given amount of time there are fewer cycles. When the ray emerges from the medium of higher refractive index it will be in a different phase of its cycle than it would have been had it not entered the more refractive substance (Fig. 2-31); it will be behind a less refracted ray, i.e., retarded. The degree to which it is retarded depends on two factors; the refractive index of the substance and its thickness. The greater the refractive index, the greater the retardation. Similarly, the greater the thickness, the longer the ray will spend in the substance and the greater will be the retardation. Differences in a refractive index will result in shifts of phase of the light rays passing through different parts of the specimen and this will, in turn, cause a shift in phase of the diffracted and direct light relative to one another. The degree to which the various parts of the specimen shift the phase difference will be reflected in a change in the way the two types of light are added to one another and, thus, a shift in the amplitude of the light from that part of the specimen. For shifts in wavelength of less than one-quarter wavelength, the interference will still be destructive and the object will be darker than the background. In the case of shifts greater than this, e.g., those caused by very refractile objects (such as lipid droplets)

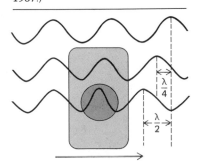

Figure 2-31. Retardations in phase in waves of light passing through a transparent living cell. The waves passing through the full thickness of the cytoplasm have been retarded by $\frac{1}{4}\lambda$, and those passing through the full thickness of a more highly refractile inclusion and some cytoplasm have been retarded by $\frac{1}{2}\lambda$. *(From K. F. A. Ross, "Phase Contrast and Interference Microscopy," St. Martin's Press, Inc., Macmillan & Co., Ltd., 1967.)*

and/or particularly thick ones, the object can actually appear brighter than the background due to constructive interference.

Phase-contrast optics for microscopes have become commonplace over the past decade or two and their cost is comparable to that for bright-field optics. Since the only special parts needed are the objectives and the condenser, ordinary microscopes can be equipped for phase contrast quite readily.

Phase-contrast optics have been most useful in the examination of intracellular components of living cells at relatively high resolution. In this capacity the dynamic motility of tiny structures, such as mitochondria, mitotic chromosomes, vacuoles, and so on, can be watched and the progress of a multitude of processes including cell division, cell migration, secretion, and cytoplasmic streaming can be followed in a manner not possible before the invention of interference-type microscopes. Simply watching the way the tiny particles and vacuoles of cells are bumped around in a random manner conveys an excitement about the living state totally unattainable from the observation of stained dead cells. The greatest benefit derived from the invention of the phase-contrast microscope has not been in the discovery of new structures, but in its every day use in research and teaching labs for observing cells in a more revealing way.

The greatest biological handicap of the phase-contrast microscope is that it is only suitable for observing single cells or thin cell layers. On the other hand, it has optical handicaps as well, which derive primarily from the lack of complete separation of the direct and diffracted rays [some diffracted rays go through the phase plate (Fig. 2-30d)]. This deficiency manifests itself in loss of resolution and various types of interfering halos and shading resulting from edges where sharp changes in refractive index occur. The phase-contrast microscope is a type of interference microscope. There are a number of other microscopes that minimize these optical artifacts and provide a truer representation, though often less dramatic, of the living cell. These various interference microscopes achieve a complete separation of direct and diffracted beams using complex light paths and prisms and are much more expensive and difficult to use. One type of interference optics, termed *Nomarski interference* after its developer, recently has become popular and more readily available. Nomarski interference provides an apparent three-dimensional quality to the image with the various parts of the object seemingly standing in relief (Fig. 2-28c). Contrast in the Nomarski system depends on the rate of change of refractive index across a specimen. As a consequence, the edges of structures, where the refractive index varies relatively greatly over a small distance, are seen with especially good contrast.

Another feature of the interference microscope is that it can be used as an instrument for measuring dry weight, i.e., the amount of material other than water present in a specimen. The basis for this capability is that refractive index is a very sensitive measure of density. The greater the weight per unit volume of a substance, the greater the decrease in velocity. Since phase retardation, which can be quantitatively measured by an interference microscope, is related to the refractive index differ-

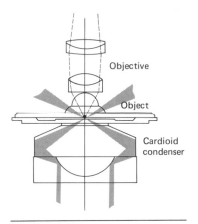

Figure 2-32. Illumination of the specimen in dark-field microscopy. The dotted line indicates diffracted light. *(Bausch and Lomb, Inc.)*

ences and thickness, the density can be determined. A technique such as this is most important because the measurement can be made on single microscopic specimens and can even provide information on the density of one part of the cell as opposed to another.

Dark-Field Microscopy

One of the difficulties with bright-field illumination is that the brightness of the background can make viewing the image of diffracted light difficult. In dark-field microscopy the condenser is constructed such that it produces a hollow cone of light in a manner analogous to that for phase contrast. The main difference is that the ring of light is emitted at a steeper angle (Fig. 2-32), one that causes it to strike the specimen but not to enter the objective lens. As a result, there is no direct light to produce the bright background but only the diffracted (or refracted) light that is deviated from its original path causing it to enter the objective lens to form the image (Fig. 2-28*d*), which appears bright against the dark background.

Fluorescence Microscopy

Fluorescence is a property of certain molecules that can be excited by radiation of invisible, short wavelengths and caused to emit radiation of longer, visible wavelengths, thereby appearing luminous. In its most widely used application, cells are stained with fluorescent dyes and the luminescence of the parts of the object are observed. Often these fluorescent molecules are chemically complexed to large macromolecules, such as antibodies. The fluorescent antibodies can then be allowed to interact with the cells in their specific manner and the location of the interaction between the antibody and the cell monitored by the location of the fluorescence. Examples are given on pages 229 and 754. As in the case of dark-field microscopy, the illuminating light is not seen.

Polarization Microscopy

The polarized light microscope serves a special function by enabling information to be gained about biological organization at a level below the limit of resolution for a light microscope. Such a function is possible because well aligned submicroscopic structures (or molecules) possess an optical property called *birefringence*. An object (microscopic or macroscopic) manifests its birefringence by appearing bright when placed between two pieces of polarizing material which are "crossed" (i.e., the two are positioned so that their polarizing properties are perpendicular). Sheets of Polaroid are generally used as the polarizing material. Light becomes plane polarized (light waves vibrating in one direction only) when passing through a single sheet of Polaroid. When two Polaroid sheets are crossed, light will not pass through the pair. A birefringent structure appears bright between crossed Polaroids because it alters plane polarized light. This alteration of light is a direct consequence of the oriented elements within the structure. The light is altered by the birefringent structure (but not rotated) in such a way that a component of it can pass through a second Polaroid. In contrast, a nonbirefringent object, i.e., one whose components have little preferred alignment, does not alter plane polarized light. Consequently, no light will pass through a second Polaroid and the object appears dark like the background.

A polarizing microscope is set up in its simplest form when one piece of Polaroid is positioned between the light source and condenser while another piece is inserted into the microscope tube, preferably above the back focal plane of the objective. Resolution is dictated by the N.A. of the objective, but the presence of cellular components of a size below this limit can be detected by a polarizing microscope provided that these components are sufficiently well aligned to be birefringent. A classic example of a birefringent cell structure is the mitotic spindle (see Fig. 15-7). The spindle's birefringence is derived from the many aligned microtubules of which it is composed. Although microtubules can only be resolved by an electron microscope (see Fig. 15-19), their presence in the spindle can be detected with a polarizing microscope. In contrast, chromosomes are not birefringent and, although easily resolvable at the light microscope level, are not nearly as distinct when viewed with a polarizing microscope as compared to an interference microscope (compare Figs. 15-11 and 15-7). Other cell structures which are birefringent include myosin filaments (A band) in striated muscle (Fig. 16-3), chloroplasts (owing to their stacks of internal membranes), and plant cell walls (oriented cellulose).

Preparation of Specimens for Light Microscopy

Specimens to be observed with the light microscope fit broadly into two categories, whole mounts and sections. Whole mounts consist of an intact subject, either living or dead, whether an entire microscopic organism such as a protozoan or a small part of a larger organism. As long as the object is sufficiently transparent it can be viewed with transmitted light. Often relatively opaque objects can be made translucent by removing the water by substitution with alcohol and immersing the object in solvents such as toluene or xylene in which they become clear.

Most tissues of plants and animals are much too opaque for microscopic analysis unless examined in the form of a very thin slice. The first step in the process is to kill the cells by immersing the tissue in a chemical solution, called a *fixative*. A good fixative is capable of rapidly penetrating into the cell and immobilizing all of its macromolecular material in such a way that the structure of the cell is maintained as closely as possible to that of the living state. It is of obvious importance, when one examines cell structure at the end of a procedure, to have confidence that what is being observed is a reflection of the true structure of the living cell rather than an artifact produced from the fixation process. The gross effect of a fixative on the appearance of a living cell can best be appreciated by watching the fixation process in the phase-contrast microscope. The most common fixatives for the light microscope are various types of formaldehyde solutions.

After fixation, the tissue is dehydrated by transfer through a series of alcohols and embedded in wax which provides mechanical support during sectioning. One of the great advantages of wax as an embedding medium is the ease with which it can be dissolved from the sections in various organic solvents. The slides are simply immersed in toluene to remove the wax, leaving the section of tissue attached to the slide and capable of being stained or treated in some manner (enzymatic, hydrolytic, etc.). Hundreds of different staining procedures have been developed over the years of light microscopy, certain of which might be best suited for one or a few types of cells and not others. A particular stain might be better because it provides the best visual contrast for a tissue or because it selectively stains one type of molecule within certain cells. After the staining procedure a coverslip is mounted over the tissue and, once the mounting medium (which has the same refractive index as the glass slide and coverslip) has dried, the preparation is ready to be examined.

Electron Microscopy

The advantage of the electron microscope over that of the light microscope is the vastly greater resolution the instrument can deliver. This capability derives from the wave properties of electrons. In the equation for the limit of resolution of a lens, $D = 0.61 \, \lambda/n \sin \Theta$, the importance of the wavelength is clearly indicated. In the case of an electron, its wavelength is not a constant distance but is related to the speed at which the particle is traveling, which is in turn dependent upon the accelerating voltage applied in the microscope. This relationship is defined by the equation $\lambda = \sqrt{150}/V$, where λ is the wavelength in angstroms and V the

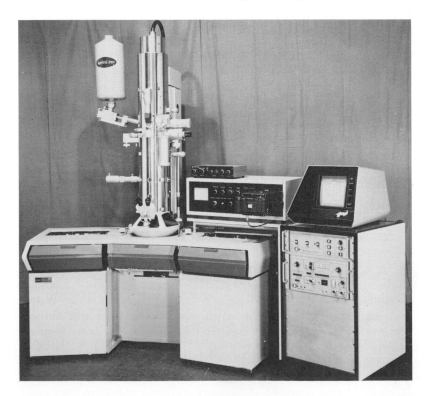

Figure 2-33. An electron microscope. *(Hitachi-Perkin-Elmer Corporation.)*

accelerating voltage (in volts). Standard electron microscopes operate with a voltage range from 10,000 to 100,000 volts. At 60,000 volts the wavelength is approximately 0.05 Å. If the numerical aperture obtainable with the light microscope were used in the equation for D, the limit of resolution would be about 0.03 Å. If one considers that the typical distance between the centers of atoms within molecules is on the order of 1 Å, the prospects become incredible. In actual fact, the resolution obtainable with the electron microscope is about two orders of magnitude less than its theoretical limit. The reason for this is the extreme problem of spherical aberration from which electron-focusing lenses suffer, requiring the numerical aperture of the lens to be made very small. The precise practical limit of resolution of present-day microscopes is debatable but is in the range of 2 to 5 Å. The actual limit when observing cellular structure is more typically in the range of 10 to 15 Å.

Electron microscopes consist essentially of a tall, hollow cylindrical column (Fig. 2-33), within which the electron beam is confined, and a console having panels of dials which electronically control the operation in the column. At the top of the column is the cathode, a tungsten wire filament which, when heated, acts as a source of electrons. Electrons are drawn from the hot filament and accelerated as a fine beam by the high voltage applied between it and the anode. Air is pumped out of the column prior to operation producing a vacuum through which the electrons travel. If the air were not removed, electrons would be prematurely scattered by collision with gas molecules and filaments would rapidly deteriorate. Electrons, being charged particles, are capable of being deviated in their paths and thus brought to focus when placed in a magentic field. There are several types of lenses in the electron microscope, all of which are powerful electromagnets located in the wall of the column surrounding the evacuated core. The strength of the magnets is controlled by the current provided them, which is determined by the positions of the various dials of the console. The condenser lenses are placed between the electron source and the specimen and are responsible for focusing the electron beam on the specimen. The specimen itself is supported on a small, thin metal grid (3-mm diameter) which is inserted with tweezers into a grid holder, which is, in turn, inserted into the middle region of the column of the microscope so that the grid is perpendicular to the electron beam.

A comparison between the lens systems of the light and electron microscopes is shown in Fig. 2-34. The objective lens, which is located below the level of the specimen focuses the electrons and is responsible for the formation of an image of the object. The objective lens is focused by changing its current which has the effect of altering the strength of its magnetic field and shifting its focal length. To overcome the spherical aberration of the objective lens, a very small aperture (approximately 50 μm in diameter) is usually inserted at the back focal plane of the objective lens. Only those electrons being focused by the center of the lens will be able to pass through the aperture to participate in forming the final stage. The numerical aperture of electron microscope objective lenses is generally between 0.01 and 0.001.

Since the focal lengths of the lenses of the electron microscope are varied by altering the current supplied to them, one lens is capable of providing the entire range of magnification rather than having to switch from lens to lens as with the light microscope. The image provided by the objective lens is magnified approximately 100 times, but unlike the light microscope, there is

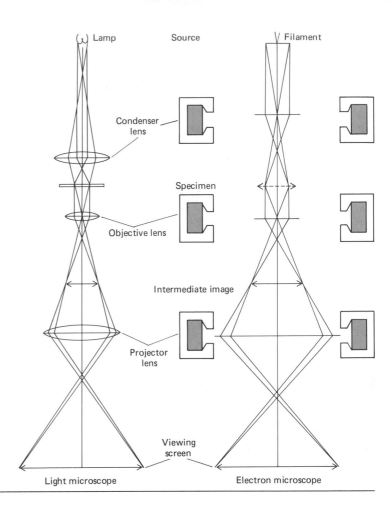

Figure 2-34. Comparison of the lens systems of a light and electron microscope. *(From A. W. Agar, "Principles and Practice of Electron Microscope Operation," Elsevier North-Holland, 1974.)*

sufficient detail present in this image for it to be magnified an additional 10,000 times to allow the eye to be able to appreciate its information. As in the case of the light microscope, the image from the objective lens serves as the object for an additional lens system. In the electron microscope there are usually two additional lenses (Fig. 2-34): the intermediate lens, which generally can enlarge the image another 20 times, and finally a projector lens, which is capable of an additional 100-fold increase. The degree of magnification desired is obtained by altering the current supplied to the intermediate and projector lenses. Together these lenses can change the magnification factor from 1000 times to 250,000 times. If the microscope is properly adjusted, the resolution is so great that even at 250,000 times there is more information in the image than can be seen by the eye. This detail is made visible by taking photographs and further enlarging the image. To obtain the very limit of resolution of present-day microscopes (≈ 3 Å), a magnification of 10^6 would be needed for the eye to resolve it. A viewing screen, coated by a fluorescent material, is placed at the bottom of the column to be bombarded by those electrons that have passed through the aperture behind the objective lens, and brought to focus on the screen. Electrons striking the screen excite the fluorescent crystals to emit their own visible light which is perceived by the eye as an image of the specimen.

The basis for the formation of an image in the electron microscope resides in the differential scattering of electrons by parts of the specimen. Consider a beam of electrons emitted by the filament and focused on the screen by the lenses. If no specimen were present the entire field would be of uniform brightness, being illuminated evenly by the beam of electrons striking the screen. When the specimen is placed in the path of the beam, a percentage of the electrons strikes atoms in the specimen and is scattered away from the specimen at some angle. Electrons that bounce off the specimen will not pass through the very small aperture at the back focal plane of the objective lens and will, therefore, be lost as participants in the formation of the image.

Contrast in the electron microscope is achieved by having some parts of the specimen serve as an obstacle to the passage of electrons while other parts allow them to penetrate without collision. The scattering of electrons by a part of the specimen is proportional to the amount of matter that is present in that part, i.e., its *mass thickness,* which is a measure of the number of atoms per unit area and their atomic density. Since the insoluble material of cells consists of atoms of relatively low atomic number—carbon, oxygen, nitrogen, and hydrogen—there is very little capacity present to scatter electrons. To obtain contrast for biological material, atomic density is introduced by fixing and staining the tissue with solutions of heavy metals (described in detail below). These metals penetrate into the structure of the cells and become *selectively* complexed with different parts of the organelles. Differential affinity for the metals is essential, for if every part bound these metal atoms to the same degree, the entire field would be of uniform density and we would be no better off than if no metals had been used. The basis for the selectivity, i.e., why particular metals interact with particular types of structures, is very poorly understood; but, whatever the basis, excellent contrast can be achieved. Those parts of cells that have the greatest concentration of these metal atoms allow the least number of electrons through to participate in the formation of the image by the lenses. The fewer the electrons per unit time that are focused on the screen at a given spot, the darker the screen at that spot, while the greater the electrons, the brighter the spot. Photographs of the object are made by lifting the viewing screen out of the way and allowing the electrons to strike a photographic plate in position beneath the screen. Since photographic emulsions are directly sensitive to electrons, much as they are to light, an image of the object can be recorded on film.

Specimen Preparation for Electron Microscopy

In most instances tissue to be examined in the electron microscope must be fixed, embedded, and sectioned as with the light microscope, though the steps taken are quite different. Fixation of tissue for electron microscopy (Fig. 2-35) is a much more critical process since the scrutiny that the sections will be subjected to is much greater. There are two

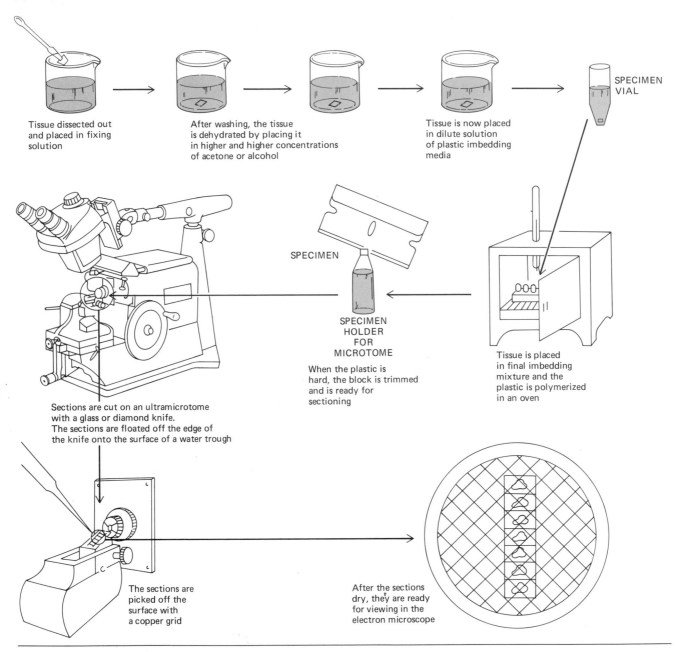

Tissue dissected out and placed in fixing solution

After washing, the tissue is dehydrated by placing it in higher and higher concentrations of acetone or alcohol

Tissue is now placed in dilute solution of plastic imbedding media

SPECIMEN VIAL

SPECIMEN

SPECIMEN HOLDER FOR MICROTOME

When the plastic is hard, the block is trimmed and is ready for sectioning

Tissue is placed in final imbedding mixture and the plastic is polymerized in an oven

Sections are cut on an ultramicrotome with a glass or diamond knife. The sections are floated off the edge of the knife onto the surface of a water trough

The sections are picked off the surface with a copper grid

After the sections dry, they are ready for viewing in the electron microscope

Figure 2-35. Preparation of a specimen for observation in the electron microscope. *(From W. A. Jensen and R. B. Park, "Cell Ultrastructure,"* © *1967 by Wadsworth Publishing Company, Inc., Belmont, California 94002. Reprinted by permission of the publisher.)*

major fixation problems. On one hand, it is desired that the fixative stop the life of the cell without altering its structure. At the level of resolution afforded by the electron microscope, relatively minor damage, such as swollen mitochondria or ruptured endoplasmic reticulum, will be very apparent. To obtain the most rapid fixation and thus the least damage to the cells, very small pieces of tissue are prepared. The other major fixation problem is the formation of structures that were not present in the living cell. Fixatives are chemicals that have a drastic effect on cell components; they denature and render insoluble the macromolecules of the cell. Chemicals capable of such action may cause the coagulation or precipitation of materials that had no structure in the living cell; a formation such as this is called an *artifact*. The best argument that a particular structure is not an artifact is the demonstration of its existence in cells fixed in a variety of different ways or, better, not fixed at all. For this latter case the tissue is rapidly frozen, rather than fixed chemically, and special techniques are utilized to cause the ultrastructure to become visible (described below). The most common fixatives for electron microscopy are glutaraldehyde and osmium tetroxide. Being a heavy metal, the latter not only fixes the cells, but complexes formed between osmium atoms and intracellular macromolecules are made electron dense and visible in the electron microscope.

Once the tissue has been fixed, the water must be removed by dehydration in solutions of increasing concentration of alcohol, and the tissue spaces filled with a material that will support being sectioned. The demands of electron microscopy require that the sections being examined be very thin. With most waxes used in embedding for light microscopy it becomes difficult to obtain sections thinner than about 5 μm, whereas sections for electron microscopy should be less than 0.1 μm (an equivalent in thickness to about four ribosomes). These sections are so small and thin that if all of the sections cut by an electron microscopist during a lifetime were stacked together, they would be equivalent to less than a 1-cm cube. Since its development as an embedding medium in 1961, an epoxy plastic called Epon has been the most widely used embedding material. A variety of ultramicrotomes are presently manufactured which are capable of sectioning this plastic material with the contained tissue at the required thickness for both light microscope observation (0.5 to 1 μm thick) on slides or electron microscope observations (0.05 to 0.1 μm) on grids. Sections are cut by bringing the block slowly down across an extremely sharp cutting edge (Fig. 2-35) made of cut glass or a finely polished diamond face. The sections coming off the knife edge float onto the surface of a trough of water that is contained just behind the knife edge. The sections are then picked up with the specimen grid and dried down onto its surface. To provide the tissue with mass thickness to scatter electrons out of the beam, the grids are floated on drops of heavy metal stains, primarily uranyl acetate and lead citrate, and stored for use. In addition to those stains which will form complexes with a wide variety of different substances, more specific procedures have been worked out whereby the location of a particular type of macromolecule can be determined. Generally, however, these procedures are more difficult to carry out and interpret than corresponding cytochemical techniques at the light microscope level and a much narrower range are available. The primary difficulty is that each electron microscope level technique requires that an electron scattering atom be deposited in the specific location being sought, which often requires a complex and unpredictable series of chemical reactions that must occur between the tissue and the reagents. Another difficulty with cytochemical techniques at the electron microscope level is that the plastic of the sections is difficult to penetrate with large molecular weight reagents, such as enzymes, which have been very successfully used as specific diagnostic tools at the light microscope level.

Negative Staining

Techniques have existed, even before sectioning methods were available, for examining particulate material in the electron microscope. The particles best suited are large molecular-level aggregates such as viruses, ribosomes, multisubunit enzymes, or long fibrous substances (collagen, microtubules, microfilaments, nucleic acids, etc.). One means of making such particles visible is to place them on grids and stain them in a positive manner, i.e., complex them with heavy metals causing them to scatter more electrons than their surroundings. Generally, however, more contrast between particle and surroundings is accomplished by negative stain procedures. In negative staining, heavy metal deposits are collected everywhere on a specimen grid except where the particles are present, and their presence stands out by their relative brightness on the fluorescent viewing screen. To carry out the procedure a drop of stain solution (uranyl acetate or potassium phosphotungstate) is placed on a grid containing the particles to be examined. After a brief period most of the drop is removed. As a result of surface tension the stain tends to surround the particle on the support film and to penetrate into any open irregularities at the surface of the particle. The remainder of the particle collects little stain. The method is particularly well suited for showing any subunit organization of which a particle may be composed. It also serves as a rapid means of following the progress of procedures attempting to isolate particulate types of materials. An example is shown in Fig. 2-36a.

Shadow Casting

Another widely used technique to make very small isolated particles visible is to use them as objects to cast shadows. The technique is described in Fig. 2-37. The grids are placed in a sealed chamber which is then evacuated. In the chamber is a filament composed of a heavy metal (usually platinum together with carbon) which is heated to a high temperature, causing it to evaporate and deposit a metallic coat over those surfaces within the chamber that are accessible to it. Since the filament and the specimens will be at some oblique angle to one another, the metal will be deposited on that surface facing the filament, while the opposite surface of the specimen and the grid space in its shadow will remain uncoated and incapable of scattering electrons. On the viewing screen the shadow will appear bright and the metal-coated regions dark, while this arrangement will be reversed on the photographic plate. The convention for illustrating shadowed specimens is to print a negative image (that of the photographic plate) in which the appearance is one where the particle is illuminated by a bright, white light (corresponding to coated surface) and a dark shadow is cast by the particle. The technique provides excellent contrast for isolated material and produces a three-dimensional effect.

Figure 2-36. Electron micrographs of a tobacco rattle virus after negative staining with potassium phosphotungstate *(a)* or shadow casting with chromium *(b)*. *(Courtesy of M. K. Corbett.)*

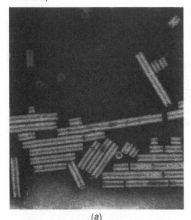

(a)

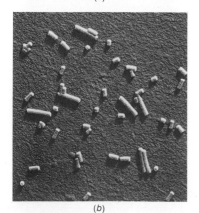

(b)

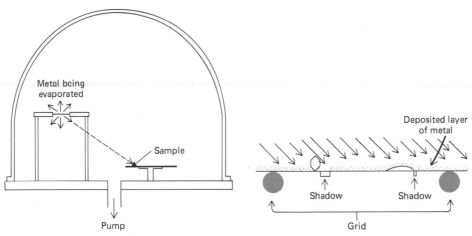

Metal being evaporated

Sample

Pump

Deposited layer of metal

Shadow

Shadow

Grid

Figure 2-37. Shadow casting for providing contrast in the electron microscope. Procedure is described in the text. *(From B. D. Davis et al., "Microbiology," 2d ed., Harper & Row, 1973.)*

Freeze-Fracture Replication

One of the concerns expressed earlier was the possibility that certain of the structures seen with the electron microscope might represent artifacts produced during the period of chemical fixation. Ideally one would like to be able to examine living, unfixed cells under the electron beam and decide if the structures are still present. Living cells cannot be observed directly with the electron microscope, but attempts have been made to place cells in a suitable medium within an airtight container which is then put in the vacuum of the microscope column and observed. So far this technique has been successful with a few bacteria using a high-voltage electron microscope, but results with cells of higher organisms are still preliminary.

An alternative means to chemical fixation for stopping the activities of the living cell is to freeze it. The effects of freezing on living tissue depend greatly on the manner in which the tissue is frozen. The most damaging aspect of the freezing process is the formation of ice crystals, which must be avoided. Consider the effect of an ice crystal growing inside a cellular organelle. As the crystal grew by conversion of liquid to solid, it would rupture the organelles it displaced. In addition, the recruitment of solvent into the solid phase leaves more and more concentrated solutes behind, effectively dehydrating the protoplasm. In the late 1940s it was found that certain compounds, such as glycerol, served as protective agents for cells during freezing, and, in conjunction with rapid freezing procedures using very low temperatures (for example, $-180°C$), cells could be frozen without forming large ice crystals and the cells could be brought back to an active life again. Under carefully controlled conditions, whole animals have been frozen and brought back to an active liquid state.

To return to the question of artifacts, how can frozen tissue be observed under the microscope? For the light microscope, tissue can be sectioned in a frozen state using a *cryostat* and the frozen sections mounted

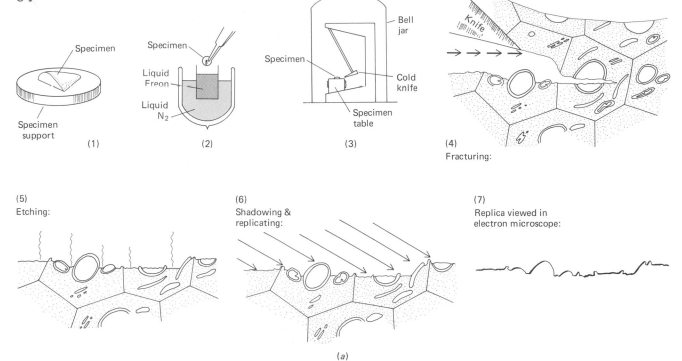

(1)

(2)

(3)

(4)
Fracturing:

(5)
Etching:

(6)
Shadowing &
replicating:

(7)
Replica viewed in
electron microscope:

(a)

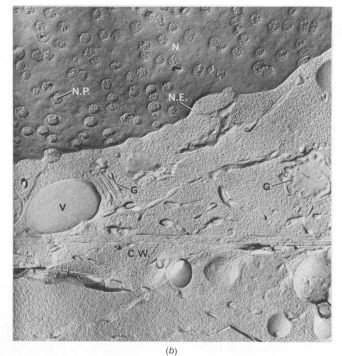

(b)

Figure 2-38. (a) Procedure for the formation of freeze-fracture replicas as described in the text. Step 5, which is discussed on page 211 in connection with membrane structure, is an optional step taken if the surfaces of the membrane are being studied as well as the interior. The step involves the sublimation of the covering ice thereby exposing the membrane surface. (See Fig. 5-17 for further information.) *(From D. Branton and D. W. Deamer, "Membrane Structure," Springer-Verlag, 1972.)* (b) Replica of a freeze-fractured onion root cell showing the nuclear envelope with its pores, the Golgi complex, a cytoplasmic vacuole, and the cell wall. *(Courtesy of D. Branton.)*

on slides for observation. This type of procedure is not suitable for the electron microscope since sections must be able to withstand the vacuum and the electron beam. An ingenious technique was developed, that of freeze-fracture replication, to combine the freezing procedure in preparation of the tissue and specimen preparation for the electron microscope (Fig. 2-38*a*). Small pieces of tissue are placed on a small metal disk and then rapidly frozen by immersion into liquid Freon or a comparable substance. The disk is then placed in a special holder and, in the cold, a knife edge is caused to strike the frozen tissue block, causing a fracture plane or fissure to spread out from the point of contact, splitting the tissue into two pieces.

Consider what might happen as a fracture plane spreads through a cell composed of a great variety of organelles of different composition. These structures tend to cause deviations, either upward or downward, in the fracture plane, causing the surfaces produced to be irregular, to have elevations, depressions, and ridges, which reflect the contours of the protoplasm traversed. In other words, the surfaces exposed by the fracture contain information about the contents of the cell; the problem is to make this information visible. The replication process accomplishes this by using the fractured surfaces as templates upon which a heavy metal layer is deposited. The heavy metal is deposited on the surfaces of the frozen tissue by evaporation in the same chamber as the fracturing was carried out. The evaporation is done at an angle to provide shadows which accentuate local topography as described in the section on shadowing (Figs. 2-37, 2-38*f*). After the metal layer is deposited, a carbon layer is evaporated on top of it, but this time from directly overhead, rather than at an angle, so that a uniform layer of carbon is formed which cements the patches of metal into a solid layer. Now that the replica of the surface is formed, the tissue that provided the template can be thawed, removed, and discarded; it is the metallic-carbon replica that is placed on the specimen grid and viewed in the electron beam. Variations in the thickness of the metal cause variations in the numbers of electrons that penetrate it to reach the viewing screen and thus contrast of the various parts of the replica is achieved. The resolution that can be achieved is on the order of 20 to 30 Å. With this technique all of the major organelles have been seen, verifying their actual existence in the living cell. An example is shown in Fig. 2-38*b*.

In the preceding discussion no indication was given that the fracture plane may spread preferentially through certain cell structures and not others. In fact there are paths of least resistance through which the fractures tend to deviate. The best of these is within the middle of membranes and it is in the study of membranes that the technique has been the most valuable. This topic will be deferred to the discussion of membranes in Chapter 5.

Scanning Electron Microscopy

The entire discussion of the electron microscope up to this point has actually considered only one of the two main types of instruments available. This has been the transmission electron microscope (TEM) in which the image is formed by those electrons that pass through, i.e., are transmitted by, the specimen. An entirely different principle is used in the other type of instrument, the scanning electron microscope (SEM). The transmission microscope has been exploited most widely in the examination of the internal structure of cells, though the negative stain and shadow-casting procedures provide information about the surfaces of very small particles. The scanning electron microscope has provided the means to examine in great clarity and detail the surfaces of objects from the size of viruses (Fig. 2-39a) to that greater than the head of an insect (Fig. 2-39b). Its construction and operation is radically different from that of the transmission electron microscope.

Specimens for the SEM are generally fixed and then dehydrated under very careful conditions. The goal of specimen preparation is to produce an object that has the same shape and surface properties of the living state, but is totally dried out. Since water constitutes such a high percentage of the weight of living cells and is present in association with virtually every macromolecule, its removal can have a very destructive effect on cell structure. Particularly destructive effects result from the surface tension properties at air-water interfaces when cells are simply air dried. There are two commonly employed drying procedures. One is to freeze-dry the specimen, much like preparing instant coffee; the cells are rapidly frozen and the water (in the form of ice) is sublimed (evaporated) under vacuum. The other procedure is termed *critical point drying*, which takes advantage of the fact that a critical temperature and pressure exist for each solvent in a closed container where the density of the vapor is equal to the density of the liquid. At this point there is no surface tension between the gas and the liquid. The technique involves replacing the solvent of the cells with a liquid transitional fluid (generally carbon dioxide) and vaporizing the fluid under pressure without any surface tension existing between the liquid and gaseous phase.

Figure 2-39. Scanning electron micrographs of *(a)* a T$_4$ bacteriophage (×275,000) and *(b)* the head of an insect (×40). *(a)* [*From A. N. Broers, B. J. Panessa, and J. F. Gennaro,* Science, **189:**635 (1975); (b) courtesy of H. F. Howden and L. E. C. Ling.]

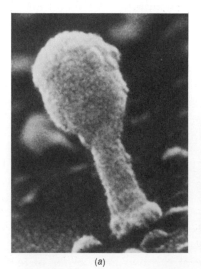

(a)   (b)

Once the specimen is dehydrated it is coated with a layer of carbon and then metal (usually gold or gold-palladium), which makes it suitable as a target for an electron beam. In the TEM, the electron beam is focused by the condenser lenses to simultaneously illuminate the entire viewing field. In the SEM, the electrons are accelerated as a fine beam (as small as 50 Å in diameter) which scans the specimen rather than simply illuminating it. In the TEM, electrons forming the image are ones that have passed through the specimen. In the SEM, image formation is based on electrons that are reflected back from the specimen (back-scattered) or secondary electrons given off by the specimen in response to being struck by the primary electron beam.

Image formation in the SEM is much more indirect than the TEM. In addition to the beam scanning the surface of the specimen, another moving electron beam is needed, one that scans the face of a cathode-ray tube. The image observed is that in the cathode-ray tube, much like the image in a television screen. These two moving electron beams are synchronized by a common electronic system so that a point-to-point correspondence exists between a spot in the specimen field and one on the cathode-ray tube screen. The image seen in the cathode-ray tube, i.e., the variations in brightness and darkness, reflects variations in the intensity of the beam as it scans along inside the tube. Brighter parts of the screen are caused by greater numbers of electrons in the beam at that point, which in turn result from a stronger signal reaching the controls of the cathode-ray tube. The electrons bouncing off the specimen are directly responsible for controlling the strength of the signal to the beam in the cathode-ray tube. The more electrons collected from the specimen at a given spot, the stronger the signal to the tube, and the greater the intensity of the beam on the screen at the corresponding spot. The result is an image of the specimen, one which reflects its surface topology, because it is this topology (the crevices, hills, and pits) that determines the number of electrons collected from the various parts of the surface.

The SEM is relatively new on the scene, the first commercially available microscopes appearing in 1965, and a great number of advances are presently being carried out in SEM technology. One of the limitations of the early microscopes was their relative lack of resolution (averaging about 100 Å). With the advent of high-resolution models (Fig. 2-39a), even this limitation has been overcome. The most remarkable properties of the SEM are its great range of magnifications and its tremendous depth of focus, approximately 500 times that of the light microscope at a corresponding magnification. It is this latter property which gives the SEM images the three-dimensional quality they possess. Most importantly, at the cellular level the SEM allows for the visualization of the outer cell surface (Fig. 6-16); its various processes and extracellular materials which are believed involved in its interactions with its environment and numerous physiological processes.

High-Voltage Electron Microscopy

The typical TEM can supply an accelerating voltage up to about 100 kilovolts. Since the penetrating power of electrons is related to their accelerating voltage, this places a limit on the acceptable thickness of the sections being examined. In recent years a number of electron microscopes have been manufactured which give much greater accelerating

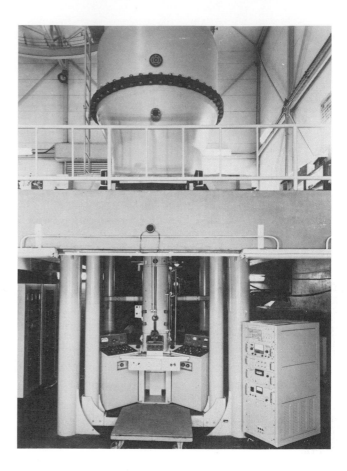

Figure 2-40. A high-voltage (1 million volts) electron microscope. *(Hitachi-Perkin-Elmer Corporation.)*

voltages (up to 3 million volts), though they remain very expensive and difficult to maintain. One of these microscopes is shown in Fig. 2-40. At higher voltages much thicker specimens can be examined under conditions of very high resolution. Since high-voltage microscopes have great depth of focus, the analysis of thick sections provides a much greater picture of the interactions among the various parts of the cell than can be achieved with thin sections, in which these insights must be arrived at by reconstructions from many different micrographs. An electron micrograph taken with a high voltage instrument is shown in (Fig. 8-1).

REFERENCES

Brock, T. D., *Biology of Microorganisms.* 2d ed., Prentice-Hall, 1974.

Campbell, A. M., Sci. Am. **235**, 102–113, Dec. 1976. "How Viruses Insert Their DNA into the DNA of the Host Cell."

Champe, S. P., ed., *Phage.* Halstead, 1974.

Costerton, J. W., Geesey, G. G., and Cheng, K.-J., Sci. Am. **238**, 86–95, Jan. 1978. "How Bacteria Stick."

Davis, B. D., et al., *Microbiology.* 2d ed., Harper & Row, 1973.

Douglas, J., *Bacteriophages.* Halsted, 1975.

Echlin, P., Sci. Am. **214,** 74–83, June 1966. "The Blue-Green Algae."

Everhart, T. E., and Hayes, T. L., Sci. Am. **226,** 54–67, Jan. 1972. "The Scanning Electron Microscope."

Fawcett, D. W., *The Cell: Its Organelles and Inclusions.* Saunders, 1966.

Glauert, A. M., *Practical Methods in Electron Microscopy.* vol. 1, Elsevier North-Holland, 1972.

————, J. Cell Biol. **63,** 717–748, 1974. "The High-Voltage Electron Microscope in Biology."

Gunning, B. S., and Steer, M., *Plant Cell Biol.* Crane, Russak, 1975.

Hartley, W. G., *How to Use a Microscope.* Natural History Press, 1964.

Haschemeyer, R. H., and de Harven, E., Ann. Rev. Biochem. **43,** 279–302, 1974. "Electron Microscopy of Enzymes."

Hayat, M. A., *Basic Electron Microscopy Techniques.* Van Nostrand, 1972.

————, ed., *Principles and Techniques of Electron Microscopy: Biological Applications.* Vol. 5, Van Nostrand, 1975.

Horne, R. W., Sci. Am. **208,** 48–56, Jan. 1963. "The Structure of Viruses."

————, *Virus Structure.* Academic, 1974.

Humason, G. L., *Animal Tissue Techniques.* Freeman, 1962.

Hündgren, M., Int. Rev. Cytol. **48,** 281–321, 1977. "Potential and Limitations of Enzyme Cytochemistry."

Jenson, W. A., and Park, R. B., *Cell Ultrastructure.* Wadsworth, 1967.

Last, J., ed., *Eukaryotes at the Subcellular Level, Development and Differentiation.* Dekker, 1976.

Lauffer, M. A., et al., eds., *Advances in Virus Research.* Vol. 21, Academic, 1977.

Ledbetter, M. C., and Porter, K. R., *Introduction to the Fine Structure of Plant Cells.* Springer-Verlag, 1970.

Lima-de-Faria, A., *Handbook of Molecular Cytology.* Elsevier North-Holland, 1969.

Lwoff, A., Science **152,** 1216–1220, 1966. "Interaction Among Virus, Cell and Organism."

Morowitz, H. J., and Tourtellotte, M. E., Sci. Am. **206,** 117–127, March 1962. "The Smallest Living Cells."

Needham, G. H., *The Practical Use of the Microscope.* Charles C Thomas, 1977.

Palade, G. E., J. Cell Biol. **50,** 5D–19D, 1971. "Albert Claude and the Beginning of Biological Electron Microscopy."

Parsons, D. F., Science **186,** 407–414, 1974. "Structure of Wet Specimens in Electron Microscopy."

Porter, K. R., and Bonneville, M. A., *Fine Structure of Cells and Tissues.* Lea & Febiger, 1968.

Roland, J. C., Szollosi, A., and Szollosi, D., *Atlas of Cell Biology.* Little, Brown, 1977.

Shillaber, C. P., *Photomicrography.* Wiley, 1944.

Sjöstrand, F. S., *Electron Microscopy of Cells and Tissues.* 2 vols., Academic, 1967–68.

Stanier, R. Y., Adelberg, E. A., and Ingraham, J., *The Microbial World.* 4th ed., Prentice-Hall, 1976.

CHAPTER THREE
The Structure of Proteins

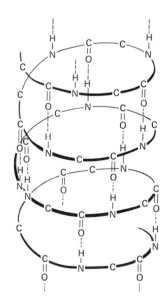

The multitude of activities that occurs within a cell can be considered a direct consequence of the variety of proteins with which the cell is endowed. Included within this class of macromolecules are proteins with the following types of functions:

1. In their best-studied role, proteins are enzymes that catalyze metabolic reactions.

2. Proteins provide structural support both within the cell and in the extra-cellular space. The best-studied structural proteins are the ones found outside of cells, for example, the collagens of connective tissue or the keratins of the skin.

3. A wide variety of regulatory functions, both within cells and between cells, are mediated by proteins. These include molecules responsible for the selection process whereby a given gene might be active in one type of cell and silent in another. Also included are many hormones, such as insulin, glucagon, and the various trophic hormones.

4. Proteins are required to specifically bind and/or transport other molecules. The transport may be within a cell, such as between the cytoplasm and nucleus; across the plasma membrane, as in the case of ion transport systems; or between cells, as happens with oxygen- or lipid-binding proteins in the blood.

5. There are many miscellaneous functions that require specific proteins. Contractility requires proteins, typically actin and myosin. Antibodies are proteins. Many toxins are proteins. Interferon, a substance with important antiviral activity, is a protein.

6. Proteins are also important in a nonspecific manner simply because they have large molecular weights, or carry a number of electrically charged groups, or because of their abundance. For example, the osmotic pressure within a cell can be strongly influenced by the protein content of the cell, as can the cell's buffering capacity, or its distribution of small molecular weight ions across the membrane.

The question to consider is how one type of molecule can have so many varied functions. The explanation resides in the virtually unlimited geometrical shapes that proteins, *as a group*, can form, with each specific protein having its own unique distribution of reactive chemical groups and ionic charge. In other words, proteins are capable of such a variety of activities because there is such a variety of different proteins. Within the group, however, each protein species is constructed in a highly ordered manner so as to carry out a highly specific function.

AMINO ACIDS

Proteins are composed either wholly of amino acids or of amino acids together with some other type of molecule. In the latter case, the protein is said to be *conjugated* (Table 3-1). Conjugated proteins include those linked to nucleic acids, the *nucleoproteins;* to lipid, the *lipoproteins;* to carbohydrate, the *glycoproteins;* or to various smaller molecular weight materials including metals and metal-containing molecules.

In certain respects, the capabilities of a protein can be explained by the properties of its constituents, the amino acids. Information on the chemistry of each amino acid can help explain the role of that amino acid in the function of the complete macromolecule. In addition, new properties and potentials emerge as individual amino acids are linked together into a higher level of organization. To begin to understand their function, we will first consider the structure of the amino acids and then the proteins they form.

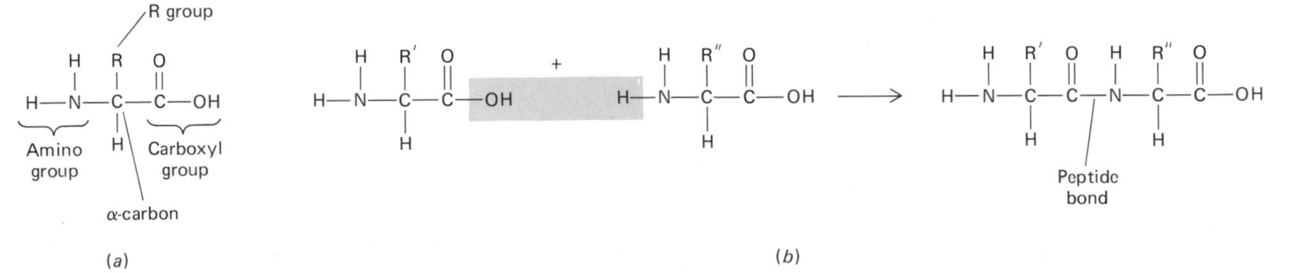

Figure 3-1. *(a)* The basic structure of an amino acid. The nature of the various possible R groups is shown in Fig. 3-4. *(b)* The formation of a peptide bond by the condensation of two amino acids. In the cell this reaction occurs on the ribosome during protein synthesis.

(a)

(b)

TABLE 3-1
Some Conjugated Proteins

Class	Prosthetic group components	Approximate percentage of weight
Nucleoprotein systems		
Ribosomes	RNA	50–60
Tobacco mosaic virus	RNA	5
Lipoproteins		
Plasma β_1-lipoproteins	Phospholipid, cholesterol, neutral lipid	79
Glycoproteins		
γ-Globulin	Hexosamine, galactose, mannose, sialic acid	2
Plasma orosomucoid	Galactose, mannose, N-acetylgalactosamine, N-acetylneuraminic acid	40
Phosphoproteins		
Casein (milk)	Phosphate esterified to serine residues	4
Hemoproteins		
Hemoglobin	Iron protoporphyrin	4
Cytochrome c	Iron protoporphyrin	4
Catalase	Iron protoporphyrin	3.1
Flavoproteins		
Succinate dehydrogenase	Flavin adenine dinucleotide	2
D-Amino acid oxidase	Flavin adenine dinucleotide	2
Metalloproteins		
Ferritin	$Fe(OH)_3$	23
Cytochrome oxidase	Fe and Cu	0.3
Alcohol dehydrogenase	Zn	0.3
Xanthine oxidase	Mo and Fe	0.4

SOURCE: A. L. Lehninger, "Biochemistry," 2d., Worth, 1975.

There are 20 different amino acids that are commonly found in proteins (see Fig. 3-4), whether they are from a virus or a man. There are two aspects of the structure of amino acids to consider, that which is common to all of them and that which is unique to each. The primary structural requirement for all of the amino acids is that they be able to couple with another amino acid on each side so as to form a long, continuous, unbranched polymer called a *polypeptide chain*. To accomplish the coupling process, each amino acid (Fig. 3-1) has both a carboxyl group and an amino group, separated from each other by a single carbon atom, the α carbon.

The α carbon of an amino acid, as with all carbon atoms, is capable of forming single bonds with four other atoms. The arrangement of the groups around a carbon atom can be depicted as in Fig. 3-2a with the carbon placed in the center of a tetrahedron and the bonded groups projecting into its four corners. If the four groups bonded to a carbon atom

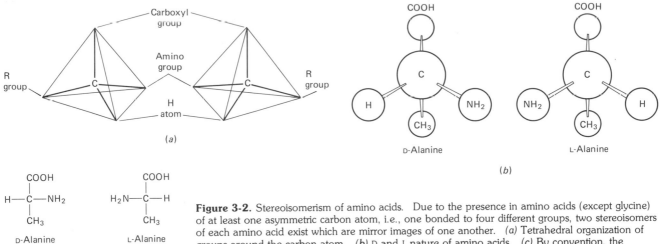

Figure 3-2. Stereoisomerism of amino acids. Due to the presence in amino acids (except glycine) of at least one asymmetric carbon atom, i.e., one bonded to four different groups, two stereoisomers of each amino acid exist which are mirror images of one another. *(a)* Tetrahedral organization of groups around the carbon atom. *(b)* D and L nature of amino acids. *(c)* By convention, the D-isomer is shown with the H atom on the left.

are all different, as in the case for the α carbon of all amino acids except glycine, then two possible configurations exist that cannot be superimposed upon one another (Fig. 3-2*a*, *b*). These two arrangements (termed *stereoisomers* or *enantiomorphs*) both represent the same amino acid, but they are mirror images of one another rather than identical molecules. They rotate plane-polarized light in opposite directions; i.e., they are "optically active." One of these configurations is the L configuration, the other the D configuration (Fig. 3-2*b*, *c*).

All amino acids isolated from all proteins regardless of the source are L-amino acids. Why proteins with D-amino acids are not found is an interesting and debatable question. Regardless, it clearly indicates that biological systems can be specific for one enantiomorph or the other. Pasteur found this out long ago when he supplied bacteria with a mixture of both D- and L-amino acids and found that, when they stopped growing, all the L-amino acids were gone and only D-amino acids remained. In other words, the enzymes responsible for selecting amino acids for incorporation into proteins are capable of distinguishing between the two forms without making mistakes.

Each amino acid in an internal position within a polypeptide chain donates both its amino group and its carboxyl group to form amide bonds (peptide bonds) to each side. As a result, polypeptide chains are characterized by having a backbone of the following nature:

Once incorporated into a polypeptide chain, amino acids are termed *residues*. The residue on one end of the chain, the C-terminal end, has a free (unbonded) carboxyl group, while at the opposite end, the N-terminal end, there is an amino acid with a free amino group.

The backbone of the polypeptide chain is composed of that part of each amino acid that is common among them. The remainder of each amino acid, the *R group* or *side chain,* is highly variable among the 20 building blocks and it is this variability (discussed below) that gives proteins their versatility. If all of the amino acids are considered together, the variety of organic reactions they can partake in and the types of bonds they can form is very great. The assorted characteristics of the R groups of the amino acids are of importance in both *intra*molecular interactions which, as discussed below, determine the structure of the molecule, and *inter*molecular interactions which determine the activities the protein can perform. What is required is that the proper reactive group be present in the protein at the proper place to give the protein the specificity required for the job at hand.

With 20 possible units, the number of different polypeptide chains that can be formed is 20^n, where n is the number of amino acids in the chain. Since most polypeptide chains have over 100 amino acids, the variety is essentially unlimited, though only a very small percentage of the possible sequences would be expected to have a structure that could perform any meaningful activity. Those that are present have evolved via natural selection and are coded within the DNA of the presently existing organisms on earth. In the following discussion of amino acids we will consider only those aspects that provide an insight into protein structure and its relation to protein function.

Ionic Properties

An acidic group is commonly defined as one that is capable of donating a proton (hydrogen ion) and a basic group as one that is capable of accepting a proton. All amino acids $H_2NCCOOH$ possess both a carboxyl (COOH) and an amino (NH_2) group, and therefore all are simultaneously acids and bases. An amino acid, however, never exists in aqueous solution in the uncharged form, but rather varies among the three ionized states:

$$^+H_3N-\overset{\overset{R}{|}}{\underset{\overset{|}{H}}{C}}-\overset{O}{\overset{||}{C}}-OH \underset{OH^-}{\overset{H^+}{\rightleftharpoons}} {}^+H_3N-\overset{\overset{R}{|}}{\underset{\overset{|}{H}}{C}}-\overset{O}{\overset{||}{C}}-O^- \underset{H^+}{\overset{OH^-}{\rightleftharpoons}} H_2N-\overset{\overset{R}{|}}{\underset{\overset{|}{H}}{C}}-\overset{O}{\overset{||}{C}}-O^-$$

The particular ionic state in which a molecule exists is determined by the pH of the medium. Suppose an amino acid, for example, glycine, is dissolved in a strongly acidic solution. In the presence of a high concentration of hydrogen ions, the dissociation of the proton from the carboxyl group will be suppressed and the amino group will be protonated

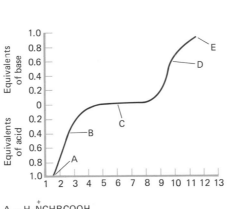

Figure 3-3. Titration curves of (a) glycine, (b) aspartic acid. The ionic nature of the amino acid at key points along the curve is shown. The isoelectric point is the pH at which there is no *net* electric charge on the molecule; i.e., it is electrically neutral.

A — $H_3\overset{+}{N}CHRCOOH$

$B - \begin{cases} H_3\overset{+}{N}CHRCOOH \\ + \\ H_3\overset{+}{N}CHRCOO^- \end{cases} = pK_1' = 2.4$

C — $H_3\overset{+}{N}CHRCOO^-$ = isoelectric point (pH of 6.0)

$D - \begin{cases} H_3\overset{+}{N}CHRCOO^- \\ H_2NCHRCOO^- \end{cases} = pK_2' = 9.6$

E — $H_2NCHRCOO^-$

(a)

A — $H_3\overset{+}{N}CHRCOOH$

$B - \begin{cases} H_3\overset{+}{N}CHRCOOH \\ + \\ H_3\overset{+}{N}CHRCOO^- \end{cases} = pK_1' = 1.9$

C — $H_3\overset{+}{N}CHRCOO^-$ = isolectric point (pH of 2.8)

$D - \begin{cases} H_3\overset{+}{N}CHRCOO^- \\ + \\ H_3\overset{+}{N}CHR^-COO^- \end{cases} = pK_2' = 3.7$

E — $H_3\overset{+}{N}CHR^-COO^-$

$F - \begin{cases} H_3\overset{+}{N}CHR^-COO^- \\ H_2NCHR^-COO^- \end{cases} = pK_3' = 9.6$

G — $H_2NCHR^-COO^-$

(b)

(Fig. 3-3a). The amino acid will have an overall positive charge. If NaOH is slowly added to the solution, the ionic nature of an increasing number of glycine molecules will change as a result of the dissociation of the proton from the carboxyl group. The pH at which half of the carboxyl groups are ionized is the pK for the carboxyl (the pK_1 of the amino acid). The stronger the acid, i.e., the more readily the proton is lost, the lower is its pK. The pK_1 for glycine is 2.4. This is quite low for a carboxylic acid; the close presence of the amino group enhances its acidity. As the carboxyls dissociate, an increasing number of the molecules are doubly charged or dipolar. They are *amphoteric*: the carboxyl group is anionic, the amino group is cationic. At physiological pH, essentially all of the molecules are in the dipolar state and are, therefore, electrically neutral. If the titration is continued into the alkaline range, protons become removed from the previously protonated amino groups causing the amino acid to lose its positive charge and to have an overall negative charge. The pH at which the amino group of glycine is half-dissociated is the pK for the amino group (the pK_2 for the amino acid), approximately 9.6 for glycine.

The ionic properties of the carboxyl and amino groups of the α carbon are of little importance in the consideration of protein structure since, with the exception of the amino acids at each end of the chain, they have disappeared in forming the peptide bonds of the primary structure. However, the nature of certain of the amino acid side chains (R groups) is such that they may be either fully or partly ionized at physiological pH, and are, therefore, of great importance in the reactivity of the polypeptide chains. The preceding discussion of the titration of the carboxyl and amino groups of the α carbon applies in principle to the R groups as well. Table 3-2 gives a list of those amino acids whose R groups contain acidic or basic groups. As in the case of the free amino acids, the degree of ionization of a particular group is dependent upon the pH, and the pK for each is given in the table. It should be kept in mind that the pK's of these groups are measured in the free amino acid state (Fig. 3-3b for example shows the titration curve for aspartic acid), which might be quite different from the pK of the same group buried deeply within a large protein. In other words, the environment in which a particular R group finds itself can be very important in the properties that it shows. The pK's of aspartic acid, glutamic acid, lysine, and arginine are sufficiently to one side or the other of physiological pH that they would be expected to be fully charged within the cell.

It is the relative balance of the negatively charged and positively charged groups in the protein that will determine the overall charge of the protein itself. At physiological pH, most proteins are anionic, i.e., they have a predominance of amino acids with a carboxylic R group. What-

TABLE 3-2
pK Values of Ionizable Groups in Proteins

Group	Acid $\rightleftharpoons$ Base + H+	Typical pK
Terminal carboxyl	$-COOH \rightleftharpoons -COO^- + H^+$	3.1
Aspartic and glutamic acid	$-COOH \rightleftharpoons -COO^- + H^+$	4.4
Histidine		6.5
Terminal amino	$-NH_3^+ \rightleftharpoons -NH_2 + H^+$	8.0
Cysteine	$-SH \rightleftharpoons -S^- + H^+$	8.5
Tyrosine		10.0
Lysine	$-NH_3^+ \rightleftharpoons -NH_2 + H^+$	10.0
Arginine		12.0

ever the relative abundance of acidic or basic R groups, a pH exists for a given protein where the overall charge is neutral; the anionic groups equal the cationic ones. The pH at which overall neutrality is obtained is termed the *isoelectric point* of the protein. For proteins with a preponderance of acidic groups, the isoelectric point will be on the acidic side since it will take a greater hydrogen-ion concentration in the medium to suppress the dissociation of a sufficient number of carboxyl groups so that a balance with the protonated amino groups of lysine and arginine is achieved. The reverse condition is true for proteins, such as histones and protamines, which have an excess of basic groups. Protamines, for example, which are found associated with DNA, contain an extremely high arginine content and have an isoelectric point of approximately 12. On the opposite side of the spectrum are the digestive proteins of the stomach which operate under very acidic conditions and have very low isoelectric points.

Amino acids are conveniently classified on the basis of the ionic potential of their side chains. They fall roughly into four categories (Fig. 3-4).

---

1. Polar, charged. In the above discussion we considered the ionic properties of those groups that become fully charged, i.e., the stronger organic acids and bases. Amino acids with R groups of this nature are lysine, arginine, aspartic acid, and glutamic acid. Histidine is generally included in this group, though in most cases it is only partially charged at physiological pH. In fact, because of its ability gain or lose a proton in physiological pH ranges, histidine is a particularly important residue in many proteins.

2. Polar, uncharged. Amino acids of this category contain side chains that are weakly acidic or basic. These groups can become charged during titration but at physiological pH's they generally exist uncharged. Even if not fully charged, these side chains would have some separation of charge producing electropositive and electronegative regions making these molecules capable of forming hydrogen bonds and associating with water. These amino acids are often quite reactive. Included in this category are asparagine and glutamine (the amides of aspartic acid and glutamic acid), threonine, serine, tyrosine, and cysteine. Cysteine contains a reactive sulfhydryl (—SH) group and is often present covalently linked to another cysteine residue, as a disulfide (—SS—) bridge (as in Fig. 3-7b). When present as the disulfide, the combined amino acid residues are given the name, cystine.

3. Nonpolar. At the other extreme from those in the first category are the amino acids whose side chains are hydrophobic and are not capable of forming hydrogen bonds or interacting with water. The amino acids of this category are valine, leucine, isoleucine, tryptophan, phenylalanine, proline, and methionine. The side chains of the nonpolar amino acids lack oxygen and nitrogen (with the exception of tryptophan).

4. The other two amino acids, glycine and alanine, have very small R groups (a hydrogen atom and a methyl group, respectively) and do not fit into any of the above categories. When present as a residue in a chain they are neutral and are accommodated in either a polar or a nonpolar environment.

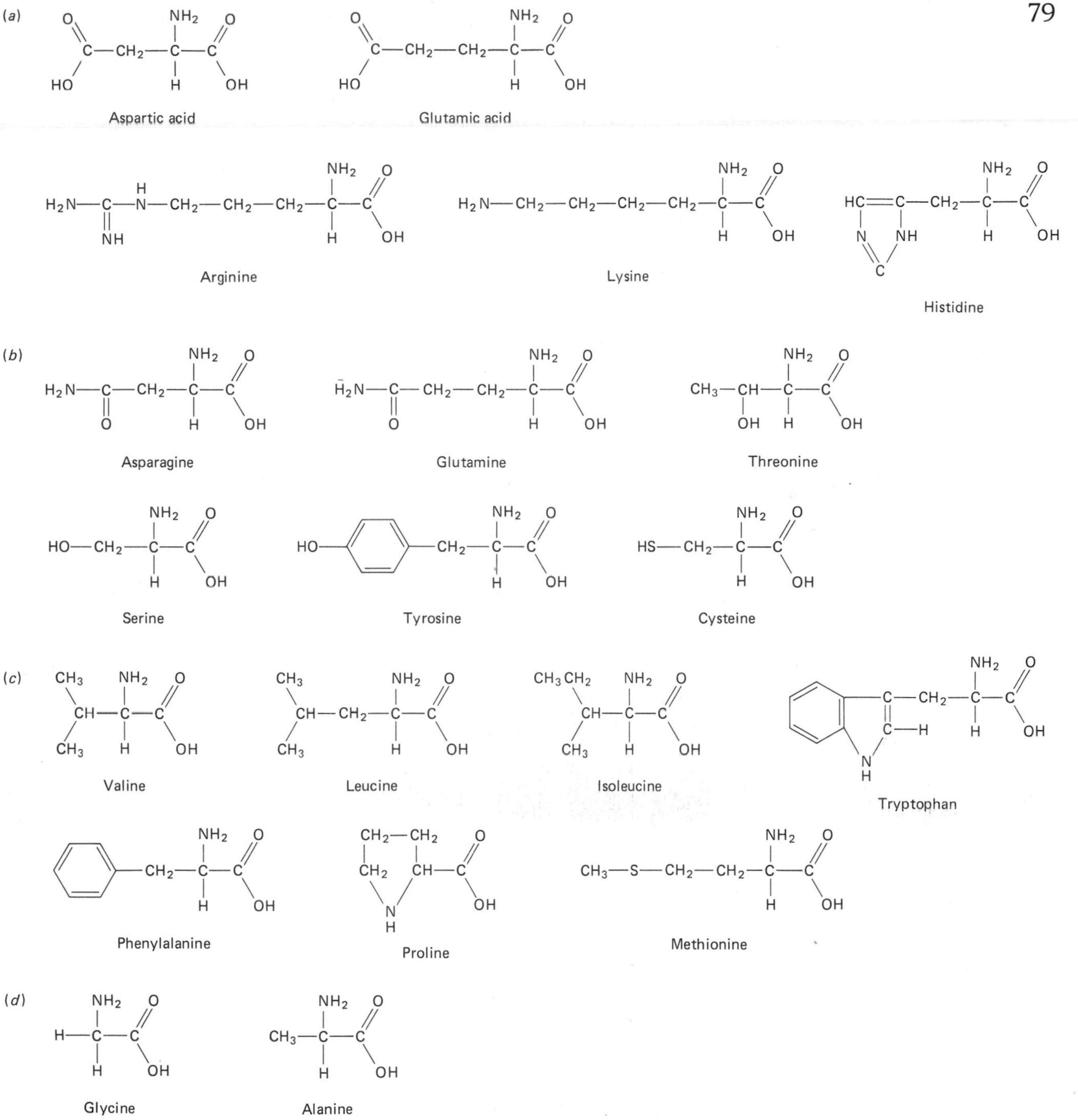

Figure 3-4. The chemical structure of the amino acids. These 20 amino acids represent those most commonly found in proteins and, more specifically, those coded for in DNA. The amino acids are arranged into four groups as described in the text. (a) polar, charged; (b) polar, uncharged; (c) nonpolar; (d) those with R groups of H or CH₃.

The effect that side chain variability has on solubility is shown in Fig. 3-5. All of the amino acids can be separated from one another simply on the basis of differences in their solubility properties. A few other amino acids have been found in proteins, for example, thyroxine, hydroxyproline, and hydroxylysine, but these arise as a result of the alteration of one of the 20 basic amino acids *after* its incorporation into the polypeptide chain. Not all amino acids are found in all proteins, nor are those amino acids that are present distributed in an equivalent manner.

The ionic properties of the side chains are very important in protein structure and function. Most proteins are organized so that the nonpolar residues are in the core of the molecule, away from the aqueous medium, and the polar groups form its outer region. When concentrated in one area of a particular protein, nonpolar residues make that area a nonaqueous island in an environment whose properties are determined by water. In a sense the interior of the protein recreates the organic solvent in which many organic reactions are best suited. Events can occur in this specialized, nonpolar medium that could not occur in the outside aqueous world. Some reactions that might proceed at an imperceptibly slow rate in water can occur in thousandths of a second within the protein.

The polar residues form various types of electrostatic bonds with one another and their environment. In many enzymes or transport proteins it is the reactive polar groups projecting into the nonpolar interior that makes these proteins work. The nonpolar environment greatly enhances electrical interactions between groups, interactions which would be competed out in the presence of water. When membrane proteins are considered in Chapter 5, the importance of the location of hydrophobic

Figure 3-5. *(a)* The technique of two-dimensional chromatography. The mixture is spotted in a lower corner, placed in a sealed chamber in a trough of solvent. The solvent moves up the paper by capillary action, carrying with it dissolved components from the mixture. The greater the solubility in the moving solvent, the greater the migration. After the first solvent has moved the desired distance, the paper is dried, rotated 90°, and placed in a second solvent system for further separation. *(b)* Two-dimensional chromatogram of a mixture of amino acids. The first solvent system is *n*-butanol-acetic acid-water (250:60:250 by volume), the second is phenol-water-ammonia (120:20:0.3). Note how the amino acids having similar structures in Fig. 3-4 migrate together (dotted lines). CySSCy represents cystine, i.e., two cysteines joined by a covalent bond. [*(b)* After A. White, P. Handler, and E. L. Smith, "Biochemistry," 5th ed.. McGraw-Hill, 1973.]

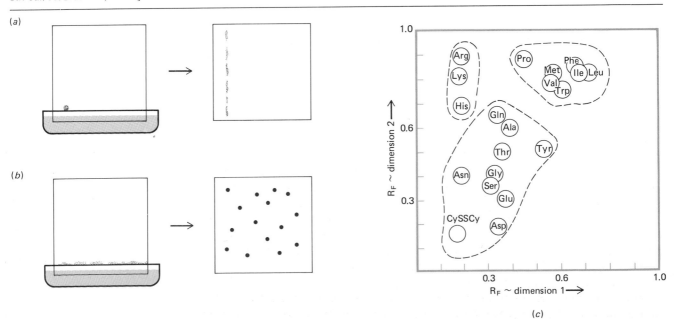

and hydrophilic residues will be very evident. The hydrophobic-inside–hydrophilic-outside arrangement is an adaptation to an aqueous environment. If this environment is markedly altered, as within biological membranes, the construction of the proteins is greatly modified.

PROTEIN STRUCTURE

Protein structure can be described at several levels of organization, each emphasizing a different aspect and each being dependent upon different types of interactions. There are generally four such levels described: *primary, secondary, tertiary,* and *quaternary.* Whereas primary structure is concerned with the particular amino acid sequence of a protein, the latter three levels are concerned with the organization in space, i.e., the conformation of parts of the molecule or the entire molecule itself.

Primary Structure

The primary structure refers to the specific linear sequence of amino acids in a given polypeptide chain. As will be described at great lengths, the information for the precise order of amino acids in every protein that an organism can produce is contained within the genetic inheritance of that organism. It is the DNA that provides the information for the amino acid sequence, but it is the amino acid sequence that provides the information for the protein's structure. As will be shown, the various three-dimensional shapes to which proteins conform are believed to be a direct consequence of their primary structure. The importance of the sequence of amino acids, therefore, is all important and changes that arise in it as a result of mutation may not be readily tolerated. The earliest and best-studied example of this is the change in the amino acid sequence of hemoglobin that causes the disease sickle cell anemia. This severe inheritable anemia is totally the result of a single change in amino acid sequence within the molecule. In this case, a valine is present where a glutamic acid is supposed to be, a replacement of a charged, polar amino acid with a nonpolar one. All of the problems concerning red blood cell shape and decreased oxygen-carrying capacity in persons with sickle cell anemia result from this change. In many other cases amino acid changes have little effect. The degree to which changes in the primary sequence are tolerated depends upon the degree to which the overall geometry of the protein or the critical functional residues are disturbed. The nonpolar residues of the core of the protein are particularly sensitive to alteration unless the substitute is an amino acid very similar to the original one.

Primary structure is maintained by covalent bonds, specifically the peptide bonds between amino acids, and, by some definitions, the disulfide bonds between cysteine residues. The primary structures of a large and growing number of proteins have been determined. The task can be an arduous one, but the results are of great importance; far more is at stake than simply knowing the names of the particular links in a long linear chain. Information obtained from amino acid sequencing is important in understanding the structure of the protein, how that structure comes about

TABLE 3-3
Sequencing the Insulin Molecule

Step 1	Oxidation (by hydrogen peroxide in formic acid) of disulfide bridges holding the two chains together.
Step 2	Separation of the two chains by selectively precipitating one chain (the B chain, having a phenylalanyl terminal), while leaving the other chain (the A chain, having a glycine terminal) in solution.
Step 3	Hydrolysis of a preparation of one or the other polypeptide chains. Hydrolysis was accomplished by (a) acid (12 N HCl at 37°C), (b) base (1 N NaOH at 100°C) or (c) enzymes (pepsin, trypsin, chymotrypsin, carboxypeptidase).
Step 4	Fractionation of the various peptide mixtures. Various fractionation procedures were used, e.g., paper chromatography, ion-exchange chromatography, and electrophoresis). Acid hydrolysates of the B chain, for example, yielded 22 dipeptides, 14 tripeptides and 12 longer fragments.
Step 5	Determination of the amino acid sequence of each of the small peptides. This step involved two types of determinations: The amino acid composition of a particular fragment, and the specific amino acid present at the N-terminal position. In the case of *di*peptides, this information automatically provided the sequence. Larger peptides were sequenced by fragmenting them further to dipeptides whose sequences were determined. Once the amino acid sequence of the peptides was determined, that of the entire chain could be deduced.

in the cell, the mechanism whereby that protein operates, and something of its evolution.

The Amino Acid Sequence of Insulin

The first protein to be sequenced was insulin, by Frederick Sanger and coworkers in the early 1950s. This feat stands as one of the most important and fundamental in the rapidly developing fields of biochemistry and molecular biology. Its primary impact was the proof that proteins, the most complex molecules of cells, had a specific, definable substructure. It indicated that protein structure and function was not outside the realm of organic chemistry, but simply involved more complex organic chemistry. The door to defining the activity of an enzyme in terms of the known principles of chemistry had been opened. Recently, the reverse task has been accomplished, that of synthesizing insulin from amino acids. No physiological difference is found between that artificially produced and that extracted from tissue. Beef insulin was chosen as the first protein to be sequenced primarily because of its availability and its small size — 2 polypeptide chains of 21 and 30 amino acids each. The general procedure utilized in obtaining the amino acid sequence of insulin is shown in Table 3-3.

Before any attempt to sequence a protein can begin, that protein must be obtainable in a purified state. Contamination of the protein under study with other proteins will obviously confuse the results. If the protein is one that is made of more than one type of polypeptide chain, these must be separated from one another so that each can be sequenced without interference from the other. Disulfide bonds, if present, must be broken. From this point the task is to convert an unmanageably large polypeptide chain into small pieces, separate the pieces from one another, and determine the amino acid sequence of each fragment. There are several points to consider.

The procedure outlined above requires techniques whereby amino acids and small peptides can be separated from one another. The chromatographic techniques needed were being developed before and during the early sequencing work. These methods utilized differences in solubility (such as is shown in Fig. 3-5) or ionic character to bring about the fractionation of a mixture of amino acids or peptides over a two-dimensional surface. An excellent example of the separation of *peptides* is shown in Fig. 3-6, in which paper chromatograms were prepared comparing digests of normal and sickle cell hemoglobin. The peptides of the two hemoglobins possess the same migratory characteristics except for that peptide in which the amino acid substitution has occurred.

Once the peptides of the insulin molecule were separated, the amino acid sequence of each one had to be determined. This required the development of reactions whereby the amino acid at the end of the fragments could be labeled and identified. Techniques have been developed since Sanger's original work that allow the amino acid at the end of the fragment to be removed for identification, leaving the remaining peptide intact so that the cycle can be repeated. Machines are now available that can automatically sequence peptides of 20 amino acids in length.

In order to sequence a particular polypeptide chain, that chain had to be fragmented in more than one pattern so that the sequence of fragments produced by different points of cleavage would overlap with one another. As a result of the overlap, Sanger and his colleagues were able to work backward from the sequences of various fragments to obtain the sequence of the complete molecule. In the determination of the sequence of the chains of insulin, Sanger used several cleavage techniques (Fig. 3-7). Several different enzymes were employed—trypsin, chymotrypsin, and pepsin—each breaking specific but different peptide bonds along the chain. Hydrolysis of the chains with acid, a procedure that releases very small peptides, was also used. The hope at the present time, as more primary sequences become available, is that it

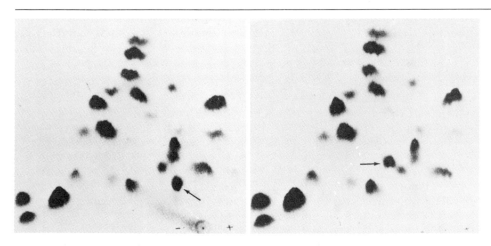

(a) (b)

Figure 3-6. Fingerprints obtained by a two-dimensional separation of the peptides isolated after treatment of normal hemoglobin *(a)* and sickle cell hemoglobin *(b)* with trypsin. The hydrolysate was subjected to paper electrophoresis in the first dimension and paper chromatography in the second. The two fingerprints are identical with the exception of one peptide (marked) which contains a single amino acid difference. [*From C. Baglioni*, Bioc. Biop. Acta, **48:**394 (1961).]

Dipeptides from acid and alkaline hydrolyzates

H-phe-val-OH H-his-leu-OH H-his-leu-OH H-ala-leu-OH H-gly-glu-OH H-thr-pro-OH
H-val-asp-OH H-leu-cySO$_3$H-OH H-leu-val-OH H-leu-val-OH H-glu-arg-OH H-lys-ala-OH
H-asp-glu-OH H-cySO$_3$H-gly-OH H-val-glu-OH H-val-cySO$_3$H-OH H-arg-gly-OH
H-glu-his-OH H-ser-his-OH H-glu-ala-OH H-cySO$_3$H-gly-OH H-gly-phe-OH

Tripeptides from acid and alkaline hydrolyzates

H-phe-val-asp-OH H-leu-cySO$_3$H-gly-OH H-ala-leu-tyr-OH H-gly-glu-arg-OH H-pro-lys-ala-OH
H-val-asp-glu-OH H-ser-his-leu-OH H-tyr-leu-val-OH
H-glu-his-leu-OH H-leu-val-glu-OH H-leu-val-cySO$_3$H-OH
H-his-leu-cySO$_3$H-OH H-val-glu-ala-OH H-val-cySO$_3$H-gly-OH

Higher peptides from acid and alkaline hydrolyzates

H-phe-val-asp-glu-OH H-ser-his-leu-val-glu-OH H-tyr-leu-val-cySO$_3$H-OH H-thr-pro-lys-ala-OH
H-phe-val-asp-glu-his-OH H-ser-his-leu-val-glu-ala-OH H-leu-val-cySO$_3$H-gly-OH
H-glu-his-leu-cySO$_3$H-OH H-ser-his-leu-val-glu-OH
H-his-leu-cySO$_3$H-gly-OH H-leu-val-glu-ala-OH
H-ser-his-leu-val-OH

Sequences deduced from above peptides

H-phe-val-asp-glu-his-leu-cySO$_3$H-gly- -tyr-leu-val-cySO$_3$H-gly- -thr-pro-lys-ala
-ser-his-leu-val-glu-ala- -gly-glu-arg-gly-

Peptides identified in peptic hydrolyzate

H-phe-val-asp-glu-his-leu-cySO$_3$H-gly-ser-his-leu-OH H-leu-val-cySO$_3$H-gly-glu-arg-gly-phe-OH H-tyr-thr-pro-lys-ala-OH
H-val-glu-ala-leu-OH
H-his-leu-cySO$_3$H-gly-ser-his-leu-OH

Peptides identified in chymotryptic hydrolyzate

H-phe-val-asp-glu-his-leu-cySO$_3$H-gly-ser-his-leu-val-glu-ala-leu-tyr-OH H-tyr-thr-pro-lys-ala-OH
H-leu-val-cySO$_3$H-gly-glu-arg-gly-phe-phe-OH

Peptides identified in tryptic hydrolyzate

H-gly-phe-phe-tyr-thr-pro-lys-ala-OH

Structure of the B (phenylalanyl terminal) chain of insulin

H-phe-val-asp-glu-his-leu- (cyS-) - gly-ser-his-leu-val-glu-ala-leu-tyr-leu-val- (cyS-) gly-glu-arg-gly-phe-phe-tyr-thr-pro-lys-ala-OH

(a)

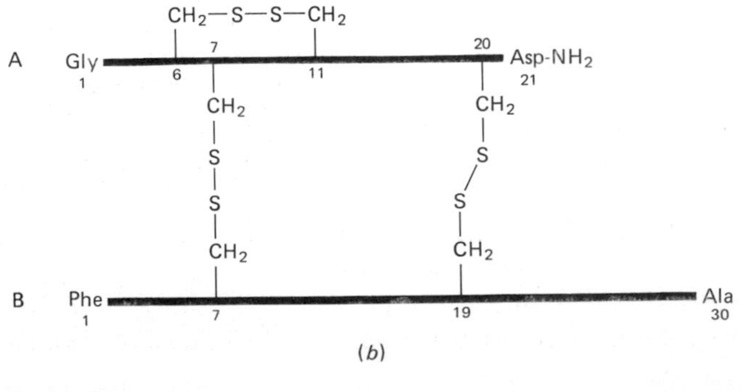

(b)

Figure 3-7. *(a)* The various peptides identified in hydrolysates of the B chain of insulin. Procedure described in Table 3-3. Deduced sequence shown at bottom of the figure. *(From J. L. Oncley, in "Biophysical Science—A Study Program," J. L. Oncley, ed., Wiley, 1959.) (b)* The structure of insulin showing the location of the three disulfide bonds. The numbers *refer* to the sequence of amino acid residues beginning from the C-terminus. *(From B. W. Low and J. T. Edsall, in "Currents in Biochemical Research," D. E. Green, ed., Wiley-Interscience, 1956.)*

will be possible to predict the higher levels of protein structure simply by knowing the order of the amino acids of the chain.

Secondary Structure

All matter exists in space and therefore has a three-dimensional expression. Proteins are large aggregates of matter, relative to others at the molecular level, and their shape is complex. The term *conformation* refers to the three-dimensional arrangement of the atoms of a molecule, i.e., to their spatial organization. Secondary structure refers to the conformation of pieces of the polypeptide chain, i.e., to the geometrical relationship of amino acids adjacent to one another in the linear sequence. The conformation of the polypeptide chain will be dependent upon the bond lengths and the bond angles formed along the backbone of the chain. The three primary bonds to consider are those between the three atoms of the backbone donated by each amino acid.

The analysis of these bonds and the consequences for secondary structure was carried out primarily by Linus Pauling and Robert Corey over many years. Measurements of the length of the peptide bond between adjacent amino acids indicated that it was intermediate in character between a single (C—N) bond and a double (C=N) bond. In other words, the peptide bond has a partial double-bond character (approximately 40%). This is generally depicted with the amide linkage existing in a resonating condition between two states.

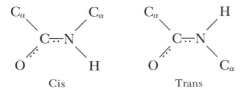

The double-bond character of the peptide bond has important consequences for the conformation of the polypeptide chain because it imposes a rigidity upon the participating atoms such that they cannot rotate relative to one another. As a consequence, the entire amide linkage is flattened in space, i.e., it is coplanar (Fig. 3-8). If the carbon and nitrogen atoms of the peptide bond cannot rotate, there are two possible configurations for the atoms on each side of the peptide bond relative to one another.

$$C_\alpha \diagdown \quad \diagup C_\alpha \qquad C_\alpha \diagdown \quad \diagup H$$
$$\quad C \cdots N \qquad\qquad C \cdots N$$
$$\diagup \quad \diagdown \qquad\qquad \diagup \quad \diagdown$$
$$O \qquad\quad H \qquad O \qquad\quad C_\alpha$$
$$\text{Cis} \qquad\qquad\qquad \text{Trans}$$

In the cis configuration the α carbons adjacent to each peptide bond would be on the same side of the bond, while in the trans configuration, they would be on the opposite sides of the bond. Due to the bulky nature of the R groups, the trans configuration is preferred (Fig. 3-8). The other two bonds of the backbone extend to either side of the peptide bond to the α carbons. These are designated ψ and ϕ in Fig. 3-9 and are single bonds and, theoretically, the atoms are free to rotate. One of the main

Figure 3-8. The planar nature of the peptide bond. The partial double-bond character of the C—O and C—N bonds causes all three atoms of the amide linkage to lie in the same plane and blocks rotation around the C—N bond. *(From R. E. Dickerson and I. Geis, "The Structure and Action of Proteins," W. A. Benjamin, Inc. Copyright 1969 by Dickerson and Geis.)*

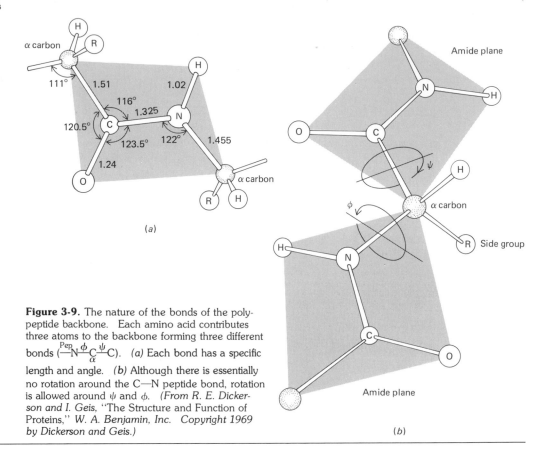

Figure 3-9. The nature of the bonds of the polypeptide backbone. Each amino acid contributes three atoms to the backbone forming three different bonds ($\overset{Pep}{-}N\overset{\phi}{-}\underset{\alpha}{C}\overset{\psi}{-}C$). (a) Each bond has a specific length and angle. (b) Although there is essentially no rotation around the C—N peptide bond, rotation is allowed around ψ and ϕ. *(From R. E. Dickerson and I. Geis, "The Structure and Function of Proteins," W. A. Benjamin, Inc. Copyright 1969 by Dickerson and Geis.)*

questions considered by Pauling and Corey was whether or not there were preferred angles for these two bonds (ψ and ϕ) along the chain which might impose some form of predicatable organization on the arrangement of the chain.

The Alpha Helix and the Beta Pleated Sheet

By studying x-ray diffraction patterns of amino acids and simple peptides and constructing appropriate models, Pauling and Corey concluded that there were preferred bond angles and that there were preferred conformations for the chain. The polypeptide chains were neither simply extended polymers nor were they randomly twisted in space. Two conformations were proposed.

In one case the form taken was a helix or twisting spiral termed the α helix, shown in Fig. 3-10. A spiral of this type occurs when the angles of the bonds ψ and ϕ are approximately 123° and 132°, respectively. These bond angles are such that, taken together with the planar nature of the peptide bond, the chain forms the outline of a tightly drawn helix of regular dimensions, enclosing a cylindrical space. The backbone is on the inside, the side chains projecting outward. There is a rotation between amino acids of 100° (3.6 amino acids per 360° complete turn) which results in the pep-

Figure 3-10. The alpha helix. *(a)* A schematic representation. [*From J. C. Dearden,* New Scientist, *21:629 (1968).*] *(b)* The molecular organization of the atoms of the backbone of the helix and the hydrogen bonds that form between every fourth amino acid. [*From L. Pauling, R. B. Corey, and H. R. Branson,* Proc. Natl. Acad. Sci. (U.S.), *37:205 (1951).*]

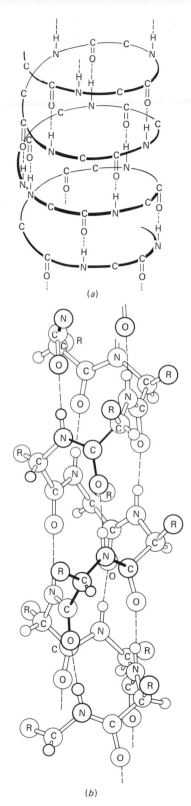

(a)

(b)

tide bonds of every fourth amino acid coming very close to one another.

The approach of the carbonyl group (C=O) of one peptide bond to the imine group (H—N) of another results in the formation of hydrogen bonds (2.8 Å in length) between them. All peptide bonds of the α helix are hydrogen bonded to another peptide bond three residues away. The hydrogen bonds are essentially parallel to the axis of the cylinder and thus serve to hold the turns of the chain together (Fig. 3-10).

It was Pauling's prediction that the favored conformation would be the one having the maximum possible formation of hydrogen bonds consistent with the appropriate bond angles that caused the α helix to be selected from among other models that could be constructed. Analysis of various peptides suggested that several amino acids were not readily accommodated into an α-helical conformation (Table 3-4). Proline, for example, which is actually an α imino acid (Fig. 3-4), since its α carbon is within a particular ring structure, will not fit into such a helix. The x-ray diffraction analysis of actual proteins during the 1950s bore out the predictions of the α helix perfectly, first in hair keratin and later in various globular proteins. X-ray diffraction pictures of keratin had, in fact, been available since the 1930s, but the reflections caused by the α helix in the pattern were not interpretable until the proposal by Pauling and Corey had been made.

The other regular arrangement of polypeptide chains proposed by Pauling and Corey was one of an entirely different nature from the helical;

TABLE 3-4
Assignment of Helical Character to Amino Acid Residues

Helix breaker	Helix former	Helix indifferent
Glycine	Leucine	Alanine
Proline	Isoleucine	Threonine
Serine	Phenylalanine	Valine
Aspartic acid	Glutamic acid	Glutamine
	Tyrosine	Asparagine
	Tryptophan	Cystine
	Methionine	Histidine
		Lysine
		Arginine

SOURCE: P. N. Lewis et al., *Proc. Natl. Acad. Sci. U. S.,* **65**:812 (1970).

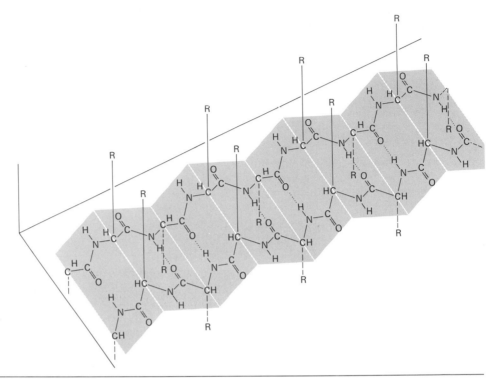

Figure 3-11. The beta pleated sheet. *(From P. Karlson, "Introduction to Modern Biochemistry," 4th ed., Academic, 1975, originally published by Georg Thieme Verlag.)*

that of the β pleated sheet (Fig. 3-11). Unlike the cylindrical or rodlike form of the α helix, the β pleated sheet is a highly flattened extended form, but is also characterized by a maximum of hydrogen bonding between peptide bonds. In this case, however, the hydrogen bonds are perpendicular to the long axis of the polypeptide chain and project across from one chain to another. X-ray diffraction analysis of several proteins has also provided evidence for the presence of the β pleated sheet conformation, as a second, naturally occurring, ordered arrangement. Since any given polypeptide chain has polarity (a C-terminal end in one direction and an N-terminal end in the other), two chains that are hydrogen bonded across to one another can be running in the same direction (parallel) or in opposite directions (antiparallel).

Fibrous Proteins

To understand the properties that these two different types of secondary structures provide, it is best to examine those fibrous proteins where they are most extensively found and best studied. For the β pleated sheet, silk protein has provided the model, while for the α helix it has been the protein of hair and wool, α keratin. The helix of α keratin provides the fiber with extensibility; it can be stretched to a moderate degree, yet its elasticity (resulting from the hydrogen bonds as well as interchain disulfide bridges) causes it to snap back. Strength is obtained by having many helices running along side by side to form long cables. A cross section through a hair is diagrammed in Fig. 3-12. The smallest unit

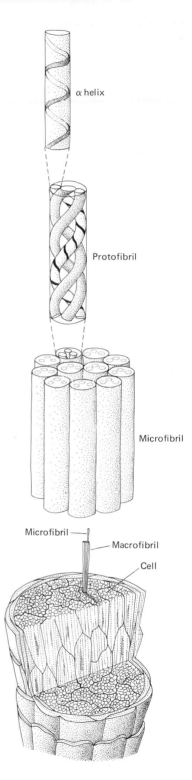

Figure 3-12. The levels of organization that exist in a single hair. *(From R. E. Dickerson and I. Geis, "The Structure and Action of Proteins," W. A. Benjamin, Inc. Copyright 1969 by Dickerson and Geis.)*

shown in the diagram is the polypeptide. The microfibril is an 80-Å cable, made up in turn of a collection of protofibrils (Fig. 3-12), which consists of several α helices coiled around each other. Even a simple inert, dead strand of hair has a complex, predictable organization upon which its properties are based. The architecture extends from the polypeptide chain made visible by techniques such as x-ray diffraction; to the fibrils, seen in the electron microscope; to the hair itself, visible with the naked eye.

Silk, in contrast, has limited extensibility and elasticity but is strong and flexible. It cannot be stretched since the β sheets are already greatly extended. Its strength is derived directly from the lengths of covalent bonds forming the backbone of the chains while its flexibility is derived from the loose manner in which the sheets are stacked one on top of another. The interrelationship between molecular structure and macromolecular function is nicely illustrated by the properties of these structural proteins. Similarly, it gives us confidence that we have some basic understanding of the role of molecular architecture. Clearly, there must be more to protein shape than these two secondary structures since, if there were not, all proteins would be very similar in construction. The analysis of tertiary structure of a variety of globular proteins has revealed that much more is involved.

Tertiary Structure

Whereas secondary structure is concerned primarily with the conformation of adjacent amino acids in the polypeptide chain, tertiary structure describes the conformation of the entire protein. In a sense, secondary and tertiary structure are closely related topics. If, for example, a polypeptide chain existed solely in the alpha helix conformation, its tertiary structure would be defined; it would be an elongate cylindrical structure with a diameter equal to that of the helix and a length proportional to its amino acid number. Most proteins, however, are globular and the polypeptide chains are folded and twisted into much more complex organizations (Fig. 3-13). Distant points on the linear sequence of amino acids are brought next to each other and a variety of types of interactions between R groups among different parts of the chain have been found to occur. Given a polypeptide chain of a hundred or more amino acids, it is easy to imagine that an unlimited number of shapes could be generated. The facts indicate, however, that just as there is a unique and predictable sequence of amino acids for a given protein, there is a unique and predictable conformation for that sequence.

How can the intricacies of protein tertiary structure be elucidated? The complex spatial organization of proteins has been analyzed primarily by the technique of x-ray diffraction, the same technique used by Watson, Crick, and Wilkens in the determination of the structure of DNA. In the discussion of the microscope in Chapter 2, the relationship between the diffraction of electromagnetic radiation, the wavelength of the illuminating source, and the resolution that can be obtained was described. The wavelength of electrons is sufficiently small to allow the resolution of intramolecular distances, but the electron microscope is not able to provide the means to achieve it. X-ray diffraction analysis of crystals cannot provide the degree of resolution theoretically obtainable by electrons; but, because its actual practical resolving power is much greater, it can provide precise information about the interatomic organization of molecules down to resolutions in the order of 1.5 Å.

The first analysis of crystals by this technique was begun by Lawrence Bragg in 1913. In 1937 Max Perutz began an examination of crystals that

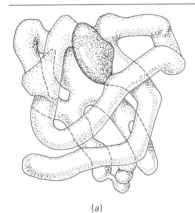

(a)

Figure 3-13. The structure of whale myoglobin. *(a)* A first relatively crude model of the protein based on x-ray diffraction analysis at low resolution. The general conformation of the molecule is visible but there is little information concerning the types of interactions among the amino acid residues. (The plane of the heme group is not correct in this diagram.) *(From A. White, P. Handler, and E. L. Smith, "Biochemistry," 5th ed., McGraw-Hill, 1975.) (b)* Front view of myoglobin showing the secondary structure of various regions of the polypeptide chain. The majority of amino acids reside within α-helical regions which comprise the linear pieces of the chain and enclose the heme. The nonhelical regions primarily occupy the corners where changes in direction occur. The position of the heme is indicated as are the locations of the two histidine residues involved in heme function (see page 93). The site at which the oxygen molecule is bound is indicated by the W. *(From R. E. Dickerson and I. Geis, "The Structure and Action of Proteins," W. A. Benjamin, Inc. Copyright 1969 by Dickerson and Geis.) (c)* A detailed model of the structure of myoglobin illustrating the enormous complexity of a protein. *(Courtesy of J. Kendrew.)*

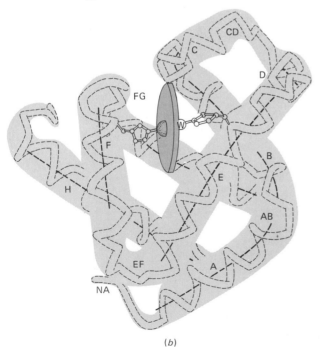

(b)

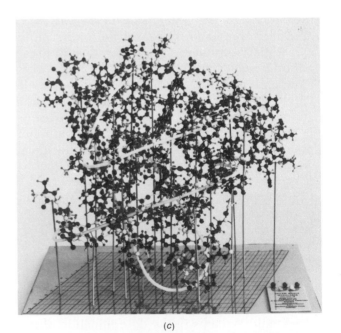

(c)

were orders of magnitude greater in complexity than those previously attempted. The crystals in this case were made of hemoglobin. It would be 23 years later before the structure of the molecule would be determined. In the meantime another investigator, John Kendrew, also at the University of Cambridge, began a similar study of a more simple, but functionally related molecule, myoglobin. In 1957, the overall structure of myoglobin was worked out. The first enzyme to be analyzed in detail by this technique was that of lysozyme, an enzyme which is capable of digesting certain polysaccharides of bacterial cell walls. The studies on lysozyme were conducted by David Phillips and coworkers in 1965. The analysis of these three proteins will be discussed in this chapter and in Chapter 4. During the past 15 years the structures of numerous other proteins have been probed by this technique, and remarkable insights into molecular activities have been obtained. The *Scientific American* articles by Kendrew, Perutz, and Phillips provide an excellent starting point into this subject.

The Structure of Myoglobin

The technique of x-ray-diffraction analysis is discussed at some length in the appendix to this chapter. The information that one obtains concerning the structure of a protein molecule depends upon the amount of information from the diffraction pattern that is processed. The early analysis of myoglobin utilized relatively less information and provided Kendrew and his colleagues with a glimpse at the overall manner in which the polypeptide chains were organized within the molecule. The profile of myoglobin at 6 Å is shown in Fig. 3-13a. It can be seen from this reconstruction that the molecule is compact, that the polypeptide chain is folded back on itself in a complex arrangement, and that there is no evidence of regularity or symmetry within the molecule. One of the questions of importance that could be answered is to what degree the α helix contributes to the structure of myoglobin. There are 8 stretches of α helix in this protein ranging from 7 amino acids to 24 amino acids in length (Fig. 3-13b). Altogether approximately 75% of the 153 amino acids in the polypeptide chain are in the α helix conformation. This is an unusually high percentage compared to that for other proteins that have been examined. In between the stretches of α helix, the chain can be considered to exist in a "random coil" signifying the absence of a typical structure. This is not to imply that these are not ordered regions or that they are not important. Just the reverse, they are likely to be very important parts with respect to the functional aspects of the structure. The α helix and β sheet conformations are more of structural importance and the other parts of the polypeptide are best suited to the unique properties of a given protein.

The higher resolution analysis provided the identification of many of the amino acid residues in the molecule. During the period when the x-ray diffraction was being carried out, the protein was also being sequenced by the analysis of overlapping fragments. Together these two approaches provided the means whereby particular amino acids could be placed at particular locations in the molecule and their roles established. For example, the factors that maintained or destabilized the pieces of α

helix could be analyzed. As expected, proline was an effective helix breaker, but certain of the places where the helix stopped and the polypeptide turned the corner did not contain a proline; other factors were responsible. Myoglobin contains no region of pleated sheet and has no disulfide bridges; the entire tertiary structure results from noncovalent interactions. All of the noncovalent bonds (Fig. 3-14) proposed to occur between side chains within proteins, i.e., hydrogen bonds, ionic bonds, and hydrophobic bonds, were found. The model of myoglobin pictured in Fig. 3-13c shows the relative position of every atom of the molecule.

Like the determination of amino acid sequences, the analysis of the three-dimensional organization of proteins has been one of the milestones in biochemistry and molecular biology. The central role of proteins in cell function has been known and appreciated for a long time; but, due to their large molecular weight and great complexity, it was not anticipated that this type of information would soon be obtained. Knowing the complete structure brings us closer to being able to explain cell functions in completely defined molecular terms. The remaining problem to consider is what happens within the protein molecule as it is carrying out its task. For example, can we tell which amino acids of an enzyme are responsible for binding a particular reactant during the reaction that it catalyzes? Can we determine if parts of the molecule are in motion relative to one another? The types of interactions that give a protein its tertiary structure are sufficient to maintain that structure in the cell but are not so rigid that flexibility within the molecule is not possible. In fact, a large body of evidence indicates that changes in conformations of proteins occur and these are critical for their functions. One of the best studied examples of a conformational change occurring during protein function is shown in Fig. 3-16. In this case, the shift in position of parts of the polypeptide chains of the hemoglobin molecule occurs upon the binding of oxygen to the protein.

Myoglobin Function

Considerable information is known, for example, about the relationship of myoglobin structure to its function. Myoglobin is a conjugated protein, made of a polypeptide chain (described above) and a heme group whose structure is shown in Fig. 9-9a and its position in the protein in Fig. 3-13b. Myoglobin functions in muscle tissue as a storehouse for oxy-

Figure 3-14. Types of noncovalent bonds involved in maintaining the native conformation of proteins. [*From H. K. Schachman,* Cold Spring Harbor Symp. Quant. Biol., **28**:418 (1963).]

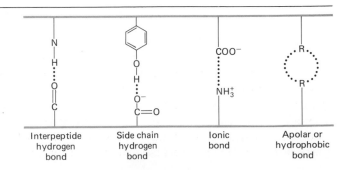

| Interpeptide hydrogen bond | Side chain hydrogen bond | Ionic bond | Apolar or hydrophobic bond |

gen, the oxygen being bound by the iron atom in the center of the heme until it is required by the tissue and is released. It is the polypeptide chain that provides the proper environment for the heme group to carry out its function. In order for the heme group to maintain its grip on molecular oxygen it must remain in the reduced form (the ferrous state) rather than become oxidized to the ferric state as would normally happen in solution as a result of an interaction with oxygen (or within the cytochrome heme of Fig. 9-9a). It is the nonpolar environment that the heme occupies in the myoglobin molecule that allows the binding of oxygen to occur in the absence of water and, therefore, without the oxidation (loss of electrons) of the iron atom. Within the nonpolar region of myoglobin, two conspicuously polar side chains, those of histidine, are also present (Fig. 3-13b). One of these histidine residues is directly bound to the iron atom, the other is nearby. Both of these amino acids are believed to play key roles in the oxygen-binding process.

Protein Folding

In this last section we have emphasized the complexity of the three-dimensional structure of proteins. An important question to consider is how such a complex, folded, twisted structure can arise in the cell. We know that the *linear sequence* of amino acids in a polypeptide chain is specified by the sequence of nucleotides in the messenger RNA, but how is the particular unique conformation characteristic of each protein generated in the cell? Since it is impossible to isolate polypeptide chains before they undergo their folding process inside the cell, some indirect procedure must be employed to study the factors involved in tertiary structure determination. The approach that has been taken is to isolate and purify intact proteins from cells, then treat them with agents that cause them to lose their secondary and tertiary structure, and then observe the capabilities that such disorganized polypeptides possess in *reforming* meaningful spatial organizations. The unfolding or disorganization of a protein is called *denaturation* and it can be brought about by a wide variety of agents including detergents, organic solvents, radiation, heat, and compounds such as urea and guanidium chloride. Though all of these are effective because they disrupt noncovalent bonds, they bring about the same result in different ways, many of which are poorly understood.

Denaturation can be either reversible or irreversible. Which of these occurs depends upon the specific protein being tested and the conditions by which it is denatured and allowed to return to its normal conformation, termed its *native state*. A commonly encountered example of irreversible denaturation is the cooking of an egg white. The transformation from the semifluid state of the raw egg to the solidified state of the cooked one illustrates an extreme case of protein denaturation and no one is ever likely to reverse this process. The first and best-studied case of *reversible* denaturation is that of *ribonuclease* studied by Christian Anfinsen and coworkers. Ribonuclease is a small enzyme which consists of one polypeptide chain of 124 amino acids with 4 disulfide bonds linking various parts of the chain (Fig. 3-15a). Denaturation is accomplished by subject-

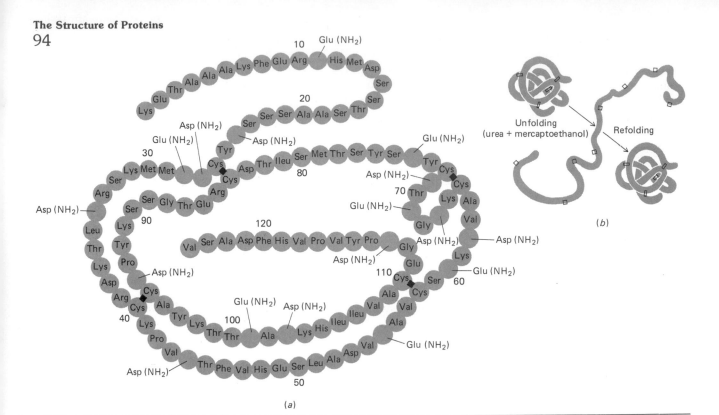

Figure 3-15. (a) The covalent structure of bovine pancreatic ribonuclease. (From C. B. Anfinsen, "New Perspectives in Biology," M. Sela, ed., Elsevier North-Holland, 1964.) (b) A native protein, with intramolecular disulfide bonds, is reduced and unfolded with β-mercaptoethanol and 8 M urea. After removal of these reagents, the protein is allowed to undergo spontaneous refolding and reoxidation. Asp (NH₂) and Glu (NH₂) represent the amino acids asparagine and glutamine. [From C. J. Epstein, R. F. Goldberger, and C. B. Anfinsen, Cold Spring Harbor Symp. Quant. Biol., **28**:439 (1963).]

ing the protein to a solution of mercaptoethanol, a reducing agent which breaks the disulfide bonds converting them to sulfhydryl (—SH) groups of cysteine, and 8 M urea, which breaks noncovalent bonds. Together these agents bring about the total disorganization of the protein. If the urea and mercaptoethanol are removed in an appropriate manner, active enzyme molecules are reformed (Fig. 3-15b) which are indistinguishable from those present at the beginning of the experiment.

In the experiment on the reversible denaturation of ribonuclease it is assumed that the urea brings about the total disorganization of the protein. There are numerous physical properties, such as viscosity, that provide a sensitive measure of the degree of denaturation, but it can always be argued that a small amount of order remains in the protein, and after the removal of the urea this residual structure allows the remainder of the folding to occur. The most unequivocal evidence that the primary structure does indeed determine the tertiary structure for ribonuclease is the more recent finding that this enzyme can be synthesized in the laboratory one amino acid at a time and will still fold up on its own into active enzyme molecules.

The results of these experiments and similar ones with other proteins, including myoglobin, suggest that the formation of the secondary and tertiary structure of a protein is a self-directed process, one that derives automatically from its primary structure. In other words, once a given linear sequence of amino acids is constructed, *given the appropriate environment*, the secondary interactions occur spontaneously to cause the protein to fold into the proper conformation; no additional information is required. As remarkable as this is, it is difficult to imagine it otherwise. Since each protein has a unique shape, if this shape did not form on its own, each would presumably require a special mechanism to accomplish it. Each mechanism would likely be as complex as the protein it folded and would have to be specified by some other mechanism, and so on, ad infinitum. Self-assembly is an elegant way out of this otherwise insoluble problem.

It would appear that of the vast number of possible conformations that could be formed, that of the native protein is thermodynamically favored above all of the others. This brings up another perplexing aspect of this phenomena. Does the polypeptide chain simply twist and fold in a random manner until the proper, thermodynamically stable conformation is arrived at? Given a polypeptide of over 100 amino acids containing over 1000 rotatable bonds, calculations indicate that it would take many years for all of the possible conformations to be visited. Yet in the cell the folding occurs very rapidly, and even in the test tube it is usually accomplished within a matter of minutes. Though conditions within the cell may speed the process somewhat as compared to folding in the test tube, it appears that the speed with which it occurs is inconsistent with it being a totally random event. It is generally believed that the folding process occurs in steps; once one structural state is reached, clues are present that direct the process to the next stage, thereby eliminating the need to search through many unstable conformations. These metastable (slightly stable) stages are called *nucleation states*. The alpha helix is believed to be an important aspect of the early nucleation states; the helical portions form spontaneously and rapidly, before the distant parts of the chain become folded.

It is believed that, for most proteins at least, the formation of hydrophobic bonds is the most important factor in directing the formation of the proper shape. Analysis of numerous globular proteins has shown that the core of the protein consists of amino acids with hydrophobic side chains that are in very close contact with one another and held together by a large number of hydrophobic bonds. Water is almost totally excluded from this region, producing what is essentially a nonaqueous environment within the protein.

The thermodynamic "force" driving the folding process is not well understood. It is generally believed that the association of hydrophobic groups results in an overall increase in the entropy (disorder) of the system. As discussed in Chapter 4, reactions proceed to maximize the entropy in the universe. As long as the hydrophobic groups are on the surface, they cannot form hydrogen bonds with the surrounding water molecules, and they will be responsible for organizing (ordering) the solvent into a cage around them. If they withdraw from the solvent, water will achieve a more disorganized state and therefore their retraction into the center of the protein will be thermodynamically favored. The situation is similar in principle to the spontaneous association of small oil droplets in an aqueous solvent to form larger drops.

Once the primary folding processes occur, the side chains are brought

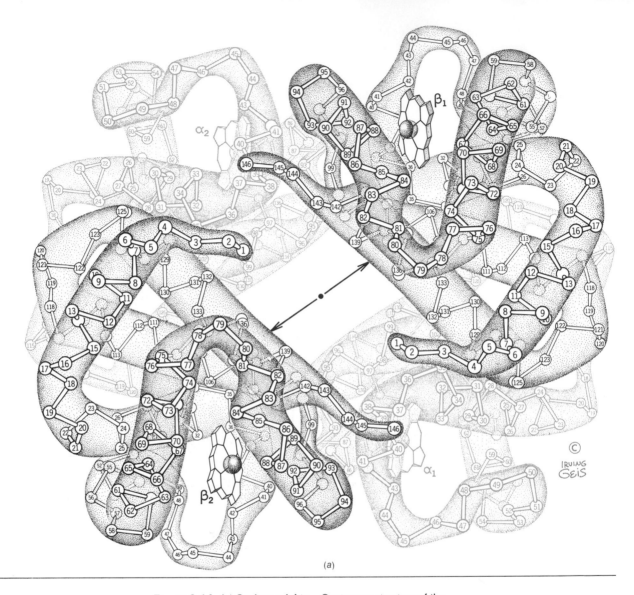

(a)

Figure 3-16. *(a)* Oxyhemoglobin. Quaternary structure of the
oxyhemoglobin tetramer. This view is looking down the true axis of
symmetry. There are two identical β chains and two identical α
chains. In this view the β chains are on top. The true axis is
represented as a dot in the center of the molecule. The double-
headed arrow represents the distance separating β chains. The
numbered circles represent positions of α carbon atoms which
define the backbone of the polypeptide chain of each subunit. *(b)*
Deoxyhemoglobin. Oxygen binds to the iron atoms (dark ball at
the center of each of the four hemes, labeled β₁, β₂, α₁, α₂).
Upon the loss of oxygen, the molecule undergoes a change in
conformation. There is some change in tertiary structure and slight
alteration in the β₂, α₂ subunit relationship. By far the greatest
change is the moving apart of the β subunits. Compare the greater
length of the arrow in deoxyhemoglobin with the arrow at the left in
oxyhemoglobin. *(Adapted from A. N. Schechter and I. Geis, "The
Molecular Biology of Hemoglobin," Elsevier North-Holland, 1979.
Hemoglobin drawings © 1979 by Irving Geis.)*

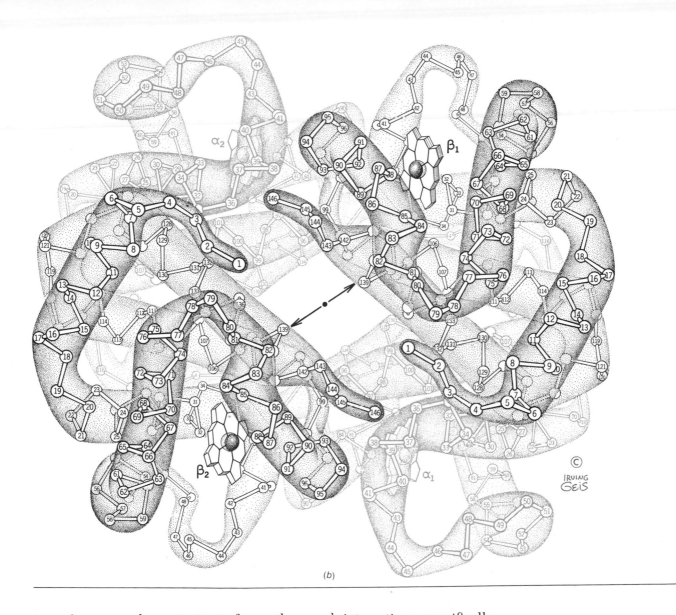

(b)

into close enough proximity to form other weak interactions, specifically hydrogen bonds and other electrostatic bonds. Disulfide bridge formation is more responsible for stabilizing a three-dimensional structure already present than for causing it to be generated in the first place. In many proteins, disulfide bridges can be gently broken and the shape of the protein, and to an extent its activity, can be retained.

Quaternary Structure and Beyond

Whereas many proteins such as ribonuclease and myoglobin are composed of only one polypeptide chain, many others are made up of more than one polypeptide chain or subunit. Proteins composed of subunits are said to have quaternary structure (Fig. 3-16). It is generally found that proteins with molecular weights above 50,000 daltons have quaternary structure. In some proteins, the polypeptide chains are identical, while in other cases there is more than one variety. In some cases isolated polypeptide chains are completely inactive, while in others they can carry out the same function as the complex, though not necessarily with the same kinetics. The best studied protein, in terms of quaternary structure analysis, is hemoglobin. Each molecule of hemoglobin present within an adult human red blood cell contains two alpha and two beta polypeptide chains. A model of the arrangement of the four subunits of the hemoglobin molecule is shown in Figure 3-16. If each of these subunits is considered individually, it is found to be very similar in overall tertiary structure to that of myoglobin. Furthermore, the oxygen-binding properties of the isolated hemoglobin subunits are somewhat similar to myoglobin. However, when intact hemoglobin tetramers are studied, new oxygen-binding properties (discussed in the following chapter) are found to emerge as a consequence of the interaction among the subunits themselves. In other words, the close association of polypeptide chains within the same protein promotes a type of communication within the molecule between distinct polypeptide units.

The subunits of quaternary level proteins are generally bound together by noncovalent bonds, either electrostatic or hydrophobic, and can be readily separated from one another. Subunits can be dissociated from one another and tested for their ability to reassociate when the dissociating agents are removed. As in the formation of the tertiary structure of ribonuclease from the denatured polypeptide chain, the quaternary structure can also be obtained spontaneously from the separated subunits. The conclusion reached once again is that the information contained in the amino acid sequence has far-reaching consequences for protein organization and, therefore, protein function. In a sense, there are two sets of information in the same stretch of amino acids. One set is responsible for the proper secondary, tertiary, and quaternary structure to be generated and the other set for providing the proper reactivity of the molecule.

The spontaneous association of polypeptide chains does not stop with the achievement of quaternary structure. There are numerous examples in which different enzymes, i.e., proteins catalyzing distinct and separate reactions, are associated together to carry out a greater cooperative function. Such aggregates are termed *multienzyme complexes* and are typically responsible for catalyzing a *sequence* of reactions. The product of one enzyme is channeled directly as a reactant for the next enzyme in the sequence without having to equilibrate in the medium of the cell. Having the enzymes associated with one another overcomes any difficulties that might result in having the steps spatially separated. An example of a multienzyme complex is shown in Fig. 3-17.

If distinct proteins can become associated with each other in a highly predictable and ordered fashion, it should be possible for different types of biological molecules to do the same. Cellular organelles are composites of various macromolecules, ribosomes consist of protein and RNA, chromosomes of DNA and protein, membranes of protein and lipid, and so forth.

Figure 3-17. Pyruvate dehydrogenase of *Escherichia coli,* an example of a multienzyme complex. Each complex contains 60 polypeptide chains comprising 3 different enzymes. Its molecular weight approaches 5 million daltons. *(a)* Electron micrograph of the negatively stained complex. *(b)* A model of the complex. Twelve pyruvate dehydrogenase dimers (black spheres) are distributed symmetrically along the twelve edges of the transacetylase cube, and the six flavoprotein dimers (small grey spheres) are placed in the six faces of the cube. *(Courtesy of L. J. Reed.)*

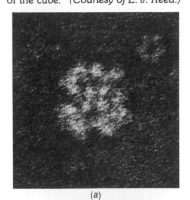

(a)

(b)

Can these structures be derived by the *spontaneous* association of less complex components, i.e., *self-assembly,* or does their association require the guidance of outside agents, for example, enzymes needed to catalyze the formation of covalent bonds? In other words, how far can subcellular organization be explained simply by having the pieces fit themselves together noncovalently to form the most thermodynamically stable arrangement? The means by which cellular organelles are assembled is one of the least understood topics in cell biology. As we consider the various organelles in later sections of the book, some attention will be paid to their formation. In the following section we will discuss a couple of the best-studied examples of molecular self-assembly.

Tobacco Mosaic Virus

The most convincing evidence that a particular assembly process is self-directed is the demonstration that the assembly can occur outside the cell (in vitro) under physiological conditions when the only macromolecular materials present are those that make up the final structure under analysis. In a "reconstitution" type experiment of this nature, the components are purified, mixed together, and the mixture analyzed for the presence of the assembled structure. In 1955, Heinz Fraenkel-Conrat and Robley Williams demonstrated, using tobacco mosaic virus (TMV), the remarkable ability that biological macromolecules have for self-assembly. Working with this virus (Fig. 3-18), whose mature particle consists of one long RNA molecule (approximately 6600 nucleotides) wound within a helical capsule made of approximately 2200 identical protein subunits, they found that, upon mixing the purified RNA and protein components in vitro, mature infectious viral particles could be formed. More recent analysis has apparently demonstrated the manner in which these viruses are assembled from the two components and illustrates the complex interactions and conformational changes that macromolecules can undergo.

The protein unit of TMV is the 4S subunit (S is roughly a measure of molecular weight, page 468), but the unit of assembly is believed to be a specific complex of 34 subunits. This complex is a disk made of two layers of subunits, 17 per layer, arranged in a circle. If a preparation of these disks is combined with the RNA, complete virus particles are formed, much more rapidly than if one starts with the purified 4S protein. The conversion from the 4S subunit to the disk, which is the intermediate in the assembly process is the slower event. Initiation of assembly begins with the association of a disk with a specific region of the RNA molecule. Although it was previously believed that assembly began at one end of the RNA molecule, more recent evidence indicates that the first disk attaches approximately 830 nucleotides from one end with assembly proceeding first in one direction and then in the other direction. The assembly of TMV results from the ordered accumulation of disks, but how can such disks form the helical structure of the virus? A pile of disks would merely form a cylinder, while a helix is a continuing spiral that requires none of the disks to have a closed shape. The answer to this puzzle is seen in

Figure 3-18. Tobacco mosaic virus (TMV). *(a)* Diagram of a portion of a TMV particle. The protein subunits are identical along the entire rod and together enclose the single helical RNA molecule shown protruding at the top (without its associated protein). [*From A. Klug and A. D. Caspar,* Adv. Virus Res., *7:274 (1960).*] *(b)* Electron micrograph of a shadowed TMV particle after a portion of the protein was removed with phenol. [*From M. K. Corbett,* Virology, *22: 539 (1964).*]

(a)

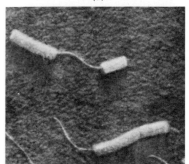

(b)

a comparison of parts (*a*) to (*b*) of Fig. 3-19, and reemphasizes the complex conformational changes that protein molecules can make.

The transition to a helical unit occurs because of a shift in the way the subunits are organized within the disk; a dislocation converts the two-layered disk into two turns of the spiral. This structure is reminiscent of a "lock washer" found in a hardware store. The most thermodynamically stable structure is the disk which minimizes the repulsive interaction of negatively charged carboxyl groups present in each subunit. If this ionic repulsion can be overcome, either by lowering the pH in the test tube or by combining with the RNA molecule inside the infected cell, then the lock-washer conformation becomes preferred and the two layers spontaneously shift over each other by about 10 degrees. The requirement for RNA in this transition ensures that particles will never form without the genetic material. An intermediate in the TMV assembly process is shown in Figure 3.19*c*. Incomplete particles of the type diagrammed in Figure 3.19*c* are seen in the electron microscope to be composed of a partial cylinder having the RNA molecule present within the central channel and its two free ends emerging from one end of the particle.

The Ribosome

Ribosomes, like TMV particles, are made of RNA and protein. Unlike TMV, ribosomes contain several different RNA molecules and a considerable collection of different proteins. It seems safe to state at the present time that the ribosome looms as the only cell organelle whose complete molecular organization is likely to be elucidated in the next few years. As a result, the ribosome holds a special place in the field of molecular biology and has been a focus of research for many years. All ribosomes, regardless of their source, are composed of two subunits of different sizes. The ribosomes of prokaryotic cells contain fewer proteins and are considerably smaller than those from eukaryotic cells. The ribosome from *E. coli* has a sedimentation coefficient (page 498 of about 70S and is composed of a large and a small subunit of 50S and 30S, respectively

Figure 3-19. Stages in the assembly of TMV. *(a)* and *(b)* The dislocation of the disks to the "lock-washer" form because of interaction with the RNA. [*Reprinted from A. Klug*, Federation Proceedings, **31**:40 *(1972)*.] *(c)* The proposed structure of a TMV particle intermediate in the assembly process. [*From G. Lebeurier, A. Nicolaieff, and K. E. Richards*, Proc. Natl. Acad. Sci. (U.S.), **74**:152 *(1951)*.]

Initiation

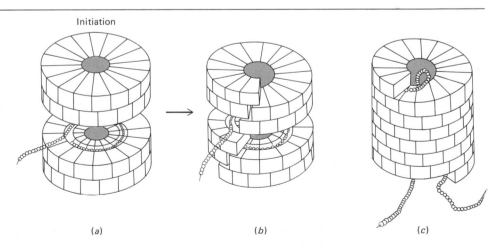

(a)   (b)   (c)

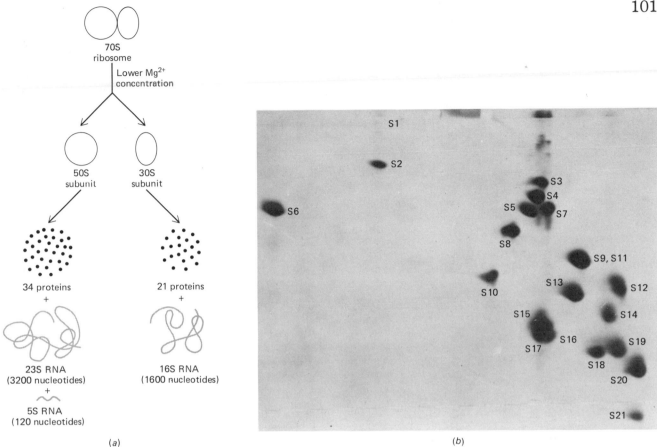

(a)

(b)

Figure 3-20. (a) The dissociation of the bacterial ribosome into its sub-units and then into its RNA and protein components. (b) Two-dimensional polyacrylamide gel electrophoresis of the proteins of the small subunit. [*After W. A. Held, S. Mizushima, and M. Nomura, J. Biol. Chem.,* **248**:5727 (1973).]

(Fig. 3-20*a*). The 50S subunit contains two molecules of RNA: the 23S ribosomal RNA (approximately 3200 nucleotides) and the 5S ribosomal RNA (approximately 120 nucleotides). In addition, the large subunit has 34 different proteins (L1 to L34). The 30S subunit contains one molecule of RNA (the 16S ribosomal RNA) of approximately 1600 nucleotides and 21 different proteins (S1 to S21). One protein of the small subunit (S20) is also found in the large subunit (L26). One protein of the large subunit exists in two different forms (L7 and L12) which differ by acetylation.

The 55 proteins of the *E. coli* ribosome can be separated from one another (Fig. 3-20*b*) and each can be purified and studied in isolation. Remarkably, if the 21 proteins of the small ribosomal subunit are mixed with the 16S RNA under the appropriate conditions, complete, fully func-tional 30S ribosomal subunits can be formed in vitro. This feat was ac-complished during the 1960s by Masayasu Nomura and his coworkers, who thereby demonstrated that the components of the 30S subunit contain all of the information for the assembly of the particle. There appears to be an approximate sequence in which the ribosomal proteins associate with the RNA molecule, a sequence which presumably reflects to some degree their assembly in vivo. For example, it was found that six or seven of

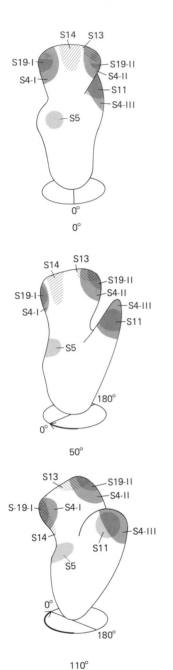

Figure 3-21. One model of the 30S ribosomal subunit showing the location of selected proteins as viewed from different positions. [*From J. A. Lake,* J. Mol. Biol., **105**:154 (1976).]

the proteins bind directly to different parts of the ribosomal RNA. The binding of this first group of primary ribosomal proteins then sets the stage for the attachment of a group of secondary ribosomal proteins. Once the secondary proteins have been added to the growing particle, the remaining proteins attach and the assembly is complete. More recent studies have led to the complete reconstitution of the 50S subunit from its components, and the stage has been set for similar accomplishments on the more complex eukaryotic ribosomes, which have a sedimentation coefficient of 80S and are made of a large and a small subunit of 60S and 40S, respectively.

The development of an in vitro reconstitution system has led to further analysis of the role of the various ribosomal proteins. For example, one can ask what the effect of omitting or modifying one or more of the component proteins has on the assembly or function of the 30S subunit. Some proteins appear to be required exclusively for the assembly of the subunit while others appear to function in some aspect of protein synthesis. The 16S ribosomal molecule is clearly required for both the assembly and function of the organelle. Chemical modification of only a few nucleotide bases is sufficient to destroy all of its activity.

A number of laboratories in recent years have been concentrating their efforts on the determination of the position of various proteins in the ribosome. Various techniques have been employed in this regard, the most successful one involving the use of antibody molecules capable of attaching to only one of the various proteins. In order to carry out this determination, 30S ribosomal subunits are mixed with a preparation of antibodies specific for one of the proteins. If this protein is present at the surface of the ribosome (and all of the 30S ribosomal proteins appear to have some part of their structure at the surface), then the antibody should be able to reach the protein and attach to it. When this complex is viewed in the electron microscope, the position of the antibody on the surface of the ribosome can be determined. A topographical map of a number of proteins of the 30S subunit is shown in Fig. 3-21. These types of studies as well as others have revealed the close association that ribosomal proteins have with one another and with the ribosomal RNA. The fact that the components are so interdependent makes it difficult to assign particular roles to one or another constituent, since its position and activity will be affected by the condition of its immediate neighbors. As a result, the emphasis in research has shifted to some degree from attempts to assign roles to specific proteins to attempts to determine the activity of particular regions (or domains) of the ribosome, knowing that each of these regions is composed of several interacting molecules.

Collagen

The formation of TMV particles appears to occur by a straightforward process of self-assembly. Within the cells of the tobacco plant, the two macromolecular components of the particle are synthesized and come together spontaneously to form mature, infectious virus. The formation of collagen fibers is a much more complex process and illustrates that structures can be built in a sequential manner in which some of the steps occur by self-assembly, while others require the intervention of an external agent. Collagen is a fibrous protein of very high tensile strength. It is estimated that a fiber of 1 mm diameter is capable of suspending a weight of over 20 pounds. Collagen is generally found outside of cells, having been secreted into the extracellular space from which it can be extracted (often with great difficulty) and studied in vitro. Collagen is a characteristic of connective tissues throughout the animal kingdom and is present in vertebrates as a major component of skin, cartilage, bone, tendons, and the cornea.

Collagen fibers are polymers constructed from a building block or monomer. Each collagen monomer is composed of three long, helical polypeptide chains (termed α chains) which are hydrogen bonded to each other in a very tightly fitting complex. Collagen molecules (the trimer) are themselves helical (a superhelix since each is composed of helical subunits) and approximately 2900 Å long and only 17 Å diameter (Fig. 3-22). The collagen molecule is therefore a thin, highly elongated structure. The amino acid composition of the α chains is highly unusual. Each chain is about 1000 amino acids in length with essentially every third residue being that of glycine. The R group of glycine is only a hydrogen atom and is a very important amino acid for just that reason. Due to its lack of a side chain, glycine residues allow the backbones of two polypeptides to approach each other very closely. The helix of the α chains of collagen is not an α helix; the two differ in many ways. The glycines, which occur every third amino acid, are located where each chain turns to the inside of the trimer, i.e., where the least amount of room is available for an amino acid side chain. In fact, glycine is the only amino acid that could fit into the triple-helical organization of collagen.

There are two other very unusual features of the amino acid composition of the α chains: they have a large amount of proline and many of the proline and lysine residues are hydroxylated. The prolines are important in generating the particular type of helix that the α chains assume and the hydroxylated amino acids are important in maintaining the stability of the triple helix by forming hydrogen bonds from one chain to another. The hydroxylation of proline and lysine occurs enzymatically after the amino acids have been incorporated into the forming polypeptide chains. Since hydroxylation occurs after protein synthesis, it is said to be a *postsynthetic (posttranslational)* modification. If the hydroxylation event is blocked, the triple helix of collagen is unstable and becomes denatured at temperatures of approximately 25°C. In other words, in direct contrast to the situation for myoglobin and other globular proteins previously discussed, the formation and maintenance of the tertiary structure of the collagen monomer depends upon changes occurring after the polypeptide chains are synthesized, not simply upon their linear amino acid sequence. The symptoms of the vitamin C deficiency, scurvy, are a consequence of an inability to hydroxylate the α chains of collagen. Ascorbic acid (vitamin C) is a required reducing agent in this enzymatic process and in its absence the collagen deposited is lacking in sufficient hydroxyl groups to maintain its necessary structure.

If the prime criterion for self-assembly is the in vitro reconstitution of the structure from the purified components, then it would appear that collagen monomers arise in some other manner, since purified α chains show very little tendency to reform the trimer when incubated under physiological conditions in vitro. As was recently discovered, the matter is more complex. The α chains present in collagen molecules prepared from extracted collagen fibers are not in the same state as the α chains that are first synthesized; they are

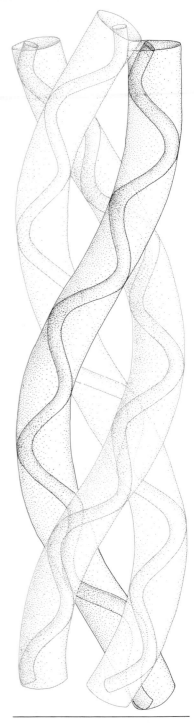

Figure 3-22. The triple helix of the collagen molecule. *(From R. E. Dickerson and I. Geis, "The Structure and Action of Proteins," W. A. Benjamin, Inc. Copyright 1969 by Dickerson and Geis.)*

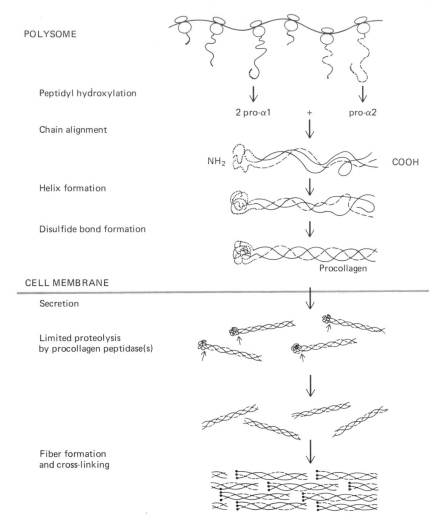

POLYSOME

Peptidyl hydroxylation

2 pro-α1   +   pro-α2

Chain alignment

NH₂   COOH

Helix formation

Disulfide bond formation

Procollagen

CELL MEMBRANE

Secretion

Limited proteolysis
by procollagen peptidase(s)

Fiber formation
and cross-linking

Figure 3-23. Stages in the formation of collagen fibers. As described in the text, recent findings indicate the presence of additional amino acid residues (not shown) at the C-terminal region of the pro-α chains which are removed during the extracellular processing steps. *(Reproduced, with permission, from P. Bornstein,* Annual Review of Biochemistry, *Volume 43.* © *1974 by Annual Reviews Inc.)*

considerably shorter (Fig. 3-23). There is an additional fragment of approximately 20,000 molecular weight at the N-terminal end and a piece of approximatley 35,000 at its C-terminus. In other words, the α chains of collagen are synthesized as a longer molecule, termed pro-α chain, pieces of which are later chopped away and discarded. This is not an unusual situation. There are numerous examples of proteins which originate from molecules larger than the final active product. A few other examples include proinsulin and insulin, chymotrypsinogen and chymotrypsin, and fibrinogen and fibrin. In the case of collagen it appears that the extra pieces are important in causing the α chains to become aligned and bound together as a trimer, procollagen. Procollagen is secreted by the cell (typically a fibroblast) to the extracellular space at a stage in which the pro-α chains are linked to one another by disulfide bridges between cysteine residues in one of the extra fragments. At a later stage, outside of the cell, the definitive collagen molecule is formed from procollagen by the action of specific proteolytic enzymes. The failure to remove the additional amino acids occurs in certain congenital diseases which are characterized by severe connective tissue problems.

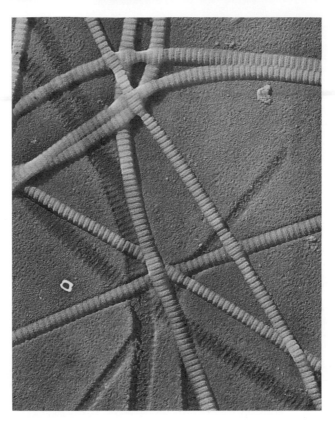

Figure 3-24. An electron micrograph of a chromium-shadowed replica of native human collagen fibrils. Note the cross-banding pattern. The bands repeat along the fibril with a periodicity of 640 to 700 Å One of the key early questions in collagen research was how a molecule 2900 Å long could polymerize to form a fiber with a banding pattern approximately one-quarter this length. Detailed mapping studies have revealed the precise manner in which the collagen molecules *overlap* to produce the observed pattern. [*From J. Gross and F. O. Schmitt*, J. Exp. Med., **88**:555 (1948).]

It would appear from the above discussion that collagen formation occurs in steps (Fig. 3-23). Some of these steps, such as the association of three pro-α chains occur by self-assembly. Other steps, such as the formation of the proper tertiary structure, require one enzymatic event, in this case hydroxylation of specific residues. Up to this point we have followed the course of events from the synthesis of the polypeptides to the formation of collagen molecules. In order to understand the organization of bundles of collagen fibers as they are found in connective tissues (Fig. 3-24) we have to consider the formation of supramolecular complexes of which the collagen molecule is simply the smallest building block. Collagen monomers become organized into micro-fibrils, which are in turn associated into fibrils and larger fibers. The assembly process does not stop with the formation of fibers, since these, in turn, must be packed in some highly specific manner within the extracellular space. Collagen fibers are organized into parallel bundles in tendons, into networks around blood vessels, into a plywoodlike matrix in the cornea (page 270), and in an essentially random manner in skin.

An unusual biochemical characteristic of collagen fibers is their continuing modification after the final polymerization steps have occurred. This modification occurs by the action of extracellular enzymes, which cause collagen molecules to become covalently cross-linked, both within the collagen molecules (between chains) and between them. As cross-linking continues, the collagen fibers become more and more insoluble until they can no longer be extracted even under acidic conditions. Cross-linking causes significant changes in the physiological properties of collagen and is believed to be of

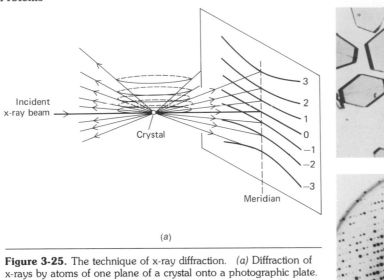

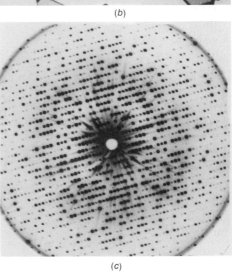

Figure 3-25. The technique of x-ray diffraction. *(a)* Diffraction of x-rays by atoms of one plane of a crystal onto a photographic plate. *(From R. F. Steiner, "The Chemical Foundations of Molecular Biology," Van Nostrand, 1965.) (b)* Myoglobin crystals such as those used for analysis. *(c)* The pattern of spots formed on a photographic plate by diffraction from a crystal of myoglobin. [*(b)* and *(c) Courtesy of J. Kendrew.*]

major importance in the aging process (Chapter 19). The picture of collagen assembly has been slowly uncovered over a number of years and many questions, particularly those relating to the packing of the smaller units to form the large fibers, remain unanswered. Regardless of the presence of unanswered questions, it is clear that the biology of collagen is a complex topic, and particularly so when one considers that most of the events occur outside the confines of the cell.

APPENDIX ON PROTEIN ANALYSIS
X-ray Diffraction Analysis

The first steps in the analysis of a particular protein by any technique is to obtain the most purified preparation of that protein that is possible (pages 109–120). X-ray diffraction analysis utilizes protein crystals, the preparation of which is one of the best guarantees of molecular purity. A crystal is composed of a regularly repeating arrangement of a unit cell, in this case an individual protein molecule. The procedure involves bombarding a protein crystal (Fig. 3-25a, b) with a thin beam of x-rays of a single (monochromatic) wavelength and allowing the radiation that is scattered (diffracted) by the electrons of the protein atoms to strike a photographic plate placed behind the crystal. Just as each protein molecule is repeated in a periodic manner within the crystal, so is each atom within the molecule. The diffraction pattern produced by the crystal (Fig. 3-25c) is determined by the nature of the structure within the protein itself; the many molecules in the crystal simply serve to reinforce the reflections, i.e., they act as one giant molecule.

As in the case of the diffraction of light by a specimen on the stage of a microscope, the diffraction of the x-rays contains information about the nature of the diffracting source. In the microscope the objective lens collects the diffracted rays and recombines them to form an image of the specimen. In x-ray crystallography, the recombination must be done by the crystallographer based on the information from (1) the position of spots (reflections), (2) their intensity on the photographic plates, and (3) the phase of the waves, which is obtained by comparison of patterns produced with and without heavy metals attached to the protein. The heavy metals provide reference points within the protein.

In other words, based on the information from the photographs, the investigator works backward to derive the structure that must have been responsible for producing the patterns. The positions and intensities of the spots can be related mathematically to the electron densities within the proteins, since it is the electrons of the atoms that produced them. The task is to determine the locations of the electron densities and, thus, the locations of the atoms from this information. The way the crystal is bombarded, each plate represents a slice of the molecule, and many plates must be analyzed before the total diffraction pattern is obtained. An example of a single photographic plate is shown in Fig. 3-25c.

The steps in going from a series of photographs such as that of Fig. 3-25c to a model of the molecule (Fig. 3-13c) is an extremely complex and laborious task. The degree of difficulty depends upon how great a resolution of the structure is sought. As it turns out, those spots closer to the center of the pattern result from x-rays scattered at smaller angles from the crystal and provide information about the grosser aspects of the protein, i.e., the long spacings within the molecule. The spots closer to the periphery of the photograph represent the closely spaced aspects of the molecules within the crystal. The resolution that is obtained is determined by the number of spots that must be analyzed.

Myoglobin has been analyzed successively at 6 Å, 2 Å, and 1.4 Å, with years elapsing between each completed determination. Considering that covalent bonds are between approximately 1 to 1.5 Å in length and noncovalent bonds between 2.8 Å and 4 Å in length, the nature of what can be learned will be greatly dependent upon the resolution obtained. A comparison of the electron density of diketopiperazine at four levels of resolution is shown in Fig. 3-26. In myoglobin, a resolution of 6 Å is sufficient to show the manner in which the polypeptide chains are folded and the location of the heme moiety, but is not sufficient to show structure within the chain. For example, to be sure of the degree to which a part of the polypeptide chain is in the α helix or some other conformation, or to try to identify which amino acids are present in which location, or to assign a particular type of bonding between two side chains, requires greater resolution and thus the analysis of many more spots.

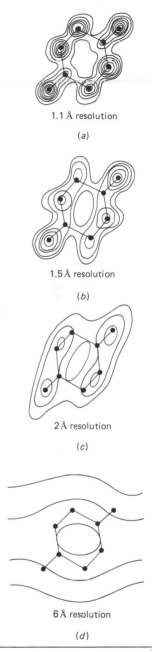

1.1 Å resolution

(a)

1.5 Å resolution

(b)

2 Å resolution

(c)

6 Å resolution

(d)

Figure 3-26. The electron density distribution of diketopiperazine calculated at various levels of resolution. At the lowest resolution, the only aspect distinguishable is the ring nature of the molecule, while at the highest resolution the electron density (indicated by the circular contour lines) around each atom is revealed. [*From D. Hodgkin*, Nature, **188**:445 (1960).]

At a resolution of 2 Å, groups of atoms can be separated from one another, while at 1.4 Å individual atoms can be seen. For resolution at 6 Å, approximately 400 reflections were analyzed, at 2 Å approximately 10,000, and at 1.4 Å approximately 25,000, essentially all of the spots present on all of the plates. The number of spots that must be examined is actually much greater than the numbers, given, since several heavy metal derivatives of the molecule must be examined at several different exposures. At 2 Å resolution, for example, the density of approximately 250,000 spots was measured. These procedures must be automated and high speed computers are utilized to analyze the data. At 1.4 Å, the electron densities for approximately a half million points in the molecule were determined.

The way in which the electron densities of the protein are displayed is shown in Fig. 3-27. The entire molecule is constructed of a stack of plastic layers. The electron density calculated for a given layer is displayed in the form of a contour map drawn on a Leucite sheet. The individual sheets are then stacked on top of one another (Fig. 3-27) to give the spatial organization of the entire molecule. The last step is to translate the electron density profiles, which show where atoms are concentrated in space, into the three-dimensional arrangement of specific molecules and bonds between molecules that they represent. The degree to which this can be accomplished depends upon the resolution that is achieved (Fig. 3-26).

Figure 3-27. The construction of the electron densities of various planes through the myoglobin molecule. *(Courtesy of J. Kendrew.)*

Isolation, Purification, and Fractionation of Proteins

In this past chapter and throughout the remainder of the book we will be considering the properties of one protein or another. Before information about the structure or function of a particular protein can be obtained, some attempt must be made to isolate and purify it prior to further analysis. It is the purpose of this section to briefly survey a few of the techniques that are most relevant to protein analysis in cell biology in order to provide a better understanding of the types of experimental approaches that have been taken to obtain some of the data presented.

For the present purpose, protein methodology can be divided into two broad categories, *preparative* and *analytical*. Preparative techniques lead to the isolation and purification of a particular molecule by a process of removal of other molecules, i.e., contaminants. The term *preparative* implies use of the material beyond simply its isolation. Analytical procedures allow one to estimate the molecular diversity present in a given preparation. The term *analytical* implies the disassembly of the whole so that its parts can be examined. For example, if one were to determine the number of different types of macromolecules present in bacterial ribosomes, the first steps would be of a preparative nature, so that a purified sample of ribosomes could be obtained from which the components could be solubilized. The second task would require that the components be spread out, i.e., analyzed, in some manner so that their diversity could be assessed. Analytical procedures require the mixture to be *fractionated* into its component ingredients. In many cases the same types of techniques accomplish both aims, since the fractionation of a mixture of materials simultaneously accomplishes a purification of each substance.

The purification process is generally performed in steps in which contaminants are removed from preparations and the desired protein is retained. Purification is generally measured as an increase in *specific activity* (the ratio of the amount of that protein to the total amount of protein present). Some identifiable feature of the specific protein must be utilized so that the relative amount of that material can be determined, i.e., *assayed*. For instance, if the protein is an enzyme, the assay consists of a test to measure the catalytic activity of the preparation. Other tests can be immunological, electrophoretic, electron microscopic, and so on. Measurements of total protein can be made in various ways. Sensitive colorimetric methods have been developed utilizing reactions specific for protein; these techniques generate a colored solution whose intensity is proportional to the amount of protein present. Certain amino acids absorb light of ultraviolet wavelengths (tyrosine and tryptophan primarily) and, assuming that these amino acids are present in normal percentages, this can provide a measure of total protein. Most often specific activity determinations are based on total nitrogen which can be very accurately measured and is quite constant at about 16% of the dry weight for all proteins. If one considers that most cells are likely to have over a thousand different proteins, purification can be a formidable task.

An important first step toward the goal of obtaining a purified preparation of a specific protein is to find an organism and tissue particularly rich

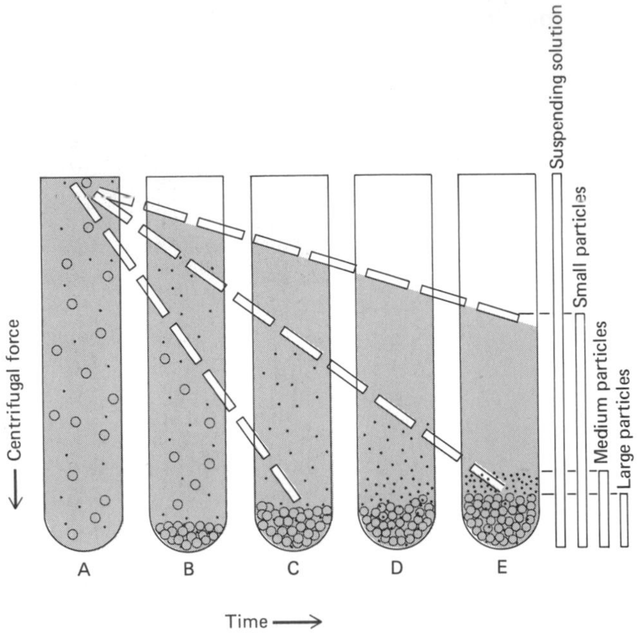

Figure 3-28. Sedimentation of a suspension of three-sized particles in a centrifugal field. The large particles sediment first but will be contaminated with smaller particles which sediment more slowly. Further purification is achieved by further centrifugation steps (not shown). [*From N. G. Anderson,* Natl. Cancer Inst. Mono., **21**:10 (1966).]

in that material. This was one of the foremost reasons that myoglobin prepared for x-ray diffraction studies was obtained from the muscle tissue of the whale, a diving mammal which relies on stored oxygen for long periods of time and provides a very rich source of this oxygen-binding protein. In some cases one must take advantage of special situations. For example, the purification of useful quantities of the β-galactosidase repressor from *E. coli* (page 590) could be obtained only after mutants had been isolated that produced abnormally high amounts of this protein. In the normal bacterial cell only a few molecules of repressor exist and such quantities do not provide a feasible source. In the following pages, the discussion of techniques utilized elsewhere in this book is set in the large typeface.

Differential Centrifugation of Subcellular Particles

The first step in the purification procedure is to determine in which part of the cell the protein resides. To accomplish this step, cells are usually broken open in an isotonic buffered solution (often containing sucrose) by one of a variety of homogenization techniques, and the contents of the cell separated into various fractions. The homogenate is fractionated most commonly by differential centrifugation (Fig. 3-28), which depends on the principle that particles of different size will travel toward the bottom of a centrifuge tube at different rates in a centrifugal field. The technique for cell fractionation is shown in Fig. 3-29a. If low centrifugal forces are applied for short time periods, only the nuclei will sediment into a pellet. At greater forces the mitochondria can be spun out, then the microsomes (the fragments of vacuolar and reticular membranes of the cytoplasm) and the ribosomes. For these last steps, the ultracentrifuge which can generate speeds to produce over 100,000 times the force of gravity is generally used. Once the ribosomes have been removed, the supernatant consists of the soluble phase of the cell's protoplasm and those particles too small to be easily removed from suspension. Each of these cell fractions can be isolated and subjected to further, more critical, fractionation procedures to isolate a particular cell organelle or ensure greater purity. In many cases, further purification is accomplished by centrifugation of the preparation through a gradient of sucrose (Fig. 3-29b) which distributes the contents into various layers. Once the fractionation of the cell's contents has been accomplished, the proteins can be extracted from each fraction and the specific activity of the desired molecule determined. Since most proteins are localized in one part of the cell or another, most of the proteins should be found primarily associated with one of the purified fractions.

Figure 3-29. Cell fractionation. (a) Homogenization and initial fractionation steps by differential centrifugation. (b) Further purification by density gradient centrifugation. In the sedimentation steps of (a) the particles will continue to sediment as long as they are subjected to centrifugation since the medium is less dense than the particles. In (b) the medium is composed of a density gradient, and the particles move until they reach a place in the tube equal to their own density where they form bands.

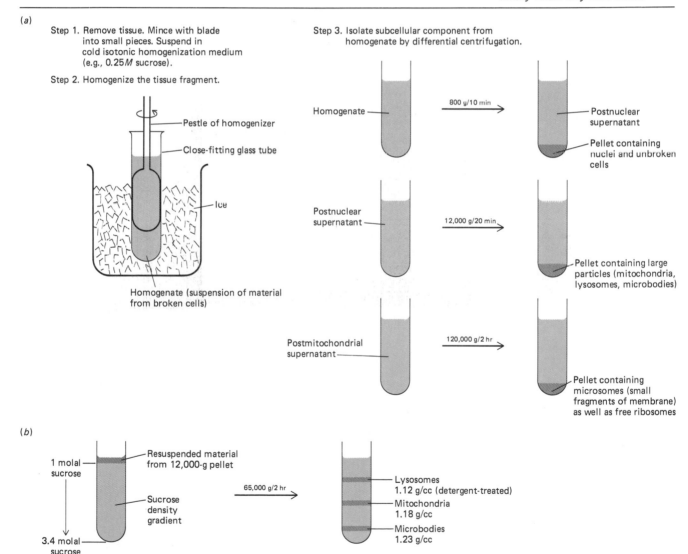

Molecular Purification and Fractionation

Selective Precipitation Two species of proteins having a different linear sequence of amino acids and a different tertiary structure will inevitably have different physical properties, particularly molecular weight, ionic charge, shape, and solubility. These differences can be utilized to purify a given molecular species. The first step in purification should be one that can be carried out on a highly impure preparation and can be accomplished with a large increase in specific activity. Usually the first step takes advantage of solubility differences among proteins by selectively precipitating the desired protein out of solution. The solubility of a protein depends upon the relative balance between protein-solvent interactions which tend to keep it in solution, and protein-protein interactions which tend to cause it to aggregate and precipitate. The ionic strength of the solution is of particular importance in determining which of these types of interactions will predominate. At low ionic strength, interactions with the solvent are enhanced and proteins are particularly soluble; they are said to be *salted-in*. However, at higher ionic strengths protein solubility can rapidly decrease, apparently as a result of a decrease in water available to hydrate the protein, due to the increased utilization of water for hydration of the ions. The precipitation of proteins at high ionic strength is termed *salting-out*, and different proteins salt-out at different ionic strengths. The most commonly employed salt for selective protein precipitation is ammonium sulfate, which is highly soluble in water and has high ionic strength. Purification is achieved by gradually adding a solution of saturated ammonium sulfate to the crude protein extract. As the addition of salt continues, precipitation of protein increases, until a point is reached at which the protein being studied comes out of solution. This point is recognized by the loss of activity in the soluble fraction when tested by the particular assay system being used.

Ion-Exchange Chromatography Proteins are large, polyvalent electrolytes, and it is unlikely that many proteins in a partially purified preparation will have the same overall ionic charge. Ionic charge is used as a basis for purification (or analytical fractionation) in a variety of techniques, primarily ion-exchange chromatography and zone electrophoresis. The overall charge of a protein is a summation of all of the individual charges of its component amino acids. Since the charge of each amino acid is dependent upon the pH of the medium (Fig. 3-3), the charge of the total protein will also be dependent upon the pH. As the pH is lowered, negatively charged groups become neutralized and positively charged groups increase in number. The opposite condition occurs upon increasing the pH. As previously mentioned, a pH exists for each protein at which the negative charges equal the positive charges. This pH is the isoelectric point at which the protein is neutral. The isoelectric point of most proteins is below pH 7, and they are therefore anionic at a neutral pH.

Chromatographic techniques are based on the distribution of proteins between a mobile and an immobile phase. The mobile phase is formed by the solvent as it flows along carrying with it a certain amount of the protein. The immobile phase can be filter paper, or charged resin, or inert porous beads, etc., any of which can provide an alternative to the solvent for protein molecules to become associated. Ion-exchange chromatography depends upon the ionic association of proteins with charged groups bound to an inert supporting material, often cellulose. The two most commonly employed ion-exchange resins are DEAE-cellulose and carboxymethylcellulose. DEAE-cellulose is positively charged and therefore acts by binding negatively charged molecules; it is an anion exchanger. Carboxymethylcellulose is negatively charged and acts as a cation exchanger. The resin is packed into a column and the protein solution allowed to percolate through it in a buffer of composition that promotes the binding of all of the proteins to the resin. Since the binding between the resin and proteins is an ionic interaction, it can be displaced in two ways, either by increasing the ionic strength of the buffer in the column or by changing its pH. Since different proteins have different ionic charges, they are displaced from the resin in the column at different ionic strength or pH. Figure 3-30 shows the way in which these features are used to purify and fractionate protein mixtures.

Electrophoresis Another powerful technique that is widely used to purify, but more commonly to fractionate proteins, is electrophoresis. There are many different variations among electrophoretic techniques, but all of them depend on the ability of charged molecules to migrate when placed in an electric field. In most cases the proteins are applied at one location and they migrate through some form of support medium in response to the applied current. Given the opportunity, proteins with an overall negative charge will move toward the positive pole, the anode, while positively charged species will move in the opposite direction. The distance that a particular protein migrates depends upon its charge and its molecular weight. The greater the charge and the less the molecular weight, the greater the distance moved per unit time. Since it is unlikely that two proteins will be present in a partially purified mixture with combinations of charge and size that cause them to migrate similarly, electrophoresis is a valuable technique to display the variety of species present in the mixture.

In several places throughout this book we will have cause to discuss results obtained by one technique, that of acrylamide gel electrophoresis

Figure 3-30. The separation of two species of protein by DEAE-cellulose. In this case a positively charged ion-exchange resin is used to bind the negatively charged protein. This diagram should not be interpreted to mean that only proteins of opposite overall charge can be separated. Any proteins which differ in their ionic properties will be eluted at different ionic strengths from the resin.

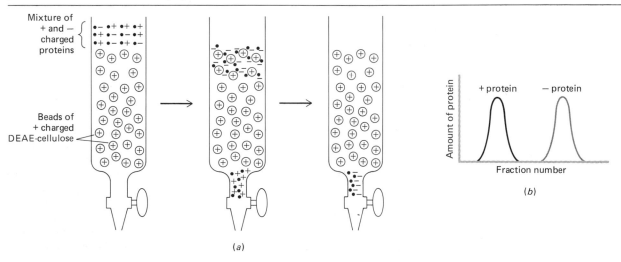

(Fig. 3-31), which has provided one of the best fractionation procedures available. To begin a fractionation on acrylamide gel, small tubes are filled with specially prepared solutions of acrylamide, which solidifies into a gel when exposed to ultraviolet light. Proteins are capable of moving through this semisolid support material, though their rate is dependent upon the porosity of the gel, which is determined by the percent acrylamide in the solution. The tube is inserted between two compartments containing buffer in which electrodes are immersed and the protein solution is layered in the tube over the gelated acrylamide. Current is applied to the buffered compartments across the column of acrylamide gel, and the proteins begin their migration as disks moving at varying rates. After the desired length of time the current is turned off, the tubes are removed, the acrylamide gels are displaced from their glass container, and the gel which emerges as one piece can be stained to facilitate location of the protein bands. Figure 3-32 shows an example of acrylamide gel electrophoresis and also illustrates the value of various steps taken during a purification process. In this case, a particular protein from blue crab serum was sought, and the purification was monitored by acrylamide gel electrophoresis. In the first gel, the variety of proteins present in unfractionated serum is shown. The second, third and fourth gels indicate the nature of the preparation after successive purification steps (ammonium sulfate precipitation, DEAE-cellulose chromatography, and sucrose density gradient centrifugation) were taken. By the end of the purification steps, a single protein is detected in the acrylamide gel; a protein which was present at such a small concentration in the original serum that it did not produce a detectable band in the first acrylamide gel of the set. With each successive step, an increase in the intensity of this band is seen with a corresponding decrease in that of the other bands.

Acrylamide gel electrophoresis can also be a valuable means for the determination of the molecular weight of migrating species when the electrophoresis is carried out in the presence of the ionic detergent, sodium

Figure 3-31. Acrylamide gel electrophoresis. *(a)* Diagram of the apparatus showing several glass tubes suspended between an upper and lower chamber of buffer. It is through these tubes, which contain the gelated acrylamide, that the proteins migrate. *(b)* Left drawing shows the protein mixture just layered over the large pore gel waiting for the power supply to be turned on. Middle drawing shows an early stage in electrophoresis. The proteins have all moved rapidly through the large pore gel (which offers little resistance) so they are all concentrated together at the "starting gate," i.e., the top of the small pore gel, which will separate the components on the basis of charge and molecular weight (right drawing).

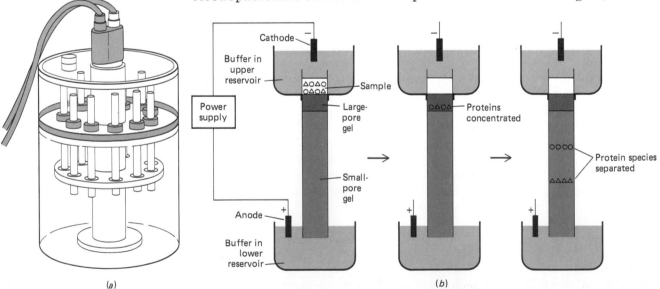

(a)

(b)

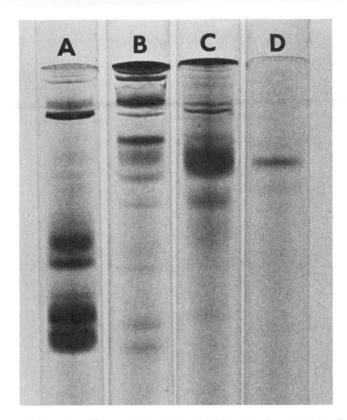

Figure 3-32. Acrylamide gel electrophoresis of successive preparations during the purification of a protein from blue crab serum that has neutralizing activity for T_2 bacteriophage. The first gel shows the fractionation of whole serum, and the second gel indicates the nature of the 40% saturated ammonium sulfate precipitate. In order to obtain material for the third gel, the precipitate was dissolved in buffer and passed through a DEAE cellulose column, with the eluant tested for phage neutralizing activity (the assay for the presence of the desired protein). The material eluted in the peak containing the activity is shown in the third gel (some contaminants remain). Final purification was obtained by sucrose density-gradient ultracentrifugation, as shown in the fourth gel. *(Courtesy of L. J. McCumber, E. M. Hoffmann, and L. W. Clem.)*

dodecyl sulfate (SDS, Fig. 5-37). In the SDS technique, the negatively charged detergent molecules become complexed with the protein to an extent which is proportional to the protein's molecular weight. Consequently, each protein species, regardless of its molecular weight, will have an equivalent charge per unit molecular weight and will tend to move at an equivalent rate in an electric field. However, as a result of the cross-linked nature of the acrylamide in the tube, the larger the molecular weight of the migrating molecule, the more it will be held up, and the slower it will move. As a result, proteins become separated on the basis of their molecular weight.

Another variation of electrophoresis, that of isoelectric focusing, has become an extremely powerful means to separate a mixture of proteins into very fine bands. In this technique the tube to be subjected to the applied current is filled with a substance that has a variable pH along its length from one end to the other. As proteins migrate through the gel in response to the electric field, they are exposed to a continually changing pH which produces a continuous change in their ionic charge. At some point within the tube each protein encounters a pH which is equal to its isoelectric point, thereby converting the protein to a neutral molecule and causing its migration to stop. Each protein comes to equilibrium to form a very sharp band at a predictable position along the tube's length.

In 1975, a new technique for protein fractionation was introduced which has already proved invaluable in the separation of large numbers of different proteins. The technique involves the separation of proteins in two dimensions across a flat gel surface. As is generally the case for two-dimensional fractionation procedures, it is most desirable that the separation in each direction take advantage of independent properties of the materials in the mixture, thereby maximizing the distribution of the components over the surface of the plate. In this technique, proteins are separated in one direction according to their isoelectric point, using isoelectric focusing. Proteins are then fractionated in a second direction according to their molecular weight by SDS gel electrophoresis. The resolution of this technique is so great that virtually all the proteins present in a cell (in detectable amounts) can be distinguished. It is estimated that up to about 5000 different proteins could theoretically be displayed by this procedure. A diagrammatic representation of the distribution of human serum proteins after two-dimensional electrophoresis is shown in Fig. 3-33 and a photograph of an actual gel containing fractionated chromosomal proteins in Fig. 13-22.

Figure 3-33. A diagram drawn from a gel after fractionation of human blood plasma by two-dimensional electrophoresis. The locations of known plasma proteins are indicated. [*From L. Anderson and N. G. Anderson*, Proc. Natl. Acad. Sci. (U.S.), **74**:*5424 (1977).*]

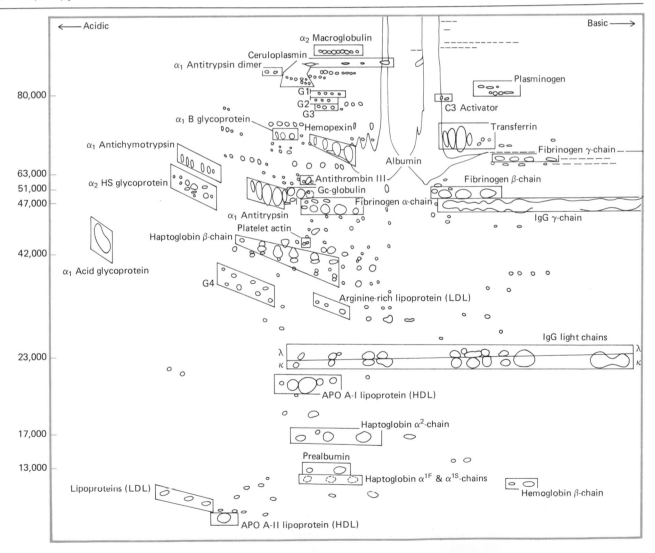

Gel Filtration Different techniques utilize different physical properties; therefore purification and fractionation procedures typically involve more than one method to ensure maximum separation. In the examples above, consider that the maximum purification by ion-exchange chromatography or electrophoresis has been obtained and the protein sought has been displaced with a contaminating species. The fraction containing the desired protein can then be subjected to another technique that depends most heavily on a different physical property of proteins, for example, molecular weight. Gel filtration is a technique that separates by molecular weight (though shape of the molecule is also important). As in the case of ion-exchange chromatography, the means for separation is a cylindrical column through which the solution of proteins slowly passes. In the case of gel filtration, the addition of the sample is followed by washing the column with large amounts of the same buffer in which the sample had been applied. While passing through the column proteins of various molecular weight are sieved in such a way that they are eluted from the column after different volumes of buffer have been added. The most commonly employed materials for gel filtration are beads made of a cross-linked polysaccharide (called Sephadex) which are perforated by holes of specific diameter. These beads allow entry of molecules up to a certain molecular weight and exclude larger molecules or ones with an inappropriate shape. A column of Sephadex beads acts like a molecular sieve.

Consider that the protein being isolated has a molecular weight of 125,000 daltons and there are two contaminating proteins, one of 250,000 daltons and the other 75,000 daltons (Fig. 3-34). In this case a column of Sephadex G150

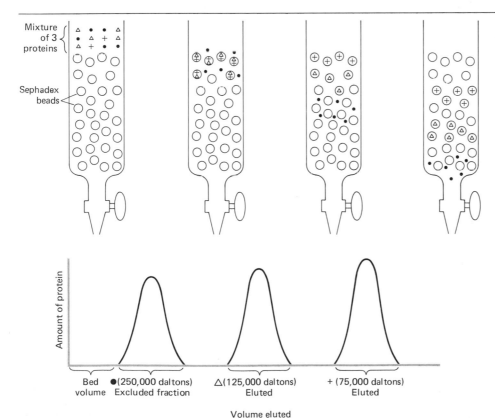

Figure 3-34. Gel filtration. The separation of three proteins on the basis of molecular weight, as described in the text.

would be used, since these beads only allow molecules of less than 200,000 molecular weight to penetrate their interiors. When a solution containing these proteins passes through the column bed, the protein whose molecular weight is 250,000 cannot enter the beads; therefore it remains dissolved in the moving solvent phase and is eluted as soon as the preexisting solvent in the column (the bed volume) has dripped out. In contrast, the other two proteins can diffuse into the interstices within the beads and are retarded in their passage through the column. As more and more solvent is poured through the column, these other proteins are carried down its length and out the bottom. Sephadex not only has the capacity to exclude larger molecular weight materials, but, within the molecular size range which does penetrate the beads, smaller species are retarded to a greater extent than larger ones. In other words, in our example the small contaminant is retarded to a greater degree than the desired protein, and thus requires a greater volume of solvent to remove it from the column. As solvent flows through the column, the 125,000 molecular weight protein is eluted in a purified state as determined by the assay procedure. Sephadex is important not only in purification and fractionation but also in the rapid determination of molecular weight.

Affinity Chromatography The techniques described up to this point are ones that utilize the bulk properties of a protein to effect a purification or fractionation. In recent years a new concept in protein purification has emerged which, if it can be applied to a particular molecule, can bring about an essentially total purification in one relatively simple, rapid step. The technique is termed *affinity chromatography* and takes advantage of the unique structural properties of a protein whereby that protein can be specifically withdrawn from solution while all other molecules remain behind in the dissolved state. The technique is best explained by example. In Chapter 5 the properties of membrane receptors, such as the insulin receptor, will be described in detail. Membrane receptors are proteins that are partially embedded in the plasma membrane and are capable of specifically combining with a particular extracellular molecule. Since the insulin receptor is specific for this particular hormone, one could imagine some means to take advantage of this interaction in the purification of the receptor in question. Affinity chromatography in this case would consist of a column containing an inert material (a matrix) to which insulin molecules are covalently attached. If an impure preparation of membrane proteins is passed through this column, the interaction between the hormone immobilized on the beads and its receptor in solution is such that the receptors will combine with the hormone and be pulled out of solution (Fig. 3-35). Once all of the contaminating proteins have passed through the column and out the bottom end, the receptor molecules can be displaced from the matrix and a one-step purification of receptor has been accomplished.

As long as a suitable molecule is available for specific interaction with a protein being sought, an affinity chromatographic procedure should be of great potential use. In other cases specific enzymes have been purified using immobilized inhibitors of the enzyme as the "capturing" molecule or specific antibodies have been used to selectively complex their respective antigen, i.e., that molecule which induced the formation of the anti-

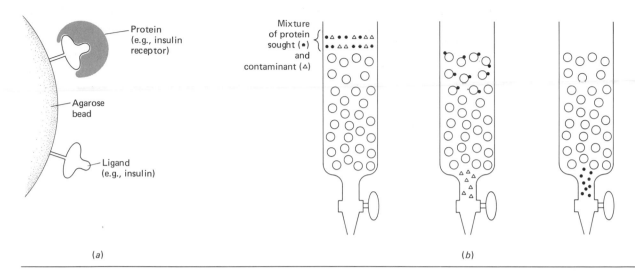

Figure 3-35. Affinity chromatography. *(a)* Schematic representation of the coated beads to which only a specific protein can combine. *(b)* Steps in the chromatography procedure.

body being used. In recent years increased interest in carbohydrate-bearing proteins (glycoproteins) has led to the development of affinity chromatographic techniques whereby lectins (carbohydrate-binding proteins, page 225) are used to complex selectively with proteins having the appropriate sugars in the molecule.

The techniques described above are the most commonly employed. Numerous other methods also exist which take advantage of one property or another that proteins can display. Once the protein is purified, various techniques are available to provide a measure of its purity and an estimation of its molecular weight and other properties. Using these methods, over 1000 different proteins have been studied in an isolated state; over 200 have been purified to the point of crystallization. In some cases these purification techniques have required large amounts of starting tissues, for example, such readily available material as beef liver, for the isolation of milligram quantities of the particular protein.

The Purification of the Insulin Receptor

To illustrate the use of preparative techniques, the steps taken in the purification of the insulin receptor by Pedro Cuatrecasas [*Proc. Natl. Acad. Sci. (U.S.)*, **69:** 1277–1281 (1972)] will be briefly described. The procedure was begun with the removal of livers from rats and their homogenization in ice-cold 0.25-M sucrose. To monitor the distribution of the receptor protein during the various purification steps, an assay system was used which measured the amount of radioactive insulin bound by each fraction, thereby providing a measure of the receptor protein present. The insulin receptor is located within the plasma membrane of the liver cells, which become purified during differential centrifugation as shown in Table 3-5. In order to solubilize the receptor protein from the membranes of the 40,000-g pellet, the detergent Triton X-100 is added. Purification, in this case, was accomplished by ammonium sulfate purification, ion-exchange chromatography on DEAE-cellulose, and affinity chromatography. The purification achieved by each step is shown in Tables 3-5 and 3-6.

TABLE 3-5
Specific Insulin-binding Activity and Fractionation of Liver Homogenates

Sample	Volume (ml)	Protein (g)	Activity (cpm/mg)†
Crude homogenate	2000	88 [100]°	1,800 [100]
600 × g Supernatant	1650	50 [62]	3,900 [122]
12,000 × g Supernatant	1400	28 [32]	5,750 [101]
40,000 × g Pellet	200	4.6 [5]	37,000 [108]

° In square brackets are percentages of the total.
† [^{125}I]—insulin-binding activity.

TABLE 3-6
Résumé of Procedures Used in the Purification of the Insulin Receptor of Liver-Cell Membranes

Preparation or procedure	Insulin-binding activity (pmol/mg of protein)	Purification
Crude liver homogenate	0.008	0
Liver membranes	0.15	20°
Triton extract of membranes	0.26	1.7†
(NH$_4$)$_2$SO$_4$ fraction 20–40%	0.75	3‡
DEAE-cellulose chromatography	14	60‡ §
Affinity chromatography¶	About 2000	About 8,000‡
		About 250,000°

° Compared to crude liver homogenate.
† Compared to liver membranes.
‡ Compared to Triton extract of liver membranes.
§ Dialysis of Triton extract results in a 3-fold purification; DEAE-chromatography results in a further purification of about 20-fold.
¶ These are tentative figures because of the difficulty in accurately determining the small amounts of protein obtained by these procedures.
SOURCE: P. Cuatrecasas, *Proc. Natl. Acad. Sci. (U.S.)*, **69**: 1277 (1972).

The greatest purification is achieved by affinity chromatography (Fig. 3-35) which illustrates the great value of this approach. The impure preparation is passed through a column to which insulin molecules are covalently linked to an agarose matrix. The receptor protein is bound, the contaminants pass through, after which the receptor protein is eluted by the addition of urea which disrupts the noncovalent bonds holding the two macromolecules together, thereby releasing the mobile receptor protein into the eluant. It is estimated that the overall purification of the receptor by these techniques is approximately 250,000-fold. Since it is estimated that a 400,000-fold purification would be required to achieve complete purity, the effectiveness of these procedures, primarily that of affinity chromatography, is evident.

REFERENCES

Adler, A. D., ed., Ann. N.Y. Acad. Sci. **244**, 1975. Symposium on "The Biological Role of Porphyrins and Related Structures."

Anfinsen, C. B., Harvey Lectures **61**, 95–116, 1965. "The Formation of the Tertiary Structure of Proteins."

————, Science **181**, 223–230, 1973. "Principles that Govern the Folding of Protein Chains."

————, Edsall, J. T., and Richards, F. M., eds., *Advances in Protein Chemistry.* Vol. 1, Academic, 1944.

Bailey, J. I., *Techniques in Protein Chemistry.* 2d ed., Elsevier, 1967.

Baldwin, R. L., Ann. Rev. Biochem. **44**, 453–476, 1975. "Intermediates in Protein Folding Reactions and the Mechanism of Protein Folding."

Bornstein, P., Ann. Rev. Biochem. **43**, 567–604, 1974. "The Biosynthesis of Collagen."

Casjens, S., and King, J., Ann. Rev. of Biochem. **44**, 555–611, 1975. "Virus Assembly."

De Duve, C., J. Cell Biol. **50**, 20D–55D, 1971. "Tissue Fractionation: Past and Present."

Dickerson, R. E., Ann. Rev. Biochem. **41**, 815–842, 1972. "X-ray Studies of Protein Mechanisms."

————, Sci. Am. **226**, 58–72, April 1972. "The Structure and History of an Ancient Protein."

———— and Geis, I., *The Structure and Action of Proteins.* Harper & Row, 1969.

Engelman, D. M., and Moore, P. B., Sci. Am. **235**, 44–54, Oct. 1976. "Neutron-Scattering Studies of the Ribosome."

Gallop, P. M., and Paz, M. A., Phys. Revs. **55**, 418–487, 1975. "Posttranslational Protein Modifications, with Special Attention to Collagen and Elastin."

Gross, J., Sci. Am. **204**, 120–130, May 1961. "Collagen."

————, Harvey Lectures **68**, 351–432, 1974. "Collagen Biology: Structure, Degradation and Disease."

Hartley, B. S., Biochem. J. **119**, 805–822, 1970. "Strategy and Tactics in Protein Chemistry."

Kendrew, J. C., Sci. Am. **205**, 96–110, Dec. 1961. "Three-Dimensional Structure of a Protein."

Kushner, D. J., Bact. Revs. **33**, 302–345, 1969. "Self-Assembly of Biological Structures."

Lehninger, A. L., *Biochemistry.* 2d ed., Worth, 1975.

Meister, A., *Biochemistry of the Amino Acids.* 2d ed., Academic, 1965.

Moore, S., and Stein, W. H., Science **180**, 458–464, 1973. "Chemical Structures of Pancreatic Ribonuclease and Deoxyribonuclease."

Morris, C. J., and Morris, P., *Separation Methods in Biochemistry.* 2d ed., Halsted, 1976.

Neurath, H., and Hill, R. L., eds., *The Proteins.* 3d ed., Academic 1975.

Nomura, M., Sci. Am. **221**, 28–35, Oct. 1969. "Ribosomes."

————, Science **179**, 864–873, 1973. "Assembly of Bacterial Ribosomes."

O'Farrell, P. H., J. Biol. Chem. **250**, 4007–4021, 1975. "High Resolution Two-Dimensional Electrophoresis of Proteins."

Okada, Y., Adv. Biophys. **7**, 1–41, 1975. "Mechanism of Assembly of Tobacco Mosaic Virus in Vitro."

Oncley, J. L., ed., *Biophysical Science: A Study Program.* Wiley, 1958.

Perutz, M. F., Sci. Am. **211,** 64–76, Nov. 1964. "The Hemoglobin Molecule."

Robinson, A. L., Science **192,** Res. News, 360, 1976. "Electron Microscopy: Imaging Molecules in Three-Dimensions."

Sanger, F., and Thompson, E. O. P., Biochem. J. **53,** 353–374, 1953. "The Amino Acid Sequence in the Glycyl Chain of Insulin."

Stein, W. H., and Moore, S., Sci. Am. **204,** 81–92, Feb. 1961. "The Structure of Proteins."

Stryer, L., *Biochemistry.* Freeman, 1975.

Symposium on "Assembly Mechanisms," J. Supramol. Struc. **2,** 81–514, 1974.

Symposium on "The Structure and Function of Proteins at the Three-Dimensional Level." Cold Spring Harbor Symp. Quant. Biol. **36,** 1972.

Thompson, E. O. P., Sci. Am. **192,** 36–41, May 1955. "The Insulin Molecule."

Traub, W., and Piez, K. A., Adv. in Prot. Chem. **25,** 243–352, 1971. "The Chemistry and Structure of Collagen."

White, A., Handler, P., and Smith, E. L., *Principles of Biochemistry.* 5th ed., McGraw-Hill, 1975.

Wittman, H. G., European J. Biochem. **61,** 1–13, 1976. "Structure, Function, and Evolution of Ribosomes."

CHAPTER FOUR
Energy, Enzymes, and Metabolism

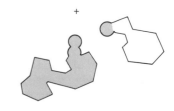

The interrelationship between structure and function is evident at all levels of biological organization from the molecular to the organismal. One of the foremost goals of molecular biology has been to fully describe the activities of complex macromolecules in terms of their molecular architecture. In the case of a few proteins this goal has been essentially realized. In the previous chapter it was shown that the complete amino acid sequence of numerous proteins has been determined as well as, in a few cases, the positions in space of all of the atoms of the molecule. This type of information has its greatest significance if it can be applied toward a better understanding of the molecular functions of the parts of the molecule as they relate to the activity of the molecule as a whole. The best understood proteins are enzymes, which we will consider in some detail below. Much more extensive treatment can be found in any of the excellent biochemistry textbooks listed in the references at the end of the chapter. Before embarking upon the analysis of the catalytic properties of enzymes, we will consider a few of the most important concepts in thermodynamics which are necessary to more fully understand and appreciate enzyme function. In addition, a brief survey of a few of the thermodynamic principles will be important in tying together a number of processes that occur in cells ranging from protein folding, to diffusion of ions across membranes, ATP hydrolysis, and membrane assembly. The thermodynamic analysis of a particular system can provide some measure of whether or not the events can occur spontaneously or not, and if not, a measure of the energy a cell must expend in order for the process to be brought about. Thermodynamic considerations will provide no help in unraveling the specific mechanism used by the cell to carry out a given process, for it is independent of the manner in which an event occurs.

ENERGY

The study of thermodynamics embodies a set of empirical concepts concerning energy by which we can predict the direction or course which events can take in the universe. The first law of thermodynamics is the law of conservation of energy. It states that energy can neither be created or destroyed; however it can be converted from one form to another. For example, electrical energy is converted to mechanical energy when we plug in a clock, or chemical energy is converted to thermal energy when fuel is burned in an oil heater. Regardless of the conversion process, the energy in the two forms is equivalent, i.e., the total amount in the universe remains constant. That living organisms conform to the first law of thermodynamics was demonstrated before the start of this century in experiments in which subjects were maintained in a controlled environment and an energy balance sheet was maintained. It was found that the energy that was taken in by these individuals in the form of nutrition was balanced by the energy released in the form of waste products and heat.

Energy is generally defined as the capacity to do work; energy and work are like two sides of the same coin and within certain limits are interconvertible. We can recognize energy in two general forms, *potential energy* and *kinetic energy*. A rock perched on the edge of a cliff has potential energy since it has the potential to perform work. It has this potential because it exists in a field of force (gravitational force); if the rock is pushed over the edge, then the force can act upon it and cause it to fall. In the process of falling it has kinetic energy and can accomplish work, for example, by lifting another object as shown in Fig. 4-1. In another case we can have potential energy by separating oppositely

Figure 4-1. The ability of a falling rock to do work. The energy available for work is proportional to the mass of the rock and the distance it falls.

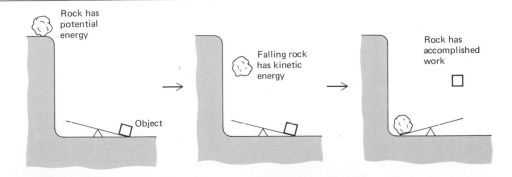

charged ions across a membrane thereby producing a voltage which serves as a different type of field of force. This potential can be converted to kinetic energy by allowing the charged particles to cross the membrane to eliminate their separation. In both of these cases, as in all cases of measurement of energy or work, there are two factors to consider: an intensity factor (potential factor) and a capacity factor (Table 4-1). In the case of the falling rock the capacity factor is its mass while the potential factor is the height down which it will fall. For the movement of the charged ions the capacity factor is the charge of the particles and the intensity factor is the voltage. The work or energy is a multiple of these two factors; as either one increases so too does the energy. Energy can be measured in calories, ergs, or joules, all of which are interconvertible.

First Law of Thermodynamics

In order to discuss energy transformations involving matter we need to divide the universe into two parts, the *system* under study and the remainder of the universe which we will refer to as the *surroundings*. A system can be defined in various ways; it may be a certain space in the universe or a certain amount of matter. For example the system might be a living cell. In most thermodynamic systems the stipulation is made that it does not exchange matter with its environment; it is a closed system. In contrast, the system is allowed to exchange energy with its surroundings. For most cases, we cannot determine the absolute amount of energy associated with a given amount of matter, and, regardless, it is not an important property to know. What is important, is the change in energy of the system that occurs during an event, and this we can measure using various types of procedures. Changes in a system's energy occurring during a process are manifested in two ways: a change in the heat content of the system and the performance of work. Even though the system may gain or lose energy, the first law of thermodynamics indicates that the loss or gain must be balanced by a corresponding gain or loss in the surroundings so that the amount in the universe as a whole remains constant. The energy of the system is termed the *internal energy* (E) and the change in the internal energy during a transformation is ΔE. One way to describe the first law of thermodynamics is that $\Delta E = Q - W$, where Q is the heat energy and W is the work energy. If heat is absorbed into the system during the event from the surroundings, then Q is positive;

TABLE 4-1
Intensive and Extensive Factors in Forms of Energy

Type of work	Intensity	Capacity
Work; gravitational	Height, h	Mass, m
Work; stretching	Tension, τ	Length, l
Work; electrical	Voltage, $\mathscr{E}$	Charge, q
Work; expansion	Pressure, P	Volume, V
Work; surface	Surface tension, γ	Area, A

SOURCE: I. M. Klotz, "Energy Changes in Biochemical Reactions," Academic Press, 1967.

if heat is lost to the surroundings it is negative. If the system does work on the surroundings (for example, by increasing in volume and compressing the surroundings), then W is positive; when work is done upon the system, W is negative. In other words, even though the amount of work performed and heat exchanged may vary during a particular transformation (such as the falling of the rock), the *difference* between these two measurable quantities does not vary. In the case of the falling rock, the shift of that amount of mass through a given height results in a specified change in the internal energy of the system; it makes no difference what happens to the rock along the way, i.e., whether it strikes the board in Fig. 4-1 or not. In a fall of this height a certain amount of work and/or heat can be exchanged with the surroundings, but the difference between the two as stated in the equation will remain constant. Energy differences are independent of the path that is taken by the system from one energy state to another (from the top of the cliff to its bottom).

Depending upon the type of process, the internal energy of the system at the end can be greater than, equal to, or less than its internal energy at the start, depending upon its relationship to its surroundings. In other words, ΔE can be positive, zero, or negative. In the case of mechanical energy changes, such as the falling rock, the internal energy change will be negative. In the case of chemical energy changes, which are more the type we will be concerned with, the outcome is less predictable. Consider a system to be the contents of a reaction vessel (like that of a living cell). As long as there is no change in pressure or volume of the contents, there is no work being done by the system on its surroundings or vice versa. In that case $\Delta E = Q$, and the energy at the end of the transformation will be greater than before if heat is absorbed (gained) and less if heat is released (lost). Under conditions of constant pressure and volume, reactions which lose heat are termed *exothermic* and ones which gain are *endothermic*. There are many reactions of both types. Since ΔE during a particular process can be positive or negative, it gives us no information as to the likelihood that a given event will occur. In order to consider the direction in which a particular spontaneous transformation is likely to occur, we need to consider some additional points.

Second Law of Thermodynamics

The second law of thermodynamics can be stated in a variety of ways, but basically it expresses the concept that events in the universe always proceed "downhill." In any energy transformation, there is a decreasing availability of energy for doing work. Rocks fall off cliffs to the ground below, and once at the bottom their ability to do additional work is reduced; it is very unlikely that they will spontaneously lift themselves back to the top of the cliff. Similarly, opposite charges spontaneously move together, not apart, and heat spontaneously flows from a warmer to a cooler body, not the reverse. The concept of the second law of thermodynamics was formulated originally for heat engines and carried with it the notion that it was thermodynamically impossible to construct a perpetual-motion machine. In other words, it is impossible for the machine to be 100% efficient; some of the energy will be lost. The energy that is unavailable for doing additional work after the transformation has an intensity factor and a capacity factor as do other energy terms. The intensity factor is temperature (in degrees) and the capacity factor is termed *entropy* and has the dimensions energy per degree (calories/degree). The unavailable energy term is $T \Delta S$, where ΔS is the change in entropy between the initial and final states.

With the development of the theories of quantum mechanics and a realization of discrete energy levels of electrons, atoms, and molecules, a new formulation of the entropy concept was developed, one of a molecular-statistical nature (see I. M. Klotz, 1967, for a readable introductory discussion). The basis for the loss of available energy during a process is a result of the tendency for there to occur an increase in the randomness or disorder of the universe. The term entropy is a measure of this disorder; it is associated with the *random* movements of the particles of matter which, because they are random, cannot be made to accomplish a *directed* work process. The greater the movements of the elementary particles, the greater the entropy of the mass. For example, when a sugar cube is dropped into a cup of hot water, there is a spontaneous shift from an ordered state of the molecules in the crystal to a much more disordered condition when the sugar molecules are spread throughout the solution (Fig. 4-2). In other words, the freedom of movement (or number of possible states in which a sugar molecule can exist) is much greater after a period of time than before the dissolving process began; the entropy has increased. It is the increase in the random movements that causes the distribution of the molecules to change from the condensed to the distributed state. The likelihood that this process will reverse itself, so that the sugar molecules will spontaneously return to one location of the cup and recrystallize, is infinitesimally small. The distribution of solute molecules after a given time is based strictly upon statistics. The state of uniform distribution is simply *the most probable one*. The greater the temperature, the more rapid will be the molecular movements and the more rapidly will the molecules spread themselves out.

As with other spontaneous events we must distinguish between the system and its surroundings. The second law of thermodynamics indicates only that the total entropy in the universe must increase; the disorder within one part of the universe (the system) can decrease at the greater expense of its surroundings. The dissolved sugar can decrease in entropy; it can be recrystallized by evaporating off the water. The consequence of this process, however, is an increase in the entropy of the surroundings (the water molecules now in the gas phase) which more than balances the decrease in the freedom of the molecules of the sugar crystals. As will be described below, this is exactly what living organisms manage to accomplish, i.e., the maintenance of a state of decreased entropy at the expense of the increase in the entropy of their surroundings (page 135).

Free Energy
Together, the first and second laws of thermodynamics indicate that the energy of the universe is constant but the entropy continues to increase to a maximum. The concepts inherent in the first two laws were

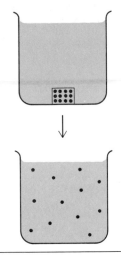

Figure 4-2. The dissolution of a sugar cube in hot water is an entropy-driven process. The random movements of the sugar molecules in the dissolved state will inevitably cause a change in their distribution. Once completely spread throughout the container there will be no further tendency for redistribution: the process is at equilibrium, entropy is at a maximum, and no work can be accomplished.

put together by Gibbs and Helmholtz into the expression $\Delta H = \Delta G + T\Delta S$, where ΔG is termed the change in *free energy,* i.e., the change in energy available during a transformation to do work; ΔH is the change in *enthalpy* or total energy content of the system (equivalent to ΔE for our purposes), T is the absolute temperature, and ΔS is the change in the entropy of the system. The equation states that the total energy change is equal to the sum of the changes in energy available and unavailable to do further work. If the equation is written as $\Delta G = \Delta H - T\Delta S$, then it takes on new importance. It allows one to predict the direction in which a spontaneous reaction will proceed and the degree to which the reaction will occur. All *spontaneous* energy transformations *within a system* must have a negative value of ΔG, i.e., the reaction must proceed from a state of higher free energy to a state of lower free energy. While the signs of ΔH and ΔS can be positive or negative (depending upon the nature of the relation between the system and its surroundings), the overall sum of the two terms, ΔH and $-T\Delta S$, must be negative in value. The magnitude of ΔG indicates the maximum amount of energy that can be passed on for use by another process.

In most cases, transformations proceed to reduce the total energy (H) of the *system,* as in the case of the falling rock. When chemical reactions are considered, the term ΔH becomes the heat of reaction and indicates whether the reaction is endothermic ($+\Delta H$) or exothermic ($-\Delta H$). Similarly, most, but not all, transformations proceed so that the entropy of the *system* increases as in the case of the dissolving sugar cube. The dissolving process is accompanied by a much greater freedom of movement of the sugar molecules, i.e., there are a great many more possible arrangements that a given molecule can occupy than in the previous condition. The more possible arrangments, the more probable the state. The same statistical properties are apparent in a flip of a pair of coins. There are two possible arrangements of the pair of coins to produce one head and one tail, and only one possible arrangement to give either two heads or two tails. Therefore, as in the case of entropy, the state for which there are more possible arrangements is the most probable state. The change in entropy (ΔS) during a reaction is proportional to the number of possible arrangements at the beginning (W_1) versus the end (W_2). This is expressed in the equation $\Delta S = k \ln W_2/W_1$ where k is Boltzmann's constant.

The counterplay between ΔH and ΔS is illustrated by the ice–water transformation. The conversion of water from the liquid to the solid state is accompanied by a decrease in entropy (ΔS is negative) and a decrease in enthalpy (ΔH is negative). In order for this transformation to occur,

TABLE 4-2
Thermodynamics of the Ice–Water Transformation

Temperature (0°C)	ΔE (cal/mol)	ΔH (cal/mol)	ΔS (cal/mol °C^{-1})	$-T\Delta S$ (cal/mol)	ΔG (cal/mol)
−10	−1343	−1343	−4.9	1292	−51
0	−1436	−1436	−5.2	1436	0
+10	−1529	−1529	−5.6	1583	+54

SOURCE: I. M. Klotz, "Energy in Biochemical Reactions," Academic Press, 1967.

ΔH must be more negative than $-T \Delta S$ is positive, a condition which occurs only below 0°C. This relationship can be seen from Table 4-2, which indicates the values for the different terms *if* one mole of water were to be converted to ice at 10°C, or −10°C. In all cases, regardless of the temperature, the energy level of the ice is less than that of the liquid (the ΔH is negative); however at the higher temperature the entropy term of the equation ($T \Delta S$) is more negative than the enthalpy term and therefore the free-energy change is in the wrong direction to occur spontaneously (ΔG is positive). At 0°C the system is in equilibrium, while at −10°C the solidification process is greatly favored, i.e., the ΔG is negative.

Biochemical Reaction Kinetics

The concept of free energy and its change during chemical reactions is of the greatest importance in understanding the rationale of metabolism and its control. In order to relate the laws of thermodynamics to biochemical reactions we must briefly consider the law of mass action and its relation to chemical equilibria. All chemical reactions within the cell are reversible and, therefore, we must consider two reactions occurring simultaneously, one forward and the other in reverse. According to the law of mass action, the rate of reactions is proportional to the concentration of the reactants. Consider, for example, the following hypothetical reaction:

$$A + B \rightleftharpoons C + D$$

The rate of the forward reaction is directly proportional to the product of the molar concentrations of A and B. This can be expressed as the rate of forward reaction $= k_1 (A)(B)$, where k_1 is a rate constant for the reaction. The rate of the backward reaction $= k_2 (C)(D)$. All chemical reactions proceed, however slowly, toward a state of equilibrium, i.e., toward a point at which the forward and backward reactions are equivalent. At equilibrium, an equivalent number of molecules of A and B will be converted into C and D, as will be formed from them. At equilibrium, therefore,

$$k_1 (A)(B) = k_2 (C)(D)$$

which can be rearranged to

$$\frac{k_1}{k_2} = \frac{(C)(D)}{(A)(B)}$$

In other words, at equilibrium there will be a predictable ratio of the concentration of products to the concentration of reactants. This ratio, which is equal to k_1/k_2 is termed the *equilibrium constant* (K_{eq}).

The equilibrium constant describes the direction (forward or reverse) in which the reaction is favored. If the equilibrium constant in the above reaction is greater than one, and we start with an equal concentration of all four substances, the reaction will proceed at a greater rate toward the formation of C and D than in the reverse direction. The opposite conditions would hold if the equilibrium constant were less than one. Once equilibrium is reached and reactions are proceeding to an equal extent in both directions, the quantities of each molecular species will remain constant. It follows from these points, that the direction in which the reaction is proceeding (which can be predicted from K_{eq}), is dependent upon the relative concentration of all molecules at any given instant.

The equilibrium ratio of reactants to products as specified by the equilibrium constant is dependent upon the relative free-energy levels of the substances on either side of the equation. In the reaction above, if the conversion of A and B to C and D is attended by a decrease in free energy of the substances, as for example in the case of the oxidation of carbohydrate, then the reaction in that direction will be thermodynamically favored. In other words, if the sum of the free energy of A and B is greater than that of C and D, then the reaction will proceed in the direction of formation of the products C and D. This does not mean that the reaction will necessarily occur at a measurable rate, only that it can occur spontaneously, i.e., without the input of energy, until the equilibrium rates are reached. As a chemical reaction proceeds and a greater and greater build-up of products occurs at the expense of reactants, the difference in free energy between the reactants and products decreases, until at equilibrium the difference is zero. All systems tend toward a minimization of their free energy.

Since ΔG for a given reaction depends upon the reaction mixture present at a given time, it is not a useful term in attempting to compare the energetics of various reactions. To place reactions on a comparable basis and to allow various types of calculations to be made, a convention has been adopted to consider the free-energy difference between the reactants and products under a set of standard conditions. The conditions are arbitrarily set using a reaction mixture at 25°C, one atmosphere of pressure, when all of the reactants and products are present at 1.0 M concentration. The standard free-energy difference ($\Delta G°$) describes the difference in free energy when one mole of each reactant is converted to one mole of each product under these conditions. It must be kept in mind, however, that standard conditions do not prevail in the cell and, therefore, one must be cautious in the use of values for standard free-energy differences in calculations of cellular energetics.

The relationship between the equilibrium constant and the standard free-energy difference is given by the equation $\Delta G° = -RT \ln K_{eq}$, where R is the gas constant and T is the absolute temperature (293 K). From this equation it follows that reactions having equilibrium constants greater than one will have negative $\Delta G°$ values. Reactions with negative $\Delta G°$ values are termed *exergonic* reactions which indicates they can occur spontaneously *under standard conditions*. Reactions having equilibrium constants of less than one will have positive $\Delta G°$ values. This latter class of reactions is termed *endergonic* indicating they cannot occur spontaneously *under standard conditions*. In other words, given the reaction written as follows: A + B $\rightleftharpoons$ C + D, if the $\Delta G°$ is negative, the reaction will go to the right when reactants and products are all present at 1.0 M concentration. The greater the negative value, the farther to the right the reaction will proceed before equilibrium is reached. Under the same conditions, if the $\Delta G°$ is positive, the reaction will proceed to the left, i.e., the reverse reaction is favored. Another way to state this is that the $\Delta G°$ is equal to the sum of the free energy of the products minus the sum of the free energy of the reactants when present at standard conditions, i.e., $\Delta G° = \Sigma G°$ products $- \Sigma G°$ reactants. The actual free energy of a molecule is determined by the nature of its atomic organization and is impossible to measure. However, the difference in free energy between molecules that are interconvertible is easy to measure once the equilibrium constant for the reaction between them is determined. For example, one of the most important chemical reactions of the cell is the hydrolysis of ATP (Fig. 4-3) into its products, ADP and P_i (inorganic phosphate). The

Figure 4-3. ATP and its hydrolysis to ADP, a highly thermodynamically favored reaction. In some cases, ATP is hydrolyzed to AMP, a compound with only one phosphate at the 5′ position of the sugar.

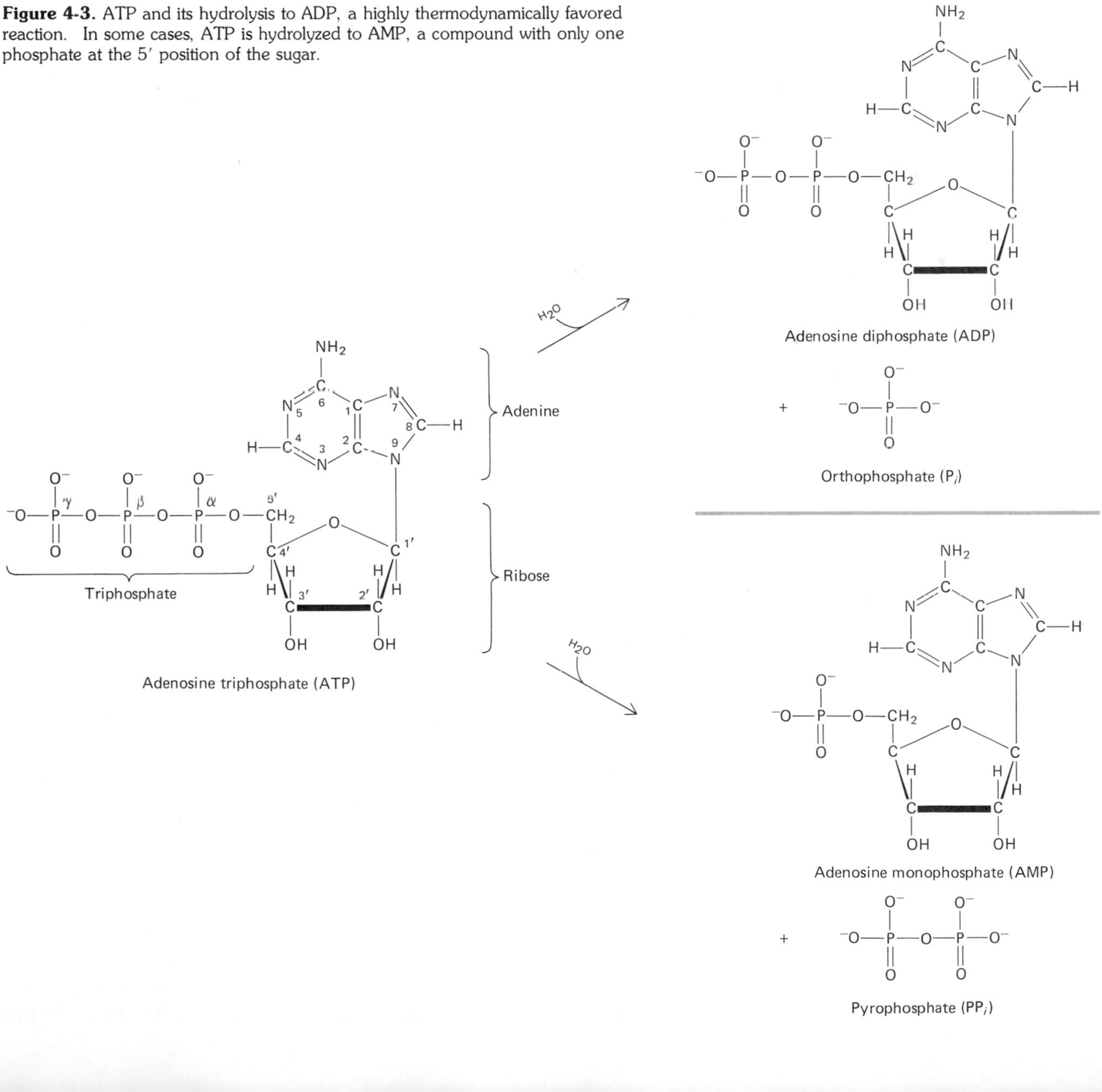

TABLE 4-3
Relationship between $\Delta G°$ and K_{eq} at 25°C

K_{eq}	$\Delta G°$ (cal/mol)
10^{-3}	4089
10^{-2}	2726
10^{-1}	1363
10^{0}	0
10^{1}	−1363
10^{2}	−2726
10^{3}	−4089

SOURCE: BIOCHEMIS-TRY by Lubert Stryer. W. H. Freeman and Company. Copyright © 1975.

standard free-energy difference between the products and reactants is −7300 cal/mol. The hydrolysis of ATP is an example of a highly exergonic (thermodynamically favored) biochemical reaction; i.e., one which tends toward an equilibrium at which the (ADP)/(ATP) ratio is very large. The relation between $\Delta G°$ and K_{eq} is shown in Table 4-3.

Free Energy in Metabolism

It is important that the difference between ΔG and $\Delta G°$ be kept clearly in mind. The $\Delta G°$ is a fixed value which describes the direction in which a reaction is proceeding when the reaction mixture is at standard conditions. Since standard conditions do no prevail within a cell, $\Delta G°$ values cannot be used to predict the direction in which a particular reaction is proceeding at a given moment within a particular cellular compartment. In order to do this one must know the ΔG which is determined by the concentrations of the reactants and products that are present in that case; information on the value of $\Delta G°$ is of no concern. Knowledge of the ΔG, however, is of great use for it reveals the direction in which the reaction in the cell is proceeding and how close the particular reaction in question is to the equilibrium state (see Fig. 4-26). For example, even though the $\Delta G°$ for the hydrolysis of ATP is −7.3 kcal/mol, the typical ΔG in the cell is considerably greater than this (over −12 kcal/mol). This large negative value reflects the high (ATP)/(ADP) ratio in the cell.

It should be kept in mind that whether ΔG is positive or negative depends solely on the manner in which the equation is written, i.e., which compounds are on the left side, and which are on the right side. A negative ΔG value indicates the forward reaction (as written) is progressing more rapidly than the reverse reaction at that moment. If ΔG is positive the reverse reaction is favored. The problem in determining ΔG is that the concentration in the cell of the reactants and products must be measured; this can be a very formidable task.

It was stated above that reactions with positive $\Delta G°$ values are endergonic reactions, i.e., they cannot occur spontaneously under standard conditions. However, there are many reactions that occur in the cell which, at first glance, appear to be endergonic, i.e., have positive $\Delta G°$ values. How can such "uphill" reactions take place? The answer is that within the cell they are not uphill at all, but rather the conditions are such that they are actually made to occur in a "downhill" fashion. There are two basic means by which reactions with positive $\Delta G°$ values can be made to occur in the cell. The first illustrates the important difference between ΔG and $\Delta G°$ and the second reveals the manner in which reactions with positive $\Delta G°$ values can be converted to reactions with negative $\Delta G°$ values by use of chemical free energy stored within the cell.

Consider the reaction (Fig. 4-26) in which dihydroxyacetone phosphate is converted to glyceraldehyde 3-phosphate. The $\Delta G°$ for this reaction is +1.8 kcal/mol, yet the formation of the product of this reaction is continually taking place in the cell. It can be assumed that within the cell this reaction proceeds because the ratio of the products to reactants is kept below that defined by the equilibrium constant, i.e., the free-energy level of the reactants is greater than that of the products (the ΔG is negative). As long as this condition holds the reaction will continue spontaneously in the direction of formation of glyceraldehyde 3-phosphate. This brings up an important characteristic of cellular metabolism, namely that specific reactions cannot be considered independently like they can in a test tube; hundreds of reactions are occurring in the cell at the same time. All of these reactions are interrelated since the products of one become the substrates for the next in the sequence and so on throughout one metabolic pathway and into the next. In order to maintain the production of glyceraldehyde 3-phosphate at the expense of dihydroxyacetone phosphate, the reaction must be placed so that the products of the reaction are removed by the next reaction in the sequence at a sufficiently rapid rate so that a favorable ratio of the concentrations of these two molecules is retained. In this reaction, the K_{eq} is approximately 0.05. In order for the forward reaction to be favored, the concentration of the reactant must be greater than 20 times that of the product. The relation between ΔG and $\Delta G°$ described above can be expressed by the equation $\Delta G = \Delta G° + RT \ln \dfrac{(C)(D)}{(A)(B)}$. If the concentrations of the products C and D are maintained sufficiently low, then the log of the ratio of products to reactants $\left(\ln \dfrac{(C)(D)}{(A)(B)} \right)$ will be sufficiently negative to cause the ΔG to be negative and the reaction exergonic.

Reactions with sufficiently large positive $\Delta G°$ values must be treated within the cell in a different manner; they must be "driven" by the input of energy. In the nonliving world endergonic reactions such as the breakdown of water into molecular hydrogen and oxygen can be driven by the addition of a large amount of energy; in this case in the form of heat or electricity (electrolysis). In the cell, however, a more gentle means must be used; endergonic reactions are driven by the use of chemical energy obtained from reactions having large differences in free energy. Consider the formation of the amino acid glutamine from glutamic acid.

$$\text{Glutamic acid} + \text{ammonia} \rightleftharpoons \text{glutamine} \qquad \Delta G° = +3.4 \text{ kcal/mol}$$

The problem the cell faces is to make favorable reactions out of what would appear to be an unfavorable energy-requiring process. It does this by coupling otherwise endergonic reactions to the exergonic hydrolysis of ATP thereby converting the reactions to exergonic ones which can occur spontaneously in the desired direction. For example, the formation of glutamine from glutamic acid is made to occur in two steps.

1st reaction: glutamic acid + ATP → glutamyl phosphate + ADP
2d reaction: glutamyl phosphate + NH_3 → glutamine + P_i
Overall reaction: glutamic acid + ATP + NH_3 → glutamine + ADP + P_i

The formation of glutamine is said to be "coupled" to the hydrolysis of ATP. As long as the hydrolysis of ATP is more exergonic than the formation of glutamine is endergonic, the "downhill" hydrolysis reaction can be used to drive the "uphill" synthesis of glutamine. An analogy can be made to the use of a falling rock (a downhill reaction) to lift another object (an uphill reaction) as in Fig. 4-1. In this case, the exergonic process is used to perform mechanical work, whereas in the case of ATP hydrolysis, chemical work is accomplished. What is required for the coupling to occur is that the product of the first reaction be used as the substrate for the second. This key molecule, glutamyl phosphate in this case, is termed the *common intermediate*, the bridge between the two reactions. What is occurring, in essence, is that the exergonic hydrolysis of ATP is taking place in two steps, the intermediate being glutamyl phosphate. In the second step, water becomes the final acceptor of the phosphoryl group and the hydrolysis is complete. Both steps of the hydrolysis process are exergonic. What is important in thermodynamic terms is the sum of the free energies of all the reactants (glutamic acid, ATP, and NH_3) and the sum of the free energies of all of the products (glutamine, ADP, P_i) of the overall reaction. If the difference between the products and the reactants of the overall reaction is negative under the conditions of the cell at that moment, then the reaction will proceed toward the formation of the products. The particular reaction path that occurs to go from original reactants to final products is not important in determining the favorability of the reaction. To determine if the reaction will be favored *under standard conditions*, one can add up the $\Delta G°$'s of the component reactions. The $\Delta G°$ of glutamine formation is +3.4 kcal/mol, that of ATP hydrolysis is −7.3 kcal/mol. If the overall value is negative, as it is in this case, then the reaction will proceed in the direction of the formation of products. Once again it should be emphasized that calculations based on $\Delta G°$'s serve only as a guide to the potential feasibility of the particular reaction; the conditions in the cell may be such that the actual feasibility, i.e., ΔG, is very different. We return to the use of ATP and its formation later in this chapter.

We have discussed the equilibrium condition at some length. Cellular metabolism is essentially nonequilibrium metabolism, i.e., it is characterized by nonequilibrium ratios of products to reactants. At equilibrium, as stated above, the difference in free energy between the products and reactants is zero. No useful chemical work can be accomplished at this state just as no useful mechanical work could have been done by the rock in Fig. 4-1 if it had been sitting on the ground when the "teeter-totter" was set up. At equilibrium there is no potential for the flow of materials in any particular direction in the reaction, simply exchange of reactants and products into one another at equal rates.

The hydrolysis of ATP can be used in the cell to drive reactions leading to the formation of molecules such as glutamine because the ATP levels are kept far higher (relative to the ADP levels) than that predicted by the equilibrium constant. As described on page 132, the $\Delta G°$ for ATP hydrolysis is −7.3 kcal/mole, while that for glutamine formation from glutamic acid and ammonia is +3.4 kcal/mole. The fact that the first value is −3.9 kcal/mole greater than the second does not mean that some of the energy from ATP hydrolysis is being wasted or that the reaction is inefficient. Rather, the more negative the overall $\Delta G°$ of the coupled reaction, the greater the rate of the forward reaction as opposed to the reverse reaction under standard conditions. Similarly, the more negative the overall $\Delta G°$, the lower the (glutamic acid)/(glutamine) ratio that can exist in the cell and still lead to the formation of glutamine.

As reactions tend toward equilibrium and their free energy available to do work decreases toward a minimum, another important consequence occurs as well: the entropy (the energy unavailable to do work) increases toward a maximum. In other words, the farther a reaction is kept from its equilibrium state, the less its capacity to do work is lost to the increase in entropy. This does not mean that there are no cellular reactions that are occurring at or near equilibrium concentrations of reactants and products (see Fig. 4-26). Those metabolic steps that occur very rapidly will inevitably tend to be close to equilibrium which simply means that reactions of this type will have important consequences for other reactions in the sequence that either precede or follow it.

The nonequilibrium conditions of metabolism have other important consequences for the cell which are less easily described and not as well understood. The basic principles of thermodynamics have been formulated using nonliving *closed* systems (no exchange of matter between the system and its surroundings) under equilibrium conditions and the unique features of cellular metabolism require a different perspective if we are to understand their significance (see A. Katchalsky, 1965). It is very clear that, just as there are thermodynamic advantages for manmade machinery over the machinery of the cell, the reverse is also true. Cellular metabolism can maintain itself at nonequilibrium conditions because, unlike the environment within a test tube, the cell is an *open* system. Materials are continually flowing into the cell from the bloodstream or culture medium and waste products are continually being removed. The extent of the input into cells from the outside becomes apparent when one considers our minute-to-minute dependence upon an external source of oxygen, a very important reactant in cellular metabolism. As a result of the continual flow of materials into and out of cells and the interrelatedness of biochemical reactions, cellular metabolism is said to exist in a "steady-state" condition (Fig. 4-4). In a steady state, the concentrations of reactants and products remain essentially constant, even though the individual reactions are not necessarily at equilibrium. Since products from one reaction are acted upon as substrates of the next reaction, the concentrations of each metabolite (chemical intermediate) can remain the same as long as new substrate is brought in from the outside and terminal products removed at a constant rate at the other end.

It is this feature of the living state that allows complex, ordered creatures, such as ourselves, to "escape," *in a sense*, the downhill tendency for all energy transformations in the universe. Living systems decrease their entropy by increasing the entropy of their environment. Just as *relatively* simple structures, such as an egg, develop into much more complex structures, such as an adult human, and relatively simple molecules, such as amino acids, become ordered into more complex structures, such as myoglobin in a muscle cell, so too does a large amount of high-energy nutritive organic molecules, such as yolk in the embryo, or glucose in the muscle cell, become degraded to lower energy states (e.g., CO_2 and H_2O). Even though living systems can maintain themselves for a period of time as a system of decreasing entropy, they are not capable of violating the laws of thermodynamics. Ultimately, all living systems on earth depend upon the radiant energy from the sun to maintain their open, steady-state energetics (Chapter 9).

ENZYMES

The study of the properties of enzymes actually began when humans first discovered the means for the formation of such products as cheese and wine, which are processes requiring the activity of microorganisms and their enzymes. The involvement of living yeast cells in the process of fermentation was first shown by Louis Pasteur. Studies at the end of the last century demonstrated that intact yeast cells were not required for the production of alcohol, but these cells could be broken and extracts prepared which could accomplish the same enzymatic conversion. By 1900, the groundwork had been laid for the further analysis of the active ingredients present in extracts of living cells that accounted for their chemical reactions.

The first direct evidence of the protein nature of enzymes was obtained by James Sumner in 1926 when he crystallized the enzyme urease from jack beans and determined its composition. Though this finding was not greeted with much positive acclaim at the time, several other enzymes

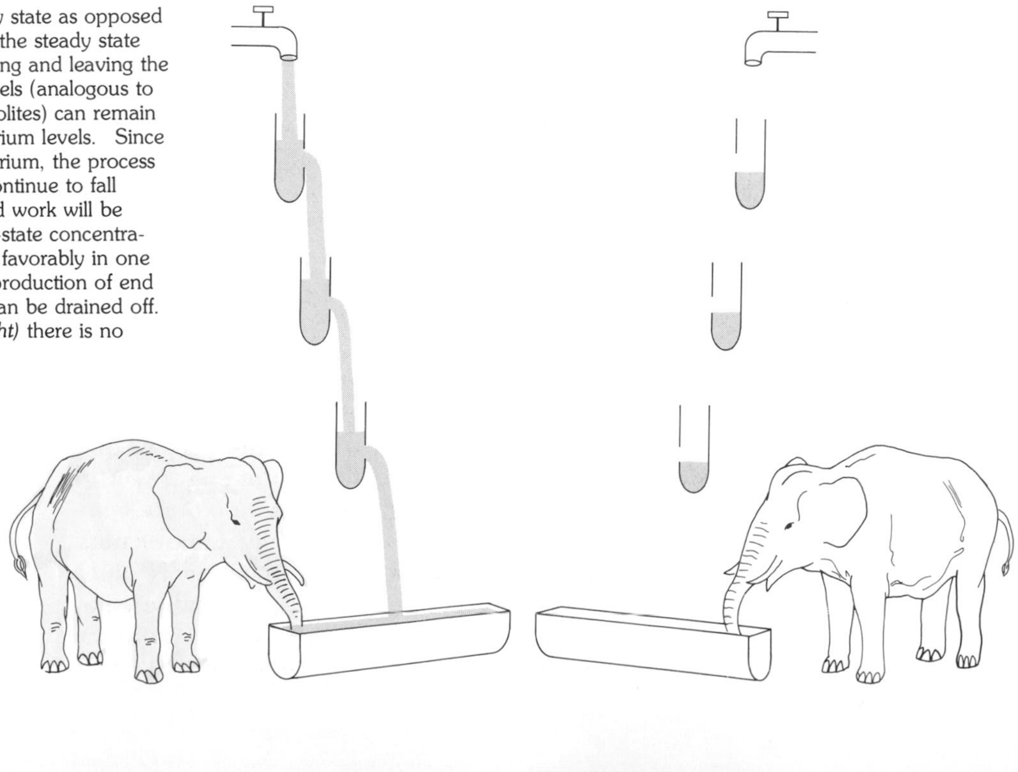

Figure 4-4. Systems in a steady state as opposed to an equilibrium condition. In the steady state *(left),* matter is continually entering and leaving the system. Consequently, fluid levels (analogous to concentrations of cellular metabolites) can remain at relatively constant nonequilibrium levels. Since the fluid levels are not at equilibrium, the process can be directed, i.e., fluid will continue to fall downhill through the system and work will be accomplished. In a cell, steady-state concentrations allow reactions to proceed favorably in one overall direction leading to the production of end products of metabolism which can be drained off. In the equilibrium condition *(right)* there is no directed movement.

were soon shown to be proteins, and it is now accepted that every enzyme is a protein. Even though all enzymes are proteins, many of them are conjugated proteins, i.e., they contain nonprotein components, called *cofactors,* which may be inorganic (metals) or organic (coenzymes) or both. When present, cofactors are important participants in the functioning of the enzyme, often carrying out activities for which amino acids are not suited. For example, in the electron transport chain of the mitochondria, it is the coenzymes that are successively oxidized and reduced by the passing electrons.

By convention of the Enzyme Commission, there are six broad categories of enzyme-catalyzed reactions which can be broken down into various subtypes as listed and described in Table 4-4. As true catalysts, the following properties can be attributed to enzymes: (1) They are present in small amounts. (2) They are not altered irreversibly during the course of the reaction, therefore each enzyme molecule can participate in the catalysis of many individual reactions. (3) They have no effect upon the equilibrium of the reaction being catalyzed. This last point is of particular importance. Enzymes do not determine whether a reaction is thermodynamically favorable or unfavorable nor do they determine what the ratio of products to reactants will be at equilibrium. These are properties of the components of the reaction as governed by the laws of thermodynamics. As catalysts, enzymes can only accelerate the rate at which a favorable chemical reaction proceeds. In fact, enzymes can catalyze the reverse reaction as well as the forward one if the conditions are such that the reverse reaction is favored, i.e., the ratio of products to reactants is greater than that expressed by the equilibrium constant. The catalysis of the reverse reaction can be demonstrated in the test tube by adding a small amount of radioactive *product* and observing the rapid appearance of radioactivity in the population of reactant molecules.

Enzymes act by increasing the rate at which favorable reactions occur, but there is no necessary relationship between the magnitude of the ΔG for a particular reaction and the rate at which that reaction takes place. The magnitude of ΔG informs us only of the difference in free energy (the *maximum* usable energy) between the initial and final or equilibrium state. It is totally independent of the pathway or the time it takes to go from one state to the other. Consider glucose for example. The oxidation of this carbohydrate is a highly favorable reaction as can be determined by its combustion in a calorimeter. However, glucose crystals can be left out for essentially an indefinite period without a noticeable conversion into less energetic materials. Even if the sugar were to be dissolved, as long as it were kept sterile, it would not rapidly deteriorate. However, if one were to add a few bacteria, within a short period of time the sugar would be taken up into the cells and enzymatically degraded.

When one considers the rate at which enzyme-catalyzed reactions are accelerated over the noncatalyzed reaction, the values are incredibly high. Catalysts of organic reactions, such as heat, acid, platinum, and magnesium, generally accelerate reactions a few orders of magnitude (e.g., 1000 times). In contrast, enzymes increase the velocity of reactions many orders of

TABLE 4-4
Classification of Enzymes According to the Recommendations of the Commission of the International Union of Biochemistry

1. Oxidoreductases

Enzymes which catalyze oxidation-reduction reactions. Reactions of this type involve the transfer of an electron (or hydrogen atom) from a donor to an acceptor. Common donors include:

$$\underset{\text{H}}{\overset{\text{H}}{\text{HC}}}-\text{OH}, \quad \text{HC}=\text{O}, \quad \underset{\text{H}}{\overset{\text{H}}{\text{HC}}}-\underset{\text{H}}{\overset{\text{H}}{\text{CH}}}, \quad \underset{\text{H}}{\overset{\text{H}}{\text{HC}}}-\text{NH}_2, \quad \underset{\text{H}}{\overset{\text{H}}{\text{HC}}}-\text{NH}, \quad \text{NADH}, \quad \text{NADPH}$$

Common acceptors include:

NAD^+, $NADP^+$, cytochromes, molecular oxygen, a disulfide

Examples include alcohol dehydrogenase (page 171), glyceraldehyde phosphate dehydrogenase (page 167), and succinic dehydrogenase (page 173).

2. Transferases

Enzymes which catalyze the transfer of some chemical group from one compound (a donor) to another (an acceptor). Common groups being transferred include alkyl groups (e.g., methyl), acyl or aminoacyl groups (e.g., acetyl, glutamyl), glycosyl groups (e.g., glucose, sialic acid), and phosphate groups.

Examples include glycosyltransferases (page 195), protein kinase (page 247), and phosphofructokinase (page 164).

3. Hydrolases

Enzymes which catalyze the splitting of C—O, C—N, C—C, and some other bonds by the addition of water.

Types of linkages hydrolyzed include esters (e.g., between fatty acids and glycerol in lipids), glycosidic bonds (e.g., between sugars in glycogen, starch, or bacterial cell walls), and peptide bonds (e.g., at specific amino acid residues in proteins).

Examples include β-galactosidase (page 585), lysozyme (page 146), and chymotrypsin (page 143).

4. Lyases

Enzymes which cleave C—C, C—O, C—N, and other bonds by elimination, leaving double bonds, or, conversely, adding groups to double bonds.

Examples included aldolase (page 164), pyruvate decarboxylase (page 171), and aconitase (page 173).

5. Isomerases

Enzymes which catalyze a redistribution of atoms or chemical groups *within* a molecule.

Examples include alanine racemase (converts L-alanine to D-alanine) and triosephosphate isomerase (page 164).

6. Ligases

Enzymes which catalyze the joining together of two molecules coupled with the hydrolysis of ATP or other triphosphate. Bonds formed can be C—O, C—S, C—N, or C—C in nature.

Examples include aminoacyl tRNA synthetases (page 497), succinyl CoA synthetase (page 173), glutamine synthetase (page 133), and pyruvate carboxylase.

magnitude greater than their nonprotein counterparts, up to 10^{10} times or more. What is even more remarkable, they accomplish this feat under the extremely mild conditions of temperature and pH that are present within the cell. In addition, unlike the inorganic catalysts, they are extremely efficient. There are virtually no side reactions occurring, i.e., the only products formed are the desired ones. A last advantage of enzymes as catalysts is their capacity to be regulated. The question of importance is how enzymes can accomplish such effective catalysis.

Activation Energy

The first question to consider is why all thermodynamically favorable reactions do not occur at noticeably rapid rates? In some cases, even at room temperature, molecules are reactive enough or sufficiently unstable that they will spontaneously undergo a chemical reaction. Generally, molecules this unstable are of little use to the cell for it cannot control their reactions. Even a molecule such as ATP, whose hydrolysis is so favorable, is basically stable in the cell until its breakdown occurs in a controlled enzymatic manner. Why do favorable reactions not occur rapidly in the cell without the need for enzymes? The reason is that an energy barrier still exists before the reactants can be converted to products. Even though a reaction is favored, certain chemical bonds (electron pairs) must be broken within the reactants and this requires that the reactants must come together with sufficient collision energy that they overcome this hurdle, termed the *energy of activation* (E_A). This is expressed diagrammatically in Fig. 4-5. The energy of activation is represented by the height of the hump. The analogy that is often made is that of an object

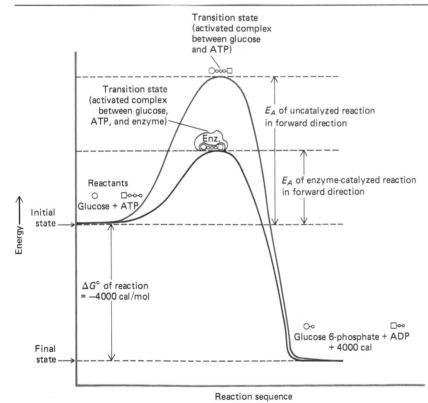

Figure 4-5. Even though the formation of glucose 6-phosphate is a thermodynamically favored reaction $(\Delta G° = -4$ kcal/mol), the reactants must have sufficient energy to attain the transition state. The energy of activation (E_A) is greatly reduced when the reactants combine (Fig. 4-7) with an enzyme catalyst.

resting on top of a cliff ready to plunge to the bottom. If left to itself, the object in all likelihood would remain there indefinitely. However, if someone were to come along and provide the object with sufficient energy to overcome the friction or some other small barrier in its way and cause it to reach the edge of the cliff, it would spontaneously drop to the bottom. The object has the potential to fall to a lower state once activated.

In a solution at room temperature, molecules are found in a state of random movement, each possessing at a given instant a certain amount of energy. Among the population of molecules, their energy is distributed in a bell-shaped curve (Fig. 4-6), some being of very low energy and others much higher. Molecules of high energy ("activated molecules") remain as such only for a brief time, losing their excess energy to other molecules by collision. If reactants collide with one another with sufficient energy to overcome the activation barrier, then the possibility exists that they will react successfully and be converted to products. In other words, the rate of the reaction depends upon the number of collisions between activated molecules that occur in a given amount of time. One way to increase the reaction rate is to increase the energy of the reactants, thereby increasing the number capable of successful collisions. This is most readily done by heating the reaction mixture.

The reactants, when they are at the crest of the hump ready to be converted to products, are said to be at the *transition state*. At this point the reactants have formed an "activated complex" which has an extremely great likelihood of being converted to products. Unlike the difference in standard free energy for a reaction, the activation energy is not a fixed

Figure 4-6. The bell-shaped curves indicate the energy content of a population of molecules present in a reaction mixture. The number of reactant molecules containing sufficient energy to undergo reaction is increased by either heating the mixture or adding a catalyst. Heat increases the rate of reaction by increasing the energy content of the molecules while the enzyme does so by lowering the energy required.

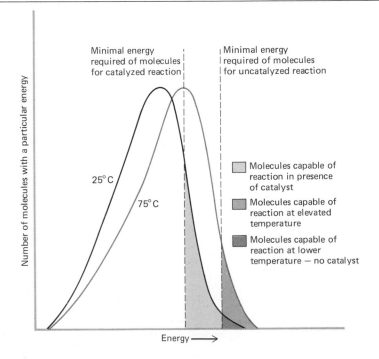

value, but rather varies with the particular reaction mechanism utilized to reach the transition state. Some pathways are less energy-requiring than others; enzymatic ones are the least requiring. In other words, unlike catalysis by heat, enzymes cause their substrates (reactants) to be very reactive without having to be raised to particularly high energy levels. Enzymes, therefore, catalyze reactions by decreasing the magnitude of the activation energy barrier. A comparison between the activation energies for the hydrolysis of urea in the noncatalyzed state, the acid-catalyzed state, and the enzymatically catalyzed state is shown in Fig. 4-6.

Mechanisms of Enzyme Action

As catalysts, enzymes accelerate bond-breaking and bond-forming processes. As one might expect, in order for a particular enzyme molecule to accomplish this task it will have to insert itself intimately into the activities that are taking place among the reactants. It was postulated at an early point in enzyme study that there must be formation of a complex between the enzyme and the reactants, i.e., an enzyme-substrate (ES) complex. Long after its proposal, direct evidence was obtained for the existence of the ES complex in a number of different reactions. In most cases the union between enzyme and substrate is noncovalent, though many examples have been discovered in which a transient covalent bond is formed. In a sense the involvement of a large macromolecule in the dealings of smaller molecular weight materials takes these substances out of solution and places them on the surface of the large catalyst molecule. Once complexed with the enzyme, the reactants will be brought very close together and presumably will be held in just the proper orientation so that the reaction is facilitated (Fig. 4-7). In contrast, when reactants are free in solution, even those possessing sufficient energy are not guaranteed that their collision will result in the successful formation of a transition-state complex. It is estimated that only 1 to 10% of such collisions are usually productive. In the case of the enzyme-catalyzed reactions, this

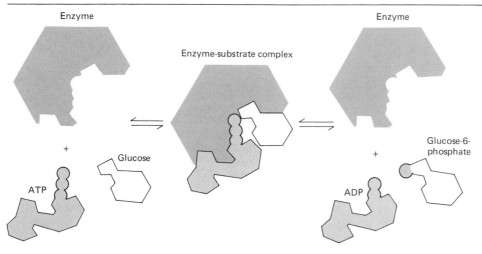

Figure 4-7. The formation of an enzyme-substrate complex followed by catalysis. *(From J. D. Watson, "The Molecular Biology of the Gene," 3d. ed., W. A. Benjamin, 1976.)*

percentage would be very high. This advantage to the reactants of being properly oriented probably accounts for some of the catalytic activity of the enzyme.

Consider the nature of an active region of an enzyme molecule with the variety of reactive amino acid side chains available to interact with the substrate. The inevitable outcome of an association of substrate with the enzyme is a change in the nature of the substrate. It is that change which is responsible, for the most part, for the increased reactivity of the substrate. The substrate has been "activated" in a sense, without the input of an external energy source, such as heat. Though it is clear that there are aspects to the activation process that are not understood, there do appear to be several general mechanisms whereby the reactivity of substrates is increased by association with enzymes. Basically, these mechanisms are of the same nature that the organic chemists have become familiar with in the analysis of organic reactions in the test tube. There is certainly no reason at this point to believe that enzymes accomplish anything either mystical, or, for that matter, not presently understood among organic reactions. The most commonly discussed means by which enzymes accelerate biological reactions fall into a few types of mechanisms. An extensive discussion of these can be found in the biochemistry texts, we will just mention them with the aim of illustrating that a considerable amount of information is available on the subject.

One of the most commonly used means of increasing the rate of organic or inorganic reactions in the test tube is to change the pH. In many cases the presence of an increased concentration of hydrogen or hydroxyl ions greatly facilitates the reaction taking place. Enzymes have numerous acidic or basic groups that are capable of donating or accepting protons (or electrons) from the substrate, thereby making it more reactive. In the discussion of protein structure in Chapter 3, the internal hydrophobic nature of globular proteins was stressed and the point was made that this region of the molecule provides a unique, nonaqueous environment within the cell. Acidic or basic amino acid side chains projecting into this hydrophobic region would be capable of very strong ionic interactions with polarized atoms in the substrate, which might greatly facilitate the reaction. These points are illustrated in the discussion of lysozyme on page 146.

Analysis of certain of the enzymes which form temporary covalent

Figure 4-8. Substitution reactions. *(a)* Nucleophilic substitution where Y^- (a Lewis base) is the nucleophilic attacking group and X^- the leaving group. The attack centers on the electropositive (δ^+) R group of the reactant, RX. *(b)* Electrophilic substitution where Y^+ (a Lewis acid) is the electrophilic attacking group and X^+ the leaving group. The attack centers on the electronegative (δ^-) R group of the reactant RX.

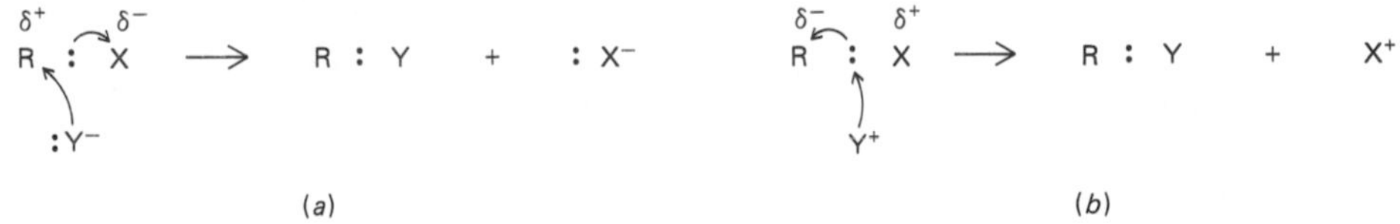

bonds with their substrates has revealed another way in which reactions are promoted. Reactive molecules are ones that are electronically polarized, i.e., they have regions of electropositive or electronegative character. Electronegative centers act as reactive nucleophilic attacking groups capable of donating electrons to electron-deficient carbon atoms (Fig. 4-8a). Electropositive centers act as reactive electrophilic attacking groups seeking complementary regions in other molecules where there is increased electron density (Fig. 4-8b). Certain of the amino acid side chains (serine, histidine, cysteine) have the potential to carry out a nucleophilic attack on a substrate thereby displacing groups from the molecule. An example of the nucleophilic displacement by a serine residue in the enzyme chymotrypsin is shown in Fig. 4-9. Enzymes that function via an electrophilic attack on the substrate do so with the aid of cofactors since there are no strongly electrophilic amino acid residues. Metal ions, being strongly electron-seeking (electropositive) make excellent electrophilic groups within enzyme molecules.

Another means by which enzymes are proposed to effect an increase in the reactivity of the substrate is to cause certain bonds in the substrate to be subjected to stress or deformation (page 147). In this proposal, the enzyme would physically distort the substrate, thereby weakening its bonds and lowering the energy need to break them. The degree to which any or all of these influences by enzymes can explain their catalytic prowess remains uncertain, though calculations generally suggest that there must be additional properties so far undiscovered.

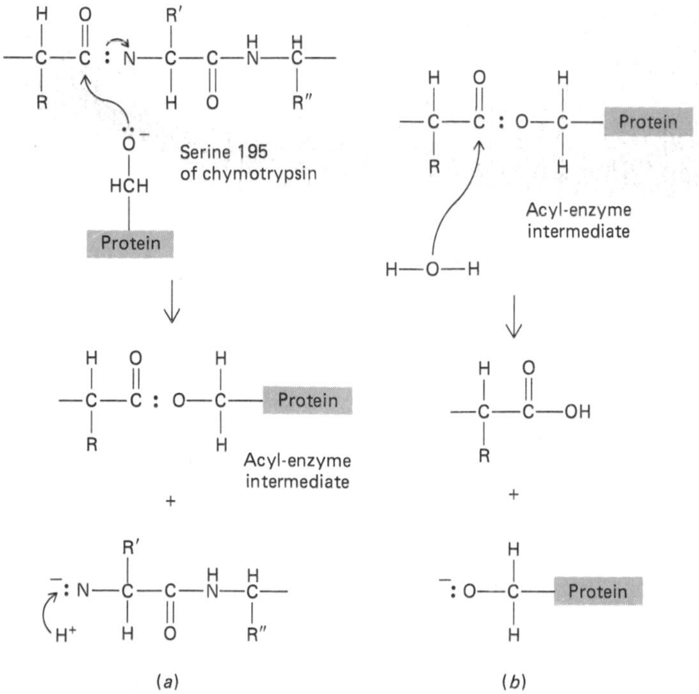

Figure 4-9. The mechanism of action of chymotrypsin. (a) The electronegative oxygen atom of a serine residue (#195) in the enzyme carries out a nucleophilic attack on the electropositive carbon atom of the polypeptide substrate, splitting the peptide bond. The electrons of the peptide bond accompany that part of the substrate molecule which is displaced and the enzyme forms a covalent bond with the remainder of the substrate molecule. (b) In the second step, the electronegative oxygen atom of a water molecule displaces the acyl group from the enzyme resulting in the release of an active catalyst molecule.

Active Site

That part of the enzyme molecule that is directly involved in the interaction with substrate is termed the *active site*. In the formation of the tertiary structure of an enzyme, a particular region appears that has the correct shape, i.e., properly disposed amino acid residues, so that substrate molecules will (1) bind to the enzyme in the necessary manner, and (2) be subjected to the necessary catalytic influences. In other words, the shape and environment of the active site is of critical importance for the catalytic activity of the enzyme. Even though the active site may contain the critical amino acid residues, this small region of the protein owes its shape to the remainder of the molecule, which should not be ignored.

The structure of the active site not only accounts for the catalytic activity of the enzyme but also for its *specificity*. The concept of specific interacting molecules is one of the most essential in cell biology and arises as a direct consequence of the complexity of cellular organization. In the case of enzymes, specificity is exemplified by the restricted number of molecules with which an enzyme can interact. For example, if the enzyme lactic dehydrogenase (LDH) is present in solution together with a hundred small molecular weight substances in addition to L(+)-lactate, its substrate, only the lactate molecules will be capable of fitting into the active site of the enzyme. For all practical purposes, the others may as well be absent. It is specificity, whether of enzymes or other biological molecules, whether in solution or attached to a membrane, which allows order to be maintained in an otherwise chaotic environment. Each event can occur independently because of the selective nature of the interactions. The precision with which interactions occur results in very efficient processes with a minimum of by-products or side effects. An example of a type of breakdown in specificity will illustrate its importance. The reactions between antibodies and antigens (Chapter 18) are remarkably specific. When a foreign substance (antigen) enters an individual, the immune system responds by the production of a protein (antibody) which is capable of specifically interacting with that antigen, thereby leading to its neutralization, removal, or destruction. In almost every case, the antibody is constructed such that the only molecule it can interact with is the antigen responsible for its production. In the case of rheumatic fever, however, a protein present in the heart tissue is similar enough in molecular structure to an antigen in the surface of a bacterial capsule that, when an infection from this bacteria (a streptococcus) occurs, the possibility arises that the antibodies produced in defense of the person will form a complex with the tissues of the heart causing tissue destruction. The complexity that is required to maintain life demands an extremely high level of precision and breakdowns in this precision can be very costly.

Not all enzymes possess an absolute specificity for their substrate, making them incapable of interaction with any other molecule. In most cases molecules related in structure to the substrate can be synthesized so that they are similar enough to make the fit and undergo reaction to a greater or lesser degree. This does not necessarily take anything away from the enzyme for these substratelike molecules (analogs) are not present in the cell and cannot interfere in the desired reaction. In many other cases, enzymes purposely have a broad specificity so that a group of related substances can all act as effective substrates. Enzymes of this type are economic for the cell since they lessen the need for additional enzymes to catalyze each specific reaction. For example, carboxypeptidase acts in the digestive tract to hydrolyze proteins, removing one amino acid at a time from the shrinking carboxyl-terminal end of the polypeptide. It makes little difference which amino acid is present at the C-terminus, carboxypeptidase will remove it. The enzyme is constructed such that the nature of the side chains of the two amino acids forming the last peptide bond in the fragment is ignored. If this enzyme were instead specific for the particular peptide bond present, 400 different enzymes (20 amino acids × 20 amino acids) would be required to replace it.

In the early concepts of enzyme-substrate interactions, it was felt that the active site was a rigid structure, a mold in which the substrate would fit. This concept was developed at the end of the last century by Emil Fischer, who showed that enzymes were constructed such that they could distinguish between stereoisomers (L form from D form) as potential substrates. It was concluded that a very tight fit between enzyme and substrate would be required to explain such a high level of specificity. The analogy was made between the enzyme-substrate contact and that of a lock and key (Fig. 4-10a). In more recent years the "lock and key" model has given way to a concept of enzymes as being more flexible and possessing the capability for internal movements of specific parts relative to one another. This concept is embodied in the "induced-fit" theory (Fig. 4-10b) proposed by Daniel Koshland in which the structure of the binding site of the enzyme is not equivalent to the structure of the catalytic site. The binding site of enzymes are constructed so as to be relatively specific in their interactions with potential substrates. However, once the proper substrate is bound, a shift in the conformation occurs so that the fit between the two is improved (an induced fit) and the proper reactive groups of the enzyme are moved into place so that the reaction can occur. The

Figure 4-10. The "lock-and-key" versus the "induced-fit" models of enzyme action. (a) In the lock-and-key model, the enzyme (by itself) is fully complementary to the substrate. (b) In the induced-fit model, the full complementarity occurs only after the initial binding event. Contact with the substrate causes a conformational change in the enzyme thereby inducing the better fit and bringing about the reaction.

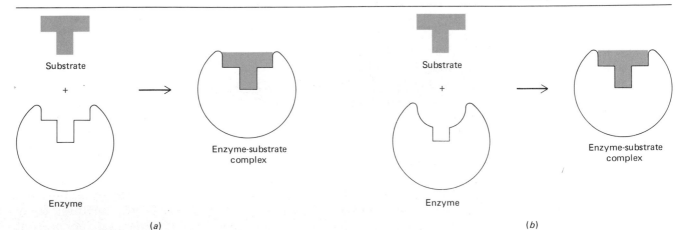

(a)

(b)

shift in conformation of the enzyme could provide a means whereby stress could be placed on the bonds of the substrate. Once the substrate is bound to the enzyme, the shift of the enzyme to the active state might serve to twist or deform the substrate.

Changes in enzyme conformation during catalysis have been detected in a variety of enzymes, providing direct evidence of their occurrence. These observations are made through x-ray diffraction analysis of the difference in conformation of enzymes in the presence of substrate (or inhibitors) and in their absence. In a number of cases, this technique has detected a shift in the relative positions of certain of the atoms of the enzyme once the binding event has occurred. Various of the points made in these last sections can be illustrated by considering an example of enzyme catalysis in some detail.

Catalysis by Lysozyme

Lysozyme is an enzyme which plays an important role in the destruction of the walls surrounding certain types of bacterial cells resulting in the cell's lysis. The enzyme is found in various tissues and secretions of the body and is particularly concentrated in egg white from which it has been purified and crystallized. It was originally discovered by Alexander Fleming, the discoverer of penicillin, when one day when suffering from a cold he found that a drop of nasal mucus, when added to a culture of bacteria, resulted in the lysis of the cells on the plate. The cell walls of sensitive bacteria are composed of an alternating copolymer of the amino sugars, N-acetylglucosamine and N-acetylmuramic acid (Fig. 4-11). The two sugars are linked to one another by a glycosidic bond which is hydrolyzed by the enzyme.

Figure 4-11. The substrate of lysozyme. Linkages between the sugar units are β (1→4) glycosidic bonds. (a) Conventional drawing. (b) Actual conformation.

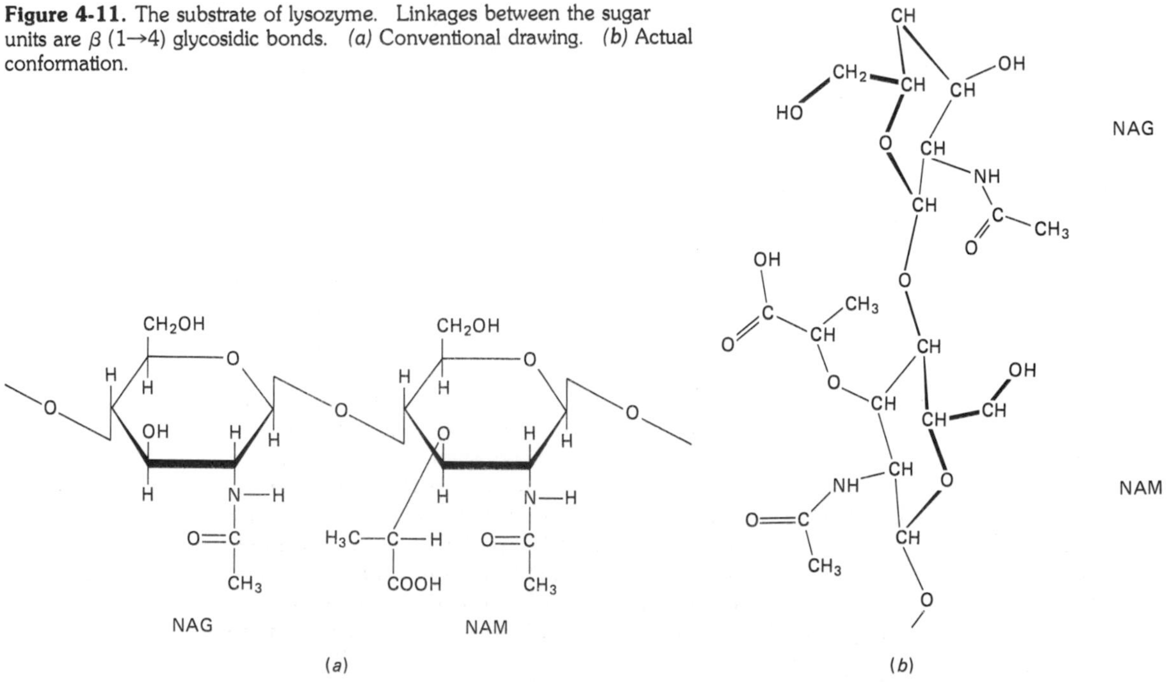

The original proposal for the mechanism of catalysis was based upon x-ray diffraction analysis of the enzyme and has been supported by a variety of subsequent studies. The normal substrate for lysozyme is a large polymeric polysaccharide. If instead of using a large molecular weight polymer, one of only three sugar units (tri-N-acetylglucosamine) is added to the enzyme, it will bind to the active site but it will not be hydrolyzed. When enzyme crystals were prepared in the presence of this compound, and their electron density profile determined by x-ray diffraction, the location of the substance within the tertiary structure of the enzyme was apparent. It was found that tri-N-acetylglucosamine sat in a cleft within the enzyme, filling about one-half of the cleft's length. It was proposed that, when a long polymer was being digested, the entire cleft would be filled by the substrate (Fig. 4-12a). Based on the construction of models of enzyme-substrate complexes and other data, it was proposed that the length of the cleft would be occupied by six adjacent sugar units of the polymer, and that the bond that would be hydrolyzed was that between the fourth and fifth sugar.

Analysis of the nature of the enzyme-substrate complex revealed several important features that suggested a mechanism to account for the hydrolytic catalysis. It was suggested that when six adjacent sugar units were bound in the cleft of the enzyme, one of the sugars (the fourth or D sugar of Fig. 4-12b) would not be readily accommodated in the space provided. In order for this sugar to fit it had to undergo a certain amount of distortion, i.e., it had to be "forced" into position.

When the tertiary structure of the enzyme in the vicinity of the glycosidic bond between the fourth and fifth sugar was analyzed, it was apparent that there were two amino acid residues that approached the bond from either side very closely (approximately 3 Å). One residue was an aspartic acid and the other a glutamic acid; both carboxyl-containing amino acids (Fig. 4-12b). When the environment of these two residues was considered, it appeared that the ionization states of the two carboxyl groups would be quite different. The environment of the glutamic acid is nonpolar which would be expected to suppress the dissociation of its proton, while that of the aspartic acid is polar which would promote the dissociation of the proton leaving the aspartyl residue with a negative charge.

In a previous section several possible means were discussed by which an enzyme might lower the energy of activation for a given reaction. It appears that the hydrolytic activity of lysozyme relies on two of these: acid catalysis and distortion of the substrate. The reaction mechanism proposed by David Phillips (additional data can be found in the text by Lubert Stryer) is as follows. The first step is the splitting of the glycosidic bond which is accomplished by the interaction of the bond with the closely applied glutamic acid. The polar nature of the oxygen atom of the bond is sufficient to draw the proton from the undissociated carboxyl of the glutamic acid, causing the acid-catalyzed hydrolysis of the bond (Fig. 4-12b). The breaking of the bond by the proton leaves the carbon atom with an excess positive charge (a carbonium ion). The formation of the carbonium ion is facilitated by the distortion the sugar had been subjected to when the enzyme-substrate complex was formed. The positively charged carbonium ion is very close to the negatively charged aspartic acid residue, which stabilizes the carbonium ion, thereby favoring the reaction in the direction of hydrolysis. The interaction of the carbonium ion with a hydroxyl ion in the solvent completes the hydrolysis.

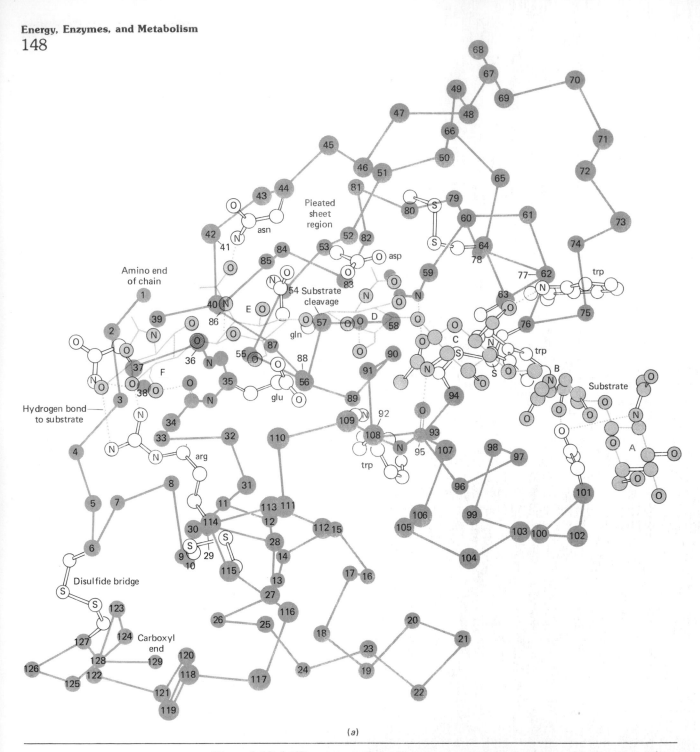

Figure 4-12. (a) The presence of six adjacent sugar units within the crevice of the lysozyme molecule. Various interactions between the substrate and enzyme are indicated as is the site of cleavage of the substrate. (b)–(d) The mechanism proposed by David Phillips to explain lysozyme action as described in the text. The dotted line indicates the undistorted conformation of the D ring. (After R. E. Dickerson and I. Geis, "The Structure and Action of Proteins," W. A. Benjamin, Inc. Copyright 1969 by Dickerson and Geis.)

Enzyme Kinetics

In several places in the preceding pages we have alluded to the rates of reaction. It has been shown that enzymes are effective because they increase reaction rates. The rates of enzyme reactions can be regulated by environmental conditions existing within the cell. We will spend a large part of this book discussing various regulatory mechanisms for many different cellular processes for the analysis of these control systems is of the greatest importance if we are to understand the lives of cells. For example, it is essential that the various metabolites of the cell are maintained in the correct concentrations and this will ultimately be determined by the reaction rates of the various enzymes. Even though we may know very little about reaction rates within the cell, a great deal is known about the capabilities of the various enzymes since each has been purified to a greater or lesser degree and the reaction it catalyzes has been studied in the test tube.

It is only when enzymes are removed from their cellular environment and studied in an isolated, purified state that the role of the enzyme can be appreciated and the mechanism of its action determined. The quantitative analysis of enzyme activity in vitro provides a means for the comparison of different enzymes (or the same enzyme acting on various substrates or under different environmental conditions). Rates (velocities) of reactions are measured as amounts of product formed (or substrate consumed) in a given amount of time. Of particular importance is the relationship between the reaction rate and the substrate concentration, when the amount of enzyme is held constant.

In 1913 Leonor Michaelis and Maud Menten reported on the mathematical analysis of the relationship between substrate concentration and the velocity of enzyme reactions. This relationship can be expressed by an equation which generates a hyperbola as shown in Fig. 4-13. Rather than considering the theoretical aspects of enzyme kinetics, we can obtain the same curve in a practical manner, as it is done for each enzyme that is studied. In order to determine the velocity of a reaction, an incubation mixture is set up at the desired temperature containing all of the ingredients required except one, which when added will initiate the reaction.

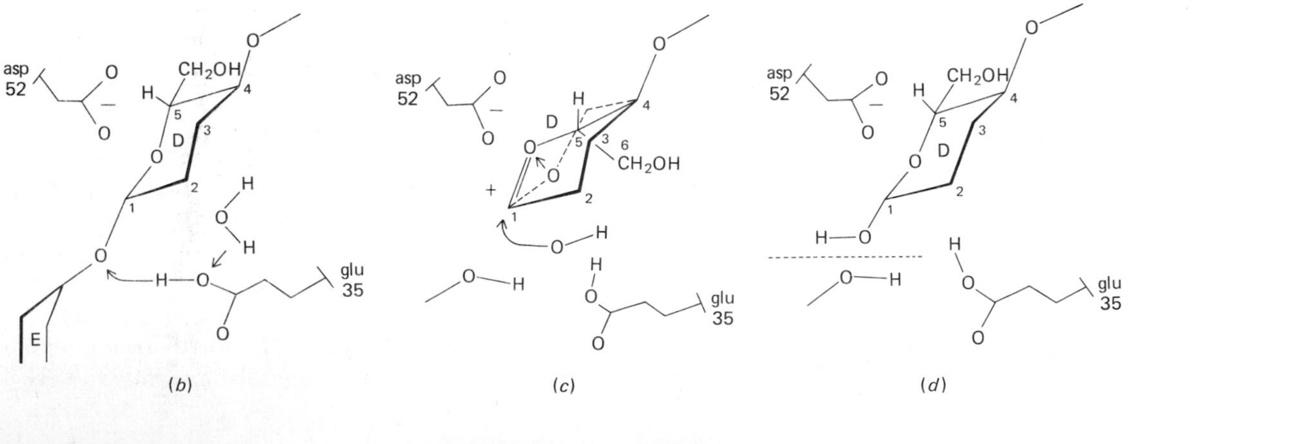

(b)

(c)

(d)

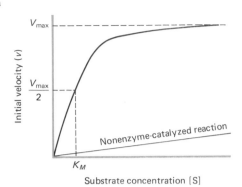

Figure 4-13. The relationship between the rate of an enzyme-catalyzed reaction and the substrate concentration.

If at the time the reaction begins, no product is present in the mixture, then the rate of appearance of product with time provides a measure of the velocity. There are complicating factors in this procedure. If the incubation time is too great, the substrate concentration becomes measurably reduced. In addition, as product appears it can be converted back to substrate by the reverse reaction which is also catalyzed by the enzyme. Ideally, what we want to determine is the initial velocity, i.e., the velocity at the instant when no product has yet been formed. In order to obtain accurate values of the initial reaction velocity, brief incubation times and sensitive measuring techniques are employed before the reaction has proceeded very far.

In order to generate a curve such as that shown in Fig. 4-13, the initial velocity is simply determined when the incubation mixture contains an increasing concentration of substrate. Since the important feature in such an experiment is that the enzyme concentration remain constant, these measurements can be made on relatively impure enzyme preparations. From this curve it is apparent that the substrate concentration has a great effect upon the initial reaction velocity. The basis for this effect resides with the capacity of each enzyme molecule. Each reaction catalyzed takes a finite amount of time which limits the number of reactions that can be catalyzed in a given time span. At low substrate concentrations, the enzyme is capable of working at a faster rate than the number of effective collisions it is subjected to. In other words, substrate molecules are rate limiting. At high substrate concentrations, the enzyme is working at its maximal capacity, being subjected to more collisions with substrate than reactions it can catalyze. At high substrate concentrations, the enzyme becomes rate limiting. As a greater and greater concentration of substrate is present, the enzyme approaches a state of *saturation*. The velocity at this theoretical saturation point is termed the *maximal velocity* (V_{max}), and essentially represents the value when the curve plateaus. If the molecular weight and the concentration of the enzyme in the reaction mixture are known, then the *turnover number* for the enzyme can be calculated from the V_{max}. The turnover number is the maximum number of molecules of substrate that can be convereted to product by one enzyme molecule/minute. A turnover number of 1000 is typical for enzymes, though values as great as 10^7 (for carbonic anhydrase) are known. It is apparent from these values that relatively few molecules of enzyme can convert a relatively large number of substrate molecules into product.

The value of V_{max} is one useful term obtained from a plot such as that shown in Fig. 4-13; another is the Michaelis constant (K_M), which is equal to the substrate concentration when the reaction velocity is $V_{max}/2$ (Fig. 4-13). In certain cases, depending upon the relative rates of formation and breakdown of the enzyme-substrate complex, the value for K_M provides a measure of the affinity of the enzyme for substrate. The greater the value, the lesser the affinity; a typical value being about 10^{-4} M. The K_M's for a variety of enzymes are given in Table 4-5, and the relative reaction rates for one enzyme with a variety of substrates is given in Table 4-6. Other factors that strongly influence enzyme kinetics are the pH and temperature of the incubation medium. For both parameters, an optimal condition exists which provides the greatest rates at a given substrate concentration (Fig. 4-14).

TABLE 4-5
K_M Values of Some Enzymes

Enzyme	Substrate	K_M
Chymotrypsin	Acetyl-L-tryptophanamide	5×10^{-3} M
Lysozyme	Hexa-N-acetylglucosamine	6×10^{-6} M
β-Galactosidase	Lactose	4×10^{-3} M
Threonine deaminase	Threonine	5×10^{-3} M
Carbonic anhydrase	CO_2	8×10^{-3} M
Penicillinase	Benzylpenicillin	5×10^{-5} M
Pyruvate carboxylase	Pyruvate	4×10^{-4} M
	HCO_3^-	1×10^{-3} M
	ATP	6×10^{-5} M
Arginine-tRNA synthetase	Arginine	3×10^{-6} M
	tRNA	4×10^{-7} M
	ATP	3×10^{-4} M

SOURCE: BIOCHEMISTRY by Lubert Stryer. W. H. Freeman and Company. Copyright © 1975.

TABLE 4-6
Rate of Oxidation of D-Amino Acids and Other Substrates by D-Amino Acid Oxidase

Substrate	Oxygen uptake°	Substrate	Oxygen uptake°
D-Tyrosine	190	D-Histidine	6.2
D-Proline	148	D-Threonine	2.1
D-Methionine	80	D-Cystine	1.9
D-Alanine	64	D-Aspartic acid	1.4
D-Serine	42	D-Lysine	0.6
D-Tryptophan	37	D-Glutamic acid	0
D-Valine	35	L-Amino acids	0
D-Phenylalanine	26	D-Peptides	0
D-Isoleucine	22	N-Acetylalanine	0
D-Leucine	14	β-Pyridyl-4-alanine	95

° Crude extract of acetone-dried sheep kidney cortex at pH 8.3 used as source of enzyme. Rate of oxygen consumption is calculated from uptake in 10 minutes, and results are given as microliter/milligram of enzyme preparation/hour.
SOURCE: A. White et al., "Principles of Biochemistry," 5th ed., McGraw-Hill, 1975.

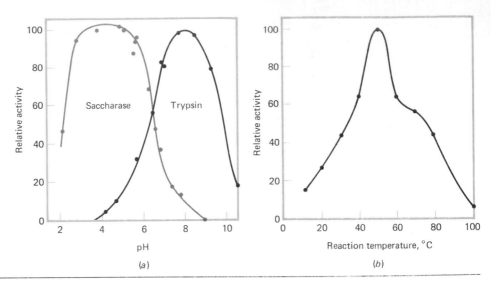

Figure 4-14. Dependence of the rate of an enzymatic reaction on (a) pH and (b) temperature. The shape of the curves and the optimal pH and temperature vary with the particular reaction. Changes in pH affect the ionic properties of the substrate and the enzyme as well as the enzyme's conformation. At lower temperatures, reaction rates rise with increasing temperature due to the increased energy of the reactants. At higher temperatures, this positive effect is offset by enzyme denaturation. [(a) From E. A. Moelwyn-Hughes, in "The Enzymes," J. B. Sumner and K. Myrback, eds., vol. 1, Academic, 1950; (b) from K. Hayashi et al., J. Biochem., **64**:93 (1968).]

In order to generate a hyperbolic curve such as that of Fig. 4-13, and make an accurate determination of the values for V_{max} and K_M, a considerable number of points must be plotted. An easier and more accurate description is gained by plotting the reciprocals of the velocity and substrate concentration against one another as formulated by Lineweaver and Burk. When this is done, the hyperbola becomes a straight line (Fig. 4-15) whose X intercept is equal to $-1/K_M$, Y intercept is equal to $1/V_{max}$, and slope is equal to K_M/V_{max}. The values of K_M and V_{max} are, therefore, readily determinable by extrapolation of the line drawn from a relatively few points.

Enzyme Inhibitors

All chemical reactions occurring in cells are believed to require enzyme catalysts. Since enzymes play such an important role in the activities of the cell it is essential that they be subject to cellular control. There are several distinct levels at which this control can be exercised. Regulatory mechanisms that determine the amount of enzyme present, i.e., the level of its synthesis and destruction, will be considered in later chapters. Regulatory mechanisms also exist that modulate the activity of enzyme molecules already present within the cell. For example, if it no longer remains in the best interests of the cell to continue to produce a particular molecule, such as an amino acid, mechanisms exist which allow that particular cellular activity to be reduced without affecting other unrelated metabolic steps. The reduction in the formation of the product is accomplished by a mechanism (page 156) which alters the activity of one or more of the enzymes responsible for the production of the product. Since the great effectiveness of enzyme catalysts is a direct consequence of the relative positions of the amino acids of the active site, any means by which these positions can be modified would be expected to alter the enzyme's activity. Changes in enzyme conformation provide the underlying means of metabolic control in the cell. Generally the activity of enzymes is modulated by small molecular weight inhibitors that may or may not be present in the cell at any given time. The study of these inhibitors in vitro

Figure 4-15. A Lineweaver-Burk plot of the reciprocals of velocity and substrate concentration from which the values for the K_M and V_{max} are readily obtained.

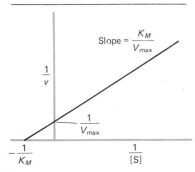

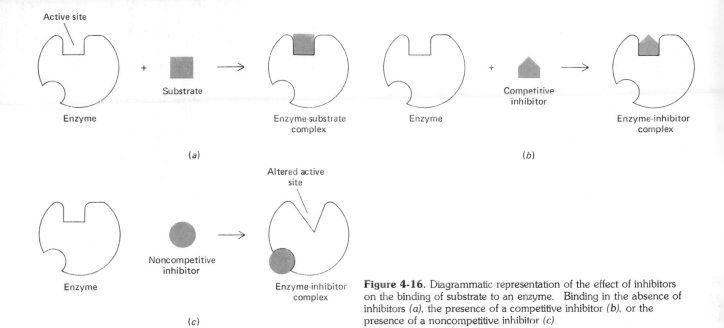

Figure 4-16. Diagrammatic representation of the effect of inhibitors on the binding of substrate to an enzyme. Binding in the absence of inhibitors *(a)*, the presence of a competitive inhibitor *(b)*, or the presence of a noncompetitive inhibitor *(c)*.

provides valuable information concerning the control of metabolism in the cell.

In other types of studies the use of artificial inhibitors of enzymes provides information about the nature of the active site and the manner in which the interaction between the natural substrate and its enzyme occurs. The basis for these studies is the concept that the active site has a complementary structure to the substrate with which it must fit. An analysis of the types of molecules that can compete with the substrate for a binding site on the enzyme, provides a better insight into the features of the substrate that are most important in its interaction with the enzyme. In this section we will discuss enzyme inhibition that can be studied with the classical Michaelis-Menten kinetics. In the next section we will consider a very important form of metabolic inhibition that does not lend itself to analysis by these more simple types of kinetics.

Inhibitors can be divided primarily into two types, competitive and noncompetitive (Fig. 4-16). In cases of *competitive inhibition,* the inhibitor competes with the substrate for access to the active site. The effectiveness of the competitive inhibitor depends upon its relative affinity for the enzyme. The inhibition can be overcome if the substrate/inhibitor ratio is great enough. In other words, if the number of collisions between the enzyme and inhibitor become insignificant relative to those between the enzyme and its substrate, then the effect of the inhibitor becomes minimal. Given a sufficient substrate concentration, it remains theoretically possible to achieve the enzyme's maximal velocity. Most intracellular competitive inhibitors are the products of a reaction since they are capable of binding to the enzyme's active site. For example, in the hydrolysis of sucrose by the enzyme invertase, its product, glucose, is a competitive inhibitor of the substrate.

In *noncompetitive inhibition* the substrate and inhibitor do not compete for an available binding site; generally the inhibitor acts at a site other than the enzyme's active site. The level of inhibition depends only upon the con-

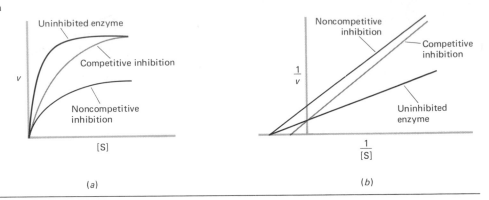

Figure 4-17. The effects of inhibitors on enzyme kinetics in a plot of velocity vs substrate concentration (a) or its reciprocal (b). The noncompetitive inhibitor reduces V_{max} without affecting the K_M, while the competitive inhibitor increases the K_M without affecting V_{max}.

centration of the inhibitor and increasing the concentration of the substrate cannot overcome it. Since, in the presence of a noncompetitive inhibitor, a certain fraction of the enzyme will necessarily be inactive at a given instant, the maximal velocity of the population of enzyme molecules cannot be reached. Examples of noncompetitive inhibitors include cyanide, which binds to metals in metalloproteins, various agents (such as PCMB), which attack the sulfhydryl group of exposed cysteine residues, and allosteric molecules (see below.) The effects upon the kinetics of enzymes in the presence of competitive and noncompetitive inhibitors is shown in Fig. 4-17. In one case the V_{max} is lowered and in the other the K_M is increased. In both types the slope (K_M/V_{max}) is increased.

Cooperative Kinetics

Though the kinetic analysis of many enzymes and enzyme inhibitors generates the types of hyperbolic curves described above, many have more com-

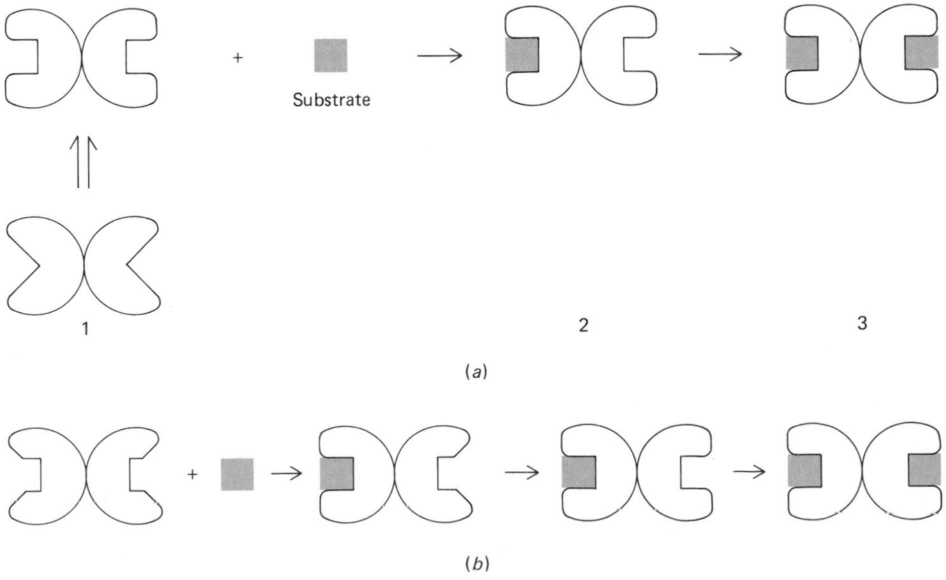

Figure 4-18. Two general models for the basis of cooperative kinetics. In (a), the enzyme (which in this case is a dimer) can exist in two different forms, only one of which has an affinity for substrate. In this model it is proposed that all subunits have the same structure (either active or inactive). As the concentration of substrate increases, the equilibrium between the two forms, is shifted toward the presence of active molecules. Since the two sites are always in the same state, the binding of substrate at one site facilitates the binding of additional substrate molecules. In (b), the binding of substrate at one site increases the tendency for the other site to change conformation and bind an additional substrate molecule. In this latter model, cooperative kinetics result from a sequential series of steps within the enzyme molecules.

plex velocity-substrate relationships. Many enzymes are oligomeric, i.e., contain several polypeptide chains, and have several catalytic sites. In many cases interactions occur between the polypeptides so that events occurring at one active site can have an effect on the events at another one. These are termed *cooperative effects* and can be positive or negative in nature. In the positive cases, when substrate is bound at one active site the affinity of the other active sites of the protein toward substrate is increased. Although the manner in which these cooperative interactions occur within a protein seem to vary among different enzymes, this type of molecular "communication" is mediated by conformational changes in protein structure (Fig. 4-18). The effect on the kinetics due to "communication" within the molecule is shown in Fig. 4-19. The curve obtained by plotting velocity vs. substrate for enzymes having cooperative kinetics is sigmoidal in shape (Fig. 4-19b) rather than hyperbolic as described above (Figs. 4-13 and 4-19a). At low substrate concentrations the likelihood is greatest that only one active site will be engaged with substrate and, therefore, the rate of reaction will be low. As the substrate concentration increases, the likelihood rises that more than one active site will have substrate bound at a given time. Since these additional catalytic sites become more active, relative to the first site, the rate of reaction rises sharply in proportion to the rise in substrate concentration. The consequence of positive cooperation is that in comparison to the hyperbolic situation (Fig. 4-19a), relatively smaller increases in substrate concentration result in disproportionate increases in enzyme activity.

The physiological significance of cooperative kinetics is best understood in the case of the binding and dissociation of oxygen from hemoglobin, which has four heme groups (Fig. 3-16) per molecule (one per polypeptide chain) and has a sigmoidal dissociation curve. The shape of the oxygen dissociation curve ensures that the optimal unloading of oxygen to needy tissues will occur since relatively small differences in oxygen pressure (analogous to substrate concentration for enzymes) cause major increases in oxygen dissociation (analogous to catalysis for enzymes). For example, in the case of hemoglobin, it takes only a threefold change in oxygen concentration to shift the binding from 10% of hemoglobin molecules complexed to oxygen all the way to 90% of the hemoglobins associated with oxygen. If the curve were hyperbolic rather than sigmoidal it would take an 81-fold increase to shift the binding from 10% to 90%. Figure 4-19c also shows the kinetics of an enzyme in which negative cooperative effects are found. In this case, as more catalytic sites are brought into action, their ability to catalyze the reaction diminishes.

Figure 4-19. Kinetics of enzymes that exhibit cooperative interactions among subunits. *(a)* The activity curve of most enzymes is hyperbolic (as in Fig. 4-13). In this case an 81-fold increase in substrate concentration is required to increase the velocity of reaction from 10% of the maximal rate to 90%. *(b)* Kinetics for an enzyme with positive cooperativity. Enzyme activity is disproportionately low at low substrate concentrations and disproportionately high at elevated substrate levels. In this case a ninefold change in substrate concentration is sufficient to increase the rate from 10% to 90%. Kinetics of negative cooperativity (a rarer situation). An increase in substrate concentration of over 6000-fold would be required for the equivalent rate increase described in *(a)* and *(b)*.

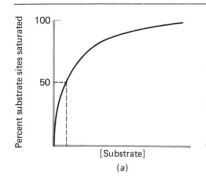

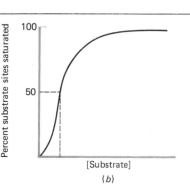

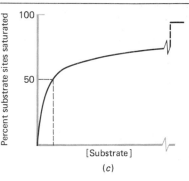

Allosteric Inhibition

Enzymes that are susceptible to *allosteric inhibition* possess, in addition to the active site, a second site for interaction with small molecular weight metabolites (as in Fig. 4-16c) which modulate (alter) the activity of the enzyme. This noncatalytic site is termed the *allosteric site* (regulatory site) and is generally as specific for the enzyme's modulators as is the active site for the enzyme's substrates. The interaction of a substance with the allosteric site causes a conformational change in the structure of the protein (Fig. 4-16c) which affects the structure of the active site and thus its catalytic properties. Allosteric inhibition is a type of noncompetitive inhibition since the inhibitor and substrate do not compete; they generally are very different in structure and are interacting with different parts of the enzyme. Allosteric inhibition once again illustrates the intimate relationship between molecular structure and function. Very small changes in the structure of the enzyme induced by the allosteric inhibitor cause marked changes in its functional activity. Allosteric inhibition provides a means for the cell to exert immediate, sensitive control over enzyme activity. It should be kept in mind that the cell also has control over enzyme production, a topic discussed at length in Chapters 11 and 13.

Another term for allosteric inhibition is *end-product inhibition*, which reflects its place in the overall metabolic scheme. Allosteric inhibitors are typically metabolites produced at the end of a sequence of enzymatic reactions, i.e., the end product of a metabolic pathway. The enzyme that is sensitive to allosteric inhibition is generally the enzyme that catalyzes the first step that is unique to that pathway (Fig. 4-20a). The best-studied

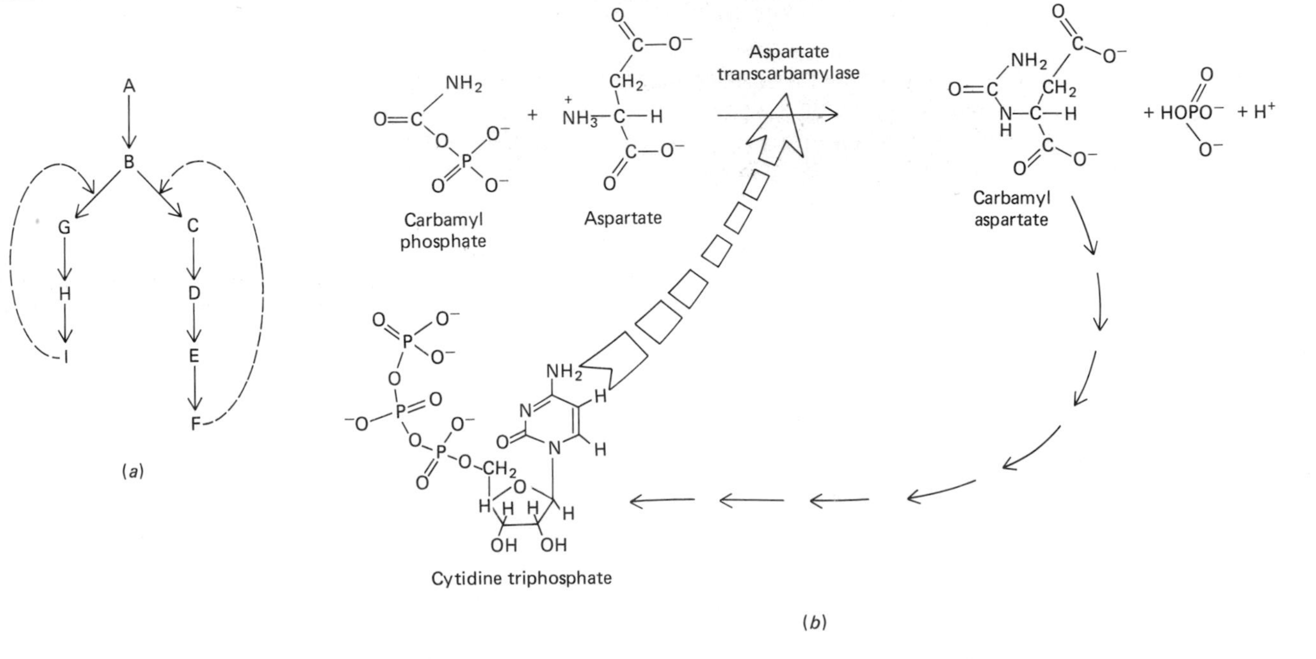

Figure 4-20. Metabolic aspects of allosteric (feedback) inhibition. *(a)* Hypothetical biosynthetic pathways leading from metabolite A to two different end products, F and I. Both F and I are capable of inhibiting the enzyme catalyzing the first reaction on the path leading only to that end product. Inhibitor loops within the cell can be much more complex. *(b)* Inhibition of aspartyl transcarbamylase by the end product CTP. [*(b) From J. C. Gerhart and A. B. Pardee, J. Biol. Chem., 237:892 (1962).*]

allosteric enzyme is aspartate transcarbamoylase which catalyzes the following reaction:

$$\text{Carbamoyl phosphate} + \text{L-aspartate} \rightleftharpoons N\text{-carbamoyl-L-aspartate} + P_i$$

This reaction is the first step of a sequence of reactions (Fig. 4-20*b*) that lead directly to the formation of cytidine triphosphate (CTP), i.e., there is no further branch point after this reaction leading to any other end product. If for some reason the concentration of CTP were to rise within the cell, there would be no reason for the cell to continue to produce unnecessary additional CTP. By having this built-in type of feedback mechanism, the cell avoids wasting valuable resources and possibly upsetting other reactions by having abnormally high CTP concentrations. Since the binding between CTP and the enzyme is a reversible one, when CTP levels in the cell drop, the dissociation of CTP from the enzyme will be favored and enzyme activity will return to the higher noninhibited level.

Allosteric inhibition is another example of homeostatic control whereby the range in concentration of a particular intermediate is regulated closely. Unlike most allosteric enzymes, the activity of aspartate transcarbamoylase is also modulated by an activating molecule, ATP. In the presence of ATP, which binds to the allosteric site in competition with CTP, the enzyme is more active. The kinetic effects of these allosteric modulators are shown in Fig. 4-21. It is the interaction between these types of regulatory molecules and enzymes (for example in Fig. 4-35) which brings order and direction to metabolism.

The presence of flexibility and potential for conformational change has been clearly shown for allosteric enzymes In this chapter the mention of conformational changes occurring within proteins has been alluded to at several points. Conformational changes occur during catalysis, during negative and positive cooperation of active sites among oligomeric proteins, and during allosteric inhibition or activation. As more becomes understood about macromolecular interactions, greater importance is placed on the role of conformational changes as the means whereby proteins accomplish many, or possibly all, of their tasks. In a few well-studied

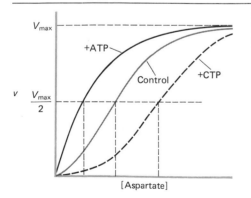

Figure 4-21. The effect on aspartyl transcarbamoylase activity of the addition of an allosteric inhibitor CTP or an allosteric activator, ATP. The apparent K_M is lowered in the presence of ATP and raised in the presence of CTP.

cases, the exact sequence of movements within the protein has been reported. Conformational changes are analogous to domino effects in which a shift in one part of the molecule will have consequences in the next part of the molecule and so on through its entire structure. In this way, the interaction of a modulator at the allosteric site in the polypeptide chain can alter the shape of the catalytic site located some distance away. In the case of cooperative effects, these conformational changes can be transmitted from one polypeptide chain to the next. The analysis of these events illustrates the manner in which proteins can effect changes in their environment by setting in motion physical changes in adjacent molecules. We will return to the consideration of conformational changes in proteins throughout this book for ultimately we will have to explain many different cellular activities with this principle.

Figure 4-22. The three stages of metabolism. The catabolic pathways (arrows downward) converge to form common metabolites and lead to ATP synthesis in stage III. The anabolic pathways (arrows upward) start from a few precursors in stage III and utilize ATP to synthesize a large variety of cellular materials. *(From A. L. Lehninger, "Biochemistry," 2d ed., Worth, 1975.)*

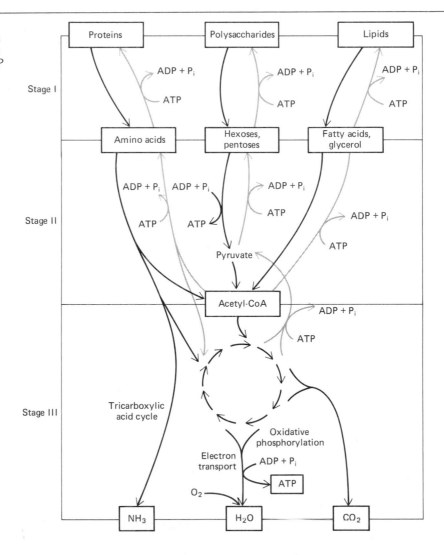

METABOLISM

It is the intent of this section to survey in a very broad manner the framework of metabolism, i.e., the sum total of the biochemical reactions that exist within a cell. We will concentrate upon those aspects of metabolism that lead to the production and utilization of chemical energy within the cell for this topic is one we will draw upon throughout the book. There are several classes of organic molecules produced within cells. There are macromolecules (DNA, RNA, protein, and polysaccharides) of very large molecular weight which are composed of covalently linked building blocks (precursors). For our purposes, we will consider lipids, which are composed of two or three long-chain fatty acids (see page 189), as macromolecules. The building blocks for the macromolecules form a second class of compounds: the nucleotides, amino acids, monosaccharides, and fatty acids. Each of these groups includes a number of specific compounds. Each amino acid, nucleotide, sugar, or fatty acid is produced by a sequence of chemical reactions, and each reaction is catalyzed by a specific enzyme. The sequence forms a metabolic pathway; the molecule at the end of the path, for example, a particular amino acid or nucleotide (as in Fig. 4-20), is termed the *end product*. The compounds produced by one reaction and consumed by the next reaction leading to the end product are *intermediates*, or *metabolites*, which as a whole represent a very large number of different organic molecules. In many cases, the metabolic pathways branch and converge so the product of one pathway may represent an intermediate on the way to the formation of a different molecule. Altogether the reactions of the cell constitute a tremendous diversity of molecular conversions represented by a significant number of different pathways, interconnected to one another at various points. All of the reactions are enzymatically catalyzed and the mechanisms whereby the cell controls all of these enzymatic activities remain the foremost challenge in the study of metabolism.

An Overview

There are generally two directions that metabolic pathways take: one direction leads to the breakdown of more complex molecules into simpler ones, while the other leads in the reverse direction. The former pathways are termed *catabolic* and are basically of a destructive nature. The breakdown of large molecular weight compounds serves two purposes; it makes available the raw materials from which other molecules can be synthesized, and it provides a large amount of chemical energy needed for the activities of the cell. In contrast, reactions that lead to the synthesis of more complex compounds compose *anabolic* pathways. Whereas catabolic pathways are exergonic and provide energy for use by the cell, anabolic pathways require the input of this chemical energy to drive otherwise unfavorable endergonic reactions.

Figure 4-22 shows a greatly simplified profile of the ways in which the major anabolic and catabolic pathways are interconnected. These metabolic pathways are present in all cellular organisms, from the simplest bacteria to the most complex eukaryotic cell. In the first chapter we made a basic distinction between the organization of the prokaryotic and eukaryotic cells, pointing out various of the morphological advances that occurred in the period between the evolution of these two types of cells. In contrast to the striking morphological differences, the basic metabolic pathways between the two groups are very similar. In fact, it is the prokaryotic cell that has remained in possession of a more sophisticated metabolism, while the eukaryotic cells have lost the ability to synthesize certain molecules as these abilities were no longer necessary. For example, many bacteria can live on very simple media as long as a single carbon and nitrogen source is provided along with a few essential minerals. In contrast, our bodies require a much more varied diet and have absolute requirements for a number of specific organic materials (vitamins).

The breakdown of materials occurs in steps. Macromolecules are first broken down into building blocks of which they are made, i.e., they are hydrolyzed (stage I, Fig. 4-22). In complex animals these hydrolytic reactions occur both extracellularly in the digestive tract and intracellularly in virtually every cell. In both cases, catabolism converts materials into usable precursors for synthesis of other compounds and the extraction of energy. In addition, intracellular digestion allows a cell to turnover, i.e., destroy and resynthesize, many of the macromolecules of which it is composed. The rate of turnover varies greatly with the particular molecule being considered and the particular cell being studied. For example, one type of RNA molecule may live only a matter of a few minutes, whereas another type in the same cell may live for weeks or longer. Similarly, a particular enzyme may be long-lived in the brain or a muscle cell and be turned over quite rapidly in a tissue such as the liver. The DNA of the cell is never turned over.

It is not intuitively obvious why a cell would go about making some molecule at the expense of considerable energy, then turn around and destroy and resynthesize it at a later time. In fact, molecular turnover is one of the less well understood phenomena, but certain explanations seem reasonable. One explanation for turnover is likely the same reason for the replacement of parts in an automobile or any other machine. As they stay in the cell they may sustain damage and become inactive or ineffective. Their destruction and resynthesis provides the cell with a continual supply of undamaged materials. More importantly, in many cells turnover provides flexibility. As long as a cell cannot rid itself of existing materials, and synthesize different ones to replace them, it cannot adapt to a new situation which may require the presence of a different set of components.

Once the macromolecules have been degraded into their components, i.e., the amino acids, nucleotides, sugars, and fatty acids, the cell can either reutilize them directly to form other macromolecules (stage I, Fig. 4-22), convert them into different compounds to make other products (stage II, Fig. 4-22), or degrade them further (stage III, Fig. 4-22) and extract a measure of their free-energy content. The course that is taken will depend upon the needs of the cell at that time. The exquisite regulatory mechanisms of cellular metabolism provide that the needs of the cell are fulfilled. Toward this end, the cell can regulate its metabolic reactions in various ways. For example, it can modify the synthesis or destruction of particular enzymes (an example of the value of turnover) or regulate the activities of the enzymes that remain in the cell (such as by allosteric inhibition or some other means).

The pathway for the degradation of the building blocks of the macromolecules varies according to the particular compound which is being catabolized. Ultimately, however, all of these molecules are converted into a rather small variety of substances (Fig. 4-22) which can then be handled in a similar manner, regardless of the nature of the macromolecule from which those compounds had orginally been derived. In other words, catabolic pathways tend to be convergent. The substances begin as macromolecules having a very different structure, but after the activity of a sequence of enzymes, all of them have been reduced to essentially the same low molecular weight metabolites.

The two most important macromolecules (Figs. 4-23 and 5-2a) serving as a supply of energy to the cell are polysaccharides (glycogen in animals and starch in plants) and fats (neutral lipids). Though both of these materials are present to some degree in nearly all cells, certain tissues of

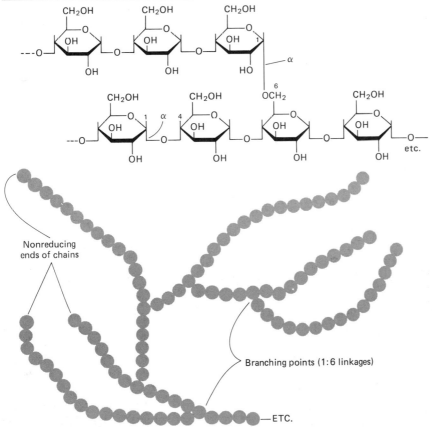

Figure 4-23. The structure of one form of starch (termed amylopectin). The top drawing illustrates the nature of the linkages between sugar units and the lower drawing shows the branched nature of a small fragment of the polymer. Each circle represents a glucose unit. *(From N. A. Edwards and K. A. Hassal, "Cellular Biochemistry and Physiology," McGraw-Hill, 1971.)*

Nonreducing ends of chains

Branching points (1:6 linkages)

—ETC.

the body are specialized for their storage, namely the liver for glycogen and adipose tissue for fat. These macromolecules are broken down into glucose and fatty acids, which can then be passed through the body in the general circulation for use by all of the cells. Fatty acids are transported in combination with certain blood proteins. These two types of molecules serve somewhat different purposes in cellular energetics. On a per weight basis, the fat contains over twice the energy content, and is used primarily as a more long-term store of energy reserves. Fat deposits are broken down only rather slowly to provide usable energy to cells, while polysaccharides are broken down and reformed at much more rapid rates to ensure that the levels of cell and blood glucose remain at the appropriate levels to keep the cellular machinery running at an optimal rate. When the rate of metabolism is high, as during an extended period of exercise, glycogen stores in the liver can be markedly depleted so that blood sugar levels can be maintained as an energy supply. When the energy demands are over, these glycogen stores can be built back up again in a relatively short period of time.

One of the important measures of the free-energy level of an organic molecule is its state of reduction (page 180). The compounds that we use as energy sources to run our machines are highly reduced organic molecules in the form of compounds such as natural gas and petroleum derivatives. In order to utilize the energy stored within these molecules, we combust them in the presence of oxygen, thereby converting the carbons to much more oxidized states as in carbon dioxide and carbon monoxide gases. Just as the degree of reduction is a measure of the energy content of a fuel, it is also a measure of its ability to perform chemical work within the cell. The more hydrogen atoms that can be stripped from a "fuel" molecule, the more ATP that can be formed. Fats contain a greater store of energy per unit weight in comparison to polysaccharides because they are considerably more reduced molecules containing long strings of $-CH_2-$ groups making up their fatty acid chains.

As will be described in Chapter 9, photosynthesis accomplishes the capture of light energy by an overall process of reduction of carbon dioxide using the electrons in water as the reducing agents. In this way, carbon dioxide, a highly oxidized form of carbon, is converted to the more complex (less entropic), more reduced (more energy-containing) organic states present in carbohydrates, fats, and other biological molecules. The extraction of this energy by catabolism is accomplished by a process which is essentially the reverse of photosynthesis, the oxidation of the substrate back to carbon dioxide and water. The large energy content of reduced biological molecules can be readily demonstrated by the combustion of a mole of glucose in oxygen to carbon dioxide and water. When this is done in a calorimeter, approximately 673,000 cal are released in the form of heat. In other words, the heat of reaction (ΔH) of glucose combustion is 673,000 cal/mol. When one large molecule such as glucose is broken into a number of smaller molecules whose possible arrangements (i.e., freedom of movement) are greater, the reaction will occur with an increase in entropy. If entropy increases, ΔS is positive and $-T\,\Delta S$ will be negative. In the combustion of glucose, the reaction is exothermic (ΔH is negative), entropy increases ($-T \Delta S$ is negative), therefore ΔG will be negative. The ΔG for the combustion of glucose is approximately 686,000 cal/mol.

For the most part, the conversion of chemical energy into heat in a cell results in the loss of that energy since, unlike a variety of machines, the cell cannot use this form of energy to perform work. In order for thermal energy to be converted to some other form, such as mechanical energy, there must be a temperature differential in the system so that heat can flow from a warmer to a cooler part of the system. No such temperature differential occurs within a cell. Since the cell cannot convert the thermal energy back into chemical energy, it will be lost for purposes other than raising the body temperature of the organism (which is useful in the case of most organisms and mandatory in the case of the warm-blooded birds and mammals). It is the responsibility of the enzymes of catabolism to prevent the conversion of the energy of the substrate into heat by conserving it in the form of an energy-rich molecule, ATP. The free energy released by glucose oxidation is very large (686,000 cal/mol *relative* to CO_2 and H_2O) while that for ATP hydrolysis is relatively modest (7300 cal/mol *relative* to ADP). In order for the energy of the carbohydrate to be stored in ATP, the breakdown of the carbohydrate occurs in many steps (Fig. 4-32). Those steps, in which the free energy difference between the reactants and products is large enough, can be used to form ATP as described below. In other words, the energy released when a molecule of glucose is oxidized in the cell to CO_2 and H_2O is conserved in "packets" in the form of ATP. Before describing some of the reactions involved we need to further briefly explore this concept of the conservation of biochemical energy.

The key word in a discussion of energy metabolism is the word *coupling* (page 133). The first law of thermodynamics states that energy can neither be created nor destroyed; but energy can be converted from one form to another. This concept is clearly depicted in the cartoons of Rube Goldberg (Fig. 1-4). Generally one refers to the conversion of energy among the different forms (light, heat, mechanical, etc.), but it also applies to the conversion of chemical energy from one molecule to another. The shift of free energy between molecules is accomplished by some type of coupling process whereby the loss of free energy that accompanies spontaneous downhill chemical reactions is used to drive otherwise unfavorable endergonic reactions as described for the formation of glutamine in an earlier section of this chapter. In other words, the loss of free energy from compounds in one reaction is used to increase the free energy of other molecules formed in the coupled reaction. Energy-yielding reactions are coupled by enzymes to energy-requiring reactions (Fig. 4-24). The intermediate between the two types of pathways is usually ATP. The catabolism of glucose provides a number of energy-yielding reactions which are coupled to one energy-requiring reaction: the formation of ATP from ADP and inorganic phosphate. As discussed earlier in this chapter, the formation of ATP is "energy-requiring" because the concentration of ATP

Figure 4-24. A mechanical analogy to the coupling of chemical reactions of large negative ΔG with ones having positive ΔG, i.e. ones that could not proceed in the absence of the coupling event. *(From L. Peusner, "Concepts in Bioenergetics," © 1974. Reprinted by permission of Prentice-Hall, Inc., Englewood Cliffs, New Jersey.)*

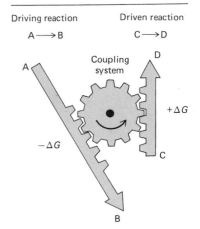

Driving reaction   Driven reaction

$A \longrightarrow B$ $C \longrightarrow D$

in the cell is maintained at relatively high levels; i.e., the (ATP)/(ADP) ratio in the cell is much greater than that found at equilibrium. As long as these nonequilibrium levels of ATP are maintained, free energy can be said to be "stored" or "conserved" in this population of ATP molecules. The free energy conserved in ATP can then be utilized by coupling its hydrolysis to some otherwise endergonic reaction, such as the formation of glutamine from glutamic acid and ammonia (page 133). In the next few pages we will consider the catabolism of glycogen, using this process to illustrate the utilization of chemical energy within the cell.

There are basically two stages in the catabolism of glucose, the building block of glycogen. The first stage occurs in the cytoplasm and is termed *glycolysis* (Fig. 4-25). There are several possible end products of glycolysis depending upon the type of organism and the conditions under which it is living; primarily whether oxygen is present or not. The second stage is the tricarboxylic acid cycle or TCA cycle (also termed Krebs cycle or citric acid cycle) in which the final oxidation of the carbon atoms

Figure 4-25. Glycolysis.

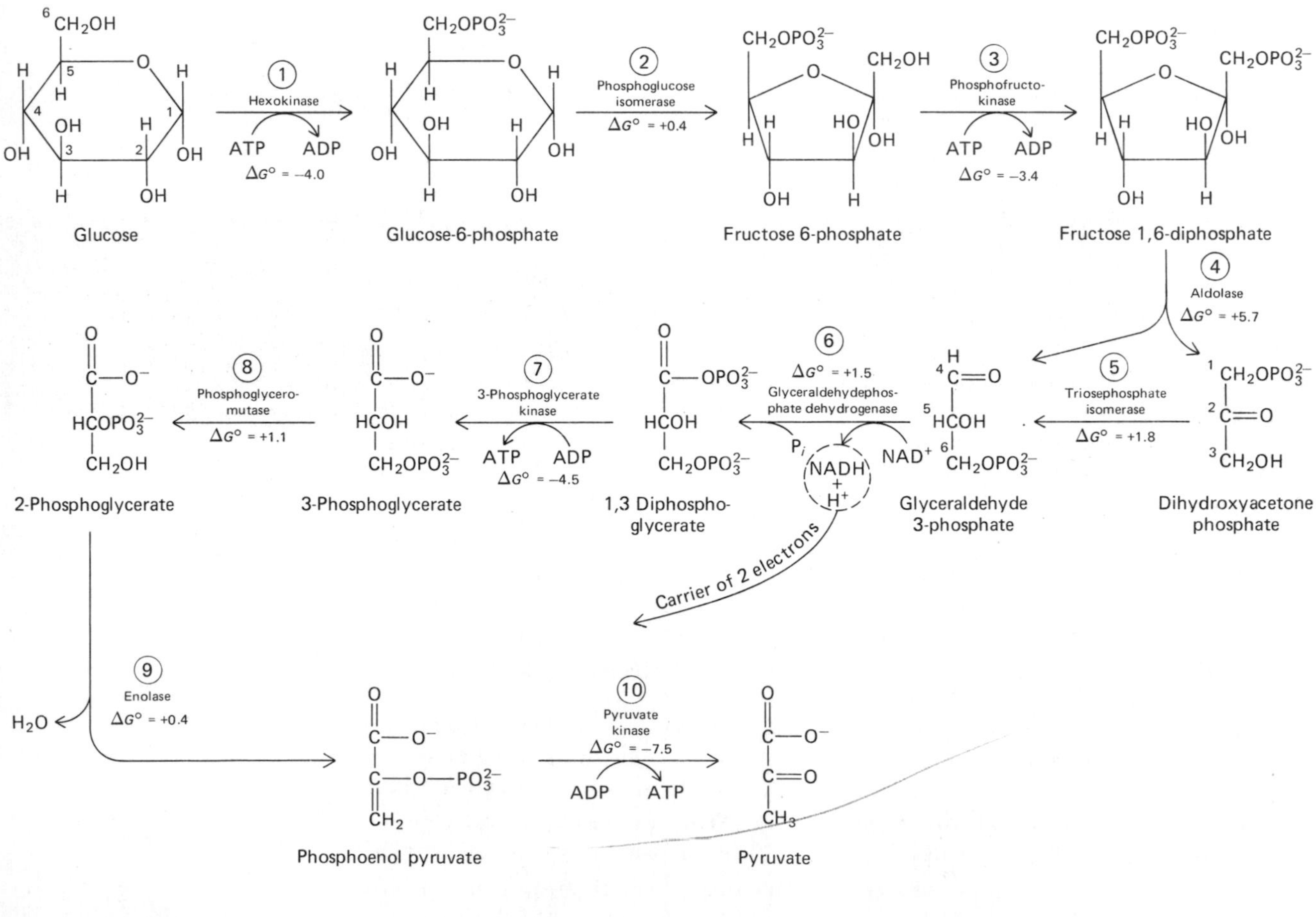

to carbon dioxide occurs. The elucidation of the reactions of these two pathways and the understanding of their importance has been another of the basic steps in the development of the science of biochemistry and cell biology.

Glycolysis and ATP Formation

The reactions of glycolysis and the enzymes that catalyze them are shown in Fig. 4-25. Before discussing the specific reactions, an important point concerning the thermodynamics of metabolism can be made. In an earlier section, the importance of the difference between ΔG and $\Delta G°$ was stressed; it is the ΔG for a particular reaction which determines its direction in the cell. Actual measurements of the concentrations of metabolites in the cell can reveal the value for ΔG of a reaction at any given time. Figure 4-26 shows the typical ΔG values for the reactions of glycolysis. In contrast to the $\Delta G°$ values of Fig. 4-25, all but three reactions have ΔG values of nearly zero, i.e., they are near equilibrium. The three reactions which are far from equilibrium, i.e., are essentially irreversible in the cell, provide the "force" which moves the metabolites through the glycolytic pathway in a directed manner.

In Fig. 4-25 glycolysis is shown as beginning with glucose, as is often the case. Alternatively, glycolysis can begin with the breakdown of glycogen by hydrolysis. Rather than simply converting glycogen to glucose, the sugar of which it is composed, the enzyme phosphorylase catalyzes the breakdown of glycogen accompanied by the phosphorylation of the hexose, forming glucose 1-phosphate. The phosphorylation of the sugar "activates" it to allow it to undergo the next several reactions, the first of which is the formation of glucose 6-phosphate. Glucose 6-phosphate is then converted to fructose 6-phosphate and then to fructose-*di*phosphate at the expense of an ATP molecule. In the next reaction, the six-carbon diphosphate is split into two, three-carbon monophosphates which sets the stage for the first exergonic reactions to which the formation of ATP can be coupled.

ATP is formed in the breakdown of glycogen in two basically different ways, both of which can be illustrated by one chemical reaction of glycolysis; the conversion of glyceraldehyde 3-phosphate to 3-phosphoglycerate. The overall reaction is an oxidation of an aldehyde to a carboxylic acid. An oxidative process is one that is accompanied by the loss of electrons (or hydrogen atoms) while a reductive process results in their gain (page 180). The oxidation of one compound is always accompanied by the simultaneous reduction of the agent responsible for its oxidation. As stated above, the oxidation of organic molecules is accompanied by the loss of free energy and the reductive process by a gain; this is precisely the means by which energy-releasing and energy-requiring processes are coupled. There are enzymes, termed *dehydrogenases*, which oxidize the substrate by removing hydrogen atoms and pass the electrons on to another molecule which is thereby reduced. The most important molecule to become reduced in dehydrogenation reactions is the pyridine nucleotide

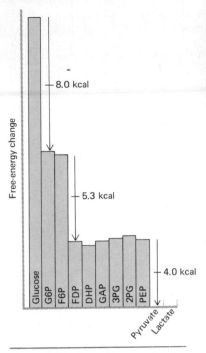

Figure 4-26. Free-energy profile of glycolysis in the human erythrocyte. All reactions are at or near equilibrium except those catalyzed by hexokinase, phosphofructokinase, and pyruvate kinase, which are characterized by large differences in free energy. In the cell all of the reactions must proceed with a decline in free energy; the slight increases in free energy depicted here for several steps must be regarded as deriving from errors in experimental measurements of metabolite concentrations. *(From A. L. Lehninger, "Biochemistry," 2d ed., Worth 1975.)*

NAD^+ whose structure is shown in Fig. 4-27. The NAD^+ acts as a loosely associated coenzyme to the dehydrogenase in position to accept the electrons (along with one proton). The reaction catalyzed by dehydrogenases is of the following type:

$$\text{Reduced substrate} + NAD^+ \rightarrow \text{oxidized substrate} + NADH + H^+$$

In order to better understand the means by which the loss of free energy of the oxidative reaction is used to drive the uphill formation of ATP, we will consider the reaction as it occurs in the cell, in steps. The first step, catalyzed by the enzyme *glyceraldehyde phosphate dehydrogenase,* involves the removal of a pair of hydrogen atoms, transferring the two electrons together with one of the protons to the acceptor NAD^+ converting it to NADH. The NADH is released from the enzyme by exchange with NAD^+. The reaction is shown by stepwise form in Fig. 4-28*a-c.* An important feature of this first step in the reaction (Fig. 4-28*a*) is that the oxidized substrate is not allowed to drop in free-energy content to the much lower level of the free acid, but is held by covalent bonds to the enzyme.

We will return to this reaction in a moment, but first we will continue with the consequences of the formation of NADH. This reduced pyridine nucleotide is a "high energy"–containing species because of the high

Figure 4-27. The structure of NAD^+ and its reduction to NADH.

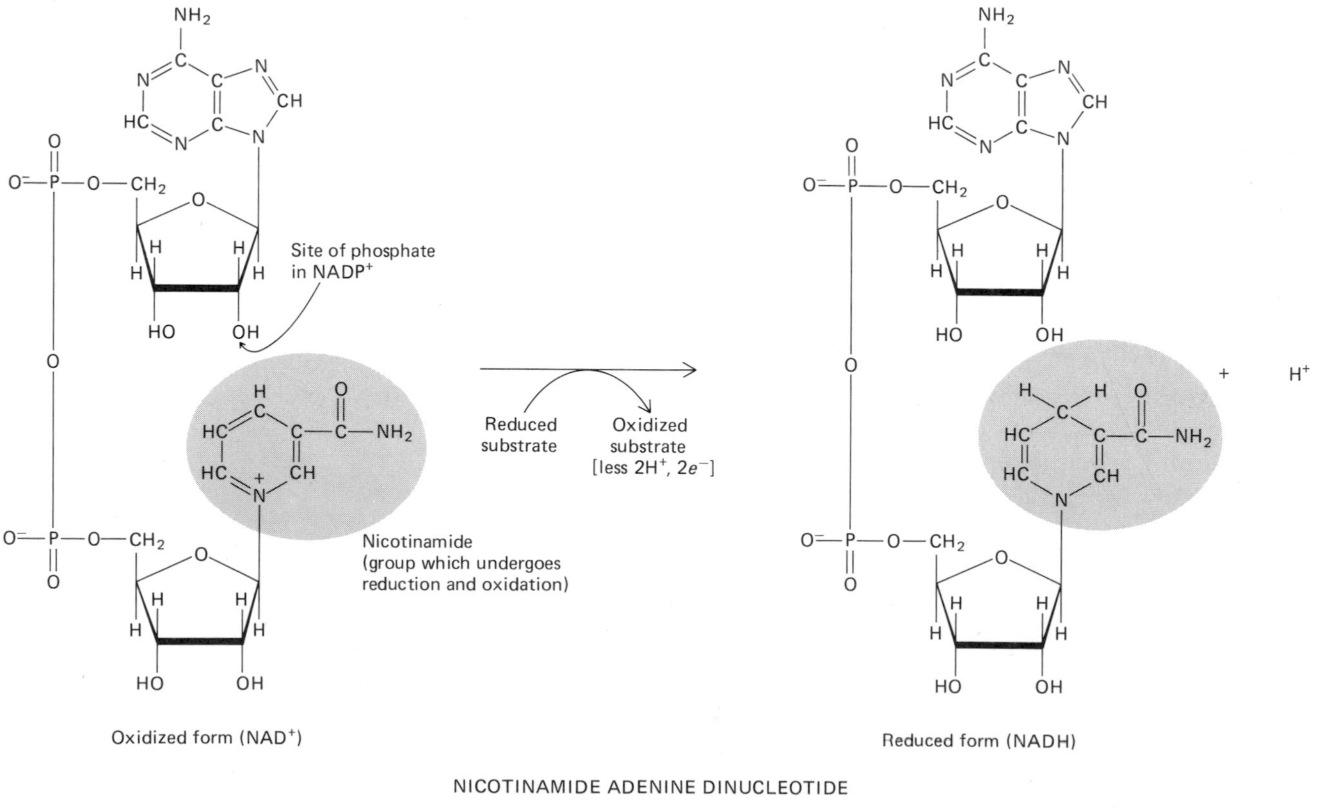

Oxidized form (NAD^+)

Reduced form (NADH)

NICOTINAMIDE ADENINE DINUCLEOTIDE

electron transfer potential that it possesses relative to the other electron acceptors in the cell (see Fig. 8-11). The most important question concerns the relationship between the formation of NADH and the production of ATP, which is one of the ultimate motives of catabolism. The process by which the energy-yielding loss of electrons from NADH is coupled to the energy-requiring phosphorylation of ADP will constitute the bulk of the discussion of Chapter 8; it is a complex and incompletely understood process. In essence, what happens is the electrons of NADH are passed to another electron acceptor which in turn passes them to another one and so on down what is called the *electron transport chain*. As the electrons move down this chain they move to a lower and lower free-energy state until eventually they are used to reduce molecular oxygen to form water (Fig. 8-11). This is the only major need for oxygen in the body, as the

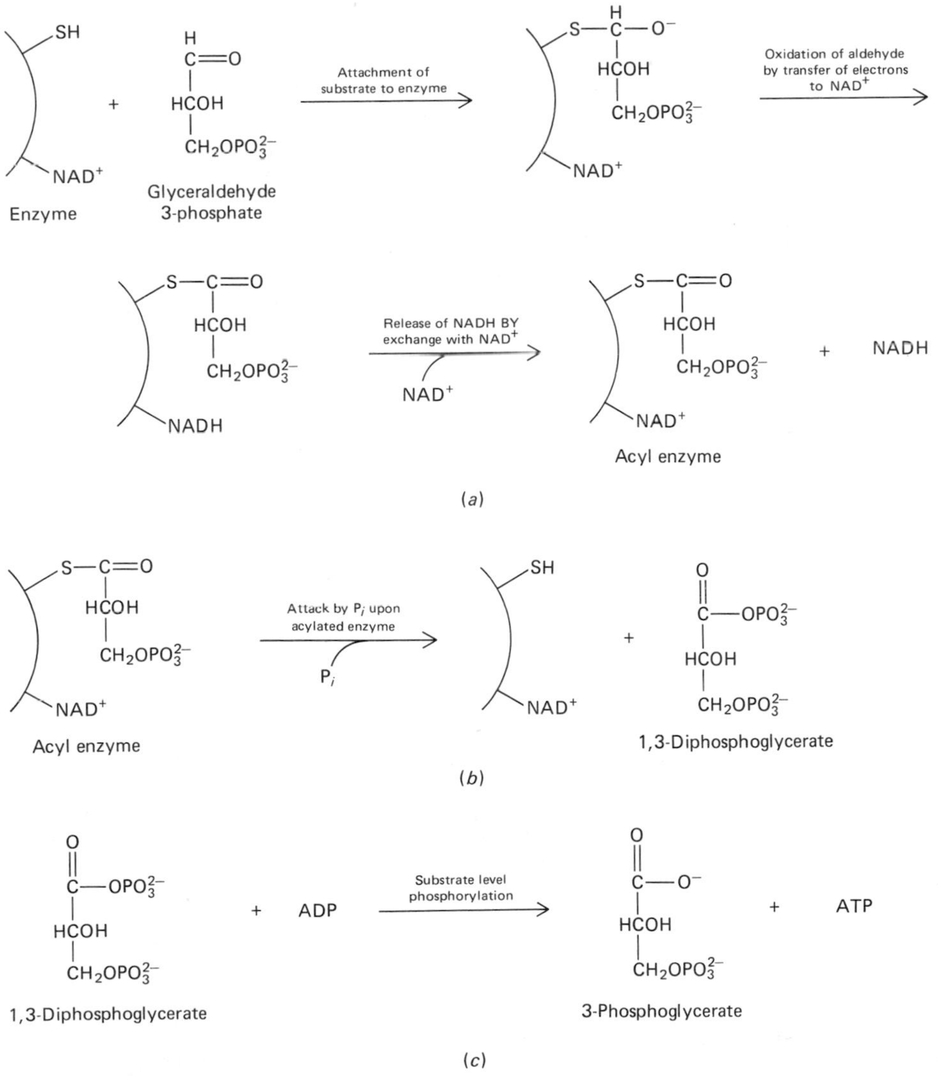

Figure 4-28. An example of substrate level phosphorylation of ADP to form ATP. Steps (a) and (b) involve the enzyme glyceraldehyde phosphate dehydrogenase, and step (c) the enzyme 3-phosphoglycerate kinase. The former enzyme has four subunits, each binding one molecule of NAD^+. Once these coenzymes are reduced to NADH, they are displaced by NAD^+ molecules from the medium.

final acceptor in the electron transport chain. The free energy lost in this downhill movement of electrons is coupled to an endergonic reaction, the formation of ATP from ADP. The coupling which occurs in the mitochondria is one between a process of oxidation and phosphorylation (termed *oxidative phosphorylation*) and forms the core of the energy-conserving mechanism in all eukaryotic cells. The reason that the details of oxidative phosphorylation are being delayed until Chapter 8 is so the topic of membranes can be explored and the information presented can be applied to this event occurring within the membranes of the mitochondria.

We can now return to the oxidized substrate still dangling by its attachment to the enzyme glyceraldehyde phosphate dehydrogenase. The formation of this acylated enzyme has stopped the process of free-energy loss midway between the free aldehyde and the free acid. The thioester linkage to the enzyme retains sufficient energy relative to the eventual product (3-phosphoglycerate) that it can be used to form an ATP by a direct *substrate-level phosphorylation*. The way this happens is as follows. The enzyme transfers its attached acyl group to an inorganic phosphate acceptor to form 1,3-diphosphoglycerate, a molecule bearing two phosphates (Fig. 4-28*b*). In the next step another enzyme (phosphoglycerate kinase) catalyzes the last step in this exergonic oxidation by transferring a phosphoryl group directly from 1,3-diphosphoglycerate to ADP to form ATP. In other words, the loss of the phosphoryl group from 1,3-diphosphoglycerate is sufficiently exergonic to drive the endergonic formation of ATP. Just as NADH has a high transfer potential for electrons, 1,3-diphosphoglycerate has a high transfer potential for phosphate, a higher one than ATP.

TABLE 4-7
Standard Free Energy of Hydrolysis of Phosphate Compounds

Compound	$\Delta G°$ (kcal/mol)	Direction of phosphate group transfer
Phosphoenolpyruvate	−14.8	
1,3-Diphosphoglycerate	−11.8	
Phosphocreatine	−10.3	
Acetyl phosphate	−10.1	
ATP	−7.3	
Glucose 1-phosphate	−5.0	
Fructose 6-phosphate	−3.8	
Glucose 6-phosphate	−3.3	
3-Phosphoglycerate	−2.4	
Glycerol-3-phosphate	−2.2	

SOURCE: A. L. Lehninger, "Bioenergetics: The Molecular Basis of Biological Energy Transformation," 2d ed. Copyright 1965, 1971 by the Benjamin/Cummings Publishing Company, Inc., Menlo Park, California.

The term *transfer potential* is a particularly important one in metabolism because it allows one to compare various reactions relative to some common standard. For example, the importance of phosphorylation, i.e., the transfer of a phosphoryl group has been stressed in the discussion. There are quite a variety of different molecules that can transfer this group and an equally large number that can accept it. It has already been pointed out that the absolute free-energy content of a compound cannot be measured and is not important. What is important is the relative free-energy differences between potential donor molecules and potential acceptor molecules for this information will determine the direction in which reactions between particular combinations of donors and acceptors will proceed. It has been stated that the $\Delta G°$ for the hydrolysis of ATP is -7300 cal/mol. By itself this information has little value for it cannot tell us which reactions can be used to reverse the hydrolysis reaction by phosphorylation of ADP, nor does it tell us about phosphorylation reactions which can be driven by ATP hydrolysis.

The information in Table 4-7 and Fig. 4-29 provides the $\Delta G°$ for the hydrolysis of several phosphorylated compounds. All of these reactions have a common acceptor, namely water, and therefore they provide a measure of the phosphoryl transfer potential for each donor. Any donor higher on the list can be used to form any molecule lower on the list and the $\Delta G°$ of the reaction will be equal to the difference between the two values given in the table. For example, the $\Delta G°$ of the transfer of a phosphate group from 1,3-diphosphoglycerate to ADP to form ATP is equal to -4.5 kcal/mol (-11.8 kcal/mol $+ 7.3$ kcal/mol). This concept of transfer potential is useful for comparing any series of donors and acceptors regardless of the group being transferred, i.e., protons, electrons, acetyls, methyls, and

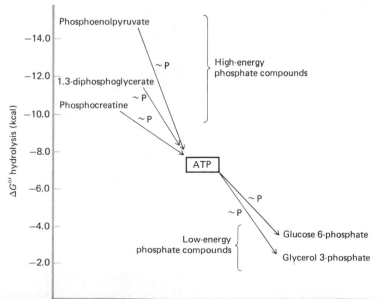

Figure 4-29. The transfer of phosphate groups from high-energy phosphate donors to low-energy phosphate acceptors. *(From A. L. Lehninger, "Bioenergetics: The Molecular Basis of Biological Energy Transformation," 2d ed., copyright © 1965, 1971 by the Benjamin/ Cummings Publishing Company, Inc., Menlo Park, California.)*

so on. Those molecules higher on the list, i.e., ones with greater free energy (higher $-\Delta G°$'s), are molecules with *less* affinity for the group to be transferred than are ones lower on the list. The less the affinity the better the donor, the greater the affinity the better the acceptor.

The oxidation of glyceraldehyde 3-phosphate to 3-phosphoglycerate illustrates the two basically different means by which ATP formation can occur in the cell: one via the formation of a reduced electron acceptor (usually NAD^+) which then passes electrons on to oxygen, and the other by direct phosphorylation of ADP using the substrate as a "high energy" phosphate donor. The former mechanism of ATP formation is strictly aerobic, i.e., occurs only in the presence of oxygen, while the latter mechanism can occur under *anaerobic* conditions. The substrate-level phosphorylation illustrates an important point about ATP: its formation is not *that* endergonic (Table 4-7). In other words, ATP is not so energetic a molecule that it cannot readily be formed by metabolic reactions, whether they be substrate-level phosphorylation or oxidative phosphorylation via electron transport. On the other hand, the reactions that use ATP to drive them (such as the formation of glutamine) are not so endergonic that the phosphate transfer potential of ATP cannot be used to accomplish the coupled reaction. Just as the overall catabolic reactions are broken up into successive smaller steps so that the energy can be conserved in "packages" of appropriate energy content, so too are the anabolic events broken up into a series of less endergonic reactions which can be driven by the hydrolysis of ATP.

An important feature of glycolysis is that it can generate a limited number of ATP molecules in the absence of oxygen. Neither the substrate-level phosphorylation of ADP by 1,3-diphosphoglycerate described above, nor a later one by phosphoenolpyruvate (Fig. 4-25), require the electron transport chain and oxygen. It is for this reason that glycolysis is considered anaerobic, indicating that it can proceed in the absence of oxygen to continue to provide ATP. Some cells, such as tumor cells or muscle cells, may rely heavily upon the anaerobic formation of ATP. If this is the case, why is it that we cannot survive for more than a few minutes in the absence of oxygen? Consider what happens to oxidative metabolism of glycogen in the absence of oxygen. In the oxidation of glyceraldehyde 3-phosphate, one of the products of the reaction is NADH. The formation of NADH occurs at the expense of one of the reactants, NAD^+. Since NAD^+ is a necessary reactant in this important step of glycolysis, it must be rejuvenated from NADH. If it is not reformed, this reaction can no longer take place, nor can any of the succeeding reactions that depend on the product. However, as long as oxygen is absent, NADH cannot be oxidized back to NAD^+ in the electron-transport chain; oxygen, the final acceptor of the electrons, must be present.

There is however a means whereby certain cells, particularly muscle cells, can oxidize NADH without oxygen. It can be used as an electron donor in the reduction of pyruvate to lactate (Fig. 4-30). This reaction can be considered a side reaction off the main path from glycolysis to the tricarboxylic acid cycle. It is a stop-gap measure to rejuvenate NAD^+ so that glycolysis can continue and ATP production can be maintained. The conversion of pyruvate to lactate is particularly important in muscle cells when they are undergoing strenuous activity. Under these conditions, the oxygen supply cannot keep pace with oxidative metabolism and the cells continue their metabolism by producing lactate which diffuses out into the bloodstream. When oxygen once again becomes available in sufficient amounts, the lactate can be converted back to pyruvate for continued oxidation. The need to convert pyruvate to lactate when insufficient NAD^+ is present illustrates how sensitive metabolism is to the concentration levels of various metabolites. These concentrations must be very carefully controlled by a complex variety of regulatory mechanisms.

If one considers that the original environment in which cellular metabolism evolved was devoid of oxygen, the analysis of pathways such as glycolysis provides a glimpse into a more primitive means for the conservation of chemical energy. There are numerous microorganisms, for example, yeast, that can live perfectly well without oxidative phosphorylation. When yeast is maintained under anaerobic conditions it uses a different reaction to rejuvenate NAD^+ from NADH from that used in the formation of lactate; it reduces pyruvate in a two-step reaction (Fig. 4-30) to ethyl alcohol. Though the evolution of this mechanism has been fortunate for ethanol consumers, the energy gained by anaerobic *fermentation* is meager when compared to the complete oxidation of the same

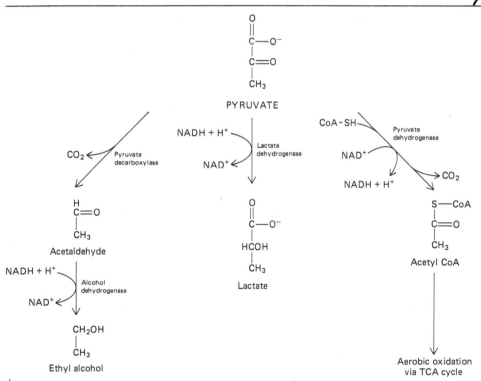

Figure 4-30. The fate of pyruvate is variable depending upon the type of organism and the conditions of growth. Under anaerobic conditions, pyruvate is reduced by NADH to rejuvenate NAD^+. In yeast, the product of pyruvate reduction is ethyl alcohol, while in animal cells it is lactate. Under aerobic conditions, pyruvate is oxidized (and decarboxylated) to form acetyl CoA which can enter the TCA cycle. This latter reaction is catalyzed by the multienzyme complex shown in Fig. 3-17.

sugar to carbon dioxide and water. Of the 686,000 cal/mol of glucose that is released when it is completely oxidized, only 57,000 cal is released when it is converted to ethanol under standard conditions, and only 47,000 cal when it is converted to lactate. In either case, only two molecules of ATP are formed per glucose oxidized (Table 4-8). It appears that the bulk of the energy obtained from glycogen is released and conserved as ATP when pyruvate is oxidized further in the tricarboxylic acid (TCA) cycle. If an organism is existing under anaerobic conditions and relying upon glycolysis for its ATP production, it will have to consume much larger amounts of food materials to extract the same amount of chemical energy in comparison to the same organism living under oxygen-rich conditions.

The TCA Cycle and ATP Formation

Pyruvate, which is produced by glycolysis in the cytoplasm, moves into the mitochondria where it is decarboxylated to form a two-carbon compound in a complex reaction carried out by a huge multienzyme complex. The two-carbon molecule becomes complexed with a large molecular weight substance, Coenzyme A to form acetyl CoA (Figs. 4-30 and 4-31), which is then fed into the TCA cycle. The first step in the cycle is the condensation of the two-carbon acetyl group with a four-carbon oxaloacetate to form the six-carbon citrate molecule. As a result of a series of reactions, this six-carbon citrate is decreased in chain length, one carbon at a time, until it is back to the four-carbon oxaloacetate molecule and ready to condense with another acetyl CoA, hence the term *cycle*. It is the two

TABLE 4-8
ATP Formation during Oxidative Metabolism

Reaction no.°	Reaction§	Number of ATPs hydrolyzed (−) or produced (+)
1	Glucose → glucose 6-phosphate	−1
3	Fructose 6-phosphate → fructose 1,6-diphosphate	−1
6	Glyceralde 3-phosphate → 1,3-diphosphoglycerate; NADH formed (Reaction forms NADH, which subsequently passes into the mitochondria. Each NADH formed in this reaction yields 2 ATPs.)	2† × 2 = +4
7	1,3-diphosphoglycerate → 3-phosphoglycerate	2 × 1 = +2
10	Phosphoenolpyruvate → pyruvate	2 × 1 = +2
11	Pyruvate → AcCoA; NADH formed	2 × 3 = +6
14	Isocitrate → α-ketoglutarate; NADH formed	2 × 3 = +6
15	α-Ketoglutarate → succinyl CoA; NADH formed	2 × 3 = +6
16	Succinyl CoA → succinate	2 × 1 = +2‡
17	Succinate → fumarate; FADH₂ formed	2 × 2 = +4
19	Malate → oxaloacetate; NADH formed	2 × 3 = +6
		36 ATPs

° Numbers in this column refer to reactions in Figs. 4-25 and 4-31.

† In this and the following reactions, the number of ATPs formed is multipled by two since each molecule of glucose is split into two, three-carbon molecules for further oxidation.

‡ In this reaction GTP is produced rather than ATP.

§ Reaction numbers 7, 10, and 16 are substrate-level phosphorylations. ATPs formed in reaction numbers 6, 11, 14, 15, 17, and 19 are from electron transport.

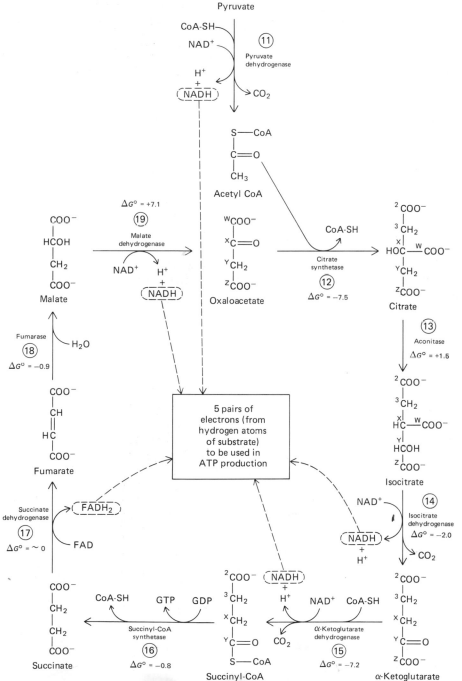

Figure 4-31. The tricarboxylic acid cycle. The cycle begins with the condensation of acetyl CoA with oxaloacetate. The carbons of these two compounds are marked with numbers or letters. The two carbons lost during passage through the cycle are derived from oxaloacetate.

carbons that are removed during the TCA cycle (which are not the same ones that were brought in with the acetyl group) that are completely oxidized to carbon dioxide. During the cycle there are four reactions whereby electrons are transferred from substrate to an electron acceptor. In three of these oxidation-reduction reactions the acceptor is NAD^+, while in a fourth it is FAD. For each pair of electrons transferred from NADH down the electron transport chain to oxygen, three ATP molecules are formed. For the pair of electrons transferred from $FADH_2$ (which has a lesser electron transfer potential than NADH), two ATP molecules are formed. If one adds up all of the ATPs formed from one molecule of glucose completely catabolized, the net gain is 36 ATPs (Table 4-8). The ATPs are formed in steps as the free energy of the intermediates is gradually lowered (Fig. 4-32).

If one considers the position of the TCA cycle in the overall metabolic scheme (Figs. 4-22 and 4-33), it can be seen that the metabolites of this cycle represent the compounds upon which most of the catabolic pathways of the cell converge. For example, acetyl CoA is a particularly important end product of a number of catabolic paths in addition to carbohydrates, such as that of the fatty acids, which are degraded by two carbon units at a time. These two carbon compounds enter the TCA cycle as acetyl CoA.

The TCA cycle is the central pathway of the cell, for not only do the catabolic paths converge upon it, but the anabolic paths diverge from it (Figs. 4-22 and 4-33). In other words, the intermediates produced in the cycle are used as the starting points from which many of the precursors of the cell's macromolecules are made. Anabolic pathways lead to the formation of more complex, reduced organic molecules and are basically energy-requiring pathways. Ultimately all of the energy expended to drive anabolic reactions is originally derived from catabolic reactions of the type described.

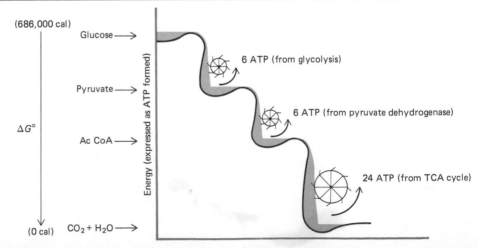

Figure 4-32. The formation of ATP during the oxidative breakdown of glucose to CO_2 and H_2O. This figure is meant to illustrate the stepwise nature of free-energy release as the intermediates fall to lower and lower free-energy states. A portion of this free energy released is conserved in ATP. Note that four of the six ATPs assigned to glycolysis are actually derived by electron transport using the electrons from NADH formed during glycolysis. *(Waterwheel concept of energy release and capture from N. A. Edwards and K. A. Hassal, "Cellular Biochemistry and Physiology," McGraw-Hill, 1971.)*

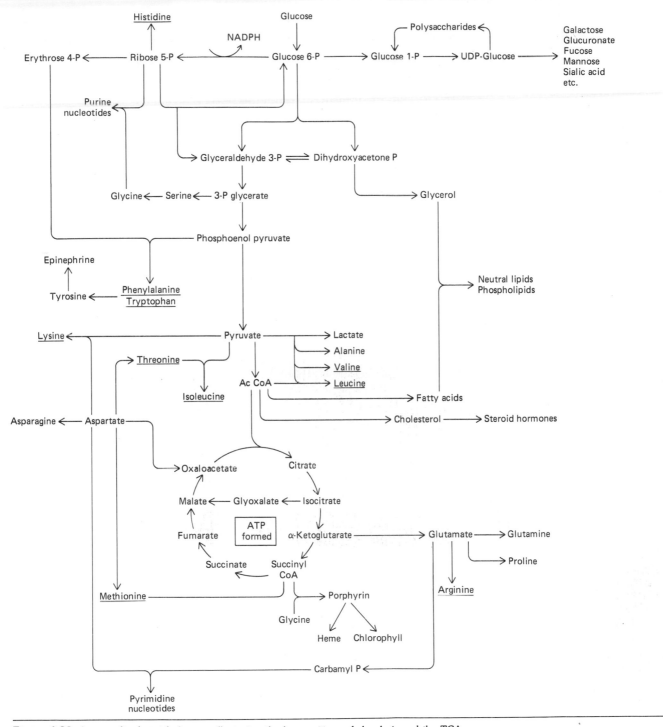

Figure 4-33. A generalized metabolic map illustrating the key positions of glycolysis and the TCA cycle in the formation of metabolites of the cell. Essential amino acids for humans, i.e., ones that must be present in the diet, are underlined. No relation exists between the lengths of any of the arrows and the number of reactions involved. This diagram is primarily meant to portray the pathways involving compounds discussed at one place or another in this book. Pathways can vary from one type of organism to another.

Reducing Power

We have spent considerable time on the formation of ATP and its use to drive endergonic reactions. This is one of the basic elements in understanding the means by which catabolism is coupled to anabolism. The other basic element by which these two types of pathways are coupled is the formation and use of "reducing power." We have shown that nearly all of the ATPs produced by the degradation of glycogen were derived from NADH. This is one example of reducing power (in the form of NADH) acting indirectly via electron transport to lead to the formation of ATP. The other use of reducing power is by the transfer of electrons (or hydrogen atoms) directly to substrate molecules thereby increasing their state of reduction. The most important compound acting in this manner is also a pyridine-nucleotide, NADPH. $NADP^+$, the oxidized form of NADPH, is formed from NAD^+ in the following reaction:

$$NAD^+ + ATP \rightleftharpoons NADP^+ + ADP$$

NADPH is formed by the reduction of $NADP^+$. The most effective catabolic pathway for the formation of NADPH from $NADP^+$ is the pentose phosphate pathway (Fig. 4-34). As in the case of NADH, NADPH is also a "high energy" compound because of its high electron-transfer potential; the loss of electrons from NADPH to an appropriate acceptor is a highly favorable event. The transfer of the free energy in the form of these electrons raises the acceptor to a more reduced, more energetic state.

The separation of "reducing power" into two distinct but related molecules, NADH and NADPH, reflects a separation of their primary metabolic role and allows the use of each to be controlled independently. At the same time, however, they are interconvertible and the relative amounts of $NADP^+$, NADPH, NAD^+, and NADH will depend to some extent upon the conditions in the cell at the particular time.

Metabolic Regulation

The amount of ATP that is present in a cell at a given moment is surprisingly small (approximately 10^6 molecules in a bacterial cell) and its half-life is very brief (in the order of one second). In other words, the average bacterial cell has only enough ATP present at one time to sustain

Figure 4-34. The first several reactions in the pentose phosphate pathway. These reactions produce reducing power in the form of NADPH as well as ribose phosphate, which is a key intermediate in the formation of numerous compounds.

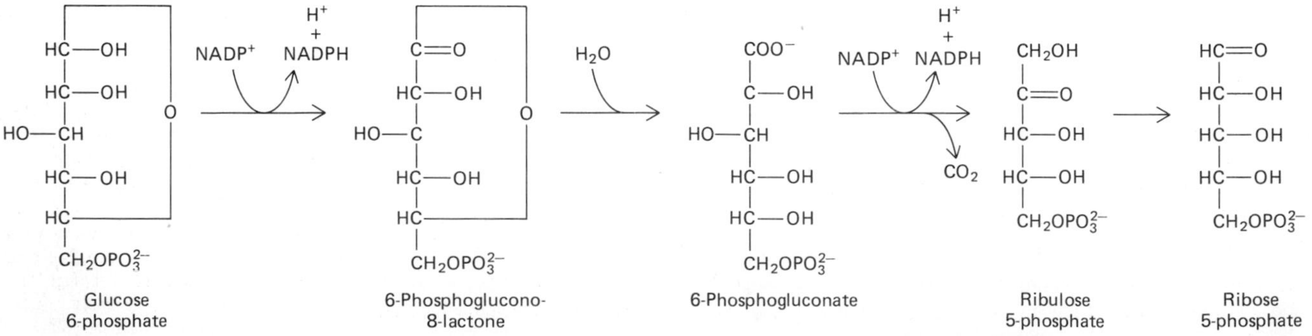

the cell's activities for a very short time. This is a remarkable condition for a molecule which is of such basic importance that it is either broken down or reformed by virtually every metabolic pathway in the cell. With so limited a supply, it is clear that ATP is not a molecule in which a large total amount of free energy is stored. It is the large organic molecules, particularly the polysaccharides and fats, which maintain the energy reserves of the cell. When the levels of ATP start to fall, it sets in motion reactions which result in its rapid formation at the expense of the energy-rich storage forms. Similarly, when its level is sufficiently high, reactions that would normally lead to its production are inhibited. For example, as will be described at more length in Chapter 8, electron transport in eukaryotic cells will not occur unless there are sufficient ADP molecules to become phosphorylated, i.e., the cell does not continue to catabolize carbohydrate to form NADH for electron transport if there is no reason for electron transport to be occurring. A number of key enzymes in the cell appear to be directly sensitive to the levels of ATP and their activity can be directly affected by interactions between the enzyme and ATP acting as a regulating molecule. One example of the effect of ATP on an enzymatic reaction was illustrated in Fig. 4-21.

For the control of glycolysis, the key regulatory enzyme is phosphofructokinase, which catalyzes the conversion of fructose 6-phosphate to fructose 1,6-diphosphate. Even though ATP happens to be a substrate of this enzyme (Fig. 4-25), ATP is also an allosteric inhibitor, while ADP is an allosteric activator. When ATP levels are high (higher than needed for its role as a substrate), the activity of the enzyme is decreased so that no additional ATP will be formed by glycolysis and the TCA cycle. When ADP levels are high relative to ATP, the activity of the enzyme is increased to promote the additional formation of ATP by oxidative metabolism. The regulation of oxidative metabolism is illustrated in Fig. 4-35. It is these types of complex regulatory mechanisms that permit all of the cellular reactions to continue in a way that serves the needs of the cell rather than leading to its destruction. The result of these feedback types of regulation is the maintenance of relatively constant ATP levels in the cell under a variety of environmental conditions. ATP levels generally do not fluctuate, but remain high in the face of great variations in demand for its use. It is important that the cell maintain high ATP concentrations relative to that of ADP and AMP, for only in this way will the $-\Delta G$ of ATP hydrolysis remain large enough to drive endergonic reactions.

A very brief consideration of the anabolic pathway leading to the formation of glucose (*gluconeogenesis*) or glycogen (*glyconeogenesis*) will illustrate a few important aspects about synthetic pathways. How can glucose (or glycogen) be synthesized from pyruvate? The first important point is that the formation of glucose cannot proceed simply by the reversal of the reactions of glycolysis. The glycolytic pathway contains three essentially irreversible reactions (Fig. 4-26), and these steps must somehow be bypassed. Even if all of the reactions of glycolysis could be reversed, it would be a very undesirable way for the cell to handle its metabolic activities since the two pathways could not be controlled independently

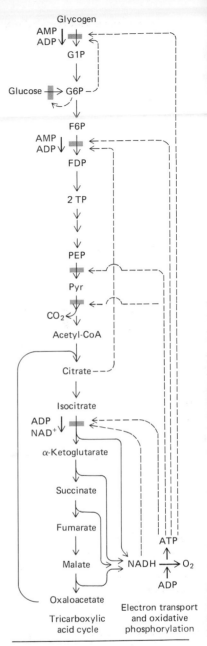

Figure 4-35. Summary of the important points of regulation of the rates of glycolysis and respiration. The dashed arrows show the origins of the feedback inhibitors (ATP, NADH, citrate, glucose 6-phosphate) and the reactions where they exert inhibition (bars across the reaction arrows). The reactions that are promoted by positive modulators (AMP, ADP, NAD+) are designated by arrows parallel to the reaction arrows. (*From A. L. Lehninger, "Biochemistry," 2d ed., Worth, 1975.*)

of one another. In other words, the cell could not shut down its synthesis of glucose and crank up its breakdown since the same enzymes would be active in both directions.

The way both of these problems are overcome, i.e., the thermodynamic difficulties and the regulatory ones, is to use at least a few different reactions when the processes are leading in different directions and to have different enzymes catalyze these reactions (Fig. 4-36). The new reactions are thermodynamically favored due to an input of energy by coupling them to ATP hydrolysis. These particular enzymes can be activated or inhibited so that either the synthesis or destruction of glucose is favored, depending

Figure 4-36. The relationship between the reactions leading to the breakdown of glucose to pyruvate, i.e., glycolysis, and those leading to the formation of glucose from pyruvate, i.e., gluconeogenesis. Three reactions of glycolysis, those with large changes in free energy (Fig. 4-26), must be bypassed in the formation of glucose.

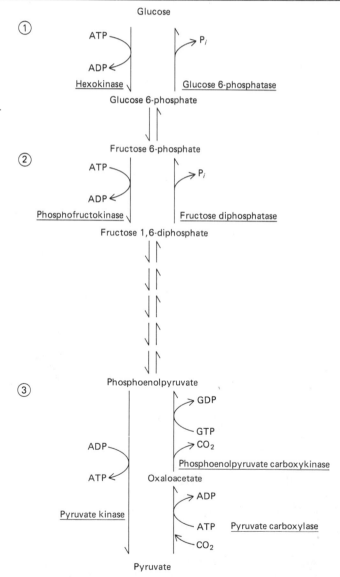

upon the needs of the cell. In the case of glycolysis vs. gluconeogenesis, when the anabolic pathway is active, certain key enzymes of the catabolic path are inhibited (page 247), while when the cell is breaking glucose down, certain enzymes required in its formation are inactivated. If this were not the case, the metabolism of the cell would simply be running around in circles, expending energy to form a molecule and then wasting that energy by destroying the molecule.

A glance at the last reaction of glyconeogenesis

$$\text{UDP-glucose} + \text{glucose}_n \rightarrow \text{UDP} + \text{glucose}_{n+1}$$
$$\qquad\text{(Glycogen)} \qquad\qquad\qquad \text{(Glycogen)}$$

illustrates another point. UDP-glucose is an activated form of glucose, i.e., one with high *glucose transfer potential*. UDP-glucose is formed in the reaction $\text{UTP} + \text{glucose 1-phosphate} \rightleftharpoons \text{UDP-glucose} + \text{PP}_i$. In this case UTP is being used in the same way as ATP was used to form glutamine (page 133). The couple intermediate in this case is UDP-glucose. In other words, ATP is not the only triphosphate used to couple reactions, though all of the other triphosphates are formed directly or indirectly from ATP. None of these other molecules (CTP, UTP, or GTP) can be formed by oxidative phosphorylation. CTP, for example, is an important reactant in phospholipid synthesis, while GTP is important in protein synthesis. We have concentrated in this chapter on the conservation of chemical energy in the form of ATP and its use in metabolism. It should be kept in mind that the energy of ATP is used in a great variety of diverse processes, all of which are coupled to ATP hydrolysis. During the remainder of the book, many of these processes will be discussed; a few of them are illustrated schematically in Fig. 4-37.

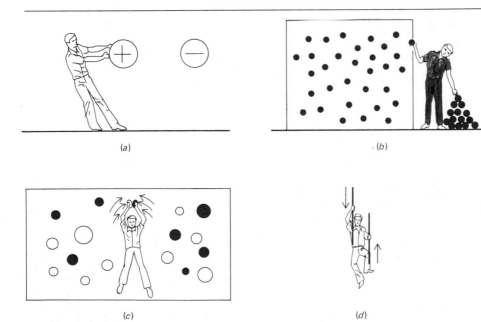

(a)

(b)

(c)

(d)

Figure 4-37. Pictorial representation of the means to which free energy stored in ATP is used by cells. *(a)* The separation of charge as occurs across a membrane. *(b)* The concentration of solutes in one or another cellular compartment. *(c)* The use of ATP to drive otherwise unfavorable chemical reactions. *(d)* The sliding of filaments across one another as occurs during contractile processes. [*(a)–(c) From L. Peusner, "Concepts in Bioenergetics,"* © *1974. Reprinted by permission of Prentice-Hall, Inc., Englewood Cliffs, New Jersey.*]

APPENDIX ON OXIDATION-REDUCTION REACTIONS

Oxidation-reduction reactions, i.e., those involving a change in the electronic state (valence) of the reactants, are one of the most basic types of chemical reactions. Changes of this type are accompanied by the gain or loss of electrons.

Consider the conversion of metallic iron (Fe^0) to the ferrous state (Fe^{2+}). This conversion involves the loss of a pair of electrons by the iron atoms which thereby attain a more positive valence; the iron atoms have been oxidized. The reaction is reversible. Ferrous ions can be converted to metallic iron, a more negative state, by the acquisition of a pair of electrons; the ferrous ions have been reduced. The reaction can be written as follows: $Fe^{2+} + 2$ electrons $\rightleftharpoons Fe^0$. Substances that can be oxidized and reduced exist in pairs or *couples* representing the two alternate states between which they can be present. Oxidation-reduction couples, which differ from each other by the presence or absence of electrons, are analogous to conjugate acids and bases (page 14), which differ from each other by the presence or absence of a proton.

It is clear that in order for metallic iron to be oxidized, there must be some substance to accept the electrons that are released and, conversely, for ferrous ions to be reduced, there must be some substance to donate the necessary electrons. In other words, the oxidation of one reactant must be accompanied by the simultaneous reduction of some other reactant, and vice versa. One possible reaction involving iron might be $Fe^0 + Cu^{2+} \rightleftharpoons Fe^{2+} + Cu^0$. The substance which loses electrons during the oxidation-reduction reaction, i.e., the one that becomes oxidized, is termed the *reducing agent* and the one that gains electrons is termed the *oxidizing agent*.

Up to this point in the discussion of oxidation-reduction we have dealt only with atoms which exist as either ions or unbonded elements, i.e., ones whose change in oxidation state is accompanied by the full loss or gain of electrons (see page 10). In contrast, the oxidations that occur to substrates during cellular metabolism are ones involving carbon atoms, atoms which are *covalently* bonded to other atoms. If we are going to discuss oxidation-reduction reactions involving covalently bound carbon in the same terms as atoms which form ions, then it is best to assign oxidation states to carbon atoms under various conditions. In order to do this we will assign the pair of shared electrons of a covalent bond to one or the other of the two atoms involved in the bond.

As discussed in Chapter 1, when a pair of electrons is shared by two different atoms, the electrons are generally attracted more strongly to one of the

TABLE 4-9
Oxidation States of Carbon in Organic Compounds

Oxidation State		Primary	Secondary	Tertiary	Quaternary
−4	CH_4				
−3		RCH_3			
−2	CH_3OH		R_2CH_2		
−1		RCH_2OH		R_3CH	
0	CH_2O		R_2CHOH		R_4C
+1		$RCHO$		R_3COH	
+2	$HCOOH$		R_2CO		
+3		$RCOOH$			
+4	CO_2				

two atoms of the polarized bond. A carbon atom can be bound to a less electronegative atom (a hydrogen atom), an equally electronegative atom (another carbon atom) or a more electronegative atom (an O, P, S, or N atom). In order to determine the overall oxidation state of a carbon atom, we will assign a value to each bond based on the direction in which the bond is polarized. If the bond is between the carbon and a hydrogen atom we will assume the carbon has possession of the electrons and assign a value of -1 to the carbon atom. If the carbon is linked to another carbon atom, it will be assigned a value of 0. If it is linked to a more electronegative atom, the carbon will be assigned a value of $+1$. Since carbon has four outer-shell electrons, its oxidation state can range from -4 when bonded to four H atoms to $+4$ when bonded totally to atoms other than C or H. The more negative the overall state of the carbon atom, the more electronegative the atom, i.e., the more reduced. These concepts are illustrated in the following examples and the accompanying table (Table 4-9), both taken from J. B. Hendrickson, D. J. Cram, and G. S. Hammond, "Organic Chemistry," 3d ed., McGraw-Hill, 1970.

$$CH_3-CH{=}CH-CH{=}O \underset{-H_2O}{\overset{+H_2O}{\rightleftharpoons}} CH_3-\underset{|}{\overset{OH}{C}}H-CH_2-CH{=}O$$

$$\underset{-3\quad -1\quad -1\quad +1}{} \qquad \underset{-3\quad 0\quad -2\quad +1}{}$$

$$\text{Total} = -4 \qquad\qquad -4$$

$$Br-CH_2-\overset{O}{\overset{\triangle}{CH----CH_2}} \xrightarrow{+2H_2O} HO-CH_2-\underset{|}{\overset{OH}{C}}H-CH_2-OH$$

$$\underset{-1\quad 0\quad\quad -1}{} \qquad\qquad \underset{-1\quad\quad 0\quad\quad -1}{}$$

$$\overset{+3\qquad 0\quad -1\quad +3}{CH_3-S-CH{=}CH-C{\equiv}N}$$

$$\overset{-1\ +1\qquad\quad +3\quad -1}{HC{\equiv}C-NH-CO-CH_2-Cl}$$

In Chapter 8 we will discuss the oxidation-reduction reactions which occur in the mitochondrion. Basically, there are two aspects of these reactions: the removal of hydrogen atoms from substrate molecules, and the transfer of the electrons from these hydrogen atoms to molecular oxygen to form water. It is important to note that (on the basis of the previous discussion) the removal of a hydrogen atom (with its electron) results in a change in the oxidation state of the carbon atom. This can be illustrated by one of the reactions of the TCA cycle, the oxidation of succinate to fumarate (Fig. 4-31). In this case, the removal of two hydrogen atoms, i.e., two protons and two electrons, results in the formation of a double bond between two carbon atoms, thereby changing the oxidation state of each of the carbon atoms from -2 to -1.

Succinic acid   Fumaric acid

A different type of oxidation-reduction involves the oxidation of pyruvate to acetyl CoA (Fig. 4-31). In this case, the removal of carbon dioxide from pyruvate (accompanied by the loss of two hydrogen atoms) results in a change in the oxidation state of one carbon atom from +3 to +4, and another from +2 to +3.

$$
\begin{array}{c}
\overset{+2}{CH_3} \\
C::O: \\
C:O:H \\
\overset{+3}{:O:}
\end{array}
+ NAD^+ + H:\ddot{S}:CoA \rightarrow
\begin{array}{c}
\overset{+3}{CH_3} \\
C::O: \\
:S:CoA
\end{array}
+ :\ddot{O}::\overset{+4}{C}::\ddot{O}: + NADH + H^+
$$

| Pyruvic acid | Acetyl CoA | Carbon dioxide |

In the discussion concerning oxidation-reduction reactions involving ions and changes in valence, we used the terms *oxidizing* and *reducing agents.* The reducing agent is the member of the pair that donates electrons; the oxidizing agent the member which receives them (thereby becoming reduced). The same terminology can be used with changes in oxidation state of covalently bonded atoms. For example, in the reaction between pyruvate and NAD^+ to form acetyl CoA, pyruvate is acting as the reducing agent and NAD^+ as the oxidizing agent. In contrast, in the reaction between pyruvate and NADH to form lactate (Fig. 4-30), pyruvate is now acting as the oxidizing agent and NADH as the reducing agent.

$$
\begin{array}{c}
CH_3 \\
C::O: \\
\overset{+2}{C}:O:H \\
\overset{+3}{:O:}
\end{array}
+ NADH + H^+ \rightarrow
\begin{array}{c}
\overset{0}{CH_3} \\
H:C:O:H \\
\overset{+3}{C}:O:H \\
:O:
\end{array}
+ NAD^+
$$

| Pyruvic acid | Lactic acid |

REFERENCES

Atkinson, D. E., Science **150**, 851–857, 1965. "Biological Feedback Control at the Molecular Level."

———, *Cellular Energy Metabolism and Its Regulation.* Academic, 1977.

Bender, M. L., and Brubacher, L. J., *Catalysis.* McGraw-Hill, 1973.

Blum, H. F., *Time's Arrow and Evolution.* Harper & Row, Torchbooks, 1962.

Boyer, P. D., *The Enzymes.* 3d ed., Academic, 1970–75.

Bray, H. G., and White, K., *Kinetics and Thermodynamics in Biochemistry.* 2d ed., Academic, 1967.

Busby, S. J. W., and Radda, G. K., Current Topics Cell Reg. **10**, 89–160, 1976. "Regulation of the Glycogen Phosphorylase System."

Changeux, J.-P., Sci. Am. **212**, 36–45, April 1965. "The Control of Biochemical Reactions."

Cohen, P., *Control of Enzyme Activity.* Chapman & Hall, 1976.

Colowick, S. P., and Kaplan, N. O., eds., *Methods in Enzymology.* Vol. 1, Academic, 1970.

Dixon, J. R., *Thermodynamics.* Prentice-Hall, 1975.

Edelstein, S. J., Ann. Rev. Biochem. **44**, 209–232, 1975. "Cooperative Interactions of Hemoglobin."

Ferdinand, W., *The Enzyme Molecule.* Wiley, 1976.

Fruton, J. S., Science **192**, 327–334, 1976. "The Emergence of Biochemistry."

Gamow, G., *One, Two, Three . . . Infinity.* Viking, 1947.

Gerhart, J. C., Current Topics Cell Reg. **2**, 275–325, 1970. "A Discussion of the Regulatory Properties of Aspartate Transcarbamylase."

Greenberg, D. M., ed., *Metabolic Pathways.* 3d ed., Academic, 1967.

Hammes, G. G., and Wu, C.-W., Ann. Rev. Biophys. and Bioeng. **3**, 1–34, 1974. "Kinetics of Allosteric Enzymes."

Hochachka, P. W., and Somero, G. N., *Strategies of Biochemical Adaptation.* Saunders, 1973.

Kalckar, H. M., "Lipmann and the Squiggle," in *Current Aspects of Biochemical Energetics*, N. O. Kaplan and E. P. Kennedy, eds., Academic, 1966.

Katchalsky, A., *Non-Equilibrium Thermodynamics in Biophysics.* Harvard, 1965.

Kirsch, J. F., Ann. Rev. Biochem. **42**, 205–234, 1973. "Mechanism of Enzyme Action."

Klotz, I. M., *Energy Changes in Biochemical Reactions.* Academic, 1967.

Koshland, D. E., Harvey Lectures **65**, 33–57, 1971. "A Molecular Model for the Regulatory Behaviour of Enzymes."

———, Sci. Am. **229**, 52–64, Oct. 1973. "Protein Shape and Biological Control."

Krebs, H. A., Current Topics Cellular Reg. **I**, 45–55, 1969. "The Role of Equilibria in the Regulation of Metabolism."

——— and Kornberg, H. L., *Energy Transformation in Living Matter.* Springer-Verlag, 1957.

Lehninger, A. L., *Biochemistry.* 2d ed., Worth, 1975.

Levinthal, C., Sci. Am. **214**, 42–52, June 1966. "Molecular Model Building by Computer."

Lienhard, G. E., Science **180**, 149–154, 1973. "Enzymatic Catalysis and Transition-State Theory."

Mahler, H., and Cordes, E. H., *Biological Chemistry.* 2d ed., Harper & Row, 1971.

Meister, A., ed., *Advances in Enzymology.* Vol. 1, Wiley, Interscience, beginning 1944.

Mildvan, A. S., Ann. Rev. Biochem. **43**, 357–400, 1974. "Mechanism of Enzyme Action."

Monod, J., Science **154**, 475–482, 1966. "From Enzymatic Adaptation to Allosteric Transitions."

———, Changeux, J.-P., and Jacob, F., J. Mol. Biol. **6**, 306–329, 1963. "Allosteric Proteins and Cellular Control Systems."

———, Wyman, J., and Changeux, J.-P., J. Mol. Biol. **12**, 88–118, 1965. "On the Nature of Allosteric Transitions: A Plausible Model."

Morowitz, H. J., *Entropy for Biologists.* Academic, 1970.

Neurath, H., Sci. Am. **211**, 68–79, Dec. 1964. "Protein-Digesting Enzymes."

Pardee, A. B., Harvey Lectures **65**, 59, 1971. "Control of Metabolic Reactions by Feedback Control."

Peusner, L., *Concepts in Bioenergetics.* Prentice-Hall, 1974.

Phillips, D. C., Sci. Am. **215,** 78–90, Nov. 1966. "The Three-Dimensional Structure of an Enzyme Molecule."

Rosing, J., and Slater, E. C., Biochem. Biophys. Acta **267,** 275–290, 1972. "The Value of $\Delta G°$ for the Hydrolysis of ATP."

San Pietro, A., and Gest, H., eds., Symposium on "Horizons of Bioenergetics: Proceedings." Academic, 1972.

Sigman, D. S., and Mooser, G., Ann. Rev. Biochem. **44,** 889–931, 1975. "Chemical Studies of Enzyme Active Sites."

Sink, J. D., *The Control of Metabolism*. Pennsylvania State University Press, 1974.

Stadtman, E. R., "Mechanisms of Enzyme Regulation in Metabolism," in *The Enzymes*, P. D. Boyer, ed., 3d ed., vol. 1, 397–459, 1970.

Stroud, R. M., Sci. Am. **231,** 74–88, July 1974. "A Family of Protein-Cutting Proteins."

Stryer, L., *Biochemistry*. Freeman, 1975.

Symposium on "Energy Transformation in Biology," Ciba Symp. 31 (new series), Elsevier, 1975.

Weber, G., ed., *Advances in Enzyme Regulation*. Pergamon, vol. 1 beginning 1963.

CHAPTER FIVE

Membrane Structure and Function

During the past few years great strides have been made in research on the structure and function of membranes. During most of the years since the beginning of molecular biology, the study of membranes remained well in the shadows, waiting for an appreciation of their complexity and importance and the development of techniques for their analysis. The fast-paced accumulation of data on informational macromolecules had for many years eclipsed most other areas, but the field of membrane study has now become wide open with the realization of how much there is to learn. One of the most exciting aspects of membrane research concerns the diversity of functions that are mediated by these extremely thin organelles. Many of the functions are quite independent of one another, yet certain of them can proceed in the same membrane at the same time. The following (illustrated in Fig. 5-1) is a list of the general types of functions that can be attributed to various membranes; in the remainder of this chapter and in parts of other chapters, the mechanisms underlying these phenomena will be considered.

A SUMMARY OF MEMBRANE FUNCTIONS

1. *Compartmentalization* Membranes are continuous unbroken sheets and as such inevitably enclose compartments. The plasma membrane encloses the contents of the entire cell. The nuclear and cytoplasmic membranes enclose various internal cellular spaces. Just as the space within a building needs to be divided so that different types of activities can proceed in their own compartments with a minimum of interference with one another, so too does the cell need to be divided. In the cell compartmentalization is particularly important since the various spaces are fluid filled, and the indiscriminate intermixing of their contents would be very chaotic. One of the most limiting aspects of the prokaryotic condition is the relative lack of intracellular membranous organelles.

2. *Regulation of the movement of materials* As just described, the plasma membrane serves as a wall or barrier between the living and nonliving worlds, the intracellular membranes between the various cytoplasmic compartments. As barriers, membranes prevent the free interchange of materials from one side to the other, but at the same time membranes provide the means of communication between the spaces. Every cell requires nutrition, water, oxygen, ions, substrates, etc., from its environment, whether that environment is the blood stream of a multicellular organism or the medium in which a culture of single-celled organisms is growing. It is the responsibility of the plasma membrane to deter-

Figure 5-1. Diagrammatic representation of the types of membrane functions. Numbers correspond to those given in the text. (1) An example of compartmentalization in which a hydrolytic enzyme, acid phosphatase, is sequestered within the membrane of a lysosome. (2) An example of the role of the membrane in the selected movement of materials. Sodium ions are transported out of the cell and potassium ions are transported in. (3) An example of the involvement of the membrane in information transfer. The glucagon binds to the outside of the membrane causing a message (cAMP) to be sent into the cytoplasm. (4) A site at the cell surface where two plasma membranes come into close contact to form a special junction. (5) An example of the role of membranes in carrying out enzymatic reactions. It is generally believed that DNA synthesis occurs in conjunction with the nuclear membrane. (6) The conversion of ADP to ATP within the mitochondrion occurs in close connection to the inner membrane.

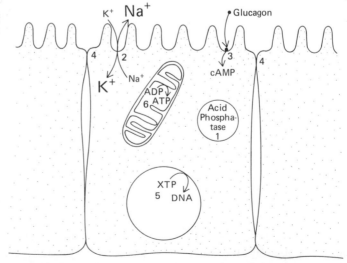

mine that the appropriate substances are allowed in and the inappropriate substances are kept out. In this capacity the plasma membrane can be considered as a *selectively permeable* wrapper.

Among the substances that are allowed to enter the cell, some do so simply by diffusion in response to a concentration difference (gradient) across the membrane. In this regard the plasma membrane acts as a gate that can be opened or closed. Other substances are carried through the membrane in order to be maintained at a higher concentration on the inner side; the plasma membrane can act as a molecular pump. Still other materials, including the fluid in which they are suspended, can be brought inside by the formation of vesicles from the plasma membrane. One important consequence of the selective permeability of the plasma membrane is its capacity to separate ions and therefore establish an electrical potential difference across itself. This potential (voltage) is critical for the so-called excitable cells, the neurons and muscle cells, but may also be of great importance in the ability of all cells to respond to their environment.

The regulation of the passage of substances across membranes is not limited to movement from the outside of the cell to the inside by the plasma membrane. The various intracellular compartments must communicate with one another as well. No one organelle of the cell is autonomous; each depends upon the exchange of materials in both directions across the membrane by which it is bounded. The problem, in this connection, is that we know less about the various intracellular membranes than we do about the plasma membrane. The intracellular membranes are much less accessible for study and must be prepared from cells that have been broken apart and the cellular compartments separated.

3. *Access to information* Another important aspect of the communication process mediated by membranes, particularly the plasma membrane, is the control of information received by and transmitted across from one side to the other. The plasma membrane, and possibly intracellular membranes as well, has receptor molecules that have the capacity to combine with specific molecules of the appropriate configuration, just as an enzyme combines with the appropriate substrate. Different types of cells have different types of receptors and, therefore, combine with different substances. Substances that combine with receptors are termed *ligands*. The best-studied ligands are hormones, growth factors, and neurotransmitters, all of which attach to the membrane but do not pass through it. The interaction of the membrane receptor with the particular ligand may cause the membrane to generate a new signal (e.g., cyclic AMP) directed to the internal compartment of the cell. In other words, the interaction between a membrane receptor and an external agent may cause a new informational stimulus to be produced by the membrane. In this way it appears that the plasma membrane has responsibility for the regulation of many different internal events. In some cases, the signal generated by the membrane tells the cell to divide or to differentiate into a new type of cell, while in the case of bacteria it may tell the cell to move toward food. Most of these processes are understood at a very superficial level or not at all; however in the case of a few hormones (particularly insulin and glucagon) quite a bit of the sequence of events has been elucidated.

4. *Intercellular interaction* At the outer edge of the living cell, the plasma membrane is responsible for mediating the interactions between cells that occur continually in a multicellular organism. Organs are usually composed of many different types of cells that must acquire and maintain a certain orientation with one another and must work together to perform an overall function. The plasma membrane allows cells to recognize one another, to adhere when appropriate, and to exchange materials and information regardless of whether the cells are fixed

in place, as in a tissue, or moving around individually within the plant or animal.

5. *Locus for biochemical activities*   Membranes provide a means to organize cellular activities.   One of the consequences of a solution of reactants is that their relative positions cannot be stabilized; instead they remain at the mercy of random collisions.   In Chapter 3, the existence of multienzyme complexes was considered. The association of enzymes catalyzing successive reactions greatly facilitates the reaction sequence since each enzyme is automatically in the right place at the right time.   Membranes provide the cell with a large-scale framework or scaffolding within which components can be ordered for effective interaction.   Within a membrane many different processes are going on simultaneously, apparently without interference with one another.   In the inner mitochondrial membrane, for example, the various components of the electron-transport chain and oxidative phosphorylation must work together in a coordinated manner.   The organization of these proteins into respiratory assemblies within the structure of the membrane allows electrons to be passed from one carrier to the next in an orderly fashion.   As a result, the energy of the electrons can be captured in the formation of ATP.   If one considers all of the enzymatic machinery within a cell, much of it would be found associated with the various membranes.

6. *Energy transduction*   The conversion of one type of energy into another (transduction) is vital in many different ways, and membranes are intimately involved in these processes.   Basic to all life is the ability of plant cells to convert light energy from the sun into the chemical energy contained in carbohydrates. Conversely, it is equally important for both plants and animals to convert the chemical energy of these nutrients into that of ATP and other "high energy" compounds.   The machinery for energy capture is contained within the membranes of the mitochondria and chloroplasts.   Light, thermal, and mechanical energy are converted by our sensory receptors into electrochemical energy for communication with the central nervous system.   The underlying mechanisms involved in these transductions are not well understood, but membranes are known to be centrally involved.

7. *Miscellaneous*   Membranes act in a variety of miscellaneous capacities. For example, they provide mechanical strength, they are involved in cell movement and secretion, they act as electrical insulators, as sites for ribosome attachment, and as conductors of nervous impulses.

MEMBRANE COMPOSITION

All membranes are lipid-protein assemblies in which the components are held together in a thin sheet by noncovalent bonds.   The proportion of lipid to protein varies considerably (Table 5-1) depending upon the type of cellular membrane (plasma vs. endoplasmic reticulum vs. Golgi, etc.), the type of organism (prokaryote vs. eukaryote), and the type of cell (cartilage vs. muscle vs. liver, etc.).   In addition to lipid and protein, certain membranes, including plasma membranes, contain carbohydrates as well. Membrane proteins often have an unusual disposition of their hydrophilic and hydrophobic residues when compared to soluble proteins since they must exist in association with a totally different environment, that of the hydrophobic lipid bilayer.   The organization of membrane proteins will be considered in detail later in this chapter.   Since protein structure has been discussed at great length in Chapter 3, no further comments on its nature will be made in this section.

Lipids

There are several basic types of membrane lipids, all of them having one point in common: they are *amphipathic*, i.e., they have hydrophilic and hydrophobic portions within one lipid molecule. Most membrane lipids contain phosphate, the main exception being cholesterol. Most of the phospholipids of membranes are phosphoglycerides; their structure is based on a glycerol backbone. Glycerol is a three-carbon polyalcohol (Fig. 5-2a) upon which substitutions for the hydroxyl groups can be made enzymatically. Storage fats (nonmembrane lipids) have all three hydroxyls replaced with long-chain fatty acids (Fig. 5-2a); they are triglycerides. In contrast, membrane glycerides are diglycerides; only two of the hydroxyls are esterified to fatty acyl chains (Fig. 5-2a), the third is esterified to the phosphate group. Without any additional modification beyond the phosphate and the two fatty acids, the molecule is phosphatidic acid, which is virtually absent in membranes. Instead, membrane diglycerides have an additional moiety linked to the phosphate. These added groups are most commonly choline, ethanolamine, serine, and inositol. All of these groups are small in molecular weight and hydrophilic and, together with the phosphate to which they are attached, this end of the phospholipid (called the *head group*) forms a highly water soluble territory within the molecule. The fatty acid chains are long hydrocarbons, i.e., they contain only

TABLE 5-1
Protein and Lipid Content of Membranes

Membrane	Protein/lipid (wt/wt)	Cholesterol/polar lipid (mol/mol)	Major polar lipids†
Myelin	0.25	0.7–1.2	Cer, PE, PC
Plasma membranes			
Liver cell	1.0–1.4	0.3–0.5	PC, PE, PS, Sph
Ehrlich ascites	2.2		
Intestinal villi	4.6	0.5–1.2	
Erythrocyte ghost	1.5–4.0	0.9–1.0	Sph, PE, PC, PS
Endoplasmic reticulum	0.7–1.2	0.03–0.08	PC, PE, Sph
Mitochondrion			DPG, PC, PE, Plas
Outer membrane	1.2	0.03–0.09	
Inner membrane	3.6	0.02–0.04	
Retinal rods	1.5	0.13	PC, PE, PS
Chloroplast lamellae	0.8	0	GalDG, SL, PS
Bacteria			
Gram-positive	2.0–4.0	0	DPG, PG, PE, PGaa
Gram-negative		0	PE, PG, DPG, PA
PPLO	2.3	0	
Halophilic	1.8	0	Ether analog PGP

* Unless otherwise indicated the data are from E. D. Korn, *Federation Proc.*, **28**:6 (1969).

† Abbreviations are: Cer, cerebrosides; DPG, diphosphatidylglycerol; GalDG, galactosyldiglyceride; PA, phosphatidic acid; PC, phosphatidylcholine; PE, phosphatidylethanolamine; PGaa, amino acyl esters of phosphatidylglycerol; Plas, plasmalogen; SL, sulfolipid; Sph, sphingomyelin.

SOURCE: E. D. Korn, *Ann. Rev. Biochem.*, **38**: 268 (1969).

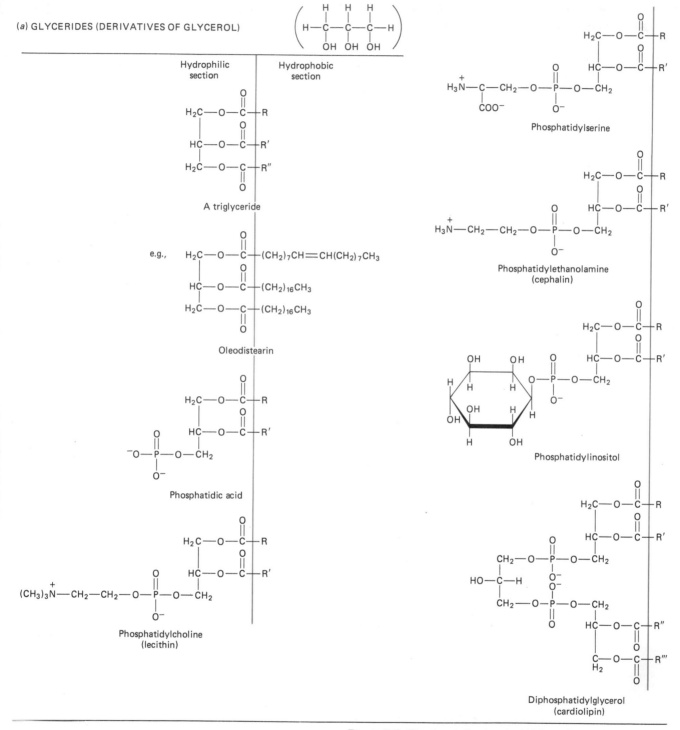

Figure 5-2. The chemical structure of biologically important classes of lipids. (*a*) Derivatives of glycerol. (*b*) Derivaties of sphingosine. (*c*) Cholesterol.

carbon and hydrogen, and are very hydrophobic, thus the asymmetric, amphipathic nature of these molecules (Fig. 5-3).

The nature of the fatty acid chains is highly variable (Table 5-2) and gives the phospholipid a wide range of properties. Fatty acid chains generally vary in number from approximately 14 to 24 carbons. The most important feature of the fatty acids is whether they are saturated (lack double bonds) or unsaturated (contain double bonds), and, if they are unsaturated, the number of double bonds that are present. Fatty acids are

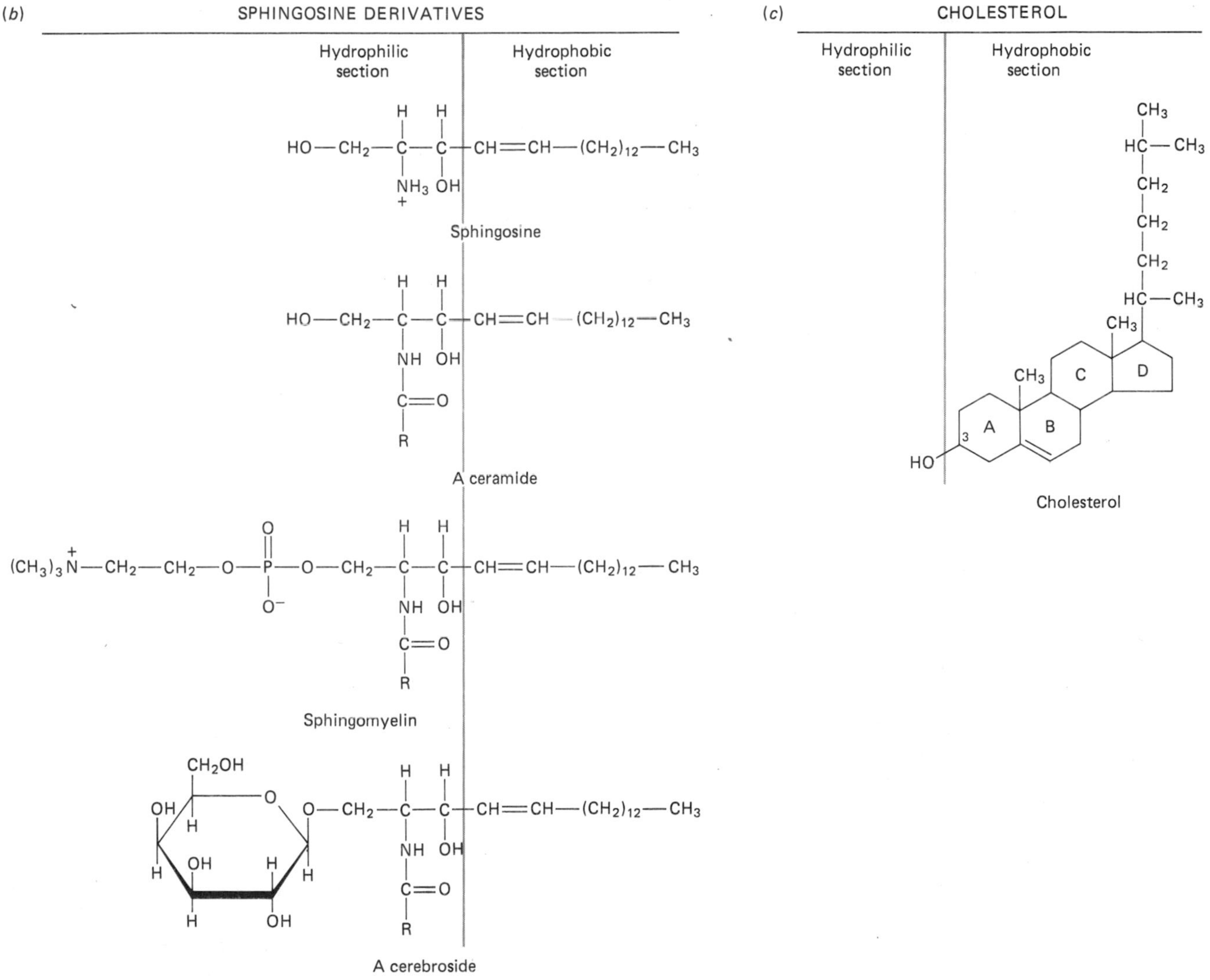
(b)
SPHINGOSINE DERIVATIVES
Hydrophilic section
Hydrophobic section
Sphingosine
A ceramide
Sphingomyelin
A cerebroside
(c)
CHOLESTEROL
Hydrophilic section
Hydrophobic section
Cholesterol

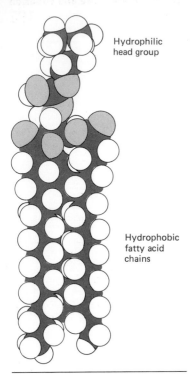

Figure 5-3. A space-filling model of a phospholipid, phosphatidylcholine.

usually described by two numbers, the first indicating the number of carbons in the chain and the second the number of double bonds. For example, linoleic acid is 18:2, oleic acid is 18:1 (the double bond having a cis configuration), and elaidic acid is 18:1 (the double bond having a trans configuration). Properties of these particular fatty acids are compared on page 220. Most of the fatty acids of the membrane are fully saturated, common ones being myristic, 14:0, palmitic, 16:0, and stearic, 18:0. The typical organization for a phosphoglyceride is one unsaturated and one saturated fatty acid, the former in the middle position of the glycerol backbone.

The other lipids of the membrane (besides cholesterol) are not glycerides, i.e., they do not have a glycerol backbone, but rather are derivatives of sphingosine, an amino alcohol that has a long hydrocarbon chain within its structure (Fig. 5-2b). When sphingosine is present as a membrane lipid, a sphingolipid, the amino group is linked to a fatty acid by an amide bond (Fig. 5-2b). This molecule is a ceramide (Fig. 5-2b). The various sphingosine-based lipids are ones with different groups esterified to the terminal alcohol. If the substitution is phosphorylcholine, the molecule is sphingomyelin, the other phospholipid of the membrane (Fig. 5-2b). If the substitute is a carbohydrate, the molecule is a glycolipid, either a cerebroside (a simple sugar) or a ganglioside (an oligosaccharide). Since all of the sphingolipids have two long hydrophobic hydrocarbon chains on one end and a hydrophilic region at the other, they are also amphipathic and basically similar in structure to the phosphoglycerides. Different membranes have widely ranging percentages of these various lipids, as well as the other major membrane "lipid," the sterol, cholesterol (Fig. 5-2c). Cholesterol is a smaller molecule than those previously described and less amphipathic. It is a four-ringed molecule, lacking a long fatty acid chain, with the hydroxyl group representing a slightly hydrophilic end. The lipid composition of a variety of membranes is shown in Table 5-3.

TABLE 5-2
Some Naturally Occurring Fatty Acids

	Saturated Fatty Acids	
$C_{12}H_{24}O_2$	Lauric acid	$CH_3(CH_2)_{10}COOH$
$C_{14}H_{28}O_2$	Myristic acid	$CH_3(CH_2)_{12}COOH$
$C_{16}H_{32}O_2$	Palmitic acid	$CH_3(CH_2)_{14}COOH$
$C_{18}H_{36}O_2$	Stearic acid	$CH_3(CH_2)_{16}COOH$
$C_{20}H_{40}O_2$	Arachidic acid	$CH_3(CH_2)_{18}COOH$
	Unsaturated Fatty Acids	
$C_{18}H_{34}O_2$	Oleic acid	$CH_3(CH_2)_7CH{=}CH(CH_2)_7COOH$ (*cis*-9-octadecenoic acid)
$C_{18}H_{34}O_2$	Elaidic acid	$CH_3(CH_2)_7CH{=}CH(CH_2)_7COOH$ (*trans*-9-octadecenoic acid)
$C_{18}H_{32}O_2$	Linoleic acid	$CH_3(CH_2)_4CH{=}CHCH_2CH{=}CH(CH_2)_7COOH$
$C_{18}H_{30}O_2$	Linolenic acid	$CH_3CH_2CH{=}CHCH_2CH{=}CHCH_2CH{=}CH(CH_2)_7COOH$
$C_{18}H_{30}O_2$	Eleostearic acid	$CH_3(CH_2)_3CH{=}CH{-}CH{=}CH{-}CH{=}CH(CH_2)_7COOH$

Carbohydrates

The membrane that has drawn the most investigation, and will be described at length in this chapter, is the plasma membrane of the red blood cell (erythrocyte). This membrane contains approximately 52% protein, 40% lipid, and 8% carbohydrate by weight. Of the 8% that is carbohydrate, only 7% is in the form of glycolipids (the sphingosine derivatives of Fig. 5-2b). The nature and functions of the glycolipids are not well understood, though changes in their structure have been found to occur upon the malignant transformation of cells. The remaining carbohydrate of the membrane is linked to proteins to form *glycoproteins*. The carbohydrate of the glycoproteins is present in short chains, *oligosaccharides*, typically having less than about 15 sugars per chain. One protein may have many such chains, attached by covalent bonds to one of five different amino acids. There are nine different sugars (Fig. 5-4) that are commonly isolated from these glycoproteins: glucose, galactose, mannose, fucose, arabinose, xylose, N-acetylglucosamine, N-acetylgalactosamine, and N-acetylneuraminic acid (sialic acid).

In contrast to most large molecular weight carbohydrates (e.g., glycogen, starch, chitin, cellulose), those attached to membrane proteins are believed to be capable of a great degree of structural variability. In other words, they have the potential to provide specificity in their interactions with one another or other molecules. The concept of biological specificity was discussed in Chapter 4 in connection with the structure of enzymes. Of the virtually infinite sequences of amino acids that could form, only a very few have evolved because of their ability to make certain-shaped proteins that can catalyze a specific reaction. As a result of this struc-

TABLE 5-3
Lipid Composition of Animal and Bacterial Membranes

	Myelin	Erythro-cyte	Mito-chondria	Microsome		Azoto-bacter agilis	Escherichia coli	Agrobac-terium tume-faciens	Bacillus megaterium
Cholesterol	25	25	5	6	°	0	0	0	0
Phosphatidyl ethanolamine	14	20	28	17	18	100	100	90	45
Phosphatidyl serine	7	11	0	0	9	0	0	0	0
Phosphatidyl choline	11	23	48	64	48	0	0	10	0
Phosphatidyl inositol	0	2	8	11	6	0	0	0	0
Phosphatidyl glycerol	0	0	1	2	0	0	0	0	45
Cardiolipin	0	0	11	0	2	0	0	0	0
Sphingomyelin	6	18	0	0	9	0	0	0	0
Cerebroside	21	0	0	0	0	0	0	0	0
Cerebroside sulfate	4	0	0	0	0	0	0	0	0
Ceramide	1	0	0	0	0	0	0	0	0
Lysyl phosphatidyl glycerol	0	0	0	0	0	0	0	0	10
Unknown or other	12	2	0	0	0	0	0	0	0

° Not analyzed.

SOURCE: E. D. Korn, *Science*, **153**: 1497 (1966). Copyright © 1966 by the American Association for the Advancement of Science.

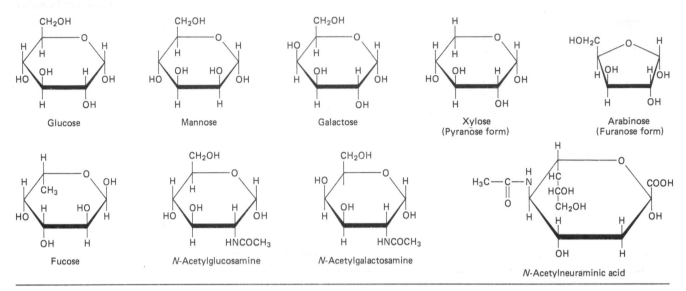

Figure 5-4. The chemical structure of the predominant sugars of the carbohydrate portion of membrane glycoproteins.

ture, only the desired reactants can fit into the active site. Since the carbohydrate portion of the membrane glycoproteins can be composed of at least nine different sugars, the possible different sequences that can be formed is also great. Although sugars lack the wide range of chemical properties that characterize the different amino acids, there is considerable structural variation among the nine sugars of Fig. 5-4.

Polymers of sugar have an additional source of structural variation that is not present among chains of amino acids since there is more than one way in which two sugars can be bonded to one another. For example, the differences among starch, glycogen, and cellulose (all glucose polymers, page 161) are determined by the ways the sugars can be attached to one another. Another feature that results from the presence of these different types of linkages is that one sugar can be bonded to three sugars, rather than just two in the case of strictly linear polymers, such as polypeptides. In other words, polysaccharides can have branch points along the chains. Though the potential to provide specific interactions based on these oligosaccharides exists, and there is good evidence that they do have such roles, very little is understood about the nature of these interactions. All of the carbohydrate of the plasma membrane, whether present as glycolipid or glycoprotein, faces outward from the membrane to the extracellular space. As such, these carbohydrates are believed to be of great importance in the manner in which the cell interacts with other cells as well as its nonliving environment. As more evidence accumulates, the realization of the importance of the cell surface grows. Some of these topics will be considered later in this chapter.

An interesting aspect of the carbohydrates of the membrane glycoproteins is their means of synthesis. There is no doubt that the sequence of the sugar components of the oligosaccharide chains is specified; if they are isolated from purified proteins, the order is consistent and predictable (Fig. 5-5a). In fact, there is a definite structural relationship among the various chains of the same molecule; those that are longer appear to be simply continuations of the shorter chain. In other words, the sequence of shorter ones tends to be the same as the more basal region of the longer ones. Shorter chains can be considered less near completion.

When a macromolecule with a specified sequence is found, one of the most important questions to be asked is the means by which the sequence is generated. In the case of the nucleic acids and the proteins, very complex events must occur in order that the precise linear sequence be produced. In these cases the essence of the process is a template, i.e., a preexisting sequence containing the information that must be passed on from one cell to the next. The formation of the carbohydrate sequences occurs without the use of a direct template. The polymerization process whereby monosaccharides are covalently linked to one another to form the oligosaccharide chains occurs by the addition of one sugar at a time to the nonreducing end of the growing chain. The addition of the sugar is catalyzed by a large group of enzymes called *glycosyltransferases*, enzymes which transfer the monosaccharide from an appropriate sugar donor to an appropriate sugar acceptor. The donor molecule in all cases is a nucleotide sugar (Fig. 5-5*b*), a molecule analogous to ATP as

---

Figure 5-5. (*a*) The structure of the branched oligosaccharide of the A blood group determinant. In the type B determinant, the terminal *N*-acetylgalactosamine residues are replaced with galactose, while in the type O determinant, neither sugar is present at these positions. (*b*) The step-by-step synthesis of the carbohydrate portion of a glycoprotein. Each sugar is added by a glycosyltransferase which is specific for both the donor nucleotide sugar and the acceptor, the latter being an incomplete oligosaccharide chain. (*Drawing by B. Tagawa, from Sugars of the Cell Membrane, by S. Roseman, Hospital Practice, Vol. 10, No. 1 and "Cell Membranes: Biochemistry, Cell Biology & Pathology," G. Weissmann and R. Claiborne, Eds., HP Publishing Co., Inc., New York, NY 1975. Reprinted with permission.)*

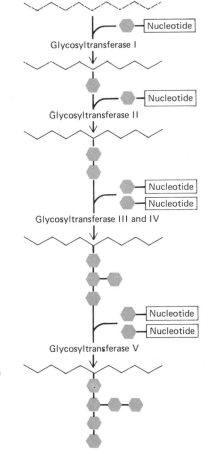

(a)

(b)

a phosphate donor, or acetyl CoA as an acetyl donor. In all of these "activated" molecules, the hydrolysis is thermodynamically favored, and the loss of the transferred group from the donor to the acceptor molecule is readily accomplished. The precise nature of the nucleotide donor varies from sugar to sugar, though for a particular type of sugar it is always the same. Examples of nucleotide sugars are CMP-sialic acid, GDP-mannose, GDP-fucose, UDP-glucose, UDP-galactose, and UDP-glucosamine.

The acceptor for the transferred sugar is the growing oligosaccharide chain (Fig. 5-5b). The glycosyltransferases are a collection of enzymes that are specific for both the sugar donor and the acceptor. This group of enzymes can be divided into families based upon the nature of the sugar transferred. For example, there is a family of sialyltransferases, which transfer sialic acid from CMP-sialic acid; a family of galactosyltransferases for galactose, and so on. Different members of *one* family recognize different acceptors to which the transferred sugar is to be added. One sialyltransferase will add sialic acid to a chain having galactose at its terminus, while another member requires some other sugar to be present. The specificity does not seem to be absolute, i.e., not every combination of donor and acceptor has its own enzyme, though the group of glycosyltransferases is known to be large.

The process by which these oligosaccharide chains are formed is, therefore, very different from that by which the much longer nucleic acids and proteins are formed. No template is used and the enzymes are specific for the various building blocks that make up the polymer. The specific sequence of sugars in the oligosaccharide chains will be determined primarily by which glycosyltransferases are present in a particular part of the cell at a given time. As the chain grows, the nature of the acceptor changes for the next sugar to be added. For example, if a galactosyltransferase for that acceptor is present in that location at that time, then a galactose will be added. The addition of the galactose would change the nature of the acceptor so that a different enzyme would take part in chain elongation. If no appropriate enzyme were present, the chain would not be continued. If a sialic acid is added the chain is automatically terminated since no further sugar can be attached to such a chain. The formation of the oligosaccharide chains of membrane glycoproteins provides an interesting example of the genetic control of cellular activities. That the sequence of sugars is genetically determined is indicated by their consistent and predictable order. It would appear that, in the absence of a direct template, the genetic influence must be manifested indirectly by a type of enzymatic assembly.

EARLY STUDIES ON THE PLASMA MEMBRANE

It is now known that the plasma membrane is an extremely thin structure; early studies on its nature were, of necessity, indirect. The first important information on the chemical nature of the outer boundary layer of a cell was obtained by E. Overton in studies during the 1890s. It was known at that time that cells would act as osmometers, i.e., they would swell due to the uptake of water when placed in various hypotonic solutions and would shrink due to the loss of water when placed in various hypertonic solutions. It appeared that living cells were covered by a membrane that had selectively permeable properties; some substances, such as water, could enter, while others could not. Overton reasoned that, by determining the nature of the chemicals that were capable of diffusing through the outer membrane, he could learn something about the nature of the membrane itself since diffusion through the membrane would require that the substance has been dissolved in it. Overton discovered a very important correlation: the more lipid soluble a compound was found to be, the more rapidly that compound would penetrate into the cell. Overton used cells of plant root hairs for his system. He found that if he placed these root hairs into a 7% solution of sucrose, no water would either enter or leave the cells. Sucrose was not able to penetrate the cells at all and, at 7%, its osmotic pressure was exactly equal to that of the cells so that equivalent numbers of water molecules were moving across the membrane in both directions. If, however, he placed the cells in a 7.5% sucrose solution, the cells were seen to rapidly shrink within the cell wall, indicating that the outer sucrose solution was hypertonic, i.e., had a greater osmotic pressure than the protoplasm.

Shrinkage due to osmotic loss of water is termed *plasmolysis* (Fig. 5-6) and Overton found that it would occur within a few seconds. To test the rate of penetration of various compounds he placed the cells in solutions of 7% sucrose that contained, in addition to the sucrose, other substances, for example, urea. If the other substance was nonpenetrating, as in the case of urea, the osmotic pressure of the medium would remain greater than that of the cells and plasmolysis would occur. If the compound could penetrate very rapidly, then it would be expected that its concentration within the cell would quickly be equivalent to its concentration outside of the cell, and no outward osmotic flow should occur. If the compound were to penetrate slowly, plasmolysis would first occur, but as the concentration of the substance inside the cell increased, its osmotic pressure would rise and water would reenter. Overton found that the more polar the solute, the less it was able to prevent plasmolysis. If, for example, instead of using urea he used a derivative of urea in which one of the hydrogens was replaced by an ethyl group (CH_3CH_2—) making it less polar, the compound would penetrate much more readily than urea, as measured by the time it took for the plasmolysis to be reversed. If two of the hydrogens of urea were replaced by methyl groups, then penetration was even more rapid, and, when three hydrogens were replaced, the concentration of the substance within the cells was almost immedi-

Figure 5-6. Plasmolysis, the shrinkage of a plant cell within the cell wall as a result of the osmotic loss of water. *(From E. C. Miller, "Plant Physiology," 2d ed., McGraw-Hill, 1935.)*

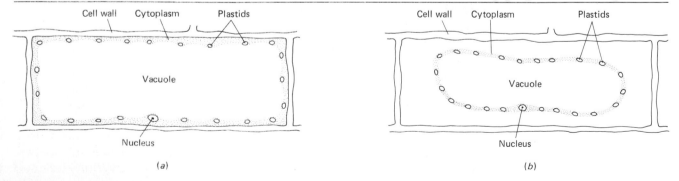

(a)   (b)

ately equal to its concentration outside. Overton concluded that the dissolving power of the outer protoplasmic layer matched that of a fatty oil. He went even further in his proposal, suggesting that the layer was most likely a mixture of lecithins (phosphatidylcholines) and cholesterol, which has turned out to be essentially the case.

As described previously, many lipids are amphipathic, they have both polar and nonpolar components. During the 1900s, several chemists, most notably Irving Langmuir, were investigating the properties of lipids and the ways in which they could be handled. It was found that amphipathic lipids, such as phospholipids, would spontaneously form layers on top of hydrophilic surfaces. If an appropriate trough of water were prepared (a *Langmuir trough* as in Fig. 5-7a), phospholipid molecules could be made to orient themselves so that their hydrophilic group was associated with the surface of the water layer and their hydrophobic chains were directed up in the air. This layer could be compressed from the sides so that all of the lipid molecules in the monomolecular layer were parallel to one another and in register (Fig. 5-7a).

In 1925 two Dutch scientists, E. Gorter and F. Grendel, extracted lipids from red blood cells and used them to prepare monomolecular layers in a Langmuir trough. By measuring the surface area of the lipid monolayer formed in the trough and calculating the total surface area of the red blood cells from which that amount of lipid was extracted, they could determine the number of monolayers of lipids that were present on the cells. This assumes that all of the lipids were from the membrane, which in the case of red blood cells is a good approximation. The ratios of the surface areas in the trough to that calculated for a variety of different red blood cells were between 1.8 and 2.2. They concluded that the actual ratio was 2, i.e., that a lipid bilayer was present. They furthermore concluded that the polar groups of each molecular layer were directed to the outside (Fig. 5-7c), the cytoplasm on one edge and the blood plasma on the other. This would certainly be the thermodynamically favored structure. In this position, the polar hydrophilic groups of the lipids could

Figure 5-7. (a) A Langmuir trough for the preparation of molecular lipid layers. (b) The lipid monolayer is confined to the smallest possible area by means of movable barriers. [(a)–(b) From R. M. Dowben, "General Physiology," *Harper & Row, 1969.*] (c) A lipid bilayer with the hydrophilic head groups facing outward and the hydrophobic fatty acid chains facing inward.

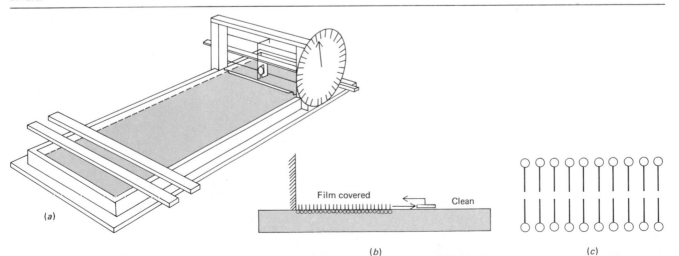

(a)

Film covered   Clean

(b)

(c)

associate with the aqueous environment. The nonpolar hydrocarbon chains would face inward, toward one another. In this orientation, the nonpolar groups of a lipid bilayer would be essentially removed from contacts with water and dissolved materials. The organization is similar to that of the globular proteins described in Chapter 3 in which the hydrophobic residues are internally disposed, being covered by a shell of hydrophilic amino acids capable of interaction with polar solvent molecules. Based on our present knowledge, Gorter and Grendel made several miscalculations, but regardless, the concept of the experiment was of great importance. The miscalculations compensated for one another such that the correct answer was obtained to the question that was asked, namely that natural membranes contain two monomolecular layers of lipid, i.e., a bilayer.

THE LIPID BILAYER

Before continuing with the historical development of our present concept of membrane structure we will briefly consider the significance of a lipid-containing membrane. The presence in the membrane of a bimolecular layer of amphipathic lipids first proposed by Gorter and Grendel has been verified by various physicochemical techniques and it has important consequences for the physiological properties of the membrane. One of the most revealing means to illustrate this is by a comparison of the properties of an *artificial* lipid bilayer with the properties of the cell membrane itself. An artificial bilayer is most readily prepared in the apparatus diagrammed in Fig. 5-8a. The lipid material to be used is picked up on the

Figure 5-8. The formation of an artificial lipid bilayer. (a) The lipid is applied across a hole in an appropriate "pot." (b) The lipid layer is initially thick but rapidly thins to form the bilayer. The electrodes in the pot and its container permit measurements of membrane conductance and capacitance. (Drawing by R. Ingle, from Models of Cell Membranes, by A. D. Bangham, Hospital Practice, Vol. 8, No. 3 and "Cell Membranes: Biochemistry, Cell Biology, & Pathology," G. Weissmann and R. Claiborne, Eds., HP Publishing Co., Inc., New York, NY 1975. Reprinted with permission.)

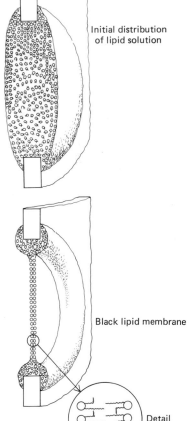

Cross-section of hole

Initial distribution of lipid solution

Black lipid membrane

Detail

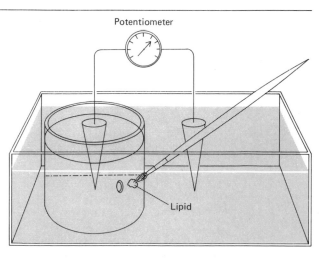

Potentiometer

Lipid

(a)

(b)

tip of a fine brush and is then applied to a small opening in a plastic sheet which is separating two aqueous chambers. The lipid forms a thin film across the opening which *spontaneously* thins to less than 100 Å (Fig. 5-8*b*). By numerous criteria, this thin film has been shown to be a lipid bilayer of the same organization as that in Fig. 5-7*c*. It appears that the bilayer is the thermodynamically preferred conformation. Table 5-4 shows a comparison between the properties of such a film and natural membranes, and it is apparent that the two are strikingly similar. In other words, much of the behavior of membranes appears to result from the presence of the lipid bilayer. Those properties not similar between the two, particularly the permeability properties and the electrical resistance, are derived from the more complex organization of the membrane as described below.

The artificial bilayer demonstrates the spontaneous nature by which lipid molecules can associate with one another to form cohesive layers of extreme molecular thinness. The situation is analogous to that of self-assembly of protein-containing structures, i.e., the potential for higher levels of organization is present within the components themselves. As in the case of proteins, this property has far-reaching structural consequences. If a structure has the capability for self-assembly, it greatly simplifies the activities of the cell in causing the structure to become organized. Whether the plasma membrane, with all of its complexity, is capable of spontaneous formation has not yet been determined. However, the ease with which many of the components can be complexed with one another in vitro to form what are essentially "reconstituted" membranes, suggests that to a large degree this may well be the case.

The capacity of lipids to spontaneously associate has physiological significance beyond the initial formation of the membrane. One conse-

TABLE 5-4
Physical Properties of Biological Membranes and Single Unmodified Lipid Bilayers

Property	Biological membranes	Bilayers
Thickness (Å):		
By electron microscopy	60–130	60–90
By x-ray diffraction	75–100	—
By optical method	—	40–80
From capacitance and dielectric constant	30–150	40–130
Resistivity (Ω cm^{-1})	10^2–10^5	10^6–10^8
Capacitance (μF cm^{-2})	0.5–1.3	0.3–1.3
Resting potential (mV)	10–90	0–140
Breakdown voltage (mV)	100	100–550
Refractive index	1.6	1.56–1.66
Interfacial tension (ergs cm^{-2})	0.03–3.0	0.2–6.0
Water permeability (10^{-4} cm sec^{-1})	0.25–58	2.3–24

SOURCE: F. Vandenheuvel, *Adv. Lipid Research*, **9**: 173 (1971).

quence of the lipid layer is that membranes are never seen to have a free edge, they are always continuous, unbroken structures. As a result, membranes form extensive, interconnected networks that ramify throughout the cell. If a membrane is ruptured, it spontaneously associates with itself or another membrane as a result of its lipoid core. If a puncture hole is made in a cell membrane, it spontaneously seals itself and the cell remains alive. The lipid bilayer facilitates the fusion or the splitting of membranes. For example, the events of secretion, in which cytoplasmic vesicles fuse to the plasma membrane, or of fertilization, where two cells fuse to form one, involve processes in which two separate membranes come together to become one continuous organelle (Fig. 7-18).

The spontaneous nature of lipid assembly has important consequences for theories of cellular evolution. A lipid-containing outer layer provides an excellent barrier for maintaining a living internal environment in a nonliving hostile world, and from our knowledge of lipids we would expect that formation of such barriers could have occurred spontaneously. Presumably the formation of such a barrier around the first primitive association of self-replicating molecules was an essential early step in allowing the next stages of evolution to occur in a much more protected condition.

THE DAVSON-DANIELLI MODEL OF MEMBRANE STRUCTURE

In the 1920s and 1930s, cell physiologists were obtaining evidence that there must be more to the structure of membranes than simply a lipid layer. Microsurgical experiments, particularly by Robert Chambers, were being carried out in which fine metal instruments were being used to puncture and tear the plasma membrane. Of particular concern were the permeability and surface tension properties of natural membranes that did not seem to be consistent with a simple lipid boundary. The permeabilities of natural membranes were selective, with the molecular size and charge being an important determining factor. Lipid solubility was not the sole determining factor as to whether a substance would penetrate the plasma membrane or be withheld. The movement of water through the membrane seemed to provide evidence in itself. Similarly, the surface tensions of membranes were believed (incorrectly) to be much lower than those of pure lipoid structures, and it was shown that the presence of a protein film over the lipid greatly lowered its surface tension.

In 1935 Hugh Davson and James Danielli proposed that the plasma membrane was a lipid layer upon which a film of globular proteins was adsorbed (Fig. 5-9a). The Davson-Danielli model was the first to attempt to explain, on a structural basis, the various physiological properties that membranes were believed to demonstrate. Over the years the model has been revised in various ways. In order to account for the permeability properties of the membrane, it was proposed that there were places where penetration of the lipid layers could occur. These openings were called "pores" and it was suggested that they were lined by protein molecules providing an internal hydrophilic surface through which polar solute molecules and ions could pass (Fig. 5-9b). Nonpolar solute would presumably

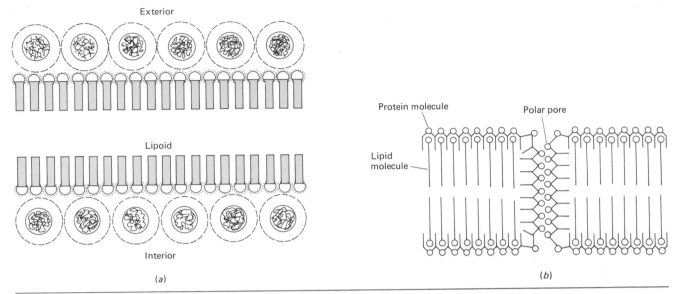

Figure 5-9. *(a)* The original Davson-Danielli model of the structure of the plasma membrane. [*From J. F. Danielli and H. Davson, J. Cell. Comp. Phys., **5**:498 (1935).*] *(b)* A revised version of the Davson-Danielli model showing the membrane containing a hydrophilic pore. [*From J. F. Danielli, Collston Papers, **7**:8 (1954).*]

dissolve in the lipid and gain entrance by directly crossing the lipid barrier. In all of these models, the lipid molecules were oriented perpendicularly to the surface of the cell.

The Davson-Danielli model, with various revisions, has served as the basis for our concept of membrane structure since the time of its proposal. It is basically a lamellar model, i.e., the components are present as layers or continuous sheets (lamellae). The sheets represent what are essentially two-dimensional planes of protein separated by a plane of lipid, a protein-lipid sandwich. During the subsequent twenty years evidence accumulated to support a lamellar model of lipid and protein. Particularly important were the techniques of polarization microscopy and x-ray diffraction which allow one to determine if there is an ordered arrangement of molecules within a given structure. These techniques, however, require specimens in which the ordered arrangement is repeated. Attention became focused on the myelin sheath which wraps the axons that comprise most nerves. Here was a lamellar organization of lipid and protein many layers thick that was amenable to examination by these physical techniques (Fig. 5-10*a*). The polarization microscopy showed myelin to possess strong positive intrinsic birefringence and weak negative form birefringence (page 54). This data was interpreted to signify that myelin was composed of layers of lipid with the molecules oriented perpendicular to the surface of the sheath alternating with layers of protein oriented parallel to the surface. The molecular dimensions were derived from x-ray diffraction data and later electron microscopy. In the 1950s it was demonstrated that the layers that make up the myelin sheath were actually layers of Schwann cell plasma membrane that were laid down around the nerve (Fig. 5-10*b*). These results provided further support for a lamellar model of the Davson-Danielli type.

THE CONCEPT OF THE UNIT MEMBRANE

By the late 1950s, resolution with the electron microscope was such that structures thinner than the plasma membrane could be resolved. The fixation of choice at the time, and remaining so today, was osmium tetroxide. Membranes of osmium-fixed tissue appeared as a thick, densely staining line. Where two cells adjoined one another, two dark lines running parallel to one another were seen (Fig. 5-11a). Presumably this represented a "double membrane" though various interpretations of its molecular organization were made.

The relationship between the electron microscopic image of membranes and their molecular structure was examined by various investigators, particularly Humberto Fernandez-Moran, J. David Robertson, and Fritiof Sjöstrand. Using a new fixative, potassium permanganate, Robertson found that each of the dense lines of Fig. 5-11b could be shown to possess a substructure having a characteristic trilaminar appearance (Fig.

Figure 5-10. *(a)* Electron micrograph of the myelin sheath showing the strict periodicity of the component layers. The basis for the alternating darker and lighter lines is indicated in the accompanying diagram. [*From E. T. Hedley-Whyte and D. A. Kirshner,* Brain Res., **113**:*493 (1976).*] *(b)* Diagram showing the mechanism of formation of peripheral nerve myelin. The Schwann cell gradually becomes wrapped around the axon and, with the accompanying removal of Schwann cell cytoplasm, the layers of plasma membrane are formed. As a result of this process the inner surfaces of plasma membrane contact one another (forming the darker periodic line), alternating with layers made up of plasma membrane whose outer surfaces have come together (forming the lighter periodic line). *(From J. D. Robertson in M. Locke, ed.,* "Cellular Membranes in Development," *Academic, 1964.)*

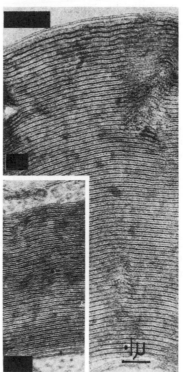

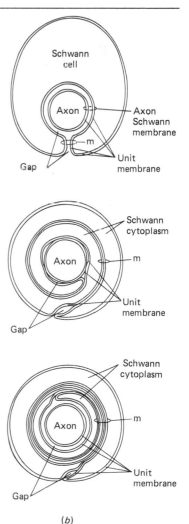

(a)

(b)

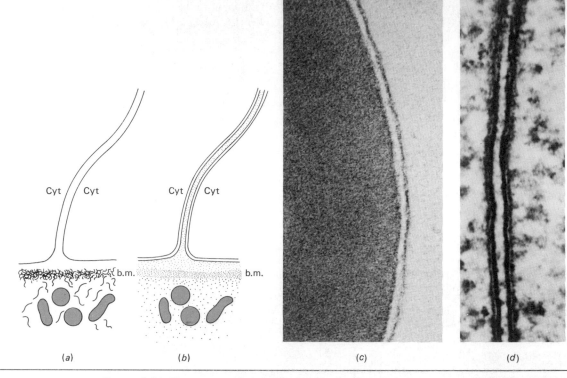

Figure 5-11. The concept of a unit membrane. *(a)–(b)* A diagrammatic comparison of the appearance of the plasma membranes of two adjacent epithelial cells using different fixation and staining procedures. In the older procedures, each membrane was seen as a single dense line *(a)*, while in the newer techniques each was seen as a triple-layered structure *(b)*. b.m., basement membrane; cyt., cytoplasm. [*From J. D. Robertson,* Prog. Biop. Biop. Cytol., **10**:*353 (1960).*] *(c)* Electron micrograph showing the trilaminar appearance of a single plasma membrane. *(From J. D. Robertson, in D. B. Tower, ed., "Basic Neurosciences," vol. I, Raven Press, p. 43, 1975.) (d)* Electron micrograph of two adjacent membranes, each showing the trilaminar nature as well as marked asymmetry between the inner and outer elements. [*From H. C. Aldrich and J. H. Gregg,* Exp. Cell Res., **81**:*407 (1973).*]

5-11*b*, *c*). The double membrane between two adjoining cells could be seen to be made of four dark lines with three separating light spaces (Fig. 5-11*d*). It appeared that each membrane was composed of two thin electron-dense lines of approximately 20 Å with an intervening space of approximately 35 Å. Even though both the inner and outer dark lines are typically in the order of 20 Å, there are often distinct differences seen in the electron microscope. In other words, the inner and outer layers are not identical; the membrane is asymmetrical with respect to its inner and outer halves. More will be said about the molecular basis of this asymmetry later in the chapter.

All membranes that were examined closely — whether they were plasma, nuclear, or cytoplasmic membranes; whether they were taken from plants, animals, or microorganisms — showed this same substructure. The appearance was not simply an artifact derived from the use of permanganate as a fixative since a similar structure was soon seen after fixation in osmium tetroxide. Robertson suggested the term "unit membrane" to describe this seemingly universal structural appearance.

The molecular interpretation of the "unit" membrane was essentially consistent with the Davson-Danielli model, which had been formulated without the aid of visual information. Based on what has been said, one would presume that the two dense lines corresponded to the inner and outer protein layers. As expected from this interpretation, extraction of the lipid from the membrane had little effect on its electron microscopic appearance; the protein layers which formed complexes with the heavy metal stains were visible in the image (Fig. 5-12a). It was also found, however, that when artificial bilayers without protein were fixed, stained, and examined with the electron microscope, they also showed the trilaminar appearance (Fig. 5-12b). This latter observation suggests that the hydrophilic ends of the lipid bilayer also stain with osmium and are rendered electron-dense in the microscope. These results suggest that, in membranes, the stain would reflect both the hydrophilic lipid head groups, all lined up in a row, and the membrane-associated protein.

Variations among Membranes

It is important to keep in mind that most of the models of membrane structure suggest that all membranes share a common structural organization. We have been discussing lamellar models; various other models

Figure 5-12. (a) Electron micrograph of a myelinated fiber in a nerve depleted of its lipid and stained with osmium tetroxide. [From L. Napolitano, F. LeBaron, and J. Scaletti, J. Cell Biol., **34**:820 (1967).] (b) Electron micrograph of an artificial lipid bilayer fixed with potassium permanganate and lanthanum nitrate showing the trilaminar appearance in the absence of protein. (Courtesy of G. L. Decker.)

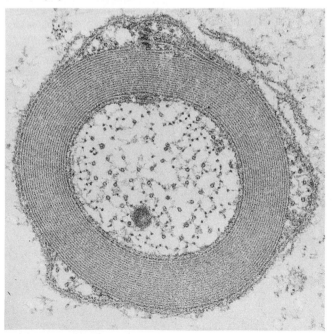

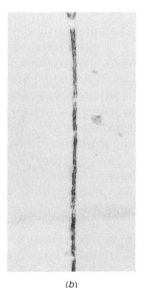

(a)   (b)

Figure 5-13. A model of membrane structure in which the lipid is organized into globular microspheres as opposed to a lipid bilayer. *(From J. D. Robertson, in A. Lima-de-Faria, ed., "Handbook of Molecular Cytology," Elsevier North-Holland, 1969.)*

have been proposed that are quite different. Other models are primarily based on the concept that membranes are composed of a linear array of globular subunits (Fig. 5-13) in which the lipids are organized into spherical micelles having their hydrophilic head groups pointing outward and their hydrophobic hydrocarbon chains pointing inward. Surrounding and encasing the lipid would be the protein, arranged in various configurations. Whether these subunit type models have any validity remains a controversial subject. Models of this type are based primarily on high-resolution electron microscopy and have not gained wide acceptance. Since some membranous organelles, such as mitochondria and chloroplasts, are the ones most often seen to be made of subunits, it may be that different membranes do show some basic differences in their organization based upon a common lipid-protein composition. Considering the tremendous differences in the functions of the various membranes, ranging from sites of energy capture in the mitochondria and chloroplasts, to electrical insulation in the myelin sheath, major differences in membranes might be expected. The question is whether the diffrences (to be described in the next paragraph) are compatible with a similar overall molecular organization. The assumption is generally made in membrane research that such is the case.

Differences among various cytoplasmic membranes are so great that they can be clearly seen with the electron microscope. There are marked differences in thickness and electron density. For example, the inner mitochondrial membrane is particularly thin (in the order of 50 Å), while the plasma membrane is relatively thick (often seen to be 90 to 100 Å). The protein/lipid ratios are extremely variable. The inner mitochondrial membrane has a very high ratio (Table 5-1) by comparison to the red blood cell plasma membrane, which is high in comparison to the membranes of the myelin sheath. To a large degree these differences can be correlated to the particular functions these membranes perform. The inner mitochondrial membrane contains the proteinaceous respiratory assemblies of the electron-transport chain. It is imperative that the proteins be in contact with one another to facilitate their coordinated activities; relative to other membranes, there is an exclusion of lipid. The myelin sheath is best described as an electrical insulation for the neurons it encloses, and this function is best carried out by a thick, unbroken lipid layer of high electrical resistance; relative to other membranes, there is an exclusion of protein.

Not only do the protein/lipid ratios of membranes vary, but the particular species of lipid and protein is characteristic for each membrane type. Among the lipids, variation can occur with respect to the nature of the phospholipid which predominates (the type of head group, the species of fatty acyl chains) and the degree to which cholesterol is present (Table 5-3). Many proteins are known to be associated with one membrane or another depending upon the catalytic activities of a given membrane. Proteins involved in communication with the environment are present within the plasma membrane, electron-transport proteins are found in the mitochondrial membranes, those involved in ribosome binding in the membranes of the endoplasmic reticulum, etc. In the following pages a number of these specific protein molecules will be considered.

THE FLUID MOSAIC MODEL

In the past few years a new concept of membrane structure has appeared, one that emphasizes the dynamic quality of membrane organization. This dynamic quality, which is difficult to capture in diagrammatic sketches, is embodied in the mobility and interaction among the various membrane components. The model that best describes this new concept is the "fluid mosaic" model (Fig. 5-14) formulated by S. J. Singer and Garth Nicolson in 1972. In certain aspects the model is based on those previously discussed. The presence of the lipid bilayer as the core of the membrane is retained, but the physical state of the lipid is emphasized and the idea of the bilayer as an unbroken lamella is rejected. The nature of the proteins and their organization within the membrane is radically different from the previous models. Rather than forming a continuous, unbroken bilayer *external* to the hydrophilic head groups of the phospholipids as in the Davson-Danielli model (Fig. 5-9), it is proposed that the proteins are present as *discontinuous* particles, many of which would penetrate deeply into, or even completely through, the lipid bilayer. In other words, the lipid bilayer is continuous in the sense that no region of lipids would be walled off from the remainder of the bilayer by a protein barrier; however, the integrity of the bilayer as an unbroken sheet would be interrupted by membrane proteins.

Membrane Proteins

In the fluid mosaic model, two categories of proteins are proposed to exist, being distinguishable by the intimacy of their relationship to the lipid bilayer. Those proteins that penetrate into the lipid are termed *integral* or *intrinsic* proteins. As might be expected, any molecule that is capable of an association with a hydrophobic environment, such as that

Figure 5-14. *(a)* The fluid mosaic model of the cell membrane as initially proposed. *(After S. J. Singer and G. L. Nicolson, copyright © 1972 by the American Association for the Advancement of Science.) (b)* A membrane "garden." This portrayal of an erythrocyte membrane focuses on the arrangement of the major proteins. The integral proteins are those which penetrate into or through the lipid bilayer while the peripheral proteins are those present outside the lipid itself, but still remaining associated with it or the integral proteins. As will be discussed later in the text, all of the proteins of the external leaflet appear to bear carbohydrate chains as indicated in the diagram. The numbers used to identify the different proteins correspond to those described in Fig. 5-37. [*From T. L. Steck*, J. Cell Biol., **62**:12 (1974).]

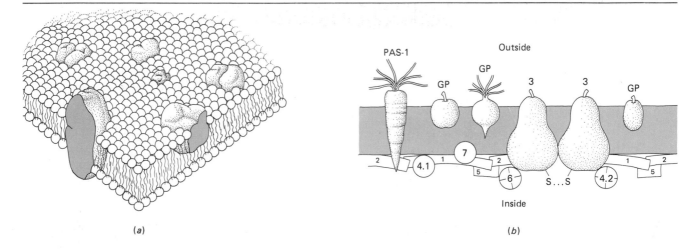

(a)   *(b)*

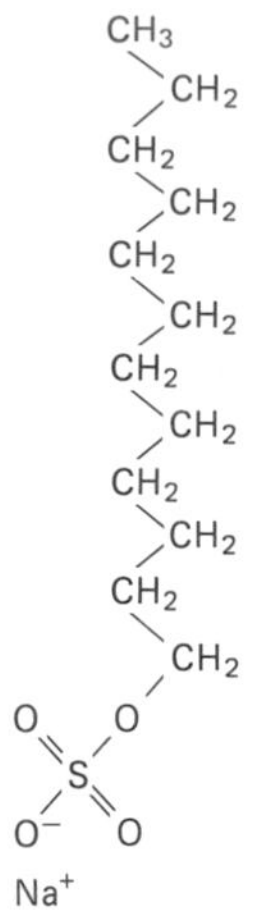

CH₃
CH₂
CH₂
CH₂
CH₂
CH₂
CH₂
CH₂
CH₂
CH₂
CH₂
CH₂

$O \quad O$
S
$O^- \quad O$
Na^+

Sodium dodecyl sulfate

Figure 5-15. The structure of sodium dodecyl sulfate (SDS).

found within the confines of the hydrocarbon chains of the fatty acids, would have to be hydrophobic in nature itself. As described in Chapter 1, hydrophobic groups tend to associate with one another by weak *noncovalent* forces with the exclusion of water and other hydrophilic substances. Numerous integral proteins have been examined and, as expected, they are characterized by having an extensive hydrophobic composition. As a result of their hydrophobic nature, integral proteins are difficult to extract and difficult to study. Removal of these proteins from the membrane requires harsh extraction procedures, usually accomplished with the aid of a detergent. Detergents, like phospholipids, are amphipathic (Fig. 5-15), being composed of a polar end and a nonpolar hydrocarbon chain. As a consequence of their structure, detergents can substitute for phospholipids in stabilizing integral proteins while rendering them soluble in aqueous solutions. Once the proteins have been solubilized by the detergent, various analytical procedures can be carried out to determine their amino acid composition (or even sequence in a couple of cases), their molecular weight, and so on. In some cases (with nonionic detergents) the protein retains its enzymatic activity while in solution.

In addition to having a capacity to form hydrophobic associations with lipids, integral proteins also possess hydrophilic portions which allow them to protrude beyond the outer surface of the lipid bilayer (Fig. 5-14). In other words, integral proteins are also amphipathic membrane components, in keeping with the amphipathic nature of the entire membrane. As a consequence, interactions with water-soluble substances (ions, small molecular weight substrates, hormones, etc.) can occur at the membrane surface while the barrier properties afforded by the continuous lipid phase can also be maintained. The hydrophobic portions of the proteins allow a means for the attachment of the proteins into the structural "wall" of the membrane much like hooks can be inserted into the holes of a pegboard. As will be discussed below, in the case of the lipoid wall of the membrane, the integral proteins are allowed an added advantage, the ability to move around within the membrane itself.

In addition to the integral proteins, there exists a class of membrane proteins that are not so tightly involved in the membrane's architecture. These are the *peripheral* or *extrinsic* proteins (Fig. 5-14b) which are capable of being solubilized by simple aqueous salt solutions and are associated with the membrane by weak bonds to either the hydrophilic head groups of the lipid or, more likely, the hydrophilic portions of the integral proteins protruding from the bilayer. Though the peripheral proteins exist entirely outside of the lipid bilayer, they do not appear to cover it uniformly, as proposed in the earlier models of membrane structure. One of the best pieces of evidence that these proteins are a discontinuous layer is the ease with which the lipids of the bilayer can be digested by the addition of phospholipid-degrading enzymes (phospholipases). When these enzymes are added to living cells, they seem to have ready access to the membrane lipid, which would not be possible were it covered by an unbroken protein layer.

Different membranes are characterized by having a different population of proteins, allowing each type of membrane to carry out the necessary functions. There are many aspects of the fluid mosaic model to be
discussed and numerous experiments to be described which reveal a whole
new panorama of membrane properties. Aspects of the membrane will be
considered piece-by-piece so that the importance of each in the overall
picture can be appreciated and the experimental approaches taken to
gather the evidence can be understood. In the following sections we will
first consider the properties of the lipid bilayer and then the nature and
properties of a few of the membrane proteins that have been examined.
However, before turning to this discussion, it is important to consider a
technique which allows one to determine (1) the distribution of the proteins
within a given region of the membrane and (2) the extent to which these
proteins penetrate deeply into the lipid bilayer.

Freeze-Fracture and Freeze-Etch Analysis

The basic procedure utilized in the preparation of freeze-fractured
replicas was described on page 65. When cells are frozen and fractured,
one of the favored paths for the fracture plane to take is that between the
two monomolecular layers of the lipid bilayer. When membranes are split
in this manner and their exposed surfaces shadowed as in Fig. 2-38, the
appearance of the replica is one of a road strewn with large pebbles, termed
membrane-associated particles (Fig. 5-16a). The typical human erythrocyte contains about 500,000 of these particles, each approximately 80Å in
diameter. Since the cells prepared for freeze-fracture are typically frozen
directly and rapidly from the living state without the use of fixatives, the
presence of these particles presumably reflect the existence of discon-
tinuities within the center of the membrane itself. Since the fracture plane
passes through the center of the bilayer (Fig. 5-16b), these particles represent membrane components that extend at least halfway through the lipid
core. When the fracture plane reaches a given particle, it apparently goes
around it (Fig. 5-16b) rather than cracking it in half. In most cases, the
particle separates with the inner (cytoplasmic) half of the membrane,
leaving a corresponding "pit" in the outer half. The face shown in Fig.
5-16 is the inner (also referred to as the A, P, or $\widehat{F}$) face. The outer face
(not shown) is termed the B, E, or $\underline{F}$ face. There is little doubt that the
intramembrane particles of the replica represent the integral proteins
that extend through or nearly through the membrane. This has been confirmed in several ways. For example, when artificial bilayers are prepared
and fractured, no particles are seen in the replicas. However, if these
bilayers are mixed with a preparation of one of these membrane proteins,
it becomes embedded in the bilayer, and particles appear in the replicas.

Not only does the freeze-fracture technique allow one to actually visualize membrane proteins that project deeply into the hydrophobic core
of the membrane, but it can also be modified to reveal those proteins or
parts of proteins that are present on either the internal or external surface.
In order to examine the surfaces of the membrane an additional step in
the procedure must be taken before the fractured surfaces are shadowed

(a)

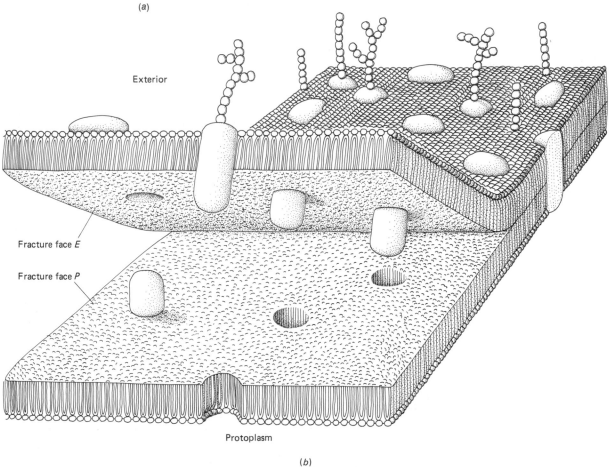

Exterior

Fracture face *E*

Fracture face *P*

Protoplasm

(b)

Figure 5-16. *(a)* Electron micrograph of a freeze-fracture replica of the cytoplasmic or P face, showing the large number of randomly-distributed membrane-associated particles. *(Courtesy of R. L. Wood.)* *(b)* Schematic illustration of the manner in which the fracture plane bisects the membrane. *(From "The Final Steps in Secretion" by B. Satir. Copyright © Oct. 1975 by Scientific American, Inc. All rights reserved.)*

(Figs. 2-38*a* and 5-17*a*). The frozen, fractured specimen, while still in place in the apparatus, is exposed to a vacuum for a brief period (e.g., one minute) during which a thin layer of water can evaporate (sublime) from above and below the surfaces of the membrane. When these surfaces are shadowed, their texture can be seen. This sublimation procedure is called *freeze-etching*. A comparison between a replica prepared with or without the etching step is shown in Fig. 5-17*b-c*.

MEMBRANE LIPIDS AND FLUIDITY

The physical state of the lipid of a membrane is best described by its fluidity, i.e., its viscosity. As with other substances, lipids can exist in a solid phase or in a liquid phase of varying viscosity, depending upon the temperature. Consider, for example, an artificial bilayer made of phosphatidylcholine (lecithin) in which the fatty acids attached to the glycerol backbone were myristic acid. If the temperature of the bilayer is held

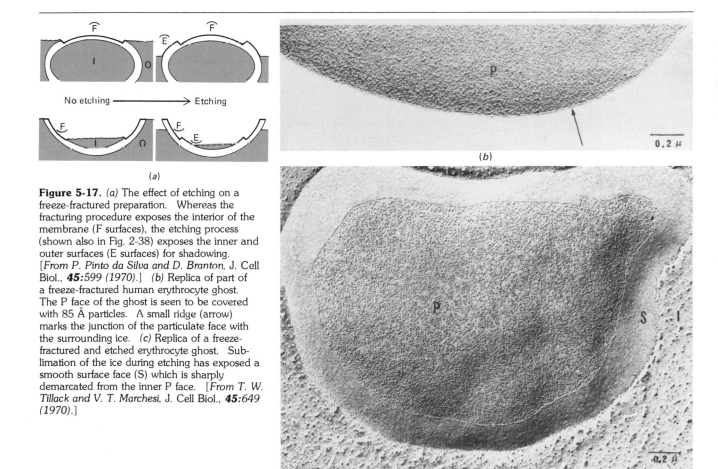

Figure 5-17. *(a)* The effect of etching on a freeze-fractured preparation. Whereas the fracturing procedure exposes the interior of the membrane (F surfaces), the etching process (shown also in Fig. 2-38) exposes the inner and outer surfaces (E surfaces) for shadowing. [*From P. Pinto da Silva and D. Branton,* J. Cell Biol., **45**:599 *(1970).*] *(b)* Replica of part of a freeze-fractured human erythrocyte ghost. The P face of the ghost is seen to be covered with 85 Å particles. A small ridge (arrow) marks the junction of the particulate face with the surrounding ice. *(c)* Replica of a freeze-fractured and etched erythrocyte ghost. Sublimation of the ice during etching has exposed a smooth surface face (S) which is sharply demarcated from the inner P face. [*From T. W. Tillack and V. T. Marchesi,* J. Cell Biol., **45**:649 *(1970).*]

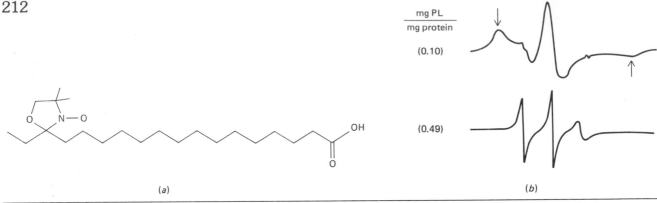

(a)

(b)

Figure 5-18. *(a)* The spin label, 16-doxylstearic acid whose unpaired electron located in the N—O group provides an electron spin resonance (esr) signal from the membrane. *(b)* The esr spectrum of 16-doxylstearic acid diffused into a model membrane made of protein-lipid of varying proportion. The spectrum is quite different when the lipid of the bilayer is immobile as in the upper trace as opposed to being fluid as in the lower trace. [*From O. H. Griffith, P. Jost, R. A. Capaldi, and G. Vanderkooi,* Annals N.Y. Acad. Sci., **222:**561 *(1973).*]

below about 25°C, the lipid will be in a gelated condition, i.e., it can be considered to be in its solid state. If the temperature is slowly raised, a point will be reached where a distinct change in the nature of the bilayer occurs. The lipid has undergone a melting process; it has been converted from a crystalline gel to a much more fluid liquid state. Above the melting temperature, termed the *transition temperature*, the lipid is best described as a liquid crystal. As in a crystal, the molecules still retain a specified orientation such that the long axes of the molecules continue to be parallel to one another.

The change in state that occurs at the transition temperature can be considered a shift from an ordered state to a much more disordered one. The increased disorder above the transition temperature is best illustrated by the increased lateral mobility that the lipids of the liquid crystal now possess. In other words, individual phospholipids are no longer frozen in place but are capable of movement across one another within the plane of the membrane. There are various physicochemical techniques in which the relative mobility within the bilayer can be measured. One of the most commonly employed techniques is electron spin resonance. In this technique a chemical group is introduced into the bilayer so that its activity can be followed when placed in an electric field. In the case of phospholipids, the chemical group is generally a nitroxide, which is a radical containing an unpaired electron that can be covalently attached to the fatty acid chains of the phospholipid (Fig. 5-18a). Once this "reporter" group is in place, its movement and orientation can be followed spectroscopically when placed in a magnetic field. The spectrum of an immobilized phospholipid is distinguishable from one that is capable of movement (Fig. 5-18b). The two traces shown in Fig. 5-18 reflect the differences in phospholipid mobility one would expect on either side of the transition temperature.

These same techniques have been applied to natural membranes as well as artificial bilayers. Since the fluidity of a membrane is related to the mobility of its phospholipid components, this and other spectroscopic methods provide a sensitive measure of this important membrane property. It is estimated, for example, that a given phospholipid molecule can

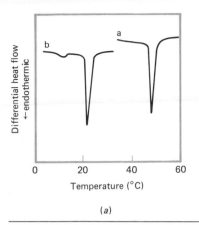

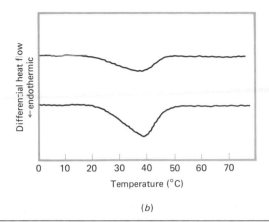

(a)   (b)

Figure 5-19. Differential scanning calorimetry. *(a)* Scans of two different phospholipids, each showing a very narrow transition temperature. *(b)* Scans of a membrane (upper) and extracted membrane lipids (lower). The melting occurs over a broad temperature range. *(Reproduced, with permission, from D. L. Melchior and J. M. Steim,* Annual Review of Biophysics, *Volume 5.* © *1976 by Annual Reviews Inc.)*

migrate from one end of a bacterium to the other in a period of one to a few seconds.

The transition temperature itself is best measured calorimetrically. The lipid-containing specimen is placed in a calorimeter, i.e., a sealed chamber in which heat generated or absorbed can be measured as the temperature is raised. As in the case of any substance undergoing a transition from a solid to a liquid state, heat is absorbed by the system in order to break the restraints, and this heat can be measured (Fig. 5-19a) by very sensitive instruments. The transition temperature of a particular bilayer is dependent upon the nature of the lipid. The most important determining factor of transition temperature is the degree to which the fatty acid chains of the phospholipid are unsaturated, i.e., contain double bonds. Transition temperature and fluidity are determined by the ability of the molecules to be packed together. Saturated fatty acids can be packed together more tightly than those with unsaturated ones, since the former are essentially straight in shape while the latter have crooks in the chain where the double bond(s) is located (Fig. 5-20). The greater the degree of unsaturation of the fatty acids of the bilayer, the *lower* will be the transition temperature (Table 5-5) and the greater the fluidity above that temperature. In other words, phospholipids of lower transition temperature will remain more fluid as the temperature is dropped and will require a lower temperature to become frozen in comparison to lipids with a higher transition temperature. In addition to unsaturation, fatty acid chain length will also, though to lesser degree, affect the temperature at which melting

TABLE 5-5
Melting Points of the Common 18-carbon Fatty Acids

Fatty acid	Double bonds	M.p., °C
Stearic acid	0	70°
Oleic acid	1	14°
Linoleic acid	2	−5°
Linolenic acid	3	−11°

SOURCE: A. White et. al., "Principles of Biochemistry," 5th ed., McGraw-Hill, 1973.

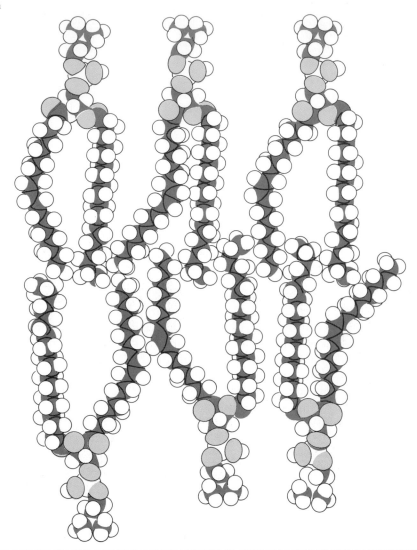

Figure 5-20. A space-filling model of a section of a highly fluid phospholipid bilayer membrane. *(From BIOCHEMISTRY by Lubert Stryer. W. H. Freeman and Company. Copyright © 1975.)*

occurs. The shorter the fatty acid chains, the lower the transition temperature. The range over which various lipids undergo their phase change is very great. For example, various phosphatidylcholines can be synthesized whose transition temperatures in an aqueous medium can range from below 0°C to temperatures in excess of 60°C.

Natural membranes contain a complex mixture of various lipids with a wide range of different fatty acids. However, preparations of natural membranes can be subjected to the same physical scrutiny as artificial bilayers, once methods to obtain a relatively pure preparation of the desired membrane have been worked out. When transition temperatures of membranes are determined, it is found that melting occurs over a rather broad range (Fig. 5-19b). Those phospholipids having lower transition temperatures melt first while those with higher transition temperatures await a further input of thermal energy before assuming the more mobile state of the liquid crystal. In other words, when complex mixtures of lipids are present, there is a range of temperatures over which different lipids may be in different phases. A condition such as this is termed a *phase separation:* some lipids are in the frozen state, others in the liquid state. A similar phase separation occurs when water freezes out of a salt solution, leaving a more concentrated solution of lower freezing point behind. The consequence of a phase separation in a membrane is that domains (regions) having different properties will exist. As lipids melt out of the frozen state they will leave behind parts of the membrane in solid patches in which lipids of higher transition temperature can be found. Within the liquid parts of the membrane, there will exist lipids of varying mobility, as well as a preponderance of membrane protein.

Another important ingredient in this question of the physical state of the membrane is the concentration and location of cholesterol molecules. Cholesterol interacts with the fatty acid chains of the phospholipids in such a manner as to compromise their fluid properties; it puts the membrane into an "intermediate" fluid state (Fig. 5-21). Cholesterol acts to abolish transition temperatures and cause phospholipids of higher transition temperature to become more mobile and phospholipids of lower transition temperature to become less mobile. Its general effect in natural membranes at physiological temperatures is to increase the viscosity and restrict the mobility of the phospholipids. Some membranes are much higher in cholesterol levels than others (Table 5-1) which must be taken into account when considering the physical state of a particular membrane.

The picture that is emerging in this discussion is that natural membranes are very complex in their organization. Complexity allows for differentiation, i.e., some parts of a membrane may be different from others. If we are examining an artificial bilayer of one phospholipid (not that more complex ones cannot be formed), we would expect a homogeneity throughout the bilayer and an ability to predict its nature across the entire structure. In a complex natural membrane the opportunity for a great deal of heterogeneity exists, such that one site in the membrane will have special properties reflecting the phospholipid composition of that region, while another site may have different properties.

It has been demonstrated that many of the integral membrane proteins, as a result of their close association with lipid, actually are surrounded by a layer of immobilized lipid molecules (Fig. 5-22). It has been suggested, and some evidence has been obtained, that specific lipids are required by specific proteins for their activity. If this were the case it would clearly

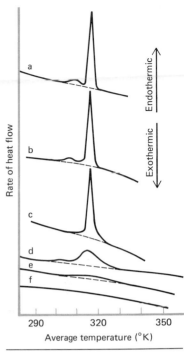

Figure 5-21. Differential scanning calorimetry curves of model membranes containing different proportions of phosphatidylcholine and cholesterol. The mole percent cholesterol in the preparations is *(a)* 0.0% *(b)* 5.0% *(c)* 12.5% *(d)* 20.0% *(e)* 32.0% *(f)* 50.0%. The transition from the gel (left side of the peak) to the liquid crystal (right side of peak) becomes blurred as the cholesterol content increases. [*From B. D. Ladbrooke, R. M. Williams, and D. Chapman,* Biochim. Biophys. Acta, *150:335 (1968).*]

Figure 5-22. Diagrammatic representation of a single protein complex and associated layer of boundary phospholipids. [*From D. C. Jost, O. H. Griffith, R. Capaldi, and G. Vanderkooi,* Proc. Natl. Acad. Sci. (U.S.), *70:483 (1973).*]

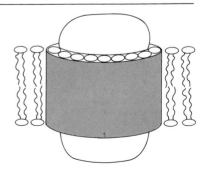

illustrate the importance of microheterogeneity of the membrane. Proteins would be associated with various lipids in various regions of the membrane. The complexity, flexibility, and need for regulation in membrane structure would be very great. The unfortunate aspect of membrane heterogeneity lies in the difficulty posed in studying it. Most biological techniques are best suited to examine the overall aspects of a specimen, i.e., its bulk properties. When the need arises to dissect sheets of 50 to 100 Å thickness into even smaller structures, many difficulties must be overcome.

The Movement of Components within the Membrane

What role does the physical state of the lipid bilayer have for the biological properties of the membrane? Most membrane biologists feel that it is of great importance, but the manner and degree of this importance has become a controversial issue. Membrane fluidity seems to provide the perfect compromise between a very rigid, ordered structure in which mobility would be lacking, and a completely fluid, nonviscous liquid in which the molecules would not be oriented and the opportunity for organization and mechanical support would be lacking. Most importantly, fluidity allows for interactions within the plane of the membrane. The interactions can be lipid-lipid, lipid-protein, or protein-protein in nature. In any case, the interacting molecules can come together, carry out the necessary reaction, and either remain together or move apart depending upon the conditions. For example, in the beginning of this chapter the role of membranes in information transfer was introduced. There are agents, such as hormones, which combine with the membrane by external receptors and cause a change in activity of an enzyme, adenylate cyclase (page 245), present on the inner side. The evidence suggests that the receptor and this enzyme are not always linked to one another, but that they interact as a result of their movements within the membrane (Fig. 5-23). The interaction occurs only when the receptor and hormone are combined. There are several different hormones that affect adenylate cyclase activity, some causing its stimulation and others its inhibition. Any one cell may have more than one receptor, and, at different times, one type of receptor or another may be interacting with adenylate cyclase

Figure 5-23. Hypothetical scheme to illustrate how interactions among different membrane proteins might occur. In this case two different hormone receptor proteins, epinephrine receptor (ER) and glucagon receptor (GR), are capable of stimulating a peripheral protein, adenylate cyclase (AC) to produce cyclic AMP. In the absence of hormone, the individual components are not associated. However, in the presence of glucagon, the GR and AC molecules interact. Had epinephrine been present, the AC would have associated with the epinephrine-bound ER component.

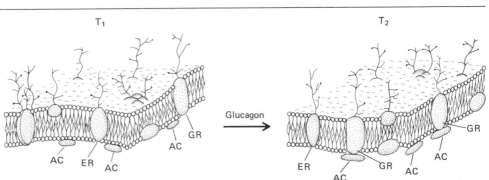

to modulate its activity. The fluidity of the membrane is important because it allows movements of integral proteins which facilitate these events.

Fluidity is also believed to be important in allowing the flow of materials to occur within membranes and possibly the flow of the entire membrane itself. In this way, materials can move from their point of insertion to other locations in the cell. It is known, for example, that the proteins of the plasma membrane are synthesized in the endoplasmic reticulum within the cell. It is believed that they remain within a membrane from the time of their formation in the endoplasmic reticulum to the time when they are found in the plasma membrane projecting to the outside of the cell. These proteins are believed to move with the membrane from place to place within the cell. In some places this process may resemble the moving belt of an assembly line in that different modifications are made to the membrane or its proteins as they move along.

The Action of Anesthetics

One of the most active current research areas in membrane biology concerns the mechanism of action of anesthetics. It has been known for a long time, based originally on the previously discussed studies of Overton, that nonspecific anesthetics, such as ethanol, were active in concentrations reflecting their membrane solubility. The greater the solubility of the substance in the lipid of the membrane, the more effective were its anesthetic properties at a given concentration (Table 5-6). Among other effects, one of the prime consequences of certain nonspecific anesthetics is that their presence in the membrane causes an expansion of the membrane (a decrease in its density) resulting in a higher degree of disorder and greater fluidity (Fig. 5-24). One of the consequences of this increased fluidity can be seen in the altered permeability of the membrane. The cells become leaky, substances diffuse out.

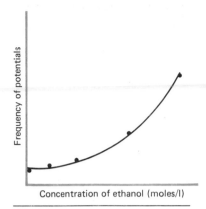

Figure 5-24. Increased fluidity of membranes caused by nonspecific anesthetics appears to facilitate release of neurotransmitter causing a potential at the synapse. With increased ethanol concentration, which is known to increase membrane fluidity, there is a corresponding rise in the frequency of such potentials. *(After P. Seeman, Neurotropic Drugs, in Hospital Practice, Vol. 9, No. 9 and "Cell Membranes: Biochemistry, Cell Biology, and Pathology," G. Weissmann and R. Claiborne, Eds., HP Publishing Co., Inc., New York, NY 1975. Reprinted with permission.)*

TABLE 5-6
Lipid Solubility and Membrane Permeability

	Narcotic conc. for tadpoles (mol/L water)	Distrib. coeff. olive oil/ water
Ethyl alcohol	0.4	0.03
Propyl alcohol	0.11	0.13
Butyl alcohol	0.13	0.18
Ethylurethane	0.04	0.03
Valeramide	0.05	0.07
Ether	0.024	2.4
Benzamide	0.005	0.44
Salicylamide	0.002	14
Carbondisulfide	0.0005	50
Phenylurethane	0.0006	150
Menthol	0.0001	250
Thymol	0.000055	600

SOURCE: R. Höber, "Physical Chemistry of Cells and Tissues," The Blakiston Co., 1945.

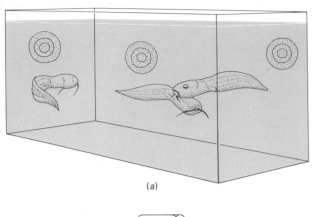

(a)

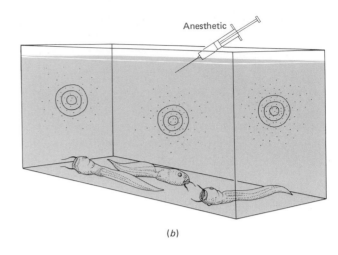

Anesthetic

(b)

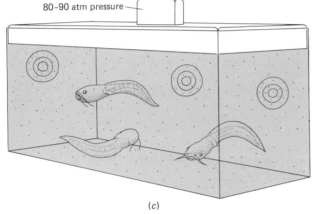

80-90 atm pressure

(c)

Figure 5-25. An experiment relating membrane integrity to the mechanism of anesthesia by ether. The addition of ether to the tank renders the tadpoles unconscious, while the application of pressure reawakens the animals. The concentric structures at the top of the tank represent enclosed membranous vesicles (liposomes) which are used in the experiment to monitor the effect of the anesthetic and the pressure on bilayer fluidity. The dots represent radioactive material originally enclosed within the liposomes (a). The addition of ether (b) disorders the liposome bilayers, releasing radioactive label to the medium. The application of pressure awakens the tadpoles (c) and restores order to the liposomes, thereby blocking the further release of label. (*Drawing by R. Ingle, from Models of Cell Membranes, by A. D. Bangham, Hospital Practice, Vol. 8, No. 3 and "Cell Membranes: Biochemistry, Cell Biology & Pathology," G. Weissmann and R. Claiborne, Eds., HP Publishing Co., Inc., New York, NY 1975. Reprinted with permission.*)

An experiment which illustrates the apparent relationship between increased fluidity and anesthetic activity is illustrated in Fig. 5-25. In this experiment, tadpoles were anesthetized by the introduction of chloroform or ether into the water in which they were swimming. It was hypothesized that if membranes of these animals could be subjected to conditions which increased the order of their lipid bilayers, it might be possible to reverse the effects of the anesthetic. When these unconscious tadpoles were subjected to hydrostatic pressures of approximately 80 to 90 atmospheres, sufficient order appeared to be restored to the membrane to cause the animals to regain consciousness despite the continued presence of the anesthetic.

Alterations in Membrane Fatty Acids

In a previous section the relationship between fluidity and the degree to which the fatty acid chains of the membrane were unsaturated was described. On this basis it would appear that membrane fluidity might be under metabolic control since the nature of the fatty acids being synthesized and/or incorporated into phospholipids may be critical in determining this membrane property. If fluidity were an important physiological parameter, it might be expected that when an organism were placed under conditions where its membrane fluidity were altered, it would respond by adjusting its activities to cause its fluidity to return to an appropriate value. Since fluidity is determined by temperature, it might be expected that alterations in the environmental temperature might trigger an alteration in the nature of its fatty acid metabolism. In the case of organisms as diverse as bacteria and cold-blooded vertebrates, it has been found that alterations in the temperature in which an organism is living is followed by adjustments of the lipid composition of their membranes so as to maintain a consistent fluidity. As the temperature is lowered, fatty acids of greater unsaturation are incorporated into membrane phospholipids to combat the effect of decreasing fluidity that accompanies a drop in temperature. There seems to exist within cells a homeostatic control system which is sensitive to a changing membrane fluidity.

Another means of studying the importance of membrane fluidity has become available with the development of techniques whereby the fatty acid composition of membranes can be manipulated. These techniques allow investigators to supply cells with a particular fatty acid which the cell will incorporate into its phospholipids. Under the appropriate conditions, the treated cells will construct membranes using these supplied materials and, therefore, these membranes will have a lipid composition dictated to a large degree by the investigator (Table 5-7). Since fatty acids of diverse melting temperatures can be provided to the cells, the effects of fluidity on various phenomena can be assessed. Various means have been employed to obtain membranes with the desired phospholipid composition. In some cases (e.g., *Acholeplasma*) the particular fatty acid can simply be added to the medium. In other cases (e.g., *E. coli*) a mutant strain of bacteria must be used that has defects in its own fatty acid metabolism such that it requires these fatty acid supplements in order to grow. Another means to "pack" the membranes of living cells is to cause the cells to pick up phospholipids from artificially prepared vesicles of the desired phospholipid composition. Most of the re-

TABLE 5-7
Fatty Acid Composition (%) of Phosphatidylethanolamines from *E. coli* Mutant Grown with Various Fatty Acids

	Fatty acid supplemented			
	cis-18:1	*trans*-18:1	19:0	18:3
16:0	22	6	27	50
14:0	6	11	17	3
12:0	...	2	1	...
cis-18:1	58	...	...	...
19:0	14	...	54	...
trans-18:1	...	81	...	...
18:3	...	...	...	46
Unidentified	...	<1	2	1
$\dfrac{16:0}{14:0}$	3.7	0.55	1.6	17
Ratio° of fatty acids	0.39	0.23	0.83	1.2

° Sum of the saturated fatty acids divided by the sum of the unsaturated and the cyclopropane derivatives. Phosphatidylethanolamine was isolated from cells grown to late exponential phase.
SOURCE: P. Overath et al., *Proc. Natl. Acad. Sci. (U.S.)*, **67**: 608 (1972).

search using these types of techniques has been carried out on microorganisms, but they have now been adapted for use with mammalian cells growing in culture and they promise to lead to important studies probing membrane function in the cells of higher organisms.

In one type of experiment attempts have been made to demonstrate that specific membrane functions are either sensitive or insensitive to the physical state of the membrane. In these experiments *E. coli* mutants with defective fatty acid metabolism were used, and the activity of specific proteins at various temperatures was determined when different fatty acids were supplied. Figure 5-26*a* shows the Arrhenius plots for transport of a sugar (a β-glucoside) when either elaidic or linoleic acid is predominant in the membrane. Figure 5-26*b* shows the Arrhenius plots for an enzyme when one of four different fatty acids has been supplied to the cells. Arrhenius plots describe the relationship between protein activity and temperature. Each of the curves of Fig. 5-26 is biphasic, i.e., there is a sharp change of slope in the rate of transport or enzyme activity as the temperature drops below a certain point. In each case, the breaking point of the curve can be correlated with the onset of the lipid phase separation of the particular fatty acid-containing phospholipid. These results suggest that the physical state of the lipids in the membrane are directly affecting the ability of a membrane protein to perform its function. As the temperature at which the bulk of the membrane becomes gelated, the activity of the protein falls off sharply. Studies with mammalian cells have also revealed the importance of lipid composition in membrane function. For example, in one study on cultured hamster cells it was shown that the presence of a high proportion of long-chain unsaturated fatty acids in the plasma membrane decreases the cell's adhesion to its substrate.

Experiments such as those described above indicate that membrane fluidity or the lack of it *can* have an effect on membrane function, but they do not prove that such is the case under normal physiological conditions where

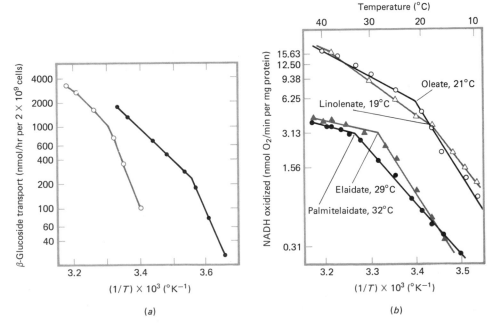

Figure 5-26. Arrhenius plots showing the relationship between membrane protein activity, the phospholipid content of the membrane, and temperature. The observed change in the slope of each plot can be correlated with the transition temperature of the particular fatty acid used as a supplement. The change in slope is believed to reflect a change in the state of the membrane which then affects the activity of membrane proteins. (*a*) The temperature dependence of glucoside transport (open circles correspond to a linoleic supplement, closed circles to elaidic supplement). [*From G. Wilson and C. F. Fox, J. Mol. Biol.,* **55:**49 *(1971).*] (*b*) The temperature dependence of NADH oxidase activities. [*From C. George-Nascimento, Z. E. Zehner, and S. J. Wakil, J. Supramol. Struc.,* **2:**646 *(1974).*]

neither fatty acid content nor environmental temperature is being manipulated. The question of importance in this connection is the degree to which frozen domains exist within membranes under physiological conditions. Studies using physical probes (e.g., electron spin resonance) suggest that phase separations do normally occur in membranes though their actual importance remains unclear. The question of fluidity will reappear below when we consider the types of movements that proteins make within natural membranes.

The discussion up to now has concerned the overall (bulk) properties of the membrane lipids. There is some evidence that different types of phospholipids may perform different functions within the membrane. It was mentioned above that membranes are capable of fusion with one another so that separate membranes can come together and become continuous structures. It appears that the ability of membranes to fuse with one another is related to their fluidity and lipid composition. One phospholipid, lysolecithin, is "fusogenic," i.e., its presence in membranes introduces an instability which promotes their fusion. Lysolecithin, a lecithin (phosphatidylcholine) molecule with one of the fatty acyl chains removed, is a naturally occurring phospholipid and the enzyme which forms it from lecithin (phospholipase A) is membrane bound. Proving that lysolecithin is actually responsible for membrane fusion within the cell is very difficult and provides an example of the problems that membrane microheterogeneity pose. It has not been established that lysolecithin is actually concentrated into tiny regions of the membrane the instant they are undergoing fusion.

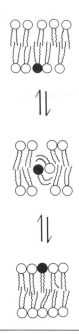

Figure 5-27. Schematic representation of one conceivable mechanism of phospholipid flip-flop. [*Reprinted with permission from R. D. Kornberg and H. M. McConnell,* Biochemistry, **10:**1119 (1971). Copyright by the American Chemical Society.]

The Asymmetry of Membrane Lipids

If different phospholipids have different properties, a question of importance is to what extent does the interchange of phospholipids occur *across* the bilayer? In other words, do lipid molecules in one of the monomolecular layers (leaflets) of the bilayer have the opportunity to move across to the other layer (Fig. 5-27)? Based on studies using various physical probes, it appears that there is a very *limited* opportunity for this type of "flip-flop" to occur. For example, measurements of the half-life of a molecule staying within one layer, as opposed to moving across to the other layer, are measured in hours. The evidence strongly indicates that flip-flop is the most restricted of all the possible motions that a phospholipid can make. This finding is not surprising since in order for flip-flop to occur, the hydrophilic head group of the lipid would have to pass through the internal hydrophobic sheet of the membrane, which makes it a thermodynamically unfavorable event. From these results it follows that the lipid bilayer can be thought of as being composed of two more-or-less stable, independent monolayers. If each monolayer should have a different phospholipid composition, we would expect that each would have different properties.

Experiments with cells indicate that there is a great degree of asymmetry with respect to the lipids in the inner and outer leaflets of the plasma membrane. For example, in red blood cells there is a greatly elevated concentration of phosphatidylcholine and sphingomyelin in the outer layer and phosphatidylethanolamine and phosphatidylserine in the inner layer. One of these phospholipids, phosphatidylserine, has a net negative charge (Fig. 5-2a) which makes it a candidate for binding to positively charged substances, e.g., Ca²⁺. All glycolipids are in the outer leaflet with the carbohydrate portion projecting out of the bilayer into the extracellular space. Another important difference between the two layers is the degree to which their phospholipids are unsaturated. The inner leaflet contains a greater percentage of unsaturated fatty acids and is, therefore, probably more fluid, though the presence of cholesterol has an effect on fluidity which must be taken into account. The importance of lipid asymmetry within the plasma membrane is not yet understood.

MEMBRANE MOBILITY

In the previous section the fluid, mobile nature of the lipid bilayer was discussed. Since it is the lipid that provides the matrix in which the integral proteins of the membrane are embedded, the physical state of the lipid will be an important determinant of the mobility of the proteins. The actual demonstration that the proteins of the membrane are in a dynamic state within the plane of the membrane has been one of the central pieces of evidence in favor of the fluid mosaic model. The diffusion of proteins within the membrane has been revealed in several ways.

The Diffusion of Proteins after Cell Fusion

The technique of *cell fusion* has been one of the most important in cell biology. Cell fusion provides the means whereby cells of two different species can be joined together to produce one cell with a common cytoplasm and a single, continuous plasma membrane. This technique utilizes the ability of certain inactivated viruses to attach to the plasma membranes of cells, making the cells "sticky" so that their membranes will associate with one another, causing the cells to adhere. Once the cells are in contact, the intervening membrane becomes interrupted, and their cytoplasms become continuous. In the experiments by L. D. Frye and Michael Edidin (Fig. 5-28), mouse and human cells were fused, and the locations of specific proteins of the plasma membrane were followed once the two membranes were continuous. In order to follow the distribution of either the mouse membrane proteins or the human membrane proteins at various times after fusion, antibodies against one or the other type of protein were prepared. Antibodies are a particular type of protein, termed an *immunoglobulin*. Antibodies are produced by an animal in response to the presence of a foreign material (an antigen), whether that material is an invading bacterium or a protein injected by an investigator.

In the present study preparations of mouse membrane proteins and human membrane proteins were injected into different rabbits, mice, and

Figure 5-28. The demonstration by cell fusion that membrane proteins (dots) are capable of lateral mobility within the plane of the membrane. Details are presented in the text.

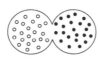

Human cell × Mouse cell — Addition of sendai (fusing) virus → → 40 minutes →

goats, and an antiserum against these two classes of membrane proteins prepared. Antibodies are capable of attaching to their respective antigens, regardless of where those antigens are located. The antibodies present in the two types of antisera (antimouse and antihuman) were then co-valently linked to fluorescent dyes so that their locations could be followed in the fluorescence light microscope. The antibodies against the mouse proteins were complexed with a dye that fluoresces green and the anti-bodies against human proteins with one that fluoresces red. These two antisera could then be used as probes to determine the location of the mouse or human proteins on the surfaces of fused cells. At the time of fusion, the plasma membrane can be considered as half human and half mouse in nature, i.e., the two proteins remained segregated in their own half. As the time after fusion increased, it was found that the membrane proteins did not remain in their original positions, but rather began to intermix with one another until by approximately 40 minutes, each species' proteins was uniformly distributed around the entire fused cell. The results indicated that the proteins of the membrane were capable of movement in lateral directions within the plane of the membrane.

The rapidity with which proteins diffuse, i.e., the magnitude of their diffusion constants, is dependent upon several factors. Most important is the consistency of the lipid matrix through which the proteins must migrate: the more fluid the bilayer, the greater the mobility. Measurements indicate that the membrane lipid has the viscosity of light machine oil. Another factor is the mass of the proteins: the greater the molecular weight of the particle, the more slowly it will diffuse. An additional factor to consider is the possibility that membrane proteins are not totally free to diffuse in a random manner; the evidence suggests that there are often restraints on these particles resulting from their interaction with materials outside of the membrane on one side or the other.

Patching and Capping

The cell that has been most intensively studied with respect to the mobility of its membrane proteins is the *lymphocyte*. Lymphocytes are cells of the immune system and are responsible for the production of antibodies against foreign materials (Chapter 18). It has been found that lymphocytes not only produce antibodies when challenged with a foreign substance, but that the membranes of these cells contain these immunoglobulins (antibodies) as well. Since the immunoglobulins (Ig's) of one animal are different from those of another, they can be used as antigens to evoke the production of antibodies, just as any other foreign protein can be used. For example, a rabbit immunized with mouse immunoglobulins (mouse Ig's) will produce its own immunoglobulins (rabbit Ig's) in response. The rabbit antiserum is said to contain antimouse Ig antibodies. These antibodies can be complexed with fluorescent markers to follow their location on the surfaces of living cells using the fluorescence microscope, or the antibodies can be complexed with ferritin (a large, iron-containing protein) and their location can be determined with the electron microscope using either ultrathin sections (Fig. 5-29c) or freeze-etch replicas (Fig. 5-29d).

When lymphocytes of one species, such as a mouse, are incubated with antimouse Ig antibodies from another species, the antibodies are at first dis-

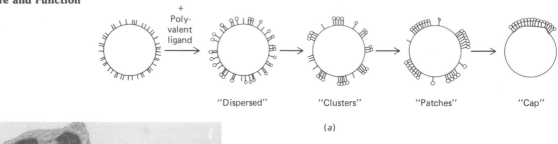

"Dispersed"   "Clusters"   "Patches"   "Cap"

(a)

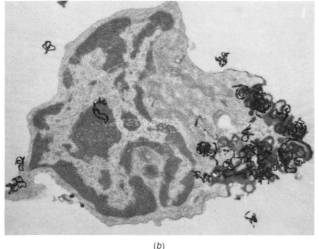

(b)

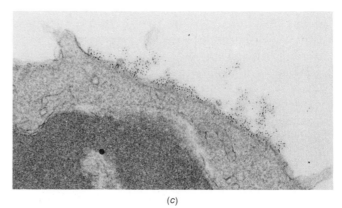

(c)

Figure 5-29. The formation of patches and caps containing membrane proteins on the surfaces of a cell treated with antibodies which serve as a polyvalent ligand. *(a)* Schematic pathway of the steps in lymphoid cell capping. [*From G. L. Nicolson,* Biochim. Biophys. Acta, **457**:*79 (1976).*] *(b–d)* The demonstration of the clustering of membrane proteins on the surfaces of lymphocytes by various techniques after the induction of patching or capping by specific antibodies. *(b)* Clustering is revealed autoradiographically when membrane proteins are radioactively labeled (in this case by use of an enzymatic procedure for radioiodination, see page 234). *(Courtesy of N. K. Gonatas.)* *(c)* Clustering is revealed by observing the location of electron dense ferritin molecules complexed to the antibodies which are, in turn, attached to the membrane proteins (in this case histocompatibility antigens). *(Courtesy of W. C. Davis.)* *(d)* Similar to *(c)* except the location of the ferritin-antibody complexes are detected by freeze-etch analysis, which reveals the presence of materials on the surface of cell membranes. *(Courtesy of M. J. Karnovsky.)*

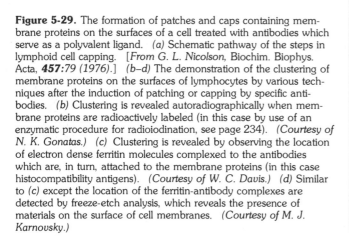

(d)

tributed rather uniformly over the cell surface (Fig. 5-29a). Within a few minutes, however, clusters of antibody molecules are seen and then patches are formed (Fig. 5-29a). The antibodies finally become distributed on the surface as a cap on one side of the cell indicating that all of the immunoglobulin molecules of the membrane have migrated over to that position (Fig. 5-29b, d). The presence of the cap is a transitory condition; it is soon removed from the surface, primarily by ingestion in the cell. The change in the distribution of

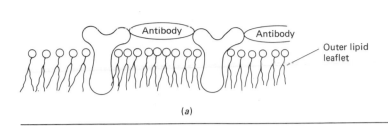

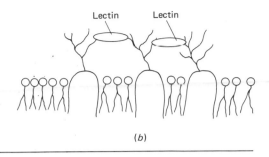

(a)

(b)

proteins within the plasma membrane provides strong evidence that integral membrane proteins can be moved. The question in this case is why these immunoglobulins should move from a random distribution into patches and then a single cap. The answer is seen most clearly when the nature of the probing antibody preparation is considered.

Antibodies are proteins that can combine with the appropriate antigen at more than one site within the antibody molecule, i.e., they are *multivalent.* As multivalent molecules, these antibodies are capable of interacting with more than one immunoglobulin in the cell's surface, thereby serving as a bridge by which these membrane immunoglobulins are brought together (Fig. 5-30a). Similarly, the immunoglobulins of the membrane are multivalent since each is capable of combining with more than one antibody molecule (Fig. 5-30a). The consequence of this type of reaction is that a network of interacting protein molecules is formed that grows in size as more and more individuals are brought together. In the unreacted condition, the membrane proteins are in an essentially random distribution, though each is capable of diffusion through the lipid bilayer. The antibodies become attached to the membrane proteins at one combining site and are then brought in contact with other antigen molecules as a result of the movement in the membrane. As the number of collisions between potentially reactive molecules increases, the size of the clusters increases until large patches are formed.

Antibodies are not the only multivalent macromolecules that can be used to visualize membrane protein clustering. Another class of multivalent interacting molecules is a collection of proteins, termed *lectins,* which are capable of combining with the carbohydrate chains projecting from the cell surface (Figs. 5-30b, 5-31). A given lectin is specific in its combining capacity for a given sugar, though different lectins have different specificities. For example, the lectin, concanavalin A is specific for α-glucosyl and α-mannosyl residues, while the lectin RCA is specific for β-galactosyl residues, and WGA for N-acetylglucosamine residues. Since each integral protein of the plasma membrane may have several oligosaccharide chains projecting outward and lectins have several combining sites capable of interaction with a specific sugar in many of these chains, the potential exists to form a cross-linked network of lectins and membrane glycoproteins in a similar manner to that described for the membrane immunoglobulins of the lymphocyte and the appropriate antibodies (Fig. 5-30b). The results of studies with lectins and with antibodies parallel one another closely.

Control of Mobility

The experiments just described suggest that the potential for migration is present among integral membrane proteins, but they do not indicate

Figure 5-30. The mechanism by which a multivalent ligand such as an antibody (a) or a lectin (b) brings about the clustering of membrane proteins.

Figure 5-31. Mouse lymphocytes bound to lectin-coated nylon fiber (diameter 135 μm). (a) Field is focused on the face of the fiber. (b) Field is focused on the edge of the fiber. *(Reproduced by permission from U. Rutishauser and G. M. Edelman in H. Bittiger and H. P. Schneble, eds., "Concanavalin A as a Tool," copyright © 1976 John Wiley & Sons Ltd.)*

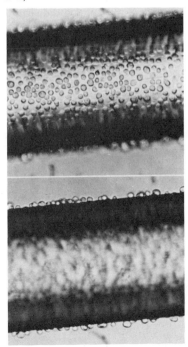

225

that all membrane proteins are always free to drift around randomly upon the lipid "sea." In fact, some types of membranes are characterized by having much greater protein mobility than others. The plasma membrane of the red blood cell is one in which patching and capping is not seen when multivalent ligands are attached to the glycoproteins of the cell surface. In this case there appears to be a network of peripheral proteins on the inner side of the membrane which has the major responsibility of keeping membrane proteins fixed in place and unavailable for diffusion through the membrane (Figs. 5-32a and 5-36b). If these inner-restraining proteins are removed from red blood cell membranes, the diffusion rates of the integral proteins are greatly increased.

Even in membranes that are characterized by extensive lateral diffusion, there are examples in which a particular organization of certain membrane proteins is maintained. Topographical differences are particularly evident in cells of organized tissues where the various cell surfaces have distinct functions. For example, in epithelia such as that lining the intestine or the kidney tubules, there are different enzymes present on the surfaces facing toward the lumen and facing away from it. In places where two cells come very close together, cell junctions are formed (page 282) in which specific proteins are segregated. Another example is the membrane of the synapse, where specific neurotransmitter substances are localized.

What mechanisms are responsible for either restraining or promoting

Figure 5-32. Restraints on membrane protein mobility. *(a)* In this case, which depicts the situation in the erythrocyte, there is transmembrane control over the glycoprotein complex. Treatment of the erythrocyte ghosts with antibodies against the inner peripheral protein, spectrin, causes the aggregation of the integral glycoproteins. *(b)* The control of membrane components by materials either outside the cell or within the cytoplasm. In this diagram, some outer surface material is controlling the distribution of the membrane glycoproteins, GP₃ and GP₄, while on the inner surface of the membrane, the cytoskeletal microfilaments and microtubules (discussed in Chapter 15) are involved in protein mobility. [*From G. L. Nicolson, Biochem. Biophys. Acta,* **457:**57 *(1976).*]

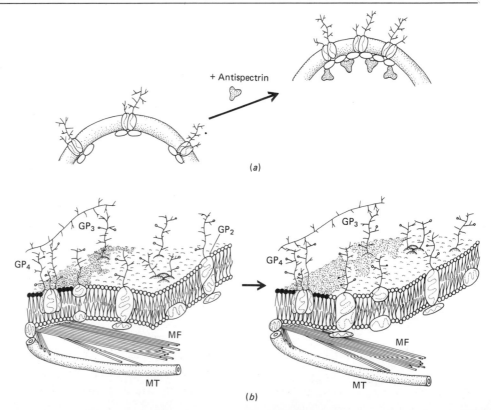

the movements of membrane proteins? In the past few years, particular attention has centered on cytoplasmic microtubules and microfilaments (and related materials) as being involved in the control of membrane protein mobility. This topic is complex and has ramifications for other areas of research in cell biology (as discussed in chapter 14 on cell growth and chapter 19 on cancer). Integrated discussions of these various topics can be found in the papers by Garth Nicolson, 1976; Gerald Edelman, 1976; J. F. Ash et al., 1977; and Theodore Puck, 1977. If the "skeletal" elements of the cytoplasm are involved in membrane protein mobility, one would expect to observe some type of physical association between the two parts of the cell as expressed diagramatically in Figure 5-32b. The evidence for microfilament association with the membrane is quite strong while that for microtubule interaction is largely indirect. The specific role of these two types of cytoplasmic organelles in membrane motility remains controversial. In one type of proposal, the microtubules and microfilaments are believed to have antagonistic functions; the microtubules being responsible for anchoring the membrane proteins in place while the microfilaments would cause them to be moved as in the formation of caps. Rather than attempt to integrate a large amount of data on this subject, a couple of experiments will be described to illustrate the type of studies that have been performed which implicate (1) microtubules and (2) microfilaments in membrane control.

One of the activities of the plasma membrane of various types of cells is the uptake of material from the environment by *endocytosis*. This process (discussed on page 323) involves the entrapment of fluid from the medium by the formation of a fold of the plasma membrane, and the subsequent formation of an internal cytoplasmic vesicle surrounding this fluid. The process of endocytosis can be induced in the appropriate cells by the addition of latex beads. In the present experiments, the activity of a cell surface transport enzyme was measured before and after stimulation of the cells to undergo endocytosis (Fig. 5-33). Under normal conditions it was found that even when a large amount of plasma membrane was being carried into the cytoplasm in the formation of these endocytotic vesicles, no loss in surface activity of this enzyme was occurring. In other words, even though preexisting plasma membrane was being removed from the surface, it appeared that essentially all of the transport proteins were remaining in the surface left behind. This result is interpreted as evidence that the cell has control over the distribution of the proteins within the membrane. Membrane involved in endocytosis appears to be cleared of some protein prior to being internalized. However, when this same experiment is performed in the presence of colchicine, a powerful microtubule-dispersing agent, a different result is observed; namely, the loss of surface transport activity. It would appear that microtubules are normally involved in maintaining endocytotic vesicles free of proteins that are not involved in this function and whose inclusion in these vesicles would simply require the cell to replace them at the surface.

As discussed at length in Chapter 16, the cytoplasm of cells, both muscle and nonmuscle, contains a variety of contractile proteins. Although the most prominent element of this contractile or cytoskeletal net-

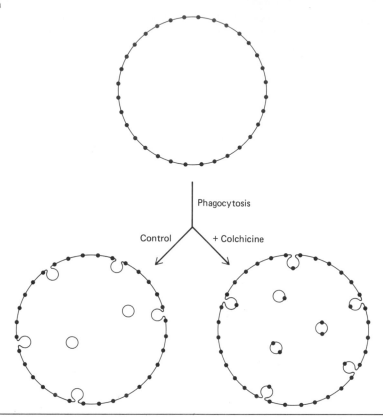

Phagocytosis

Control   + Colchicine

Figure 5-33. Diagrammatic representation of an experiment implicating microtubules in the control of membrane protein motility. Under normal conditions, a particular transport protein (represented by dots) is not lost from the cell surface, while in the presence of colchicine (a microtubule disrupter) the protein is lost from the plasma membrane to the phagocytotic vesicles.

work is the actin-containing microfilament, other proteins (such as myosin and tropomyosin) are also involved. If any or all of these proteins are involved in limiting or enhancing membrane protein mobility, then it should be theoretically possible to correlate their location within the cell with the position of integral proteins within the membrane, as illustrated diagramatically in Fig. 5-32*b*. Recent experiments in S. J. Singer's laboratory have succeeded in demonstrating this topographical relationship. In these studies, cells are treated *simultaneously* with pairs of fluorescent reagents that bind specifically to intracellular cytoskeletal materials (actin, myosin, or tropomyosin) as well as to externally exposed membrane proteins. The membrane proteins that have been examined include the Con A receptors, histocompatibility antigens, and specific enzymes. In all cases, a close topographical relationship is found between microfilaments (or the other proteins) on the inner side of the membrane and the distribution of one or another of the various membrane proteins. As might be expected, the specific results obtained depend in some degree upon the type of cell being examined. We will briefly compare and contrast the results obtained using cultured mouse A10 cells (a smooth musclelike cell line) with that from lymphocytes. In the case of the A10 cell (shown in Fig. 5-34), the membrane proteins are initially found to be distributed randomly over the cell surface, as revealed by fluorescent Con A (Fig. 5-34*b*) or fluorescent antibodies. At this same initial time, the distribution of contractile proteins (Fig. 5-34*a*) shows no obvious relationship to

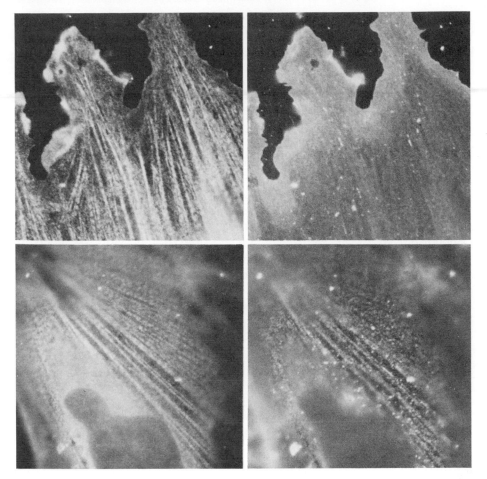

Figure 5-34. Demonstration of the relationship between surface membrane receptors and internal cytoskeletal elements in the mouse A10 cell. The upper row shows a portion of a single cell that was fixed and *then* treated with Con A. The lower row shows a portion of another cell that was treated with Con A for 15 minutes and then fixed. The left pair of photographs shows the distribution of tropomyosin-containing fibers within the cell as revealed by a two-step fluorescent antibody technique. In the first step, the cells are treated with rabbit antitropomyosin antibodies which combines with the tropomyosin of the cell. In the second step, the cells are treated with fluorescent goat anti-rabbit IgG, which complexes with the previously bound rabbit antibody and thereby reveals the distribution of tropomyosin. The right pair of photographs shows the distribution of surface Con A receptors as revealed by fluorescent Con A molecules. These photographs indicate that the Con A receptors become redistributed after Con A treatment into lines which correspond to the underlying tropomyosin-containing cables. *(Courtesy of A. Gotlieb, J. F. Ash, and S. J. Singer.)*

that of the membrane proteins. However, as the multivalent lectin or antibody begins to cause the membrane receptors to become clustered (Fig. 5-34*d*), the clusters become localized directly over the positions occupied by the contractile proteins (Fig. 5-34*c*) or microfilaments. Based on this type of evidence, it has been proposed that the clustering of external membrane proteins *induces* a linkage of the protein cluster with underlying microfilaments and related materials. In the case of the mouse A10 cell, the protein clusters do not go on to form large patches and caps. It is suggested that the linkage of the microfilaments to the membrane proteins actually serves to prevent further aggregation. Similar experiments on lymphocytes, which do undergo capping, also reveal a close relationship between the location of clustered membrane proteins and underlying contractile proteins. However, in the plasma membrane of the lymphocyte, the association of microfilaments and membrane protein may facilitate the capping phenomenon rather than preventing it. Regardless of the precise role of the cytoskeletal materials in the control of membrane mobility in one cell versus another, some type of relationship seems clearly established.

THE ANALYSIS OF MEMBRANE PROTEINS

Whereas the lipid bilayer seems to act primarily as the structural backbone of the membrane, it is the membrane proteins which are responsible for the specific functions which the membrane must carry out. From a structural point of view, the best studied membrane has been the plasma membrane of the erythrocyte (red blood cell), particularly that of our own species (Fig. 5-35). There are several reasons for the popularity of this membrane. The cells are inexpensive to obtain and readily available in great numbers from whole blood. They are already present as single cells and need not be dissociated from a complex cellular tissue. The cells are extremely simple by comparison with other cell types, the cytoplasm being composed essentially of a solution of one protein, hemoglobin; there are no other intracellular membranes present to interfere in the analysis. In addition, there is a very simple procedure available to obtain preparations of intact erythrocyte plasma membranes (Fig. 5-36a) completely uncontaminated by other materials. Intact plasma membranes (Fig. 5-36b) are called "ghosts" and are obtained from erythrocytes by placing them in a hypotonic salt solution. The cells respond to this osmotic shock by taking up water and swelling, a phenomenon termed *hemolysis*. As the surface area of the cells increase, a point is reached where the cells become leaky and the contents, composed almost totally of the dissolved hemoglobin, flows out. If the cells are centrifuged into a pellet, washed with additional saline, and the process repeated, a preparation of erythrocyte ghosts is obtained. At this point the ghosts are *unsealed* and are permeable to materials of large molecular weight. Examination of these ghosts in the electron microscope or by other techniques shows them to be composed solely of the plasma membrane. The remarkable aspect of this procedure is that the membrane seems undamaged. It can be returned to an isotonic solution and *resealed* so that it once again acts as a selectively permeable membrane with the same permeability properties that it originally possessed, and in numerous other ways acts as if it was unchanged from the living cellular state.

Figure 5-35. Scanning electron micrograph of human erythrocytes. [*From F. M. M. Morel, R. F. Baker, and H. Wayland*, J. Cell Biol., **48**:91 (1971).]

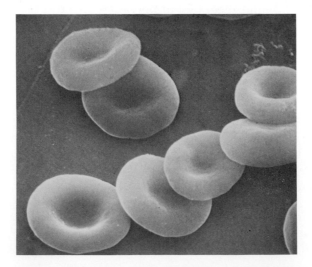

The opportunity to easily obtain purified, isolated plasma membrane preparations has so far caused the erythrocyte membrane to dominate the analysis of membrane structure, though other cells, particularly lymphocytes and various cultured cells, are receiving more and more attention. The lymphocyte membrane has special features (described in Chapter 18) that make it the most suitable for physiological studies. The disadvantage in working with the erythrocyte membrane, is the relative lack of activities that it carries out. The red blood cell is one that is produced and released to the blood stream for one purpose: to transport oxygen for a very limited length of time. It has no nucleus, is virtually without metabolism, and as such is not representative of growing, metabolizing, dividing cells. At the same time, however, its relative simplicity makes it a good membrane to start with, and there are membrane proteins of the erythrocyte that are believed to be present in other cells.

Given the opportunity to obtain purified erythrocyte plasma membranes, one of the most important analyses to perform is the solubilization of the membrane proteins and their fractionation. Once the populations of membrane proteins are separated from one another, some idea of their variety is immediately apparent and the opportunity exists to try to study each protein without interference from all the others. One of the difficulties in dealing with the integral membrane proteins is their insolubility in aqueous solutions. This difficulty has been overcome with the use of detergents, particularly sodium dodecyl sulfate (SDS), in both the extraction medium and during the fractionation procedure. Fractionation of membrane proteins has best been accomplished using acrylamide

Figure 5-36. (a) Procedure for obtaining erythrocyte ghosts. (b) Scanning electron micrograph of a torn and flattened ghost showing the reticular network on the inner surface of the membrane. (From T. L. Steck and J. F. Hainfeld in B. R. Brinkley and K. R. Porter, eds., "International Cell Biology, 1976–1977," Rockefeller University Press, 1977.)

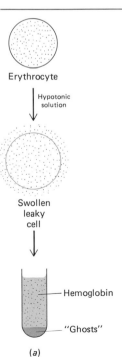

Erythrocyte

Hypotonic solution

Swollen leaky cell

Hemoglobin

"Ghosts"

(a)

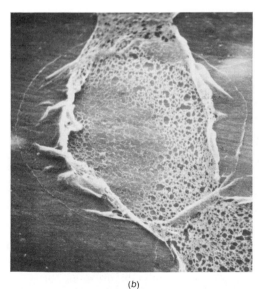

(b)

gel electrophoresis (page 114) with SDS present in the gels. The presence of SDS keeps the integral proteins soluble, and in addition adds a large number of negative charges to the proteins with which it associates. Since the number of SDS molecules per unit weight of protein tends to be relatively constant, the molecules no longer separate from one another on the basis of charge, but instead separate by molecular weight. The largest proteins move the slowest through the molecular sieve of the gel.

The most informative procedures with respect to assigning functions to the various proteins has been to carry out an experiment with the intact cell or the ghost using compounds that radioactively label specific proteins. The identity of these proteins can be determined by locating the radioactive band(s) in the gel. For example, transport proteins can be labeled using radioactive inhibitors which remain associated with the protein during the separation. In other experiments, specific proteins have been labeled with radioactive lectins, or phosphorylated using radioactive ATP to donate the phosphate.

The subject of acrylamide gel electrophoresis of membrane proteins is a complex research topic and there are many technical matters of importance, which bear upon possible interpretations, that will not be discussed. The acrylamide gel profiles of Fig. 5-37 are typical of those obtained from erythrocyte membranes; somewhat different profiles are described among the various laboratories carrying out this type of research. The following points should be kept in mind. Not all proteins are present in sufficient quantity to appear in these profiles, yet the minor species may have very important roles in membrane function. Some bands could represent cytoplasmic proteins that adhere to the membrane during its preparation, and conversely some of the loosely bound membrane proteins might be lost. Some of the bands that appear to be single proteins may represent several proteins (particularly band 3 of Fig. 5-37) and, in one case at least (PAS-1 and PAS-2), the same protein appears in more than one band.

The photographs of Fig. 5-37a, b show acrylamide gels that have been stained in different ways, the one on the left with Coomassie blue and the one on the right with periodic acid Schiff (PAS). Coomassie blue stains all of the proteins, though some of the glycoproteins (e.g., PAS-1) with very large amounts of sugars are not readily visible. Periodic acid Schiff stains carbohydrates, particularly sialic acid, and therefore the positions of heavily glycosylated proteins are best observed. Glycoproteins with small amounts of carbohydrate (e.g., band 3) do not show up in the PAS profile. Figure 5-37c, d shows the stain-density profile when gels such as those of Fig. 5-37a, b are scanned. The more stain present in a band, the taller the peak. The bands of Fig. 5-37 are given numbers which are used in the following discussion and in the schematic diagrams in Figs. 5-14 and 5-38 which indicate the apparent positions of the various proteins in the membrane.

It is apparent from these profiles that there are a significant number of proteins (approximately 15 to 20) in the erythrocyte membrane. Before discussing what is known of their function, we will consider techniques that allow one to determine the location of these proteins on one side of the membrane or the other. Two major approaches have been taken to

Figure 5-37. SDS-acrylamide gel electrophoresis of erythrocyte membrane proteins. *(a)* and *(b)* are photographs of actual gels that are stained for protein by Coomassie blue or for carbohydrate by PAS, respectively. The information in *(a)* indicates which of the bands correspond to extrinsic (peripheral) proteins and which are integral proteins. [*(a) Reproduced, with permission, from V. T. Marchesi, H. Furthmayr, and M. Tomita,* Annual Review of Biochemistry, *Volume 45.* © 1976 by *Annual Reviews Inc.; (b) From T. J. Mueller and M. Morrison,* J. Biol. Chem., ***249**:7571 (1974).*] *(c) and (d)* Represent a densitometric scan (measuring the amount of stain present) of gels stained with Coomassie blue and PAS, respectively. The numbers identifying the various peaks refer to one accepted terminology for erythrocyte membrane proteins. These numbers are used in Figs. 5-14*b* and 5-38. [*(c) and (d) From T. L. Steck,* J. Cell Biol., ***62**:3 (1974).*]

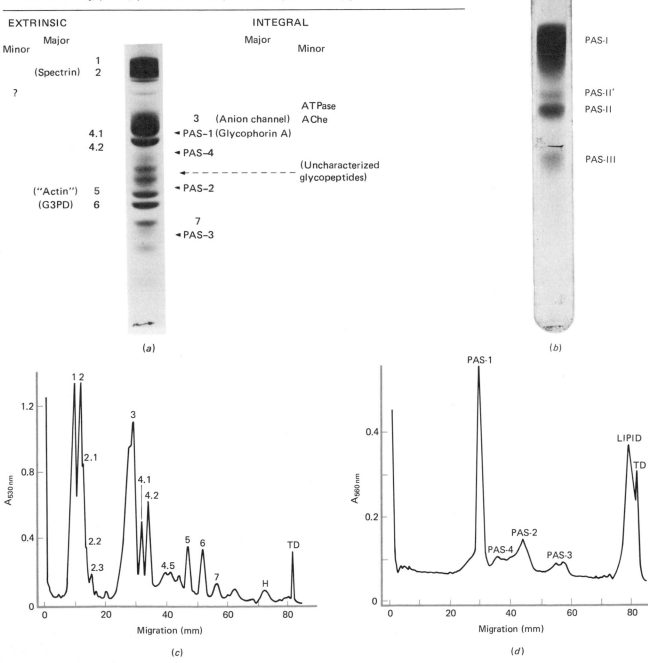

233

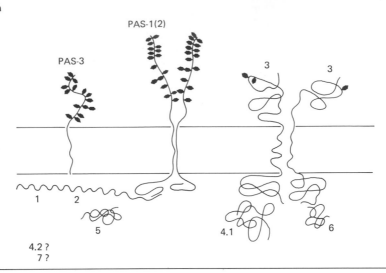

PAS-1(2)

PAS-3

3   3

1   2

5

4.1   6

4.2 ?
7 ?

Figure 5-38. Schematic representation of the arrangement of the major polypeptides of the human red cell membrane in relation to the lipid bilayer and to each other.   Dimeric complexes are proposed for the two major transmembrane proteins, PAS-1 and band 3.   Some of the major peripheral proteins may be close to (perhaps in contact with) the C-terminal portions of the integral proteins.   *(Reproduced, with permission, from V. T. Marchesi, H. Furthmayr, and M. Tomita, Annual Review of Biochemistry, Volume 45. © 1976 by Annual Reviews Inc.)*

assign proteins to one leaflet or the other; both involve modifying the proteins of the membrane before extraction, and then fractionating the proteins and observing to see which bands are affected.   In one type of experiment, membranes are incubated with proteolytic enzymes (Fig. 5-39*a*) (trypsin, pronase, papain are commonly used), and it is determined which bands are shifted to a new location as a result of the removal of a part of an exposed polypeptide.   The other approach is also an enzymatic one, but, instead of removing a part of the membrane protein, the enzyme attaches a radioactive label which can later be identified as being associated with particular bands.

Techniques for Labeling Membrane Proteins

Two enzymatic labeling procedures are most commonly employed, one which places a radioisotope in an amino acid (tyrosine) of an exposed polypeptide (Fig. 5-39*b*) and the other which places the label in a sugar (galactose or galactosamine) of an exposed oligosaccharide (Fig. 5-39*c*).   In the first case the enzyme is lactoperoxidase which catalyzes a reaction in the presence of peroxide in which a radioactive iodine atom (^{125}I or ^{131}I) can be attached to the tyrosine residue.   The second procedure utilizes the enzyme, galactose oxidase, which oxidizes galactose and galactosamine residues in carbohydrate chains.   When these oxidized sugars are incubated with a radioactive reducing agent (tritiated borohydride), they are returned to the reduced state by incorporating radioactive hydrogen atoms (^{3}H).

The enzymes galactose oxidase and lactoperoxidase can be used to determine whether a given protein is present on the inner surface or the outer surface of the membrane because these enzymes are too large to penetrate a sealed plasma membrane.   Since only those proteins and glycoproteins that are accessible to the enzymes will be radioactively labeled, their location within the membrane can be determined.   The autoradiograph (technique described on page 297) of Figs. 5-40 and 5-29*b* illustrates that the label transferred to the cell by the enzyme is only present at the surface, i.e., the enzyme has not been able to penetrate into the cell to label cytoplasmic proteins or the inner surface of the plasma membrane.   The following examples will serve to illustrate the use of these labeling techniques.

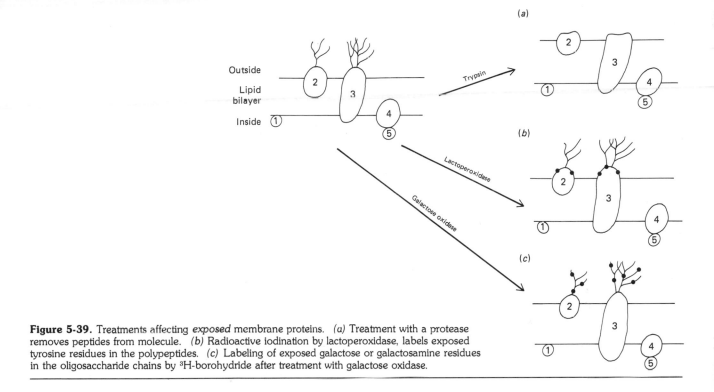

Figure 5-39. Treatments affecting *exposed* membrane proteins. *(a)* Treatment with a protease removes peptides from molecule. *(b)* Radioactive iodination by lactoperoxidase, labels exposed tyrosine residues in the polypeptides. *(c)* Labeling of exposed galactose or galactosamine residues in the oligosaccharide chains by ³H-borohydride after treatment with galactose oxidase.

Which of the proteins of the erythrocyte membrane are present facing outward from the cell membrane? Which proteins face the cytoplasm? Which, if any, penetrate the entire lipid bilayer? In order to answer these questions, various types of erythrocyte membrane preparations must be made. Some of the manipulations that can be performed on the RBC ghost are shown in Fig. 5-41. When techniques of membrane preparation are combined with the appropriate labeling procedure, the location of the major components of the membranes can be determined. Determination of the components of the external surface is the most straightforward of the labeling procedures. Intact cells (or resealed ghosts) are incubated with either lactoperoxidase or galactose oxidase, together with the appropriate labeled compounds. When membrane proteins are extracted and fractionated after either of these enzymatic labeling procedures, several of the bands are radioactive. Most importantly, the same major bands are labeled by both of these enzymes, namely band 3 and the PAS-stained bands (PAS-1, PAS-2, and PAS-3). All of the proteins that are known to contain carbohydrates are labeled by the galactose oxidase–³H-borohydride procedure when applied to intact cells (Fig. 5-42a, 3d box). Since the enzyme cannot penetrate into the cytoplasm to label the inner side, all of the carbohydrate-bearing proteins, i.e., glycoproteins, must be on the outside. Conversely, none of the proteins of the membrane that are lacking carbohydrate are labeled by the lactoperoxidase–¹²⁵I procedure, indicating that there are no major proteins exposed to the outside that are not glycoproteins.

The remaining bands seen in the gel that are not labeled by procedures designed for external proteins can be presumed to be either entirely within the membrane or on its inner side. There are several ways in which labeling reagents can gain access to the interior (cytoplasmic) side of the membrane.

Figure 5-40. Electron microscopic autoradiograph of a mouse L cell iodinated by lactoperoxidase. The radioactive iodine atoms are largely confined to the cell surface. [*From A. L. Hubbard and Z. A. Cohn,* J. Cell Biol., **64**:445 (1975).]

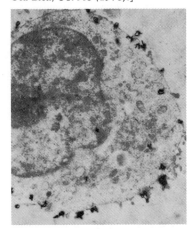

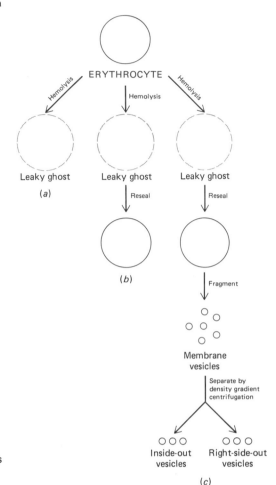

ERYTHROCYTE

Hemolysis

Hemolysis

Hemolysis

Leaky ghost

Leaky ghost

Leaky ghost

(a)

Reseal

Reseal

(b)

Fragment

Membrane
vesicles

Separate by
density gradient
centrifugation

Inside-out
vesicles

Right-side-out
vesicles

(c)

Figure 5-41. Procedure for obtaining various types of erythrocyte membrane preparations. The first step in all procedures is subjecting the erythrocyte to hemolysis in which all of the internal contents leak out leaving one with the ghost. The preparation of ghosts (after being washed several times) can be resealed so as to essentially regain its initial permeability properties. If one wants a preparation in which the internal membrane surface is exposed and the external surface is protected, inside-out vesicles can be prepared by fragmentation and density-gradient centrifugation.

During the preparation of erythrocyte ghosts by hemolysis (Fig. 5-41), the membrane is made leaky enough for all of its internal contents to escape. In this state the membrane is open for passage of macromolecules into the ghost as well as in the reverse direction. Enzymatic labeling procedures on unsealed ghosts would be expected to label proteins or glycoproteins that were exposed on either membrane face, and this is just the results that are seen. Lactoperoxidase labeling of unsealed ghosts renders essentially all of the major membrane proteins and glycoproteins radioactive.

In order to specifically label the cytoplasmic surface *without* the simultaneous labeling of the exterior proteins, special tricks must be employed. There are several ways in which the selective labeling of the internal proteins can be carried out. In one technique, *unsealed* ghosts are incubated with the enzymes (lactoperoxidase or galactose oxidase) such that these labeling agents will penetrate inside the ghosts. Once the enzymes are inside, the ghosts can be sealed so as to prevent their exit, washed to remove enzyme molecules on the outside of the ghost, and then the sealed ghosts can be incubated in the radioactive labeling reagents (radioactive borohydride or iodine) which are of small molecular weight and capable of penetrating within the sealed ghosts through the intact membrane. Once inside, these labeled molecules can be used as substrates for the sealed-in enzymes, and membrane proteins of the interior surface can be labeled.

In the other technique, erythrocyte ghosts are first prepared, then these ghosts are fragmented into smaller pieces by homogenization in dilute phos-

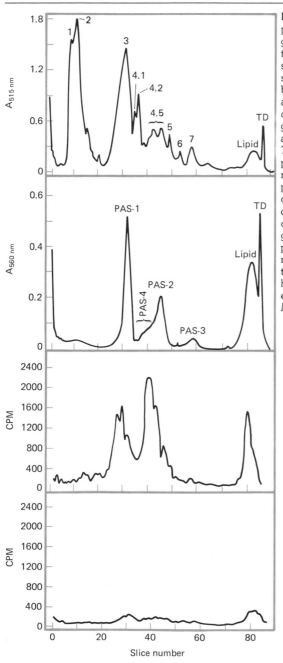

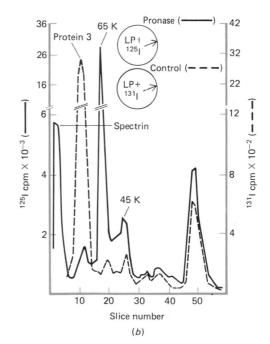

Figure 5-42. Acrylamide gel profiles of membrane proteins after various labeling procedures. *(a)* Upper profile shows the positions of the membrane proteins after gel electrophoresis as revealed by Coomassie blue staining. Second profile shows the positions of the major carbohydrate-bearing proteins as revealed by PAS stain. Third profile shows the distribution of radioactivity in the gel after labeling sealed ghosts externally by the galactose oxidase procedure. The relative similarity between the second and third profiles suggests that all major glycoproteins are accessible to external labeling by galactose oxidase. Fourth profile shows the distribution of radioactivity in the gel after labeling inside-out vesicles with galactose oxidase. The essential lack of incorporated isotope indicates the absence of carbohydrate groups on the inner surface of the membrane. [*From T. L. Steck and G. Dawson,* J. Biol. Chem., ***249:**2139 (1974).*] *(b)* The effect of pronase digestion on the distribution of iodine label in the proteins isolated from resealed ghosts that had previously been labeled internally by the lactoperoxidase procedure. This profile represents the radioactivity in the gel after electrophoresis of a combined membrane protein preparation. In one experiment, i.e., the control, resealed ghosts were simply labeled internally with ^{131}I by the lactoperoxidase procedure and the proteins extracted. In the other case, the resealed ghosts were labeled internally with ^{125}I and then subsequently treated with the proteolytic enzyme externally, prior to protein extraction. The two preparations of membrane protein were then mixed, subjected to electrophoresis, and the distribution of the two isotopes of iodine determined. The fact that band 3 protein has an altered electrophoretic pattern after being labeled internally and cleaved externally indicates that it spans the membrane. [*From T. J. Mueller and M. Morrison,* Biochemistry, ***14:**5514 (1975).*]

phate buffer. Since membranes are not stable as pieces with free edges, the fragments spontaneously seal themselves and form tiny vesicles, which are completely membrane bound. Two types of vesicles can be formed: those in which the original orientation of the membrane is preserved so that what was originally facing the outside of the cell is now facing the outside of the vesicle; this type is termed a *right side-out vesicle*. In the other type of vesicle, the membranes have fused so that what was originally the cytoplasmic surface of the membrane is now facing the outside environment of the vesicle; these are termed *inside-out vesicles*. These two types of vesicles have different properties and they can be separated from each other. Once each of the two types of vesicles has been purified and sealed, they can be subjected to the same enzymatic labeling procedures previously described for intact cells and ghosts. In the case of sealed inside-out vesicles, the enzyme can only label the exposed proteins (or glycoproteins if such were present), which actually represent the proteins of the cytoplasmic membrane surface. When the galactose oxidase–borohydride procedure is carried out on inside-out vesicles, no labeled bands appear on acrylamide gels (Fig. 5-42a, 4th box) confirming the hypothesis that all membrane carbohydrate is on the external surface. When sealed inside-out vesicles are labeled with the lactoperoxide–^{125}I procedure, the Coomassie blue–stained bands, which were not previously labeled when intact cells were treated, are now radioactive. The newly labeled bands correspond to 1, 2, 2.1, 2.2, 2.3, 4.1, 4.2, 5, 6, 7, and several trace polypeptides. These newly labeled bands represent those proteins present on the interior side and absent on the exterior side. On this basis it has been determined that the majority of proteins of the erythrocyte plasma membrane are associated with the inner leaflet.

One of the most important findings by these techniques is that two of the proteins (band 3 and PAS-1) are labeled from the external side (by either enzyme) and are also labeled from the internal side by lactoperoxidase. The conclusion that was drawn when these results were obtained was that these two proteins span the entire thickness of the membrane. In other words, it was suggested that these proteins project out of the membrane on either side and are therefore accessible to labeling procedures from either direction. In keeping with the results of the galactose oxidase procedure, it would seem that these proteins have carbohydrates attached at the membrane's outer surface and a naked polypeptide projecting from the inner surface. This proposal has been confirmed by several different procedures. The following experiment will illustrate the type of approach that has been taken to verify the existence of transmembrane (spanning) proteins.

The interior surface of the membrane can be labeled by the first of the above techniques, that in which lactoperoxidase is trapped inside the ghost before the radioactive iodine is added. In the next step, these sealed, labeled ghosts are treated with a proteolytic enzyme which will clip off polypeptides projecting out from the surface of the membrane (Fig. 5-39a). Since this type of enzyme, like other macromolecules, cannot penetrate the sealed membrane, interior proteins are protected from destruction. When acrylamide gels are prepared, the labeled proteins of these protease-treated ghosts migrate in a different manner from that of ghosts that were not protease-treated. When a piece or pieces of the protein has been removed its molecular weight and migratory properties are altered. This type of experiment (Fig. 5-42b) shows that a protein labeled from the inside can be affected by a protease treatment applied to the outside. The protein must span the membrane.

The Function of Membrane Proteins

The results of these types of labeling procedures have indicated a striking asymmetry with respect to the proteins and glycoproteins of the erythrocyte membrane (Fig. 5-14b); it appears that none of the proteins present on one side are also found on the other, though at least two, and possibly all of the glycoproteins of the external leaflet span the membrane. Results from the labeling studies suggest that all of the carbohydrate projects outward from the membrane. This can be confirmed by examination of the inner or outer surface of erythrocyte ghosts after treatment with agents that combine specifically with carbohydrates (Fig. 5-43). If one considers the difference in the properties of the medium on either side of the plasma membrane, this total asymmetry may not be surprising. The presence of the carbohydrate is believed to have very important functions in the interactions of a cell with its environment (discussed below); but, in addition, its hydrophilic nature serves as an excellent lock to prevent these glycoproteins from tumbling in the membrane. It is believed that these proteins are always maintained with their hydrophilic portions projecting out of the lipid bilayer.

A closer look at the proteins of the erythrocyte plasma membrane will illustrate the types of functions that such proteins can have. Among these proteins whose functions are known are a variety of enzymes (glyceraldehyde 3-phosphate dehydrogenase, acetylcholinesterase, and protein kinase), transport proteins (for ions, amino acids, and sugars), and structural proteins (spectrin). A few of these proteins will be considered below.

The best-studied of the peripheral proteins are those representing bands 1, 2, and 5 of Fig. 5-37. Two of these proteins (bands 1 and 2) have extremely large molecular weights, correspond to about 30% of the total membrane protein, are bound to one another, and are given the name *spectrin*. Spectrin is believed to be associated with the protein of band 5 (Figs. 5-14b and 5-38) to form a structural network beneath the membrane. Band 5 is composed of a protein that is very much like muscle actin, and spectrin has properties in common with muscle myosin. This inner protein network can be visualized in the electron microscope as a filamentous reticular network beneath the membrane (Fig. 5-36b) and is believed to be of importance in maintaining the shape of the biconcave erythrocyte. If these proteins are removed from erythrocyte ghosts, the membrane becomes fragmented and forms small vesicles, indicating that the protein network is required to maintain the integrity of the membrane. As discussed previously, these proteins have also been implicated as restraining influences upon the movement of integral membrane proteins.

The transmembrane proteins, band 3 and PAS-1 (also known as *glycophorin*), have been the most intensively studied, but are still not well understood. One of the most important techniques in the study of these proteins has been freeze-fracture analysis described earlier in this chapter. One of the great features of the freeze-fracture technique is that it takes advantage of microheterogeneity of the membrane rather than being defeated by it. In other words, localized differences in parts of the membrane stand out in these replicas and can be identified, as shown in the

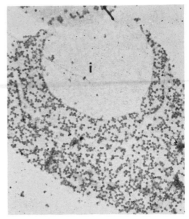

Figure 5-43. An erythrocyte ghost was prepared so a part of both its inner and outer surface was exposed. The mounted ghost was then labeled with purified influenza virus which attaches specifically to carbohydrate portions of glycoproteins. The ghost was then stained for electron microscopy after which the location of the carbohydrate receptor on the outer membrane surface becomes apparent. The lack of adsorbed virus on the exposed inner portion of the membrane indicates the absence of carbohydrates on this surface. [*From G. L. Nicolson and S. J. Singer, J. Cell Biol.,* **60:**241 (1974).]

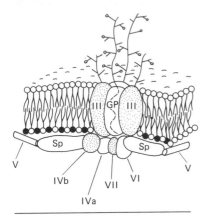

Figure 5-44. Hypothetical model of the human erythrocyte "permeaphore." Solid lipid head groups indicate asymmetric distribution of phosphatidylserine and phosphatidylglycerol. GP, glycophorin; Sp, spectrin; III, band 3 component; other proteins identified by roman numerals. [*From G. L. Nicolson,* Biochem. Biophys. Acta, **457**:74 (1976).]

replica of a gap junction in Fig. 6-21. Unlike biochemical analysis, microscopical observations do not average out all of the individualities, but show them up to be appreciated.

The transmembrane nature of the two integral proteins, band 3 and PAS-1, makes them uniquely suited to transmit materials and/or information from one side of the membrane to the other. In a transport capacity, there is evidence that these types of proteins are responsible for the movement of ions and sugars across the membrane. Permeability will be considered in more detail in a later section, but for now it will suffice to point out the potential for movement across the membrane that these proteins afford the cell. There is evidence that these proteins do not exist as monomers, i.e., individual polypeptide chains, but rather as oligomers, where several have come together (Fig. 5-44). It is these aggregates that are seen as particles in freeze-fractured membrane preparations and it has been suggested that one of the consequences of their oligomeric structure would be the formation of pores between the individual proteins through which water, ions, and dissolved substances could pass (see Fig. 5-54a). The concept of information transfer across the membrane has been most extensively studied with respect to the action of certain hormones and neurotransmitters. These molecules are known to interact with specific receptor molecules on the outer surface of the membrane and stimulate events on the membrane's inner surface (see page 245 for a detailed discussion). The mechanism by which this stimulation crosses the membrane is not yet understood, but the presence of transmembrane proteins makes them likely candidates in some of these activities.

The integral glycoprotein, band 3, has an apparent molecular weight of 100,000 and has a relatively small amount of carbohydrate (approximately 6 to 8%). This membrane protein appears to exist as a dimer and is believed responsible for the transport of anions (including HCO_3^-) and glucose across the membrane; agents that block these transport activities bind to this glycoprotein. The movement of water through the membrane may also be routed through the band 3 glycoprotein. The specific function of the carbohydrate portion of this protein is not known, though certain properties have been associated with it. It is known for example, that one of the most extensively employed lectins, concanavalin A, binds to sugars of this protein. The function of the other major glycoproteins, PAS-1 or glycophorin, is also very poorly understood, though there are also properties that can be identified with it, particularly ones associated with the extensive carbohydrate portion which makes up about 60% of the molecule (Fig. 5-38). Of particular significance, glycophorin was the first integral membrane protein whose complete amino acid sequence has been elucidated, in this case by Vincent Marchesi and coworkers. The determination of the sequence of 131 amino acids making up the single polypeptide chains of glycophorin has revealed some very interesting aspects of its amphipathic structure. The amino terminal portion is hydrophilic in its amino acid composition and projects outward into the environment. Attached to this portion are 16 oligosaccharide chains, each containing from 4 to 15 sugars. The next section of the *linear sequence* of the polypeptide contains hydrophobic residues and corresponds to that part within the lipid bilayer. The carboxyl terminal section is once again hydrophilic and projects through the membrane into the cytoplasm. The carbohydrates of glycophorin determine the blood group antigens that have been used for so long in blood typing. Both the ABO and MN systems are borne by this protein. The ABO antigens are branched oligosaccharide chains of 11 to 13 sugars. A person having blood type A has N-acetylgalactosamine at the end of the chains (Fig. 5-5a), while a person with type B has these terminal sugars replaced with galactose. The O blood type lacks both of these sugars.

Within the carbohydrate chains of glycophorin are the sites of attachment for certain viruses and the receptors for certain lectins (phytohemagglutinin and wheat germ agglutinin). Analysis of the interaction between phytohemagglutinin (PHA) and the cell surface is of particular interest because in the case of lymphocytes this lectin is *mitogenic*, i.e., it causes them to divide. When a population of nongrowing lymphocytes of the appropriate variety is incubated with PHA, the lymphocytes begin to proliferate and eventually differentiate into mature antibody-producing cells. Though the nonnucleated mammalian red blood cell obviously cannot be stimulated to divide, the analysis of the lectin-glycophorin interaction in the erythrocyte is presumably of general interest, since a similar protein may also be present on the lymphocyte surface. The importance of this type of interaction is that it clearly illustrates the role that transmembrane proteins can have in the transfer of information across the membrane. The cell surface is believed to be of great influence in determining whether cells will grow and divide or remain in a stationary growth phase. It would appear that the interaction of certain lectins with receptor molecules in the outer surface of the membrane is translated through the membrane as information for the cell to crank up its machinery for growth and division. The transmembrane structure of these proteins makes them likely candidates in the transfer of information from a great variety of environmental stimuli to the cytoplasm through the membrane.

A particularly important property of glycophorin is the high levels of sialic acid (Fig. 5-4) that are present. Sialic acid (also called N-acetylneuraminic acid), when present in an oligosaccharide chain, is always the last sugar in the sequence; no further monosaccharides can be added to it. Sialic acid carries a net negative charge and along with the carboxyl groups of the polypeptides it is responsible for the net negative charge of the outer surface of the membrane. These terminal sialic acid residues can be selectively removed by an enzyme, *neuraminidase*, and the effects of their removal studied. The best-studied effect of desialyzation of erythrocytes is the change in the recognition of these cells by the other tissues of the body. Normally if red blood cells are withdrawn from an animal, radioactively labeled so that their presence can be followed, and then reinjected back into the animal, the labeled cells will continue to circulate for quite a period of time as they would have done had they not been withdrawn. The normal lifetime of erythrocytes is approximately 120 days. If, however, labeled cells are treated with neuraminidase prior to being reintroduced into the blood stream, these same cells are rapidly removed from the body's circulation by the spleen. In other words the loss of sialic acid earmarks the cells for destruction. Correlated with this is the observation that as the length of time that red blood cells *normally* circulate increases, the sialic acid content of their surfaces decreases, suggesting that this mark of the aging process of these cells is the means by which their circulation time is regulated.

The experiment just described touches on another very important topic in cell biology, that of cell-cell interaction and recognition, for which the carbohydrates of the plasma membrane glycoproteins are intimately involved. In the case of the erythrocyte, the sialic acid can be considered as having a protective or blocking role. Once this sugar is removed, the cell surface can be *recognized* by cells of the spleen, in this case, and the erythrocytes destroyed. The most commonly found sugar beneath the sialic acid in the oligosaccharide chain is galactose. It appears to be the presence of the terminal galactose in the neuraminidase-treated cells that is most responsible for causing the erythrocyte's demise, since if the galactose residues are also removed, in this case by a β-galactosidase, the labeled red blood cells are no longer picked out of the blood stream. It would appear that the galactose recognition site is masked by the terminal sialic acids. This type of balance between recognition and antirecognition or adhesion and antiadhesion may be very common in cell-cell interactions. Of one point there is no doubt, that the carbohydrate moieties of the cell surface play extremely important roles in the communication and interactions of cells and their cellular and noncellular environments. Besides the interactions that normal cells must make, these properties are believed to be of basic importance in the variant behavior that malignant cells demonstrate, and, as such, the roles of surface carbohydrate are being intensively analyzed for the clues that they might provide on the nature of the malignant transformation.

SURFACE RECEPTORS AND MEMBRANE-MEDIATED CONTROL

Communication within an organism is generally accomplished by the release of a chemical from one cell and its interaction with another cell at some distance. The two cells may be very close to one another as in the case of two nerve cells separated by a narrow synaptic cleft, or they may be at a great distance as generally occurs when a hormone is involved in the communication. In the latter case, the chemical is secreted directly into the blood by the cells of the endocrine gland where it is circulated through the body to interact with target cells to produce a particular response. It has become increasingly apparent that a great variety of chemical "messengers" act on their target cells as a result of an interaction with the target cell membrane. Included within this group of messengers one finds neurotransmitters and other chemicals of the central nervous system, hormones, and growth-promoting substances. We will concentrate upon hormones in this section, for they have provided the best information as to how these membrane-mediated messages are transmitted.

The Control of Glucose Metabolism

The best-studied hormones that act at the cell surface are ones whose main function is to control blood sugar levels by regulating carbohydrate metabolism and the transport of sugars across cell membranes. Before discussing the action of these specific hormones it is necessary to briefly consider hormone action and sugar metabolism in general. As stated above, hormones are secreted directly into the blood, and thereby provide all cells with equal access to the substances. Why do particular cells react to a particular hormone while others do not? The basis for a cell being a target of a given hormone is the presence in that cell of a specific receptor for that hormone. Cells that respond to epinephrine (adrenaline), or insulin, or progesterone (page 621) have proteins which specifically bind the hormone in question. Cells lacking a particular receptor are incapable of serving as a target for the corresponding hormone. The presence of a particular receptor determines whether or not a cell will respond to a hormone, but not necessarily the nature of the response. In many cases the same hormone will produce different effects in different target cells within the same organism. The nature of a cell's response depends upon the components present in the cell at the time of hormone-receptor interaction. This aspect of hormone action will be illustrated later in the discussion.

Blood sugar levels are determined primarily by the blood levels of three hormones: epinephrine secreted by the adrenal medulla, and glucagon and insulin secreted by the pancreas. Although each of these hormones has a number of metabolic effects, for the sake of simplicity we will restrict the discussion to glucose metabolism, which will illustrate the points. There are two main directions in which a molecule of glucose is utilized; it can be oxidized by the brain, heart, skeletal muscles, etc., for the generation of energy, or it can be polymerized into glycogen (or converted to fat) and be stored. Epinephrine, which is secreted as a response to stress, and glucagon, which is secreted when blood sugar levels fall, stimulate the breakdown of glycogen to glucose (primarily in muscle and liver cells). In contrast, insulin, which is secreted in response to high blood sugar levels, stimulates the polymerization of glucose to glycogen in the liver and conversion to fat in adipose tissue (fat cells). As discussed in Chapter 4, the enzymes responsible for glycogen breakdown and glycogen synthesis are phosphorylase and glycogen synthetase, respectively (Fig. 5-45). Epinephrine and glucagon cause a marked activation of phosphorylase and inhibition of glycogen synthetase thereby promoting glycogen breakdown *(glycogenolysis)*. Insulin, in contrast, stimulates glucose uptake and activates glycogen synthetase leading to glycogen formation. The means by which these hormones bring about changes in enzyme activity has been uncovered gradually over a number of years. The most prominent figures in this area have been Earl Sutherland and his co-workers.

Sutherland began his studies by incubating slices of liver tissue with epinephrine and glucagon, and measuring the increased output of glucose from the tissues into the medium. The effect of the hormones was soon traced to an increased activity of the enzyme phosphorylase, a protein

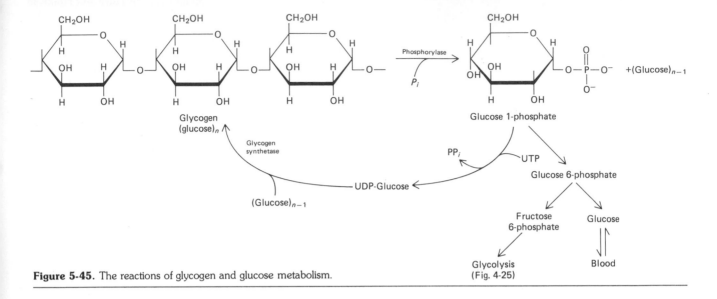

Figure 5-45. The reactions of glycogen and glucose metabolism.

which had been studied in great detail by Carl Cori and Gerty Cori. The question of importance was the means by which the hormones brought about the activation of phosphorylase. Further studies revealed that this enzyme could exist in two states, an active phosphorylated state and an inactive dephosphorylated one. If liver tissue was incubated with radioactive phosphate, radioactivity was rapidly incorporated into phosphorylase in the presence of either hormone. As it turns out, two enzymes exist in liver cells which function to control the level of phosphorylase activity. One of the enzymes, phosphorylase kinase, adds the phosphates, while the other enzyme, phophorylase phosphatase, removes them.

In order to understand the chain of events from hormone reception to glycogen breakdown, it was important to develop some type of in vitro system in which the physiological response to a hormone could be obtained. After considerable effort, Sutherland developed a system in which he could incubate epinephrine or glucagon with a preparation of *broken* cells and demonstrate an activation of phosphorylase. This broken cell preparation could be divided into a particulate fraction and a supernatant fraction by centrifugation. Even though phosphorylase was in the supernatant fraction, the particulate material had to be present if the hormonal stimulation was to be obtained. Subsequent experiments indicated that the response to the hormone occurred in at least two distinct steps. If the particulate fraction of a liver homogenate was isolated and incubated with the hormone, some new substance was produced which could then activate the soluble phosphorylase when subsequently added to the supernatant fraction. The substance produced by the particulate fraction was later identified as cyclic AMP (Fig. 5-46), a compound produced in the reaction (Fig. 5-46) catalyzed by the enzyme adenylate cyclase, a component of the liver cell membrane.

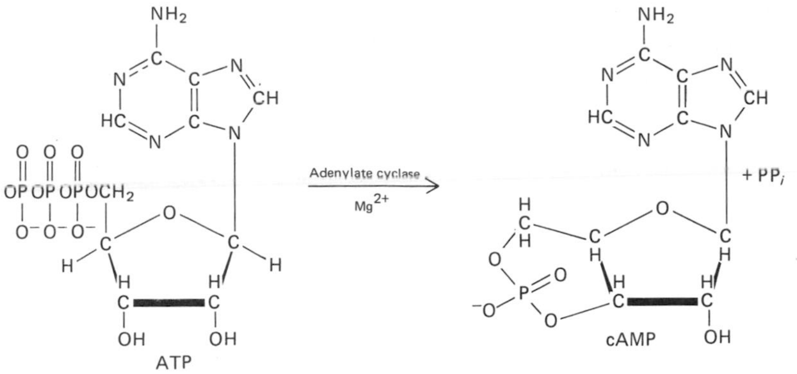

Figure 5-46. The formation of cyclic AMP from ATP catalyzed by the enzyme adenylate cyclase (or adenyl cyclase). The breakdown of cAMP (not shown) is accomplished by a phosphodiesterase, which converts the cyclic nucleotide to a 5′ monophosphate.

The analysis of the mechanism by which glucagon and epinephrine stimulated glycogen breakdown in liver cells led to a new concept in cell biology, that of the *second messenger*. The second messenger is that compound released within the cell as a result of the interaction between cell and hormone. In this case the hormones do not penetrate the cell membrane but instead react with a specific membrane receptor on the external surface of the outer leaflet. The interaction between receptor and hormone apparently activates adenylate cyclase (Fig. 5-23), a protein on the inner membrane surface, which then begins to synthesize cyclic AMP. It is the cyclic AMP which then diffuses into the cell to bring about a specific response, such as the activation of phosphorylase. Although this is the least understood step in the hormone response, it is believed that the interaction between receptor and adenylate cyclase is mediated by the lateral mobility of each protein within the membrane. If this is the case, a given adenylate cyclase molecule may be susceptible to activation by more than one type of receptor, depending on which particular receptor is in the hormone-bound form at any given time (Fig. 5-23). In this way, different types of stimuli, such as epinephrine or glucagon, can produce the same effect. Similarly, different cells may have different second messengers in response to the same hormone, or the same second messenger might produce different responses in two cell types due to the nature of the reacting systems within the cells. For example, cyclic AMP has been shown to produce a wide variety of effects in different cells (Table 5-8). Regardless of the situation, the formation of a new signal at the cell membrane adds considerable flexibility to the reacting systems. Another value of this type of mechanism is that it greatly amplifies the message. The concentration of hormones in the blood is generally very low. Considering the volume of blood in the organism, this low concentration makes for a very economical communication signal. Given the very low concentration of blood hormone levels, the membrane receptors must have a very great affinity for the hormone to complex it. Studies on the binding properties of the receptors indicate that they do have the necessary binding constants.

TABLE 5-8
Effects of Cyclic AMP

Enzyme or process affected	Tissues organism	Change in activity or rate
Protein kinase°	Several	Increased
Phosphorylase	Several	Increased
Glycogen synthetase	Several	Decreased
Phosphofructokinase	Liver fluke	Increased
Lipolysis	Adipose	Increased
Clearing factor lipase	Adipose	Decreased
Amino acid uptake	Adipose	Decreased
Amino acid uptake	Liver and uterus	Increased
Synthesis of several enzymes	Liver	Increased
Net protein synthesis	Liver	Decreased
Gluconeogenesis	Liver	Increased
Ketogenesis	Liver	Increased
Steroidogenesis	Several	Increased
Water permeability	Epithelial	Increased
Ion permeability	Epithelial	Increased
Calcium resorption	Bone	Increased
Renin production	Kidney	Increased
Discharge frequency	Cerebellar Purkinje	Decreased
Membrane potential	Smooth muscle	Increased
Tension	Smooth muscle	Decreased
Contractility	Cardiac muscle	Increased
HCl secretion	Gastric mucosa	Increased
Fluid secretion	Insect salivary glands	Increased
Amylase release	Parotid gland	Increased
Insulin release	Pancreas	Increased
Thyroid hormone release	Thyroid	Increased
Calcitonin release	Thyroid	Increased
Histamine release	Mast cells	Decreased
Melanin granule dispersion	Melanocytes	Increased
Aggregation	Platelets	Decreased
Aggregation	Cellular slime molds	Increased
Messenger RNA synthesis	Bacteria	Increased
Synthesis of several enzymes	Bacteria	Increased
Proliferation	Thymocytes	Increased
Cell growth	Tumor cells	Decreased

° Stimulation of protein kinase is known to mediate the effects of cyclic AMP on several systems, such as the glycogen synthetase and phosphorylase systems, and may be involved in many or even most of the other effects of cyclic AMP.

SOURCE: E. W. Sutherland, *Science*, **177**:405 (1972). Copyright © 1972 by The American Association for the Advancement of Science.

The presence of a small number of bound hormone molecules (a few hundred in the case of insulin) is capable of producing a maximal response which, for glucagon, involves the activation of several orders of magnitude more phosphorylase molecules. The production of the second messenger is one of the important amplification steps in the process. If one receptor-hormone complex is capable of activating one adenylate cyclase molecule, a large number of cyclic AMP molecules can be produced in a very short time. Each cyclic AMP molecule can then carry out the next step in the chain of response. The rapidity of the stimulation is best seen in the response by the heart to an injection of epinephrine, the compound which is used in attempts to revive heart arrest cases. Within less than five seconds, the level of cyclic AMP in cardiac muscle cells is greatly increased as is the strength of the contraction.

The chain of events in the response by a liver cell to glucagon or epinephrine is shown in Fig. 5-47. The cyclic AMP released by the mem-

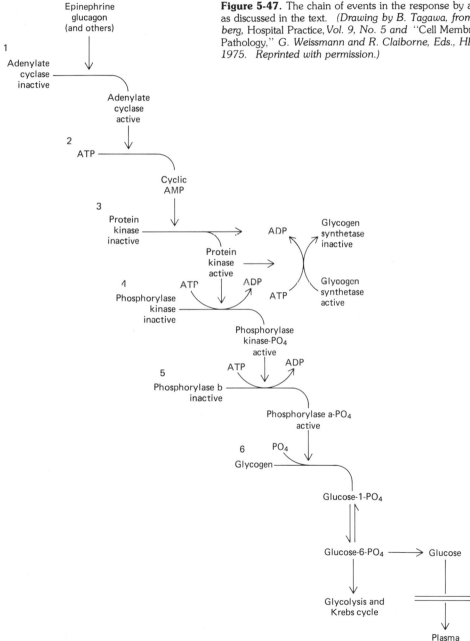

Figure 5-47. The chain of events in the response by a liver cell to glucagon or epinephrine as discussed in the text. *(Drawing by B. Tagawa, from Cyclic Nucleotides, by N. D. Goldberg,* Hospital Practice, *Vol. 9, No. 5 and "Cell Membrane: Biochemistry, Cell Biology, & Pathology," G. Weissmann and R. Claiborne, Eds., HP Publishing Co., Inc., New York, NY 1975. Reprinted with permission.)*

brane into the cytoplasm acts allosterically to activate an enzyme, protein kinase which, as the name implies, catalyzes the phosphorylation of other proteins. In this case, protein kinase acts to add phosphate groups to two enzymes, glycogen synthetase and phosphorylase kinase. The phosphorylation of glycogen synthetase serves to inhibit its catalytic activity and thus to prevent the conversion of glucose to glycogen. In contrast, the phosphorylation of the other enzyme activates it, and it begins to phosphorylate phosphorylase. The addition of phosphates to phosphorylase converts it into the active form, and glycogen breakdown is stimulated.

As one might expect, some type of mechanism must exist to reverse these steps, otherwise the cell would remain in the activated state from that time on. There are several enzymes that act to reverse these modifications. For example, there is an enzyme, phosphodiesterase, which catalyzes the destruction of cyclic AMP. As long as the hormone is bound to the receptor, cyclic AMP is produced and phosphorylase is activated. However, in the absence of hormone, the existing cyclic AMP is rapidly degraded and its effect stops. Similarly, there are enzymes, termed *phosphatases,* in the cell which function to remove the phosphate groups added by the kinases. In the absence of cyclic AMP, the phosphatases deactivate phosphorylase and reactivate glycogen synthetase and direct glucose metabolism in the opposite direction.

Even though the insulin receptor has been better characterized than that of any other hormone, less is understood about its mechanism of action. Since in many ways the insulin effect is a reversal of the effects of glucagon or epinephrine, one might expect that insulin would produce some type of an antagonistic response to the other hormones. Measurements on changes in cyclic nucleotide concentrations after administration of insulin have been very controversial. Some reports suggest that insulin causes a decrease in cyclic AMP levels and an increase in cyclic GMP levels. Cyclic GMP is another cyclic nucleotide which has also been implicated as a second messenger. In those systems where cyclic GMP has been best studied (such as in the control of cell division, page 645) it has been found to produce an opposite type of response to cyclic AMP, which is in keeping with it being a mediator of insulin action. Other investigators believe that the insulin response is mediated by cations whose permeability is altered upon insulin-receptor interaction. The cation which has received the most attention as a second messenger is calcium. A number of different responses, including secretion, muscle contraction, cell movement, etc., have been shown to result from a sudden increase in the Ca^{2+} concentration within the cytoplasm. Similarly, it has been shown that a number of hormones are capable of initiating a particular response as a result of the increased permeability of the plasma membrane to calcium. It may also be possible that certain of the effects of insulin result from the delayed entrance of the hormone into the cell. This latter possibility has gained recent support.

Research on membrane receptors has greatly expanded in the past few years with the discovery that receptors play an important role in the response by the central nervous system to addictive drugs. After considerable effort it was finally demonstrated by several laboratories that opiates, such as morphine and heroine, become specifically bound at the surfaces of nerve cells in certain brain regions. Further research indicated that these opiate receptors were normally involved in interactions with hitherto undiscovered brain substances that have been termed *endogenous opiates,* i.e., opiumlike substances normally produced within the brain. Although the role of these substances is not yet understood, some evidence exists which suggests that their activity is mediated by cyclic nucleotides. Recent experiments by Marshall Nirenberg and coworkers have provided an in vitro system for the study of the action of opiates on nerve cells as well as a theory on the biochemical basis for drug tolerance (the need for increasing amounts of the drug with time) and withdrawal symptoms. Using a special type of nerve cell which grows in culture, it has been found that initially these cells respond to the presence of opiates by decreasing their levels of cyclic AMP. After two to three days in the presence of morphine the cyclic AMP levels are back to normal, but the maintenance of the normal levels is dependent upon the drug in the medium. If the drug is removed, the cyclic AMP concentrations appear to rise, a phenomenon which may be involved in the withdrawal symptoms.

THE MOVEMENT OF SUBSTANCES
ACROSS CELL MEMBRANES

Since the contents of the cell are completely surrounded by the cell membrane, all communication between the cell and the extracellular compartment must be mediated by this organelle. In a sense, the plasma membrane has a dual function to perform. On one hand, it must retain the dissolved materials of the cell so that they do not simply leak out into the environment, while on the other hand it must allow the necessary interchange of materials into and out of the cell. The lipid bilayer of the membrane is ideally suited to prevent the loss of the dissolved materials of the cells, but some provisions must be made for the influx of nutrients and the efflux of waste products that might otherwise be blocked by the relatively impermeable lipid layer. As will be described below, the plasma membrane is a selectively permeable barrier, i.e., it is not equally permeable to all types of molecules, and it is this selectivity that allows the distribution of substances inside the cell to be totally different from that outside the cell.

To understand the selectively permeable nature of the membrane we need to consider the means by which individual molecules can traverse the membrane. There are basically two means for such movement: passively by diffusion or actively by some energy-coupled transport process. In either case it is possible for there to exist a directed movement, i.e., net flux, of a particular type of molecule. The term *net flux* indicates that the movement of the substance into the cell (influx) and out of the cell (efflux) is not balanced, but that one exceeds the other. We will consider diffusion and active transport in turn, but first we will describe the energetics when the net flux of a substance is occurring.

The Energetics of Solute Movement

Consider the events that occur when a cube of sugar is dropped into a container of hot water. At time zero, all of the sugar molecules are localized within the solid, none are in solution. After a sufficient period of time, the distribution of sugar molecules will change as a result of diffusion to one in which they are distributed in a uniform manner throughout the container. Each molecule in solution possesses a level of kinetic energy that causes it to bounce around randomly such that it is as likely to move in one direction at a given instant as another. The inevitable outcome of such motion is that concentration gradients originally present in the solution will not be maintained. The length of time it takes to achieve an equilibrium condition of uniform distribution will depend upon the temperature of the solution and the size of the molecule. The movement of molecules from a more concentrated to a less concentrated state is a shift from a more ordered to a more disordered state. Diffusion is therefore a spontaneous process that is driven by an increase in entropy. The reverse process, in which molecules are concentrated, results in a decrease in entropy and is a thermodynamically unfavorable event which must be driven by an input of energy.

The free-energy change that occurs during diffusion across a membrane is dependent upon the extent of the concentration gradient, i.e., the difference in concentration between the two compartments. As the concentration gradient decreases, the free-energy difference decreases until at equilibrium it is zero. The standard free-energy difference ($\Delta G°$) refers to that which occurs when one mole of solute is moved at 20°C from one side of the membrane at concentration C_1 to the other side at concentration C_2. When the solute is a nonelectrolyte (uncharged) the following relationship is found:

$$\Delta G° = RT \ln \frac{C_2}{C_1}$$

If the movement of a nonelectrolyte occurs from high concentration to low concentration by diffusion, the sign of $\Delta G°$ will be negative, indicating the process is exergonic. If the movement of a nonelectrolyte is to occur in the reverse direction, it is endergonic and must be coupled to an appropriate energy-yielding process. In other words, diffusion is accompanied by a loss in free energy while active transport is accompanied by a gain in free energy for the transported molecules at the expense of free energy of the energy-yielding reaction to which the transport must be coupled.

If the solute is a charged species another factor must be considered, namely the overall charge present in the two compartments. As a result of the mutual repulsion of electrolytes of like charges, it will be thermodynamically unfavorable for a charged species to move from one compartment to another compartment having an overall charge of the same nature. Conversely, if the charge of the moving substance is opposite in sign to the compartment into which it is moving, the process will be thermodynamically favored. The difference in free energy between the two compartments depends upon the magnitude of the difference in charge (potential difference or voltage). When one is considering the diffusion of an electrolyte there are two gradients which must be taken into account to determine the direction in which diffusion will occur. These are the chemical gradient, which depends upon the concentration difference of the substance between the two compartments, and, secondly, the electrical difference, which depends upon the difference in charge. Together these differences are referred to as the *electrochemical gradient*. The standard free-energy difference for an electrolyte is

$$\Delta G° = RT \ln \frac{C_2}{C_1} + ZF \Delta \psi$$

where Z is the valence of the solute, F is the faraday (96,493 C/g-equiv), and ψ the potential difference between the two compartments. The presence of a charge on a dissolved substance adds a complicating factor that has important consequences for the distribution of materials across cell membranes.

The Donnan Equilibrium

The major complexity in the distribution of charged substances across the membrane by diffusion is that the membrane is not equally permeable to all species. For example, membranes are generally most permeable to chloride and potassium ions, much less permeable to sodium ions, and impermeable to the ionic charges of the amino acid residues (mostly anionic) in the proteins. The effect of a nonpermeating ionic charge on the distribution of the diffusible ions was described by Frederick Donnan in 1911. To illustrate the effect, we will consider the following hypothetical situation of placing a "cell" into an isotonic solution of potassium chloride (Fig. 5-48). Within the "cell" there is potassium chloride plus a fixed number of nondiffusible charges borne by molecules of protein (one charge per molecule). The number of negative charges (both chloride and protein) is balanced by an equivalent number of potassium ions so that the "cell" is electrically neutral. Outside of the "cell" there is only potassium chloride. Since the total number of particles on each side of the membrane is initially equivalent, the osmotic activity of the two sides is equal.

First consider the concentration gradients. Within the "cell" there is a higher concentration of a nondiffusible protein than is present on the outside. However, as long as these molecules cannot penetrate the membrane, this gradient remains undisturbed. In contrast, chloride ions are initially more concentrated on the outside of the "cell" and therefore we would expect a *net* diffusion of chloride into the "cell." Since the movement of chloride ions will be accompanied by an equivalent movement of potassium ions in the same direction (in order to maintain the electrical neutrality of each compartment), the concentration of K^+ inside will increase while the K^+ concentration outside will decrease. As we will see below, the tendency for the ions to move in response to a concentration gradient will result in the establishment of a voltage.

Since the internal compartment is electrically neutral when

$$(K^+)_i = (Prot^-)_i + (Cl^-)_i$$

the concentration of K^+ inside will invariably exceed that of Cl^-. The greater the protein concentration, the greater $(K^+)_i > (Cl^-)_i$. In contrast, $(K^+)_o$ will always equal $(Cl^-)_o$. Donnan predicted that in this situation at *equilibrium* the product of the concentrations of *diffusible* ions on each side of the membrane should be equal to one another, i.e.

$$(K^+)_i(Cl^-)_i = (K^+)_o(Cl^-)_o$$

This equation follows from the fact that at equilibrium, there should no net movement of any ionic species, i.e., the energies (which are determined by the concentrations of the diffusible ions) on both sides of the membrane should be equal. According to the Nernst equation

$$\pi = \frac{RT}{ZF} \ln \frac{(K^+)_i}{(K^+)_o} = \frac{RT}{ZF} \ln \frac{(Cl^-)_o}{(Cl^-)_i}$$

where π is the potential difference, R the gas constant, F the faraday constant,

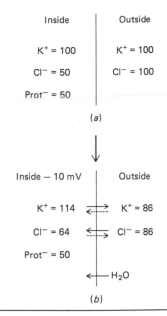

Figure 5-48. An example of Donnan equilibrium, which arises when a nonpermeating ion is present on one side of a semipermeable membrane. The relative numbers of the various ions are shown at the beginning of the experiment in *(a)* and after equilibrium in *(b)*. At equilibrium the tendency for the ions to move down their concentration gradient (solid arrows) is just balanced by their tendency to move in the opposite direction in accordance with the electrical gradient (dotted arrows). *(From B. Katz, "Nerve, Muscle, and Synapse," McGraw-Hill, 1966.)*

Z the valency of the ion, and T the absolute temperature. It is apparent that this equation simplifies to the previously stated Donnan equilibrium.

Since there is a difference in the concentration of diffusible ions on both sides of the membrane (Fig. 5-48b), a voltage will be established across the membrane at equilibrium. The tendency for ions to move down this electrical gradient will be exactly countered by their tendency to move in opposite directions down their concentration gradient. In Fig. 5-48b, the concentration of K^+ inside will be greater than that of K^+ outside, and that of Cl^- outside will be greater than Cl^- inside. The tendency for K^+ to move outside and Cl^- to move inside is exactly balanced by the voltage (internal negative relative to the outside environment). The establishment of a Donnan equilibrium in Fig. 5-48b has also resulted in the net movement of particles into the "cell," thereby raising its osmotic pressure. Where initially the two sides of the membrane were isotonic to each other, the inside is now hypertonic to its environment and water will enter the "cell" increasing its volume.

As we will see below, when the actual concentrations of diffusible ions on either side of cell membranes are measured, their values (particularly that of Na^+) do not fit that predicted by the Donnan equilibrium, i.e., the energies of the two sides are not equal. In order to explain actual ion concentrations within a cell, new mechanisms will have to be invoked.

Diffusion of Substances through the Membrane

We will begin by considering the diffusion of a nonelectrolyte through the plasma membrane into a cell. There are two qualifications that must be met for diffusion to occur. The substance must be present at higher concentration outside of the cell and the membrane must be permeable to the substance. The permeability of the membrane to a given substance can be a complex property. Given the presence of a lipid bilayer within the core of the membrane, it is not surprising that the lipid solubility of the permeating substance can be a very important determining factor in its rate of penetration. As previously discussed (page 197), the relationship between the hydrophilic nature of molecules and their relative lack of penetration into cells was one of the earliest parameters measured in the study of the nature of cell surfaces. In order for a hydrophilic molecule to diffuse into the cytoplasm of a cell, it must have sufficient energy to break the noncovalent bonds it has formed with the aqueous solvent

TABLE 5-9
Relationship between Partition Coefficient and the Rate of Penetration of a Substance

	Part. coef. ($\times 10^2$)	Relative rate
Methyl alcohol	0.78	0.99
Glycerol ethyl ether	0.74	0.077
Propylene glycol	0.57	0.087
Glycerol methyl ether	0.26	0.043
Ethylene glycol	0.049	0.043
Glycerol	0.007	0.00074
Erythritol	0.003	0.000046

SOURCE: R. Collander, *Trans. Faraday Soc.*, **33**:985 (1937).

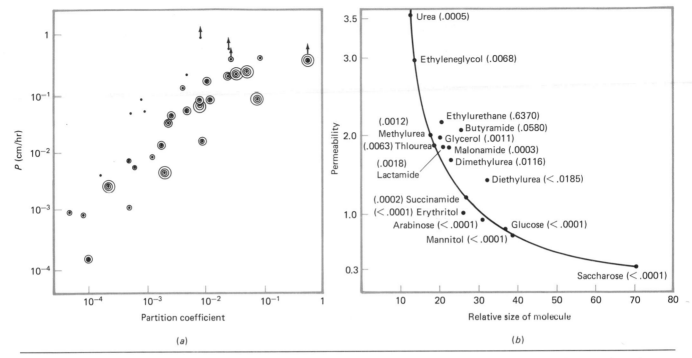

Figure 5-49. (a) The relationship between the partition coefficient of a molecule, i.e., its solubility in oil/ solubility in water, and its penetration (P) into the cells of the alga *Chara*. The size of the circle provides a relative measure of the molecular weight of the compound. The overriding importance of lipid solubility is apparent. [*From R. Collander*, Trans. Faraday Soc., **33**:986 (1937).] (b) The relationship between molecular size and permeability in *Beggiatoa*. Partition coefficients in ether-water are given in parentheses. [*From A. C. Giese*, "Cell Physiology," 4th ed., Saunders, 1973; data based on Ruhland and Hoffman, Planta, **1**:1 (1925).]

and move through the lipid layer with which it cannot interact. The more polar the molecule, the greater must be its energy content to be moved through a hydrocarbon layer of very low dielectric constant.

One simple measure of the polar or nonpolar nature of a substance is its partition coefficient, i.e., the ratio of its solubility in oil as opposed to water. The partition coefficient for a compound is measured by dissolving the substance in the oil or water, then shaking the two immiscible solvents together, allowing them to separate into two phases, and measuring the concentration of the solute in each phase. The partition coefficient is its concentration in oil over that in water. The importance of lipid solubility in permeability is shown in Table 5-9, where the penetration through the erythrocyte membranes of a series of alcohols of decreasing partition coefficient is compared. Figure 5-49a shows the relationship between partition coefficient and permeability for a number of molecules. There are several important consequences apparent from this figure. Since the small molecular weight intermediates present in a cell are generally of a polar nature, the membrane will provide an effective barrier in keeping them from diffusing out. Similarly, since it is generally these same types of molecules that must enter cells from the bloodstream (sugars, amino acids, etc.), it is apparent that they will not do so by simple diffusion, but will require special mechanisms to effect their penetration. The presence of these special mechanisms allows the cell to control the movement of substances across its surface barrier. We will return to these features of membranes in a later section.

Even though the relationship expressed in Fig. 5-49*a* is a very basic one, there are a number of small hydrophilic molecules that penetrate much more rapidly than would be expected on the basis of their partition coefficient. The most surprising of all of these is water itself, which moves very rapidly through the plasma membrane. Consider the deplasmolysis experiments of Overton described on page 197. It was found that when root hair cells were placed into solutions that were hypertonic, there would occur an almost instantaneous shrinkage of the cell due to the exit of water from the cell into its environment. Hypertonic solutions are ones which have a greater number of solute particles in a given volume of solvent, i.e., they have a greater osmotic pressure, which will cause water to move by diffusion (osmosis) into that compartment. Virtually all the evidence from a very large number of studies indicates that the movement of water through cell membranes is always a passive process. It appears that membranes are generally very permeable to diffusion by water. Since cell volumes will ultimately depend upon the relative movement of water into and out of the cell, it is imperative that the osmotic pressure of a cell be maintained at just the right level relative to its surroundings. This is an important function of the membrane. An exception to this requirement occurs in the case of plant and bacterial cells bounded by tough extracellular walls which maintain the cell's volume in the face of a hypotonic environment.

The rapid diffusion of water and other small hydrophilic molecules (urea, methanol, certain ions) was considered for a long time as evidence that the membrane must contain hydrophilic pores, for it did not seem possible that such highly polar molecules could move through the nonpolar lipid layer. Based on various types of measurements it was proposed that the membrane pores were in the range of about 8 Å diameter. The presence of membrane pores has been one of the most controversial topics in this area of research, and the matter has still not been completely decided. One important point in this regard is the finding that artificial lipid bilayers are readily penetrated by water molecules (Table 5-4). In fact, these nonprotein-containing membranes show approximately the same penetration by water as do natural membranes. It has also been demonstrated that small ions can also penetrate artificial membranes, particularly if small amounts of charged detergent molecules are included in the membrane. It would appear that small hydrophilic molecules can get through the lipid layers without invoking special, water-filled pores.

In addition to the overriding importance of lipid solubility to the permeability of most organic molecules, the molecular size is also a determining factor in their rate of penetration. If two molecules have approximately equivalent partition coefficients, the smaller molecule tends to penetrate more rapidly than the larger one (Fig. 5-49*b*). The other major determining factor in the diffusion of a substance is whether or not it bears a charge. Since polar molecules have more difficulty passing through the plasma membrane than nonpolar ones, it might be expected that charged molecules would be even less likely to penetrate. This prediction can be tested by comparing the penetration of weak electrolytes at a pH where dissocia-

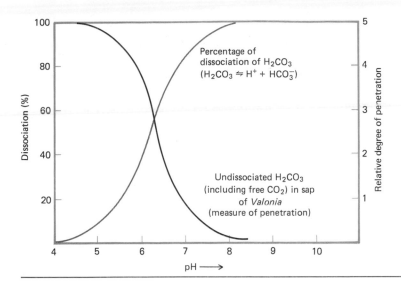

Figure 5-50. Relationship between pH and penetration of a weak electrolyte in *Valonia*. [*From A. C. Giese, "Cell Physiology," 4th ed., Saunders, 1973; based on data from W. J. V. Osterhout and M. J. Dorcas, J. Gen. Phys., 9:259 (1926).*]

tion is suppressed and a pH where the molecules exist in the charged state (Fig. 5-50). When this is done, the uncharged molecule invariably diffuses through the membrane more readily. The best-studied charged particles are small ions (Na^+, K^+, Cl^-, H^+, Mg^{2+}, Ca^{2+}) which can be present at relatively high concentrations both in the cell and the extracellular medium and are of such importance in the physiological activities of cells. Generally, small ions are capable of diffusing through membranes, but rather slowly. In the red blood cell, for example, anions generally penetrate more rapidly than cations; monovalent ions more rapidly than divalent ions; smaller ions more rapidly than larger species. As will be discussed at length below, movement of ions across membranes is much more complex than can be accounted for by simple diffusion and various types of transport mechanisms have evolved.

It should be kept in mind when one is considering the diffusion of molecules through a membrane that regulatory mechanisms exist by which these properties can be changed. Membranes of different types of cells can have considerably different permeabilities and the same membrane can vary greatly under different physiological states. For example, the basis of the formation of a nerve impulse is a sudden change in the permeability of the membrane to sodium ions. In another example, the diffusion of water through the membranes of the distal, convoluted tubule of the kidney is regulated by the presence or absence of antidiuretic hormone (vasopressin) secreted by the pituitary gland.

Facilitated Diffusion

The diffusion of a substance across a membrane always occurs from a region of high concentration on one side to a region of lower concentration on the other side, but it is not always the case that the substance diffuses by itself. Numerous examples have been uncovered where a protein (permease) exists within the membrane that facilitates the diffusion process. *Facilitated diffusion* is particularly common in the movement of sugars and amino acids across particular cell membranes, for example the band 3 protein of the erythrocyte (Fig. 5-37). The advantages afforded by the proteins are similar to the advantages provided a catalytic reaction by an enzyme. In both cases the protein has no effect on the direction of the reaction (or movement through the membrane), these are determined by the thermodynamic aspects of the system. However, in both cases the energy that the molecules have to possess to react or move through the membrane is greatly lowered, thereby accelerating the rate at which the reaction or movement occurs. This should not be surprising. The presence of a permease in the membrane offers a pathway through the membrane which is an alternative to the lipid layer, whose penetration would require much greater energy levels than the solute molecules would likely possess.

As in the case of enzymes, proteins that facilitate diffusion show saturation-type kinetics (Fig. 5-51), are specific for the molecules they bind, are capable of being inhibited (both competitively and noncompetitively), and are not changed irreversibly in the process. As with enzymes, permeases will facilitate the transfer process equally well in both directions, depending upon the relative concentration of the substance on the two sides of the membrane. In addition to accelerating the movement, these proteins provide an opportunity to regulate the flow of traffic through the membrane since the synthesis, degradation, and/or activity of the proteins can be controlled. For example, many hormones can have an important role in regulating the facilitated movements of substances across membranes.

Active Transport

If one measures the concentration of the two most important cations, Na^+ and K^+, their distribution across the membrane is not that expected from the Donnan electrochemical equilibrium. The typical K^+ concentration inside the cell is 100 mM, while that outside the cell is only 5 mM. The direction of the potassium gradient is that expected from the presence

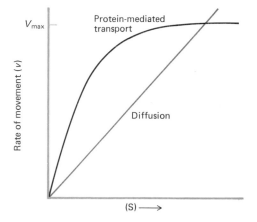

Figure 5-51. A comparison of the kinetics of transport when mediated by a carrier as opposed to simple physical diffusion.

of nondiffusible anions (primarily proteins and phosphates) inside the cell, but the magnitude of the gradient is unexpectedly high. The distribution of Na^+ is opposite from that expected (150 mM on the outside and only 10 to 20 mM on the inside). The ability of a cell to develop such extreme concentration gradients across its plasma membrane can only occur by active transport. One might suppose that the cell can form the gradients in the first place and then simply maintain them by causing its membrane to be impermeable to the diffusion of these ions. In other words, once the ionic distribution is established by active transport, the cell would not have to expend additional energy to maintain it. As can be seen by a simple experiment, this is not the case. If cells are placed in a medium containing radioactive sodium ions, it can be shown that labeled ions move into the cell down the concentration gradient. The membrane is clearly not impermeable to sodium ions, they are continually leaking inside. Since the overall concentration of sodium ions does not increase inside the cell, the sodium ions entering must be subsequently moved back in the opposite direction against the concentration gradient. This outward movement of sodium ions by active transport can also be demonstrated using radioactively labeled sodium ions. It would appear that the distribution of ions across the membrane is a steady-state condition, in which sodium ions are continually leaking in by diffusion and being pumped out by active transport. Similarly, the movements of potassium ions occur in the reverse directions (Fig. 5-52) by the same types of processes.

Since the large concentration gradients of the cations are generated by active transport, it would be expected that, if the energy input into the transport process were cut off, the pumping activity would cease. This is exactly what occurs when ATP formation is brought to a halt by the addition of metabolic poisons acting to inhibit oxidative metabolism (Fig. 5-53a). When a cell is poisoned in this manner, the influx of sodium and the efflux of potassium ions continue by diffusion, but the reverse process

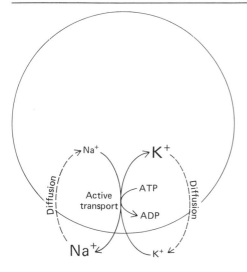

Figure 5-52. The movements of sodium ions and potassium ions into and out of the cell. The relative size of the letters indicates the direction of the concentration gradients. ATP hydrolysis is required for the movements of the ions against their gradients.

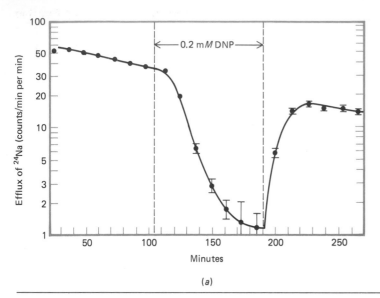

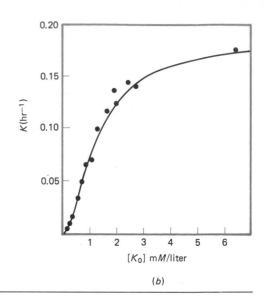

(a)

(b)

Figure 5-53. *(a)* Sodium ion efflux from a squid axon before, during, and after treatment with dinitrophenol, a substance which blocks the formation of ATP. [*From A. L. Hodgkin and R. D. Keynes,* J. Physiol., **128:** *34 (1955).*] *(b)* The dependence of Na^+ efflux on the external K^+ concentration. Sodium ion efflux is plotted on the ordinate as the "rate constant" for the potassium-dependent component of the sodium efflux. [*From J. R. Sachs and L. G. Welt,* J. Clin. Invest., **56:** *72 (1967).*]

whereby each ion is pumped back cannot occur. As a consequence, the steep ionic gradient for each cation is lost and the osmotic balance is disturbed so that the cells take up water and swell.

The Nature of the Na^+–K^+ Pump

The analysis of the mechanism whereby the sodium and potassium gradients are generated has been one of the most intensively studied in cell biology. In the past few years, along with the increase in our understanding of membrane structure and function in general, considerable advances have been made in unraveling the molecular mechanisms of these ionic pumps. These studies have concentrated on the ionic pumps of the red blood cell, certain giant nerve cells, and various specialized cells that function to secrete concentrated salt solutions. The most important feature of the ion pumps in all of these cells is that the same mechanism is responsible for the movement of both sodium and potassium ions. The pump is a reciprocal one so that the transport of one type of ion cannot occur without the simultaneous movement of the other species in the opposite direction. One of the most important findings was made by J. C. Skou in 1957 when the relationship between the ion pump of a large nerve of a crab and an ATPase activity was discovered. It was found that this ATPase was active only in the presence of both Na^+ and K^+ (as well as Mg^{2+} which acts as a cofactor). It was proposed at that time, and correctly so, that the enzyme, a (Na^+–K^+)–ATPase, responsible for the hydrolysis of ATP was the same protein that was active in transporting the two ions. The means by which the coupling of energy metabolism to active transport had been uncovered.

Another series of experiments that demonstrated some of the important properties of the (Na^+–K^+)–ATPase was carried out using hemolyzed, resealed erythrocyte ghosts. In a previous section, the preparation of ghosts was described using hypotonic solutions that cause the erythrocyte to swell until it becomes permeable to its dissolved contents which leak out. This technique has been used in a number of ways to understand various properties of red cell plasma membranes. In the present experiments, the hemolyzed erythrocyte ghosts were placed into solutions having defined concentrations of ions and then allowed to shrink back to their original size and resealed. These resealed ghosts could then be placed into various types of media and the transport activities of the membrane assessed. It was found that transport of either Na^+ or K^+ would occur only if both were present on the appropriate side of the membrane. Potassium ions had to be present on the outside (Fig. 5-53b) and Na^+ present on the inside so that both could be pumped across the membrane. ATP had to be present and it had to be available inside the resealed ghost; externally applied ATP could not be used to drive the pump. Similarly, it was determined that a well-known inhibitor of cation transport, ouabain, was effective only if applied externally to the resealed ghosts. Ouabain is a digitalislike compound, and the effect of digitalis on strengthening heart beat is believed to be due to its inhibition of the cation pump.

On the basis of a large number of studies the evidence suggests that the ratio of Na:K pumped is not 1:1, but 3:2. In other words, for each ATP hydrolyzed, 3 sodium ions are pumped out to only 2 potassium ions pumped in. One consequence of this pumping ratio is that the pump can be *electrogenic*, i.e., contribute directly to the separation of charge across the membrane. This latter statement would be the case as long as the cations are not accompanied by chloride ions as they traverse the membrane since chloride ions would cause the cation movements to be electrically neutralized. There is considerable evidence that the cation pump does contribute directly to the potential difference of the membrane along with the diffusion potential already discussed on page 251.

The experiments with red blood cell ghosts clearly illustrate the asymmetrical nature of the Na^+–K^+ pump. This asymmetry is one of the most important features of all active transport systems. Unlike the protein-mediated movement of the facilitated diffusion systems, which will carry the substance equally well in either direction, the active transport systems are coupled to their energy sources in such a way as to drive the reaction in only one direction.

The mechanism by which sodium and potassium are transported is not fully understood, but it is believed to occur in steps and involve the phosphorylation of the transport protein and conformational changes in its structure. Phosphorylation occurs as ATP is hydrolyzed, so that the energy of the phosphate bond is "captured" by the formation of a bond to an aspartic acid residue in the protein. The phosphorylation requires the presence of sodium ions and the dephosphorylation requires the presence of potassium ions. If red blood cells containing radioactively labeled ATP are incubated in a medium lacking potassium, the transport protein can be radioactively labeled by phosphorylation. Since, in the absence of potassium, the dephosphorylation cannot occur, the protein will remain labeled and its presence in an electrophoretic gel can be detected. When this is done, the radioactivity migrates with one of the minor glycoproteins (not band 3 or one of the prominent PAS bands of Fig. 5-37). It is estimated that there are only about 200 to 300 pumps (ouabain binding sites) per cell, and each pump can move approximately 6000 K^+ per minute. It is believed that the pump consists of two types of subunits, a catalytic and noncatalytic one.

The conformational changes that are believed to occur in the transport protein during the process would seem to be necessary as a means whereby its affinity for the two cations can change. Consider the situation confronting the protein. It must pick up the sodium or potassium ions from a region where they are at low concentration, which means that it must have a relatively high affinity for the ions. Then it must somehow release the ions on the other side of the membrane in the face of the pressure of a much greater concentration of each ion; its affinity for that species must decrease. Somehow its affinity on the two sides of the membrane must be different and this is best explained on the basis of a change in shape of the protein molecule.

The topic of conformational changes of transport proteins brings up another important and controversial topic. Up until a few years ago it was believed that the movement of a substance from one side of the membrane to the other was accompanied by the movement of the transport protein across the membrane as well. If not an actual movement of the protein, it was thought that at least it would rotate so that the binding site was turned from one side to the other. The term *carrier* has generally been used to suggest the physical movement of the transport protein. Recent evidence indicates, though not conclusively, that these types of movements do not occur. With the development of the fluid mosaic model and the demonstration of the presence of transmembrane proteins, new evidence for the existence of *fixed pores* has been suggested. The fixed pore would be represented by the association of transmembrane proteins (oligomeric complexes) to form channels (Fig. 5-54a) from which lipid molecules would be excluded. The binding sites for the transported substances would be within this pore and the molecule or ion would move from one side to the other (Fig. 5-54b) via a conformational change which would

Figure 5-54. (a) Water-filled channel for transport of specific ions and hydrophilic molecules through the membrane may be formed by groupings of four (or more) protein subunits. *(Drawing by B. Tagawa, from Architecture and Topography of Biologic Membranes, by S. J. Singer, Hospital Practice, Vol. 8, No. 5 and "Cell Membranes: Biochemstry, Cell Biology, and Pathology," G. Weissmann and R. Claiborne, Eds., HP Publishing Co., Inc., New York, NY 1975. Reprinted with permission.)* (b) A schematic representation of a mechanism for the translocation of a hydrophilic molecule across a membrane. In one state of the protein complex, the active site is accessible from the outside. With the appropriate energy input, the aggregate rearranges and, in the process, the molecule is translocated to the inside of the membrane. [A. Dutton, E. D. Rees, and S. J. Singer, Proc. Natl. Acad. Sci. (U.S.), **73**:1535 (1976).]

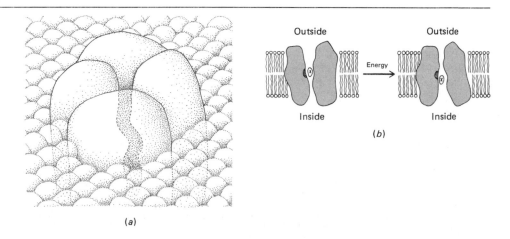

(a)

(b)

also lower the affinity of the protein for the substance so that it could be released. Another possibility is that the substance might be passed through the pore by conformational changes to successively weaker and weaker binding sites as it traversed the membrane.

One of the most recent advances in this area of research has been the purification of the Na⁺–K⁺ transport system and the study of its properties when isolated from other proteins of the membrane. One of the most important techniques that has been developed is the inclusion of integral membrane proteins into artificially prepared membrane vesicles. In a sense, the "reconstitution" of a membrane system involves taking it apart and then putting it back together again under a more defined set of conditions in which it can be studied with a minimum of interference. These "reconstituted" transport systems are capable of the same activities that were present in the cell and have the same requirements. The primary difference between the native and reconstituted state is that in the latter case the pump is oriented in the reverse direction. In other words, it pumps sodium into the cavity of the vesicle and potassium to the outside, in both cases against a gradient. To carry out this active transport, sodium and ATP must be present outside of the vesicle and potassium present inside. Though the presence of phospholipids is required for transport activity, it appears to make little difference which phospholipids are associated with the pump to cause it to transport cations.

The establishment of concentration gradients, such as those of sodium and potassium, provides a means by which free energy can be stored in the cell. The potential energy present in such a gradient can be utilized in various ways to perform work for the cell. This concept will be explored in detail in later chapters when it will be demonstrated how ionic gradients are believed to be coupled to the formation of chemical energy in the mitochondrion and chloroplast. For the present discussion we will show how this energy can be tapped to drive the movement of other molecules against a concentration gradient.

The Movement of Sugars across the Intestinal Wall

It has been shown that various types of sugar molecules can move across eukaryotic cell membranes in several ways; by simple diffusion, by facilitated diffusion in many cells such as the erythrocyte, and by a special type of active transport in cells of the epithelial wall of the small intestine and kidney. Consider the physiological activity of the intestine. Within its lumen, enzymes are present which break down large molecular weight polysaccharides into simple sugars. These sugars then move across the length of the epithelial cells of the intestine (across both outer and inner plasma membranes) and into the bloodstream as the capillaries flow beneath the basal end of the cells. Experiments have shown that the movement of sugars across the outer membrane (the brush border in contact with the lumen) of these cells against a concentration gradient, occurs by cotransport with sodium ions. The driving force is the difference in concentration of sodium on the two sides of the membrane. This arrangement is illustrated in Fig. 5-55. The Na⁺ concentration is kept very low within the cells by the action of an active sodium transport system which pumps sodium ions out of the cell against a concentration gradient.

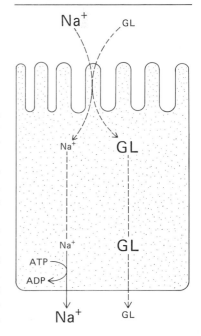

Figure 5-55. The movement of sugars across the outer membrane of the intestinal epithelial cells is coupled to the diffusion of sodium ions down a concentration gradient. The sodium ion gradient at the outer membrane is maintained by actively transporting the ions out of the cell at its inner surface. The dotted lines indicate diffusion, the solid line active transport. The relative size of the letters indicates the direction of the concentration gradients.

As a result of this pump, a Na^+ gradient is established across the membrane by which sodium will tend to move by diffusion. In the intestine this sodium gradient is "tapped" to drive the cotransport of sugar molecules *against* a concentration gradient of lesser magnitude than the gradient of sodium ions in the reverse direction. On the outside of the cell the transport protein binds both sodium and the sugar. When the sodium ion is released on the inside of the cell into a solution of lesser sodium concentration, the conformation of the protein is apparently changed so that it loses its affinity for the sugar molecule which is then released into the cell against its concentration gradient. Once inside, the sodium ions can be pumped back across the membrane to maintain the ionic gradient while the sugar can diffuse across the basal membrane of the cell and into the bloodstream in response to the lowered sugar concentration on the other side of the basal plasma membrane.

From this last example it can be seen that the means by which energy is coupled to the uphill movement of a substance can be quite different in various cases. In the case of the (Na^+-K^+)-ATPase, the coupling occurs via a phosphorylated membrane protein. In the movement of sugars (and also amino acids) in the intestine and kidney, the energy is expended in an earlier step, that in which the formation of the ion gradient occurs. In other cases, the actual state of the transported molecule is changed. For example, in certain cases sugars are phosphorylated as they move through the membrane, a step which requires energy but places the sugar in a new chemical state which cannot diffuse back through the membrane. In other cases, the sugar molecule (or other substance) might be polymerized. In any case, whether the molecule is phosphorylated, or polymerized, or precipitated (as occurs for calcium in some cases), the effect is to remove it on the inside of the membrane so that the concentration gradient favoring its diffusion across the membrane continues to be maintained.

Before leaving the subject of transport, it should be mentioned that some of the most intensive investigations have been carried out on bacterial systems and these have been ignored in this section. Since the same basic principles apply in both eukaryotic and bacterial cells, we have chosen to concentrate on the former in keeping with the overall focus of this book. One bacterial system will be explored in the next chapter, others can be found in various reviews (e.g., Boos, 1974).

REFERENCES

Ash, J. F., Louvard, D., and Singer, S. J., Proc. Natl. Acad. Sci. (U.S.) **74**, 5584–5588, 1977. "Antibody-Induced Linkages of Plasma Membrane Proteins to Intracellular Actomyosin-Containing Filaments in Cultured Fibroblasts."

Askari, A., ed., Ann. N.Y. Acad. Sci. **242**, 1974. Symposium on "Properties and Functions of $(Na^+ + K^+)$-Activated Adenosinetriphosphatase."

Bergelson, L. D., and Barsukov, L. I., Science **197**, 224–230, 1977. "Topological Asymmetry of Phospholipids in Membranes."

Biomembranes, Plenum. Vol. 1 beginning 1971.

Boos, W., Ann. Rev. Biochem. **43**, 123–146, 1974. "Bacterial Transport."

Bradshaw, R. A., and Frazier, W. A., Current Topics in Cellular Reg. **12**, 1–37, 1977. "Hormone Receptors as Regulators of Hormone Action."

Branton, D., Proc. Natl. Acad. Sci. (U.S.) **55**, 1048–1056, 1966. "Fracture Faces of Frozen Membranes."

—— and Deamer, D. W., *Membrane Structure*. Springer-Verlag, 1972.

—— and Park, R. B., *Papers on Biological Membrane Structure*. Little, Brown, 1968.

Bretscher, M. S., Science **181**, 622–629, 1973. "Membrane Structure: Some General Principles."

—— and Raff, M. C., Nature **258**, 43–49, 1975. "Mammalian Plasma Membranes."

Brinkley, B. R., and Porter, K. R., eds., symposium on "Plasma Membrane Organization" in *International Cell Biology 1976-1977*, pp. 5–28. Rockefeller University Press, 1977.

Bronner, F., and Kleinzeller, A., eds. *Current Topics in Membranes and Transport*. Academic. Vol. 1 beginning 1970.

Capaldi, R., Sci. Am. **230**, 26–33, March 1974. "A Dynamic Model of Cell Membranes."

Carraway, K. L., Biochem. Biophys. Acta **415**, 379–410, 1975. "Covalent Labeling of Membranes."

Catt, K. J., and Dufau, M. L., Ann. Rev. Phys. **39**, 529–557, 1977. "Peptide Hormone Receptors."

Chapman, D., Biomembrane **7**, 1–9, 1975. "Fluidity and Phase Transitions of Cell Membranes."

Christensen, H. N., *Biological Transport*. 2d ed., W. A. Benjamin, 1975.

Cuatrecasas, P., Ann. Rev. Biochem. **43**, 169–214, 1974. "Membrane Receptors."

—— and Greaves, M. F., eds., *Receptors and Recognition*. Halsted, 1977.

Czech, M. P., Ann. Rev. Biochem. **46**, 359–384, 1977. "Molecular Basis of Insulin Action."

Dahl, J. L., and Hokin, L. E., Ann. Rev. Biochem. **43**, 327–356, 1974. "The Sodium-Potassium Adenosinetriphosphatase."

Davson, H., *A Textbook of General Physiology*. 4th ed., Churchill, 1970.

Edelman, G. M., Science **192**, 218–226, 1976. "Surface Modulation in Cell Recognition and Growth."

Edidin, M., Ann. Rev. Biophys. and Bioeng. **3**, 179–203, 1974. "Rotational and Translational Diffusion in Membranes."

Eisenberg, M., and McLaughlin, S., Bioscience **26**, 436–443, 1976. "Lipid Bilayers as Models of Biological Membranes."

Fox, C. F., Sci. Am. **226**, 30–38, Feb. 1972. "The Structure of Cell Membranes."

Frye, L. D., and Edidin, M., J. Cell Science **7**, 319, 1970. "The Rapid Intermixing of Cell Surface Antigens after the Formation of Mouse-Human Heterokaryones."

Giese, A. C., *Cell Physiology*. 4th ed., Saunders, 1973.

Glynn, I. M., and Karlish, S. J. D., Ann. Rev. Physiol. **37**, 13–55, 1975. "The Sodium Pump."

Gurr, M. I., and James, A. T., *Lipid Biochemistry: An Introduction*. Cornell, 1971.

Helmreich, E. J. M., Zenner, H. P., and Pfeuffer, T., Current Topics Cell Reg. **10**, 41–87, 1976. "Signal Transfer from Hormone Receptor to Adenylate Cyclase."

Hers, H. G., Ann. Rev. Biochem. **45**, 167–190, 1976. "The Control of Glycogen Metabolism in the Liver."

Hughes, R. C., Essays Biochem. **11**, 1–36, 1975. "The Complex Carbohydrates of Mammalian Cell Surfaces and Their Biological Roles."

Juliano, R. L., and Behar-Bannelier, M., Biochem. Biophys. Acta **375**, 249–267, 1975. "An Evaluation of Techniques for Labelling the Surface Proteins of Cultured Mammalian Cells."

Kahn, C. R., J. Cell Biol. **70**, 261–286, 1976. "Membrane Receptors for Hormones and Neurotransmitters."

Kimmich, G. A., Biochem. Biophys. Acta **300**, 31–78, 1973. "Coupling between Na^+ and Sugar Transport in Small Intestine."

Kolata, G. B., Science **196**, 747, 1977. Research News. "Hormone Receptors: How Are They Regulated?".

Kotyk, A., and Janocek, K., *Cell Membrane Transport.* 2d ed., Plenum, 1975.

Lee, A. G., Endeavour **34**, 67–71, 1975. "Interactions Within Biological Membranes."

Lloyd, C. W., Biol. Revs. **50**, 325–350, 1975. "Sialic Acid and the Social Behaviour of Cells."

Machtiger, N. A., and Fox, C. F., Ann. Rev. Biochem. **42**, 575–600, 1973. "Biochemistry of Bacterial Membranes."

Maddy, A. H., ed., *Biochemical Analysis of Membranes.* Chapman & Hall, 1976.

Marchesi, V. T., Furthmayr, H., and Tomita, M., Ann. Rev. Biochem. **45**, 667–698, 1976. "The Red Cell Membrane."

Markham, R., et al., eds., Phil. Trans. R. Soc. (London) **B268**, 1–159, 1974. Symposium on "The Electron Microscopy and Composition of Biological Membranes and Envelopes."

Meister, A., Science **180**, 33–39, 1973. "On the Enzymology of Amino Acid Transport."

Morgan, C., Int. Rev. Cytol. **32**, 291–326, 1972. "The Use of Ferritin-Conjugated Antibodies in Electron Microscopy."

Mühlethaler, K., Int. Rev. Cytol. **31**, 1–19, 1971. "Studies on Freeze-Etching of Cell Membranes."

Nicolson, G. L., Int. Rev. Cytol. **39**, 90–190, 1974. "The Interactions of Lectins with Animal Cell Surfaces."

———, Biochem. Biophys. Acta **457**, 57–108, 1976. "Transmembrane Control of the Receptors on Normal and Tumor Cells."

Pardee, A. B., Science **162**, 632–637, 1968. "Membrane Transport Proteins."

Parsons, D. S., ed., *Biological Membranes: Twelve Essays on Their Organization, Properties, and Functions.* Oxford University Press, 1975.

Pastan, I., Sci. Am. **227**, 97–105, Aug. 1972. "Cyclic AMP."

Pinto da Silva, P., and Branton, D., J. Cell Biol. **45**, 598–605, 1970. "Membrane Splitting in Freeze-Etching."

Poo, M., and Cone, R. A., Nature **247**, 438–441, 1974. "Lateral Diffusion of Rhodopsin in the Photoreceptor Membrane."

Poste, G., and Nicolson, G. L., eds., *Cell Surface Reviews.* Vol. 1, Elsevier, 1976.

Postma, P. W., and Roseman, S., Biochem. Biophys. Acta **457**, 213–257, 1976. "The Bacterial Phosphoenolpyruvate: Sugar Phosphotransferase System."

Puck, T. T., Proc. Natl. Acad. Sci. (U.S.) **74**, 4491–4495, 1977. "Cyclic AMP, the Microtubule-Microfilament System, and Cancer."

Quinn, P. J., *The Molecular Biology of Cell Membranes.* Macmillan, 1976.

Raff, M. C., Sci. Am. **234**, 30–39, May 1976. "Cell Surface Immunology."

Robertson, J. D., Sci. Am. **206**, 64–72, April 1962. "The Membrane of the Living Cell."

Rothman, J. E., and Lenard, J., Science **195**, 743–753, 1977. "Membrane Asymmetry."

Rothstein, A., Cabantchik, Z. I., and Knauf, P., Federation Proc. **35**, 3–10, 1976. "Mechanism of Anion Transport in Red Blood Cells: Role of Membrane Proteins."

Schwartz, A., Lindenmayer, G. E., and Allen, J. C., Current Topics Membrane and Transport **3**, 1–82, 1972. "The Na+, K+–ATPase Membrane Transport System: Importance in Cellular Function."

Segal, H. L., Science **180**, 25–32, 1973. "Enzymatic Interconversion of Active and Inactive Forms of Enzymes."

Sharon, N., Sci. Am. **230**, 78–87, May 1974. "Glycoproteins."

———, *Complex Carbohydrates.* Addison-Wesley, 1975.

———, Sci. Am. **236**, 108–119, June 1977. "Lectins."

Silbert, D. F., Ann. Rev. Biochem. **44**, 315–340, 1975. "Genetic Modification of Membrane Lipid."

Singer, S. J., Ann. Rev. Biochem. **43**, 805–834, 1974. "The Molecular Organization of Membranes."

——— and Nicolson, G. L., Science **175**, 720–731, 1972. "The Fluid Mosaic Model of the Structure of Cell Membranes."

Skou, J. C., "The (Na+–K+)-Activated Enzyme System," in *Perspectives in Membrane Biology,* S. Estrada and C. Gilter, eds., Academic, 1974.

Snyder, S. H., Sci. Am. **236**, 44–56, March 1977. "Opiate Receptors and Internal Opiates."

Sutherland, E. W., Science **177**, 401–408, 1972. "Studies on the Mechanism of Hormone Action."

Shamoo, A. E., ed., Ann. N.Y. Acad. Sci. **264**, 1975. Symposium on "Carriers and Channels in Biological Systems."

Symposium on "Cell Membrane Biophysics," J. Gen. Phys. **51**, 1s–391s, 1968.

Symposium on "Insulin—Its Synthesis, Release, Interaction with Receptors and Mechanism of Action," Federation Proc. **34**, 1537–1568, 1975.

Symposium on "Membranes and Mechanism of Hormone Action," Federation Proc. **32**, 1833–1858, 1973.

Thomas, R. C., Phys. Revs. **52**, 563–594, 1972. "Electrogenic Sodium Pump in Nerve and Muscle Cells."

Van Heyningen, W. E., Nature **249**, 415–417, May 1974. "Gangliosides as Membrane Receptors for Tetanus Toxin, Cholera Toxin, and Serotonin."

CHAPTER SIX
The Cell Surface

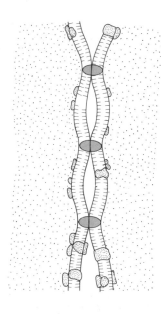

THE EXTRACELLULAR SPACE

Multicellular organisms are composed of two compartments, the cellular and the extracellular. Though the cells of an organism are obviously the basis for its being alive, the "nonliving" parts generally have very important functions to serve. If we consider the cell surface as the outer boundary of a living cell, we find ourselves in some difficulty since the plasma membrane is not necessarily the outer edge of the cell surface. For the present discussion we will consider three categories of cell surface materials (a) the cell coat, (b) the "fuzzy" layer, and (c) extraneous structures. The literature on the structure and function of the materials just outside the cell is more confusing than most and a clear integrated picture based on both electron microscopic and molecular techniques is generally unavailable.

The Cell Coat and "Fuzzy" Layer

The cell coat is generally considered to be equivalent to the carbohydrate chains attached to the proteins of the outer leaflet of the plasma membrane. This layer is visualized in the electron microscope by treating the cells or sections of the cells with heavy metal stains that form

(a)

(b)

Figure 6-1. *(a)* Freeze-etch replica showing a portion of the luminal surface of an epithelial cell of the rat rectum. Surface coat filaments extend from the tips of the microvilli into the lumen. [*From J. G. Swift and T. M. Mukherjee,* J. Cell Biol., *69:492 (1976).*] *(b)* Electron micrograph of a negatively stained isolated amoeba surface showing a thick layer of external filaments. [*From H. J. Allen, C. Ault, and J. F. Danielli,* J. Cell Biol., *60:26 (1974).*]

complexes with the carbohydrates of the coat. Even though these materials are actually a part of the structure of the membrane, they are on the outside of the cell and they can be removed without dealing the cell a mortal wound. The analogy has been made by John Luft between the plasma membrane–cell coat complex and the pelt of an animal. The skin of the pelt is analagous to the membrane—it forms a cover for the living contents which cannot be removed, though it can heal if damaged. The fur is like the cell coat: both are useful and necessary under many environmental conditions, but can be removed without fatal consequences, and both can be resynthesized by the underlying layer.

In many cells, there is a separate, "fuzzy" layer of material which exists outside of the cell coat and is composed primarily of carbohydrate-containing materials which are secreted by the cell into the extracellular space. The two types of cells in which the "fuzzy" layer is most prominent and has been best studied are the cells that compose the intestinal epithelium of mammals (Fig. 6-1a) and that of several single-celled amoebae (Fig. 6-1b). The "fuzzy" layer is generally visualized by the same types of stains that reveal the cell coat, and therefore it becomes impossible to see where one ends and the other begins. In other words, when stains such as ruthenium red are used on cells (Fig. 6-2), a thick dense layer at the cell surface is usually seen, but it is usually impossible to determine

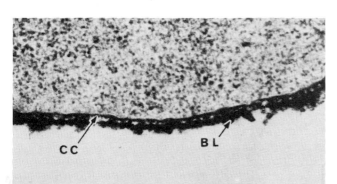

CC   BL

Figure 6-2. The basal surface of an ectodermal cell of an early chick embryo. The ruthenium red stain reveals two distinct surface structures, an inner cell coat and an outer basal lamina, i.e., an organized extracellular structure independent of the cell membrane. [*From A. Martinez-Palomo,* Int. Rev. Cytol., *29:64 (1970).*]

how much of this material is actually covalently bonded to membrane components (as either glycoproteins or glycolipids) and how much is an independent extracellular layer. The term *glycocalyx* is often used to refer to the complex of material closely adherent to the outer edge of the cell membrane. The function of this extracellular material is not always evident. In addition to its role in mediating cell-cell interactions, it has been suggested that it provides mechanical protection to cells, acts as a lubricant, or serves as a barrier to particles reaching the membrane, i.e., as a sieve.

The third category of cell surface materials includes a great variety of structures ranging from jelly coats and thick capsules around eggs to the basement membranes upon which epithelia rest and the materials in which cartilage cells and bone cells are embedded. These structures, and many other ones, are outside the limits of a text in cell biology, but two features of these extracellular materials bear a special note. One is the remarkable self-assembly properties that these materials seem to possess and the other is the major influences they can have upon the activities occurring within the cells that they surround.

Extracellular Materials: Mucopolysaccharides and Collagen

The two most prominent materials secreted by cells into the extracellular space are collagens (page 102) and mucopolysaccharides (Table 6-1). Mucopolysaccharides (or *glycosaminoglycans*) are large molecular weight sugar polymers of a sort very different from the glucose polymers (cellulose, starch, glycogen) discussed earlier. The carbohydrate chains of mucopolysaccharides exist as repeating disaccharides (Table 6-1) in which one of the two sugars is always an amino sugar, either *N*-acetylglucosamine or *N*-acetylgalactosamine (Fig. 5-4), hence the term glycosaminoglycan. Glycosaminoglycans are extremely acidic molecules due to the presence of carboxyl and/or sulfate groups (Table 6-1) on one or the other of the disaccharide sugars. In most cases, mucopolysaccharides exist

TABLE 6-1
Some Mucopolysaccharides and Their Repeating Disaccharide Units

Mucopolysaccharide	Monosaccharides in the disaccharide units	Linkages[°]
Hyaluronic acid	D-Glucuronic acid; *N*-acetyl-D-glucosamine	$\beta, 1 \to 3; \beta, 1 \to 4$
Chondroitin	D-Glucuronic acid; *N*-acetyl-D-galactosamine	$\beta, 1 \to 3; \beta, 1 \to 4$
Chondroitin sulfate A	D-Glucuronic acid; *N*-acetyl-D-galactosamine 4-sulfate	$\beta, 1 \to 3; \beta, 1 \to 4$
Chondroitin sulfate C	D-Glucuronic acid; *N*-acetyl-D-galactosamine 6-sulfate	$\beta, 1 \to 3; \beta, 1 \to 4$
Dermatan sulfate (formerly chondroitin sulfate B)	L-Iduronic acid; *N*-acetyl-D-galactosamine 4-sulfate	$\alpha, 1 \to 3; \beta, 1 \to 4$
Keratosulfate	D-Galactose; *N*-acetyl-D-glucosamine 6-sulfate	$\beta, 1 \to 4; \beta, 1 \to 3$

[°] The linkages are given in the same order as the names of the monosaccharide units, e.g., for chondroitin the linkages, as given, indicate that the linkage of glucuronic acid to acetylgalactosamine is β, $1 \to 3$, and that of acetylgalactosamine to the next glucuronic acid is β, $1 \to 4$, etc.

SOURCE: A. White et al., "Principles of Biochemistry," 5th ed., McGraw-Hill, 1973.

in combination with proteins, the complex being termed a *mucoprotein* or *proteoglycan*. Mucoproteins, like glycoproteins, contain both proteins and carbohydrates and both are commonly found on the outer surfaces of cells. However, these two types of macromolecules differ in both their chemical composition and their extracellular location (described below). Unlike the glycoproteins, which contain short carbohydrate chains (oligosaccharides) of a variable nature, mucoproteins contain a greater number and much longer polysaccharide chains composed of the repeating disaccharides just described. Unlike the extracellular collagen molecules which are fibrous (Fig. 3-24), the mucopolysaccharides are amorphous in their appearance, extremely viscous (as in mucous secretions) and occupy huge volumes for a given weight of material. In many cases the collagen and mucoprotein molecules interact with one another to form a variety of extracellular structures (Fig. 6-3). Once these materials find themselves outside of the cell they appear to be self-organizing; there is no evidence of enzymes or other materials which bring about their polymerization, interaction, or orientation.

The cornea of the eye, i.e., the transparent outer layer, provides a well-studied example of a structure that consists primarily of extracellular materials. The formation of the cornea has been studied as a model system for interactions among extracellular components and several interesting observations have been made which illustrate the complexity and importance of these materials. Figure 6-4a shows a section across the cornea. Basically there are three layers: an outer cellular epithelium, an inner cellular endothelium, and a greatly thickened stroma. The stroma consists primarily of collagen fibers embedded in a mucopolysaccharide matrix with flattened fibroblasts scattered throughout the layer. The formation of the stroma has been best studied in the chick embryo. The epithelium of the cornea is derived from the outer ectodermal layer (epidermis) of the embryo. By day 3 of incubation this layer secretes collagen and chondroitin sulfate, a mucopolysaccharide, on its basal surface. These materials interact to produce the primary stroma of the cornea, a totally noncellular structure composed of approximately 20 layers of collagen fibers, the fibers in each layer being perpendicular to those of the layers on either side (Fig. 6-4b). In some manner this orthogonal arrangement (per-

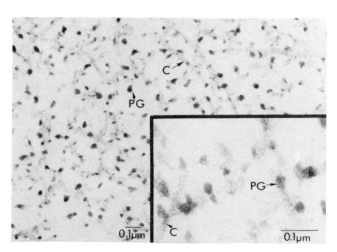

Figure 6-3. Electron micrograph of the extracellular matrix of cartilage. This material is composed primarily of collagen fibrils (C) and associated amorphous deposits of proteoglycan (PG). Inset shows a portion of the matrix at higher magnification. [*From R. W. Orkin, B. R. Williams, R. E. Cranley, D. C. Poppke, and K. S. Brown*, J. Cell Biol., **73**:295 (1977).]

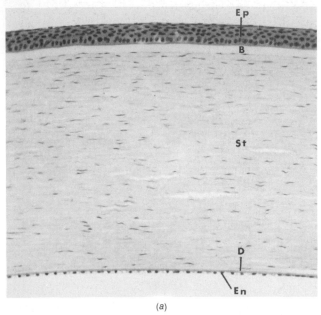

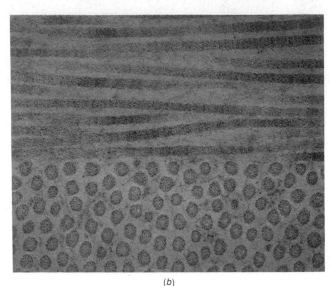

(a)　　　　　　　　　　　　　　　　(b)

Figure 6-4. (a) Cross section of the cornea. Ep, epithelium; B, Bowman's membrane; St, stroma; D, Descemet's membrane; En, endothelium. (b) Lamellar arrangement of the corneal fibers. All of the fibers are arranged parallel to the surface of the cornea, but those in adjoining layers are perpendicular to one another. (From D. G. Cogan and T. Kuwabara, R. O. Greep and L. Weiss, eds., "Histology," 3d ed., McGraw-Hill, 1973.)

pendicular layers) arises by self-assembly. On day 6, the primary stroma swells to a much greater thickness. This swelling occurs simultaneously with the secretion of hyaluronic acid (a nonsulfated mucopolysaccharide) into the stroma, and the uptake of water is believed to result from its presence.

The sudden swelling of the primary corneal stroma sets the stage for the next step of the process, the invasion of the primary stroma by wandering fibroblasts (connective tissue cells). These cells have been studied as a model system for cell movement in vivo (page 757), and it has been shown that the primary corneal stroma acts as a scaffolding to provide the appropriate substrate for these cells to move across. Once these cells have migrated into the swollen primary stroma, they begin to synthesize and secrete their own collagen and chondroitin sulfate, which becomes organized into the secondary or adult stroma. Most importantly, the bundles of collagen fibers in the layers of the secondary stroma are organized identically to the orthogonal arrangement of the primary stroma, i.e., the primary stroma is a miniature version of the final layer. Either the orientation of the fibers of the primary stroma is directly involved in some template process in orienting the new fibers, or it exercises this role indirectly by orienting the cells in the proper directions before they secrete the materials. Beginning on day 10, after the secondary stroma has been laid down, an enzyme is secreted into the layer and the hyaluronic acid produced prior to swelling is destroyed. In the wake of the removal of hyaluronic acid, the stromal layer shrinks by loss of water to its characteristic final composition.

Cell Walls

Since a lipoprotein plasma membrane of approximately 100 Å thickness can be expected to offer only minimal protection for a cell's contents, it is not surprising that cells are extremely fragile structures. When cells of a mammal, for example, are removed from the body and cultured in vitro, elaborate precautions must be taken to ensure that the temperature, pH, osmotic pressure, nutrient content, etc., are precisely adjusted or else the cells cannot survive and grow. Within the body of the animal, the extracellular conditions are maintained at the appropriate settings by the homeostatic mechanisms of the organism. The organism itself is covered by a thick, primarily nonliving hide which serves to protect the animal as a whole. Considering the fragility of cells, it becomes apparent that those organisms which exist as single cells stand in particular need of some type of extracellular protection since they interface directly with their environment.

Protective outer envelopes occur in various ways among different organisms: protozoa have a thickened outer coat (Fig. 6-1), bacteria and plants have distinct cell walls. In this section we will briefly consider the nature of the bacterial and plant cell walls and the types of functions they serve. Bacteria are generally divided into two groups on the basis of their appearance after treatment with the Gram stain, a procedure which involves the deposition of a colored complex between crystal violet and iodine. Gram-positive bacteria (such as *Streptococcus*) are ones which retain the crystal violet-iodine complex, while gram-negative bacteria (such as *E. coli*) include those in which the complex is extracted from the cells in a poststaining alcohol treatment. The difference in the extractability of the colored complex is related to the nature of the extracellular wall. The cell wall of gram-positive bacteria consists of a single thick (generally about 500 Å) layer of macromolecular material. The gram-positive wall consists of two materials, peptidoglycan and techoic acid. Peptidoglycans vary somewhat in their chemical nature, but all share a basic plan (Fig. 6-5a). The backbone of the peptidoglycan portion of the wall (the major portion), consists of alternating residues of N-acetylglucosamine (NAG) and N-acetylmuramic acid (NAM) joined together by β, 1-4 linkages. This repeating disaccharide should be remembered as being the substrate for the enzyme lysozyme (page 146).

In addition to the repeating disaccharide, peptidoglycan contains short chains of amino acids (4 to 5 in number) which are covalently cross-linked to one another. One unusual aspect of the amino acids of bacterial cell walls is the presence of D-amino acids which are never found in proteins. The structure of the peptidoglycan of *E. coli* is shown in Fig. 6-5b; additional components are present in many other species. The cross-linking within the peptidoglycan molecule is so extensive that the entire wall of a bacterium can be considered as one giant, bag-shaped macromolecule (Fig. 6-5c), which acts as a molecular sieve for material reaching the cell itself. It is the cross-linking step in cell wall formation which is sensitive to penicillin. The other component of the gram-positive wall, techoic acid, exists at the outer surface of the wall and consists of a polymer of ribitol or glycerol with the units joined by phosphodiester bonds. The cell wall of gram-negative bacteria is thinner but more complex. In these organisms, the wall consists of a thin inner peptidoglycan layer (approximately 25 Å) and a thicker outer layer, which in many respects resembles the plasma membrane. The outer layer of the wall is believed to contain a continuous lipid bilayer, together with associated proteins and lipopolysaccharides.

The cell walls of bacteria perform numerous functions. They give the cell its characteristic shape (spherical, rod-shaped, etc.), they protect the cell against mechanical or osmotic damage, they mediate cell-cell interactions, they act as a primary barrier to the penetration of large molecular substances, and they help in cell movement and cell division. In addition, they provide the surface at which bacterial viruses attach or to which antibodies produced by a host's defense system are directed. If the cell wall is removed from a bacterium, for example, by treatment with lysozyme in the case of gram-positive species, the living cell itself, termed a *protoplast*, is released for further study. As expected, these protoplasts are found to be extremely sensitive to disruption. Even slight differences in the osmotic pressure between the cell and its environment cause the protoplast to either swell and rupture or shrink.

Plant cell walls, although very different in chemical composition and

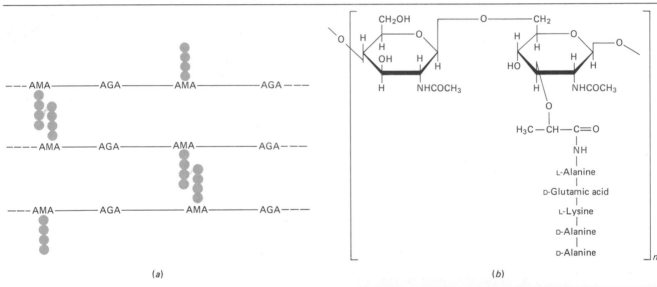

(a)

(b)

Figure 6-5. The bacterial cell wall. *(a)* General structure of peptidoglycans showing the alternating *N*-acetylmuramic acid (AMA) and *N*-acetylglucosamine (AGA) residues to form linear chains interconnected by small peptides. *(b)* The structure of the repeating unit shown in *(a)*. *(c)* An isolated cell wall of *Bacillus*. [*(a)* From M. J. Pelczar, R. D. Reid, and E. C. S. Chan, "Microbiology," 4th ed., McGraw-Hill, 1977. *(b)* From A. White, P. Handler, and E. L. Smith, "Biochemistry," 5th ed., McGraw-Hill, 1973. *(c)* From N. Sharon, Sci. Am., **220**:93 (April 1969).]

(c)

molecular organization from those present among bacteria, carry out similar protective and supportive functions. Plant cell walls are often likened to fabricated materials such as reinforced concrete or glass fiber-reinforced plastic, in that they contain a fibrous element embedded in a nonfibrous gel-like matrix (Fig. 6-6). In the case of the cell wall, the tightly packed fibers (termed *microfibrils*) are in the order of 100 Å diameter and composed of bundles of cellulose molecules. Cellulose is an unbranched polymer of glucose in which the sugar units are joined by β, 1-4 linkages. In addition to cellulose, materials that make up cell walls include hemicellulose (a pentose polymer), pectin (a hexuronate polymer), lignin (a phenylpropanoid polymer), protein, and water. The percentages of these various materials in cell walls is highly variable, depending upon the type of plant, the type of cell, and the stage of the wall.

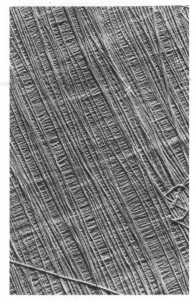

Figure 6-6. Metal shadowed replica of the side walls of a filament of the alga, *Chaetomorpha*, showing the orientation of the cellulose microfibrils. *(Courtesy of E. Frei and R. D. Preston.)*

Cell walls arise with the formation of a *cell plate* (described on page 712) between the cytoplasm of newly formed daughter cells. Once the cell plate has formed, materials are secreted into the extracellular space where they become adsorbed to the two surfaces of the plate to form the more complex cell wall. The nature of the cell wall and the orientation of the components are determined by the activities of the cell secreting them; however the materials themselves are capable of self-assembling into a relatively complex structure. In addition to providing mechanical support and protection, a cell wall must be able to grow in connection with the orderly growth of the cell it surrounds. The walls of growing cells are referred to as *primary* walls and are characterized by an extensibility which is lacking in the *secondary* walls of mature plant cells. Secondary walls are characterized by having an increased cellulose content and, in many cases, a large amount of lignin (the major component of wood) which is present in little or no quantity in primary walls. The surface of the wall of many plant cells (those of the epidermal cells of the leaves or other arial organs) contain a waxy substance for additional protection.

INTERCELLULAR RECOGNITION AND ADHESION

In the previous chapter considerable attention was paid to the types of proteins that are found in the plasma membrane, and the nature of the carbohydrate portions projecting from the membrane to the environment. The evidence suggests that particular membrane proteins have specific roles in the communication of the cell with both the cellular and noncellular components of its environment. One of the most important responsibilities of the cell surface is to *recognize* other cells, for it must know which ones to interact with and which to ignore. In this sense recognition is an identification or screening process. *Self-recognition* is of particular importance, since cells of a common type are most often found working together, but recognition of other types of cells in the surrounding tissue can be of equal importance. Unfortunately, the best-studied membranes, those of erythrocytes and lymphocytes, are of cells that are basically antisocial; they wander through the body as isolated single cells without forming permanent types of contacts. The transitory, highly specialized types of interactions in which lymphocytes engage will be discussed in Chapter 18.

Though we know less about the membranes of cells that form coherent tissues, it is clear that their surfaces possess a great number of proteins (probably glycoproteins) which might act as recognition sites for proteins on the surfaces of other cells. Some of these specific proteins are found in essentially all cells of the body. Included in this group are the histocompatibility antigens (page 799), which can serve to identify the normal cells of an individual from "foreign" cells, such as those that compose a transplanted organ. In other cases it appears that there are surface proteins which are shared by cells of the same type or those having a common embryonic origin. In at least some cases, there is evidence that each individual cell in a tissue may have its own unique surface identification. This latter condition seems to exist with the nerve cells of the neural retina and the brain cells to which the neurons of the eye project. Even though we are relatively ignorant of the nature of the specific proteins involved and the means by which they interact with one another across intercellular spaces, a great deal of experimental work has been performed in this area.

In order to construct a laboratory situation in which the processes of cell recognition and adhesion can be studied, the favored approach has been to take an existing tissue, dissociate it into its single-component cells, and observe these cells reaggregate. Surprisingly, dissociated cells are capable of coming back together and rearranging themselves to form similar counterparts with normal tissues of the organism. Though this type of laboratory experiment has no equivalent occurring within the body, it lends itself to a wide variety of experimental manipulations and has been very informative. It is generally believed that tissue reconstruction in vitro is based on the same types of properties of cell surfaces that are responsible for the formation of the embryonic tissues in the first place. There have been two favored systems for carrying out dissociation-reaggregation type experiments: adult sponges and embryonic organs. Both have been selected for the ease with which the cells can be separated from one another with a minimum of cell damage.

Sponge Cell Adhesion

The history of this approach goes back to 1907 to experiments on sponges by H. V. Wilson. Sponges are ideal for these types of experiments for they are composed of only a few cell types. The cells are relatively loosely attached to one another, and once dissociated they rapidly reassociate in simple isotonic salt solutions. Wilson found that whole sponges could be forced through a silk mesh and converted to a suspension of cells in sea water. The cells are said to be *mechanically dissociated*. If these cells were allowed to remain together they would aggregate to form large clusters, which would ultimately become new sponge individuals. Several sponges could form from what were originally the cells of a single organism. Further research on the phenomenon revealed that the reformation of new individuals depended upon the movement of cells within the cluster so that their appropriate position within the reaggregated mass was regained. In other words, cells that were lining the outer surface of the original sponge would take up a corresponding residence in the aggregate. Although not interpreted as such, Wilson had discovered a type of "cell self-assembly" comparable to the molecular self-assembly discussed in Chapter 3. The component cells of the organism seem to possess the information to construct the whole; no outside information or guiding force is necessary. It should be pointed out that this degree of self-direction is not present in more complex metazoans and not even in many other sponges. However, it illustrates the capabilities that cells have, and the property can be readily demonstrated in the reconstruction of vertebrate *tissues* discussed below.

In subsequent experiments, Wilson found a different principle of equal importance in cell biology, that of self-recognition. He found that if he dissociated the cells from two different species of sponge and mixed the cells together, that cells of each species would become grouped with their own type. The sponges that reformed were found to be exclusively of one or the other species' cells. Since the cells of the two species used were of different colors, one red and the other yellow, Wilson could actually watch like cells associate with one another and "sort-out" of the mixed cell mass.

The simplicity of the sponge system has made it of continuing value in the study of recognition and adhesion, and various investigators have continued to explore it. An important finding was made when the means of dissociating the cells was changed. Instead of using natural sea water, sponges were placed in artificial sea water (made in the lab from ordinary salts), in which Ca^{2+} and Mg^{2+} had been left out. When this was done, the sponges fell apart into their component cells, just as if they had been mechanically dissociated. However, there is an important difference between the "chemically" dissociated cells and a mechanically prepared suspension. If both are washed and placed in normal sea water (contains Ca^{2+} and Mg^{2+}) and the temperature kept very low (e.g., $4°C$), only the mechanically dissociated cells will reaggregate; the chemically dissociated cells remain as single cells or small, loose clusters. If, however, the supernatant in which the chemical dissociation took place is added back to the cells, then their reaggregation can also occur. It appears that in the process of chemical dissociation in Ca^{2+}-Mg^{2+}-free sea water, something is lost from the cells which is required for their reaggregation. As long as the temperature is kept low, the cells cannot resynthesize this "aggregation factor" (their metabolism is too sluggish at cold temperatures) and they cannot adhere. Analysis of the supernatant demonstrated that the factor responsible for promoting reaggregation was a large molecular weight glycoprotein composed of over 50% carbohydrate. Destruction of either the protein part or the carbohydrate part of this molecule renders the aggregation factor inactive.

To determine if the presence of the aggregation factor was involved in the species-specific nature of reaggregation, the supernatant in which a red sponge was chemically dissociated was added to a mixture of chemically dissociated red and yellow sponge cells kept at low temperatures.

In the presence of the factor from red sponges, only the red *Microciona* sponge cells would aggregate, leaving dissociated *Haliclona* yellow sponge cells behind.

The results on the sponge system suggest that species-specific intercellular adhesion is due to the presence of an intercellular cement (aggregation factor) which is complexed to the negatively charged cell surface by divalent calcium bridges. If these bridges are removed by suspending the cells in Ca^{2+}-Mg^{2+}-free sea water, the cementing material can no longer remain attached to the cell surface, and the cells come apart. The fact that the "cement" isolated from one species will not promote the re-aggregation of the cells from another species, indicates that the association mechanism is more intricate than simple Ca^{2+} bridges. In order for specificity to be present, complexity must be built into the structural association. Even though calcium ions may play an important role, some more complex aspect of the cell surface is required to explain the data. One recent view of the arrangement of the components is shown in Fig. 6-7.

Vertebrate Cell Adhesion

Studies of reaggregating cells from vertebrate tissues have revealed certain similarities with the results on sponges and certain basic differences. As might be expected, not all of the results on different vertebrate tissues performed in different laboratories using assorted assay procedures are in agreement. Research in this field was begun primarily by Johannes Holtfreter in the late 1930s and by A. A. Moscona in the 1950s. Moscona, working with various chick tissues, developed a new type of dissociation procedure based on a treatment with the proteolytic enzyme, trypsin, followed by gentle mechanical agitation. If one considers the use of trypsin to destroy the glycopeptides projecting from the external leaflet of the plasma membrane as described in the last chapter, it is not surprising that cells so treated would come apart. The use of trypsin has raised certain questions concerning the state of the surface of the dissociated cells. Generally, there appears to be a period of a few hours in which the cells interact in a less specific manner. This is believed to reflect the time during which the surface is being repaired. Whatever the effects might be, there does not seem to be any irreversible damage to the cell or its surface as a consequence of the enzyme treatment.

In the original assay for reaggregation, the suspension of single cells (Fig. 6-8*a*) was rotated in flasks, causing the cells to collide with one another and thereby providing the opportunity for them to adhere (Fig. 6-8*b,c*). Aggregates

Figure 6-7. A tentative model for specific cell-cell interaction in tissue construction. Two macromolecular aggregation factors (AF) are illustrated, each consisting of at least two subunits. The black termini at each pole carry the carbohydrates, which are recognized by the baseplate (BP) anchored in adjoining cell surfaces. [*From G. Weinbaum and M. M. Burger,* Nature, **244**:510 (1973).]

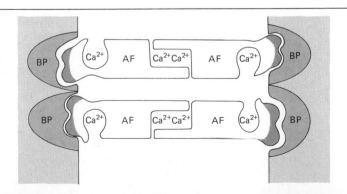

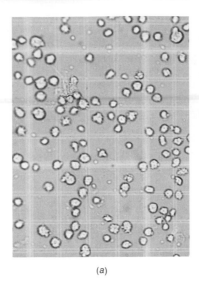
(a)

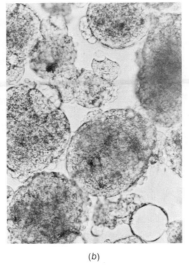
(b)

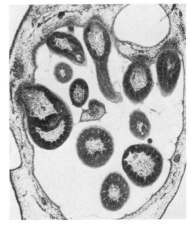
(c)

Figure 6-8. (a) Suspension of skin cells obtained by trypsin dissociation of 8-day chick embryo skin. (b) The appearance of an aggregate of dissociated skin cells after 24 hours of incubation. (Courtesy of M. H. Moscona and A. A. Moscona.) (c) Scanning electron micrograph of an aggregate of 8-day chick embryo neural retina cells originally dissociated by trypsin. (Courtesy of Y. Ben-Shaul and A. A. Moscona.)

were observed at various times and their number and size recorded. The potential for dissociated cells goes far beyond their ability simply to aggregate. If cell aggregates such as those of Fig. 6-8b and c are placed under the appropriate culture conditions, the potential for differentiation and morphogenesis can be revealed. Figure 6-9 shows a cross section through a feather primordium that has developed from a mass of reaggregated embryonic skin cells.

As in the studies of sponge cells, the cells prepared from vertebrate tissues were found to bear specific substances on their surface that allowed them to recognize certain cells and reject others. In the experiments of Moscona and coworkers, the major difference from the results on sponge cells is that vertebrate cells bear cell- and tissue-specific markers, as opposed to species-specific markers, that govern the nature of their interactions. For example, Moscona found that if he mixed cells dissociated from two tissues, such as embryonic chick liver and cartilage, these cells would initially form mixed aggregates but would then proceed to "sort out" so that each cell type became associated only with the same type of cell (Fig. 6-10). When cells of different tissues taken from two species were mixed, cells sorted out by tissue, apparently ignoring species differences. For example, when embryonic mouse cartilage and liver are mixed with embryonic chick cartilage and liver, the cartilage cells sort out together to form a chimaeric (mixture of two species) structure, as do the liver cells.

More recent studies by other laboratories have approached these same types of problems in different ways. Systems for the assay of cell adhesion can be quite different and thereby result in the measurement of different parameters. For example, in one assay (Fig. 6-11a) very different from that of Moscona's, *radioactively labeled cells* in suspension are added to dishes that are completely covered with a layer of cells (either the same or ones of a different type), and the radioactivity adhering to the cells in the dish is measured. In these measurements it is often found that labeled cells adhere to a much greater extent when the cell layer is from the same tissue than when it is from a heterologous tissue (Fig. 6-11b). Results of a wide variety of studies have come to similar conclusions, namely, that cells bear tissue-specific substances on their surfaces, presumably glycoproteins, that are responsible for self-recognition processes and cell adhesion. Beyond this point,

Figure 6-9. Light micrograph of a section through an aggregate of skin cells from an 8-day chick embryo that had been grafted for 6 days to the chorioallantoic membrane of a chick egg, a site on which pieces of tissue are routinely cultured. Feather formation has occurred within the aggregate. (Courtesy of M. H. Moscona and A. A. Moscona.)

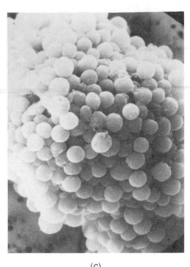

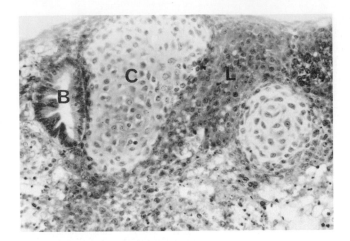

Figure 6-10. A light micrograph of a section through a mixed aggregate containing 8-day embryonic chick liver and cartilage cells. After an initial period of aggregation, the two types of cells have sorted out according to tissue and each has undergone recognizable differentiation. B, bile duct; L, liver; C, cartilage. *(Courtesy of A. A. Moscona.)*

Figure 6-11. *(a)* In this assay, radio-labeled cells in suspension are shaken above a cell layer. The rate of adhesion is determined by incubating for various times, washing off the nonadhered cells, and counting those remaining attached to the cell layers. *(From S. Roseman in E. Y. C. Lee and E. E. Smith, eds., "Biology and Chemistry of Eukaryotic Cell Surfaces," Academic, 1974.) (b)* Kinetics of cell adhesion to homologous and heterologous monolayers. Dark line shows the interaction between homologous single cells and monolayer, light line shows the heterologous interaction. The experiment shown compares inter-actions of mouse teratoma cells and mouse kidney cells in the two reciprocal combinations. [*From B. T. Walther, R. Öhman, and S. Roseman, Proc. Natl. Acad. Sci. (U.S.), **70**:1569 (1973).*]

however, there is considerable disagreement. Much of the controversy stems from the difficulty in attempting to analyze the basis for very complex inter-cellular interactions using readily measured properties under in vitro conditions. One can observe reaggregation and measure rates or strengths or selec-tivities of adhesion or sizes of aggregates, etc., but which of these assays is most relevant to cell-cell recognition and adhesion within the organism? The answer is unclear. Evidence for several types of recognition and adhesion factors has been obtained. In some cases, these cell-surface materials appear to promote general adhesion between a cell and its cellular and/or noncellular environment. In other cases, aggregation factors have been isolated which promote the selective interaction of a cell with other members of the same cell or tissue type. For example, if neural retina cells are grown in tissue cul-ture (outside of the organism in dishes in a controlled environment), they release substances that, when added to dissociated cells of the same tissue, cause larger aggregates to form more rapidly. In some cases at least, super-natant from one tissue has no effect upon the aggregation of cells from a *heter-ologous* (different) tissue, even when the tissues are quite closely related. A question of importance concerns the degree to which these isolatable factors are responsible for the sorting out activities discussed above, particularly when one considers the extreme complexity and precision with which sorting out can occur. For example, when retinal cells are allowed to reaggregate, dif-ferent types of component cells gradually become positioned and aligned in correct locations relative to one another until eventually an essentially intact organ is reconstructed.

Another important question concerning the aggregation factors is the degree to which these materials are actually part of the plasma membrane

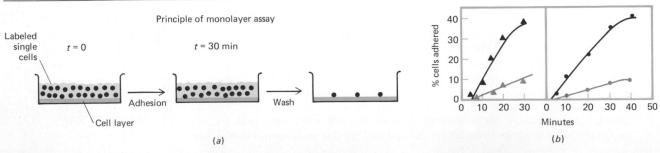

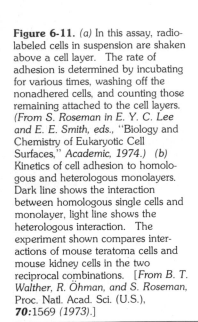

(a)

(b)

itself as opposed to being intercellular (between cells, but separable from each); this is still a matter of debate and may vary from cell to cell. In some cases, it has been shown that isolated, purified membranes can specifically interfere with aggregation of cells of the same type from which the membranes were prepared. For example, the presence of plasma membranes from neural retina cells will inhibit the reaggregation of neural retina cells but will not interfere with the aggregation of cerebellum cells. In further experiments, butanol extraction of retinal cell membranes has led to the purification of a glycoprotein of approximately 50,000 daltons, which mimics the properties of the neural retina cell-aggregation factor that had been previously purified from the supernatant medium of cell cultures. Results such as these provide strong evidence that tissue-specific aggregation factors do exist as components of plasma membranes.

The results described above and others to be presented below have implicated carbohydrates as being of major importance in determining the specific nature of intercellular interactions. In certain experiments it has been shown that if the synthesis or interacting capacity of these cell surface carbohydrates is disturbed, the ability of cells to adhere to one another can be greatly inhibited. For example, in one study it was found that cells of a mouse embryonic tumor would aggregate in a complex tissue culture medium but would not do so in a simple, glucose-balanced salt solution. The complex medium contained 51 ingredients not present in the simpler solution. By a process of elimination through adding various of these substances and testing for aggregation, it was found that only one of the 51 ingredients was required for cells to adhere, L-glutamine. Glutamine is an important molecule in the formation of the amino sugars (glucosamine, galactosamine, sialic acid) as it provides the amino group in the following reaction:

Fructose 6-phosphate + glutamine →
glucosamine 6-phosphate + glutamic acid

If trypsinized cells are not provided with glutamine (or the amino sugars), they cannot repair the damage to their surface resulting from enzymatic digestion of the membrane glycopeptides and *they cannot stick together.*

In a few other cases specific sugars have been singled out as being particularly important in the interactions of one type of cell or another. For example, in one experiment, tiny inert beads were prepared with different sugars attached to their surfaces. In this experiment it was found that if the beads were coated with galactose, cells would temporarily stick to them (Fig. 6-12a), whereas beads coated with other sugars did not provide the proper substrate for attachment to occur (Fig. 6-12b). Considering the structural variability that can occur among oligosaccharide chains, their involvement in selective adhesion is generally accepted, though the mechanism by which the intercellular connections are made is not understood. As in the sponge system, adhesion is believed to be mediated by complementary, interacting molecules, one of which is a carbohydrate and the other a carbohydrate-binding protein. Selectivity in the adhesion event would result from different cells having different interacting macromolecules.

In the previous chapter, the role of glycosyltransferases in the construction of the oligosaccharide chains of glycoproteins and glycolipids was discussed. These enzymes are present *within cells* as a collection of transferases that remove specific sugars from a nucleotide donor and transfer them to specific acceptors on a growing oligosaccharide chain (Fig. 5-5). The finding that these enzymes are *also* present on the outer surface of the plasma membrane has

Figure 6-12. Adherence of virally transformed fibroblasts to galactose-coated beads *(a)* as opposed to glucose-coated beads *(b)*. *(Courtesy of S. Roseman.)*

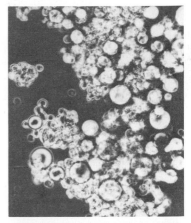

(a)

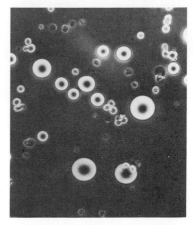

(b)

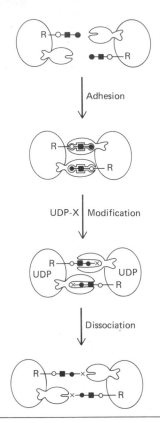

Figure 6-13. Schematic representation of the hypothesis that glycosyltransferases on the cell surface are involved in intercellular adhesion. The complex formed between the enzyme and the oligosaccharide acceptor holds the two cells together. The addition of the nucleotide sugar (UDP-X) results in the elongation of the acceptor chain and an alteration in adhesion. [*From S. Roseman*, Chem. Phys. Lipids, **5**:295 (1970).]

led to the proposal by Saul Roseman and coworkers that these enzymes are involved in the recognition and adhesion process. The existence of these enzymes on the cell surface provides a ready-made protein that would be able to *selectively* associate with one type of oligosaccharide as opposed to another. In the cell, these enzymes would be expected to transfer the sugar from a donor to the specific acceptor, whereas on the cell surface they would be expected, more-or-less, to remain attached to the specific acceptor (Fig. 6-13). In other words, this theory uses enzymes, presumably the same molecules that had earlier functioned enzymatically in making surface glycoproteins (Fig. 7-10) in a nonenzymatic role. The particular interactions that different cells could have with one another would depend on the variety of enzymes and acceptors on their surfaces. The catalytic potential of these transferases adds another dimension to this proposed role. If the appropriate nucleotide sugar were to be available to these surface enzymes, the reaction might be catalyzed whereby the sugar is transferred covalently to the acceptor, thereby changing the nature of the acceptor and ending its contact with the enzyme (Fig. 6-13). The potential to finish the reaction and release the acceptor gives this interaction a dynamic quality, more compatible with the reversibility of cell adhesion that is seen in many cases where cells come together, make temporary contacts with one another, and move on to some other interaction.

Up to this point we have considered only the qualitative aspect of cell adhesion—whether two interacting cells have complementary surfaces which allows them to combine. It was pointed out that cell aggregation in vitro leads to more than just a clump of like cells. Rearrangements within cell aggregates causes different types of cells to take up positions relative to one another that resemble their positions in the organ prior to dissociation. This was clearly shown in 1955 by the experiment of P. L. Townes and Johannes Holtfreter on cells from very early amphibian embryos. In the gastrula and neurula stages upon which this work was done, the embryo is composed essentially of three types of undifferentiated cells: outer ectodermal, inner endodermal, and mesodermal cells in between. It was found that embryonic cells could be dissociated (by high pH in this case) and mixed together in various combinations and the cells would reorganize themselves such that the same approximate positions they held in the embryo would be restored in the culture dish (Fig. 6-14). As in the case of the reformation of a whole sponge in vitro, cells from complex vertebrate tissues also possess the capability for self-assembly into structures of higher order. To explain the formation of specific patterns formed by mixtures of more than one cell type, it would appear that a new dimension to this topic must be considered.

The differential adhesion theory of Malcolm Steinberg attempts to explain the relative positions that masses of cells occupy on the basis of the strengths of the contacts that cells make with one another. The theory is not concerned with the specific chemical interactions involved in adhesion, but is compatible with any molecular theory. Rather it concerns itself with the physical aspects of cells capable of moving around and sorting out into predictable configurations. Steinberg found that if he mixed two types of cells together, one of the types tends to move inward forming a central core surrounded by the cells of the other type. The final configuration that the cells maintain is the thermodynamically favored one, a type of equilibrium configuration, with the cells having the strongest contacts being located in the interior. For example, if the precartilage cells of the embryonic limb bud are mixed with cells taken from embryonic heart ventricle, at equilibrium the latter cells will have surrounded the former (Fig. 6-15a). If these same ventricle cells are mixed with cells of the embryonic liver, they will move to the interior of the combined mass, becoming surrounded by the liver cells (Fig. 6-15b). The differential adhesion theory explains this predictable relationship by proposing that in any combination of cells, those with the greater cohesiveness will be found in the interior, having squeezed out the less cohesive cells to the periphery. Each type of cell-cell interaction can be assigned a relative cohesiveness and a scale or hierarchy of adhesiveness can be established.

In the final analysis, no one theory can explain the many different forces that are needed to bring about the complex organization of vertebrate tissues and organs. Consider *morphogenesis,* the development of form and structure. Complex forces must act to shape the hundreds of bones of the body, or the lobulated salivary gland, or the intricately sculptured kidney tubule. In the case of the retina, each of the several layers must be arranged and interconnected. Most formidable of all is the cell interactions that lead to the complex circuitry of the nervous system. In the optic nerve, for example, it appears that each neuron from the retina extends itself so as to hook up with a precise cell or group of cells in the brain. The study of cell recognition and cell adhesion is central to our understanding of all of these phenomena, but much more seems to be involved.

Other Systems

Another perspective on the surface interactions of cells is provided by the examination of systems in which single, independent cells come into contact with one another and adhere. In contrast to the reaggregation of a dissociated tissue, this type of event is a naturally occurring one, though the fact that it involves the union of single cells raises questions as to the degree to which results can be generalized to complex, multicellular vertebrate tissues. There are two types of systems that have been primarily studied in this regard. In one case, the union of two cells is a reproductive process, whether it be between an egg and sperm from a multicellular animal, or two yeast cells, or two *Chlamydomonas,* a unicellular phytoflagellate. In the other case, the cells are amoebae which live as single-celled organisms for most of their life cycle but come together under certain conditions to form a multicellular construction. We will begin with a brief discussion of the latter type.

The cellular slime molds, such as *Dictyostelium,* have been one of the most important organisms in the study of cell and developmental biology. Conveniently, these organisms provide an evolutionary and physiological bridge between the single-celled and multicellular state. The life cycle of *Dictyostelium* was shown in Fig. 1-9. The *vegetative* stage is spent as free-living individual amoebae, growing in the laboratory on bacteria provided in the culture medium. If the supply of bacteria becomes exhausted, a new series of events is set into motion: the aggregation of the amoebae into masses of cells that lift themselves off the floor of the culture dish (Fig. 1-10). The change from the single-celled to the aggregated stage is not simply a behavioral one but is accompanied by changes in the surfaces of the cells that facilitate the aggregation. Within a few hours after removal of the bacteria the cells acquire a cohesiveness not present in the vegetative state. If the plasma membranes are isolated from cells that have acquired the ability to aggregate, a protein can be extracted which is capable of causing the agglutination of sheep erythrocytes. It appears that the membranes from cohesive slime mold cells contain a multivalent carbohydrate-binding protein, termed *discoidin,* that

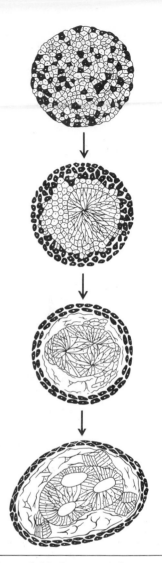

Figure 6-14. Sorting out of embryonic amphibian cells to form a composite structure with an outer epidermal cover enclosing neural and mesodermal tissues. [*From P. L. Townes and J. Holtfreter,* J. Exp. Zool., **128**:53 (1955).]

Figure 6-15. *(a)* Section through an aggregate formed by intermixed precartilage and heart ventricle cells. The precartilage is reassembled internally to the heart tissue. *(b)* Section through an aggregate formed by intermixed heart ventricle and liver cells. Heart ventricle tissue is reassembled internally to liver tissue. [*From M. S. Steinberg,* J. Exp. Zool., **173**:*411 (1970).*]

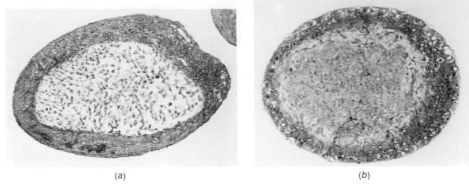

(a)   *(b)*

can combine with the oligosaccharide chains projecting from the surfaces of the erythrocytes and hold the erythrocytes together (Fig. 6-16*a*). The agglutination of red blood cells provides a convenient assay for the presence or absence of this slime mold lectin. Membranes from vegetative amoebae do not have such a protein; it is produced by the cells only after the absence of food has triggered the change toward the aggregative state (Fig. 6-16*b*). Recent experiments by the author and coworkers indicate that the change to the aggregation-competent state is dependent upon the synthesis of glycoproteins. If slime mold amoebae are exposed to tunicamycin, a drug which inhibits the addition of carbohydrate residues to proteins, these cells do not adhere and the aggregation event is blocked.

Utilization of these easily cultured single-celled organisms has advantages stemming from the simplicity of the system and the opportunity to extract and purify the specific macromolecules involved in the phenomena. Similar progress has been made in the purification and analysis of specific, interacting molecules present on the surfaces of yeast cells and *Chlamydomonas*, that are responsible for holding cells together during mating activities. These concepts have also been extended to interacting gametes in mammalian fertilization and it appears that similar types of molecules may be responsible for allowing the attachment of the sperm to the egg or egg membranes during conception. For example, various of the lectins (such as wheat germ agglutinin, page 225), when added to unfertilized mammalian eggs, block the attachment of the sperm. It would appear that as long as the wheat germ agglutinin molecules are associated with the carbohydrates, the binding sites for the sperm to attach to are blocked. In all of the systems described in the last few pages, research is just beginning to provide an understanding of the means by which cells can recognize one another, and, when appropriate, adhere.

CELL JUNCTIONS

In the previous section we considered the experimental and molecular aspects of cell adhesion, focusing primarily upon the initial events whereby cells determine if their surfaces are mutually compatible, i.e., have cell surface components which can recognize and combine with one another. In this section we will examine some of the information on the morphological nature of intercellular interaction, focusing primarily upon results obtained with the electron microscope. The initial contacts that reaggregating cells make have been examined with the scanning electron microscope (SEM). In the SEM, the surfaces of cells generally appear very irregular with large blebs and various projections. These surface protrusions presumably reflect the dynamic nature of the cell surface in the living state. When neural retina tissue is dissociated with trypsin, the surfaces of the cells possess randomly oriented filopodia (long thin processes). It appears that the initial contacts between reaggregating cells are mediated by these filopodia (Fig. 6-17a). As the cells approach each other more closely, the filopodia connecting the cells become shortened, giving the impression that they are pulling the cells together. Within a short time the membranes of the two cells are intertwined (Fig. 6-17b).

Examination of reaggregating cells has provided information on the initial nature of intercellular contacts, but we need to turn to studies on complex tissues of the body to describe the cell junctions that have been observed between permanently associated cells. The term *cell junction* refers to specializations in the plasma membrane formed in places where two cells make contact. There are several types of junctions which differ primarily with respect to (a) the closeness with which the plasma membranes come together, i.e., the width of the *extracellular space* between the membranes; (b) the extent of the surfaces of the cells that are involved; (c) the appearance of material between the membranes; (d) the function which is served. We will consider the three main types of cell junctions found in vertebrate cells: the *desmosome*, the *tight junction*, and the *gap junction*. Discussions of more specialized types can be found in the reviews cited at the end of the chapter.

Figure 6-16. *(a)* Scanning electron micrograph of fixed sheep erythrocytes clustered around living slime mold cells. *(Courtesy of C-M. Chang, S. D. Rosen, and S. H. Barondes.)* *(b)* Experiment illustrates the change in the nature of the cell surface of the slime mold amoebae with the stage in their life cycle (Fig. 2-10). The lighter line indicates the cohesiveness of the amoebae toward one another, while the darker line indicates the presence of the agglutination factor on the cell membranes. It would appear that the ability of amoebae to aggregate (Fig. 2-10a) prior to the formation of a fruiting body, depends on the appearance of the surface factor. *[From S. D. Rosen, J. A. Kafka, D. L. Simpson, and S. H. Barondes, Proc. Natl. Acad. Sci. (U.S.), **70:**2555 (1973).]*

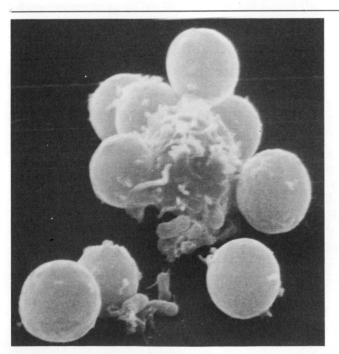

(a)

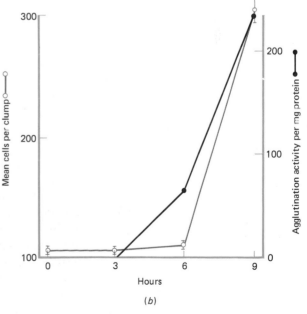

(b)

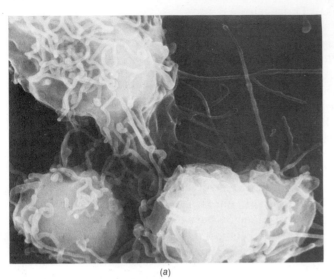

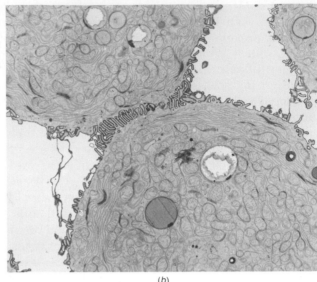

(a)

(b)

Figure 6-17. *(a)* Scanning electron micrograph of a small retina cell aggregate showing the interaction of filopodia between cells. [*From Y. Ben-Shaul and A. A. Moscona, Exp. Cell Res.,* **95**:*193 (1975).*] *(b)* Electron micrograph of hepatocytes incubated for four hours. The first cell contacts are established consisting of numerous intertwined microvilli. The cell coat is intensely stained by ruthenium red. [*From J-C. Wanson, P. Drochmans, R. Mosselmans, and M-F. Ronveaux, J. Cell Biol.,* **74**:*863 (1977).*]

Desmosomes

If one examines a cross section through a vertebrate organ and looks for the free spaces, such as the outer edge of the organ or the lining of a lumen or small duct, the cells bordering the space are closely pressed up against one another forming an unbroken layer, an *epithelium*. Epithelial cells are notoriously tough to mechanically separate from one another, for example, by pulling on the cells with a pair of fine needles. However, treatment of the tissue with trypsin will usually do the job. One of the bases for the tight mechanical adhesion of cells is the presence of desmosomes between them. Desmosomes are of two major types: ones which are present as patches *(macula adherens)*, or the less common ones which encircle the cells like a belt *(zonula adherens)*. Desmosomes provide adjacent cells with a very tight hold on one another and are particularly numerous in tissues that are subjected to mechanical stress, such as the skin and the uterine cervix. Certain diseases of the skin, in which the cells are readily shed, result from a deficiency of desmosomes. Desmosomes are elaborate in their appearance (Fig. 6-18), having modifications both within the extracellular space between the plasma membranes and the cytoplasmic region on each side; the membrane itself appears unaltered. The space between the membranes is between 200 to 350 Å, a typical width for closely applied cells, and is filled with a fine filamentous material which is believed to act as an intercellular cement. On the basis of its sensitivity to enzymes (e.g., trypsin) and its staining properties (e.g., positive for ruthenium red), the material of the extracellular space appears to contain protein and carbohydrates, probably present as glycoproteins and mucopolysaccharides.

The electron-dense material within the cytoplasm of the spot desmosome (the *macula adherens*) appears to be an amorphous matrix within which bundles of filaments of approximately 100-Å diameter, termed *tono-filaments*, approach the inner side of the membrane and loop back into the cytoplasm. In certain cases the tonofilaments can be seen to run across the cytoplasm, attaching the inside of one desmosome to the inside of another one *in the same cell*. In some cases, forces impinging upon one of the desmosomes can be transmitted to the others. Belt desmosomes (*zonula adherens*) have a simpler appearance (Fig. 6-21), and instead of having 100-Å tonofilaments in the adjoining cytoplasm, they have 70-Å filaments shown to be made of actin.

Tight Junctions

The tight junction brings the plasma membranes of adjacent cells together in their closest contact; there is no space whatsoever between the two plasma membranes at the site of contact (Fig. 6-19*a,b*). As a result, tight junctions are occluding junctions, i.e., they block off the extracellular space on one side of the junction from that on the other. Since they occur as belts around the circumference of the cells (Fig. 6-19*c*) they are termed *zonula occludens*. The physiological importance of tight junctions can be seen in the following example. The function of the kidney is to produce a urine whose composition is different from that of the blood and tissue fluids. The fluid produced is stored in the bladder, a thin-walled structure situated among other tissues and organs of the body. It is imperative that the contents of the bladder are not able to exchange with the fluid surrounding it. Ignoring for the present the possibility that water or solute might move across the bladder wall by passing through the cells themselves, what is to prevent these substances from crossing the wall by diffusing *between* its cells? If one examines the apical edge (one facing

Figure 6-18. (*a*) Electron micrograph of a desmosome from newt epidermis. [*From D. E. Kelly, J. Cell Biol. **28**:51 (1966).*] (*b*) Diagrammatic representation of the structure of a desmosome. (*Drawing by B. Tagawa, from Junctions Between Cells, by G. D. Pappas, Hospital Practice, Vol. 8, No. 8 and "Cell Membranes: Biochemistry, Cell Biology & Pathology," G. Weissmann and R. Claiborne, Eds., HP Publishing Co., Inc., New York, NY 1975. Reprinted with permission.*)

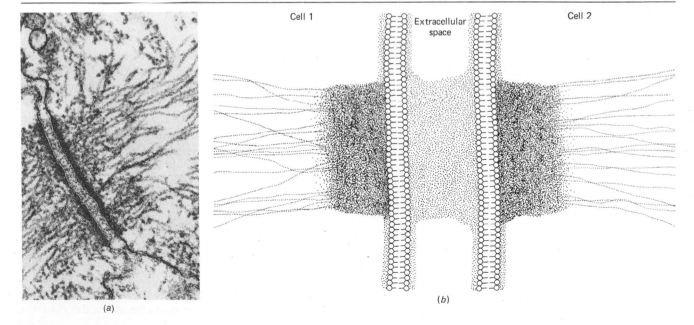

(*a*) (*b*)

Cell 1   Extracellular space   Cell 2

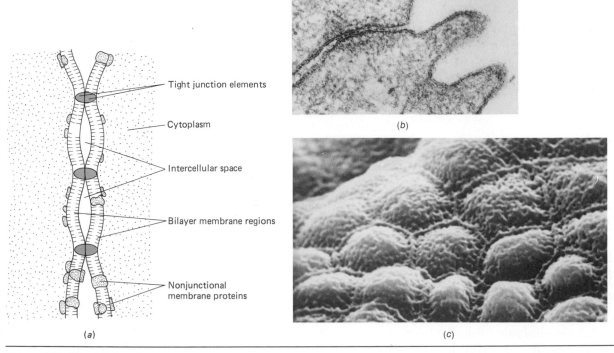

Tight junction elements

Cytoplasm

Intercellular space

Bilayer membrane regions

Nonjunctional membrane proteins

(a)

(b)

(c)

Figure 6-19. The tight junction. (a) Model of a tight junction indicating that the membranes are held together along lines of attachment composed of proteinaceous particles. [*From L. A. Staehlin,* Int. Rev. Cyt., *39:207 (1974).*] (b) Electron micrograph of this type of junction. (*Courtesy of D. Friend.*) (c) Scanning electron micrograph of the apical surface of an epithelium showing the encircling nature of the junctions. (*Courtesy of D. Tarin.*)

the cavity) of the epithelial cells of the bladder wall, tight junctions can be seen to encircle the cells. If the resistance across the bladder wall of the amphibian is measured, an extremely high value (possibly 2,000 ohm-cm²) is found, indicating that the passage of ions (current) across the wall is virtually nonexistent. It is the function of the tight junction to seal off the extracellular space of the lumen from the extracellular space between the epithelial cells. Tight junctions perform similar services in other places, though in most cases their sealing ability is not so great; i.e., they are either "leaky" to a varying degree, or they restrict only large molecular weight materials, not ions and water.

If one examines electron micrographs of sections cut perpendicular to the planes of the adjoining membranes, it appears that the membranes in the region of the junction come together at intermittent points (Fig. 6-19a, b). In other words, the outer surfaces of the membranes are not in contact over a large surface area, but only at specific sites. Freeze-fracture analysis is ideal for examining the structure of the junctions, since the fracture planes tend to traverse the membranes, revealing the distribution of the integral membrane proteins. Freeze-fracture replicas of the tight junction (Fig. 6-20) are seen to contain a network of ridges which run parallel to the apical surface of the epithelium. These ridges (or grooves in the opposing fracture face) represent sites within the membranes where the individual membrane particles have become tightly aligned next to one another to form a continuous linear stretch of integral membrane protein. The ridge of protein in one membrane projects outward into the extracellular space to meet the corresponding row of proteins projecting from the adjacent membrane. It is the presence of these aligned rows of proteins, which are shared between adjacent membranes and encircle each cell, that serve to seal and divide the extracellular space. Most importantly, there appears to be a correlation between the number of parallel rows that make up the junction and the tightness of the seal (Fig. 6-20*a, b*). For example, the frog bladder epithelium discussed above, which is a particularly effective barrier, has an average of eight ridges in the junction. Similarly, the distal convoluted tubule of the mammalian kidney, which is also very restrictive (300 to 600 ohm-cm²) has four or five interconnecting strands. In contrast, the proximal convoluted tubule of the same kidney, which is very leaky (approximately 6 ohm-cm²) has only one. In certain epithelia, such as the intestine and that of various glands, there is an organized junctional complex (Fig. 6-21) with an apical tight junction, an adjacent belt desmosome *(zonula adherens)*, and a spot desmosome *(macula adherens)*.

Gap Junctions

As in the case of the desmosome and the tight junction, the gap junction (or *nexus*) also has a distinctive morphology and a special function to perform. Gap junctions are present as patches on which adjacent membranes approach each other very closely (20 to 40 Å), though they are never seen to be in contact as in the tight junction. Electron micrographs of cells cut perpendicular to the planes of the adjacent membranes show that there are very fine connections present between the membranes in the region of the junction (Fig. 6-22*a*). The organization of the components of the junction is best seen if the tissue is first immersed in a colloidal suspension of lanthanum salts before being fixed and sectioned. If the sections of lanthanum-treated tissue are made parallel to the membranes *within the extracellular space*, the appearance of the gap junction is seen as shown in Fig. 6-22*b*. The electron-dense regions represent the extracellular space within which the lanthanum had circulated prior to fixation. Within the electron-dense background, the outlines of electron-translucent particles of about 70-Å diameter can be seen which represent the connections between the two plasma membranes. In other words, the electron-translucent particles represent structures into which the lanthanum cannot

Figure 6-20. Freeze-fracture replicas showing tight junctions. *(a)* Mouse proximal convoluted tubule, a very leaky epithelium, consisting in most places of a single strand of membrane particles (arrows). Portions of the inner (A) face and outer (B) face are shown. *(b)* Mouse distal convoluted tubule, a tight epithelium, consisting of four or more anastamosing strands. [*From P. Claude and D. A. Goodenough,* J. Cell Biol., **58:**390 *(1973).*]

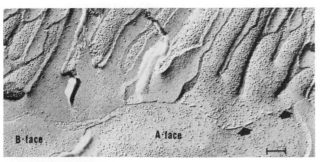

B-face   A-face

(a)

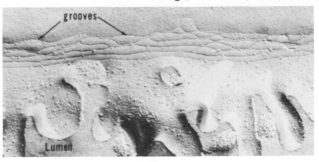

grooves

Lumen

(b)

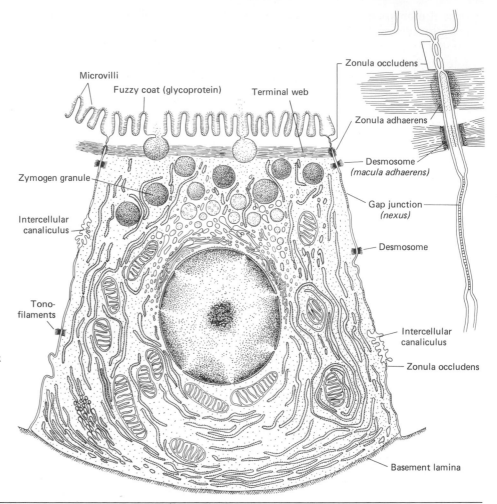

Figure 6-21. Diagram illustrating the principal specializations that would be found on a simple columnar epithelial cell. The junctional complex at the luminal surface consists of a zonula occludens (tight junction) and zonula adherens (belt desmosome) and a row of desmosomes. Under the row of desmosomes, gap junctions occur. Other desmosomes and gap junctions are located deeper in the tissue. *(From E. D. Hay in R. O. Greep and L. Weiss, "Histology," 3d ed., McGraw-Hill, 1973.)*

penetrate; structures which bridge the extracellular space. Within the center of these particles, a pit of about 15-Å diameter can often be seen.

Freeze-fracture replicas of gap junctions clearly show that the 70-Å-diameter connections represent membrane-associated particles. A patch of these particles is shown in Fig. 5-22c). The current view is that these particles represent proteins that span the membranes (Fig. 6-22d), and project out into the extracellular space. The protein complexes of both membranes come together within the extracellular space to form a continuous structure which is not penetrated by the lanthanum. The most important structural aspect of these membrane proteins is that they do not form a solid complex, but one with a pore or pipeline running through it as seen represented by the 15-Å spot in the center when cut in cross section. In other words, particles seen in the electron microscope are actually doughnut shaped and embedded in the membrane on each side of the extracellular space.

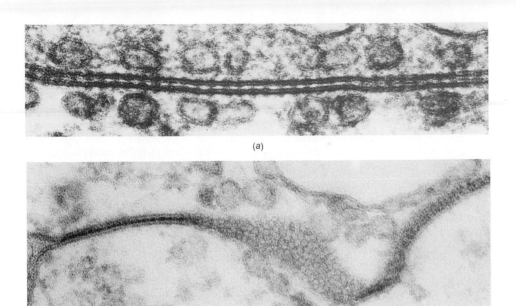

(a)

(b)

(c)

Cytoplasmic
membrane
surface

Gap junction particles

Lipid bilayer
membrane

Hydrophilic channels

Cytoplasm

Intercellular
space

External
membrane
surface

Integral membrane proteins

(d)

Figure 6-22. The gap junction. (a) Electron micrograph of a section through a gap junction perpendicular to the plane of the membrane. The beaded nature of the profile is due to rows of intramembrane particles which are in register and protrude from both membrane surfaces. (b) Electron micrograph of a section cut parallel to the membranes forming the gap junction, and between them. The tissue was immersed in a heavy metal (lanthanum hydroxide) before and during fixation and this material has penetrated between the membranes into the open spaces of the junction. As a result of the presence of lanthanum, the membrane particles appear negatively stained against the more dense background. [(a) and (b) From C. Peracchia and A. F. Dulhunty, J. Cell Biol., **70**:419 (1976).] (c) Freeze-fracture replica through a gap junction showing the concentrated presence of membrane particles. (Courtesy of D. F. Albertini.) (d) Three-dimensional reconstruction of a gap junction. [From L. A. Staehlin, Int. Rev. Cyt., **39**:249 (1974).]

On the basis of both morphological evidence (described above) and physiological evidence (described below), the gap junction is believed to represent a region where open communication exists across the extracellular space. Since the pipelines span the membranes, the openings on each side of the channels would be on the inner side of each membrane, thereby providing an open means of communication from the cytoplasm of one cell directly across to the cytoplasm of the adjacent cell. Cells connected in this manner are said to be "coupled." There are two principal means to demonstrate the direct exchange of material from one cell to another. In one technique, materials of different molecular weight are injected into one cell and are seen to diffuse rapidly and directly into the adjoining cell (Fig. 6-23a). If the materials injected are fluorescent compounds, such as fluorescein, their spread from cell to cell can be followed directly using the fluorescence microscope. A variety of tracer studies of this sort indicate that many substances, particularly of small molecular weight (below 1000), readily cross from the cytoplasm of one cell to that of another, without being seen to diffuse into the extracellular space. The size of the molecules that can pass through these channels varies considerably with the types of cells that are joined and presumably reflects the diameter of the particular channels.

In the other method of analysis, electrodes are placed within two cells in a line of cells and the passage of ionic current (a measure of the ability of ions to move between cells) is determined (Fig. 6-23b). Since the gap junctions contain spots where there is no membrane present between the two adjoining cells, there is much less resistance in those regions than across a lipid bilayer. As a result, gap junctions (also called *low resistance junctions*) will readily transmit a pulse of current from one cell to the next. Cells between which a flow of current can be measured are said to be electrically coupled. Both tracer flow and electrical coupling have been convincingly correlated with the presence of gap junctions.

The function of certain of the gap junctions was discovered a number of years before its structure was seen in the electron microscope. The

Figure 6-23. *(a)* Dark-field photomicrograph showing the passage of fluorescein from one cell into which it was injected (X) to the surrounding cells. [*From R. Azarnia and W. R. Loewenstein,* J. Memb. Biol., **6:**378 *(1971).*] *(b)* Measurement of intercellular electrical communication through gap junctions. The photograph shows microelectrodes inserted into living cells of the Malphigian tubule of an insect. *(Courtesy of W. R. Loewenstein.)*

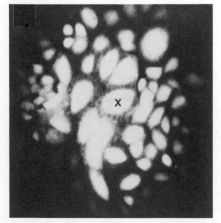

(a)

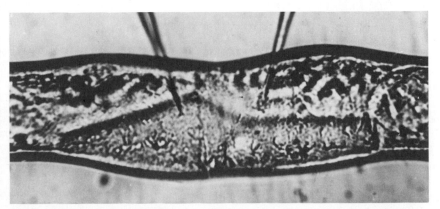

(b)

first indication of their existence was the finding that an electrical potential could spread directly between adjacent neurons in a crayfish motor nerve; no chemical transmission of the impulse across a synapse was necessary. The neurons of the nerve were electrically coupled across a low resistance junction, and propagation of the nerve impulse from cell to cell did not have to be delayed as occurs when chemical neurotransmitter substances are required. In this case, the nerve operates the crayfish tail muscle causing it to contract rapidly leading to the animal's escape response. In vertebrates the primary use of electrical coupling in excitable tissue is in cardiac and smooth muscle. In contrast to striated (skeletal) muscle, where each cell is directly innervated by neurons, smooth muscle cells are connected to each other by gap junctions through which a flow of potassium ions occurs. This ionic flow successively activates adjacent muscle cells leading to a slow, sustained wave such as that found in the walls of the digestive tract.

The role of these junctions in nonexcitable tissue, such as the malphigian tubule epithelium of Fig. 6-23b is much less clear, though their existence in many tissues is widespread. The fact that gap junctions can put a large number of cells into intimate cytoplasmic contact with one another provides what is, in essence, one giant compartment with respect to those molecules small enough to pass through the communicating channels. This condition would be expected to have very important physiological consequences since various regulatory molecules, such as cyclic AMP and many others, might be expected to rapidly diffuse among all the cells causing their activities to be coordinated. Recent studies suggest that the degree to which gap junctions allow intercellular diffusion depends upon the local concentration of calcium ions. Experiments in which very small volumes of calcium-containing solutions are microinjected into cells indicate that the greater the free Ca^{2+} concentration in the region of a junction, the less its permeability. As will be apparent in the following chapters, the calcium ion (like the cyclic nucleotide) is a highly active substance with numerous effects within cells. Under normal conditions, the calcium ion level is kept low by the action of intracellular calcium pumps which sequester the ion within membranous containers. However, if the concentration of this ion should rise to above approximately $5 \times 10^{-5} M$ in the general cytoplasmic compartment, one of the effects that might be expected is the closure of the cell's gap junctions. Communication, in general, is one of the most important activities carried out by cells, and gap junctions provide a means for the direct, short-range communication between adjacent cells.

The presence of the various types of specialized junctions raises interesting questions concerning cell cooperation and membrane activities. In all of the junctions discussed, there is a striking plane of symmetry running down the middle of the extracellular space; the appearance on one side (in one cell) is the same as that on the other side (in the other cell). How this cooperation is accomplished is not understood, but it is most evident in the formation of these junctions. There are two means by which their formation has been observed. In one type of study, dissociated cells are allowed to reaggregate and samples of the forming tissues are fixed at various intervals. In the other case, examination of early embryonic stages are made in an attempt to observe the initial formation of these junctions in the life of an organism. Both types of observations agree and show evidence of cell-cell cooperation.

Gap junction and tight junction formation have been best studied using the freeze-fracture technique. At first there is an adjacent region in both membranes that is essentially devoid of particles, termed a *formation plaque*. It is within this zone of the membrane that particles collect to become organized into a particular arrangement. In the tight junction the particles become aligned closely together and appear to fuse with one another into a growing strand. In a gap junction the particles form small aggregates which grow in size by the recruitment of additional particles or the fusion of smaller aggregates into larger patches. The process is not understood but it clearly illustrates the fluid properties of the membrane described in Chapter 5. Presumably these particles represent integral membrane proteins which are moving laterally through the plane of the membrane to the site where they are needed. The rapidity with which these junctions form in reaggregating cells (generally within a few minutes) suggests that the proteins are already in the membrane before the actual contact takes place. That these junctions which form in vitro are functional is elegantly shown in experiments where single, pulsating embryonic heart cells are observed in culture. When two of these cells come together, within minutes of the formation of a stable contact, these two cells are beating in perfect synchrony with one another; they are electrically coupled via low-resistance junctions. The formation of cell junctions, and their dissolution as well, serves as an excellent illustration of the dynamic nature of cell membranes.

REFERENCES

Albersheim, P., Sci. Am. **232**, 80–95, April 1975. "The Walls of Growing Plant Cells."

Ashwell, G., and Morrell, A., Trends Biochem. Sci. **2**, 76–78, 1977. "Membrane Glycoproteins and Recognition Phenomena."

Bennett, M. V. L., Federation Proc. **32**, 65–75, 1973. "Function of Electrotonic Junctions in Embryonic and Adult Tissue."

Bosmann, H. B., Int. Rev. Cytol. **50**, 1–23, 1977. "Cell Surface Enzymes: Effects on Mitotic Activity and Cell Adhesion."

Brinkley, B. R., and Porter, K. R., eds., symposia on "Cell-to-Cell Interactions and Communication," in *International Cell Biology 1976-1977*, pp. 31–100, Rockefeller University Press, 1977.

Cox, R. P., *Cell Communication*. Wiley, 1974.

Ito, S., Federation Proc. **28**, 12–25, 1969. "Structure and Function of the Glycocalyx."

Loewenstein, W. R., Sci. Am. **222**, 78–86, May 1970. "Intercellular Communications."

————, Federation Proc. **32**, 60–64, 1973. "Membrane Junctions in Growth and Differentiation."

Luft, J. H., Int. Rev. Cytol. **45**, 291–382, 1976. "The Structure and Properties of the Cell Surface Coat."

Marchase, R. B., Vosbeck, K., and Roth, S., Biochem. Biophys. Acta **457**, 385–415, 1976. "Intercellular Adhesive Specificity."

Martinez-Palomo, A., Int. Rev. Cytol. **29**, 29–75, 1970. "The Surface Coats of Animal Cells."

Marx, J. L., Science **196**, 1429, 1977, Research News. "Looking at Lectins: Do They Function in Recognition Processes?"

McNutt, N. S., and Weinstein, R. S., Prog. Biophys. and Mol. Biol. **26**, 45–101, 1973. "Membrane Ultrastructure at Mammalian Intercellular Junctions."

Moscona, A. A., ed., *The Cell Surface in Development.* Wiley, 1974.

———— and Hausman, R. E. in *Cell and Tissue Interactions,* J. W. Lash and M. M. Burger, eds., Raven, 1977.

Nelson, G. A., and Revel, J. P., Dev. Biol. **42**, 315–333, 1975. "Scanning Electron Microscopic Study of Cell Movements in Corneal Endothelium of the Avian Embryo."

Northcote, D. H., Ann. Rev. Plant Phys. **23**, 113–132, 1972. "Chemistry of the Plant Cell Wall."

Poste, G., and Nicholson, G. L., eds., *Cell Surface Reviews.* Elsevier, vol. 1 beginning 1976.

Rambourg, A., Int. Rev. Cytol. **31**, 57–114, 1971. "Morphological and Histochemical Aspects of Glycoproteins at the Surface of Animal Cells."

Sharon, N., Sci. Am. **220**, 92–98, May 1969. "The Bacterial Cell Wall."

Staehelin, L. A., Int. Rev. Cytol. **39**, 191–283, 1974. "Structure and Function of Intercellular Junctions."

Tanzer, M. L., Science **180**, 561–566, 1973. "Cross-Linking of Collagen."

Trelstad, R. L., and Coulomb, A. J., J. Cell Biol. **50**, 840–858, 1971. "Morphogenesis of the Collagenous Stroma in the Chick Cornea."

Weismann, G., and Claiborne, R., eds., *Cell Membranes*, Hospital Practice, 1975.

Winzler, R. J., Int. Rev. Cytol. **29**, 77–125. "Carbohydrates in Cell Surfaces."

Woodhead-Galloway, J., and Hukins, D. W. L., Endeavour **35**, 73–78, 1976. "Molecular Biology of Cartilage."

CHAPTER SEVEN
Cytoplasmic Membrane Systems

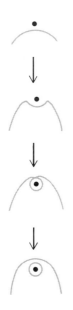

As is readily apparent from a quick examination of an electron micrograph of almost any eukaryotic cell, the cytoplasm is characterized by an extensive series of membranous structures. There are membrane-bound vesicles of varying diameter, containing material of different electron density. There are often long channels bounded by membranes which ramify through the cytoplasm to form an interconnected network of canals. In some cases there are stacks of flattened membrane-bound sacs, termed *cisternae*. The existence of certain of these structures was known from the examination of stained tissue sections with the light microscope, but an appreciation of the diversity and organization of these membranous structures awaited the development of the electron microscope. During the last 30 years a great deal of attention has been focused on the morphological, biochemical, and physiological aspects of these organelles. The study of the structure and function of the cytoplasmic membranes has required a coordinated effort between biochemists and microscopists as well as the development of new techniques. We will begin this discussion by considering the methods that were developed during the early period of study of these organelles and then consider the nature of the organelles themselves. We will be particularly concerned with one cellular process, that of *secretion,* which has been demonstrated to tie many of these structures together into an integrated functioning unit. It should be kept in mind that the topics considered in this chapter are relevant only to eukaryotic cells, since prokaryotes lack a corresponding cytoplasmic membrane system.

TECHNIQUES FOR THE STUDY OF CYTOMEMBRANES

Electron Microscopy

Progress in the morphological analysis of the cytoplasmic membranes paralleled progress in the development of techniques in electron microscopy. The situation in the early 1940s was one in which cytologists had within their means an instrument capable of delivering much greater resolution than the available techniques of tissue preparation would permit. In order to take advantage of the potential offered by the electron microscope, the means had to be developed whereby very thin, resistant sections could be cut and stains had to be prepared to provide the necessary contrast (see Chapter 2). A particularly important area of electron microscopic research was the development of stains that could localize specific molecular components, for example, a particular enzyme, type of sugar, or secretory product. The availability of a specific cytochemical method for a substance allows one to perform a biochemical test and visualize it with a microscopical method. As with autoradiographs (described below) this type of technique allows us to directly observe a physiological process. The ability to identify these substances cytochemically has been invaluable in assigning various functions to the cytoplasmic organelles.

Although the electron microscope can provide exquisitely detailed pictures of the parts of the cell, they are inevitably pictures that are frozen in time. The processes are dynamic ones, yet our means of observing them portray the scenes in a totally static manner. As a result, we must attempt to put these various structures into action in our own mind's eye and formulate a theory as to what they might be doing. Compounding the difficulty, most steady-state cellular processes occur in a continuous manner and therefore all stages in the process are present at the same time. This can make it very difficult to determine the sequence of events in time, particularly when the materials being processed (such as secretory products) are below the limits of resolution of the instrument. In some cases the difficulties can be eased by analyzing the event under conditions where it is not continuously occurring. For example, it might be possible to examine the process in the cells of the embryo in which it is just beginning for the first time. Or perhaps initiation of the process can be inhibited for a period of time so that what has already been initiated goes to completion but a new cycle cannot begin. If the inhibition is then removed, the sequence of events can be followed in succession by fixing pieces of tissue at various intervals after reversal of the inhibition. In other cases, the process may be one that normally does not occur but can be induced by some treatment, for example, a hormone or drug. As in the previous case, the sequence of events can be illuminated by examining the cells at successive intervals after the onset of the activity.

Differential Centrifugation

Simultaneous with the development of electron microscopic techniques, the methodology of differential centrifugation (page 111) was being worked out, primarily by Albert Claude, whereby the membranous elements of the cytoplasm could be isolated in vitro and hopefully separated from one another. When the cell is ruptured by homogenization, the cytoplasmic membranes become fragmented and the ends of the pieces of membrane fuse and form small vesicles, termed *microsomes*. The microsomal fraction usually contains a heterogeneous collection of small vesicles derived from a variety of the cell organelles. More recent procedures have been developed whereby the different types of vesicles can be at least partially separated from one another. For example, vesicles with attached ribosomes (representing the rough endoplasmic reticulum) can be caused to aggregate by the addition of cations such as Ca^{2+} or Mg^{2+} to the suspension, causing them to sediment more rapidly than smooth vesicles in a centrifuge tube. A comparison of the ultrastructure of a smooth and rough microsomal fraction is shown in Fig. 7-1. In other cases, the microsomes can be centrifuged through a gradient of increasing sucrose concentration and the membranes of the Golgi complex can be separated from those of the endoplasmic reticulum. Once these separations have been achieved, each fraction can be isolated independently of other fractions containing different organelles. It has been found that membranous organelles retain a remarkable degree of their cellular activity under these in vitro conditions; not only can their composition be determined but their functional capacities, for example, their enzymatic activities, can be as well. The most serious limitation to the study of cellular fractions is the uncertainty of the changes that might have occurred during the isolation procedure. In other words, to what extent do these fragments being studied in the test tube represent their state in the living cell?

Figure 7-1. *(a)* The electron microscopic appearance of a section through a smooth microsomal fraction in which the membranous vesicles lack ribosomes. *(b)* A rough microsomal fraction containing ribosome-studded membranes. [*From J. A. Higgins and R. J. Barnett,* J. Cell Biol., **55**:293 (1972).]

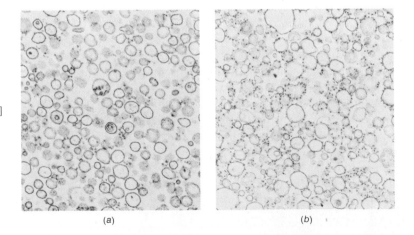

(a)   (b)

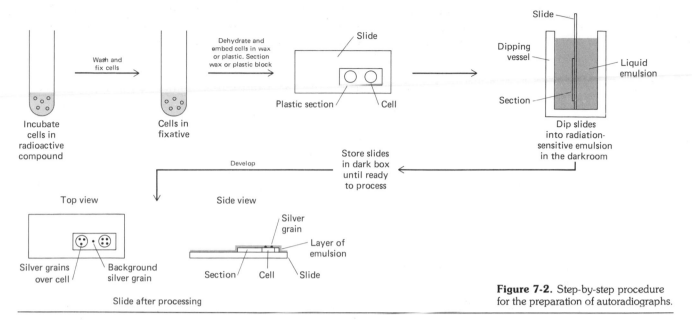

Figure 7-2. Step-by-step procedure for the preparation of autoradiographs.

Autoradiography

A third technique of great importance in the analysis of cytomembrane function is that of *autoradiography* (or *radioautography*) which was applied to these problems after the initial use of electron microscopy and cell fractionation. The steps in the technique are illustrated in Fig. 7-2. Basically, autoradiography takes advantage of the ability of a particle emitted from a radioactive atom to activate a photographic emulsion, much like light or x-rays activate the emulsion that coats a piece of film. If the photographic emulsion is brought into close contact with a radioactive source, the particles emitted by the source leave tiny, black silver grains in the emulsion after processing (Fig. 7-3). Autoradiography is used to localize radioisotopes within sections of cells and tissues that have been

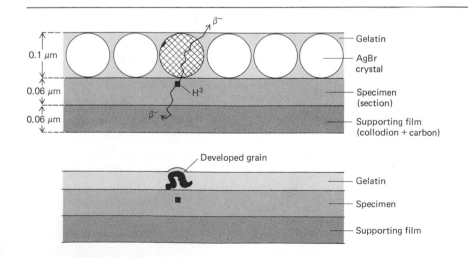

Figure 7-3. Diagrammatic representation of an electron microscope autoradiograph preparation. Top: during exposure, the silver halide crystals, embedded in a gelatin matrix, cover the section. A β-particle, from a tritium point source in the specimen, has hit a crystal (crosshatched), causing the appearance of a latent image on the surface (black speck). Bottom: After processing, the exposed crystal has developed into a silver grain; the nonexposed crystals have been dissolved. *(From L. Caro, in D. M. Prescott, ed., "Methods in Cell Physiology," vol. 1, p. 337, Academic, 1964.)*

immobilized on a slide or EM grid. The emulsion is applied to the sections on the slide or grid as a very thin overlying layer and the specimen is put into a lightproof container to allow the film to be exposed by the emissions. The longer the specimen is left before processing, the greater will be the number of silver grains. When the slide or grid is examined in the microscope, the location of the silver grains in the layer of emulsion just above the tissue indicates the location of the radioactivity in the cells. Examples of autoradiographs are shown on pages 224 and 315.

Autoradiography provides a means to visualize biochemical events. For example, if one wants to know where in the cell proteins are synthesized, the tissue can be immersed in a radioactive precursor to proteins, i.e., an amino acid, and the cell allowed a brief period of time to take the labeled amino acid up from the medium and incorporate it into the proteins that are being synthesized at that time in the cell. As far as the cell is concerned, there is no difference between a radioactive molecule and that same molecule in its nonradioactive state; once the labeled amino acid is in the cell it mixes with the unlabeled ones already present and is as likely to be selected for incorporation as an unlabeled molecule. If the cell is fixed very soon after the isotope is administered, the only sites within the cell that should contain *incorporated* radioactivity are the sites where the protein is being assembled. If the tissue is fixed, embedded, sectioned, and prepared for autoradiography, the location of the silver grains should indicate the location of the synthetic activity. There is one dilemma to consider. How can one distinguish radioactive amino acids in the cells that have not been incorporated yet, i.e., they are still present in the *precursor pool*, from those that have already been included in either a growing or completed polypeptide chain? The means to accomplish the distinction between these two types of labeled amino acids, i.e., incorporated from unincorporated, is to selectively remove the unincorporated ones. Since free amino acids are of small molecular weight, they are easily extracted from the cells prior to the embedding procedure while the larger molecular weight macromolecules remain in a precipitated state within their intracellular location.

One of the most important features of autoradiography, besides making visible an otherwise invisible biochemical process, is the technique's ability to get around the problems inherent in the analysis of steady-state events. Since the time at which the isotope is administered acts as a zero time for the process, it makes no difference that the process is occurring continuously and all stages are present at the same time. For example, in the following pages we will consider the events of secretion in the pancreas. We can define the process as beginning with the synthesis of the material to be secreted and ending with the discharge of the material outside the cell. There is a time sequence of events from the start to the finish of this process. How can we determine the steps in this sequence? If the secretion product is a protein, for example, an enzyme, we can administer radioactive amino acids to a number of pieces of tissue at time zero. If one piece of tissue is fixed very soon after the isotope is given, then the autoradiograph will reveal the sites where the incorporation of the amino acid into protein has occurred. If one waits for a period of time before another piece of tissue is fixed, the radioactive protein made at the beginning of the incubation period will have had the opportunity to move on from its site of synthesis to places where later stages in the process are occurring. The longer we wait, the farther along the radioactive protein will have moved before the cell was killed.

To ensure that the maximum amount of information is gained from the experiment, the pieces of tissue are generally exposed to the radioactive precursor for a brief period of time, the tissue is quickly washed free of excess isotope, and the medium is replaced with one containing unlabeled amino acids. An experiment of this type is called a *pulse-chase;* the pulse refers to the brief incubation with radioactivity and the chase refers to the period when the tissue is exposed to the unlabeled medium. Under these types of conditions, the supply of labeled amino acids in the cell will be rapidly exhausted and the subsequent proteins being synthesized will be made from unlabeled precursors and will no longer be evident in the autoradiographs. In this way one can ideally follow the sequence of events by observing a wave of radioactive material moving through the cell from one location to the next until the process is complete (see Fig. 7-10).

The most serious limitation of the autoradiographic technique for the type of studies to be discussed in this section is one of resolution. The purpose of the technique is to determine the location of the radioactivity within the tissue from the position of the silver grains. However, if the silver grain itself is too large relative to the organelle from which the radioactivity is originating, it will be impossible to determine if the radioactivity is in the organelle or some surrounding structure or solution. Similarly, the particle emitted by the radioactive source can travel a certain distance from its point of origin before it strikes the emulsion (from 0.5 μm using ^{3}H to greater distances for other commonly employed isotopes such as ^{14}C and ^{35}S). As a result, the location of the silver grain, i.e., where the particle interacted with the emulsion, may be some distance from the source within the tissue. Since at the electron microscope level the objects being examined are often very small, these factors can be of importance in limiting the resolution of the techniques.

ENDOPLASMIC RETICULUM

Under the light microscope the cytoplasm of most living cells appears relatively empty except for the presence of various types of large granules or droplets. However, before the start of the century examination of stained sections of cells, such as those of the pancreas, had indicated the presence of some type of extensive cytoplasmic structure, which was termed the *ergastoplasm*. It was noted that the presence of this rather vague stainable region of the pancreas cell was dependent upon the physiological state of the animal; it would tend to disappear upon extended starvation and reappear if the animal were fed. It was proposed that the ergastoplasm was somehow involved in the preparation of digestive juices by the pancreas. The morphological nature of the ergastoplasm was brought to light in the 1940s with the early studies of tissue sections in the electron microscope, originally by Keith Porter, who renamed the structure the *endoplasmic reticulum* (ER). The endoplasmic reticulum is generally divided into two broad categories (Fig. 7-4): the rough endoplasmic reticulum (RER), and the smooth endoplasmic reticulum (SER). In both cases the organelles are composed of a system of membranes which are always seen to enclose a space. The fluid content of the cytoplasm is accordingly divided by the ER into two compartments: that space enclosed within its membranes, i.e., generally termed the *cisternal* space, and that region outside of the membranes, i.e., the actual cytoplasmic space.

The morphological distinction between the RER and the SER is the presence of attached ribosomes in the former and their absence in the latter. When present, the ribosomes are always on the outer surface of the membranes, i.e., that surface facing the cytoplasmic space. The rough endoplasmic reticulum generally appears as an extensive membranous organelle composed of vesicles, tubules, and flattened sacs *(cisternae)* as shown in Fig. 7-5*a*. The cisternae are present one on top of the other with a cytoplasmic space between the sacs (Fig. 7-5*b*). In contrast, the elements of the smooth ER are more tubular in nature (Figs. 7-4 and 7-6) and form an interconnecting system of pipelines curving through the cytoplasm in which they occur. In cross section the SER tubules appear as

Figure 7-4. Electron micrograph of part of a bat pancreas secretory cell showing both smooth and rough endoplasmic reticulum. *(Courtesy of K. R. Porter.)*

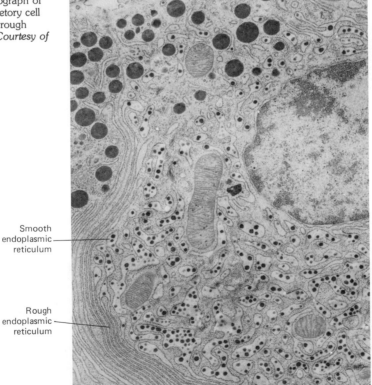

Smooth endoplasmic reticulum

Rough endoplasmic reticulum

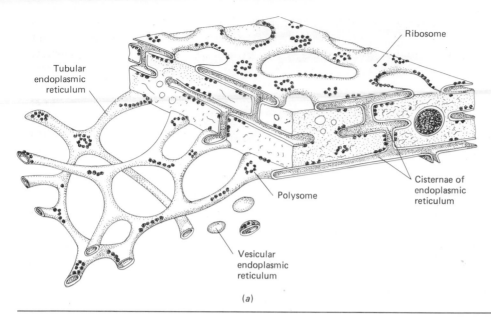

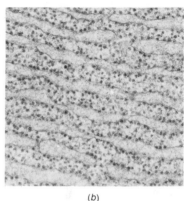

(a)

(b)

Figure 7-5. The rough endoplasmic reticulum. *(a)* Schematic representation showing the stacks of flattened cisternae as well as a portion of tubular rough ER. The ribosomes are seen to be present on the membranes in linear arrays which represent polysomes. *(From W. M. Copenhaver, R. P. Bunge, and M. B. Bunge, "Bailey's Textbook of Histology", 16th ed., Williams and Wilkins, 1971.)* *(b)* Electron micrograph of a portion of the rough endoplasmic reticulum of a human secretory pancreatic acinar cell. Note the manner in which the cell contents is divided into an intracisternal space (devoid of ribosomes) and an external "cytoplasmic" space. *(Courtesy of S. Ito.)*

vesicles, elongated to varying degrees depending upon the plane of the section relative to the axis of the tubule. In many cells the two types of channels appear to be interconnected (Fig. 7-6b) and studies indicate that materials can be transported from the cavities of the RER into the smooth surfaced channels. Different types of cells are characterized by having markedly different amounts of one type of ER or the other, depending upon the activities of the cell. For example, cells that are active in secreting proteinaceous materials, such as the cells of the pancreas, have extensive regions of RER, while the SER in these cells is generally restricted to certain edges of the RER cisternae as shown in Fig. 7-6b. We will return to the function of the rough ER below.

In many types of cells, such as those found in skeletal muscle, the tubules of the kidney, or steroid-producing endocrine glands, there is an extensive network of SER tubules with a greatly reduced RER. In all probability, the nature of the smooth ER, i.e., its specific enzymatic content, varies considerably from cell to cell, even though its appearance may be fairly similar. The heterogeneous nature of the SER can be seen from the variety of functions that it can carry out. Specialized functions of the SER include:

1. The synthesis of steroids in a variety of steroid-producing cells including those in the gonad and adrenal gland.

2. In the liver the smooth ER is specialized for the detoxification of various drugs, for example, barbiturates. At least two complex series of enzymes which act to oxidize these substances are known to occur in SER membranes. If an animal is injected with a large dose of phenobarbitol, for example, there is a rapid increase in the smooth ER of the liver cells (Fig. 7-6a) and in their content of these proteins.

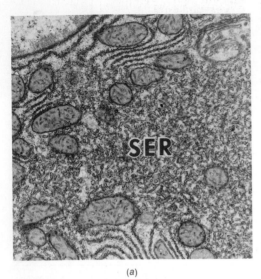

(a)

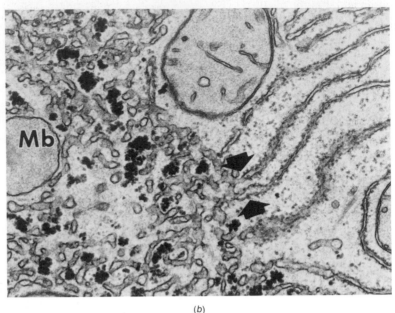

(b)

Figure 7-6. *(a)* The smooth ER of a rat liver cell after treatment of the animal with phenobarbitol, a drug which causes a tremendous increase (hypertrophy) in this type of membranous organelle. *(b)* Electron micrograph of a portion of a normal rat liver cell showing the continuity between the rough and smooth surfaced endoplasmic reticulum membranes (large arrows) as well as the intimate association of glycogen granules (G) with the smooth ER. A portion of a microbody (MB) is shown. [*From A. L. Jones and D. L. Schmucker, Gastroenterology, **73**:847 (1977).*]

3. In liver cells one of the characteristic enzymes of the SER is glucose 6-phosphatase, which catalyzes the dephosphorylation of glucose 6-phosphate to form glucose. As a result of this reaction, energy stored in glycogen can be made available to the body. Large reserves of glycogen are usually found in cells of the liver in the form of small granules attached to the outside of the SER membranes (Fig. 7-6*b*). When the need for chemical energy arises in the body, the glycogen is broken down under hormonal control (page 243) by the enzyme phosphorylase into glucose 1-phosphate, which is then converted to glucose 6-phosphate in the cytoplasm. Since glucose phosphates are intermediates in glycolysis (Chapter 8), one might expect that they could be used directly for ATP formation as a substrate of oxidative metabolism. The problem is that in the phosphorylated state they cannot leave the liver cell, nor can they be taken up by the cells of the body; the membranes are impermeable to sugar phosphates. It is the function of the glucose 6-phosphatase in the SER membrane to catalyze the removal of the phosphate group and transfer the glucose formed into the space within the SER. From there the glucose can somehow gain entrance to the bloodstream where it can be transported to needy cells. Once within the cells the sugar can be rephosphorylated and metabolized.

4. In striated muscle cells there is an extensive, highly specialized smooth ER, called the *sarcoplasmic reticulum*, which is involved in carrying the stimulus from the nerve cell into the cytoplasm of the muscle fiber. We will consider the function of this organelle in Chapter 16.

The degree to which the membranes of the smooth ER and rough ER differ in composition is a matter of some debate and probably varies with the cell type. In many cases where both SER and RER are found together the same proteins are found in both smooth-surfaced and rough-surfaced microsomal vesicles, though the proportions of each may vary. Taken together, the membranes of the ER are characterized by a relatively high protein/lipid ratio. Up to 70% of the weight of the ER membrane can be accounted for by its protein; this represents approximately 23 phospholipid molecules to each protein molecule. Approximately 30 to 40 different enzymes have been found within ER membranes (Table 7-1). Even though there are both integral and peripheral proteins, as in the case of the plasma membrane, the greatly increased levels of protein relative to that of the plasma membrane may require a somewhat different organization. At the least, the structure of the ER membrane would be expected to have a greater dependence upon protein-protein interactions, as opposed to protein-lipid or lipid-lipid interactions.

THE SECRETORY CELL

The appearance and function of the rough endoplasmic reticulum appears to be intimately linked to the production of proteins for export out of the

TABLE 7-1
Transverse Localization of Various Enzymes of the Endoplasmic Reticulum

Enzyme	Localization	Criteria
Cytochrome b_5	Cytoplasmic surface	Release with protease Inhibition with antibodies EM of ferritin-labeled antibody Lack of latency
NADH-cytochrome b_5 reductase	Cytoplasmic surface	Release with protease Inhibition with antibodies Lack of latency (page 327)
NADPH-cytochrome c reductase	Cytoplasmic surface	Release with protease Inhibition with antibodies Lack of latency Localization of products
Cytochrome P-450	Cytoplasmic surface	Denaturation with protease ^{125}I-labeling Localization of products
	Luminal surface	Denaturation with protease in presence of deoxycholate
ATPase	Cytoplasmic surface	Denaturation with protease Localization of products
Nucleoside pyrophosphatase	Cytoplasmic surface	Release with protease
GDPmannosyl transferase	Cytoplasmic surface	Denaturation with protease
Nucleoside diphosphatase	Luminal surface	Lack of protease effect Lack of antibody inhibition Latency
Glucose 6-phosphatase	Luminal surface	Protease denaturation only in presence of deoxycholate Localization of products
Acetanilide-hydrolyzing esterase	Luminal surface	Lack of effect of various proteases No adsorption of specific antibodies
β-Glucuronidase	Luminal surface	Lack of protease effect Release with low concentrations of deoxycholate Latency

SOURCE: J. W. Depierre and G. Dallner, *Bioc. Biop. Acta,* **45**: 454 (1974).

cell. A considerable variety of secretory cells have been studied using the techniques described at the beginning of this chapter. These studies include endocrine cells which secrete polypeptide hormones, salivary gland cells which secrete enzymes and mucoproteins, plasma cells which secrete antibodies, pancreatic acinar cells which secrete digestive enzymes, intestinal goblet cells which secrete mucoproteins, cartilage cells which secrete collagen and mucoprotein, liver cells which secrete serum proteins, and plant cells which secrete material for their cell wall. In all of these cells a great number of similarities have been found (including the presence of considerable rough ER) and the same organelles appear to carry out roughly the same functions. The details, however, vary considerably from one type of cell to the next as does the morphology of the various organelles. In the following discussion we will draw most heavily upon the studies of the guinea pig pancreas and the rabbit parotid salivary gland carried out in the laboratory of George Palade, and the intestinal goblet cell based upon the work of Charles Leblond and colleagues. References to these extensive studies can be found at the end of the chapter. The subcellular organization of secretory cells in shown in Fig. 7-7.

As is clear from the diagrams of the secretory cells of Fig. 7-7, the internal organelles are very asymmetrically placed. The contents of these glandular epithelial cells are polarized along the tall axis of the cell, i.e., from their basal to their apical end. Toward the basal surface of the cell, i.e., that facing the blood supply, is the nucleus and the parallel cisternae of RER. In the central region of the cell one finds the Golgi complex, while toward the apical end of the cell, i.e., that edge facing the lumen (cavity) of the duct into which the secretions will be discharged, are a large number of membrane-bound vesicles, termed *secretory granules*. As will be described shortly, this polarity from one end to the other parallels the flow of secretory products through the cell.

In order to determine the site of synthesis of the polypeptide chains comprising the various secretory products, pieces or slices of tissue are immersed in radioactive amino acids for the brief "pulse" (Fig. 7-8). If these cells are fixed directly after incubation in isotope without being "chased," virtually all of the silver grains are localized over the rough ER (Fig. 7-8a). The basis for parts b and c of Fig. 7-9 are considered in the following section. The same site of incorporation is seen in all of the secretory cells and the concept has emerged from these studies that all proteins produced for export from the cell are synthesized on *membrane-bound ribosomes*. Even though the ribosomes upon which the proteins are manufactured are located on the *outer* surface of the ER membrane, the protein is released into the space of the cisternae (Fig. 7-9). It has been proposed that the polypeptide chain as it is being formed by the ribosome, passes through very small open channels across the membrane into the cavities of the ER. Once inside the cisternae, the protein would undergo its folding, causing it to assume its tertiary structure and become too large in diameter to pass back through the membrane.

If proteins to be exported from the cell, i.e., secretory proteins, are produced on membrane-bound ribosomes and proteins destined to remain

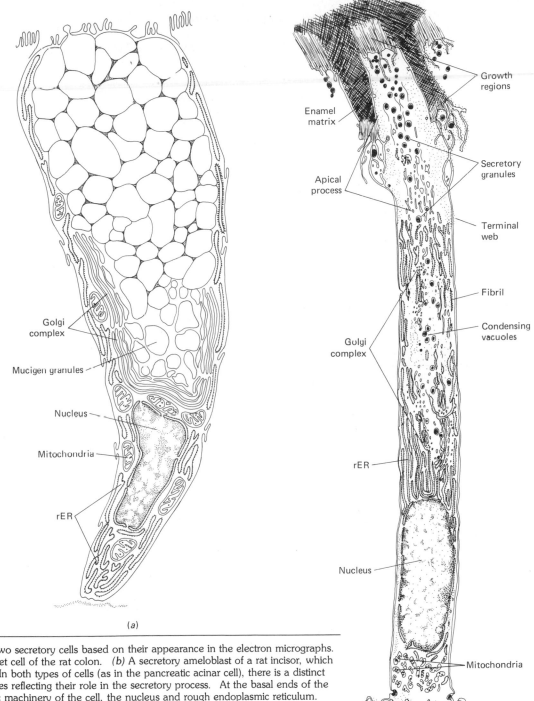

Labels in figure (a): Golgi complex, Mucigen granules, Nucleus, Mitochondria, rER

Labels in figure (b): Growth regions, Enamel matrix, Apical process, Secretory granules, Terminal web, Fibril, Condensing vacuoles, Golgi complex, rER, Nucleus, Mitochondria

(a)

(b)

Figure 7-7. Diagrams of two secretory cells based on their appearance in the electron micrographs. (a) A mucus-secreting goblet cell of the rat colon. (b) A secretory ameloblast of a rat incisor, which produces enamel matrix. In both types of cells (as in the pancreatic acinar cell), there is a distinct polarization of the organelles reflecting their role in the secretory process. At the basal ends of the cell, one finds the synthetic machinery of the cell, the nucleus and rough endoplasmic reticulum. Proteins synthesized in the RER move into the closely associated Golgi complex and from there into condensing vacuoles in which the final secretory product is concentrated. The apical region of the cells are filled with secretory granules containing the mucus or enamel matrix glycoprotein, ready for release. [(a) From M. Neutra and C. P. Leblond, J. Cell Biol., **30**:119 (1966); (b) from A. Weinstock and C. P. LeBlond, J. Cell Biol., **51**:26 (1971).]

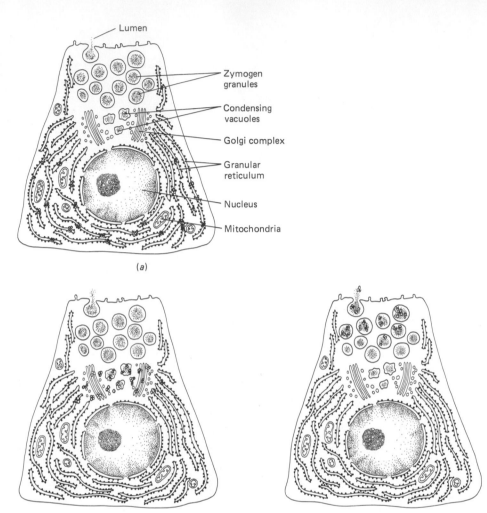

Lumen

Zymogen
granules

Condensing
vacuoles

Golgi complex

Granular
reticulum

Nucleus

Mitochondria

(a)

(b)   (c)

Figure 7-8. Diagrams showing the
organization of a guinea pig pancreatic acinar cell and the localization
of radioactivity after injection of
³H-leucine. *(a)* After a 3-minute
pulse, the rough endoplasmic reticulum is labeled. *(b)* After a 3-minute
pulse and 17-minute chase, the rough
ER near the Golgi complex and the
condensing vacuoles are labeled.
*(c) After a 3-minute and 117-minute
chase, only the zymogen granules at
the apical pole are labeled. (From
P. Favard, in A. Lima-de-Faria, ed.,
"Handbook of Molecular Cytology,"
Elsevier North-Holland, 1969, after
J. D. Jamieson and G. E. Palade.)*

in the soluble phase of the cytoplasm (such as the enzymes of glycolysis
in a liver cell or hemoglobin in a reticulocyte) are produced on free ribosomes, the question arises as to how the cell identifies these two classes
of proteins in order to separate their site of synthesis. In response to this
dilemma, a proposal termed the *signal hypothesis* has been formulated
which has continued to gain experimental support. The basic points of
the signal hypothesis are illustrated in Fig. 7-9. The synthesis of the polypeptide would actually begin in connection with a ribosome that was not
attached to a membrane. It is proposed that the polypeptide chain, as it
is formed, is directed through a tunnellike region of the large ribosomal
subunit. Furthermore, it is proposed that the initial portion of the polypeptide chain (the amino terminus) contains a sequence of amino acid
residues which play a special role in the compartmentalization of secretory
proteins. In this hypothesis, this stretch of amino acids function as a
"signal" by which the mRNA-ribosome complex becomes attached to the

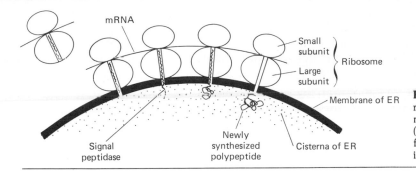

Figure 7-9. The concept of the interrelation between the ribosomes and membrane of the rough ER. The attachment of the ribosome, presumably by the signal peptide (dotted section) of the forming polypeptide chain, facilitates the movement of the newly synthesized protein into the cisterna of the ER as described in the text.

membrane of the endoplasmic reticulum. It is proposed that the amino terminal region of the growing polypeptide interacts specifically with one or more membrane protein receptors to which it becomes attached. The attachment of the signal region of the polypeptide to the membrane would lead to the further attachment of the ribosome to the membrane as well as to the formation of a tunnel across the membrane through which the polypeptide can pass into the internal space of the ER. Once the polypeptide begins to enter the rough ER cisterna, it is proposed that a proteolytic enzyme, the signal peptidase, removes the "address" which is no longer needed. Recent experiments with in vitro systems have confirmed that secretory proteins are synthesized with an extra 15 to 30 amino acids (primarily hydrophobic) at their amino end. If the in vitro synthesis of the protein occurs in the presence of microsomal membranes, then the newly synthesized protein passes into the membrane of the vesicles and the signal peptide is cut from the remainder of the polypeptide. Further experiments indicate that a very similar mechanism operates in bacteria for the exportation of external (periplasmic or outer membrane) proteins. In this case, the proteins destined to be expelled from the cell are synthesized by ribosomes bound directly to the plasma membrane itself; there are no internal membranes homologous to the endoplasmic reticulum of the eukaryotic cell.

The endoplasmic reticulum is perfectly constructed for its role in secretory cells. The membrane provides a large surface area upon which the ribosomes can attach (approximately 13 million per liver cell) and the cisternae provide a compartment into which the secretory product can be segregated and then transported through its channels. As indicated above it is important that the cell distinguish between the proteins produced for secretion and the proteins to be kept within the cytoplasm since the two have such different fates. The secretory products remain in a membrane-bound form throughout the secretory process and this packaging begins in the RER. The segregation of newly synthesized proteins in the ER cisternae essentially removes them from the cytoplasm. This process provides a clear illustration of the role of membranes in compartmentalizing the space within the cell so that different types of activities can be separated. Since the cisternae of these secretory cells exist as large flattened cavities, they can act as transport channels for secretory proteins from their site of synthesis to a more apical position in the cell where they are believed to leave the endoplasmic reticulum heading for the Golgi complex.

It is apparent from a number of studies that the trip through the ER cisternae and into the Golgi complex involves more than just the movement of the products. As the proteins move along they are chemically acted upon and modified by enzymes within the wall of the containers. For example, there appears to be an enzyme that promotes the formation of disulfide bridges between appropriate cysteine residues in the polypeptide. In fibroblasts, the hydroxylation of proline and lysine residues of collagen (Chapter 3) are catalyzed by ER enzymes. In the cells of the islets of Langerhans of the pancreas (its endocrine cells rather than its enzyme-secreting cells), the proinsulin molecule is split by a proteolytic enzyme into its two chains. In other words, it appears that the membranes of the transport channels through which the secretory product moves act as a type of assembly line which sequentially interact with the contents of the channel and modify it.

If one considers the composition of various secretory materials, it becomes apparent that a large percentage of them contain carbohydrates. In fact, an earlier proposal suggested that carbohydrate was an essential component of all products synthesized for export and that the carbohydrate provided a means for the cell to identify materials for this purpose. Though it is now clear that there are secretory proteins that contain no carbohydrate (for example, serum albumin of the liver or numerous of the pancreatic enzymes), a large number of these substances are glycoproteins, mucoproteins, or simply complex polysaccharides. The carbohydrate may be present as a relatively small percentage of the total product, as in the case of antibodies or a few of the pancreatic hydrolases, or it can account for the bulk of the product as in the case of chondromucoprotein of the cartilage matrix. In some cases, the material being secreted is totally of a polysaccharide nature as in the case of the hyaluronic acid of the synovial fluid of joints, or the components of the plant cell wall.

A question of importance is where in the cell the carbohydrate portion is formed and/or attached to the protein. The answer generally depends upon the position of the sugar in the carbohydrate chain (Fig. 7-10). If the sugar is one that is very close to the protein, then, at least in some cases, it is attached to the polypeptide while it is still in the ER. If, however, it is one of the terminal sugars in the chain, such as is usually the case for galactose and always the case for fucose and sialic acid, the sugar is added within the Golgi complex (as described below).

Secretory proteins are not the only important glycoproteins of the cell. A great deal of attention was paid in Chapter 5 to glycoproteins that constituted the integral proteins of the outer leaflet of the plasma membrane. The synthesis of the glycoproteins of the various cellular membranes and the glycoproteins of various secretory products is believed to be a closely related process. The major steps that occur in the formation of both secretory and membrane glycoproteins is illustrated in Fig. 7-10. We have already discussed the signal hypothesis by which polypeptides destined for secretion become attached to the membrane of the rough ER. This same proposal is believed to apply to the formation of membrane proteins as well. As discussed in Chapter

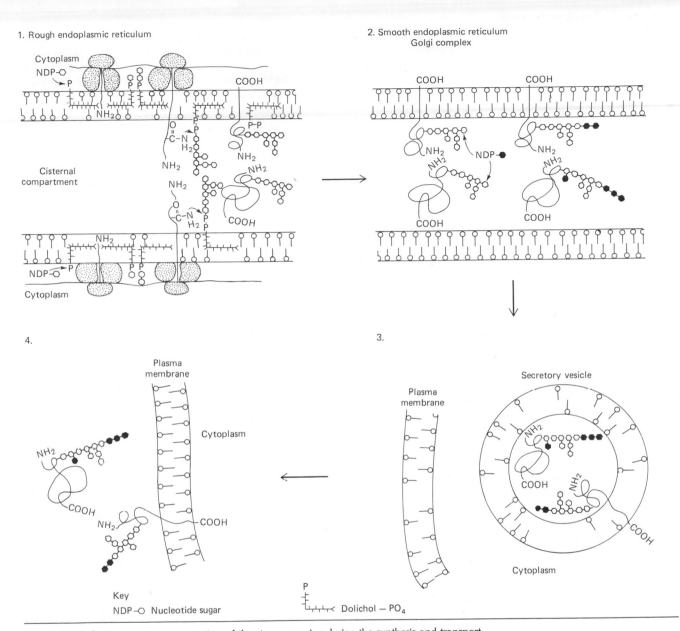

1. Rough endoplasmic reticulum

2. Smooth endoplasmic reticulum
Golgi complex

Cytoplasm
NDP-O
P
Cisternal
compartment
NDP-O
Cytoplasm

COOH
NH₂
C-N
H₂
NH₂
NH₂
P-P
NH₂
C-N
H₂
COOH
NH₂

COOH
NH₂
NDP-●
NH₂
COOH

COOH
NH₂
NH₂
COOH

4.

Plasma
membrane

Cytoplasm

NH₂
COOH
NH₂
COOH

Key
NDP-O Nucleotide sugar

3.

Plasma
membrane

Secretory vesicle
NH₂
COOH
NH₂
COOH
Cytoplasm

P
Dolichol — PO₄

Figure 7-10. Diagrammatic representation of the steps occurring during the synthesis and transport of both secretory and membrane glycoproteins. Open hexagons represent sugar residues that were initially added to the dolichol phosphate carrier shown in the first diagram while closed hexagons represent sugar residues added to incomplete oligosaccharide chains already attached to proteins. It should be noted that the finding of membrane proteins (such as band 3 of the erythrocyte) with the carboxyl terminals on the outside of the plasma membrane and amino terminals on the inner side suggests that additional factors may be involved in membrane protein metabolism. (*Courtesy of W. J. Lennarz.*)

5, the site of attachment of the carbohydrate to the protein, as well as the sugar composition, is a very important property of a particular glycoprotein. In other words, it is important that the proper sugar residues be incorporated into the glycoprotein at the proper location in the molecule. The construction of particular oligosaccharides by the action of individual glycosyltransferases was discussed at some length in Chapter 5 (page 196), although no mention was made as to where in the cell these reactions took place. As indicated in Fig. 7-10, the initial sugars are added to the protein in the rough ER, probably as the polypeptide chain is being formed. Recent studies indicate that the first few sugars of the oligosaccharide chains are not added directly to the protein, but rather are transferred, as a group, from a lipid carrier. This lipid carrier, termed *dolichol-phosphate*, is an extremely long hydrophobic molecule believed to occur within the membrane itself (part 1, Fig. 7-10). The first sugars are added to the dolichol phosphate molecules, one at a time in the proper order by the action of glycosyltransferases. In every case, the sugar donor is a nucleotide sugar. Once a certain block of sugar residues are linked, this block is then transferred as a unit to the protein (indicated by the arrow in part 1, Fig. 7-10). From this point, individual sugars (indicated by blackened sugar groups in Fig. 7-10) are added one at a time by glycosyltransferases as described in Chapter 5. As mentioned above, secretory proteins and membrane proteins are treated in a very similar manner. The main difference between the two classes of proteins is the release of the secretory protein into the cisterna and the retention of the membrane protein by the membrane itself (shown in Fig. 7-10). Presumably, the membrane proteins consist of polypeptide chains containing stretches of hydrophobic amino acids that cause them to remain in the membrane from the time of their formation. In the next part of this chapter, we will follow the movement of secretory materials produced in the endoplasmic reticulum along their path through the cell, a path which takes them through the Golgi complex, into secretory vesicles, and out via fusion with the plasma membrane. As seen in Fig. 7-10, those proteins (and glycoproteins) initially present in the cisternae of the ER are ejected from the cell, while many of those present within the membranes of the ER eventually become part of the membrane surrounding the cell itself.

THE GOLGI COMPLEX

The Golgi complex has had a stormy history since its first description in 1898 by the 1906 Nobel laureate, Camillo Golgi. Working with nervous tissue that had been impregnated for several days in osmium-containing stain, he discovered a dark yellow network that could be visualized near the cell nucleus and which was later identified in a variety of cell types (Fig. 7-11a). Up until the 1950s when the Golgi complex was first identified in the electron microscope, the existence of this structure had been the center of much controversy. Its presence in unfixed, freeze-fractured specimens has indicated its existence within the *living* cell beyond doubt. Though the appearance of the Golgi complex varies somewhat with the nature of the cell, it has a characteristic structure composed of flattened disklike cisternae and associated vesicles (Fig. 7-11b–d). The cisternae, whose diameters are typically from 0.5 to 1.0 μm. are arranged in an orderly stack, much like a stack of pancakes, and are curved in a manner so as to form a shallow cup. Typically, there are less than eight cisternae per stack, and there may be anywhere from one or a few stacks to several thousand per cell, depending upon the cell type.

In the diagram in Fig. 7-7a, the cisterna closest to the ER is said to be at the outer edge (or face) of the stack, while the cisterna facing toward the apical end of the cell is said to be at the inner face. In many cases the edges of the cisternae, particularly the ones toward the inner face, are dilated and these are associated with vacuoles which are believed to be budded from the dilated tips. Since the membranes of the Golgi complex are devoid of ribosomes, they are classified as smooth membranes and after homogenization these membranes can be found in the smooth microsome fraction (Fig. 7-1). Techniques have been devised to partially purify Golgi membranes and certain enzymes are known to be particularly rich in the membranes of the Golgi fraction.

The precise relationship between the endoplasmic reticulum and the Golgi complex is not often clear, and where it is evident seems to vary with the type of cell. In the cells of the pancreas, it appears that the apical edges of the RER cisternae (Figs. 7-6b, 7-12, 7-18), which are devoid of ribosomes (and therefore considered SER), break into small vesicles which act as transport structures carrying materials from the RER to the Golgi region. In other words, smooth-surfaced vesicles are budded from the endoplasmic reticulum as a type of *transitional element* between the two organelles. In some cells there is evidence of direct communication between the cavities of the ER and those of the Golgi complex. These are most evident in high-voltage electron micrographs where thicker sections can be examined.

A variety of studies have indicated that the Golgi complex is not of uniform composition from one end of its stack to the other. In some cases there are striking cytochemical differences between cisternae at either end of the stack. For example, osmium impregnation occurs preferentially at the outer face (Fig. 7-13b), while certain enzymes of the Golgi complex (thiamine pyrophosphatase or nucleoside diphosphatase) are found preferentially at the inner face (Fig. 7-13a). In a few cases at least, the membranes of the outer face are considerably thinner than those of the inner face and the difference in thickness across the Golgi stack is similar to the difference in thickness between the membranes of the endoplasmic reticulum (50 to 60 Å) and the plasma membrane (75 to 100 Å). The Golgi complex is generally considered to be a transitional organelle between the ER and plasma membrane, undergoing a progressive differentiation from one membrane type to the other. In support of this contention, the protein and lipid composition of the Golgi complex tends to be intermediate between that of the other two types of membrane. For example, in comparison to the endoplasmic reticulum, the membranes of the Golgi complex have an elevated cholesterol and sphingomyelin level and a lowered percentage of unsaturated fatty acids. In these respects, the Golgi membranes are more like the plasma membrane.

The formation of the Golgi complex is envisioned to occur by the fusion of smooth-surfaced vesicles that had previously been budded off the endoplasmic reticulum. These vesicles would fuse to form the outermost cisternae, which are often termed the *forming face*. At the other end of the stack, the *maturing face*, the cisternae are proposed to disperse into a number of smooth-surfaced vesicles. This dispersal process is believed to occur, at least partially, by the budding of vacuoles at the tips

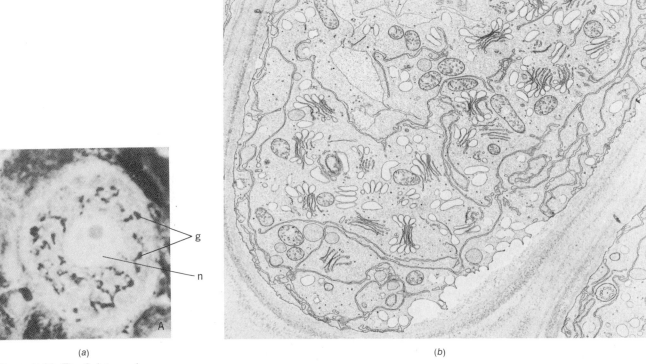

(a)

(b)

Figure 7-11. The Golgi complex. (a) Light micrograph of a spinal ganglion cell stained to best reveal the Golgi complex. (b) Low-power electron micrograph of the cytoplasm of a plant cell showing the distribution of the various membranous organelles. (c) Electron micrograph of a portion of a single dictyosome (a stack of Golgi cisternae) of an animal cell showing the presence of vesicles near the forming face and the budding of larger secretory vesicles from the ends of the cisternae of the maturing face. (d) Electron micrograph of a dictyosome of a plant cell depicting the maturation of cisternae from one face to the other. [(a) from G. Rasmussen and H. Elftman, in W. F. Windel, "Histology," 5th ed., McGraw-Hill, 1976; (b and d) courtesy of H. H. Mollenhauer; (c) courtesy of O. Kiermayer.]

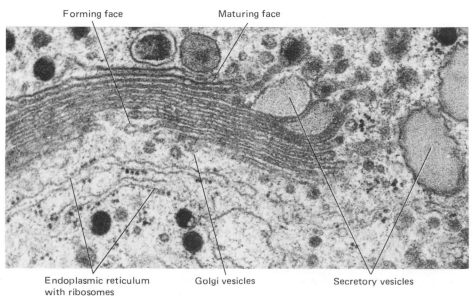

Forming face   Maturing face

Endoplasmic reticulum
with ribosomes   Golgi vesicles   Secretory vesicles

(c)

of the inner cisternae. It is estimated that in the intestinal goblet cell it takes 20 to 40 minutes for the movement of membrane through the stack of cisternae from the outer to the inner face. In other words, it is estimated that the Golgi complex is completely renewed (turned over) in this period of time. Whether this flow of membrane through the Golgi complex from end to end is widespread among different cell types is not known, but there do seem to be variations. Regardless of the manner in which the Golgi complex becomes involved in the transport of secretory products through the cell, pulse-chase autoradiographic experiments indicate labeled proteins move from the ER into that region of the cell. For example, in the pancreas, when the chase period is extended to approximately 10 to 20 minutes, the bulk of the silver grains are found over the Golgi complex (Fig. 7-8*b*).

The involvement of the Golgi complex in polysaccharide metabolism has long been suggested by the cytochemical demonstration of the presence of these materials within that organelle. The role of the Golgi complex in glycosylation was first shown autoradiographically with intestinal goblet cells, which secrete large quantities of mucin, a mucoprotein of

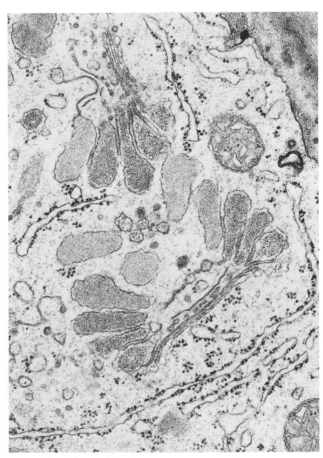

(*d*)

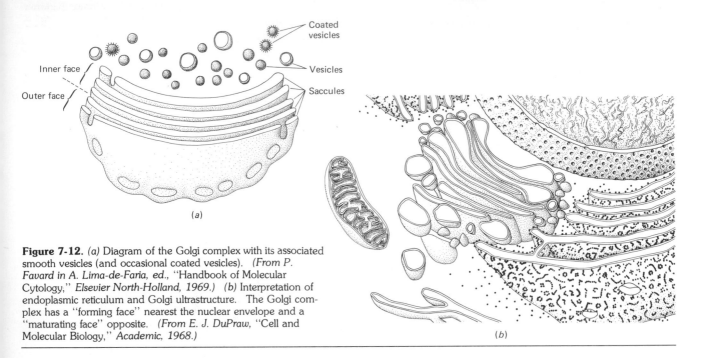

Figure 7-12. *(a)* Diagram of the Golgi complex with its associated smooth vesicles (and occasional coated vesicles). *(From P. Favard in A. Lima-de-Faria, ed., "Handbook of Molecular Cytology," Elsevier North-Holland, 1969.) (b)* Interpretation of endoplasmic reticulum and Golgi ultrastructure. The Golgi complex has a "forming face" nearest the nuclear envelope and a "maturing face" opposite. *(From E. J. DuPraw, "Cell and Molecular Biology," Academic, 1968.)*

high sugar content. When these cells were exposed to radioactively labeled sugars, the incorporation was first seen to occur within the Golgi complex (Fig. 7-14*a-c*). Since this early study, a large number of different cells have been tested and the incorporation of the terminal sugars, such as fucose, always occurs in the Golgi complex. In addition to carbohydrate, a number of secretory products also contain sulfate esterified to certain sugars, as is the case for the chondroitin sulfate (page 268), of car-

Figure 7-13. Cytochemical demonstration of the regional differences in the nature of the Golgi complex of a nerve cell. *(a)* The reaction product of the enzyme inosine diphosphatase, which forms the basis of this stain, is concentrated within the inner cisternae of the organelle. *(b)* Reduced osmium tetroxide preferentially impregnates the outer cisternae. [*From R. S. Decker, J. Cell Biol., **61:**603 (1974).*]

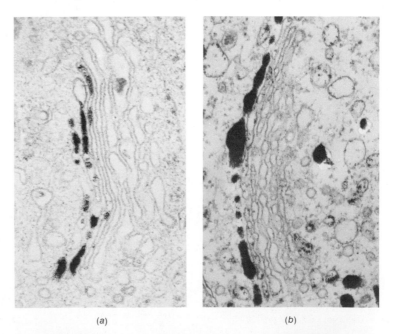

tilage matrix. It has been shown in a variety of cells that the sulfation process, which requires that the activity of sulfotransferases, which transfer sulfate from an activated donor to the appropriate acceptor, also occurs within the Golgi complex (Fig. 7-14d).

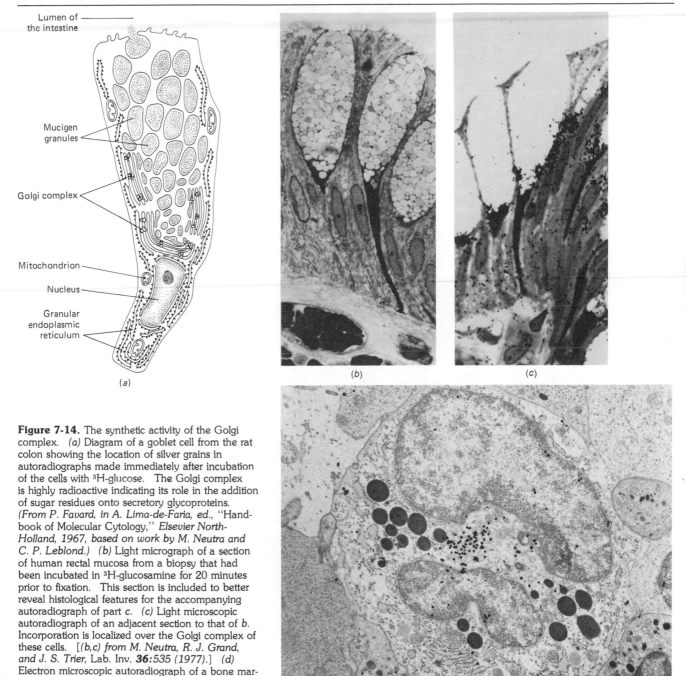

Figure 7-14. The synthetic activity of the Golgi complex. (a) Diagram of a goblet cell from the rat colon showing the location of silver grains in autoradiographs made immediately after incubation of the cells with ³H-glucose. The Golgi complex is highly radioactive indicating its role in the addition of sugar residues onto secretory glycoproteins. (From P. Favard, in A. Lima-de-Faria, ed., "Handbook of Molecular Cytology," Elsevier North-Holland, 1967, based on work by M. Neutra and C. P. Leblond.) (b) Light micrograph of a section of human rectal mucosa from a biopsy that had been incubated in ³H-glucosamine for 20 minutes prior to fixation. This section is included to better reveal histological features for the accompanying autoradiograph of part c. (c) Light microscopic autoradiograph of an adjacent section to that of b. Incorporation is localized over the Golgi complex of these cells. [(b,c) from M. Neutra, R. J. Grand, and J. S. Trier, Lab. Inv. 36:535 (1977).] (d) Electron microscopic autoradiograph of a bone marrow cell incubated in ³⁵SO₄ for five minutes and immediately fixed. Sulfate incorporation is seen to be strikingly localized in the Golgi complex. [(From R. W. Young, J. Cell Biol., 57:177 (1973).]

As described above, it appears that as the secretory products move along their route, they are successively modified by membrane-bound enzymes including those present within the walls of the Golgi complex. The presence of membrane-bound glycosyltransferases in the Golgi complex can be verified directly by isolating the fragments of the Golgi membranes from the microsomal vesicles and testing them for their enzyme activity. A number of different transferases can be identified within the membranes, particularly those responsible for attaching the sugars characteristic of the terminal portions of the carbohydrate chains. The involvement of the Golgi complex in the elaboration of extracellular polysaccharides holds true for plant cells as well as animal cells. For example, the pectin and hemicellulose components of the cell wall are assembled in the Golgi cisternae.

THE FORMATION OF SECRETORY GRANULES AND THEIR RELEASE

In numerous cells the trip through the Golgi complex results in the concentration of the secretory product. The concentration process is reflected

Figure 7-15. Scanning electron micrograph of the interior of a rabbit pancreatic acinar cell showing the nucleus, membranous cisternae, and the large numbers of secretory vesicles. *(Courtesy of K. Tanaka and T. Naguro.)*

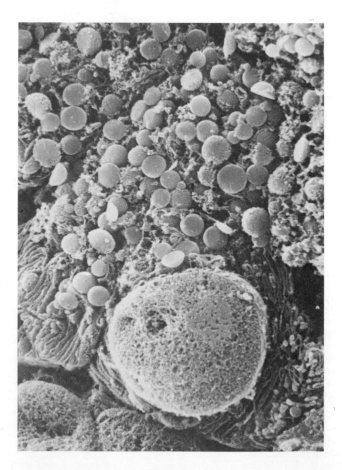

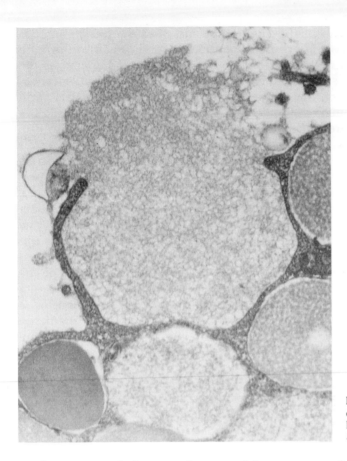

Figure 7-16. A secretory granule discharging its contents after fusion has been completed. *(Courtesy of S. J. Bunnen.)*

in the increased electron density of the contents of the inner cisternae as compared to the outer ones or the ER. When the secretory product is seen in the inner Golgi cisternae, it extends to the end of the space and fills the dilated tips from which vacuoles containing the product appear to be pinched. Once these vacuoles are released, they are termed *condensing vacuoles* which can fuse with one another to form larger vacuoles within which the concentration process continues. Since the membranes of the condensing vacuoles originate in the Golgi complex, the organelle can be considered as a source of membranous containers in which secretory products can be stored. In the pulse-chase experiments with radioactive amino acids administered to the pancreas, radioactivity is found predominantly in the condensing vacuoles when the chase period has been extended to about 40 minutes. As the concentration process continues, the membrane-bound vacuoles become *secretory granules,* which move closer and closer to the apical surface of the cell. The accumulation of secretory granules within the pancreatic acinar cell is clearly revealed in the scanning electron micrograph of Fig. 7-15. When the secretory granules have finished their final maturation process, they are ready for fusion with the plasma membrane and the discharge of their contents (Fig. 7-16). In the pulse-chase experiment with the pancreas illustrated in Fig. 7-8, radioactivity appears in the lumen outside of the cell approximately 2 hours after the chase has begun. This time period indicates the duration of the entire process from the time the polypeptide is synthesized to the time the finished product is released from the cell. An overall summary of the entire secretory process is illustrated in Fig. 7-17.

The degree to which the secretory material is stored in mature granules within a cell depends upon the cell type. In some cells, for example those of the liver, there is a relatively continuous release of serum proteins into the bloodstream and there is little opportunity for the secretions to be stored within the cell before their release. In other cases, for example the pancreatic acinar cell, secretion is not continuous but intermittent, and, when needed, large amounts of material must be released. In cells that secrete upon demand, as in the pancreas, a considerable quantity of secretory product is kept on hand within the cells until needed (Fig. 7-15). In the pancreas, the secretory granules are termed *zymogen granules*, and their release is triggered by the presence of elevated blood hormone levels. The exact mechanism by which the interaction of the hormone (pancreozymin) with the pancreas cell leads to the discharge of the granules is not known, but the process (exocytosis) is known to be triggered by an elevated calcium ion concentration and to be mediated by cyclic nucleotides (page 245). Discharge of the mature secretory granules can be initiated by the microinjection of calcium ions into the cell.

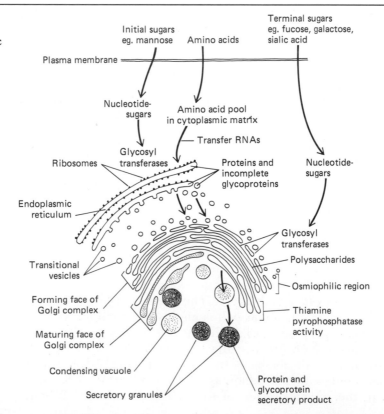

Figure 7-17. Schematic representation of the structure and relations of the Golgi complex to the endoplasmic reticulum. *(After W. Bloom and D. W. Fawcett, "Textbook of Histology," 10th ed., Saunders, 1975.)*

The underlying event in the discharge process is membrane fusion. Information concerning membrane fusion has been obtained primarily by freeze-fracture analysis of fusing membranes of ciliated protozoa. In ciliates, there exist membrane-bound structures (*mucocysts* and *trichocysts*) situated just beneath the plasma membrane. When the appropriate stimulus is received, these vesicles fuse with the plasma membrane, discharging their contents into the medium. When freeze-fracture replicas of the plasma membrane and the vesicle membranes are examined, a particular arrangement of particles in the plasma membrane (Fig. 7-18a) and particle-free areas in the mucocyst membrane is seen at the sites where the first contacts will be made. It appears that there exists a complementary organization within the two members of a fusing pair that provides both the specificity for the process and the basis for the association of the two lipid bilayers (Fig. 7-18b). There is some evidence that the membranes of secretory granules of vertebrate cells also contain some type of distinguishable organization that provides the means for the fusion event.

(a)

Figure 7-18. The fusion of the plasma membrane with that of the underlying mucocysts in *Tetrahymena.* (a) A freeze-fracture replica showing the arrangement of particles in a rosette formation within the plasma membrane before fusion. (b-e) Diagram of steps during the fusion of the two membranes. A and B refer to freeze-fracture faces, Mc to mucocyst membrane, and PM to plasma membrane. [*From B. Satir, C. Schooley, and P. Satir, J. Cell Biol.,* **56:***174 (1973).*]

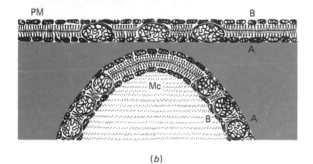

(b)

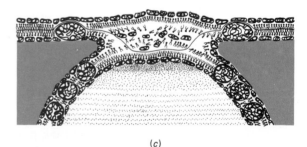

(c)

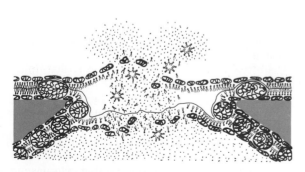

(d)

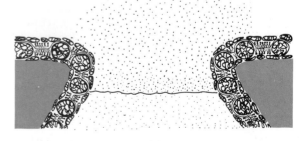

(e)

MEMBRANE FORMATION AND TURNOVER

During the discussion of secretion it was pointed out that the secretory proteins were not only materials traveling in a directed manner through the cell. In addition, it appears that the membranous containers surrounding the secretory product are also moving from the endoplasmic reticulum through the Golgi complex to the plasma membrane. If membrane is actually flowing through the cell in this manner, and not everyone agrees that this is the case, then these membranous organelles are conceived of as being transitory in nature, i.e., the materials of which they are made are continually arriving and departing, though the organelle itself persists. The most serious difficulty with this proposal is that each cytomembrane has its own identity, i.e., its own unique composition. How can these different types of membranes transform themselves into one another when each has a distinct phospholipid content and enzymatic activities? If membrane flow occurs, membranes must have the ability to undergo changes in their composition to become adapted to their new position. Presumably this is what is observed in the stack of Golgi cisternae from one end to the other, i.e., they are gradually becoming modified from an ER type of membrane to a plasma membrane as they move through the organelle.

If the rough endoplasmic reticulum is the source of membrane, then we would expect it to be the site of synthesis of the membrane components, just as it is the site of synthesis of secretory materials. If one provides the cell with precursors of phospholipids, for example, ^{3}H-choline which is incorporated into phosphatidylcholine (Fig. 5-2), it is incorporated into phospholipid in the rough ER as expected. This is corroborated by examination of the enzymatic activities of the various cell fractions; only the RER has all of the enzymes needed for the complete synthesis of the various phospholipids. Similarly it is believed that the protein components of membranes are synthesized on membrane-bound ribosomes of the RER. If the components of the membrane move along the same route as that followed by the secretory products, then a pulse-chase experiment using a labeled membrane precursor should yield the same type of progression of radioactivity through the cell. When a labeled precursor of membrane glycoproteins, such as ^{3}H-fucose, is taken up by various cells, it is first incorporated into the Golgi complex and then eventually finds its way to the cell surfaces. The time it takes to move through the cell can vary considerably. For example, in the cells of the distal region of the kidney tubule, labeled glycoproteins are maximally present at the surface after 30 minutes, while it takes 4 hours for peak surface labeling in the proximal section of the same tubule. However, if one does a pulse-chase experiment with a phospholipid precursor such as ^{3}H-choline, there is no orderly time course. Even though the lipids are synthesized in the RER, they are rapidly found in all of the other cytomembranes, including both the inner and outer membranes of the mitochondria and the plasma membrane.

How can the newly synthesized phospholipids move so rapidly from one membrane to another? There are two main possibilities. One is that lipids are diffusing laterally through the membranes of the cell, from one to the next, *within the plane of the membrane*. In order for this to be the case it would require all of the membranes to be interconnected, another controversial topic. Numerous electron micrographs have been published showing contacts between different types of membranes, for example, between the nuclear membrane and ER (Fig. 7-19), or the ER and outer mitochondrial membrane. On the basis of these types of observations, it has been proposed that cytoplasmic membrane systems are interconnected. However, most cytologists feel that

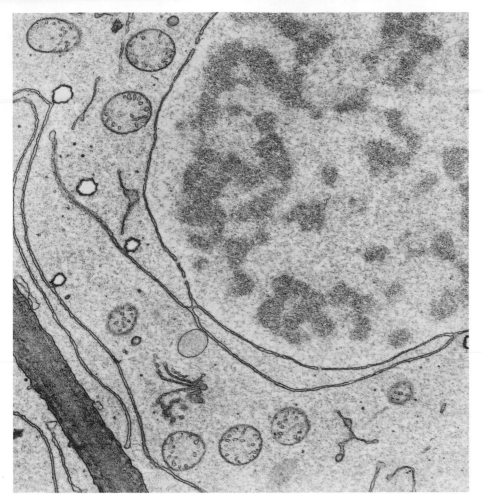

Figure 7-19. Electron micrograph of a portion of a meristematic root-cap of corn showing the continuity between the outer membrane of the nuclear envelope and that of the endoplasmic reticulum. [*From W. G. Whaley, H. H. Mollenhauer, and J. H. Leech*, Am. J. Botany, **47**:425 (1960).]

the interconnections are of such a limited nature that the organelles are essentially morphologically independent structures. This is certainly the case between the outer and inner membranes of the mitochondria. In other words, it would not generally be acceptable that lipids could rapidly diffuse from one organelle to another within the membrane.

The other possible explanation for the rapid appearance of newly synthesized phospholipids in all of the membranes of the cell is that they move from the RER where they are formed into the other membranes by passing through the cytoplasm itself. Since there would be definite solubility problems for lipid substances to move in the aqueous cytoplasm, this may seem at first glance an unlikely event. However, there is convincing evidence for the widespread presence of *phospholipid-exchange proteins*, molecules which function by transporting phospholipids through the cytoplasm from one membrane in one part of the cell to another membrane at some other location. These proteins, which are present in the soluble phase of the cytoplasm, have been purified and they have been shown to possess a hydrophobic region to which the membrane lipid is bound during the transport process. It appears that a great variety of membranes can donate lipids to these proteins or accept lipids from them. Consider the following experiment. If one prepares a

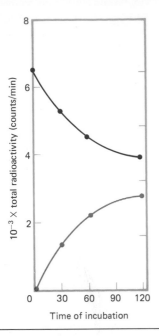

Figure 7-20. Incubation of mitochondria containing ³H-choline–labeled membranes with unlabeled microsomes in an unlabeled supernatant. Radioactivity is seen to be transferred from the labeled mitochondria (dark line) to the unlabeled supernatant (light line) until both are in equilibrium. [*From W. C. McMurray and R. M. C. Dawson, Biochem. J., **112**:103 (1969).*]

purified ³H-choline–labeled microsome fraction and adds a purified, unlabeled mitochondrial preparation, there is no significant transfer of radioactivity from the microsomes to the mitochondria. However, if a preparation of these exchange proteins is added, then label is rapidly transferred into the membranes of the mitochondria until an equilibrium is reached among the phospholipids of all membranes (Fig. 7-20).

We have discussed at some length in this chapter the movement of membrane from the endoplasmic reticulum to the plasma membrane. The expulsion of the contents of the secretory vesicle during exocytosis leads to the incorporation of vesicle membrane into that of the plasma membrane. During periods of active secretion, one would expect that the total surface area of the plasma membrane would increase in a disproportionate manner and that some means of removing excess membrane would have to occur. The results of various studies indicate that excess plasma membrane can be taken back into the cell by endocytosis, a subject discussed in the following section. In some cases, at least, this process of membrane retrieval leads to the return of surface membrane to the Golgi complex, where its fate becomes difficult to follow.

The analysis of membrane flow within the cell is complicated by another process, that of membrane turnover; i.e., the removal of existing membrane components and their replacement by new ones. The way in which turnover is generally studied is to grow the cells in a radioactive precursor for a period of time, then wash the cells free of isotope and chase the cells with unlabeled precursor for increasing amounts of time. The amount of radioactivity in the particular component is measured at the beginning of the chase and at subsequent intervals. The result of such an experiment is shown in Fig. 7-21, for several membrane proteins. There are two major points to be drawn from this figure. The first is that different proteins are destroyed at different rates, i.e., the destruction process appears to be selective. Secondly, the radioactivity present in *each* protein drops off at a linear rate indicating that a newly synthesized protein (a radioactive one) is just as likely to be destroyed as a protein that has been in the membrane for a period of time.

Regardless of the manner in which membrane components are assembled into membrane, it appears that this process occurs within preexisting membrane. In other words, the evidence suggests that membrane formation is

Figure 7-21. Rate of replacement of membrane components varies from one to another. Note differing slopes of curves. Rate in no case depends upon the age of individual molecules but is essentially random, as shown by linear plots on semilogarithmic scale. (*Drawing by B. Tagawa, from Dynamics of Intracellular Membranes, by P. Siekevitz, Hospital Practice, Vol. 8, No. 11 and "Cell Membranes: Biochemistry, Cell Biology & Pathology," G. Weissmann and R. Claiborne, Eds., HP Publishing Co., Inc., New York, NY 1975. Reprinted with permission.*)

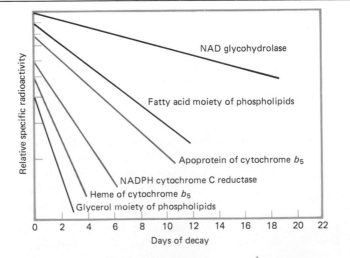

accomplished by the incorporation of lipids and proteins into existing membrane, presumably causing its expansion. To counteract the proliferation of membrane, the cell can destroy excess membrane either by removal of molecules, one at a time, or by the excision and destruction of intact membrane fragments. The balance between formation and destruction depends upon whether the cell is growing and dividing, and is therefore in need of additional membrane, or rather is maintaining a constant cell volume, in which case additional membrane is not needed.

ENDOCYTOSIS

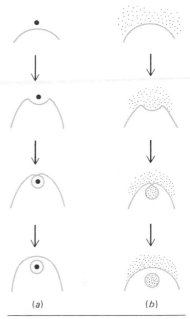

In the previous chapter we considered the movement of small molecular weight materials directly through the plasma membrane, but what of the uptake of materials that are too large to penetrate this membrane, regardless of its permeability properties? To meet this need, special mechanisms have evolved which bring about the enclosure of material from the environment within special vesicles derived from folds or channels of the plasma membrane. The phenomenon is generally termed *endocytosis* and is usually divided into two related events, *phagocytosis* and *pinocytosis*. Phagocytosis refers to the intake of particulate matter, and pinocytosis the intake of fluid and dissolved materials (Fig. 7-22). The distinction between the two processes is vague; both are believed to reflect the same type of underlying events. In this section we will consider the ingestion process. The digestion of the enclosed material occurs by fusion of the vesicles with lysosomes which will be discussed in the following section.

Figure 7-22. Diagrammatic representation of phagocytosis and pinocytosis. *(From A. Roller, "Discovering the Basis of Life," McGraw-Hill, 1974.)*

Phagocytosis is carried out extensively by a few types of cells specialized for the uptake of particles from the environment. Single-celled heterotrophic organisms, such as amoebae and ciliates, make their livelihood by trapping food particles and smaller organisms in various manners, enclosing them within food vacuoles (phagosomes) pinched off from the plasma membrane, and digesting them intracellularly. Some of the primitive metazoans, the sponges, coelenterates, and flatworms, also gain their nutrition by the intracellular digestion of small food particles in plasma membrane-derived vesicles. Within higher organisms, specialized cells have evolved, such as the macrophages and leucocytes (white blood cells), whose function is to wander through the tissues and blood phagocytizing invading organisms or debris.

In the best-studied case, that of the amoeba, the initiation of the event depends on the contact of the membrane with an appropriate object, usually a small ciliate, but in experimental situations often a latex bead. The surface of many amoebae has a considerably different appearance from most single cells within a multicellular organism. Since the cell surface of the amoeba is at the same time the surface of the entire animal, special modifications are present to facilitate the necessary exchange with the outside environment. Many amoebae have a surface covered with fine filamentous hairs (Fig. 6-1) and/or a type of slime coat composed of mucopolysaccharides. One of the prime functions of these extracellular materials is to adsorb potential food particles from the environment, as is readily apparent by watching a ciliate try to untangle its cilia from the amoeba's slime coat. It appears that maintained contact of an object with the amoeba's surface initiates phagocytosis. The response to contact by the amoeba is the appearance of undulations or folds of the membrane around the particle and its engulfment (Fig. 7-23). The indentation formed by these folds is referred to as a food cup. In the continued presence of food particles, as many as 100 food cups have been recorded as having been formed by an individual amoeba within its generation time ("lifetime") of about 24 hours. Since each cup requires approximately 10% of the surface area of the organism, it appears that it consumes about 10 times its surface per generation. In other words, the endocytotic activities that a cell engages in appear to be quite costly in terms of its surface membrane, which must be replaced. This does not mean that the vacuole membrane or its components cannot be reutilized in some way for future membrane assembly; insufficient information about membrane formation makes it difficult to describe the pathways taken by the specific materials of the ingested membrane. Similarly, attempts to measure membrane synthesis before and during periods of rapid endocytosis have failed to indicate the need for new membrane formation.

The term *pinocytosis*, meaning "cell drinking," reflects the early concept that this process represented a means for the cell to take up fluid in a bulk manner from its environment. Later observations have shifted the emphasis from uptake of solvent to uptake of solute. The diagram shown in Fig. 7-24 is taken from one of the early studies of pinocytosis in amoebae; the morphology varies somewhat from cell to cell. It was seen in the early studies that when a culture of amoebae was engaged in pinocytosis, long thin channels formed into which the medium and its dissolved materials could pass. At the bottom of the channel, vesicles are pinched off forming membrane-bound vacuoles (pinosomes) with their enclosed fluid. Approximately 20 years after the early morphological observations, an experimental analysis of the underlying mechanisms was begun both on amoebae and various vertebrate cells which also can be seen to undergo pinocytosis. It was found that pinocytosis in amoebae could be induced by various substances while others had no effect. Particularly effective were basic dyes, proteins, cations, viruses, and amino acids. Carbohydrates and nucleic acids were not effective. It appeared that substances that initiated pinocytosis were ones that would bind to the amoeba's surface. Since the surface coat is negatively charged, the most effective inducing agents were ones having positively charged groups. In other words, uptake of solute by pinocytosis is selective to a degree but not specific. For example, in one study using fluorescent antibody to locate the presence of a particular protein during pinocytosis in amoebae, this protein was found to form a dense coating on the cell's surface as well as a lining on the pinosomes. Binding sites on the surfaces of these cells allows the ingested materials to be concentrated to a far greater level than that found in the medium.

Using radioactively labeled protein it was shown that in five minutes amoebae could take up approximately 50 times the amount of protein

Figure 7-23. Scanning electron micrograph of an amoeba showing the phagocytosis of latex beads. P, profile of internalized bead; C, food cup. [*From R. J. Goodall and J. E. Thompson,* Exp. Cell Res., **64**:4 *(1971).*] Transmission electron micrograph of a phagocytic vesicle in the process of formation around latex beads. [*From E. D. Korn and R. A. Weisman,* J. Cell Biol., **34**:219 (1967).]

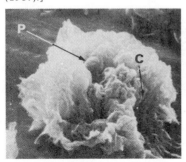

(a)

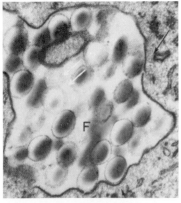

(b)

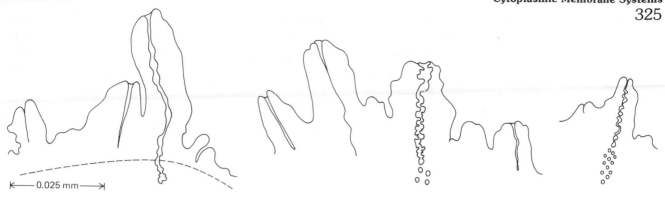

← 0.025 mm →

Figure 7-24. The formation of pinocytotic vesicles in an amoeba. [*From S. O. Mast and W. L. Doyle,* Protoplasma, **20**:556 (1933).]

present in a volume of medium equivalent to their own volume. In other words, if no concentration process occurred at the cell surface, an amoeba would have to take up an amount of medium equivalent to its own volume every 6 seconds, a rate orders of magnitude greater than its capability.

A wide variety of cells of higher organisms have also been shown to undergo pinocytosis. Unlike the case of the amoeba, no channels are formed, but, rather, droplets of fluid are trapped by undulations of the surface membrane. Whether pinocytosis exists in higher plants remains an unsettled question. In animals, it appears that this process represents the principal means by which blood-borne macromolecules gain entry to the various needy cells. In the transfer of dissolved materials from the blood to the cytoplasm of cells, the pinosomes formed are usually very small, requiring the electron microscope for their observation. In such cases, the process is termed *micropinocytosis*. To illustrate the types of materials that can be transferred from the blood by this means, two examples of pinocytosis in vertebrates will be considered.

The egg of most vertebrates including fish, frogs, and chickens is a giant cell packed with sufficient nutritive materials to support the needs of the developing embryo until it is sufficiently advanced to be able to obtain nutrition for itself. The nature of the material varies somewhat from animal to animal, but the term *yolk* refers to a proteinaceous material, often high in lipid and/or phosphate, which serves as the primary source of material with which the embryo builds its own macromolecules. Yolk is packaged into the egg while it is in the ovary (termed an *oocyte* at this stage) but it is not synthesized within the oocyte, or even within the ovary. The yolk materials are produced by the liver, secreted into the blood and taken up into the oocyte by micropinocytosis.

Another example of the uptake of important substances from the blood is that of a class of proteins that transport cholesterol. The synthesis and utilization of cholesterol in mammalian cells has complex and interesting properties. Cholesterol is at the same time an essential component of many cell membranes and a potentially hazardous molecule in causing deposits within the walls of major arteries. The capability for cholesterol synthesis is widespread among various mammalian cells, but normally the liver is primarily responsible for its production and release into the bloodstream for distribution to the tissues. Sterols, such as cholesterol, must be carried through the blood as a complex with other molecules, the *low-density lipoproteins* (LDL).

These important transport proteins contain a lipid region with which the cholesterol (esterified to a fatty acid) is associated.

The uptake of cholesterol-bearing proteins has been best studied in mammalian cells grown in culture. Recent studies have indicated that *specific* receptor molecules are present on the surfaces of cells to which these proteins must bind prior to their entry by pinocytosis into the cell. Furthermore, the receptors appear to be concentrated in patches on the surface, regions which have a coated appearance in the electron microscope (Fig. 7-25). Once inside the cell the cholesterol is split off from the protein carrier for use within the cell. One of the important responsibilities of the cholesterol split off from the LDL is the suppression of the synthesis of cholesterol by the cell and the stimulation of its own conversion to an esterified storage form. These effects of the transported cholesterol maintain the levels of free cholesterol needed in the cell for membrane synthesis without raising its extracellular levels to a damaging degree. It appears that a particular human disease state, that of familial hypercholesterolemia, which is characterized by greatly elevated cholesterol levels and resultant circulatory disease, results from a defect in the pinocytosis of low density lipoproteins. The evidence suggests that cells from persons with this disease lack LDL cell surface receptors. The cells lacking uptake of these proteins have no counterbalance to their own cholesterol synthesis and excessive levels of the sterol are produced.

LYSOSOMES

The discovery of lysosomes as a cytoplasmic organelle containing digestive enzymes occurred through an unusual course of events. In about 1950,

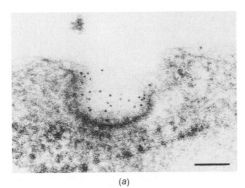

(a)

Figure 7-25. The uptake of ferritin-conjugated low-density lipoprotein (LDL) by cultured human fibroblasts. Bar is equivalent to 1000 Å. (a) A typical coated pit containing the electron dense LDL-ferritin. (b) A coated pit being transformed into an endocytic vesicle. (c) Formation of a coated vesicle containing LDL-ferritin. Some of the LDL-ferritin is excluded from the interior and left on the surface of the cell (arrow). [*From R. G. W. Anderson, M. S. Brown, and J. L. Goldstein,* Cell, **10**:356 (1977).]

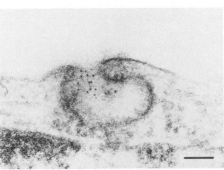

(b)

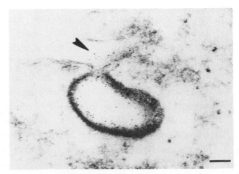

(c)

Christian de Duve and his colleagues were studying the distribution of various enzymes of carbohydrate metabolism (such as glucose 6-phosphatase) in the cell fractions obtained from homogenates by differential centrifugation. One of the enzymes being studied was acid phosphatase, which splits the phosphate group from a variety of substrates containing phosphate esters. This particular phosphatase was known to have its maximal activity at an acidic pH (approximately 5), hence its name. Since these studies were aimed at the isolation of cell organelles in an intact state, very gentle homogenization methods were used to minimize mechanical damage. When the acid phosphatase activity of a gently prepared homogenate was measured, a particular value was obtained. If, however, the same assay was carried out on a more roughly prepared homogenate or one that had remained in the icebox for a number of days, it was found that the acid phosphatase activity was increased approximately 10 times. It appeared that the enzyme showed some type of *latency* in that its full activity could not be demonstrated until the preparation was harshly treated.

When subsequent observations were made, the reason for the latency became apparent: the enzyme was normally enclosed within a membrane, which prevented it from reaching the substrate (Fig. 7-26). Once the membrane was ruptured, the full activity of the enclosed enzyme could be measured. Further analysis indicated that it was the large particle fraction obtained by differential centrifugation in which the enzyme was located. This was the fraction known to contain the mitochondria, however the particle being sought was not a mitochondrion, but one which sedimented in a gradient a little more slowly.

In 1955 the appearance of the acid phosphatase-containing particle was observed by Alex Novikoff in electron microscopic examination of isolated fractions that had been enriched for the presence of these particles (Fig. 7-27b), which by then had been termed *lysosomes*. Once their morphology had been revealed, the presence of these organelles in thin sections of various tissues could be sought and some idea of their distribution and quantity might be determined. However, the identification of lysosomes in various types of tissues proved difficult. There are many different types of granules and vesicles present in various cells, and lysosomes (Fig. 7-27a) are neither distinctive nor uniform in appearance. In fact, lysosomes range in size from relatively large structures (over 1 μm diameter) to very small vesicles (250 to 500 Å diameter) and their irregular shape and variable electron density makes them even less distinctive. Even though acid phosphatase is also known to occur within the cell in nonlysosomal locations, it is generally agreed that the demonstration of this enzyme within a membrane-bound organelle is sufficient grounds for calling the organelle a lysosome. With the development of a cytochemical procedure for the localization of acid phosphatase, the means for identifying the organelles within sections became available. The procedure that is most commonly employed is to incubate a section with a suitable substrate in the presence of lead ions. When the inorganic phosphate is split off from the substrate by the enzyme, the formation of an insoluble lead phosphate precipitate forms at the site which can then be directly identified in the electron micro-

Figure 7-26. Schematic representation of the structure-linked latency of lysosomal hydrolases. *(From R. Wattiaux, in A. Lima-de-Faria, ed., "Handbook of Molecular Cytology," Elsevier North-Holland, 1969.)*

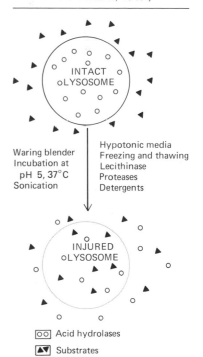

INTACT LYSOSOME

Waring blender Incubation at pH 5, 37°C Sonication

Hypotonic media Freezing and thawing Lecithinase Proteases Detergents

INJURED LYSOSOME

Acid hydrolases

Substrates

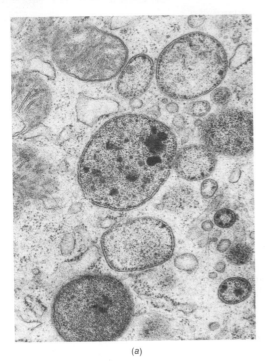

(a)

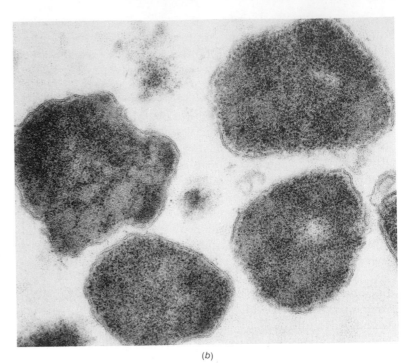

(b)

Figure 7-27. *(a)* Portion of a Kupffer cell (a type of macrophage) containing at least ten lysosomes of highly variable size, all showing the presence of "ferritin granules" in their matrix. *(b)* High magnification electron micrograph from an isolated liver lysosomal fraction. The lysosomes are bordered by a triple-layered membrane and contain abundant electron dense ferritin granules. [*From H. Glaumann, H. Jansson, B. Arborgh, and J. L. E. Ericsson, J. Cell Biol.*, **67**:887 (1975).]

scope. In addition to acid phosphatase, approximately 40 different enzymes have been localized within lysosomes from various sources (Table 7-2) and several of these enzymes can be identified by specific cytochemical tests.

If one examines the variety of enzymes present in lysosomes, there are two basic similarities: they are all hydrolytic enzymes, and they all have their optimal activity at an acidic pH. Among the group are hydrolases capable of digesting almost every known type of macromolecule. In many ways lysosome formation (Fig. 7-28) resembles that of the secretory granules discussed earlier in this chapter, though there may be more than one route in which the packaging of lysosomal enzymes can occur. In one of the best-studied cases, that of the polymorphonuclear leukocyte (a white blood cell), the enzymes are produced in the rough endoplasmic reticulum and transported to the Golgi complex where they can be seen in the inner cisternae and in the vacuoles being pinched from the dilated tips. In other cells it has been suggested that the Golgi complex is bypassed and the packaging occurs via a network of smooth-surfaced tubules (called the *GERL*) adjacent to the Golgi membranes. Regardless of the precise manner in which the packaging process occurs, it would seem that the cell must have some control over the fate of its various membrane-bound materials. For example, if a cell is involved in secretion, as well as lysosome formation, it must be able to direct the flow of the different types of proteins into different types of vesicles, particularly since the lysosomal enzymes have the potential to destroy the contents of the secretory package. That the separation is not always complete is suggested by the occasional finding of some acid phosphatase activity in secretory granules.

The presence within a cell of what is, in essence, a bag of destructive enzymes suggests a number of possible functions. The best-studied role of the lysosome is in the breakdown of materials brought into the cell from the outside by endocytosis. In single-celled organisms these endocytotic vesicles contain food particles. In the cells of multicellular organisms, endocytotic vesicles are not used for nutritive purposes but for various other functions. For example, white blood cells and macrophages are important scavengers within the body, collecting debris or potentially dangerous microorganisms by taking them up in phagocytotic vesicles. In other cases, proteins are taken up by cells within pinocytotic vesicles (pinosomes). For example, the proteins of the blood serum are broken down by being taken up by pinocytosis in cells of the liver. In another example, the protein thyroglobulin is stored in an extracellular space within the thyroid gland and when needed taken up by the thyroid cells by pinocytosis.

TABLE 7-2
Some Acid Hydrolases Which Have Been Located in Lysosomes

Enzyme	Natural substrate	Source of lysosomes
Phosphatases		
Acid phosphatase	Most phosphomonoesters	Many tissues of
Acid phosphodiesterase	Oligonucleotides and phosphodiesters	animals and plants; protists
Nucleases		
Acid ribonuclease	RNA	Many tissues of
Acid deoxyribonuclease	DNA	animals and plants; protists
Polysaccharide and Mucopolysaccharide Hydrolyzing Enzymes		
β-Galactosidase	Galactosides	Animals, plants, protists
α-Glucosidase	Glycogen	Animals
α-Mannosidase	Mannosides	Animals
β-Glucuronidase	Polysaccharides and mucopolysaccharides	Animals; plants
Lysozyme	Bacterial cell walls and mucopolysaccharides	Kidney
Hyaluronidase	Hyaluronic acids; chondroitin sulfates	Liver
Arylsulfatase	Organic sulfates	Liver; plants
Proteases		
Cathepsin(s)	Proteins	Animals
Collagenase	Collagen	Bone
Peptidases	Peptides	Animals; plants; protists
Lipid Degrading Enzymes		
Esterase(s)	Fatty acid esters	Animals; plants; protists
Phospholipase(s)	Phospholipids	Animals; plants?

SOURCE: D. Pitt, "Lysosomes and Cell Function," Longman, 1975.

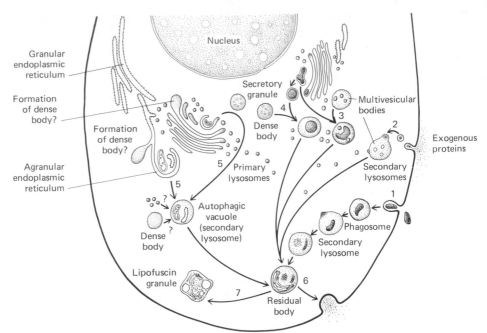

Granular
endoplasmic
reticulum

Formation
of dense
body?

Formation
of dense
body?

Agranular
endoplasmic
reticulum

Nucleus

Secretory
granule

Multivesicular
bodies

Exogenous
proteins

Dense
body

Primary
lysosomes

Secondary
lysosomes

Autophagic
vacuole
(secondary
lysosome)

Phagosome

Dense
body

Secondary
lysosome

Lipofuscin
granule

Residual
body

Figure 7-28. Summary of pathways in the lysosomal system. Most of the interrelationships among the various components are discussed in the text. *(From W. M. Copenhaver, R. P. Bunge, and M. B. Bunge, "Bailey's Textbook of Histology", 16th ed., Williams and Wilkins, 1971.)*

In all of these examples of endocytosis, the material is degraded by lysosomal enzymes to smaller molecular weight substances. The manner in which this is accomplished was first shown by Werner Straus in 1964 on experiments with an injected foreign protein. The protein used was horseradish peroxidase, an enzyme of relatively small molecular weight (40,000 daltons), whose enzymatic activity is very resistant to fixatives, and for which excellent cytochemical methods are available to localize its presence. When horseradish peroxidase is injected into an animal, it is small enough to pass out of the bloodstream through the glomerulus of the kidney and into the lumen of the kidney tubules. Once inside the lumen, it is reabsorbed by the cells of the tubule via the formation of pinocytotic vesicles. When sections of the kidney are made from an animal killed 30 minutes after injection of horseradish peroxidase, the enzyme as detected cytochemically is seen within pinosomes at the apical surface of the cell. In the particular cytochemical procedure used in these experiments, the peroxidase-containing vacuoles stain bright blue in color. If the sections are stained by a different technique for acid phosphatase activity, it appears in small granules (lysosomes) located toward the base of the cell which stain bright red. In other words, at 30 minutes after injection the protein has been taken up by the cells but is located within the cell in a different site from the lysosomes. When the animal was killed at later periods following the injection and the sections stained by both of these techniques, instead of finding blue and red granules at separate locations, there were purple granules located near the base of the cell where the lysosomes had previously been seen. The purple color indicated the combined presence of both enzymes in the same particle. The results of this experiment suggested that the peroxidase-containing vacuoles were being carried to the base of the cell where they were fusing with the lysosomes. After a day or two, there was no longer any cytochemical evidence of the peroxidase within the cells; it had been digested by the lysosomal enzymes.

As in the case of the fusion of a secretion granule with the plasma membrane, the fusion of lysosomes with other membranous structures is highly selective. Lysosomes fuse only with membrane-bound vesicles containing material to be digested. It would appear, as in the case of the discharge of secretory granules, that the membranes of lysosomes and endocytotic vesicles contain some type of recognition system which facilitates their fusion. Examples are known in which the interactions between lysosomes and phagocytotic vesicles are aberrant. There are certain bacteria, for example, the tubercle bacillus, that will be taken up by a macrophage, but the phagocytotic vacuole will not fuse with a lysosome as ordinarily occurs. It appears that this virulent organism somehow inhibits the fusion which would lead to its destruction and instead multiplies within the cell. An experimental procedure has also been discovered that inhibits the fusion process. It has been found that if cells are stimulated to undergo pinocytosis in the presence of the lectin concanavalin A (page 225), the pinosomes that form remain in the cytoplasm and do not fuse with lysosomes. For some reason the presence of the lectin attached to the cell coat which then becomes the *inner* lining of the pinosome, inhibits the fusion process presumably by interfering with recognition between the two types of membranes.

Once a lysosome has fused with a vesicle containing material to be digested, the structure is termed a *secondary lysosome* (Fig. 7-28). Secondary lysosomes can be recycled, i.e., they still possess the capacity to fuse with other endocytotic vesicles in order to digest additional material. Since different secondary lysosomes within a cell contain different materials undergoing degradation and are at different stages in the process, these organelles are seen to be heterogeneous in size and appearance. As digestion continues within the vesicle, low molecular weight precursors are generated which are small enough to penetrate the lysosome membrane and will diffuse into the cytoplasm for use by the cell. Generally there is a certain amount of material that is not digestible and this residue remains within the lysosome. Once the digestive process has been completed, the organelle is termed a *residual body* (Fig. 7-28), which in protozoa is merely eliminated by a process of exocytosis. In the cells of vertebrates, there appears to be no mechanism for the elimination of residual bodies and they accumulate within the cytoplasm. Often these remnants of lysosome activity are termed *lipofuscin granules,* which increase in number as the individual increases in age; this is particularly marked in long-lived cells such as neurons where it is considered a major characteristic of the aging process.

Considering all of the various processes occurring in different cell types throughout a complex multicellular organism, it is not surprising that a need exists for a great variety of destructive activities. We have discussed the phenomenon of turnover in previous chapters, focusing primarily upon molecular turnover whereby individual macromolecules are selected for destruction. Molecule-by-molecule turnover is reflected in the markedly different half-lifes of the various components of an organelle. For example, the half-life of the protein of the inner soluble compartment of the mitochondria is three times that of the outer mitochondrial membrane. There is also another type of turnover, that involving whole cellular organelles, which is accomplished by lysosomes. In this case the organelle, whether a mitochondria, chloroplast, or some other structure, is surrounded by a membrane donated by the endoplasmic reticulum and digested by the activity of lysosomal enzymes. The process of destruction of a cell's own cytoplasmic contents is termed *autophagy* (Fig. 7-29), and the structure in which it occurs is called an *autophagic vacuole*, a type of secondary lysosome. It is not uncommon when surveying an electron micrograph to see a mitochondria or some other organelle trapped in a membranous vacuole being digested (Fig. 7-29). It is calculated that in a liver cell one mitochondrion undergoes autophagy approximately every 10 minutes. The number of autophagic vacuoles seen within a cell can vary considerably with the physiological state of the cell. If the cell is placed under various types of stressful conditions, such as starvation, a

Figure 7-29. (a) Electron micrograph showing an autophagic vacuole of a rat liver cell in which a mitochondrion is undergoing digestion. [*From U. Pfeifer,* Verh. Dtsch. Ges. Path., **60:** *32 (1976).*] (b) Electron micrograph of a portion of a cell of the mouse kidney proximal convoluted tubule showing numerous lysosomes. The lysosomal nature of these bodies is demonstrated by the presence of the end product of acid phosphatase activity in one of them (L_1). The fusion of the mitochondrial membrane (M) with that of L_2 is apparent. *(From V. Maggi, in E. E. Bittar, ed., "Cell Biology in Medicine," Wiley, 1973.)*

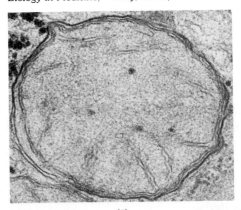

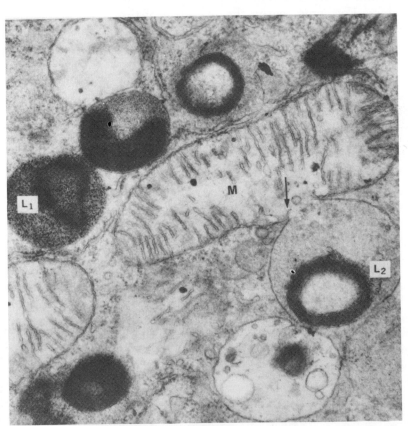

(a)

(b)

marked increase in autophagy is noted. In the case of a starving cell, it is presumed that autophagy is providing the cell with energy to maintain its life by cannabilizing its own material.

A survey of the digestive functions carried out by lysosomal enzymes includes the following cases: the release of the thyroid hormone, thyroxine, from the thyroglobulin molecule in which it is contained; the destruction of the extracellular matrix of bone and cartilage during normal turnover or when the need arises (as is often the case in the embryo) for a remodeling of the structure; and the transformation of an erythroblast (a red blood cell precursor) into the mature red blood cell which is accompanied by the destruction of virtually all of the intracellular contents. During fertilization, the head of the sperm releases lysosomal enzymes that act to digest certain barriers surrounding the egg so that the sperm can gain entrance to the egg surface. During metamorphosis in an amphibian tadpole or the sculpturing of the toes and fingers of a mammalian or avian limb, a considerable number of cells must die, and it is the lysosomal enzymes that are responsible for their death.

Considering that hydrolytic enzymes are capable of the digestion of the entire contents of a cell, it is obvious that special precautions must be taken when such enzymes are present. In the case of the secretory cells of the pancreas, a number of potentially hazardous enzymes are produced in an inactive precursor form, which normally becomes activated only in the intestine. For example, the inactive proteins, pepsinogen and trypsinogen, are the precursors of the active proteolytic enzymes, pepsin and trypsin. In addition, in the case of trypsin, an inhibitor of the enzyme is also produced in the cell presumably as an added precaution to prevent the enzyme from becoming active within the cell. In a rare circumstance, a condition can occur in which the ducts leading from the pancreas cells to the intestine become blocked, causing the enzymes to build up within the organ. When this happens, the tissue can become rapidly damaged by self-digestion, illustrating the risk involved in a cell's handling these proteins. In the case of the lysosome, protection is afforded the cell by the presence of the surrounding membrane which acts to compartmentalize the hydrolytic enzymes. For some unknown reason, the membranes of the lysosome are not affected by its enzymatic contents. Under normal conditions, the intracellular digestion accomplished by lysosomes is always conducted within membranes, which protects the cell.

Almost from the time of the discovery of lysosomes, it was proposed that their malfunction might be a major cause of various disease conditions. The widespread occurrence of lysosome-related disease has been difficult to verify, though a few examples are known. For example, there is a miner's disease known as *silicosis*, which results from the uptake of silica fibers by macrophages in the lungs. The fibers become enclosed within secondary lysosomes, but they cannot be digested, and instead cause the lysosomal membrane to become leaky, spilling its contents into the cell. A similar result occurs when asbestos fibers are taken up by macrophages and in both cases the condition can be debilitating. Certain types of inflammatory diseases, such as rheumatoid arthritis, are believed to result in part from the release of lysosomal enzymes from the cell into the extracellular space, causing damage to materials in the joints. The hormones cortisone and hydrocortisone, which act as anti-inflammatory agents owe their activity, at least partially, to their ability to stabilize the lysosome membrane against breakdown. Other agents, such as excessive amounts of vitamin A, can have the opposite effect of making the lysosome membrane less stable and can lead to connective tissue damage.

Just as there are disease conditions resulting from excessive lysosome activity, there are serious problems in individuals lacking one or another of the lysosomal enzymes. These rare genetic disorders are called *storage diseases* and are characterized by the buildup within tissues of the particular macromolecules which the missing enzyme would normally use as its substrate. For example, one of these diseases results from the inherited absence of α-fucosidase, an enzyme required to remove fucose residues of the carbohydrate chains of glycoproteins. Another storage disease results in the accumulation of glycogen due to the inability of the cell to catalyze its hydrolysis. Two of the best studied storage diseases (Hurler's and Hunter's syndrome) result from an inability of the cells to degrade acid mucopolysaccharides. Figure 7-30 shows an electron micrograph through a skin biopsy of a person with Hurler's disease. Cells of the sweat glands of this individual are seen to be filled with vacuoles not found in corresponding cells of normal individuals. These vacuoles, which stain for both acid phosphatase (a lysosomal enzyme)

Figure 7-30. Electron micrograph of a section through a portion of a sweat gland of a patient with Hunter's disease. Of the three types of cells shown, myoepithelial cells (ME), clear cells (CC), and dark cells (DC), the latter types are characterized by the presence of large numbers of vacuoles believed to represent the accumulation of aberrant secondary lysosomes. [*From S. S. Spicer, A. J. Garvin, H. J. Wohltmann, and J. A. V. Simson*, Lab. Inv., **31**:494, © 1964 by US-Canadian Division of the International Academy of Pathology.]

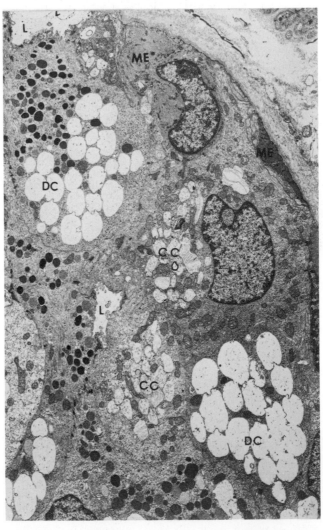

and acid mucopolysaccharide (an undegraded substrate), are believed to represent secondary lysosomes whose contents could not be degraded. The presence of these very fatal storage diseases (of which about 30 are known) illustrates the importance of these acid hydrolases in cell metabolism.

MICROBODIES (PEROXISOMES AND GLYOXISOMES)

In 1954 electron microscopic examination of kidney tubule cells revealed a membrane-bound particle of approximately 0.5 to 1.0 μm diameter with a dense granular appearance. The structure was termed a *microbody*. Two years later a similar organelle was described in liver tissue, though in these cells some type of crystalline core was seen within the vesicle. When cells are homogenized and subjected to differential centrifugation, microbodies sediment with the small particle fraction containing the lysosomes and mitochondria. Owing to their difference in density from these other two organelles, microbodies can be purified by centrifugation through density gradients (Fig. 3-29b). The availability of isolated microbodies has led to the assignment of a variety of enzymatic activities to these organelles and the realization that they are involved in particular types of oxidative reactions.

There are generally two types of enzymes in microbodies which act in sequence. The first step in the reaction sequence is the oxidation of one of a variety of substrates using molecular oxygen as the oxidizing agent (Fig. 7-31), resulting in the formation of hydrogen peroxide (H_2O_2). These reactions are catalyzed by various flavin-containing oxidases including urate oxidase, glycolate oxidase, and amino acid oxidases. It is urate oxidase which is present in the crystalline core of the microbodies of rat liver. Since one of the products of the first reaction is hydrogen peroxide, a very reactive and toxic oxidizing agent, some means must be available in the organelle to destroy this compound so that it cannot accumulate. The breakdown of H_2O_2 occurs in the second step (Fig. 7-31) in the sequence which is catalyzed by the enzyme catalase, present in high concentrations in these vesicles. The formation of peroxides in these organelles has led to the use of another term for these structures, namely *peroxisomes*. Since the development of a cytochemical technique at the EM level for the identification of catalase, peroxisomes have been identified in a great variety of cells in both plants (Fig. 7-32a) and animals (Fig. 7-6b). The close morphological association between peroxisomes and endoplasmic reticulum suggests that the membrane of these vesicles has its origin in the membrane of the ER.

A variety of studies on the peroxisomes of leaf cells have revealed a striking example of interdependence among different organelles. The electron micrograph of Fig. 7-32a shows a peroxisome of a leaf cell closely apposed to the surfaces of two adjacent chloroplasts and in close proximity to a nearby mitochondrion. This distribution is not fortuitous but is believed to reflect an underlying biochemical interaction whereby the products of one organelle serve as the substrate for another. Although the

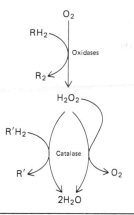

Figure 7-31. Peroxisomes contain enzymes which carry out the accompanying reaction, i.e., the two-step reduction of molecular oxygen to water. In the first step, one of a variety of oxidases removes electrons from a variety of substrates (RH_2), such as uric acid or amino acids. In the second step, the enzyme catalase converts the hydrogen peroxide formed in the first step to water using electrons donated by one of a number of molecules. [*From C. de Duve and P. Baudhuin, Physiol. Revs.*, **46**:325 (1966).]

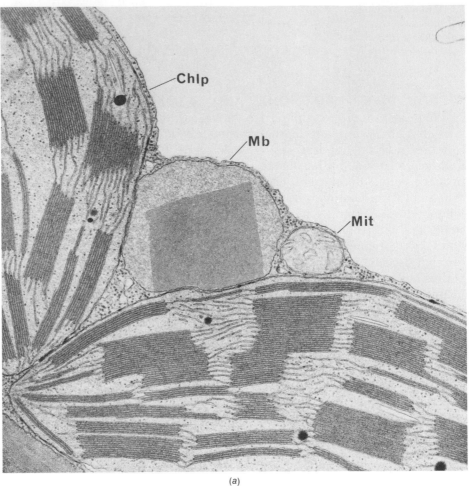

(a)

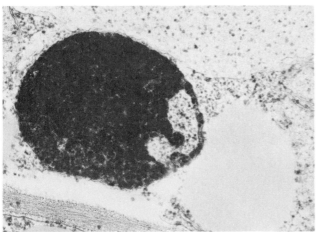

(b)

Figure 7-32. *(a)* Electron micrograph of a portion of a leaf mesophyll cell of a tobacco plant showing the microbody with its crystalline core pressed up against a pair of chloroplasts and close to a mitochondrion. The arrangement of these organelles is believed to facilitate their metabolic interactions. [*From S. E. Frederick and E. H. Newcomb, J. Cell Biol., **43**:350 (1969).*] *(b)* Electron micrograph of a glyoxisome in a cell of a cucumber cotyledon after treatment to reveal the presence of catalase cytochemically. [*From R. N. Trelease, W. M. Becker, and J. J. Burke, J. Cell Biol., **60**:491 (1974).*]

metabolic interactions are complex and not fully understood, a superficial examination of the reactions wll serve to illustrate the cooperative activities. Depending upon the plant and its environmental conditions (light, oxygen availability, etc.), chloroplasts have the potential to produce a two carbon compound, glycolate, by the following series of reactions:

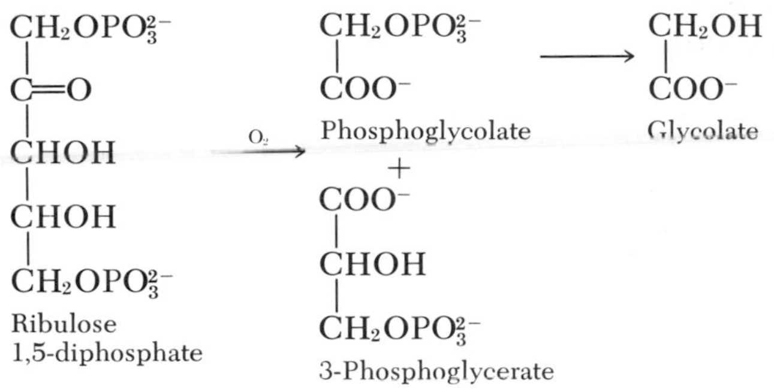

$$
\begin{array}{ccc}
\underset{\text{Ribulose}}{\underset{\text{1,5-diphosphate}}{\begin{array}{c}CH_2OPO_3^{2-}\\|\\C=O\\|\\CHOH\\|\\CHOH\\|\\CH_2OPO_3^{2-}\end{array}}}
& \xrightarrow{O_2} &
\underset{\text{Phosphoglycolate}}{\begin{array}{c}CH_2OPO_3^{2-}\\|\\COO^-\end{array}} \xrightarrow{} \underset{\text{Glycolate}}{\begin{array}{c}CH_2OH\\|\\COO^-\end{array}}
\end{array}
$$

$$
\begin{array}{c}
+\\
\underset{\text{3-Phosphoglycerate}}{\begin{array}{c}COO^-\\|\\CHOH\\|\\CH_2OPO_3^{2-}\end{array}}
\end{array}
$$

Once formed, the glycolate of the chloroplast can diffuse into the peroxisome, where the enzyme glycolate oxidase converts it to glyoxylate in a reaction of the type shown in Fig. 7-31:

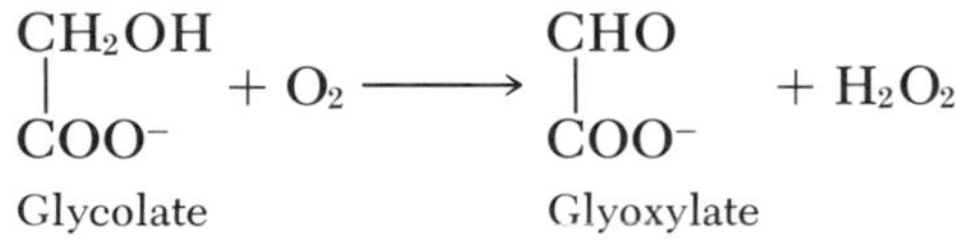

$$
\underset{\text{Glycolate}}{\begin{array}{c}CH_2OH\\|\\COO^-\end{array}} + O_2 \longrightarrow \underset{\text{Glyoxylate}}{\begin{array}{c}CHO\\|\\COO^-\end{array}} + H_2O_2
$$

The glyoxylate of the peroxisome can be metabolized in several ways, one of which involves the acceptance of an amino group to form glycine,

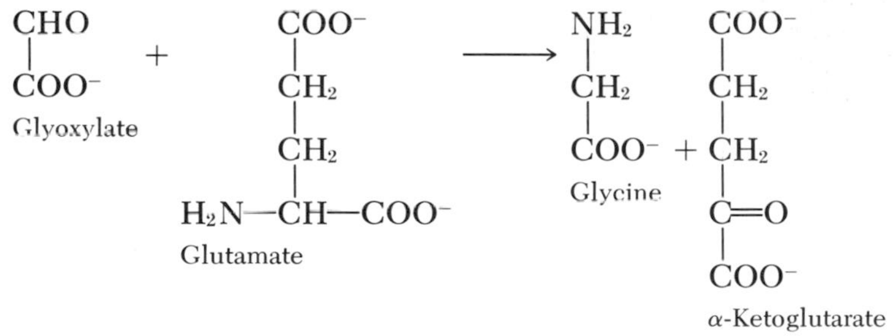

$$
\underset{\text{Glyoxylate}}{\begin{array}{c}CHO\\|\\COO^-\end{array}} + \underset{\text{Glutamate}}{\begin{array}{c}COO^-\\|\\CH_2\\|\\CH_2\\|\\H_2N-CH-COO^-\end{array}} \longrightarrow \underset{\text{Glycine}}{\begin{array}{c}NH_2\\|\\CH_2\\|\\COO^-\end{array}} + \underset{\alpha\text{-Ketoglutarate}}{\begin{array}{c}COO^-\\|\\CH_2\\|\\CH_2\\|\\C=O\\|\\COO^-\end{array}}
$$

which is believed to be transferred primarily to the mitochondrion, where it is converted to serine:

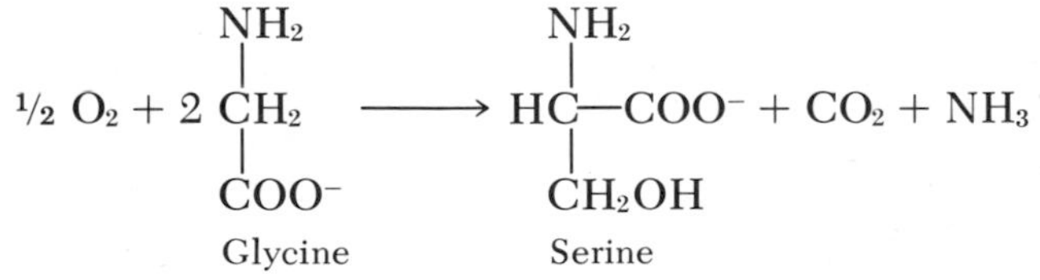

$$
\tfrac{1}{2} O_2 + 2 \underset{\text{Glycine}}{\begin{array}{c}NH_2\\|\\CH_2\\|\\COO^-\end{array}} \longrightarrow \underset{\text{Serine}}{\begin{array}{c}NH_2\\|\\HC-COO^-\\|\\CH_2OH\end{array}} + CO_2 + NH_3
$$

The term *photorespiration* is used to refer to the process described in the above reactions which in the light serve to utilize molecular oxygen (three of the reactions indicated incorporate O_2) and release carbon dioxide. The role of photorespiration in the overall metabolism of the plant is not well understood. The serine produced in the peroxisome may be converted to glycerate in a two-step peroxisomal reaction and the glycerate shuttled back to the chloroplast to be utilized in carbohydrate synthesis via the formation of 3-phosphoglyceric acid (Fig. 9-17).

Along with the type of peroxisome described above, plants contain another organelle which in addition to glycolate oxidase and catalase, also possess a number of enzymes that are not found in animal cells. These organelles, called *glyoxisomes* (Fig. 7-32b), are most prominent in plant seedlings which rely upon stored fatty acids to provide them with the energy and material to begin the formation of a new plant. One of the primary activities in these germinating seedlings is the conversion of stored fatty acids to carbohydrate. Since the product of fatty acid hydrolysis is acetyl coenzyme A (AcCoA), one might expect that glucose formation (gluconeogenesis, page 178) could occur by shuttling the acetate through the TCA cycle. However, if the reactions of the TCA cycle are examined (page 173), it can be seen that after the condensation of acetate with oxaloacetate (OAA), two decarboxylation reactions take place which defeat the purpose of building a larger molecule from the acetate originally introduced. Whereas the TCA cycle provides a route for the oxidation of

Figure 7-33. The glyoxylate cycle. The malate synthetase and isocitrase are special enzymes of this cycle. *(From S. J. Edelstein, "Introductory Biochemistry," Holden-Day, 1973.)*

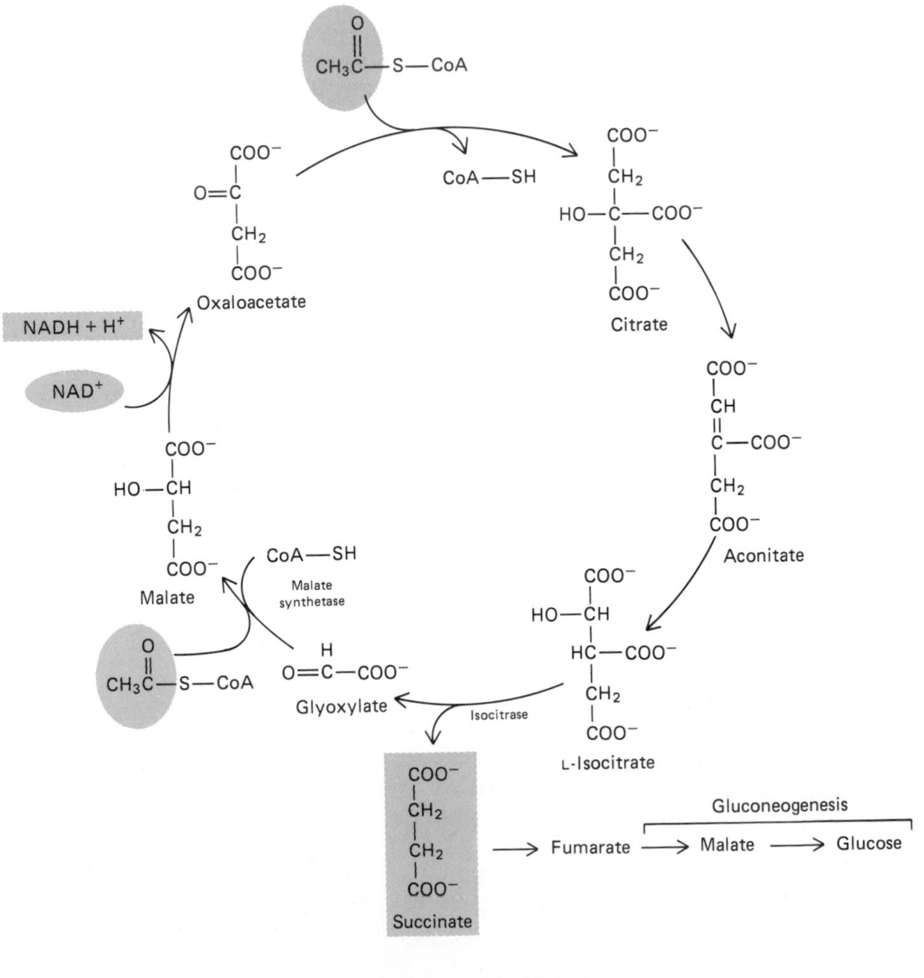

acetate, and the conservation of its free energy in ATP, it is not suitable for the conversion of acetate into more complex molecules.

An alternate pathway for acetate in plant cells is afforded by the glyoxylate cycle shown in Fig. 7-33, which occurs within glyoxisomes. The enzymes of the glyoxylate cycle contain several enzymes present in the TCA cycle as well as additional ones unique to this particular pathway. The first step is the condensation of AcCoA with OAA as occurs in the TCA cycle, but in the subsequent reactions the two decarboxylation steps of the TCA cycle are bypassed, and succinate is formed along with glyoxylate. The succinate can then be utilized in gluconeogenic reactions while the glyoxylate can serve to regenerate OAA for an additional round of the cycle. The presence of glycolate oxidase and catalase in the glyoxisome provides the means for other compounds to be converted to glyoxylate which can then enter the cycle.

REFERENCES

Allison, A., Sci. Am. **217**, 62–72, Nov. 1967. "Lysosomes and Disease."

Bennett, G., Leblond, C. P., and Haddad, A., J. Cell. Biol. **60**, 258–284, 1974. "Migration of Glycoproteins from Golgi Apparatus to the Surface of Various Cell Types as Shown by Radioautography after Labeled Fucose Injection into Rats."

Birnie, G. D., *Subcellular Components: Preparation and Fractionation.* University Park Press, 1972.

Brinkley, B. R., and Porter, K. R., eds., symposium on "Endoplasmic Reticulum, Golgi Apparatus and Cell Secretion" in *International Cell Biology 1976-1977*, pp. 267–340, Rockefeller University Press, 1977.

Brown, M. S., and Goldstein, J. L., Science **191**, 150–154, 1976. "Receptor-Mediated Control of Cholesterol Metabolism."

Budd, G. C., Int. Rev. Cytol. **31**, 21–56, 1971. "Recent Developments in Light and Electron Microscope Radioautography."

Cardell, R. R., Jr., Int. Rev. Cytol. **48**, 221–279, 1977. "Smooth Endoplasmic Reticulum in Rat Hepatocytes during Glycogen Deposition and Depletion."

Castle, J. D., Jamieson, J. D., and Palade, G. E., J. Cell Biol. **53**, 290–311, 1973. "Radioautographic Analysis of the Secretory Process in the Parotid Acinar Cell of the Rabbit."

Claude, A., Science **189**, 433–435, 1975. "The Coming of Age of the Cell."

Cook, J. S., ed., Symp. Soc. Gen. Physiol. **31**, 1976. Symposium on "Biogenesis and Turnover of Membrane Macromolecules."

Dauwalder, M., Whaley, W. G., and Kephart, J. E., Subcellular Biochem. **1**, 225–275, 1972. "Functional Aspects of the Golgi Apparatus."

Dean, R. T., and Barrett, A. J., Essays Biochem. **12**, 1–40, 1976. "Lysosomes."

De Duve, C., Harvey Lectures **59**, 49–87, 1963. "The Separation and Characterization of Subcellular Particles."

———, Sci. Am. **208**, 64–72, May 1963. "The Lysosome."

———, Science **189**, 186–194, 1975. "Exploring Cells with a Centrifuge."

Depierre, J. W., and Dallner, G., Biochem. Biophys. Acta **415**, 411–472, 1975. "Structural Aspects of the Membrane of the Endoplasmic Reticulum."

———— and Ernster, L., Ann. Rev. Biochem. **46**, 201–263, 1977. "Enzyme Topology of Intracellular Membranes."

Holtzman, E., *Lysosomes: A Survey*. Springer-Verlag, 1976.

Koatz, F. N., et al., Proc. Natl. Acad. Sci. (U.S.) **74**, 3278–3282, 1977. "Membrane Assembly in Vitro: Synthesis, Glycosylation, and Asymmetric Insertion of a Transmembrane Protein."

Kolodny, E. H., New Eng. J. Med. **294**, 1217–1220, 1976. "Lysosomal Storage Diseases,"

Lennarz, W. J., Science **188**, 986–991, 1975. "Lipid Linked Sugars in Glycoprotein Synthesis."

Maggi, V., "Lysosomes," in *Cell Biology in Medicine*, E. E. Bittar, ed., Wiley, 1973.

Morré, D. J., Ann. Rev. Plant Phys. **26**, 441–481, 1975. "Membrane Biogenesis."

———— and Oltracht, L., Int. Rev. Cytol. Supp. **5**, 61–188, 1977. "Dynamics of the Golgi Apparatus: Membrane Differentiation and Membrane Flow."

Neufeld, E. F., Lim. T. W., and Shapiro, L. J., Ann. Rev. Biochem. **44**, 357–376, 1975. "Inherited Disorders of Lysosomal Metabolism."

Neutra, M., and Leblond, C. P., Sci. Am. **220**, 100–107, Feb. 1969. "The Golgi Apparatus."

Northcote, D. H., "The Golgi Complex," in *Cell Biology in Medicine*, E. E. Bittar, ed., Wiley, 1973.

———— Phil. Trans. Roy. Soc. (London) **B268**, 119–128, 1974. "Complex Envelope System. Membrane Systems of Plant Cells."

Novikoff, A. B., and Allen, J. M., eds., J. Histchem. and Cytochem. **21**, 941–1020, 1973. Symposium on "Peroxisomes."

Palade, G. E., Science **189**, 347–358, 1975. "Intracellular Aspects of the Process of Protein Synthesis."

Pitt, D., *Lysosomes and Cell Function*. Longman, 1975.

Rothman, S. S., Science **190**, 747–753, 1975. "Protein Transport by the Pancreas."

Rustad, R. C., Sci. Am. **204**, 120–130, April 1961. "Pinocytosis."

Satir, B., Sci. Am. **233**, 28–37, Oct. 1975. "The Final Steps in Secretion."

Siekevitz, P., Ann. Rev. Phys. **34**, 117–140, 1972. "Biological Membranes: The Dynamics of Their Organization."

Silverstein, S. C., Steinman, R. M., and Cohn, Z. A., Ann. Rev. Biochem. **46**, 669–722, 1977. "Endocytosis."

Stossel, T. P., New Engl. J. Med. **290**, 717–723, 1974. "Phagocytosis."

Straus, W., J. Cell Biol. **21**, 295–308, 1964. "Occurrence of Phagosomes and Phagolysosomes in Different Segments of the Nephron in Relation to the Reabsorption, Transport, Digestion, and Extrusion of Intravenously Injected Horseradish Peroxidase."

Symposium on "Cytopharmacology of Secretion." Adv. Cytopharm. **2**, 1–379, 1974.

Tolbert, N. E., Ann. Rev. Plant Phys. **22**, 45–74, 1971. "Microbodies — Peroxisomes and Glyoxysomes."

Waechter, C. J., and Lennarz, W. J., Ann. Rev. Biochem. **45**, 95–112, 1976. "The Role of Polyprenol-Linked Sugars in Glycoprotein Synthesis."

Weissmann, G., and Claiborne, R., *Cell Membranes*. Hospital Practice, 1975.

Whaley, W. G., *The Golgi Apparatus*. Springer-Verlag, 1975.

———, Dauwalder, M., and Kephart, J. E., Science **175**, 596–599, 1972. "Golgi Apparatus: Influence on Cell Surfaces."

Wirtz, K. A., Biochem. Biophys. Acta **344**, 95–117, 1974. "Transfer of Phospholipids Between Membranes."

Young, R. W., J. Cell Biol. **57**, 175–189, 1973. "The Role of the Golgi Complex in Sulfate Metabolism."

CHAPTER EIGHT

Mitochondria and the Conservation of Chemical Energy

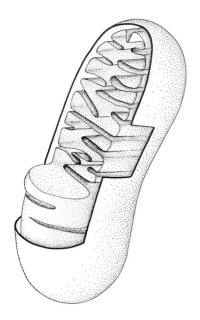

The overriding function of the mitochondria (singular *mitochondrion*) is to provide chemical energy in the form of ATP for use by the cell in virtually all of its energy-requiring activities. Whereas in the prokaryotic cell, the plasma membrane bears the responsibility for the conservation of chemical energy, eukaryotic cells possess a special organelle, a miniature "power plant," to carry out these specialized biochemical functions. Our understanding of mitochondrial function has emerged very slowly with many important findings occurring periodically during the past hundred years. Despite a great deal of intensive study, we are still uncertain of the precise mechanism by which mitochondria accomplish the transfer of energy from reduced substrates into ATP. However, there have been many recent new findings and the definitive answers may be forthcoming. The greatest difficulty in understanding the underlying mechanisms of mitochondrial activity relates to the fact that these events occur within membranes by the action of integral membrane proteins, and it has been only recently that techniques to study hydrophobic membrane interactions have been developed. We will consider some of the experiments that have been performed and some of the questions still to be answered later in the chapter, but first we need to describe the organelle itself and its role in metabolism.

MITOCHONDRIAL STRUCTURE AND FUNCTION

Unlike most of the cytoplasmic organelles, mitochondria are large enough to be clearly resolved with the light microscope (Fig. 8-1*a*) and their presence within cells has been known for over a hundred years. Before the turn of the century, mitochondria had been isolated from tissues by teasing the cells apart with a fine needle and a few of their properties were discovered. For example, it was realized that mitochondria were osmotically active, i.e., they would swell in hypotonic media and shrink in hypertonic ones. It was soon found that mitochondria were capable of carrying out oxidation-reduction reactions with various dyes and this property was taken advantage of in the development of stains that would reveal the presence of these organelles. However, the role of mitochondrial oxidation-reduction capacity in cell metabolism would not be unraveled for many years.

As with most other organelles, mitochondria possess recognizable morphological characteristics while demonstrating a considerable variability in appearance. The typical mitochondrion is a sausage-shaped structure (Fig. 8-1*b, c*) approximately 0.2 to 1.0 μm in cross-sectional diameter and 2 to 8 μm in length; dimensions which place them in the size range of bacteria. The fact that mitochondria are similar in size to bacteria may be more than coincidental, since a body of evidence (page 785) suggests that these organelles may have evolved from bacteria that had come to live symbiotically within other cells. In some cells such as those of early embryos, mitochondria are almost spherical in shape, while in others such as fibroblasts, they are elongate, threadlike structures. Since the mitochondria are involved in the extraction of energy from chemical substrates and the production of ATP, their number and intracellular location varies with the nature of the cell. In the typical liver cell there are approximately 500 to 1000 mitochondria distributed in an essentially random manner throughout the cell. In muscle cells there are typically a greater number of mitochondria, which reflects the elevated need for chemical energy in contractile cells, and they tend to be distributed in association with the contractile filaments (Fig. 8-2*a*) that require ATP. In other cells the mitochondria tend to be associated with fatty acid-containing oil droplets from which they derive the raw materials to be oxidized. A particularly striking arrangement of mitochondria (Fig. 8-2*b*) is seen in sperm cells in which they are located in the middle of the cell (the midpiece) behind the anterior nucleus. The movements of the sperm are powered by ATP production in these mitochondria.

With the development of the technique of differential centrifugation (page 296), isolated preparations of mitochondria became available and with them the opportunity for more intensive study. Electron microscopic examination of mitochondria, originally by George Palade, Fritiof Sjöstrand, and others, revealed the structure of the organelles as shown in Fig. 8-1*c*. Before describing the various features of mitochondrial ultrastructure, it is important to keep in mind that in the living cell these organelles are very active and are seen in phase-contrast movies to be in continual motion within the cell and are apparently capable of rapid acts of splitting and fusing. In addition, electron microscopic observations suggest that the

Mitochondria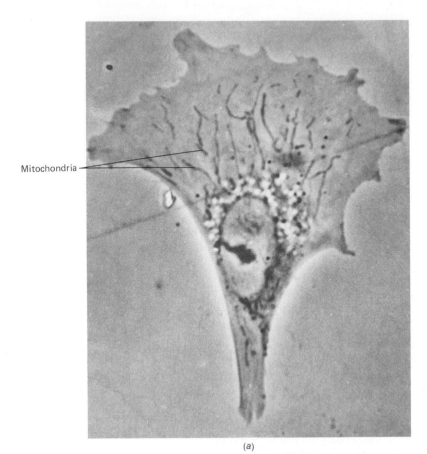

(a)

Figure 8-1. *(a)* A living fibroblast as seen with the phase-contrast microscope. Mitochondria are seen as elongated, dark bodies. *(Courtesy of N. K. Wessels.)* *(b)* High-voltage electron micrograph of the edge of an unsectioned, cultured mouse cell showing the highly elongated mitochondria. [*From J. J. Wolosewick and K. R. Porter,* Am. J. Anat., **149:** *(1977).*] *(c)* Electron micrograph of a thin section through a mitochondrion revealing the internal structure of the organelle, particularly the large number of cristae. *(Courtesy of K. R. Porter.)*

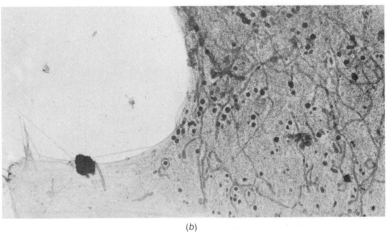

(b)

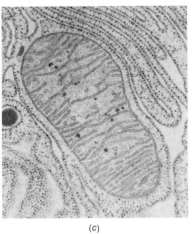

(c)

picture shown in Fig. 8-1 is not necessarily an accurate portrayal of the mitochondrion as it exists within an actively metabolizing cell, but rather is a consequence of the anaerobic conditions accompanying the fixation procedure. (The different morphological states of the mitochondrion of rat liver cells are shown in Fig. 8-22.)

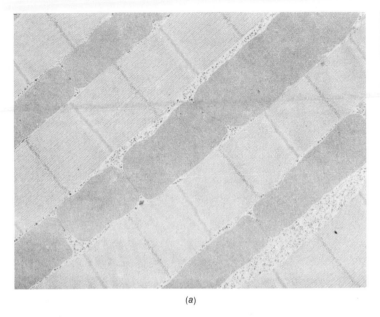

(a)

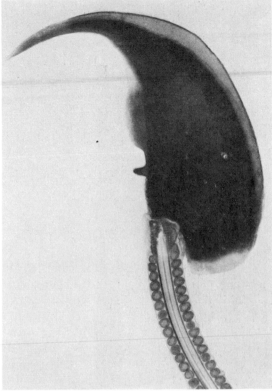

(b)

Figure 8-2. *(a)* An electron micrograph of a section through the flight muscle of a fruit fly. The contractile elements (termed *microfibrils*) are surrounded by mitochondria and glycogen granules which together supply the required ATP. *(Courtesy of S. J. O'Brien.)* *(b)* The localization of mitochondria around the flagellum of a mammalian sperm cell. *(Courtesy of D. W. Fawcett.)*

Careful studies with the electron microscope have indicated that mitochondria contain two different membranes, an outer one and an inner one. The outer membrane completely encircles the mitochondrion, serving as its outer boundary. The inner membrane forms a continuous sheet which is present partly as a layer just within the outer membrane and partly as a series of folds which penetrate the interior of the organelle. This organization is seen in the electron micrograph of Fig. 8-1*b* and the diagram of Fig. 8-3. The infolds of the inner membrane are termed *cristae* and their connections with the inner membrane at the periphery of the mitochondrion are not always apparent. However, the continuity between the membrane of each crista and the peripheral inner membrane can be seen when serial sections through an entire mitochondrion are examined. In some cases, such as in rat liver, the cristae are wide sheets that cut across the entire diameter of the mitochondrion, while in other cases, as in most plant cells, they are more tubular in nature. The membranes of the mitochondrion divide the organelle into two main compartments, that within the center of the mitochondrion and that between the inner and outer membranes. The central space is termed the *matrix* which is rather gellike due to the presence of a high concentration of soluble proteins. The space between the two membranes is represented by two continuous compartments, that within the folds of the cristae (the *intracristal space*) and that thin space between the outer and inner membranes at the periphery of the

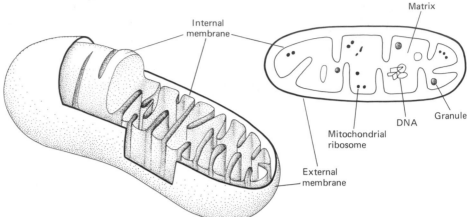

Figure 8-3. Diagrammatic representation of the structure of a mitochondrion. *(From J-C. Roland, A. Szollosi, and D. Szollosi, "Atlas of Cell Biology," Masson and C^re, 1974.)*

organelle. This latter space appears to be rather limited in the micrographs and diagrams of Figs. 8-1 and 8-3, but under conditions of active respiration this space can be quite extensive (Fig. 8-22) and the matrix space can be considerably reduced.

Mitochondrial Membranes

The outer and inner membranes of the mitochondrion appear similarly in the electron microscope, but their properties are very different. The outer membrane is composed of approximately 50% lipid, including a large amount of cholesterol, and is a considerably more permeable membrane allowing molecules up to about 10,000 daltons to pass into the intracristal space. The inner membrane, on the other hand, is highly impermeable, allowing only small uncharged molecules, such as water and pyruvic acid (free acid form) to penetrate. Larger molecules and ions require special transport systems to gain entrance to the matrix. The ability of mitochondria to shrink and swell osmotically is a reflection of the restrictive permeability of the inner membrane. When mitochondria are placed in water, the inner membrane holds back the loss of the contents of the matrix so the organelle takes up water and swells. As a consequence, the cristae unfold and the outer membrane is ruptured (Fig. 8-4). The protein/lipid ratio of the inner membrane is particularly high (about 4:1), with at least 60 different proteins being present within its structure. Among the lipids of the inner membrane is an unusual one, cardiolipin (see Fig. 5-2), which is found associated with certain of the integral membrane proteins. As will be described, the composition and organization of the inner membrane are the key to the activities of the organelles. It is the architecture of the inner membrane which provides the capacity for the successive interactions of the proteins to occur. Since the bulk of the inner membrane is present in the cristae, it might be expected that more active cells would have a greater number of these infoldings. This has been found to be the case. The cristae provide the mitochondria with a great increase in membrane surface in which its reactions can occur. In heart muscle, for example, the density of the cristae is very high.

A particularly important electron microscopic observation was made by Humberto Fernández-Morán when isolated mitochondria were placed on grids and examined after negative staining (page 62). It was seen that, attached to the inner side (matrix side) of the inner membrane, was a layer of spheres projecting from the membrane and attached to it by stalks (Fig. 8-5). The finding of these "lollipoplike" structures foreshadowed a new phase of mitochondrial research dealing with the molecular nature of the membrane and the molecular mechanisms it facilitated. Though the nature of these spheres is now fully established (page 360), their precise position with respect to the membrane itself remains uncertain. The inner membrane spheres are most prominent in negatively stained preparations which involve an air-drying process that is believed to enhance their protrusion from the membrane.

The Mitochondrial Matrix

In addition to various enzymes, the contents of the matrix include a variety of materials which are not found in other cytoplasmic organelles. These include ribosomes of a considerably smaller nature than those found within the cytoplasm of the cell and circular threads of DNA (Fig. 8-6). The mitochondria have their own, albeit limited, genetic material and their own capacity to produce RNA and proteins. This nonchromosomal DNA is important in coding for a small number of mitochondrial proteins which are integrated into the structure of the mitochondrion together with proteins coded for by nuclear DNA. Mitochondria are semiautonomous, self-replicating organelles that arise by some type of splitting process from preexisting mitochondria. The dual nature of mitochondria, their assembly, and the function of the mitochondrial genetic system will be considered in detail in Chapter 17. In addition, the matrix often contains filaments and dense granules. One type of granule contains a store of calcium ions in the form of precipitates of calcium phosphate. The mitochondrion is an active accumulator of calcium ions, though the function of this activity is not yet fully understood. The calcium ion is a very important substance in regulating numerous biochemical activities within the cell and the concentration of the ion in the cytoplasm may be regulated via its storage and release from the mitochondria.

The Role of Mitochondria in Metabolism

In Chapter 4 we considered the oxidation of carbohydrates. Beginning with glucose or glycogen, the first steps of the oxidation process are carried out by the enzymes of glycolysis which are located in a soluble state in the cytoplasm outside of the mitochondria. There are two products of interest from glycolysis, NADH and pyruvate. The NADH contains a relatively small amount of energy stored in the two electrons that it picked up from the substrate, glyceraldehyde 3-phosphate (see Fig. 4-25). In contrast, pyruvate still contains a great deal of energy. The energy of these compounds must be transferred into a usable form, that of ATP, which must occur within the mitochondrion. Since mitochondria are not

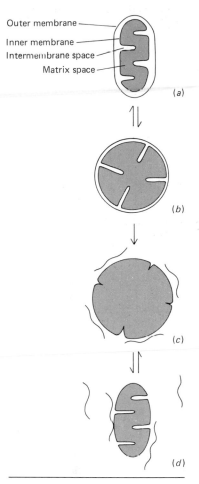

Outer membrane
Inner membrane
Intermembrane space
Matrix space

(a)

(b)

(c)

(d)

Figure 8-4. Schematic diagram of mitochondrial swelling. *(From P. Borst, in A. Lima-de-Faria, ed., "Handbook of Molecular Cytology," North-Holland, 1969.)*

347

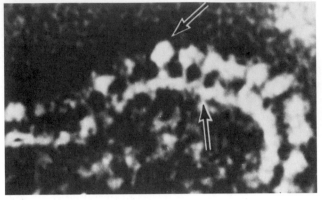

Figure 8-5. Electron micrograph of a small portion of a beef heart mitochondrion that has been negatively stained. At magnifications of approximately one-half million, spherical particles are seen attached by a thin stalk to the inner surface of the cristae membranes. [*From H. Ferrández-Morán, T. Oda, P. V. Blair, and D. E. Green, J. Cell Biol., 22:71 (1974).*]

Figure 8-6. An electron micrograph of a circular molecule of amphibian mitochondrial DNA. [*From D. R. Wolstenholme and I. B. Dawid, J. Cell Biol., 39:222 (1968).*]

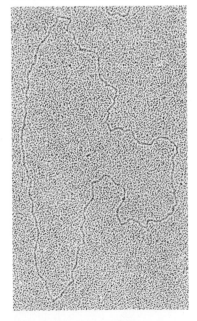

permeable to NADH, some special means must be employed to make the energy of its electrons available within the mitochondria. This is accomplished by using the electrons of NADH to reduce a small molecular weight metabolite (Fig. 8-7) which can enter the mitochondrion. Once inside the organelle the reduced metabolite (glycerol 3-phosphate) can pass its electrons to a flavoprotein of the inner mitochondrial membrane. These electrons (now present in $FADH_2$) can then be used for ATP formation.

The pyruvate from glycolysis enters the mitochondrion directly and is converted to acetyl CoA (see Fig. 4-30) in the matrix by the activity of the giant multienzyme complex, pyruvate dehydrogenase. The acetyl CoA is then ready to be circulated through the TCA cycle as described in Chapter 4. With the exception of succinate dehydrogenase, which is located within the inner membrane, all of the enzymes of the TCA cycle are present in a soluble form within the matrix. Fatty acid oxidation (Fig. 8-8a), which also generates acetyl CoA molecules, occurs within the mitochondrion itself, so there is no need in this case for the entrance of this TCA cycle metabolite from the outside. The proximity of mitochondria to fat droplets in the cytoplasm reflects the mitochondrial localization of the catabolic enzymes of fatty acid degradation. So far we have accounted for the role of the mitochondria in the catabolism of polysaccharides and fats; they are also of importance in the metabolism of amino acids. Even though the amino acids as a group represent a heterogeneous collection of molecules, their breakdown (Fig. 8-8b) also generates metabolites of the TCA cycle, namely acetyl CoA, succinyl CoA, α-ketoglutarate, fumarate, and oxaloacetate.

It is apparent from this discussion that all of the energy-providing macromolecules (polysaccharides, fats, and proteins) are broken down into metabolites of the TCA cycle, and therefore the mitochondrion becomes the focus for the final energy-conserving steps in metabolism regardless of the nature of the starting material. The importance of mitochondria in oxidative metabolism was determined primarily by Albert Lehninger and colleagues around 1950. Working with mitochondria that had been isolated by differential centrifugation from rat liver homogenates, they found that these organelles were capable of oxidizing substrates such as pyruvate or fatty acids, or the electron carrier NADH, at the expense of molecular oxygen with the accompanying formation of ATP from ADP and P_i. When NADH was used as the source of electrons, the mitochondria were treated with hypotonic medium to facilitate NADH penetration.

OXIDATION-REDUCTION POTENTIALS

The focus of attention in the study of mitochondrial function is on the mechanism by which this organelle converts the energy obtained from the oxidation of substrate into the phosphorylation of ADP. Before turning to observations on this process we need to briefly reconsider the nature of oxidation-reduction reactions (see Appendix to Chapter 4) and their relation to free-energy release. If one compares a variety of oxidizing agents, they can be ranked in a series according to their affinity for electrons: the greater the affinity, the stronger the oxidizing agent. Similarly, reducing agents can also be ranked according to their affinity for electrons: the less the affinity (the more easily electrons are released), the stronger the reducing agent.

The terms *affinity, donor,* and *acceptor* are the same ones used in Chapter 4 when the phosphate-transfer potential of various compounds was discussed. The same concepts apply to oxidation-reduction reactions, the difference being that what is transferred is not a phosphate or an acetyl or some other chemical group, but an electron, or pair of electrons, or hydrogen atoms. In the case of reducing agents we can use the term *electron-transfer potential,* signifying that those substances having a high transfer potential (such as NADH) are strong reducing agents while those with a low transfer potential (such as H_2O) are weak reducing agents. The other member of the couple is in a reversed position. For example, NAD+ (of the NADH–NAD+ couple) is a weak *oxidizing* agent, while the other member of the couple from H_2O, namely oxygen (actually $\frac{1}{2}O_2$) is a strong oxidizing agent.

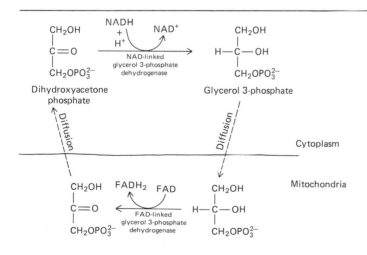

Figure 8-7. The transfer of electrons from NADH molecules formed in the cytoplasm to the electron-transport chain of the mitochondria occurs in an indirect manner. Due to the impermeability of the mitochondrial membrane to NADH, the electrons are passed to dihydroxyacetone phosphate which shuttles them into the mitochondrion where they are used to reduce FAD.

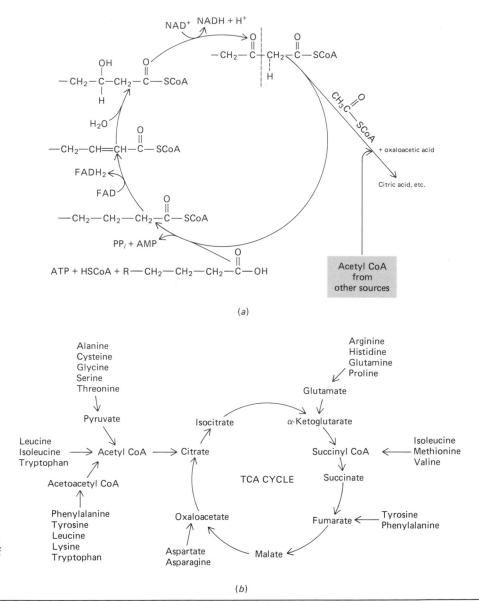

Figure 8-8. (a) The steps that occur in each cycle of fatty acid oxidation leading to the removal of acetyl groups. (*After A. White, P. Handler, and E. L. Smith, "Biochemistry," 5th ed., McGraw-Hill 1975.*) (b) Input of amino acids into the TCA cycle. (*From S. J. Edelstein, "Introductory Biochemistry," Holden-Day, 1973.*)

Since the movement of electrons involves the separation of charge, the affinity of substances for electrons can be measured by instruments that detect voltages. The means by which these measurements are made will not concern us here; discussions can be found in textbooks of chemistry, biochemistry, and physics. What is measured for a given couple is an oxidation-reduction (redox) potential relative to the potential for some standard couple. The standard couple has arbitrarily been chosen as hydrogen. Just as in the case of free energy changes when a *standard*

$\Delta G°$ was used, a similar assignment is used for redox couples. The standard redox potential for a given couple is designated as the voltage produced by a half-cell (only members of one couple present) in which each member of the couple is present at a standard concentration under standard conditions. Standard concentrations are 1.0 M for solutes and ions and 1 atm pressure for gases (e.g., H_2) at 25°C. The standard redox potential for the oxidation-reduction reaction involving hydrogen ($2H^+ + 2$ electrons $\rightleftharpoons H_2$) is 0.00 volt. The assignment of sign (positive or negative) to couples other than the hydrogen couple is *arbitrary* and *varies* among different sources. We will consider the assignment in the following way. Those couples having reducing agents which are better donors of electrons than H_2, i.e., have less affinity for electrons than H_2, will have negative redox potentials. For example, the standard redox potential for NADH is -0.32 volt (Table 8-1). Those couples having oxidizing agents which are better electron acceptors than H^+, i.e., have greater affinity for electrons than H^+, will have positive redox potentials. Table 8-1 gives the redox potentials of some biologically important couples. The value for the hydrogen couple in the table is not 0.00 but -0.42. This number represents the value when the concentration of H^+ is 10^{-7} M (pH of 7.0) rather than 1.0 M (pH of 0.0) which is of little physiological use.

Just as any other spontaneous reaction is accompanied by a loss in free energy, so too are oxidation-reduction reactions. The standard free-energy change during a reaction of the type $A_{ox} + B_{red} \rightleftharpoons A_{red} + B_{ox}$ can be calculated from the standard redox potentials of the two couples involved in the reaction according to the equation $\Delta G° = -nF \Delta E°$, where n is the number of electrons transferred, F is the caloric equivalent of the

TABLE 8-1
Standard Oxidation-Reduction Potentials of Some Conjugate Redox Pairs Expressed on the Basis of Two-Electron Transfers at pH Near 7.0 and Temperature 25 to 30°C °

Electrode equation	E_0' volts
Acetate + $2H^+$ + $2e^-$ $\rightleftharpoons$ acetaldehyde	-0.58
$2H^+$ + $2e^-$ $\rightleftharpoons$ H_2	-0.421
α-Ketoglutarate + CO_2 + $2H^+$ + $2e^-$ $\rightleftharpoons$ isocitrate	-0.38
Acetoacetate + $2H^+$ + $2e^-$ $\rightleftharpoons$ β-hydroxybutyrate	-0.346
NAD^+ + $2H^+$ + $2e^-$ $\rightleftharpoons$ NADH + H^+	-0.320
$NADP^+$ + $2H^+$ + $2e^-$ $\rightleftharpoons$ NADPH + H^+	-0.324
Acetaldehyde + $2H^+$ + $2e^-$ $\rightleftharpoons$ ethanol	-0.197
Pyruvate + $2H^+$ + $2e^-$ $\rightleftharpoons$ lactate	-0.185
Oxaloacetate + $2H^+$ + $2e^-$ $\rightleftharpoons$ malate	-0.166
Fumarate + $2H^+$ + $2e^-$ $\rightleftharpoons$ succinate	-0.031
Ubiquinone + $2H^+$ + $2e^-$ $\rightleftharpoons$ ubiquinol	$+0.10$
2 cytochrome $b_{K(ox)}$ + $2e^-$ $\rightleftharpoons$ 2 cytochrome $b_{K(red)}$	$+0.030$
2 cytochrome c_{ox} + $2e^-$ $\rightleftharpoons$ 2 cytochrome c_{red}	$+0.254$
2 cytochrome $a_{3(ox)}$ + $2e^-$ $\rightleftharpoons$ 2 cytochrome $a_{3(red)}$	$+0.385$
$\frac{1}{2}O_2$ + $2H^+$ + $2e^-$ $\rightleftharpoons$ H_2O	$+0.816$

° The standard potentials of the cytochromes and of ubiquinone vary somewhat with their state, i.e., whether isolated or present in the mitochondrial membrane; values given are for the latter case.

SOURCE: A. L. Lehninger, "Biochemistry," 2d ed., Worth, 1975.

faraday (23.062 kcal/volt mol^{-1}), and $\Delta E°$ is the difference in volts between the standard redox potentials of the two couples. Exergonic reactions $(-\Delta G°)$ are ones having positive $\Delta E°$ values. The greater the difference in redox potential between two couples, the farther the reaction will proceed to the formation of products before an equilibrium state is reached.

In the case of the reduction of NADH by molecular oxygen according to the reaction $NADH + \frac{1}{2}O_2 + H^+ \rightleftharpoons H_2O + NAD^+$, the $\Delta E°$ is 1.14 volts. In other words, if the standard redox potentials of the two couples

$$\frac{1}{2}O_2 + 2H^+ + 2 \text{ electrons} \rightleftharpoons H_2O \qquad E° = +0.82 \text{ volt}$$

$$NAD^+ + 2H^+ + 2 \text{ electrons} \rightleftharpoons NADH + H^+ \qquad E° = +0.32 \text{ volt}$$

are considered, the difference between the two $E°$ values is 1.14 volts. Substituting this value into the equation $\Delta G° = -nF \, \Delta E°$, the standard free energy difference is -52.6 kcal/mol. As in the case of other reactions, the actual ΔG values will depend upon the relative concentrations of reactants and products present in the cell at a given instant. Regardless, it is apparent that the drop in free energy of a pair of electrons as they pass from NADH to molecular oxygen is sufficient to drive the formation of several molecules of ATP ($\Delta G°$ of $+7.3$ kcal/mol). The transfer of this energy from NADH to ATP within the mitochondrion will be the topic of concern in the remainder of this chapter. Before considering the details of the process, a brief overview may be useful to put the various pieces into perspective.

As previously described, the catabolic pathways of the cell convert the large molecular weight materials to a small number of low molecular weight organic compounds (two to four carbon backbones) that are interrelated by the reactions of the TCA cycle. Up to this point in the process relatively little oxidation has occurred; most of the carbons remain in a reduced state, i.e., covalently bonded to carbons and hydrogens. During passage through the TCA cycle, two important changes occur to particular carbons: the hydrogen atoms are stripped from the carbons and the carbons are removed from the rest of the molecule in the form of CO_2 (see Fig. 4-31). The carbon atom in carbon dioxide can be considered to have an oxidation state of $+4$, its maximally oxidized (therefore, least energy-containing) state. It is this increase in positive state (relative to the reduced substrates, such as pyruvate), associated with the loss of electrons, that is the essence of the oxidation process. The two products of oxidation, CO_2 and H_2O arise in very different ways. Carbon dioxide is formed from simple decarboxylation reactions, while water is formed only after a very indirect route.

The substrates of the TCA cycle, i.e., isocitrate, α-ketoglutarate, malate, and succinate, have redox potentials of relatively high negative values — sufficiently high to transfer electrons to the appropriate electron acceptors, NAD^+ or FAD, *under conditions in the cell*. Table 8-1 gives the standard redox potential for several couples involved in the reactions of the TCA cycle. Certain of these couples (such as α-ketoglutarate-isocitrate) have more negative E_0' values than the NAD$^+$-NADH couple. In these cases, electrons would be transferred from substrate to NAD$^+$ when reactants and products were present at equimolar concentrations, i.e., standard conditions. In contrast, the oxaloacetate-malate couple has a more positive E_0' value than the NAD$^+$-NADH couple. The oxidation of malate to oxaloacetate can proceed to the right only when the ratio of products to reactants is kept below that of standard conditions. In other words, the ΔG of this reaction is kept negative by maintenance of elevated malate and/or NAD$^+$ levels relative to oxaloacetate and/or NADH levels. The situation is analogous to that concerning the formation of glyceraldehyde 3-phosphate from dihydroxyacetone phosphate (page 132). In contrast, the oxidation of succinate to fumarate (E_0' of -0.031) proceeds by the reduction of FAD, a coenzyme of lesser electron affinity than NAD$^+$.

With the exception of the enzyme succinate dehydrogenase, which is membrane-bound, the enzymes of the TCA cycle are free within the fluid matrix of the mitochondrion. However, the electrons that are removed from the substrate are soon bound to carriers associated with the inner mitochondrial membrane and the electrons are then passed from one carrier to the next until the final acceptor becomes reduced. The final acceptor of this electron "bucket brigade" is molecular oxygen which becomes reduced to water, forming the other end product of metabolic oxidation. Each electron carrier along the line has a more positive redox potential than the previous one, and with each successive transfer the electrons lose additional free energy. The loss of free energy accompanying electron transport is coupled to the phosphorylation of ADP. The means by which the energy released by electron transport is used to drive the endergonic formation of ATP remains one of the central problems in biochemistry and molecular biology and one which has inspired a great number of interesting experiments.

ELECTRON TRANSPORT

The components of the electron-transport chain comprise a heterogeneous collection of molecules all of which are capable of undergoing oxidation and reduction. The atoms that actually undergo changes in oxidation state include iron, copper, nitrogen, and carbon. All but one of the components, coenzyme Q, are believed to be associated with a protein, and all but one of the proteins are integral components of the mitochondrial membrane. The exception is a peripheral membrane protein, cytochrome c. The fact that electron carriers represent membrane proteins has made their analysis more difficult than that of other proteins of the cell. The best-studied carriers are the cytochromes, which are heme-containing page 90) proteins (Fig. 8-9a), the flavin nucleotide FMN (Fig. 8-9b), which is a component of NADH dehydrogenase, and coenzyme Q (Fig. 8-9c), a lipid-soluble quinone. The order in which these particular carriers are present within the transport chain is shown in Fig. 8-10. The various cytochromes differ from one another by substitutions within the heme group as well as within the protein. These substitutions are respon-

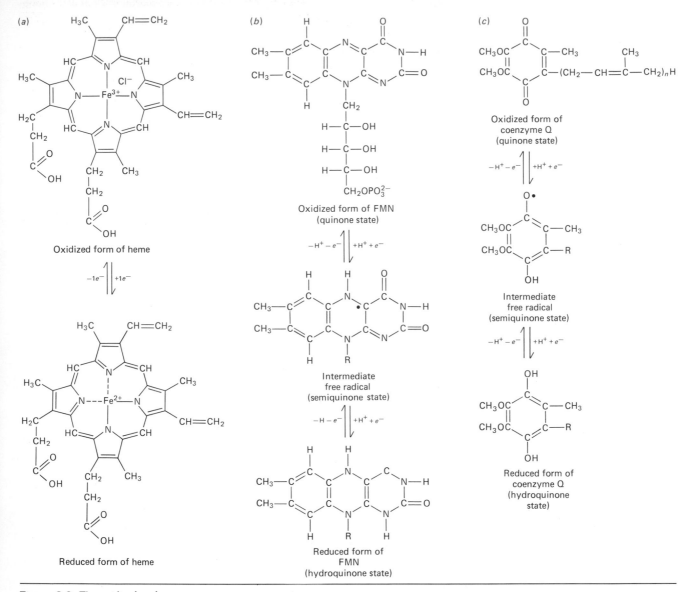

Figure 8-9. The oxidized and reduced forms of the major electron carriers. *(a)* Heme, of the cytochromes. *(b)* FMN of NADH dehydrogenase (FAD has an adenosine bonded to the phosphate group). *(c)* Coenzyme Q.

sible for the differences in redox potential among the cytochromes. Each of the carriers along the path is capable of existing in two states, one oxidized and the other reduced. Each carrier is successively reduced by the gain of electrons from the preceding carrier in the line and is subsequently oxidized by the loss of electrons to the carrier following it. In order for transport to occur, the carriers must be organized into a series of increasing (more positive, more electron affinity) potential. The drop in the free energy of the electrons during transport is shown in Fig. 8-11.

The sequence of carriers shown in Fig. 8-10 indicates the pathway for a pair of electrons when NADH is the donor of the electrons to the transport chain. When FADH$_2$ is the electron donor, as occurs when the elec-

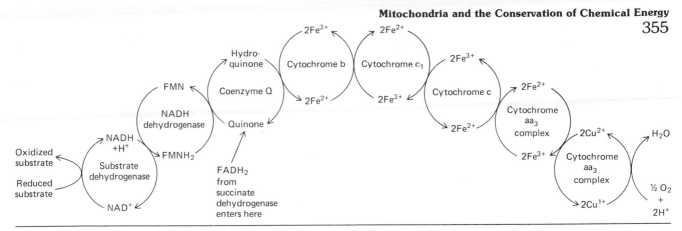

Figure 8-10. The electron-transport chain.

trons are stripped from succinate by succinate dehydrogenase (see Fig. 4-31) or the electrons are taken from glyceraldehyde 3-phosphate and then enter the mitochondrion via the shuttle (Fig. 8-7), an alternate route is taken. The electrons from $FADH_2$ are passed to coenzyme Q, bypassing the left end of the chain which has too negative a redox potential to accept the "less energetic" electrons of the flavin nucleotide. In other words, $FADH_2$ lacks sufficient electron-transfer potential to enter the chain at a higher level. Not all of the carriers transport the same particles; each cytochrome is reduced by the acceptance of only one electron while coenzyme Q and FMN are reduced by the acceptance of two hydrogen atoms (two electrons plus two protons). The last carrier in the sequence is cytochrome oxidase, a term which denotes that it is the carrier that transfers its electrons to oxygen. In fact, it is the only carrier in the chain that will react with molecular oxygen. Cytochrome oxidase is a large oligomeric protein containing two different heme-containing cytochromes (u and a_3) as well as other subunits. In addition to the hemes, cytochrome oxidase contains two atoms of copper. Electrons are passed from cytochrome a_3 to oxygen, a step which is believed to involve the reduction and oxidation of the copper atoms. It is not known how the formation of water, which requires one-half an O_2 molecule and two electrons, is accomplished. Since each cytochrome transfers only one electron, four molecules are probably re-

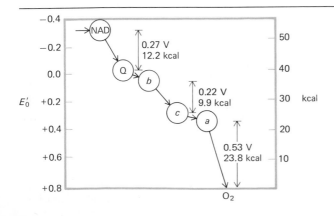

Figure 8-11. The decline in free energy as electron pairs flow down the respiratory chain to oxygen. Each of the three segments denoted yields sufficient energy to generate a molecule of ATP from ADP and phosphate. *(From A. L. Lehninger, "Biochemistry," 2d ed., Worth, 1975.)*

quired to deliver four electrons to oxygen which together with four protons can produce two molecules of water.

If one examines the redox potentials of successive carriers of Fig. 8-10, it is apparent that there are three places in which the transfer of electrons is accompanied by a major release of free energy (Fig. 8-11). These are the three sites (sites I, II, and III) that are responsible for the release of sufficient energy to drive the phosphorylation reaction forming ATP. When $FADH_2$ is the donor of the pair of electrons, rather than NADH, the first large drop in redox potential (site I) is bypassed (Fig. 8-10) and only two molecules of ATP are formed. Since each pair of electrons from NADH is responsible for the reduction of one oxygen atom ($\frac{1}{2}O_2$), it would be expected that three ATPs would be formed per O atom. This relationship, termed the P:O ratio, was first shown to have a value of 3 by Severo Ochoa in 1943. When $FADH_2$ is the donor, only two ATPs are formed per atom of oxygen so the P:O ratio is 2.

If one determines the relative amounts of each of the various components of the electron-transport chain, it is found that a precise stoichiometric relationship exists among them. These and other findings suggest that the various components of the electron-transport chain are organized into a predictable, geometrical relationship within the inner mitochondrial membrane; a relationship that facilitates the transfer of electrons from one carrier to another. It has been found that when the components of the electron-transport chain are prepared from fragmented mitochondria, certain of the carriers remain associated with one another in a complex. It was found by David Green that there were four complexes that could be identified, each containing two or more components. The complexes are numbered I, II, III, and IV. In addition there are two components of the chain, cytochrome c and coenzyme Q, which are not part of the four complexes. The components of each of the four complexes and their interrelationship is shown in Fig. 8-12. The three sites providing the energy for phosphorylation are related to complexes I, III, and IV. Complex II contains succinate dehydrogenase whose carrier is FAD, which, once reduced, passes its electrons on to coenzyme Q, bypassing complex I and the first site of phosphorylation. The precise spatial organization of the various complexes within the membrane is not understood, though the term *respiratory assembly* is often used to denote the collection of components that would be necessary to carry out both electron transport and

Figure 8-12. The four complexes of the respiratory chain. The components of each of these complexes are firmly integrated into the inner mitochondrial membrane. The lipid-soluble coenzyme Q and the peripheral membrane protein cytochrome c are readily separated from the other components. *(From N. A. Edwards and K. A. Hassal,* Cellular Biochemistry and Physiology, *McGraw-Hill, 1972.)*

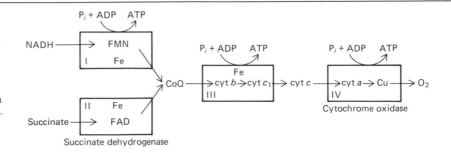

oxidative phosphorylation by a pair of electrons. It is estimated that there are approximately 4000 respiratory assemblies per square micron of inner membrane in a liver mitochondrion.

Before one receives the impression that the entire series of electron-transport carriers is well established, it is important to point out that the oxidation-reduction centers shown in Fig. 8-11 represent only a fraction of those now known to exist. In the early 1960s, an entirely new group of oxidation-reduction centers in the inner membrane were discovered. These new components were termed *iron-sulfur proteins* and were detected using the technique of electron-spin resonance. The iron atoms of these proteins are not located within a heme group but are instead directly associated with the sulfur atoms in cysteine residues (Fig. 8-13) of the protein, hence the term iron-sulfur. As in the case of the iron-containing heme groups of the cytochromes, the nonheme irons also undergo a change in valence from the ferrous (Fe^{2+}) to the ferric (Fe^{3+}) state. There are at least nine nonheme iron centers distributed through the electron-transport chain and their precise function remains largely unknown. By some estimates there may be up to 30 different centers within the chain that can undergo reduction and oxidation. The number of these that are directly involved in the transfer of electrons remains unknown, but they are all likely to have some role in the process. Even though the basic function of the electron-transport chain is well understood, there are numerous questions that remain to be answered.

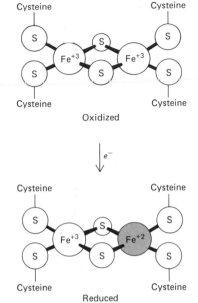

Figure 8-13. Model of the iron-sulfur group of adrenodoxin. *(From A. White, P. Handler, and E. L. Smith, "Biochemistry," 5th ed., McGraw-Hill, 1975, after R. Cammack et al.,* Biochem. J., **125**:849, *1971.)*

ATP FORMATION

In the previous section we considered the means by which the free energy stored in the electrons of NADH and $FADH_2$ is released. Before turning to examine the mechanism by which this energy is made available to the phosphorylation process, we will examine some of the important aspects of the phosphorylation apparatus itself. First we need to consider where the actual phosphorylation of ADP occurs. Fortunately for the study of oxidative phosphorylation, it was found that it was not necessary to use intact mitochondria in order to form ATP from the energy stored in NADH. It was shown in the 1950s that mitochondria could be broken into pieces by mechanical treatments or ultrasonic vibration (*sonication*), and the fragments, termed *submitochondrial particles* (Fig. 8-14a), were capable of transferring electrons from NADH to molecular oxygen, using the energy released to form ATP.

Electron microscopic examination of the submitochondrial particles revealed that they represented fragments of the inner membrane, which, in typical membrane fashion, had their ends fused together to form small closed vesicles (Fig. 8-15a). Since these vesicles could carry out the entire sequence of activities from the removal of electrons from NADH to the phosphorylation of ADP, it was clear that all of the components necessary for electron transport and phosphorylation were associated with the inner mitochondrial membrane. Electron micrographs revealed one very important piece of information: the lollipoplike structures, which in the intact

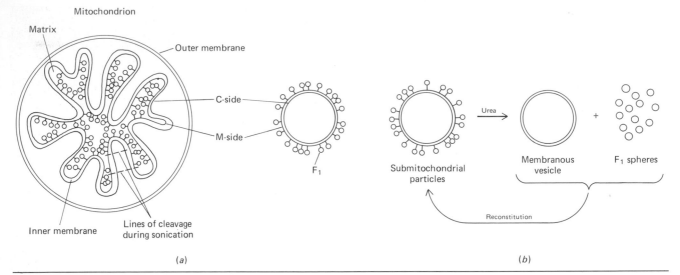

Figure 8-14. (a) Formation of submitochondrial particles by ultrasonic
oscillation. (From E. Racker,
Biochem. Soc. Trans., **3:**789, 1975.)
(b) Separation of the membranous
component from the F₁ spheres by
treatment with urea. Whereas the
intact submitochondrial particle is
capable of oxidative phosphorylation,
the isolated membranes will oxidize
substrate but cannot phosphorylate
ADP. The isolated F₁ spheres act as
an ATPase.

In the figure, the following labels appear: Mitochondrion; Matrix; Outer membrane; C-side; M-side; F₁; Inner membrane; Lines of cleavage during sonication; (a); Urea; Submitochondrial particles; Membranous vesicle; F₁ spheres; Reconstitution; (b).

mitochondrion are facing the matrix side of the inner membrane, were now facing outward from the submitochondrial particles into the medium (Fig. 8-15a). In other words, the mechanical disruption of mitochondria produces inner membrane vesicles with the reverse orientation from that in intact mitochondria, i.e., they are inside-out vesicles. The availability of submitochondrial particles provided an excellent opportunity to study the function of all of the components of oxidative phosphorylation, but the nature and function of the spheres projecting from the vesicles were particularly accessible to analysis. It was soon found that the spheres could be detached from the remainder of the vesicle (Figs. 8-14b and 8-15b, c) (for example, by urea treatment) and the function of the two parts could be determined in isolation.

It was found that, in the absence of the spheres, the vesicles would carry out the oxidation of NADH with the formation of water, but they would not phosphorylate ADP. It appeared from this finding that the entire electron-transport chain from NADH dehydrogenase to cytochrome oxidase was present in the membrane itself and remained active without the spheres. In contrast, the isolated preparation of spheres had no electron-transport activity but instead catalyzed the hydrolysis of ATP; it was an ATPase. At first glance this seems like a peculiar situation. Why should mitochondria possess an enzyme that hydrolyzes the substance it is supposed to produce? If one considers that the hydrolysis of ATP is the reverse reaction from its formation, the function of the sphere becomes more apparent: it is the site at which ATP formation normally occurs. Since enzymes do not affect the equilibrium constant of the reaction they catalyze, and enzymes are capable of catalyzing both the forward and reverse reactions, the direction the reaction takes depends totally upon the conditions under which the enzyme is placed. This is nicely shown in experiments with other ATPases, such as that catalyzing the uptake of K⁺ and ejection of Na⁺ against their respective concentration gradients

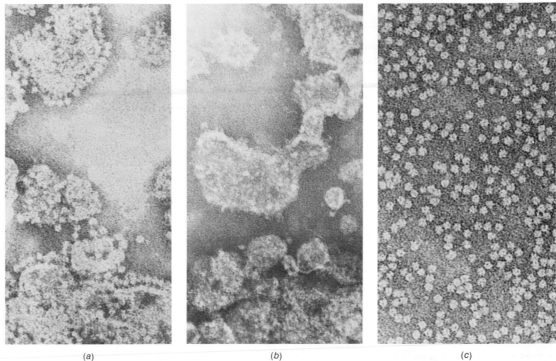

(a)   (b)   (c)

Figure 8-15. Electron micrographs of the preparations described in the previous figure. *(a)* Submitochondrial particles prepared from mitochondria by sonication. *(b)* Isolated membranous vesicles without F_1 spheres. *(c)* The F_1 particles in a purified state. *(d)* The reconstituted submitochondrial particles. *(Courtesy of E. Racker.)*

(page 257). When this enzyme was discussed in Chapter 5, it was considered strictly as an enzyme that used the energy obtained from ATP hydrolysis to actively transport these two ions across the plasma membrane. In the cell this *is* the only function of the enzyme. However, if the conditions are constructed so that the reverse reaction is favored (Fig. 8-16), this enzyme can catalyze the formation of ATP rather than its hydrolysis. In this case, the conditions are obtained by preparing red blood cell ghosts and sealing them in such a way as to have a *very* high K^+ concentration within the cell and a *very* high Na^+ concentration on the outside. Under these conditions, each ion moves in the opposite direction it normally would, though in this case the ions are moving down their respective gradients rather than against them. As a result of the ion movement, ATP is synthesized (if ADP and P_i are present within the ghost) rather than being hydrolyzed. Experiments such as this illustrate the reality of what would be theoretically expected from information about the reversibility of enzyme-catalyzed reactions.

When the isolated ATPase spheres are added back to the vesicles from which the spheres had been removed, they reattach to one another (Figs. 8-14*b*, 8-15*d*) and the reassembled submitochondrial particles are once again capable of phosphorylation coupled to electron transport. In the past few years considerable information has been obtained about the

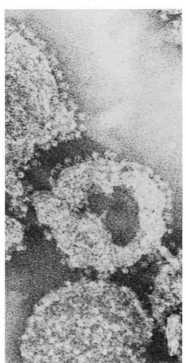

(d)

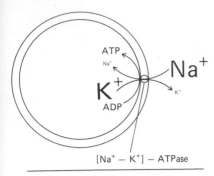

Figure 8-16. The activity of membranous vesicles reconstituted with the (Na⁺-K⁺)-ATPase. By causing these vesicles to have very high internal K^+ concentrations and very high external Na^+, the reaction can be run in the reverse direction from its normal course in the plasma membrane. In the process, ATP is formed from ADP. The size of the letters indicates the direction of the concentration gradient.

Figure 8-17. The morphology of the mitochondrial ATPase complex. The arrangement of individual subunits drawn in this figure is totally speculative; the actual arrangement is not known. *(From A. E. Senior,* Biochim. Biophys. Acta, **301**:265, 1975.)

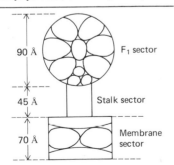

nature of the ATP-forming machinery. The spheres are obviously very large structures in macromolecular terms, measuring about 85 Å in diameter. Each sphere has a molecular weight of about 360,000 daltons and contains a number of polypeptides representing what is believed to be six different species. Five of the six are firmly complexed to one another, while the sixth, which acts as an inhibitor of the enzymatic activity of the complex, is more readily dissociated. This complex of proteins is distinct from the membrane itself and, unlike the membrane components, is soluble in aqueous solutions. The sphere is generally referred to as coupling factor 1, or F_1. The F_1 sphere is attached to the membrane by a stalk (Fig. 8-17), which is also protein. The stalk is believed to include the OSCP (oligomycin sensitivity-conferring protein, see below).

The study of mitochondrial function has been greatly aided by the discovery of a variety of inhibitors which act at different sites in the electron-transport and phosphorylation process. For example, inhibitors played a very important role in working out the sequence of the carriers of the electron-transport chain (see review by B. Chance and G. Williams, 1956, for a discussion). Oligomycin is an inhibitor which acts to block ATP formation by interaction with the ATP-synthesizing complex. It does not, however, act on the F_1 sphere, but rather on a membrane protein at the base of the complex (Fig. 8-17). When the sphere is detached from the stalk, it is no longer connected to the site of oligomycin action and therefore its activity, i.e., its ATPase ability, is no longer sensitive to this antibiotic. However, once it is reattached to the vesicle by the stalk, its activity is once again sensitive to oligomycin. This result, as well as others, indicates that there are three components to the ATP-forming machinery: the sphere, the OSCP, and the piece of the membrane (the *membrane sector*) which forms the base of the entire structure. More recently, extraction procedures employing detergents have been developed and the entire complex, together with its associated membrane lipids, has been isolated. Since all three of the elements of the complex remain together in these detergent preparations, the ATPase of the sphere retains its sensitivity to oligomycin. For this reason the entire complex is generally referred to as the *oligomycin-sensitive ATPase*. In the remainder of this discussion this tripartite complex (Fig. 8-17) will be referred to as the *ATP synthetase*. Though this term is not a widely used one, it has the advantage of brevity and it stresses the role of the enzyme in the mitochondrion in forming ATP rather than breaking it down. There is a striking similarity in the structure of the ATP synthetase of the membranes of bacteria, chloroplasts, and mitochondria. It appears that the basic mechanisms of energy conservation are very similar among all living organisms.

The isolation of the entire ATP-synthesizing complex by Efraim Racker and his colleagues has opened the door to a wide range of in vitro studies aimed at uncovering the underlying mechanisms of oxidative phosphorylation. The most informative studies have involved the disassembly of the respiratory apparatus and its reassembly under defined, in vitro conditions, a procedure that is termed *reconstitution*. It has been mentioned previously that the electron-transport chain can be broken into four complexes (I to IV), cytochrome c, and coenzyme Q, and that the ATP synthetase can be isolated independently from the components of the electron-transport chain. The development of these isolation procedures has allowed investigators to recombine the various pieces of the oxidative phosphorylation apparatus in any way desired. One example of this approach was described on page 261 in connection with the isolated (Na^+-K^+)-ATPase. Basically, the reconstitution is accomplished by sonicating phospholipids together with the detergent-containing preparations of the membrane protein complexes, followed by the removal of the detergent. The result of this procedure is the formation of an artificial phospholipid vesicle (a *liposome*) containing the proteins of the original mixture. When all of the components of the inner membrane are present, the reconstituted vesicles that form are capable of the entire respiratory process, i.e., the oxidation of NADH by molecular oxygen and the simultaneous formation of ATP. Using these techniques, each of the three sites of phosphorylation can be reconstituted separately. For example, if the cytochrome oxidase complex, cytochrome c, and the ATP synthetase are combined, the reconstituted vesicles are capable of site III phosphorylation when the appropriate electron donor is provided. Interestingly, it appears that the nature of the phospholipids used in forming the lipid bilayer of the vesicle is an important factor in determining the vesicle's activity. Maximal activity is obtained when a 4:1 mixture of phosphatidylethanolamine/phosphatidylcholine is used. It appears that the amino group of phosphatidylethanolamine (see Fig. 5-2) may carry out some specific role in the process. These experiments illustrate the extent to which subcellular components can be manipulated experimentally without destroying their activity. Were biological elements more fragile, as might have first been expected, we would know much less about the workings of cells than we do.

COUPLING OF OXIDATION AND PHOSPHORYLATION

Oxidative phosphorylation was first proposed by V. A. Englehardt in the early 1930s and it was demonstrated to be a major means of energy recovery in association with the TCA cycle by V. Belitser and H. Kalckar in 1938. The question that has proved to be the most insoluble, and certainly the most controversial, has concerned the means by which the oxidation of NADH or $FADH_2$ by molecular oxygen is coupled to the phosphorylation of ADP. The question can be restated—how does the ATP-synthesizing machinery "feel" the passage of electrons down the electropotential gradient? Since the early 1960s there have been three major hypotheses to explain the coupling of oxidation to phosphorylation. All three of them provide a means by which the free energy released by electron transport can be stored in some high-energy state. Each of them proposed a different type of high-energy intermediate that is used to drive the endergonic phosphorylation reaction. An important point to keep in mind is that the theories are not mutually exclusive, i.e., any two or all three may have some role in the overall coupling process. In defense of one or another of these theories, a great deal has been learned about mitochondrial physiology and membrane function in general. One of the most important features of a good theory is its ability to stimulate experiments directed to gather evidence either in its behalf or against it; all of these theories have accomplished this goal.

Chemical-coupling Hypothesis

The first proposal to explain the coupling of phosphorylation to oxidation was based upon the experience gained from research on other high-energy reactions. In 1953 E. C. Slater proposed that the energy released by the electron-transport chain was trapped in the formation of a high-energy chemical intermediate which could be used to drive the unfavorable phosphorylation of ADP. The theory was a logical extension of the work on the other means by which ATP is formed in the cell, i.e., substrate-level phosphorylation (page 168), whose mechanism had previously been unraveled. Consider the substrate-level phosphorylation of ADP during the oxidation of glyceraldehyde 3-phosphate to 3-phosphoglycerate (see Fig. 4-28). The first step is the removal of the electrons from the substrate and the formation of a nonphosphorylated, high-energy intermediate, the acyl-enzyme complex. The term *nonphosphorylated* refers to the 1 carbon position and ignores the phosphate at the 3 carbon position which is not involved in this reaction. The second step in the reaction is the formation of a second high-energy intermediate by the transfer of the acyl group to a phosphate to form 1,3-diphosphoglycerate. The formation of this phosphorylated, high-energy intermediate sets the stage for the formation of ATP since the phosphate now present at the 1 carbon has a high transfer potential. In the last step the phosphate at the 1 position is transferred to the acceptor, ADP, to form the final product.

The chemical-coupling hypothesis involves the formation of common intermediates between the electron-transport carriers and ATP in a manner similar to that just described for substrate-level phosphorylation. The chemical-coupling theory has been presented in several ways but the most common series of proposed reactions is as follows:

$$1. \quad AH_2 + B + I \rightleftharpoons A \sim I + BH_2$$
$$2. \quad A \sim I + A \rightleftharpoons X \sim I + A$$
$$3. \quad X \sim I +_i \rightleftharpoons X \sim P + I$$
$$4. \quad X \sim P + ADP \rightleftharpoons ATP + X$$

In the first reaction, A and B are successive electron carriers and I is some nonphosphorylated group. In this reaction, electrons (and protons) are being passed from A to B with the simultaneous formation of the high-energy ($\sim$) intermediate A $\sim$ I, which is analogous to the acyl-enzyme of the 3-phosphoglyceraldehyde oxidation. Since there are three different sites of phosphorylation, it would be expected that there would be three different intermediates (A $\sim$ I and two others) representing the three different carriers involved. In the second reaction, these three different intermediates would transfer their high-energy group ($\sim$ I) to form some common high-energy nonphosphorylated intermediate (X ~ I), regenerating the electron-transport carriers in the process. The third reaction converts the nonphosphorylated intermediate (X ~ I) to a phosphorylated one (X ~ P) which can act as a common intermediate in ATP formation in a manner analogous to 1,3-diphosphoglycerate. The fourth reaction shows the transfer of the phosphate group to ADP.

Despite the original likelihood of a mechanism involving a chemical intermediate, the theory is no longer held in much favor. There are several reasons for its general abandonment, the primary one being the lack of evidence for the existence of any of the proposed high-energy intermediates. Despite over 20 years of search, including the investigation of several unconfirmed claims, no intermediates have been found. However, negative evidence such as this is always unconvincing, particularly since such intermediates might be expected in very small amounts and possibly present within the membrane. Additional evidence against the theory will be presented below.

Chemiosmotic-coupling Hypothesis

In 1961 a radically different hypothesis was proposed by Peter Mitchell to explain the coupling of phosphorylation to electron transport. In essence Mitchell proposed that the free-energy released by electron transport was enlisted to produce an ionic gradient across the inner membrane. The free energy stored in this gradient would then be used to drive the phosphorylation reaction. In other words, in the chemiosmotic theory the membrane of the mitochondrion acts as an energy transducer, converting the electrochemical energy of a gradient into the chemical energy of ATP. The theory is complex and remains very controversial though many mitochondriacs believe that the major proof for the theory has been obtained. This belief is reflected in the award of the Nobel prize to Mitchell in October 1978. We will divide the topic into two parts: first, the proposed mechanism and, second, the predictions that have been made together with some of the evidence that has been gathered.

The details of the theory have undergone some revision since its original proposal in 1961, but the basic elements have remained unchanged. In the first part of the theory, Mitchell has proposed that the movement of electrons down the transport chain results in the formation of a *proton* gradient. The gradient is formed because protons are picked up on one side of the membrane and dumped off on the other side. If the nature of the electron carriers is considered, it is apparent that there are two types: one that transports both protons and electrons and one that transports only electrons. The establishment of the proton gradient would be accomplished by the absorption of protons on one side of the membrane by carriers that transported both protons and electrons, and the deposition of the protons on the other side when the electrons are passed to carriers that do not accept the accompanying protons. Figure 8-18 shows a hypothetical arrangement of carriers to accomplish this condition. Most importantly, whereas the electrons from NADH pass along the entire transport chain, the protons do not (Fig. 8-10). For example, the two

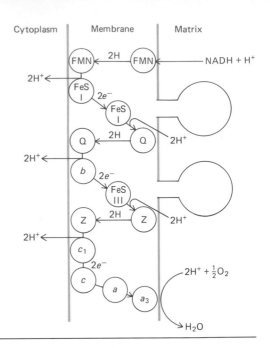

Figure 8-18. Pathway of proton and electron transfer according to the chemiosmotic hypothesis. Q, coenzyme Q,; Z, hypothetical hydrogen carrier; FeS, nonheme iron proteins; *b, c, c₁, a, a₃* are cytochromes. In this diagram there are three loops for the movement of protons across the membrane. The carriers FMN, Q, and Z are each reduced by combining with a pair of electrons and protons to form FMNH₂, QH₂, and ZH₂. The pair of hydrogen atoms (2H) are then moved across the membrane after which the electrons are passed to the next carrier in line while the protons are forced out into the cytoplasm, a region of relatively high proton concentration. The release of protons into the cytoplasm is driven by the voltage difference ($-\Delta G$) between the two adjacent carriers. The electrons are then passed back across the membrane to a hydrogen atom carrier (or to molecular oxygen) at the matrix side. It should be noted that there is no evidence to indicate that the hydrogen atom carriers FMN, Q, or the hypothetical Z actually move through the membrane. Rather the hydrogen atoms may be passed between carriers as they move across the membrane. (*From F. M. Harold,* Bacteriol. Rev., **36**:177, 1972.)

protons of FMNH₂ in Fig. 8-18 are not transferred to the iron-sulfur (FeS) protein (not shown in Fig. 8-18); only the electrons are transferred while the protons are discarded on the *outer* side of the membrane.

The formation of an excess of protons on one side of the membrane (the cytoplasmic side), relative to their concentration on the other (matrix) side, constitutes a storage of free energy that can be tapped to do work. The same principle applies when the storage of free energy in the sodium gradient across the membrane of the intestinal epithelial cell (page 261) accomplishes the uphill movement of sugars into those cells. In the present case, the gradient is an electrochemical one and as such there are two components to consider. One of the components is the concentration difference between the hydrogen ions on one side of the membrane versus the other; this is a pH gradient (ΔpH). The other component is the electrical one resulting from the separation of charge across the membrane; this is an electrical gradient ($\Delta\psi$), i.e., a voltage. The energy present in both components of the electrochemical gradient is termed the *protonmotive force* (ΔP). The steeper the gradient, the greater the protonmotive force, and the greater the free energy to be utilized.

$$\Delta P = \Delta \psi + 2.3 \frac{RT}{F} \Delta \mathrm{pH}$$

Up to this point we have established that the purpose of electron transport in the chemiosmotic theory is to generate an electrochemical gradient which can be considered as a high-energy *state,* just as the previous theory proposed the existence of a high-energy chemical intermediate. In the chemiosmotic theory the free energy released by electron transport is used to pump protons into a region in which they are already present at

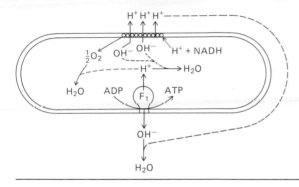

Figure 8-19. Diagrammatic illustration of that part of the chemiosmotic theory concerned with the formation of ATP. The movement of protons to the cytoplasm by electron transport establishes a proton gradient. In contrast, the spatial organization of the active site of the F_1 sphere is such that protons formed during ATP synthesis are routed into the matrix (a region of high OH^- concentration) and hydroxyl ions are routed into the cytoplasm (a region of high H^+ concentration). Consequently, the gradients formed during electron transport can be used to drive the reaction catalyzed by the F_1 sphere in the direction of the formation of ATP.

high concentration, i.e., against a gradient. It was mentioned in an earlier section that there are three places within the electron transport chain where the transfer of electrons is accompanied by a major release of free energy. These have been designated sites I, II, and III of phosphorylation. In the chemiosmotic theory these sites are only indirectly linked to phosphorylation. Rather, they represent the events of electron transport whose energy release drives the movement of protons out of the membrane. The greater the drop in voltage between two carriers, the more negative the ΔG and the farther the reaction can be driven toward the formation of products. In other words, the relative attraction for electrons by a subsequent carrier (e.g., cytochrome b) is sufficiently strong as to remove the electrons from the previous carrier (e.g., $CoQH_2$) despite the fact that the associated protons must be forced out into a region of very high proton concentration. The second part of the theory is concerned with translating the high-energy state into a driving force for phosphorylation.

Consider the reaction whereby ATP formation occurs:

$$ADP + P_i \rightleftharpoons ATP + H_2O$$

An important component in this reaction is the molecule water; its concentration is of the utmost importance in determining the ΔG of the reaction at any given moment. We are used to thinking of cellular reactions occurring in an aqueous environment in which the concentration of water is approximately 55 M. However, within various hydrophobic environments, such as within the core of a globular protein or within a membrane, its concentration may be very low. Since it can be difficult to determine the concentration of water in a particular microenvironment within the cell, it is difficult to be certain of how endergonic the formation of ATP might actually be. This does not mean that ATP can be formed from ADP without the expenditure of energy, rather it indicates that the energy may not be required for the actual phosphorylation itself. The actual endergonic steps may be located somewhere else in the mechanism, for example, in the movement of the hydrophilic reactants into the hydrophobic environment, or the release of the product from the enzyme. Regardless of the thermodynamics of phosphorylation, one of the components of the chemiosmotic theory is the limited access by water to the active site of the ATP synthetase.

In order to understand how a proton gradient can drive the phosphorylation reaction, we need to consider the reaction once again. The reaction is essentially one of dehydration and it has been shown using radioisotopes that the OH^- is removed from the phosphate and the H^+ is removed from the ADP. In other words, the reaction is most properly written

$$ADP + P_i \rightleftharpoons ATP + OH^- + H^+$$

Since the formation of ATP is accompanied by the formation of a proton and a hydroxyl ion, the opportunity exists to pull the reaction in the direction of ATP formation by removing the proton and hydroxyl ion formed. It is proposed by the chemiosmotic theory that the hydroxyl ion formed by the reaction is displaced to the cytoplasmic side of the membrane where it can react with the excess protons in that location to form water (Fig. 8-19). Conversely, the proton formed by the reaction would be displaced to the matrix side containing the lowered proton (and thus elevated hydroxyl) concentration. In this way the proton gradient established by electron transport would drive phosphorylation, and the gradient would be neutralized in the process. Continuing electron transport would be required to maintain the proton gradient.

A number of additional pages will be devoted to a discussion of the chemiosmotic theory; the reasons for this extensive treatment include the following. The theory is one of the most important and provocative to emerge in cell and molecular biology in the past few decades and it appears to explain many of the mysteries of a critically important biological process. Its importance goes beyond mitochondria, extending to chloroplast and bacterial membrane energetics as well. Equally important for our purposes, it provides a good example of the importance of a theory in making predictions that can be tested.

Predictions and Evidence

The chemiosmotic theory is relatively complex and demands that a number of conditions within the mitochondrion be met. Since it makes specific predictions about the physiological activities of the inner membrane, the theory can be put to the test in a number of ways. Since its proposal in 1961, a substantial body of evidence has accumulated in support of the chemiosmotic theory, though disagreement exists over some of the data and controversy over its implications. In this section, the predictions made by the theory will be presented as well as some of the evidence in its behalf.

1. One of the most basic predictions is that *electron transport should be accomplished by the ejection of protons from the cytoplasmic side of the inner membrane* (Figs. 8-19, 8-20). When isolated mitochondria are actively engaged in the oxidation of substrates, the medium in which they are suspended becomes acidic, as expected from the theory. Similarly, when inside-out submitochondrial particles are engaged in oxidative metabolism, the interior of the vesicles becomes acidic relative to the surrounding medium, as expected. The ejection of protons by mitochondria during their activity is an accepted fact, but the number of protons ejected is a controversial subject. In its strict sense, the theory predicts that two protons should be ejected per phosphoryla-

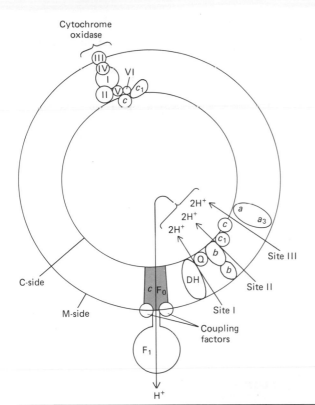

Figure 8-20. The topography of a submitochondrial particle. The location of the various components of the respiratory chain within the membrane is determined by using various probes, particularly antibodies prepared against purified mitochondrial proteins. *(From E. Racker, Biochem. Soc. Trans., 3:790, 1975.)*

tion site. If this were the case, six protons should be ejected per molecule of NADH oxidized and four protons per molecule of $FADH_2$. Numerous measurements have been made and numbers such as these have been reported by several laboratories. However, other investigations, particularly some recent ones, indicate that there may be three or four protons released per site. If there are more than two protons ejected per site, this would add definite complications to the theory, though would not invalidate it.

2. A second basic prediction of the chemiosmotic theory is that *in order for protons to be picked up on the inner side of the membrane and released on the outer side, the electron carriers must be organized asymetrically within the membrane* (Fig. 8-18). Certain carriers that bound both protons and electrons would be present on the inner side so as to remove protons from the matrix. In order to release these protons on the other side, the electrons must be passed to carriers on the outer side of the membrane that will not accept the protons. As long as the proper organization existed within the membrane, electrons could be passed along the entire chain to oxygen, while protons would simply be shuttled (*vectorially translocated*) across the membrane. Each passage of an electron from a carrier on the inner side, to one on the outer side, and back to the inner side again would constitute a "loop." The chemiosmotic theory predicts that *there should be three such loops* (as shown in Fig. 8-18), one per phosphorylation site.

At the present time there is considerable evidence that the components of the electron-transport chain and the ATP synthetase are organized asymmetrically within the membrane. The evidence for asymmetry comes from studies in which either intact mitochondria or the inside-out submitochondrial particles are treated with agents that attach to specific components they can reach. The basic approach is similar to that described in the studies of the asymmetric localization of proteins of the plasma membrane (Chapter 5). The agents used are ones that cannot penetrate the inner membrane, for example, antibodies specific for particular cytochromes or the F_1. The results show, for example, that cytochrome c is on the outer surface of the membrane, that the F_1 complex is on the inner surface (as is apparent from the electron micrographs), and that cytochrome oxidase spans the membrane. The apparent location of the best-studied carriers is shown in Fig. 8-20. So far the results indicate that asymmetry, which is required by the theory, does exist, but there is as yet no evidence for three successive loops correlated with the three successive sites of phosphorylation. The matter is complicated by the presence of so many oxidation-reduction centers within the membrane whose precise function is not yet understood.

3. Another basic prediction of the chemiosmotic theory is that *the membrane be capable of establishing a proton gradient*. There are several aspects to this prediction. (a) If the membrane is to hold a gradient, it should be impermeable to protons, which has been shown to be the case. This is not unexpected since even artificial lipid bilayers are extremely resistant to the passage of protons. (b) Not only does the inner membrane have to be impermeable to protons, but it must also block the diffusion of other ions across the membrane which would lead to the collapse of the voltage component of the proton gradient. In other words, if the pump is to be electrogenic (voltage producing, page 259), the ejection of protons to the outside of the membrane cannot be accompanied by the movement of anions in the same direction or the movement of some other cation in the opposite direction. Either of these shifts in charged species would tend to neutralize the separation of charge required to maintain the voltage. The evidence suggests that the membrane is very impermeable to all charged species. (c) Since the maintenance of an ion gradient requires an intact membrane surrounding an enclosed space, it would be expected that any damage to the membrane would destroy the gradient and terminate oxidative phosphorylation. Though claims have been made for oxidative phosphorylation in the absence of an intact membrane, it is generally accepted that this does not occur and that damage to the membrane abolishes the capacity for coupled phosphorylation. (d) If the membrane were to become permeable to protons, the electrochemical gradient would collapse and oxidative phosphorylation would cease. There appears to be a class of inhibitors, called *uncouplers,* whose precise action is to cause the membrane to become permeable to protons. Inhibitors of this type are termed uncouplers because they detach the oxidative end of the process from the phosphorylation end. In the presence of an uncoupler, such as 2,4-dinitrophenol, respiration continues, i.e., substrate is oxidized at the expense of molecular oxygen, but no ATP is produced. It appears that the protons ejected from the membrane by the action of electron transport simply diffuse back through the membrane as a consequence of the uncoupler-induced permeability.

4. The preceding predictions together lead to a fourth prediction, one of the basic tenets of the theory, namely that *a difference in the concentration of protons occurs across the membrane*. As discussed above, the theory predicts that the proton gradient manifests itself as both a pH gradient and a voltage. Both of these have been measured, but a major controversy exists over the magnitude of the gradients and, equally as important, just how steep a gradient would have to exist to be able to drive the phosphorylation. It

is generally believed that the formation of a mole of ATP under the conditions existing within the cell requires up to 17 kcal. If this is an accurate measure, and there is disagreement on the value, then it is estimated that, under physiological conditions, the electrochemical gradient would have to be quite substantial, well over 200 millivolts. Measurements by Mitchell's laboratory, and others, suggest that a potential difference of this magnitude does exist across the inner membrane, while others find much lower values. Mitchell estimates that approximately 80% of the free energy of the gradient is represented by the voltage component of the proton gradient and the other 20% by the concentration difference (approximately one pH unit difference). Since disagreement exists both on the energy required for phosphorylation in the mitochondrion as well as the energy generated by the proton gradient, it is difficult at the present time to decide if the chemiosmotic theory can account for coupled phosphorylation in the cell.

5. Another prediction of the chemiosmotic theory concerns *the ATP-synthesizing machinery, which must be oriented in such a manner as to displace the protons formed upon ATP synthesis to the inside of the membrane and the hydroxyl ions formed to the outside of the membrane—the opposite from that produced by the electron-transport chain.* As a consequence of the asymmetry of the ATP synthetase, the proton gradient produced by electron transport can serve to neutralize the products of ATP formation and pull the reaction forward. Verification of the asymmetry of the active site of the ATP synthetase has been accomplished from the analysis of the reverse reaction. If mitochondria or submitochondrial particles are supplied with added ATP in the absence of oxygen, the ATP synthetase will act as an ATPase and hydrolyze the ATP. Just as protons and hydroxyl ions are products of ATP synthesis, they must be reactants for ATP hydrolysis. Therefore, if protons and hydroxyl ions are selectively released to opposite sides of the membrane during ATP synthesis, they should be selectively absorbed from those respective sides during ATP hydrolysis. In other words, if protons are normally released to the matrix when ATP is being synthesized, protons should be absorbed from the matrix during ATP hydrolysis. Under conditions of ATP hydrolysis, the matrix should become more alkaline while the surroundings should become more acidic. When inside-out submitochondrial particles are used, the reverse changes should occur. Measurements on both intact mitochondria and submitochondrial particles have confirmed these expectations.

6. At the heart of the chemiosmotic theory is the proton gradient: its formation by electron transport and its use during phosphorylation. Based on the theory, *one would expect that if an* artificial *proton gradient could be imposed across the membrane, one should observe the formation of ATP in the absence of electron transport.* This is a rather daring prediction, seemingly one that places a proponent of the theory far out on the limb. This experiment has been performed and once again the predictions are confirmed. How does one go about constructing an artificial proton gradient across a membrane? The experiment was first performed successfully by André Jagendorf and coworkers using chloroplasts, whose function in phosphorylation is also covered by the chemiosmotic theory. The artificial proton gradient was imposed by rapidly changing the medium in which the chloroplasts were suspended (Fig. 8-21); a change to a medium of a different pH.

As described in the next chapter, chloroplasts are composed of membranous disks, termed *thylakoids*, which contain an ATP-synthesizing complex as well as an electron-transport assembly. Although the same types of ele-

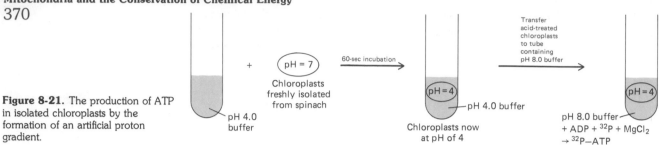

Figure 8-21. The production of ATP in isolated chloroplasts by the formation of an artificial proton gradient.

ments are present in the membranes of both chloroplasts and mitochondria, their orientation is reversed in the two organelles: the internal compartment of the thylakoid contains an elevated hydrogen ion concentration, while the F_1 spheres are located on the outer surface of the membrane (see Fig. 9-26). In order to impose an artificial gradient across the membranes of the thylakoids, Jagendorf prepared isolated chloroplasts from spinach cells and suspended them in a tube containing a pH 4 buffer for about 60 seconds (Fig. 8-21), the time required for the protons of the medium to cross the membrane of the thylakoid so as to lower the pH within that space. After the interior chamber had achieved a pH of approximately 4, the acidified chloroplasts were injected into a second tube containing a buffer of pH 8, along with the substances necessary to make radioactive ATP. Jagendorf found that within a few seconds of the time that the chloroplasts were transferred to the alkaline medium, newly synthesized ATP could be detected. The results indicate that the pH gradient established when the chloroplasts were suspended in the alkaline buffer was capable of driving the phosphorylation of ADP. More recently, similar types of experiments have been performed with submitochondrial particles which also respond to artificially imposed gradients by phosphorylating ADP.

The results of these experiments are important in a number of ways. Not only do they demonstrate that an electrochemical potential can provide the free energy for ATP formation, but they indicate that electron transport and ATP formation need not be *directly* coupled. In this experiment (and one described on page 375), no electron transport is taking place, therefore one can conclude that phosphorylation can occur independently from electron transport. Electron transport normally generates a high-energy state which is used to drive phosphorylation, but there are no particular conditions of that high-energy state that cannot be met by an artificially imposed gradient.

Conclusion

It is generally agreed by mitochondrial physiologists that the inner membrane does generate a proton gradient and that the ATP-synthesizing machinery *can* use this gradient to form ATP. Even if one accepts all of the evidence presented in behalf of the chemiosmotic theory, there remain several primary reservations. One is that the electrochemical gradient may be a prerequisite for phosphorylation without being the direct force that drives the reaction. The gradient may be simply one of the steps of the coupling mechanism, but not necessarily the last step. In other words, that part of the chemiosmotic theory dealing with the generation of the gradient may be correct, but those points that are concerned with the formation of ATP may be incorrect. There is much more controversy over the precise means by which the ATP synthetase is able to phosphorylate

ADP than over the existence of a protonmotive force. The other reservation is that the electrochemical gradient may exist within the mitochondria for some other function beside phosphorylation, though it can be used for this purpose when electron transport is lacking and the proton gradient is artificially imposed. At first glance this latter reservation may seem far-fetched; however, the proton gradient does appear to provide the energy for certain other functions in the mitochondrion (described later in this chapter) and it is possible that some other high-energy intermediate is normally responsible for phosphorylation.

Conformational-coupling Hypothesis

In its original state, the conformational theory of oxidative phosphorylation suggested that the transport of electrons by the carriers caused conformational changes within the proteins of the transport chain. These structural modifications would then be passed on to the ATP-synthesizing apparatus, changing its conformation and thereby favoring the phosphorylation reaction. In other words, the energy released by electron transport would be passed on to the ATP synthetase, converting it to a reactive high-energy state. The condition would be somewhat analogous to using energy in pulling back on a bowstring: when the string is released, the energy is directed into propelling the arrow. In the conformational theory, the energy released by electron transport is stored as a displacement in the position of various weak bonds within the enzyme. The release of this stored energy would accompany the phosphorylation.

Two types of conformational changes have been observed to occur within mitochondria. One of these is the major change in the structure of the mitochondria between the inactive and active state (Fig. 8-22). This is certainly a major reorganization of the conformation of the entire organelle, but it is believed to occur more slowly than would be expected by the conformational-coupling theory and is more likely involved in the control of mitochondrial activation and inactivation. In other words, the ultrastructural differences seen in a mitochondrion when it is under anaerobic as compared to aerobic conditions may reflect a mechanism to switch respiration on or off in response to cellular needs. The other type of conformational change that has been recorded occurs within the various proteins of the inner membrane as they carry out their activities. For example, the fluorescent compound aurovertin, which specifically binds to the F_1 sphere, undergoes an increase in fluorescence during the phosphorylation reaction. This type of observation is indicative of a structural change occurring within the enzyme during catalysis. This is the type of conformational change that is predicted by the theory, but there is no evidence it is involved in the release of stored free energy. Conformational changes accompany many enzymatic reactions but they are not generally a means whereby exergonic reactions are coupled to endergonic ones. Most importantly, the demonstration that ATP synthesis can occur in the absence of electron transport (such as that illustrated in Figs. 8-21 and 8-23b) indicates that the phosphorylation reaction need not be driven by conformational changes passed on to the ATP synthetase from electron-transport proteins.

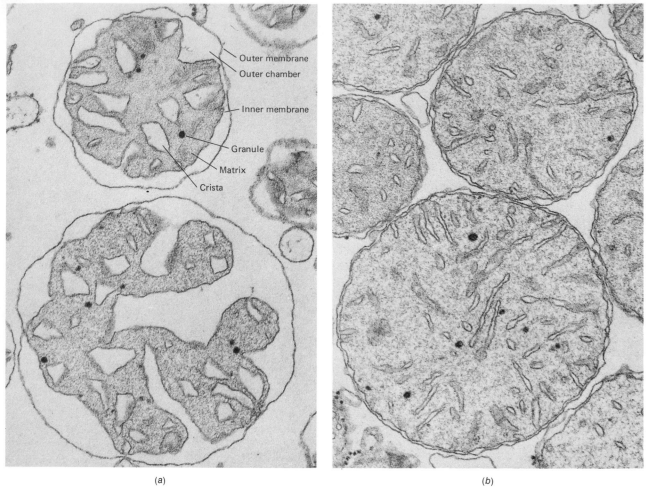

Outer membrane
Outer chamber
Inner membrane
Granule
Matrix
Crista

(a)

(b)

Figure 8-22. Rat liver mitochondria exist in two very different structural and functional states. In the condensed conformation *(a),* which is characteristic of mitochondria engaged in electron transport, the inner membrane is more randomly folded and the outer chamber, i.e., the space between the two membranes, is greatly enlarged. The matrix volume of condensed mitochondria is correspondingly reduced. In the orthodox conformation *(b),* which is characteristic of inactive mitochondria, the inner membrane is folded into cristae as shown in most electron micrographs. *(From C. R. Hackenbrock, J. Cell Biol., 37:345, 1968.)*

In the past few years the conformational-coupling theory has been restated in several different ways. In one form, proposed by Paul Boyer and E. C. Slater, conformational changes occurring in the ATP synthetase are not responsible for the formation of ATP but for its release from the enzyme. Though the space is not available to consider the complex biochemistry that has led to this proposal, briefly it suggests that the actual formation of ATP is not endergonic under the conditions present at the active site of the enzyme, but that the release of the newly formed ATP molecule from a tight-binding site on the enzyme is the energy-requiring step. In this revision of the conformational-coupling theory, no mention is made of how the energy from electron transport is delivered to the enzyme to cause the release of the ATP molecule and there is no reason to exclude the proton gradient as an intermediate in the process. In other words, electron transport might establish the electrochemical gradient which in turn would provide the energy for the necessary conformational changes within the enzyme. In fact, there is considerable support for the concept that protons move in a controlled manner from a region of high concentration outside the mitochondrial membrane *through* the ATP synthetase assembly into the matrix. As they moved through this specialized passage (in an otherwise impermeable membrane), they would effect conformational changes in the F_1 particle that would drive the phosphorylation of ADP. As stated before, the theories are not mutually exclusive and the mechanisms may be interrelated. A more complex proposal based on conformational changes has been proposed by David Green (see Green, 1974).

CONTROL OF RESPIRATORY ACTIVITY

We have considered the respiratory activities of the mitochondria in some detail but have ignored the means by which they are activated or suppressed. Since the primary function of the mitochondria is to provide ATP for the cell, it would be expected that ATP levels may be an important influence on mitochondrial activities at any given instant within a cell. It was pointed out in Chapter 4 that ATP levels play an important controlling role in glycolysis via the regulation of one of the key enzymes, phosphofructokinase. In a similar manner, a key enzyme of the TCA cycle, isocitrate dehydrogenase, is also modulated by the level of ATP. When the concentration of ATP is high, the activity of the enzyme is inhibited at an allosteric site by ATP. When ATP levels start to drop (and ADP levels start to rise), the activity of the enzyme is stimulated. However, in the mitochondria, ATP levels are only indirectly responsible for the control of respiration; the direct responsibility belongs to ADP, the molecule to be phosphorylated. When ADP levels are low, ATP levels are correspondingly high and there is no need for additional substrate to be wasted in order to provide electrons for transport. When ADP levels rise, an abrupt increase in oxygen consumption is noted along with a change in the organization of the mitochondria toward the condensed state (Fig. 8-22). Under normal conditions the level of ADP, the level of electron transport, and the level of phosphorylation are all tightly coupled to one another reflecting their common purpose. The result of this tight coupling is that under normal conditions electron transport never occurs in the absence of phosphorylation. The simplest way to uncouple the components is to add a substance, such as 2,4-dinitrophenol, which causes the membrane to be permeable to protons thereby destroying the high-energy state generated by electron transport. As a consequence of adding an uncoupler of this nature, phosphorylation cannot occur; ADP levels remain high and electron transport (and oxygen consumption) remains high in a vain attempt by the mitochondria to make ATP.

Since the inner membrane is a highly impermeable one, special transport mechanisms must be present for ADP and P_i to pass into the matrix and for ATP to pass out into the cytoplasm where it is needed. These movements are facilitated by special transport proteins present within the membrane. One of these proteins carries out an exchange of ATP and ADP in opposite directions across the membrane. Another important enzyme in the reactions of the adenine nucleotides (AMP, ADP, and ATP) is adenylate kinase, which is present in the space between the inner and outer membrane. This enzyme catalyzes the following reaction:

$$AMP + ATP \rightleftharpoons 2ADP$$

The reaction provides a means whereby AMP, a product of a number of very important reactions, can be recycled back into the energy metabolism of the cell via the formation of ADP which can then pass across the inner membrane to be phosphorylated.

THE ENERGIZED MEMBRANE

Certainly for our purposes the most important activity of the mitochondrion is the production of ATP, but there are other physiological activities occurring in this organelle beside oxidative phosphorylation that also require energy. One of these, for example, is the uptake of calcium ions. It is beyond the scope of this book to discuss the evidence, but it appears that these other energy-requiring activities are not necessarily coupled to ATP hydrolysis, but rather are driven directly by the free energy stored in the proton gradient. In other words, as a result of electron transport a high-energy state exists within the mitochondria in the form of a proton gradient which can be tapped to do work. That work can be the phosphorylation of ADP if the chemiosmotic theory is correct, or it can be the uptake of calcium ions or some other energy-requiring process. Surprisingly, if mitochondria are given a choice between Ca^{2+} to transport, or ADP to phosphorylate, they will preferentially accumulate the calcium over synthesizing ATP. Though the importance of this type of activity is not understood, it shows that, unlike other cell organelles, the mitochondria have another source of stored energy to call upon beside ATP, the energy currency for all of the other endergonic activities in the cell. This other source is the energized membrane, itself.

The fundamental importance of the energized state of a membrane is best seen in bacteria. Bacteria, being prokaryotic organisms, lack mitochondria and have their electron-transport activity and oxidative phosphorylation machinery associated with their plasma membrane, the only membrane they possess. The plasma membrane, in essence, substitutes for the inner membrane of the mitochondrion and, if mitochondria are actually distant relatives of bacteria, the two membranes would represent homologous structures. Regardless of the evolutionary relationship between the energized inner membrane of the mitochondria and the energized plasma membrane of the bacteria, it appears that both use electron transport to establish a proton gradient. In bacteria this gradient is of utmost importance to many of the activities of the membrane beside phosphorylation. Most particularly, numerous bacterial transport systems rely upon this gradient for their energy supply. It is believed that the difference in H^+ concentration across the bacterial membrane is used in cotransport mechanisms for the accumulation of other substances (ions, amino acids, some sugars) by the bacterium. Again, the process is similar to the use of the sodium gradient to drive sugar transport through intestinal epithelial cells (page 261). In both cases the transport proteins have binding sites for the energy-supplying ion (Na^+ or H^+) as well as the substance to be carried against the gradient. The downhill movement of the Na^+ or H^+ drives the uphill movement of the other molecule. The similarity in mechanisms between a plasma membrane and a mitochondrial membrane, of a bacterium and a higher animal, indicates its basic biological importance. The study of the bacterial system not only provides a simplified model for this type of membrane activity, it also provides a glimpse into what is probably an earlier evolutionary condition.

The Membrane of the Purple Bacterium

Before leaving the subject of proton gradients and phosphorylation, one last system will be described briefly for it provides some important additional evidence for the role of electrochemical gradients in phosphorylation as well as serving as a link with the next subject, photosynthesis. This last topic concerns recent work by Walter Stoeckenius and his colleagues on a bacterium of the genus *Halobacterium* that lives in extremely salty environments, such as that found in the Great Salt Lake. In fact, the bacteria will not survive at salt concentrations *below* about 3 *M*. When cultures of this bacterium are grown under anaerobic conditions, its plasma membrane is seen to have patches that are purple in color. The purple membrane results from the presence of one particular conjugated protein termed *bacteriorhodopsin*, which contains a prosthetic group (retinal) of identical structure to that of rhodopsin, the conjugated protein of the rods of the vertebrate retina. Bacteriorhodopsin is a transmembrane protein (Fig. 8-23a) making it ideally suited for its single-handed role in generating a proton gradient across the membrane. Just as in the case of the electron-transport chain, the purple protein absorbs protons from one side of the membrane and releases them on the other, i.e., translocates them, a process which can be detected by the acidification of the medium in which the bacteria are growing. One of the important clues in determining the function of the purple membrane came when it was shown that the acidification of the medium was light induced, i.e., it did not occur when

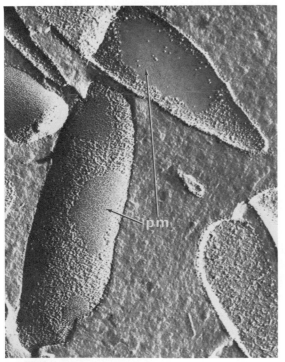

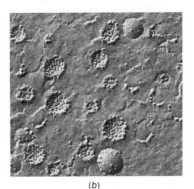

Figure 8-23. Freeze-etch replica of *Halobacterium* cells. The patches of purple membrane (pm) have a smooth appearance and contain bacteriorhodopsin. *(From W. Stoeckenius, in R. K. Clayton and W. R. Sistrom, eds., "The Photosynthetic Bacteria," Plenum, 1978.)* (b) Freeze-fracture replica of lipid vesicles reconstituted with bacteriorhodopsin. In the presence of light, the protein pumps hydrogen ions from the medium into the vesicle. *(From S.-B. Hwang and W. Stoeckenius, J. Memb. Biol., **33:**325, 1977.)*

(a)   (b)

the bacteria were kept in the dark. It was proposed that it is the light energy absorbed by the pigment of the protein which is used to drive the protons against the electrochemical gradient. The energy stored in the gradient is then used to drive the phosphorylation of ADP by a membrane-bound ATP-synthesizing system. *Halobacterium*, unlike most organisms, has two options available to ensure ATP synthesis. In the dark and the presence of oxygen, the bacteria can use electron transport and engage in oxidative phosphorylation. When the cells are growing anaerobically in the light, they can capture the energy of the light and engage in *photophosphorylation*. When neither oxygen or light is available, they must rely on substrate-level phosphorylation (page 165) and their ATP levels fall drastically.

In addition to serving as an unusual example of photophosphorylation, studies on *Halobacterium* have provided important supporting evidence for the role of the proton gradient in *mitochondrial* oxidative phosphorylation. The experiments have utilized the technique of constructing reconstituted phospholipid vesicles containing bacteriorhodopsin. When the purple protein was the only protein present in the vesicles (Fig. 8-23b), a light-dependent proton gradient could be detected across the artificial membranes. It was clearly established that the purple protein was a light-sensitive proton pump. It was soon found that reconstituted vesicles containing both the purple protein and the *mitochondrial* ATP synthetase (the oligomycin-sensitive ATPase) were capable of light-induced ATP formation. In other words, using the energy absorbed from light, the proton gradient generated by bacteriorhodopsin was capable of driving the phosphorylation of ADP utilizing the enzymatic machinery of the mitochondrion.

REFERENCES

Azzone, G. F., ed., symposium on "Biochemistry and Biophysics of Mitochondrial Membranes: Proceedings." Academic, 1972.

Boyer, P. D., chairman, Federation Proc. **34**, 1699–1722, 1975. "Biological Energy Transduction."

———, Fed. Europ. Biochem. Soc. **50**, 91–94, 1975. "Energy Transduction and Proton Translocation by Adenosine Triphosphatases."

———, Chance, B., Ernster, L., Mitchell, P., Racker, E., and Slater, E. C., Ann. Rev. Biochem. **46**, 955–1026, 1977. "Oxidative Phosphorylation and Photophosphorylation."

Chance, B., and Williams, G. R., Adv. in Enzymol. **17**, 65–134, 1956. "The Respiratory Chain and Oxidative Phosphorylation."

Ciba Foundation Symposium **31**, "Energy Transformation in Biological Systems." Elsevier, 1975.

Ernster, L., Estabrook, R. W., and Slater, E. C., eds., *Dynamics of Energy Transduction.* Elsevier, 1974.

Green, D. E., Biochim. Biophys. Acta *346*, 27–78, 1974. "The Electromechanicochemical Model for Energy Coupling in Mitochondria."

———, ed., Ann. N.Y. Acad. Sci. **227**, 1974. Symposium on "The Mechanism of Energy Transduction in Biological Systems."

Harold, F. M., Bacteriol. Rev. **36**, 172–230, 1972. "Transformation of Energy by Bacterial Membranes."

Henderson, R., Ann. Rev. Biophys. and Bioeng. **5**, 87–110, 1976. "The Purple Membrane from *Halobacterium Halobium.*"

———— and Unwin, P. N. T., Nature **257**, 28–32, 1975. "Three-Dimensional Model of Purple Membrane Obtained by Electron Microscopy."

Hinkle, P. C., and McCarty, R. E., Sci. Am. **238**, 104–123, March 1978. "How Cells Make ATP."

Kozlov, I. A., and Skulachev, V. P., Biochim. Biophys. Acta **463**, 29–89, 1977. "H+-Adenosine Triphosphatase and Membrane Energy Coupling."

Lehninger, A. L., *Bioenergetics*, 2d ed. W. A. Benjamin, 1971.

Luria, S. E., Sci. Am. **233**, 30–37, Dec. 1975. "Colicins and the Energetics of Cell Membranes."

Mitchell, P., Nature **191**, 144–148, 1961. "Coupling of Phosphorylation to Electron and Hydrogen Transfer by a Chemi-Osmotic Type of Mechanism."

————, Fed. Europ. Biochem. Soc. **43**, 189–194, 1975. "A Chemiosmotic Molecular Mechanism for Proton Translocating Adenosine Triphosphatases."

————, Biochem. Soc. Transact. **4**, 399–430, 1976. "Vectorial Chemistry and the Molecular Mechanics of Chemiosmotic Coupling: Power Transmission by Proticity."

Munn, E. A., *The Structure of Mitochondria*. Academic, 1975.

Palmer, J. M., Ann. Rev. Plant Phys. **27**, 133–157, 1976. "The Organization and Regulation of Electron Transport in Plant Mitochondria."

———— and Hall, D. O., Prog. Mol. Biol. **24**, 125–176, 1972. "The Mitochondrial Membrane System."

Papa, S., Biochim. Biophys. Acta **456**, 39–84, 1976. "Proton Translocation Reactions in the Respiratory Chains."

Racker, E., Biochem. Soc. Transact. **3**, 785–802, 1975. "Reconstitution, Mechanisms of Action, and Control of Ion Pumps."

————, *A New Look at Mechanisms in Bioenergetics*. Academic, 1976.

San Pietro, A. and Gest, H., eds., symposium on "Horizons in Bioenergetics: Proceedings," Academic, 1972.

Senior, A. E., Biochim. Biophys. Acta **301**, 249–277, 1973. "The Structure of Mitochondrial ATPase."

Simoni, R. D., and Postma, P. W., Ann. Rev. Biochem. **44**, 523–554, 1975. "The Energetics of Bacterial Active Transport."

Slater, E. C., Harvey Lectures **66**, 19–42, 1972. "The Mechanism of Energy Conservation in the Mitochondrial Respiratory Chain."

Stoeckenius, W., Sci. Am. **234**, 38–46, June 1976. "The Purple Membrane of Salt-Loving Bacteria."

———— and Oesterhelt, D., Federation Proc. **36**, 1797–1839, 1977. Symposium on "Light Energy Transduction by the Purple Membrane of Halophilic Bacteria."

Tedeschi, H., *Mitochondria: Structure, Biogenesis and Transducing Functions*. Springer-Verlag, 1976.

Thayer, W. S., and Hinkle, P. C., J. Biol. Chem. **250**, 5330–5335, 1975. "Synthesis of ATP by an Artificially Imposed Electrochemical Proton Gradient in Bovine Heart Submitochondrial Particles."

Tribe, M., and Whittaker, P., *Chloroplasts and Mitchondria*. Crane-Russak, 1972.

CHAPTER NINE
Photosynthesis and the Chloroplast

EVOLUTIONARY SPECULATIONS

In Chapters 4 and 8 we considered the processes whereby organic molecules are oxidized and dismembered. In this chapter we will start by turning to the beginning of the story and considering the means whereby these molecules came to be available in the first place. There are obvious difficulties in attempting to conceive of the conditions existing on earth several billion years ago when the first "living" structures must have arose. It is generally believed that "in the beginning" the earth was surrounded by a reducing atmosphere containing such gases as hydrogen (H_2), ammonia (NH_3), and methane (CH_4), while in the seas there were water, minerals, and dissolved gases. As has been shown in closed-reaction vessels in the laboratory, these compounds are capable of reacting together under the influence of electrical energy (in the form of lightning) and electromagnetic radiation to form a variety of small, stable, organic molecules, including sugars, amino acids, and precursors of nitrogenous bases (page 446). We suspect, therefore, that the prebiotic period was marked by the very slow buildup of organic compounds that we recognize today as the building blocks of cellular macromolecules. Presumably, some of these lower weight compounds became associated with one another to form higher molecular weight polymers. Over extremely long periods, macro-molecular complexes arose that became organized into the earliest form of structure capable of self-replication and some primitive level of inde-pendence from its environment. Presumably, some primitive type of metabolism evolved whereby these structures could take advantage of substances present in their environment for use in various ways inside the organism.

Heterotrophy versus Autotrophy

In the discussion of thermodynamics in Chapter 4, living systems were characterized as representing a bastion of negative entropy, i.e., a state in which organization and complexity was maintained by increasing the entropy of the chemicals in the environment. If there is any feature of the earliest life forms that is most assured, it is that they must have ob-tained their substance and energy from the organic materials in their environment—materials that came to be present as a result of the chemical reactions occurring in the primeval seas under the influences of the prevailing conditions. In other words, just as we survive by assimilating nutrients from out environment, so too must have the original life forms. Organisms that must depend upon an external source of organic com-pounds are termed *heterotrophs*.

Just as certain as the proposition that the first organisms were hetero-trophs is the proposition that such a means of livelihood, on a global scale, is a very limited one. As long as all living organisms were dependent upon nonliving reactions to provide them with their organic needs, the oppor-tunities for the increase in biomass on Earth was severely restricted, since the spontaneous production of these molecules does not occur rapidly. Somewhere along the way, a new type of life style had to evolve, one in which organisms could survive on molecules that were much more abun-dant on earth than the sparsely present, complex organic variety. This new type of organism became capable of manufacturing its own reduced organic nutrients from the most simple types of molecules, such as carbon dioxide and water or hydrogen sulfide. Organisms capable of surviving on CO as a principal carbon source are termed *autotrophs*. The appear-ance of autotrophs meant the availability of more nutrients for heterotrophs and the opportunity for much greater evolutionary diversification. Just as the breakdown of complex molecules into carbon dioxide and water is accompanied by the release of relatively large amounts of energy, the manufacture of more complex molecules from carbon dioxide requires the input of comparable amounts of energy. During the course of evolution, two main types of autotrophs have evolved which are distinguished from each other by their source of energy (Table 9-1). *Chemoautotrophs* are capable of using the energy stored in inorganic molecules and *photo-autotrophs* of using the radiant energy from the sun.

Leaving evolutionary considerations behind until the last section of this chapter, we can briefly survey the types of organisms that exist today with regard to their source of matter and energy. All chemoautotrophs are bacteria, though the particular energy source used to convert CO_2 into

TABLE 9-1
Major Nutritional Types of Bacteria

Type	Source of energy for growth	Source of carbon for growth	Example of genus
Phototroph			
Photoautotroph	Light	CO_2	*Chromatium*
Photoheterotroph	Light	Organic compound	*Rhodopseudo-monas*
Chemotroph			
Chemoautotroph	Oxidation of inorganic compound	CO	*Thiobacillus*
Chemoheterotroph	Oxidation of organic compound	Organic compound	*Escherichia*

SOURCE: M. Pelczar et al., "Microbiology," 4th ed., McGraw-Hill, 1976.

organic nutrients can vary considerably. In all cases the energy of a reduced inorganic molecule is conserved during an oxidation process, just as heterotrophs conserve energy from the oxidation of organic molecules. Whereas CO_2 serves as the *carbon source,* the *energy source* for chemoautotrophs can be ammonia, nitrite, hydrogen gas, hydrogen sulfide, ferrous iron, etc. These remarkable bacteria are often capable of growing in strictly mineral media in the dark.

The other type of autotroph, the photoautotroph, includes all higher plants and eukaryotic algae, various flagellated protozoa, the prokaryotic blue-green algae, as well as a variety of other types of bacteria. Photoautotrophs are those organisms that carry out *photosynthesis,* i.e., the conversion of CO_2 to carbohydrate (or other metabolites), utilizing the energy of sunlight. In keeping with this definition, the purple bacterium *Halobacterium,* discussed at the end of Chapter 8, is not a photosynthetic organism, nor is it an autotroph. It uses light energy to establish a proton gradient and subsequently to form ATP, but it does not convert CO into organic compounds, i.e., "fix" CO. *Halobacterium* can be classified as a photoheterotroph (Table 9-1), indicating its use of sunlight for an energy source but its dependence upon organic compounds for its reducing power (page 176). In all heterotrophs and autotrophs, the energy from oxidation is stored in ATP and the reducing power stored in reduced pyridine nucleotides (NADH and NADPH). It is strikingly clear how little change has occurred with respect to the basic aspects of metabolism during the entire period of bioevolution on this planet.

In Chapter 4, we contrasted two types of thermodynamic domains, the closed system and the open system. In both cases energy can be exchanged between the system and its environment. However, in the closed system there is no exchange of matter between the system and its environment, while in the open system, such exchange is allowed. Whereas a living cell is an example of an open system, the earth for all practical purposes is an example of a closed one. In other words, all matter on earth must be reutilized in a cyclic manner, whereas the flow of energy is non-cyclic—energy is derived from the sun and lost to the remainder of the universe. During the passage of energy through the earth, a certain amount is conserved for a period of time (a considerable period in the case of fossil fuels) by the photosynthetic organisms. Due to the activity of photoautotrophs, a huge quantity of light energy, which would otherwise have passed through the system performing little work, is converted to chemical energy to be exploited by all organisms on earth.

Figure 9-1. Schematic illustration of the experiment by Priestley which revealed that plants can sustain the life of an animal present in a sealed chamber.

EARLY STUDIES AND THE PRINCIPLES OF PHOTOSYNTHESIS

As is the case for most phenomena, our understanding of photosynthesis has come about slowly and remains far from complete. Study in the field began in 1771, when, in one simple elegant experiment, the British clergyman Joseph Priestley found a fundamental relationship between plants and animals (Fig. 9-1) that had previously gone undiscovered. In Priestley's own words:

> I have been so happy as by accident to hit upon a method of restoring air which has been injured by the burning of candles and to have discovered at least one of the restoratives which Nature employs for this purpose. It is vegetation. One might have imagined that since common air is necessary to vegetable as well as to animal life, both plants and animals had affected it in the same manner; and I own that I had that expectation when I first put a sprig of mint into a glass jar standing inverted in a vessel of water; but when it had continued growing there for some months, I found that the air would neither extinguish a candle, nor was it at all inconvenient to a mouse which I put into it.

Priestley had discovered the production of oxygen by plants. The importance of light in the process was soon discovered by Jan Ingenhousz, a Dutch physician who came to London to spend the summer of 1779 and carried out some 500 experiments on the subject during the brief period and had written a book before the year's end. By 1850 the overall reaction of photosynthesis was known

$$CO + H\,O \xrightarrow[\text{chlorophyll}]{\text{light}} O_2 + \text{carbohydrate}$$

as well as the understanding that light energy had been converted to chemical energy in the process. Near the end of the nineteenth century, the site of photosynthesis within the cell was identified when it was observed that it is the chloroplasts from which oxygen was evolved. This was first shown by Theodor Engelmann who found that, when certain algal cells were illuminated, small active bacteria would collect outside of the cell near the site of the large ribbonlike chloroplast.

The Role of CO_2 and H_2O in Photosynthesis

A major advance in our understanding of the chemical reactions of photosynthesis came with a proposal by C. B. van Niel in the early 1930s.

The prevailing belief at the time held that the energy of light was used to cause a breakdown of carbon dioxide from which molecular oxygen was released and the carbon atom transferred to a molecule of H_2O. This scheme fits well with the overall equation of photosynthesis

$$CO_2 + H_2O \xrightarrow{\text{light}} (CH_2O) + O_2$$

where (CH_2O) represents a unit of carbohydrate of variable backbone length. In 1931 van Niel proposed an alternate scheme based upon his work with sulfur bacteria. It was conclusively demonstrated that these organisms were able to produce carbohydrate using the energy of light without the simultaneous production of molecular oxygen. During this process, hydrogen sulfide is oxidized to elemental sulfur. Here was an example of an entirely different photosynthetic reaction, one which utilized CO_2 but did not release O_2. It was proposed that the reaction for sulfur bacteria was

$$CO_2 + 2H_2S \xrightarrow{\text{light}} (CH_2O) + H_2O + 2S^0$$

Recognizing the basic similarity in the photosynthetic processes of all organisms, van Niel proposed a general reaction to include all of these activities.

$$CO_2 + 2H_2A \xrightarrow{\text{light}} (CH_2O) + H_2O + 2A$$

For the production of a hexose, such as glucose, the reaction would be

$$6CO_2 + 12H_2A \xrightarrow{\text{light}} C_6H_{12}O_6 + 6H_2O + 12A$$

In this scheme, H_2A is an electron donor (a reducing agent) and can be represented by H_2O or H_2S or some other reducing substance. In the plant cell, the formation of glucose would be

$$6CO_2 + 12H_2O \xrightarrow{\text{light}} C_6H_{12}O_6 + 6H_2O + 6O_2$$

In this reaction scheme, each molecule of oxygen is formed from the breakdown of two molecules of H_2O. The precise molecular mechanism by which this occurs is still not certain and will not concern us. Note that, if both atoms of molecular oxygen are derived from water, the conversion of CO_2 (with its C:O ratio of 1:2) to glucose (with its C:O ratio of 1:1) must be accompanied by the formation of water, which appears among the products of the above equation. The van Niel proposal placed photosynthesis in a different light; it becomes, in essence, the reverse of respiration. Whereas oxidation in the mitochondria reduces oxygen to form water, photosynthesis in the chloroplast oxidizes water to form oxygen. We will consider the obvious thermodynamic problems that arise from this reaction a little later.

The Hill Reaction

Experimental support for water as the source of molecular oxygen during photosynthesis in plants came in 1937 in a landmark experiment by Robert Hill, which opened the door to the study of the biochemistry of photosynthesis. One of the benefits of working on photosynthesis is the large size of the organelles in which it occurs. The isolation of chloroplasts is a very easy task, but, as will be pointed out, the early isolation procedures resulted in preparations of damaged chloroplasts with the consequent loss of particular components. One of the great advantages of using *isolated* chloroplasts in physiological studies is the separation of photosynthesis from aerobic respiration. When whole cells are employed, mitochondria are present and oxygen uptake during oxidative metabolism in the mitochondria confuses the determination of oxygen *production* by the chloroplast. Hill simply crushed leaves in a sucrose solution, strained the suspension through glass wool, and collected the chloroplasts which passed through this porous sieve. Hill showed that, even though these isolated chloroplasts were not capable of fixing CO_2 (because certain of the components had leaked out of the organelle), they were able to produce molecular oxygen. In the Hill experiment, oxygen production by isolated chloroplasts was found to *occur in the absence of CO_2* and was strictly dependent upon *illumination* and the presence of an *electron acceptor*. This experiment by Hill revealed several basic features of photosynthesis:

1. Since oxygen was being formed in the absence of carbon dioxide, the oxygen atoms could not be derived from that source, but must be provided by water as proposed by van Niel. This was confirmed a few years later when radioactive oxygen gas was recovered in experiments using radioactive water and nonradioactive carbon dioxide as reactants.

2. The reactions involving water must be separated from the reactions involving carbon dioxide since one can occur without the other.

3. According to the van Niel reaction, in order for an oxygen atom to be released from water, some electron acceptor (oxidizing agent) must be present to take the electrons.

$$2H_2O \longrightarrow 4H^+ + 4e^- + O_2$$

In the van Niel reaction CO_2 is the electron acceptor, and we now know that ultimately this is the case. However, there are many steps between the removal of electrons from water and the acceptance of those electrons by carbon dioxide. In order to remove the electrons from water, Hill had to provide some acceptor — he used ferric salts (ferric oxalate or ferricyanide). In the presence of these iron salts, isolated choloroplasts would carry out the photosynthetic production of oxygen and the simultaneous reduction of the acceptor to the ferrous state. The reaction was

$$4Fe^{3+} + 2H_2O \xrightarrow[\text{chloroplasts}]{\text{light}} 4Fe^{2+} + 4H^+ + O_2$$

A more general reaction (the Hill reaction) can be written

$$H_2O + A \xrightarrow[\text{chloroplasts}]{\text{light}} AH_2 + \tfrac{1}{2}O_2$$

where A can be any number of possible electron acceptors in the test tube and some other (yet to be described) acceptor in the cell. Since CO_2 was not necessary for

O_2 production, it was virtually eliminated as the direct acceptor of electrons in the chloroplast.

4. The Hill reaction brings the focus of photosynthesis upon the removal and acceptance of electrons, which, as we will see, is the essence of the phenomenon. In Chapter 8 the role of oxygen as the terminal electron acceptor was described. Oxygen is capable of this function because it is such a strong oxidizing agent, i.e., it is so electropositive. Electrons normally flow downhill, i.e., from electronegative to electropositive acceptors and one would expect that oxygen, with its high affinity for electrons, would not easily relinquish them to some other electron acceptor. Since the redox potential of the iron oxalate couple (Fe^{3+}–Fe^{2+}) of approximately 0.0 is not nearly as electropositive as the O_2–H_2O couple (approximately +0.8), under normal circumstances ferric ions would not be capable of attracting electrons from water. However, under conditions that exist in the chloroplast when illuminated, the reduction of the ferric salt goes nearly to completion. The focus of the study of photosynthesis (as reflected in the following pages) has been concerned with the mechanism by which light can cause electrons to flow from an electropositive couple (O_2–H_2O) to a much more electronegative couple (CO_2–CH_2O or Fe^{3+}–Fe^{2+}) against an electropotential gradient (Fig. 9-2).

The Formation of Reducing Power

One of the central questions following the Hill experiment concerned the nature of the electron acceptor in the chloroplast. Presumably, whatever the acceptor was found to be, it must be involved in the subsequent reduction of CO_2 to carbohydrate. By 1950 research on the oxidative breakdown of carbohydrates, amino acids, and fats and their synthesis in animal cells had revealed the most important aspects of cellular metabolism. One of the most apparent features of biochemistry was the use of pyridine nucleotides in oxidation-reduction reactions. NADH (see Fig. 4-27), then called DPNH) was known to be of particular importance in catabolic metabolism and NADPH (then called TPNH) of particular importance in anabolic metabolism. Since photosynthesis in plant cells involved the formation of carbohydrates similar to those made in the livers or other tissues of any mammal, it was suspected that the same type of reducing agent would be involved.

The formation of NADPH in photosynthesis was uncovered by three different laboratories in 1951. Working with isolated chloroplasts, they

Figure 9-2. The overall energetics of photosynthesis and aerobic respiration.

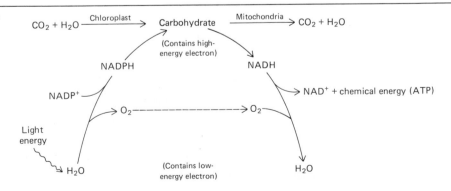

found that NADP$^+$ would serve as the acceptor of electrons from photosynthesis, i.e., would replace ferric salts in the Hill reaction. Subsequent experiments confirmed that NADP$^+$ is normally the stable intermediate in carrying electrons from water to carbon dioxide. The two phases of photosynthesis, the production of NADPH and its subsequent use in CO_2 reduction, are normally separated both spatially and temporally from one another. In 1954, it was shown by Daniel Arnon and coworkers that isolated chloroplasts were capable of the entire photosynthetic process from the release of O_2 and the formation of NADPH to the conversion of CO_2 to glucose. Overall, there is a flow of electrons during photosynthesis from water to pyridine nucleotide to carbohydrate, which is an exactly opposite flow to that which occurs during oxidation (Fig. 9-2). As will be shown below, the electrons removed from water are passed through many hands within the chloroplast before reaching carbon dioxide, their ultimate acceptor.

AN OVERVIEW OF THE MECHANISM OF PHOTOSYNTHESIS

Photosynthesis is a phenomenon that includes a large number of reactions, not all of which are known, and embodies several important concepts. Rather than discuss the components piece-by-piece, it seems advisable to first summarize the overall process to provide a framework for the subsequent description of individual reactions. The events of photosynthesis can be divided into three types of reactions: those involving photons of light, those involving transfer of electrons, and those involving chemical modification of metabolites. We will begin with the first of the three, the photochemical reactions.

The first step in photosynthesis is the absorption of a photon of light by pigment molecules in the chloroplast and the subsequent excitation of an electron. This is the mechanism by which light energy is trapped; it is used to boost an electron to a more energetic state, one from which the electron is readily lost. In other words, the energy of light is used to cause a molecule of chlorophyll of relatively poor electron-donating ability to become a molecule of relatively high electron-donating ability. Consider what happens when a pigment molecule (Y) loses an electron to an acceptor (X). X is termed the *primary acceptor* because it is the direct receiver of electrons from the excited pigment. Each of these molecules becomes charged, the pigment becomes Y$^+$ and the primary acceptor becomes X$^-$. The function of the electron donor (Y) and the electron acceptor (X) becomes more apparent when the charged state is considered. Y$^+$ is electron deficient and will seek electrons, i.e., it is an oxidizing agent. In contrast, X$^-$ has an extra electron that it will readily lose, i.e., it is a reducing agent. This event, i.e., the formation of an oxidizing agent and a reducing agent (which takes less than one-billionth of a second) is the essence of photosynthesis. The oxidizing agent formed by light absorption is sufficiently electropositive to extract electrons from water while the reducing agent simultaneously formed is sufficiently electronegative to be able to pass electrons to NADP$^+$.

We can put this event into more quantitative terms by considering the energetics of what we have just described. The O_2–H_2O couple has a redox potential of approximately +0.82 volt while the $NADP^+$–$NADPH$ couple is approximately −0.32 volt. The difference between these two values (approximately 1.2 volts) represents the minimal extent to which the energy of an electron must be raised. If the reactions are to proceed to a reasonable extent, we would expect the oxidizing agent to be considerably more electropositive than water and the reducing agent to be considerably more electronegative than $NADP^+$. If we assume there is approximately a 0.1 volt margin on each side of the scale (which may be quite an underestimate for the reducing agent), we would need an energy boost equivalent to 1.4 volts, which is considerably more energy than one photon (one quantum) of red light can deliver (page 388). It appears that the way the plant has gotten around this problem is to have two entirely different photochemical reactions involving two different and spatially separated photosystems. Each photosystem is responsible for boosting electrons part of the way up the energy hill (Fig. 9-3). One photosystem boosts electrons from below water at the bottom of the trough to a midway point, while the other photosystem raises electrons from a midway point to the top above $NADP^+$. The two photosystems are said to act in "series," i.e., one after the other (Fig. 9-4). In both of these photosystems (termed *PS II* and *PS I*) the energy of light is used to raise an electron to a higher energy state to be removed by an electron acceptor.

The absorption of light by PS II results in the formation of a very strong oxidizing agent, Y'^+, and a weak reducing agent, X'^- (Fig. 9-4). The oxidizing agent formed is able to remove electrons from water. In contrast, the absorption of light by PS I results in the formation of a strong reducing agent X^- and a weak oxidizing agent Y^+. The reducing agent formed is able to donate electrons to $NADP^+$. Each of these electrons shifts, from H_2O to Y'^+ and from X^- to $NADP^+$, proceeds by passage via electron carriers (see Fig. 9-11). Up to this point, we have the two ends of the process, electrons are taken from water and given to $NADP^+$. We need only con-

Figure 9-3. An overview of the energetics of photosynthesis as described in the text. An electron of low energy in water is boosted in two light-dependent steps to a level at which it is capable of reducing NAD^+. The characteristics of the two photosystems will be discussed later in the chapter. [*From "Molecular Physics in Photosynthesis" by Roderick K. Clayton, © Copyright, 1965, by Ginn and Company (Xerox Corporation). Used with permission.*]

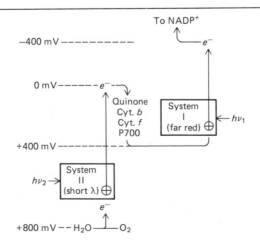

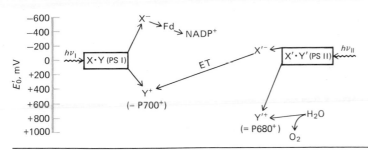

Figure 9-4. The two-photosystem scheme of green-plant photosynthesis. The generation of a primary oxidant and reductant during the photochemical reaction($h\nu_I$ and $h\nu_{II}$) in photosystem I (PS I) and photosystem II (PS II), respectively, is indicated. At left is the redox-potential scale. Fd, ferredoxin; ET, electron transport. *(From B. Ke. Biochim. Biophys. Acta, **301**:2, 1973.)*

nect the two photosystems to allow a flow of electrons from one end to the other. The two photosystems are connected by having the weak reducing agent from PS II (X'^-) reduce the weak oxidizing agent from PS I (Y^+) as shown in Fig. 9-4.

To summarize, the absorption of one quantum of energy by each of the two photosystems results in the formation of two high-energy electrons complexed to the two electron acceptors, X and X'. The high-energy electron from PS I is passed downhill to $NADP^+$ while the high-energy electron from PS II is passed downhill to fill the "hole" in Y^+ of PS I left by the loss of its electron. Meanwhile, electrons move from water to PS II to fill the hole left in its pigment molecule and the chain is complete. Once the primary electron acceptor of each photosystem has gained its electron, the need for illumination has ended. The remainder of the electron exchanges can occur in the dark, even though the term *light reactions* is often used to include the entire process up to the formation of NADPH.

In the preceding discussion, we placed the blame for the need of two photosystems on the excessive energy requirement in boosting an electron from its position in water to that in NADPH. Corroborating evidence that this is the primary reason comes from the analysis of bacterial systems which do not use water as the electron source and do not produce molecular oxygen. These bacteria are found to contain only one photosystem (analogous to PS I) because their photosynthesis is governed using an electron donor, such as H_2S, which is considerably more electronegative than H_2O. The voltage difference between H_2S and $NADP^+$ is only about 0.1 volt and is easily bridged with the energy obtained from the absorption of one quantum of light.

An important consequence of the transport of electrons downhill between the primary electron acceptor of PS II and the positively charged chlorophyll of PS I is the formation of ATP. Though this phosphorylation reaction, termed *photophosphorylation*, is not as well understood as oxidative phosphorylation in the mitochondria, the same principles apply and the same mechanism is believed responsible. The last step in photosynthesis utilizes the NADPH and the ATP formed in the early stages in the conversion of CO_2 to carbohydrate or some other organic molecule. We are now in a position to consider each of the stages in photosynthesis in more detail.

PHOTOCHEMISTRY

The Photoelectric Effect

The presence of various colored objects in our environment reflects the absorption by molecules of light of varying wavelengths. As is the case for all electromagnetic radiation, its energy content depends upon its wavelength according to the equation

$$E = h\nu = h\frac{c}{\lambda}$$

where h is Planck's constant (1.58×10^{-34} cal-s), ν is the frequency of the radiation, c is the speed of light in a vacuum, and λ is the wavelength. The longer the wavelength, the less is its energy content.

The process of photosynthesis centers on the absorption of light energy by the pigments of the chloroplast and its conversion to the kinetic energy of an electron. In order to better understand this process we need to consider briefly certain classical experiments in photoelectricity carried out near the turn of the century. The experiments are particularly useful for their clear-cut demonstration of the quantum properties of electromagnetic radiation. In 1888, after some preliminary observations, it was noted that the illumination of a zinc plate caused the metal to become positively charged. In 1900, three years after the discovery of the electron, the basis of the "photoelectric effect" was shown to be the result of the ejection of electrons from the metal as a result of illumination. A very important finding was soon made—the kinetic energy of the photoelectron is inversely related to the wavelength of the exciting radiation, while the *number* of photoelectrons ejected is directly related to the intensity of the beam. The explanation for these relationships was revealed by Albert Einstein when he proposed the dual nature of radiation (page 39) and explained the photoelectric effect as the absorption of the energy of one light particle, i.e., a quantum of energy contained in one photon.

The ejection of each electron results from the absorption of one quantum of energy resulting in the obliteration of that photon. When a photon is absorbed, its entire energy content must be accepted by the absorbing substance; photons are indivisible and parts of their energy cannot be absorbed. The more energetic the radiation absorbed, i.e., the *shorter* the wavelength, the greater is the energy (velocity) of the electron ejected. In contrast, the greater the intensity of the radiation, the greater the number of electrons released.

The Absorption of Light

What happens to an atom when a quantum of light energy is absorbed? The electrons surrounding the positively charged atomic nucleus exist in *discrete orbitals,* each with its own energy level. Those orbitals closer to the nucleus contain electrons of lower energy than orbitals farther away.

When a photon is absorbed, an electron becomes sufficiently energetic to be pushed from an inner to an outer orbital. The molecule is said to have shifted from the *ground state* to an *excited state*. The difference between the two states is termed the *energy of transition*. Since there are a limited number of orbitals within which an electron can exist, and each orbital has a specific energy level, it follows that any given atom can only absorb specific amounts of energy. Since the energy of the absorbed photon must exactly match the energy of transition between the two states, only specific wavelengths can be absorbed. A plot of the amount of absorption versus the wavelength of light provides an *absorption spectrum* for that substance. When one examines the absorption spectrum for individual atoms, one finds that the energy that can be accepted falls into very narrow ranges, or lines. However, when complex molecules such as chlorophyll are considered, the absorption bands are quite broad, having peaks of varying height (Fig. 9-5). The collection of bands forms a set of excited states that is characteristic for that particular molecule. Of the four excited states shown in Fig. 9-5, only the lowest state (that in the red) is of use in photosynthesis. When photons of higher energy are absorbed, the electron falls to the lower state (with the release of heat) before it can be used.

The excited state of a molecule is an unstable one which can be expected to last only about 10^{-9} seconds. There are several consequences that can be met by the excited electron depending upon the circumstances. If it drops back to its lower orbital, the energy it has absorbed must be released. The energy can be released in the form of heat or light (fluorescence or phosphorescence). In the case of a chlorophyll molecule within a chloroplast, if the excitation energy is released in the form of heat or light, the chlorophyll has returned to the original ground state and the energy of the absorbed photon has been wasted. This is precisely what is observed when a preparation of *isolated chlorophyll* in solution is illuminated—one finds that the solution is strongly fluorescent since the ab-

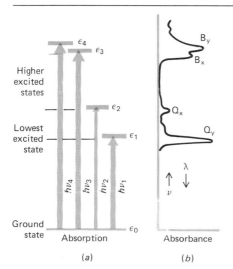

Figure 9-5. (a) Energy-level diagram, showing the spectral transitions (vertical arrows) for bacteriochlorophyll. The illustration applies qualitatively to chlorophyll a and b as well. The energy levels are broadened (shading) by vibrational sublevels that are not usually resolved in solution spectra. Electrons boosted to the higher energy levels fall to the lowest excited state before being passed to the primary acceptor. (b) Absorption spectrum corresponding to energy levels of (a). This spectrum is turned 90° from the usual orientation in order to show the relationship to the energy levels. (*From K. Sauer, in Govindjee, ed., "Bioenergetics of Photosynthesis," Academic, 1975.*)

sorbed energy is being reemitted at longer wavelength. However, if one does the same experiment on a preparation of *isolated chloroplasts* containing a comparable amount of chlorophyll, one finds only a very faint fluorescence indicating that very little of the absorbed energy is being dissipated. Instead, as previously described, the energetic electron is transferred to some acceptor before it has had a chance to drop back to fill the hole in its inner orbital. It is estimated that the transfer of the electron can occur within about 10^{-12} seconds.

PHOTOSYNTHETIC PIGMENTS

The green appearance of plant cells result from the presence within the chloroplasts of the pigment chlorophyll, which absorbs most strongly in the blue and red, leaving the intermediate green wavelengths to pass through to our eyes. The basic structure of chlorophyll is shown in Fig. 9-6. Each molecule consists of two main parts, a hydrophilic porphyrin ring and a hydrocarbon lipid-soluble chain. Unlike the red, iron-containing porphyrins of hemoglobin and myoglobin, that of the green chlorophyll contains magnesium, as well as other substitutions on the ring itself. Each porphyrin is composed of four pyrrole rings connected to one another by a one carbon link. The basis for the absorption of light by a chlorophyll molecule is seen from an analysis of its electronic structure. The presence of the alternating single and double bonds along the porphyrin ring forms a conjugated bond system characterized by the presence of a cloud of π electrons around the ring (Fig. 9-6). Conjugated systems of this type are

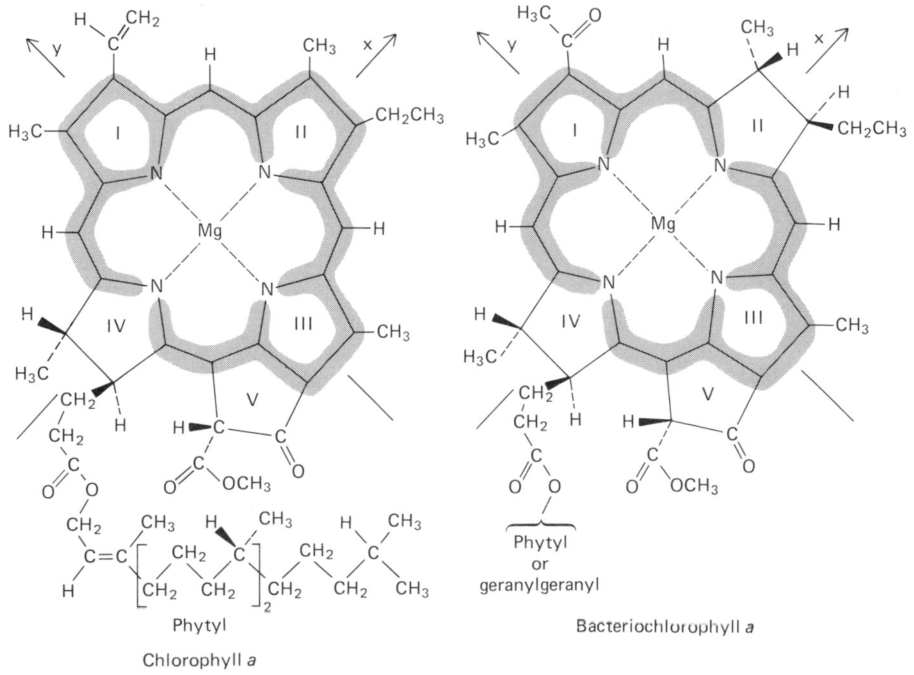

Figure 9-6. Molecular structures of chlorophyll *a* and bacteriochlorophyll *a*. The porphyrin π electrons are indicated by the shading. Chlorophyll *b* differs from chlorophyll *a* in the replacement of —CH₃ on ring II by —C. *(From K. Sauer, in Govindjee, ed., "Bioenergetics of Photosynthesis," Academic, 1975.)*

strongly absorbing, with the absorption causing a redistribution of the electron density of the molecule and, in the case of chlorophyll, the loss of an electron to an acceptor.

Several classes of chlorophyll, differing from one another in the chemical nature of the porphyrin ring, are known to occur among photosynthetic organisms. Chlorophyll *a* (Fig. 9-6) is present in all oxygen-producing photosynthetic organisms, but is absent in the various sulfur bacteria. In addition to chlorophyll *a*, chlorophyll *b* is present in all higher plants and green algae, while a third variety, chlorophyll *c*, is present in brown algae, diatoms, and certain protozoa. Prokaryotic organisms that do not evolve oxygen (all except blue-green algae) contain only one chlorophyll, termed *bacteriochlorophyll*, which is slightly different in structure from the above types (Fig. 9-6). Even though the chlorophylls are indispensable pigments in all photosynthetic cells, they are not the only pigments involved in photosynthesis. Two groups of accessory pigments, the carotenoids and the phycobilins, have also been implicated in light-absorbing reactions. Examples of these pigments are shown in Fig. 9-7. Carotenoids, which are yellow to orange in color, are widespread among photosynthetic organisms, while the phycobilins are restricted to the red and blue-green algae. The roles of these various light-absorbing molecules will be considered in the next section.

PHOTOSYNTHETIC UNITS

An important experiment carried out by Robert Emerson and William Arnold in 1932 suggested that not all of the chlorophyll molecules in the chloroplast were actively engaged in the conversion of light energy into chemical energy. Using flashing lights of increasing duration, they determined the amount of light required to saturate a given number of chlorophyll molecules, i.e., the minimal amount of light needed to produce maximal oxygen production. Based upon the number of chlorophyll molecules present in the preparation, they calculated that one molecule of oxygen

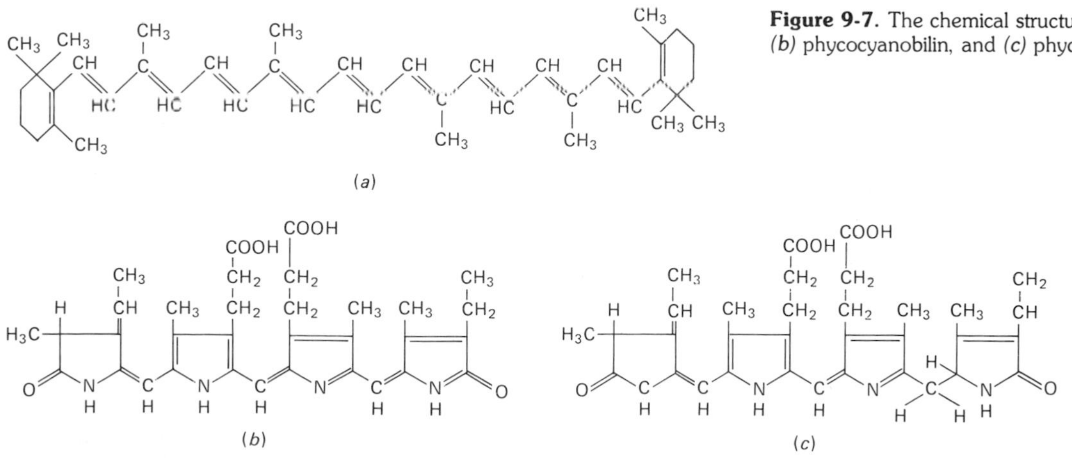

Figure 9-7. The chemical structures of (a) β-carotene, (b) phycocyanobilin, and (c) phycoerythrobilin.

was being released for approximately every 2500 molecules of chlorophyll present. Some factor was limiting the ability of the chloroplasts to respond to light. The results suggested that for each 2500 chlorophyll molecules present there was only one set of dark-reaction enzymes. Since it was later shown that a minimum of eight quanta of light are absorbed, i.e., that eight photoacts were required to produce one molecule of oxygen gas (page 399); the ratio of chlorophyll molecules present to the maximum number of quanta that could be absorbed at a given time was about 300:1. One possible interpretation of this ratio is that 299 of the 300 chlorophyll molecules are not involved in photosynthesis at all, but this is not the cause. Rather, the 300 or so chlorophyll molecules all act together as one *photosynthetic unit* in which only one member of the group is actually capable of transferring electrons to the primary electron acceptor. The small percentage of chlorophyll molecules capable of transferring the electrons are called *reaction centers*.

Why are there so few reaction centers and what is the role of all of the other chlorophyll molecules in the photosynthetic unit? We will begin to answer this question by considering the role of 99% of the pigment molecules and then return to the less than 1% acting as reaction centers. Even though the bulk of the pigment molecules are not directly involved in the conversion of light energy into chemical energy, they are responsible for light absorption; they are light "harvesters" collecting photons which are bombarding the cell. Consider the nature of a photosynthetic unit. Within the group are pigment molecules having a great variety of absorption spectra. There are two reasons for this variation. First, there are different types of pigments within each photosynthetic unit (Fig. 9-8): chlorophylls *a* and *b* and carotenoids in the plant cell. Each of these types has a considerably different absorption spectrum. Secondly, the absorption properties of pigment molecules are affected by the environment in which the particular molecule resides. If one examines the absorption properties of a purified preparation of chlorophyll *a* molecules, all of them would show exactly the same absorption profile; they all have exactly the same structure. However, within the chloroplast their maximal absorption occurs at a variety of different wavelengths (Fig. 9-8). Even though the pigment molecules are not covalently bound to other substances, they are tightly complexed to a variety of different proteins, and the nature of the association affects the nature of the excitation. In addition, these pigments are present within membranes of the chloroplast (page 413) and are therefore subject to a variety of environmental influences (fluidity, aggregation, etc.) that can also affect their interactions with a light beam. Presumably, the pigment molecules of the reaction centers exist within an environment which promotes their oxidation.

Since the light falling upon a leaf is composed of a great variety of wavelengths, the presence of this diverse group of receptor pigments ensures that a greater percentage of the incoming photons will have the opportunity to stimulate photosynthesis. This can be seen by examining an *action spectrum,* a plot of the efficiency of photosynthesis at all of the various wavelengths (Fig. 9-9*a*). The difference between an action spectrum and an absorption spectrum is an important one. Whereas the

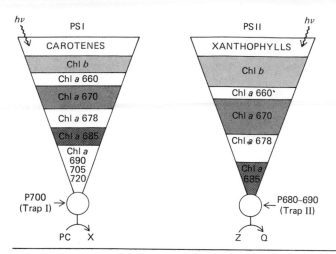

Figure 9-8. An hypothesis for the approximate distribution of the various pigments in the two pigment systems in green plants. PS I, pigment system I: PS II, pigment system II. The numbers indicate the approximate absorption maxima in the far red end of the spectrum. Z, the primary electron donor of PS II, i.e., that substance which passes electrons (originally obtained from water) to the electron-deficient reaction center of PS II: Q, the primary electron acceptor of PS II (referred to as X' in the text): PC, plastocyanin which may serve as the electron donor of PS I, i.e., that substance which passes electrons (originally obtained from the reaction center of PS II) to the electron-deficient reaction center of PS I; X, the primary electron acceptor of PS I. In diatoms and brown algae, chlorophyll c replaces chlorophyll b. In red and blue-green algae, phycobilins replace chlorophyll b. (*After Govindjee and R. Govindjee, in Govindjee, ed., "Bioenergetics of Photosynthesis," Academic, 1975.*)

absorption spectrum measures the wavelengths capable of providing the proper transition energy to a specific substance to raise electrons to higher orbitals, the action spectrum indicates the wavelengths that are actually effective in bringing about a particular physiological response. The action spectrum for photosynthesis, shown in the dotted line of Fig. 9-9*a*, indicates that a considerable fraction of the visible spectrum is capable of promoting oxygen production and therefore a variety of different pigment molecules must be responsible for absorbing light. In higher plants there is a deficiency in pigments which absorb light in the middle regions of the visible spectrum (Fig. 9-9*b*), and this is reflected in the drop in efficiency in this part of the action spectrum. In contrast, those species containing phycobilins, which absorb in the middle region, are more effective in utilizing light of the yellow and green wavelengths (Fig. 9-10).

The problem then is to transfer the energy from the light-gathering pigment molecules, which act like "antennae" in receiving light from the environment, to the pigment molecule at the reaction center which

Figure 9-9. (a) Correspondence between the wavelengths of light absorbed by cells, i.e., the absorption spectrum, of the green alga *Ulva* and the wavelengths of light which promote photosynthesis, i.e., the action spectrum. (*From F. T. Haxo and L. R. Blinks, J. Gen. Phys., **33**:404, 1950.*) (b) The absorption spectra of several isolated chlorophylls. [*From "Molecular Physics in Photosynthesis" by Roderick K. Clayton, © Copyright, 1965, by Ginn and Company (Xerox Corporation). Used with permission.*]

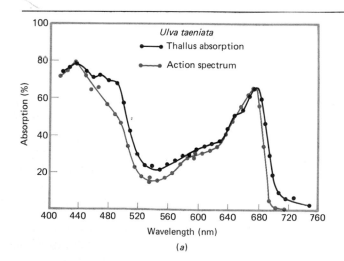

(a)

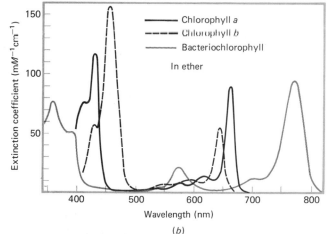

(b)

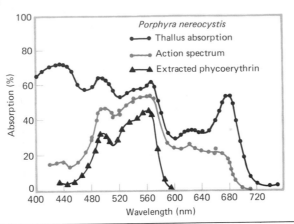

Figure 9-10. Absorption and action spectra of a red alga (in which phycoerythrin is the principal phycobilin pigment). The action spectrum corresponds more closely to the water extract of phycoerythrin than to the absorption curves of chlorophylls and carotenoids. *(From F. T. Haxo and L. R. Blinks,* J. Gen. Physiol., **33**:411, 1950.)

is involved in transferring the excited electron to the acceptor. The close presence of all of the pigments of the photosynthesic unit provides the means for the very rapid migration of the energy of excitation throughout the unit. The means by which excitation energy can be passed from one molecule to another is very complex and not fully understood. There are basically two types of transfers possible: one in which the electron of the receiving molecule requires the same energy as the donating molecule, and one in which the electron of the receiving molecule requires less energy. In the latter case, the receiving pigment molecule is one which absorbs light of longer wavelengths (less energetic), and the energy lost in the transfer is released as heat.

As the energy wanders randomly through a photosynthetic unit, each time it is transferred to a receptor absorbing at a longer wavelength, the nature of the subsequent transfers becomes restricted. The reason for this restriction is that the energy of an electron can never be utilized to excite a more energy-requiring molecule. The molecule of the reaction center is the one having the absorption peak at the longest wavelength (Fig. 9-8), which causes it to act as a sort of "trap" or "sink" into which the energy harvested by all of the pigment molecules of the unit will inevitably travel. Once the energy is received by the reaction center, which is a chlorophyll *a* molecule of particularly long wavelength absorption, the excited electron can be transferred to the waiting acceptor. It is estimated that the reaction center of PS II has an absorption maximum at 680 nm while that of PS I has a maximum at 700 nm. For this reason, these two pigment molecules are termed P680 and P700.

It is easy to understand the presence of a variety of receiving molecules to ensure that a greater variety of wavelengths will be effective in promoting photosynthesis, but it is less evident why there should be so few reaction centers relative to the total pigment population. Why should each pigment molecule in the unit not have the opportunity to transfer its excited electron, rather than simply passing on the energy to some other molecule? It would seem that, if all of the pigment molecules were hooked up to their own electron-acceptor and electron-transport system, a much greater rate of photosynthesis would result from a given amount of illumination. One might expect that, under the existing conditions, something of an "energy traffic jam" would occur with all of the harvesting molecules trying to shuttle energy into the reaction center at the same time. The reason that a traffic jam does not occur can be seen when the time course of events is considered. When one measures the number of quanta of light that each chlorophyll molecule can absorb per unit time, it is found that, even in direct sunlight, each pigment molecule is limited to about ten absorptions per second. In dim light, the number becomes much less—on the order of 0.1 absorption per second. In contrast, the reaction centers can use energy at much faster rates; transfer of energy to the reaction center and the subsequent transfer of the electron to the acceptor can occur in a very small fraction of a second. In other words, even if there were only one reaction center for each photosynthetic unit of approximately 300 pigment molecules, transfer of the energy to the reaction center pigment and out to the electron acceptor may still not be rate limiting.

PHOTOSYSTEMS I AND II

In the overview of a previous section, the existence of two photosystems was indicated. The effect in both systems of the absorption of light is the separation of charge between a pigment molecule (namely the chlorophyll a of the reaction center) and a primary electron acceptor molecule. The positively charged pigment produced acts as an oxidizing agent (Y^+ of Fig. 9-4), while the negatively charged electron acceptor acts as a reducing agent (X^- of Fig. 9-4). The nature of the primary electron acceptors of both photosystems remains uncertain. In addition to the presence of the P680 or P700 pigments and their corresponding electron acceptors, each reaction center is also believed to contain a primary electron donor, i.e., a molecule which supplies electrons to the positively charged pigments from which an electron had previously escaped in response to the absorption of light.

As illustrated in Fig. 9-4, the two photosystems are believed to act in series. Photosystem II operates at a more electropositive level than photosystem I. The response by PS II to the absorption of light is the production of a strong oxidizing agent capable of removing electrons from water, whereas the response by PS I is the production of a strong reducing agent capable of donating electrons to NADP$^+$. Though the "Z" scheme, as first proposed by Robert Hill and D. S. Bendall, depicted in Figs. 9-4 and 9-11, is still only a hypothesis, it is supported by a fairly substantial body of evidence. The greatest uncertainties reside in the chemical nature of the primary electron acceptors within each photosystem and the nature and position of all of the pieces of the electron-transport chains. An even more basic problem is the organization of the entire collection of pigments and proteins within the membranes of the chloroplast (page 412). The nature of the Z scheme will be discussed in detail after the following digression into the experimental basis for two photosystems.

The Importance of the Wavelength of Light

Before discussing the photosystems themselves, it is worthwhile to consider the types of experiments that led to their discovery in the first place.

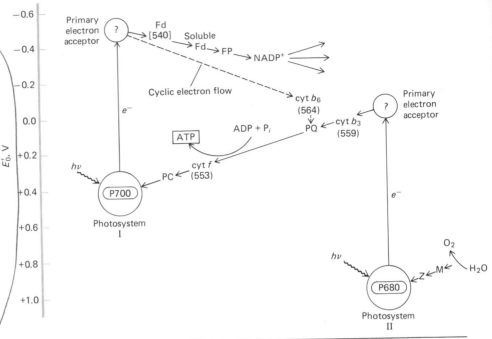

Figure 9-11. The "Z" scheme postulated to describe the flow of electrons during photosynthesis. The standard redox potentials E_0' of the various redox couples are indicated on the scale to the left. The question marks at the sites of the primary electron acceptors of the two photosystems reflect the uncertainty as to the identity of these molecules. In noncyclic transport, electrons derived from water are passed through both photosystems and ultimately to $NADP^+$ with ATP being formed by noncyclic photophosphorylation. In cyclic transport, electrons from PS I are cycled back to the PS I reaction center as indicated by the dotted line. In this case ATP is formed by cyclic photophosphorylation. These events are discussed fully in later sections of this chapter.

Keep in mind from the previous discussion, that, in order for photosynthesis to occur, *both* photosystems must participate. There were two main findings made by Robert Emerson that led to the proposal of a dual pigment system: the first was the discovery of the so-called "red drop" and the second the discovery of "enhancement." Both were uncovered in experiments measuring the effects of light of different wavelengths upon photosynthesis. First, we will describe the red drop. Emerson found that the ability of light to support photosynthesis dropped off markedly when wavelengths above 680 nm were used, even though the absorption spectrum showed that some chlorophyll could still absorb in that region (Fig. 9-9a, 9-12a). In other words, when an action spectrum and an absorption spectrum were compared, there was some discrepancy between the two curves in the far end of the spectrum. Another way to state the "red drop" is that the quantum yield, i.e., the number of oxygen molecules produced to the number of quanta of light absorbed, dropped from about 12% (1:8) to much lower values when wavelengths above 680 nm were employed (Fig. 9-12b). This can be explained by placing all of the chlorophyll molecules absorbing above 680 nm into one of two required pigment systems, leaving the other system unable to respond to far red light, thereby inhibiting the entire process.

The second finding, that of enhancement, occurred when Emerson supplemented a beam of far red light (above 680 nm) with one of shorter wavelength (about 650 nm). Together the two beams resulted in greater photosynthesis (more O_2 released) than the sum of the two given independently (dotted line of Fig. 9-12a). In the presence of both wavelengths the quantum yield for the combined beams once again approached 12%. It appeared that the cells could use light of greater than 680 nm, but only when light of shorter wavelength was also available. Again the result suggested the existence of two separate, light-driven reactions. Both of the reactions could be driven by

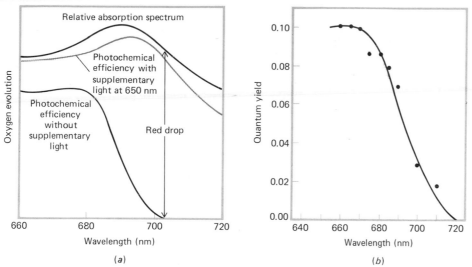

Figure 9-12. *(a)* Red drop in *Chlorella,* a green alga. A comparision of the absorption spectrum, which is due primarily to chlorophyll, with the action spectrum in the far red region of the visible range. It is apparent that the efficiency of photosynthesis, as measured by oxygen evolution, drops off sharply at wavelengths that are still being absorbed by the cells. However, if a beam of far red light (above 680 nm) was supplemented with a beam of light at 650 nm, an enhancement in oxygen release was noted, above that found with either beam alone. These results suggested the existence of two separate light-driven reactions. *(From A. L. Lehninger, ''Biochemistry,'' 2d ed., Worth, 1975.) (b)* Quantum yield of oxygen evolution as a function of wavelength in *Navicula. (From Govindjee and R. Govindjee, in Govindjee, ed., ''Bioenergetics of Photosynthesis,'' Academic, 1975.)*

shorter wavelength light, but only one could be driven by the longer wavelength beam. Subsequent experiments showed that not only would light above 680 nm supplement light of longer wavelength, it could still do so if presented to the cells at a different time. It was found, for example, that, if algae were illuminated with light of 710 nm and then switched to light of 650 nm, a short burst of oxygen production was noted (Fig. 9-13). The short

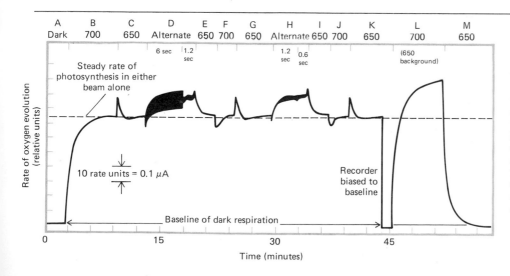

Figure 9-13. Enhancement of photosynthesis between two light beams of 650 nm and 700 nm. A reconstructed tracing from a single experiment. The wavelength of the illumination during each time period is indicated at the top of the record. Each time the illumination is switched from 700 nm to 650 nm, a burst of oxygen release occurs indicating that the energy of the previous light beam had been absorbed and can still be used if followed closely enough with light of shorter wavelength. *(From J. Myers and C. S. French, Plant Physiol., **35:**966, 1960.)*

burst represents a fleeting enhancement during the second or so that the cells still "remember" the light of longer wavelength. The fact that enhancement occurs between light flashes that are not applied simultaneously indicates that the two photosystems are connected to each other by reactions that can proceed in the dark. If all of the reactions involved in the release of oxygen were strictly of a photochemical nature, the memory of the longer wavelength light would be extinguished a billion times faster than the results that were found. Examination of Fig. 9-11 indicates that the dark reactions connecting the two photosystems comprise the electron-transport chains leading out of PS II and into PS I.

The difficulty in studying the properties of the reaction center pigments, P680 and P700, is their presence in such low concentration compared to the other molecules in the unit, and the fact that they retain these special properties only when present within their membrane environments. In order to focus on the reaction centers in a way that their properties would not be obscured by the other molecules, special techniques had to be devised. For example, in order to observe the absorbance properties of the reaction centers (Fig. 9-14), a special ultrasensitive spectrophotometer had to be built. One technique that has been very important in the analysis of both the primary oxidizing and reducing agents is that of electron-spin resonance, particularly when studying the operation of the photosystems at extremely low temperatures. Electron-spin resonance (page 212) detects the presence of molecules with unpaired electrons, a situation which occurs when the reaction center loses an electron and the primary acceptor gains it.

The use of low temperatures, such as that provided by liquid nitrogen (77°K) allows one to separate the photochemical events from the subsequent steps such as electron transfer or enzymatic modification. Only the photochemical reaction is independent of molecular diffusion, which is greatly decreased at very low temperatures. In addition, when chloroplasts are illuminated at temperatures of 77°K or so, not only does the formation of Y^+ and X^- still occur, but they become relatively stable as compared to their fleeting existence at normal temperatures. Recent studies have been further aided by the opportunity to separate fragments of the chloroplast membrane (see

Figure 9-14. These curves represent the difference between the absorption spectrum of extracted chlorophyll in the light versus dark or oxidized versus reduced state. The values obtained in the dark or reduced state are subtracted from the values in the light or oxidized state. The differences are attributed to the absorption of light by the reaction centers. For example, one of the peaks (at 698 nm) can be attributed to the primary oxidizing agent, i.e., the positively charged P700 chlorophyll, of PS I. *(From B. Kok, Biochim. Biophys. Acta., **48**:529, 1961.)*

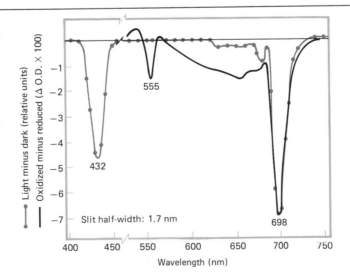

page 412) containing the two photosystems allowing each to be studied independently.

The "Z" Scheme

Even though the nature of the primary electron acceptors remains controversial, a number of the pieces of the three parts of the "Z" scheme (Fig. 9-11) have been identified. The least understood part of the entire chain extending from H_2O to $NADP^+$ is the first piece, i.e., that between H_2O and PS II. The splitting of water would be expected to be a highly endergonic event due to the stable association of the hydrogen and oxygen atoms. The formation of one molecule of oxygen requires the loss of four electrons (whose redox potential is $+0.82$ volt) from two molecules of water according to the reaction

$$2H_2O \longrightarrow 4H^+ + O_2 + 4e^-$$

The first aspect to consider is the nature of the steps believed to be responsible for the extraction of four electrons from two molecules of water by the PS II complex.

Step 1 $chl + h\nu \longrightarrow chl^*$

Step 2 $chl^* + P680 \cdot X \longrightarrow chl + P680^* \cdot X$

Step 3 $P680^* \cdot X \longrightarrow P680^+ \cdot X^-$

Step 4 $Z \cdot P680^+ \cdot X^- \cdot R \longrightarrow Z^+ \cdot P680 \cdot X^- \cdot R$

Step 5 $Z^+ \cdot P680 \cdot X^- \cdot R \longrightarrow Z^+ \cdot P680 \cdot X \cdot R^-$

Step 6 $\begin{cases} M^0 \cdot Z^+ \longrightarrow M^{1+} \cdot Z \\ M^{1+} \cdot Z^+ \longrightarrow M^{2+} \cdot Z \\ M^{2+} \cdot Z^+ \longrightarrow M^{3+} \cdot Z \\ M^{3+} \cdot Z^+ \longrightarrow M^{4+} \cdot Z \end{cases}$

Step 7 $2H_2O + M^{4+} \longrightarrow 4H^+ + O_2 + M^0$

In step 1, light is absorbed by an antenna chlorophyll molecule. In step 2, the excitation energy is transferred to the chlorophyll molecule of the reaction center (P680), which is associated with the primary electron acceptor, X (a molecule generally referred to as Q in the literature). In step 3, a charge separation occurs between the pigment molecule and the primary electron acceptor, thereby producing a strong oxidizing agent, $P680^+$ (or Y^+ of Fig. 9.4) and a weak reducing agent, X^-. In step 4, two additional components, Z and R are indicated. Z is the electron donor of the photosystem, i.e., the molecule which supplies electrons to the deficient reaction center pigment. R is the secondary electron acceptor, i.e., the molecule which receives an electron from X^-. The nature of R, which may be a plastoquinone or a cytochrome, is not known for certain. In step 4, Z donates an electron to the positively charged molecule, while in step 5 the electron from the primary acceptor is passed to R. In step 6, another component, M, is introduced. The molecule M refers to a protein (believed to contain manganese) which is capable of accumulating four positive charges. The four charges are obtained by four successive transfers of electrons to Z^+ which, in turn, passes the electrons to the pigment (step 4). Once M has accumulated these four charges, it is capable of catalyzing the reaction (step 7) whereby four electrons are transferred from two water molecules, resulting in the formation of a molecule of O_2.

The piece of the "Z" connecting PS II to PS I is known to contain at least two different cytochromes as well as plastoquinone and a copper-containing protein, plastocyanin, which appears to be the direct supplier of electrons to the positively charged reaction center of PS I. It is during the passage of electrons from PS II to PS I, that ATP is formed. The third and last piece of the "Z" leads from the primary electron acceptor of PS I, i.e., the strong reducing agent, to $NADP^+$. As stated previously, although the primary electron acceptor of PS I is uncertain, one of the components near the beginning of this segment of the chain is an iron-sulfur protein. Iron-sulfur proteins are those in which the iron is not contained within a heme but is directly associated with sulfur atoms of cysteine (see Fig. 8-13). Several different iron-sulfur proteins have been identified, one of which is a small one present in both plants and bacteria, termed *ferredoxin*. It appears that the first iron-sulfur protein is a membrane-bound ferredoxin, while farther on along the chain is another ferredoxin, though this latter one is not bound to a membrane and is less electronegative. It is this soluble ferredoxin which is readily lost during chloroplast isolation that caused Hill and others in their early investigations to be unable to detect CO_2 fixation even though oxygen production occurred in the presence of artificial acceptors. The loss of ferredoxin meant that electrons could no longer flow to $NADP^+$, and the reducing power to convert CO_2 to carbohydrate was lost with it. The last component in this third segment is a flavin-containing enzyme called NADPH-ferredoxin oxidoreductase, which catalyzes the transfer of the electrons to $NADP^+$. Since it takes two electrons to form NADPH and a given ferredoxin molecule can accept only one, two ferredoxins must act together in the reduction.

$$2 \text{ Ferredoxin}_{red} + 2H^+ + NADP^+ \xrightarrow{\text{Ferredoxin-NADPH} \atop \text{oxidoreductase}}$$
$$2 \text{ Ferredoxin}_{ox} + NADPH + H^+$$

Not all electrons passed to ferredoxin inevitably end up in NADPH; there are two other routes that can be taken depending upon the particular organism and the circumstances at the time. One route that an electron can follow is a return to the electron-deficient reaction center of PS I (shown by the dotted line in Fig. 9-11) with the simultaneous formation of ATP. This will be discussed further in the next section. Another fate which can befall electrons from PS I is their passage to various inorganic acceptors which are thereby reduced. This path for electrons can lead to the eventual reduction of nitrate (NO_3^-) to ammonia (NH_3) or sulfate (SO_4^{2-}) to sulfhydryl (—SH). In other words, the reducing power made available by photosynthesis can be put to uses other than the fixation of carbon dioxide.

PHOTOPHOSPHORYLATION

The conversion of 1 mol CO_2 to 1 mol carbohydrate (CH_2O) requires the input of considerable energy, specifically 3 mol ATP and 2 mol NADPH (see Fig. 9-19). With the finding in 1951 that photosynthesis was accompanied by the reduction of $NADP^+$, it appeared that the source of the high-energy molecules needed for the formation of carbohydrate from carbon dioxide had been found. Photosynthesis would be expected to produce sufficient NADPH to serve in the reduction of CO_2 as well as in the formation of ATP via the electron-transport chain. In the latter case, the NADPH would be used to generate NADH which could then be oxidized in the mitochondria with the formation of ATP. The problem with this proposal was the relatively scarce number of mitochondria that are present in many of the specialized plant cells that are particularly active in carbon dioxide fixation. It did not seem feasible that the high demand for ATP could be met by these few mitochondria.

In 1954 Daniel Arnon found that isolated chloroplasts were themselves capable of the entire photosynthetic process and, therefore, presumably could manufacture their own ATP. It was soon demonstrated that isolated chloroplasts could, in fact, phosphorylate ADP and could even do so in the absence of added CO_2 or $NADP^+$. In other words, it seemed that chloroplasts had some special means for ATP formation without involving most of the photosynthetic reactions which would have led to oxygen production, CO_2 fixation, or $NADP^+$ reduction. All that was necessary was illumination, chloroplasts, ADP, and P_i. The process that had been found was later termed *cyclic photophosphorylation* and is illustrated by the dotted line in Fig. 9-11. Cyclic photophosphorylation is a process carried out by PS I independently of PS II, as described briefly in the last section. The process begins with the absorption of a quantum of light by PS I and the transfer of a high-energy electron to the primary acceptor. From there the electron is passed to ferredoxin as is always the case, but, rather than being passed on to $NADP^+$, the electron is passed back to the electron-deficient reaction center to complete the cycle. During the flow of the electron around this course, sufficient free energy is released so that at least one molecule of ATP can be formed.

The other means available to the chloroplast for photophosphorylation occurs during the normal course of events when oxygen is being released and NADPH being formed. In this case ATP formation results from the loss of free energy accompanying the transport of electrons between PS II and PS I, a process termed *noncyclic photophosphorylation*. Since the voltage drop between the weak reducing agent formed by PS II (approximately 0.0 volt) and the weak oxidizing agent formed by PS I (approximately +0.4 volt) is 0.4 volt, sufficient energy ($\Delta G'^0 = -9.2$ kcal/mol) may be available along this noncyclic path to form one molecule of ATP per electron transported. If not, then one ATP molecule would be formed for each *pair* of electrons transported. Presumably, the preponderance of cyclic vs. noncyclic photophosphorylation in a particular chloroplast would reflect the needs of the cell for ATP, NADPH, and carbohydrate at the time. Although NADPH levels are believed to be of particular importance, the means by which the cell determines the relative levels of one or the other of these electron pathways is not yet known.

Though less is known about the formation of ATP during photosynthesis as compared to oxidative phosphorylation in the mitochondria, the same mechanisms are believed to operate in both organelles. The same three hypotheses discussed in Chapter 8 have been put forward for the chloroplast as well. At the present time, the evidence points to the

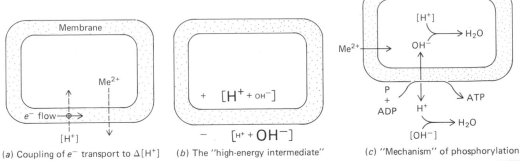

(a) Coupling of e^- transport to $\Delta[H^+]$ (b) The "high-energy intermediate"   (c) "Mechanism" of phosphorylation

Figure 9-15. The chemiosmotic model of ATP formation by the thylakoid membrane. *(a)* Electron transport, either cyclic or noncyclic, by carriers of the thylakoid membrane results in the formation of a proton gradient and an electrical gradient across the membrane. *(b)* The storage of energy in the electrochemical gradient. *(c)* The use of the gradient to drive the phosphorylation of ADP as described in Chapter 8. [*Reprinted from A. Jagendorf, Federation Proceedings,* **26**:*1362 (1967).*]

chemiosmotic mechanism (Fig. 9-15), though as in the mitochondria, there are indications of conformational changes which may or may not be involved in phosphorylation. As in the case of an active preparation of mitochondria, isolated chloroplasts also change the pH of the medium in which they are suspended, though the changes are in the opposite direction. Whereas isolated, active mitochondria eject protons into the medium, isolated, active chloroplasts absorb protons from the medium, raising its pH. The alkalinization of the medium is strictly dependent upon illumination. Measurements of proton gradients suggest that concentration differences up to 10,000 (ΔpH of 4) can occur during maximal ATP synthesis. The greater the gradient, the more rapidly ADP is phosphorylated. Analysis of ion movements suggest that the movement of protons in one direction is compensated by the movement of other ions so that no large membrane potential is built up. In other words, the pH gradient may not be accompanied by an electrical gradient. A most significant experiment in the study of chloroplast phosphorylation was performed by André Jagendorf and alluded to in Chapter 8. Jagendorf found that, when he imposed an artificial proton gradient across the membranes of the chloroplast (see Fig. 8-21), the phosphorylation of a significant amount of ADP would occur *in the dark*. The gradient was generated by keeping the chloroplasts in an acidic medium and then abruptly switching them to an alkaline medium in the presence of ADP and P_i.

CARBON DIOXIDE FIXATION AND THE FORMATION OF CARBOHYDRATE

The C3 Cycle

In the 1940s, Melvin Calvin, A. A. Benson, James A. Bassham, and their coworkers began what was to be approximately a ten-year study of the enzymatic reactions by which carbon dioxide was assimilated into the organic molecules of the cell. Armed with the newly available, long-lived radioactive isotope of carbon (^{14}C) and a new technique, two-dimensional paper chromatography, they began the task of identifying all of the labeled molecules that were formed when cells were allowed to utilize $^{14}CO_2$. The studies began with plant leaves but soon shifted to a more simple system, the alga *Chlorella*. Algal cultures were grown in closed chambers in the presence of unlabeled CO_2, and then radioactive CO_2 (in the form of dissolved gas) was introduced by injection into the culture medium. After the desired period of incubation with the labeled CO_2, a valve at the bottom of the incubation vessel was opened and the algal suspension drained into a container of alcohol, which served to immediately kill the cells, stop the enzymes, and extract the soluble molecules. Extracts of the cells were then placed as a spot on the chromatographic paper and subjected to two-dimensional chromatography (Fig. 3-5). In order to identify the location of the radioactive compounds at the end of the procedure, a piece of x-ray film was pressed against the chromatograph and the plates kept in the dark until sufficient time had passed to expose the film. The principle of the technique is no different from that of microscopic autoradiography described on page 297. After photographic development, the location of the radioactive compounds was apparent. The task that remained was to identify the nature of the molecules responsible for each darkened spot on the film by removal of the material from the paper chromatogram and subjecting it to the appropriate analytical examination.

It was found that the conversion of labeled carbon dioxide to reduced organic compounds occurred very rapidly. For example, if the cells were exposed to isotope for as long as 30 to 60 seconds, the chromatogram contained a large number of spots (Fig. 9-16b). However, if the incubation period was very short (a fraction of a second to a few seconds), one radioactive spot predominated (Fig. 9-16a). The compound responsible for this particular spot proved to be 3-phosphoglycerate (PGA), one of the intermediates in glycolysis. With longer incubation periods a greater number of spots appeared and eventually all of them were identified. Knowing the nature of these compounds, and, most importantly, the position of the labeled carbon atoms within each of the intermediates, the pathway for the conversion of CO_2 to carbohydrate was determined.

One of the most troublesome pieces in the determination of the chain of reactions was the nature of the molecule to which the CO_2 was first being attached. It was initially suspected that a two-carbon compound was responsible since the first intermediate seemed to be PGA. But for various reasons this proposal was eliminated, as were certain other possibilities. After an extensive search, it became apparent that the initial acceptor was a five-carbon compound, ribulose 1,5-diphosphate which, when condensed with CO_2, formed a six-carbon molecule (Fig. 9-17). This six-carbon compound was immediately split into two molecules of PGA, one of which contained the recently added carbon. Both the condensation and the splitting (Fig. 9-17a, b) were carried out by a huge multisubunit enzyme, ribulose diphosphate carboxydismutase. The role of ribulose diphosphate in the process was revealed in the following experiment.

It was known that, if labeled CO_2 were given to the cells, each of the spots became maximally radioactive after a certain time period, i.e., they became saturated with radioactivity. It was found that, if the cells were exposed to the $^{14}CO_2$ in the light for sufficient time to saturate the radioactivity of the intermediates and then the lights were turned off, there was an immediate rise in the label present in PGA and an immediate drop

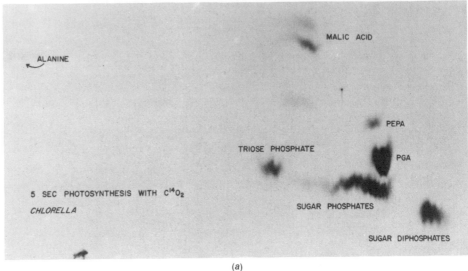

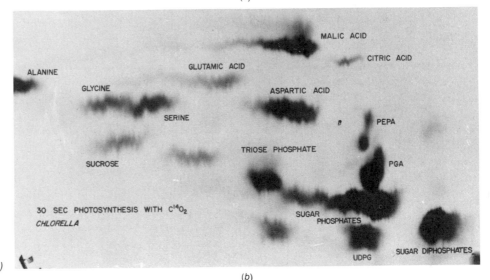

Figure 9-16. The location of radioactivity on chromatograms of organic compounds extracted from algae incubated with $^{14}CO_2$ for 5 seconds *(a)*, or 30 seconds *(b)*. *(From the work of J. A. Bassham and M. Calvin.)*

in the label present in ribulose diphosphate (Fig. 9-18). The reason for this result becomes apparent when one considers the need for ATP and NADPH formed by the light reactions in the formation of ribulose diphosphate (Fig. 9-19). In the absence of ATP and NADPH, the existing labeled PGA cannot be converted to other intermediates, nor can new ribulose diphosphate molecules be formed. In contrast, the condensation of CO_2 with existing ribulose diphosphate and the subsequent conversion to PGA can continue, since this reaction is not dependent upon high-energy products of the light reaction. Consequently, radioactivity in PGA will rise, while radioactivity in ribulose diphosphate will fall.

The nature of the reactions involved in carbon dioxide reduction are shown in Fig. 9-17. As the nature of the intermediates was determined together with the positions of the labeled carbon atoms, it became apparent

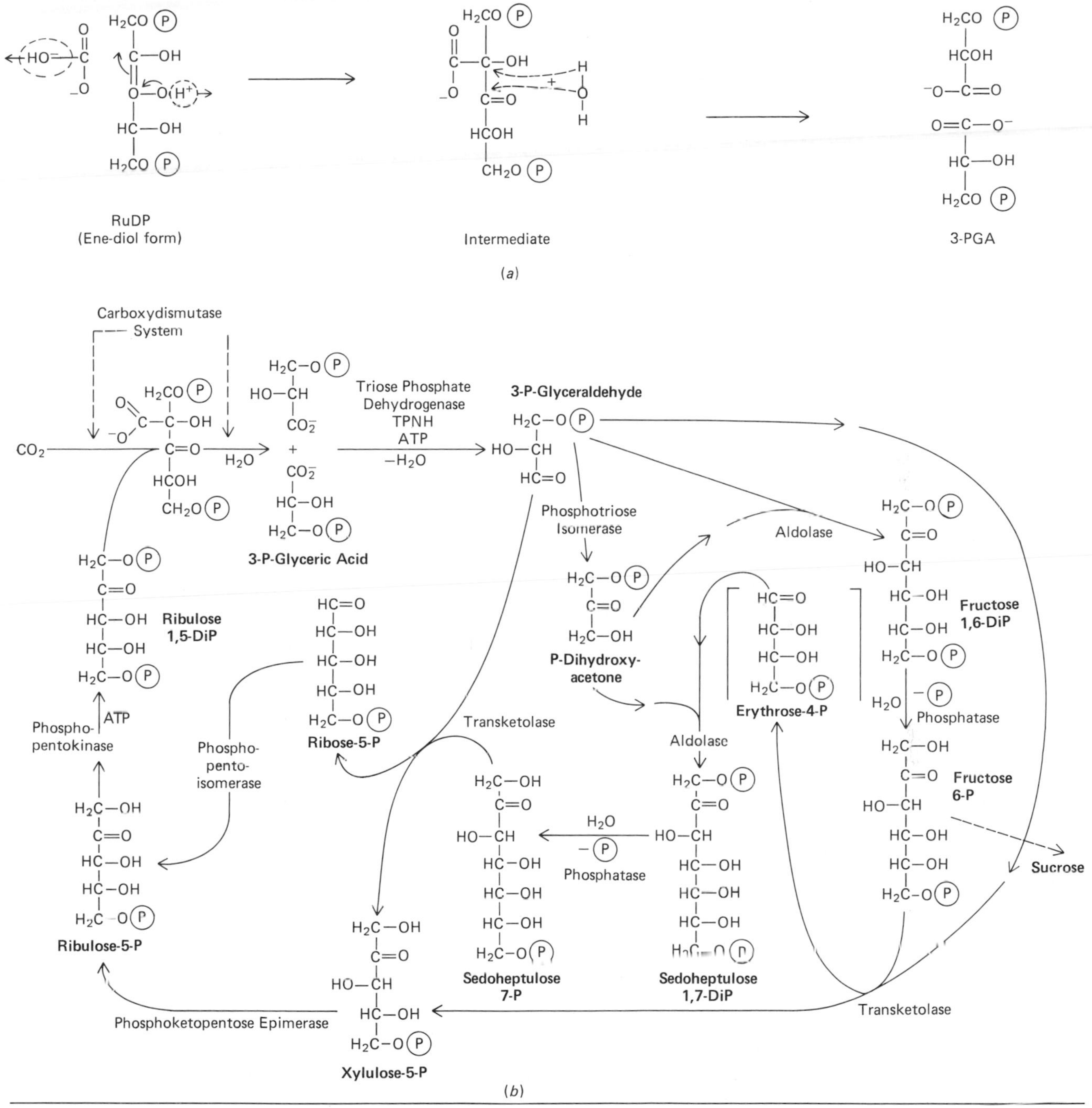

(a)

RuDP
(Ene-diol form)

Intermediate

3-PGA

(b)

Carboxydismutase
System

Triose Phosphate
Dehydrogenase
TPNH
ATP
−H₂O

3-P-Glyceraldehyde

CO_2

H_2O

3-P-Glyceric Acid

Phosphotriose
Isomerase

Aldolase

P-Dihydroxy-
acetone

Fructose
1,6-DiP

Ribulose
1,5-DiP

Erythrose-4-P

Ribose-5-P

Aldolase

Phospho-
pentokinase

ATP

Phospho-
pento-
isomerase

Transketolase

H_2O
− P
Phosphatase

Phosphatase

Fructose
6-P

Sucrose

Ribulose-5-P

Sedoheptulose
7-P

Sedoheptulose
1,7-DiP

Transketolase

Phosphoketopentose Epimerase

Xylulose-5-P

Figure 9-17. (a) Mechanism of the carboxydismutase reaction. (b) The reactions of the C₃ cycle. (From J. A. Bassham and M. Calvin, "The Path of Carbon in Photosynthesis," 1957. Reprinted by permission of Prentice-Hall, Inc., Englewood Cliffs, N.J.)

that the pathway basically formed a circle (Fig. 9-19). Ribulose diphosphate acts as the acceptor of carbon dioxide and, by virtue of one or another series of reactions, is regenerated to allow its participation in another round. In this way there is no need for the cell to manufacture these com-

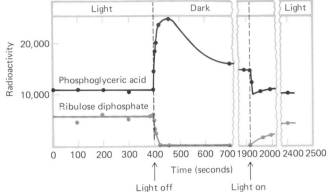

Figure 9-18. Light-dark changes in concentration of phosphoglyceric acid and ribulose diphosphate. *(From M. Calvin, in J. L. Oncley, ed., "Biophysical Science; A Study Program," Wiley, 1959.)*

pounds by other pathways. For each six turns of the cycle, a hexose can be drained away without any effect upon the levels of all the other intermediates. Alternatively, intermediates of the cycle can be siphoned off to form various molecules (amino acids, carbohydrates, nucleotides, etc.) that might be in demand at the time.

The enzymes required for the complete activity of the Calvin (or C3) cycle are not unique to photosynthetic organisms but include enzymes of the glycolytic pathway as well as the pentose phosphate shunt. Once again, the importance of membranes in cell compartmentation can be illustrated. Even though many of the intermediates formed in the Calvin cycle are also substrates in other pathways, their formation within chloroplasts insulates them from the enzymes of the cytoplasm.

The C4 Cycle

Although the Calvin cycle appears to be the most common reaction sequence leading to the formation of carbohydrate in photosynthetic organisms, it is not the only path. The existence of an alternate means by

Figure 9-19. A simplified diagram of the Calvin cycle showing the utilization of ATP and NADPH. *(From Govindjee and R. Govindjee, in Govindjee, ed., "Bioenergetics of Photosynthesis," Academic, 1975.)*

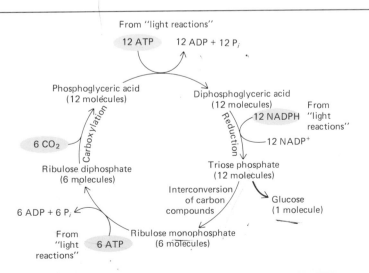

which plants can fix atmospheric CO_2 was uncovered by Hugo Kortschak in the early 1960s in sugar cane. Kortschak surprisingly found that when these plants were given $^{14}CO_2$ to fix, radioactivity first appeared in organic compounds containing four carbon skeletons rather than three. Further analysis revealed that these four-carbon compounds (malic acid and aspartic acid) resulted from the combination of CO_2 with phosphoenolpyruvate (structure shown in Fig. 4-15). Before considering the fate of this newly fixed carbon atom, it is important to examine the reason for this alternate means of CO_2 fixation.

If one places a typical plant in a closed chamber and follows its photosynthetic activity with time, one finds that, once the plant reduces the CO_2 levels in the chamber to approximately 50 ppm, photosynthesis slows and soon stops. In contrast, a plant carrying out fixation via phosphoenolpyruvate (PEP) will continue to photosynthesize until CO_2 levels are practically negligible (1 to 2 ppm). The enzyme, phosphoenolpyruvate carboxylase, which catalyzes the reaction between CO_2 and the highly reactive PEP molecule, is capable of operating at much lower CO_2 levels than ribulose diphosphate carboxydismutase. The question then centers on the value to a plant of being able to fix CO_2 at such low levels when the atmosphere invariably contains CO_2 at levels well above 50 ppm. The value becomes even more questionable when it is realized that (1) the use of PEP requires an additional expenditure of ATP and (2) there is no direct biosynthetic route leading from malic and aspartic acid to carbohydrate.

The value of the C4 pathway (also termed the Hatch-Slack pathway (Fig. 9-20) after the investigators who uncovered the component reactions) can be seen if one considers the environment in which the C4 plants live. Most C4 plants tend to live in hot dry areas (such as the deserts of the southwestern United States) under conditions of high temperatures and high light intensity. The most serious problem faced by these plants is the loss of water, termed *transpiration*, which inevitably accompanies CO_2

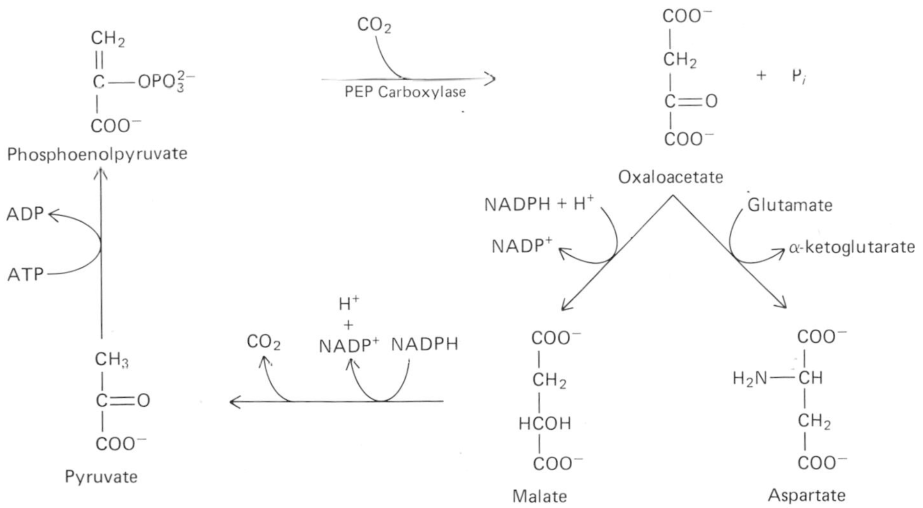

Figure 9-20. The reactions of the C_4 (Hatch-Slack) cycle.

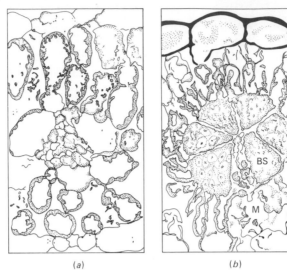

Figure 9-21. Diagrammatic representations of cross sections of leaves of the C$_3$ species *Atriplex patula* (left) and the C$_4$ species *Atriplex rosea* (right). M, mesophyll cells; B.S., bundle-sheath cells.

(a)   (b)

uptake. Carbon dioxide enters the leaves of plants through openings termed *stomata* by which water can also escape. The C4 plants can manage in hot arid environments because they are capable of closing down their stomata to prevent water loss, yet are still capable of maintaining sufficient CO$_2$ uptake to fuel their photosynthetic activity at a maximal rate. They are said to be "high efficiency" photosynthesizers due to their high levels of photosynthesis per unit of water lost.

If one follows the fate of the CO$_2$ fixed in the Hatch-Slack pathway, one finds that the CO$_2$ group is soon released only to be trapped by the ribulose diphosphate carboxydismutase system and converted to metabolic intermediates via the Calvin (or C3) cycle. As stated above, there is no direct pathway from malic or aspartic acid to carbohydrate. Since the C4 plants are very successful in their hostile environments, one might assume that there is some marked advantage in this two-step fixation process as opposed to the direct conversion of CO$_2$ to PGA. In order to un-

Figure 9-22. The separation of the reactions ot the C4 cycle between the bundle-sheath cells and those of the mesophyll. *(From O. Bjorkman, in A. C. Giese, ed., "Photophysiology," **8**:25, 1973.)*

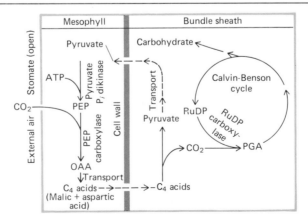

derstand this process, we need to briefly consider the anatomy of the leaf of a C4 plant (Fig. 9-21). Unlike the C3 plants, the leaves of C4 plants contain two concentric cylinders of cells. The outer cylinder is made of *mesophyll cells* and the inner cylinder of *bundle-sheath* cells. These two types of cells cooperate with one another in CO_2 fixation (Fig. 9-22). The fixation of CO_2 via reaction with PEP occurs in the outer mesophyll cells and the C4 products are then transported to the bundle-sheath cells which are sealed off from atmospheric gases. Once in the bundle-sheath cells the CO_2 can be split from the C4 carrier in such a way as to cause a high CO_2 level in the latter cells, a level suitable for reactions via ribulose diphosphate carboxydismutase. Carbon dioxide levels in the bundle-sheath cells may be 100 times that of the mesophyll. It appears, therefore, that the C4 system is necessary to drive CO_2 fixation by the less efficient C3 pathway. Once the CO_2 is split from the four-carbon compound, the pyruvate formed returns to the mesophyll cell to be recharged as PEP (Fig. 9-22).

CHLOROPLAST STRUCTURE

Chloroplasts, like most organelles, come in various shapes and sizes. In higher plants, they are generally lens shaped (Fig. 9-23), approximately 2 to 4 μm wide and 5 to 10 μm long, and numbering approximately 20 to 40 per cell. These dimensions place chloroplasts as giants among organelles, as large as the entire red blood cell discussed at length in Chapter 5. The outer covering of the chloroplast consists of an envelope composed of two membranes separated by a thin space (Fig. 9-24a). In addition to these outer membranes, there is an extensive internal membrane system with a distinctive organization. Unlike the situation in mitochondria, the internal membrane of the chloroplast is organized into flattened membrane sacs, called *thylakoids*, which are arranged in very orderly stacks called *grana*, which resemble a stack of coins. The space within a thylakoid is the *loculus*, the space outside of the thylakoids and within the outer envelope is the *stroma* (or *matrix*). Membrane-bound channels called *stroma lamellae* connect the thylakoids of one granum with that of another (Fig. 9-24b). Reconstructions of chloroplasts from serial sections indicate that the internal chambers of all of the thylakoids of a typical plant chloroplast *may* be interconnected, thereby forming one giant compartment. As in the mitochondria, the matrix of chloroplasts contains small DNA molecules and ribosomes which together form a store of genetic information and a means to utilize it. Like mitochondria, chloroplasts are semiautonomous and self-replicating (their genetics will be considered together with that of mitochondria in Chapter 17).

The Nature of the Thylakoid Membrane

The thylakoid membranes of the chloroplast contain all of the components necessary for the absorption of light and the transfer of electrons from water to $NADP^+$. Though it is certain that the geometry within the membrane of the interacting enzymes, electron carriers, and pigment

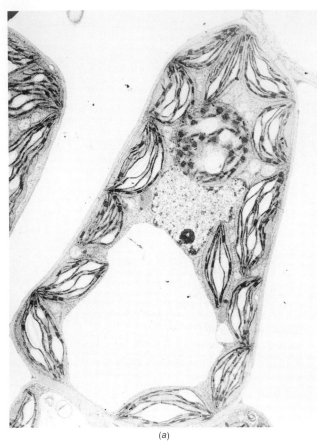

(a)

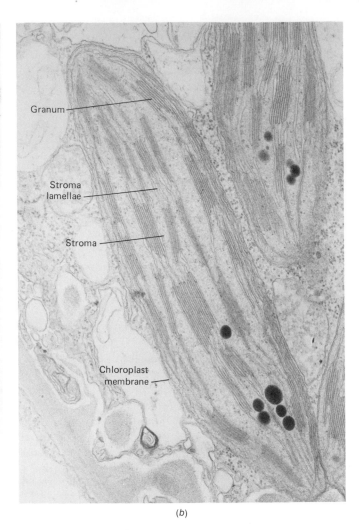

Granum

Stroma
lamellae

Stroma

Chloroplast
membrane

(b)

Figure 9-23. *(a)* Electron micrograph through a soybean leaf mesophyll cell showing the distribution of chloroplasts around the large central vacuole. *(b)* Higher power micrograph of a pea leaf chloroplast showing the stacked thylakoid membrane and stroma lamellae. *(Courtesy of H. Berg.)*

molecules is of critical importance, very little is yet known of the nature of the organization. Several approaches have been taken in attempts to determine the location within the membrane of the photosynthetic machinery. The presence of various types of particles either on or within the membrane (Fig. 9-25) has been revealed using both freeze-fracture analysis and negative staining. Negative-staining procedures are particularly valuable for indicating the presence of structures attached to one or the other *surface* of a membrane, but it must be kept in mind that the steps in which the specimen is air-dried may introduce distortions in the final preparation to be examined.

Negative-staining procedures have revealed the presence of two types of structures attached to the outer surface of the thylakoid membrane facing into the stroma. One of these particles is believed to be the large multisubunit enzyme ribulose 1,5-diphosphate carboxydismutase which catalyzes the fixation step of carbon dioxide and the immediate formation of PGA. The enzyme is loosely attached to the membrane and easily re-

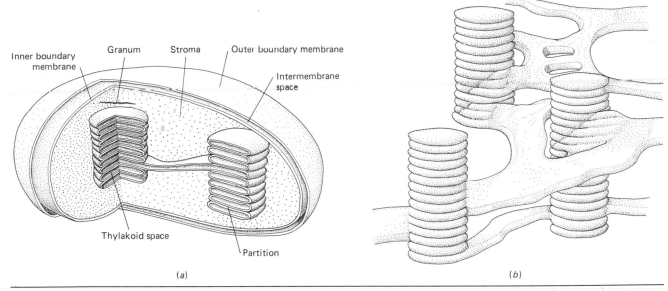

Inner boundary membrane · Granum · Stroma · Outer boundary membrane · Intermembrane space · Thylakoid space · Partition

(a)

(b)

moved. The other particle (Fig. 9-26) is believed to be the chloroplast ATPase (ATP synthetase), the counterpart to the F_1 sphere (coupling factor) of the inner mitochondrial membrane. The phosphorylation machinery of the chloroplast is believed to be essentially identical in structure to its mitochondrial counterpart, though it faces outward from the membrane rather than inward, in keeping with the reversed orientation of the proton gradient. With the exception of the ribulose 1,5-diphosphate carboxydismutase, all of the enzymes of the Calvin cycle are located within the stroma of the chloroplast in an analagous manner to those of the TCA cycle of the mitochondria.

Figure 9-24. *(a)* Diagram of the overall structure and internal contents of the chloroplast. *(From "Biology of the Cell" by Stephen L. Wolfe. © 1972 by Wadsworth Publishing Company, Inc., Belmont, California 94002. Reprinted by permission of the publisher.) (b)* Three-dimensional view of the grana with interconnecting stroma lamellae. *(From T. E. Weier, C. R. Stocking, and L. K. Shumway, Brookhaven Symp. Biol., **19**:371, 1961.)*

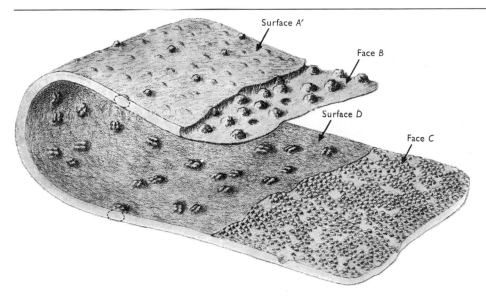

Surface A' · Face B · Surface D · Face C

Figure 9-25. Diagrammatic representation of the deep-etch surfaces (A' and D) and fracture faces (B and C) commonly seen in thylakoids. The fracture faces reveal the presence of particles of two different size classes. The large particles of the B face are only found in the region of a grana stack, while smaller particles are found in the comparable face of the stroma lamellae. The structures on the outer surface (A') are believed to represent the carboxydismutase and the coupling factor. *From R. B. Park and A. O. Pfeifhofer, J. Cell Sci., **5**:304, 1969.)*

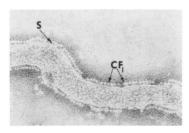

Figure 9-26. Negatively stained preparation of a thylakoid revealing the presence of 90-Å particles believed to represent the coupling factor (CF₁). As in the case of the mitochondria, a stalk (S) can be seen connecting the particle to the membrane. *(From M. P. Garber and P. L. Steponkus, J. Cell Biol., **63**:25, 1974.)*

Freeze-fracture analysis (Fig. 9-27) is generally found to reveal the presence of various types of intramembrane particles, which can be divided into smaller particles (less than 140 Å) and larger particles (greater than 140 Å). When the freeze-fracture plane passes through the thylakoid membrane, it tends to separate the two types of particles into opposite halves of the membrane. The half of the membrane representing the inner leaflet contains predominantly the larger particle, while the smaller particles predominate in the outer leaflet. Conveniently, other techniques, such as digitonin treatment, have been discovered which will separate thylakoid membranes into fragments containing one or the other of the two size classes of particles. Unlike freeze-fracture preparations, digitonin extracts can be prepared in sufficient quantity to carry out biochemical analyses on the materials. The finding that membrane fragments enriched in the larger particles were correspondingly enriched in PS II, while those containing the smaller particles were characterized by a preponderance of PS I, has led to the concept of a binary structure of membrane organization in which the two photosynthetic units are located in spatially distinct structures.

In other attempts to obtain information on the possible asymmetry of the chloroplast membranes, various types of nonpenetrating probes have been used to attach specifically to one component or another. Antibodies, trypsin, electron acceptors, and radioactive labeling agents are among the probes that have been employed. Taken as a whole, there is considerable

Figure 9-27. Freeze-fracture replica of a thylakoid. Since the membrane of the thylakoid is quite different in a region in which the thylakoids are arranged in a grana stack as opposed to being unstacked, four distinct fracture faces can be distinguished, all representing a view within the inner hydrophobic plane of the membrane. PF faces are views of the half of the membrane on the inner leaflet, i.e., facing the thylakoid lumen. EF faces are those revealed by the half of the membrane facing the cytoplasm. The designators s and u refer to the stacked and unstacked region. *(From K. R. Miller, G. J. Miller, and K. R. McIntyre, J. Cell Biol., **71**:627, 1976.)*

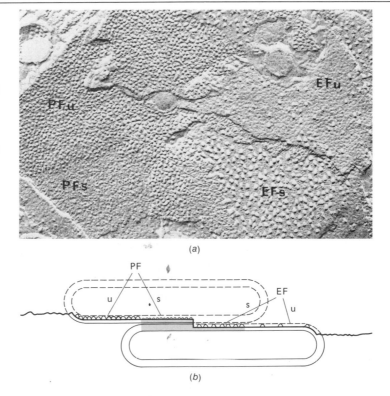

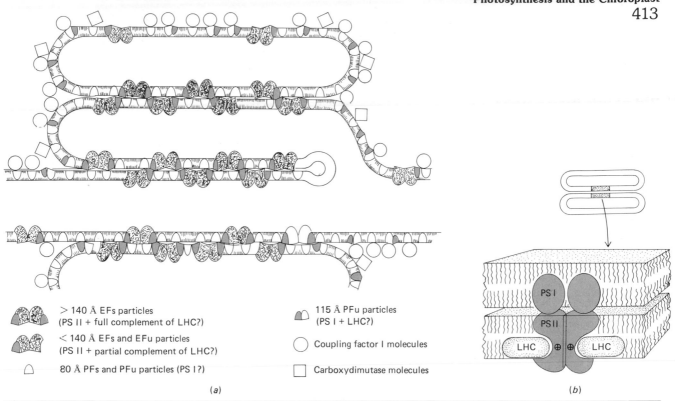

> 140 Å EFs particles
(PS II + full complement of LHC?)

< 140 Å EFs and EFu particles
(PS II + partial complement of LHC?)

80 Å PFs and PFu particles (PS I?)

115 Å PFu particles
(PS I + LHC?)

Coupling factor I molecules

Carboxydimutase molecules

(a)

(b)

Figure 9-28. (a) A recent model of the organization of components of the thylakoid membrane. Note the distribution of the different particle categories between stacked and unstacked membrane regions, the disposition of the particles with respect to the lipid bilayer continuum, and the spatial association of the particles in the stacked membrane regions. It should be noted that more recent investigations suggest that the particles containing PS II are composed of a considerably higher proportion of light harvesting pigments than indicated in this diagram (see the paper by L. A. Staehelin and C. J. Arntzen in Ciba Foundation Symposium #61 entitled "Chlorophyll Organization and Energy Transfer in Photosynthesis," for an updated diagram). The terms PFs, PFu, EFs, and EFu are defined in the legend to Fig. 9-27. *(From L. A. Staehelin, J. Cell Biol., **71:**155, 1976.)* (b) A closer look at the possible organization of the stacked membrane region. In this model the PS II-containing particle spans the membrane and makes contact with small PS I-containing particles on the opposite membrane. LHC refers to the light-harvesting chlorophyll-protein complex of PS II. One possible location of the PS II (⊕) reaction center is shown. *(From K. R. Miller, G. J. Miller, and K. R. McIntyre, J. Cell Biol., **71:**637, 1976.)*

evidence of the localization of various of the components of the photosynthetic reaction chain, though specific assignments remain inconclusive. Two recent models of the thylakoid membrane are shown in Fig. 9-28.

Though an understanding of the precise organization of the membrane is far in the future, it is clear that the components must be localized in such a manner as to eject protons on the inside of the thylakoid and absorb them from the outside. The two photosystems must be arranged so that electrons can pass from one to the other via the electron carriers. It has also been proposed that the membrane serves to separate the primary electron acceptor from the primary electron donor of each photosynthetic unit. If this were not the case, these two molecules with their opposing redox potentials would be expected to be very reactive toward one another.

The fact that the positively and negatively charged donor and acceptor remain in existence at low temperatures supports this contention. Although chloroplasts are totally absent in prokaryotic autotrophs, the photosynthetic apparatus is membrane bound, generally associated with infoldings of the plasma membrane. Although algae possess true chloroplasts, the organization of the membrane into grana is lacking.

EVOLUTION OF PHOTOSYNTHETIC MECHANISMS

Since we began this chapter with evolutionary speculations, it is fitting to close with additional ones, in this case based upon some of the information presented in the intervening pages. It is likely that the earliest biological use of sunlight was to harness its energy for the production of ATP. We saw one example of a mechanism of this type in *Halobacterium* (page 374), in which light absorption by the purple membrane is converted into the chemical energy of ATP. If one examines the process of cyclic phosphorylation by itself, a similar use becomes apparent. Both are presumably mediated by the storage of energy in a proton gradient. The only consequence of the cyclic flow of electrons from an excited PS I center back to that center is the phosphorylation of ADP; no O_2 or NADPH is formed. It may well be that this part of the entire photosynthetic process is the most primitive. Certainly any organism with the ability to produce ATP in this manner would have had a great advantage over those obtaining all of their ATP by anaerobic oxidation. The next step may well have been the utilization of a compound such as H_2S or H_2 as an electron donor. Now, rather than having to recycle its excited electron back to its oxidized reaction center, it could pass it on to $NADP^+$ and obtain reducing power.

The last step in this hypothetical chain of events was the shift to water as an electron donor. Once water could be utilized, an unlimited supply of electrons became available and with it the opportunity for photoautotrophs to expand their habitats around the world, rather than being tied to local sulfur holes. In order to utilize such an electropositive source of electrons as water, a second photosystem was required; therefore the switch to water must have involved complex evolutionary change. Among organisms living today, the blue-green algae are the only prokaryotes capable of oxidizing water. It may be that an organism of this type was the first to evolve this capability. In this same vein, blue-green algae are the only prokaryotes having a cytoplasmic membranous photosynthetic apparatus, one which is very similar to the thylakoids of the plant chloroplast. The primary structural difference is found in the organization of the thylakoids of the prokaryotic cell. Unlike the plant cell, the membranes of the blue-green algae are not stacked as disks and are not separated from the cytoplasm by a membrane. An intriguing hypothesis is that chloroplasts of plant cells are direct descendants of some primitive, oxygen-producing prokaryote, possibly similar to blue-green algae, that originally began some symbiotic relationship inside a heterotrophic cell.

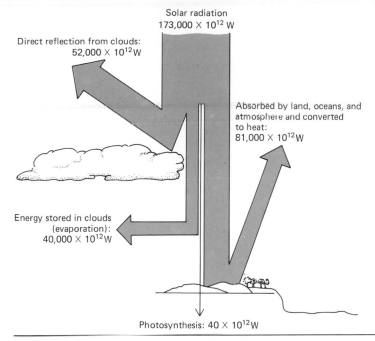

Solar radiation
173,000 × 10^{12} W

Direct reflection from clouds:
52,000 × 10^{12} W

Absorbed by land, oceans, and atmosphere and converted to heat:
81,000 × 10^{12} W

Energy stored in clouds (evaporation):
40,000 × 10^{12} W

Photosynthesis: 40 × 10^{12} W

Figure 9-29. Only a small amount of the total solar energy reaching the earth is fixed by photosynthesis. The thickness of the arrows represents the amount of energy absorbed, reflected, or stored per unit time in units of watts (W). (The U.S. energy consumption per unit time is approximately 3 × 10^{12} W.) *(From L. Peusner, "Concepts in Bioenergetics," 1974, reprinted by permission of Prentice-Hall, Inc., Englewood Cliffs, N. J.)*

Once the means had evolved, the utilization of water was inevitably accompanied by the release of oxygen into an atmosphere previously devoid of this gas. With the accumulation of oxygen came the opportunity for much greater metabolic efficiency by a rapidly increasing variety of heterotrophs.

A different aspect of photosynthesis concerns the percentage of light falling on a plant that is actually absorbed and utilized for CO_2 fixation. It is generally estimated that, under the most optimum conditions such as those existing in a field of cultivated plants, approximately 1 to 2% of the incident light energy actually finds its way into chemical energy, the remainder being lost from the system (Fig. 9-29). Eugene Rabinowitch estimated that approximately 0.24% of the radiant energy reaching the earth's surface is actually trapped in biological materials. It is also estimated that plant life on earth fixes approximately 600 billion tons of CO_2 and releases about 400 billion tons of O_2 each year. The majority of this activity is accomplished by the phytoplankton, the single-celled algae living in the thin upper layer of the world's sensitive and increasingly more polluted oceans.

REFERENCES

Anderson, J. M., Biochim. Biophys. Acta **416**, 191–235, 1975. "The Molecular Organization of Chloroplast Thylakoids."

Arnon, D. I., Sci. Am. **203**, 105–118, May 1960. "The Role of Light in Photosynthesis."

Barber, J., ed., *The Primary Processes in Photosynthesis.* Elsevier, 1977.

Bassham, J. A., Sci. Am. **206,** 88–100, June 1962. "The Path of Carbon in Photosynthesis."

———, Science **172,** 526–534, 1971. "The Control of Photosynthetic Carbon Metabolism."

Bearden, A. J., and Malkin, R., Quart. Rev. Biophys. **7,** 131–177, 1974. "Primary Photochemical Reactions in Chloroplast Photosynthesis."

Bolton, J. R., and Warden, J. T., Ann. Rev. Plant Phys. **27,** 375–383, 1976. "Paramagnetic Intermediates in Photosynthesis."

Calvin, M., Science **135,** 879–889, 1962. "The Path of Carbon in Photosynthesis."

Clayton, R. K., *Light and Living Matter.* 2 vol. McGraw-Hill, 1970–1971.

———, Ann. Rev. Biophys. and Bioeng. **2,** 131–156, 1973. "Primary Processes in Bacterial Photosynthesis."

Govindjee, ed., *Bioenergetics of Photosynthesis.* Academic, 1975.

———, and Govindjee, R., Sci. Am. **231,** 68–82, Dec. 1974. "The Primary Events of Photosynthesis."

Gregory, R. P. F., *The Biochemistry of Photosynthesis,* 2d ed. Wiley, 1977.

Hall, D. O., and Rao, K. K., *Photosynthesis.* Crane-Russak, 1972.

Hatch, M. D., and Slack, C. R., Ann. Rev. Plant Phys. **21,** 141–162, 1970. "Photosynthetic CO_2-Fixation Pathways."

Hill, R., Proc. Roy. Soc. (London) **B127,** 192–210, 1939. "Oxygen Produced by Isolated Chloroplasts."

Katz, J. J., and Norris, J. R., Jr., Current Topics in Bioenerg. **5,** 41–75, 1973. "Chlorophyll and Light Energy Transduction in Photosynthesis."

Ke, B., Biochim. Biophys. Acta **300,** 1–73, 1973. "The Primary Electron Acceptor of Photosystem I."

Kelly, G. J., Latzko, E., and Gibbs, M., Ann. Rev. Plant Phys. **27,** 181–205, 1976. "Regulatory Aspects of Photosynthetic Carbon Metabolism."

Krogmann, D. W., *The Biochemistry of Green Plants.* Prentice-Hall, 1973.

Laetsch, W. M., Ann. Rev. Plant Phys. **25,** 27–52, 1974. "The C4 Syndrome: A Structural Analysis."

Levine, R. P., Sci. Am. **221,** 58–70, June 1969. "The Mechanism of Photosynthesis."

Miyami, S., et al., eds., *Photosynthetic Organelles.* Center for Academic Publications, Japan, 1977.

Nelson, N., Biochim. Biophys. Acta **456,** 314–338, 1976. "The Structure and Function of Chloroplast ATPase."

Park, R. B., and Sane, P. V., Ann. Rev. Plant Phys. **22,** 395–430, 1971. "Distribution of Function and Structure in Chloroplast Lamellae."

Photosynthesis Bicentennial Symposium, Proc. Natl. Acad. Sci. (U.S.) **68,** 2875–2897, 1971.

Rabinowitch, E. I., and Govindjee, Sci. Am. **213,** 74–83, Jan. 1965. "The Role of Chlorophyll in Photosynthesis."

Racker, E., *A New Look at Mechanisms in Bioenergetics.* Academic, 1976.

Radmer, R., and Kok, B., Ann. Rev. Biochem. **44**, 409–433, 1975. "Energy Capture in Photosynthesis: Photosystem 11."

Reeves, S. G., and Hall, D. O., Biochim. Biophys. Acta **463**, 275–297, 1977. "Photophosphorylation in Chloroplasts."

Siedow, J. N., Yocum, C. F., and San Pletro, A., Current Topics Bioenerg. **5**, 107–123, 1973. "The Reducing Side of Photosystem I."

Simonis, W., and Urbach, W., Ann. Rev. Plant Phys. **24**, 89–114, 1973. "Photophosphorylation in Vivo."

Thornber, J. P., Ann. Rev. Plant Phys. **26**, 127–158, 1975. "Chlorophyll-Proteins: Light-Harvesting and Reaction Center Components of Plants."

Trebst, A., Ann. Rev. Plant Phys. **25**, 423–458, 1974. "Energy Conservation in Photosynthetic Electron Transport of Chloroplasts."

Whittingham, C. P., *The Mechanism of Photosynthesis*. Elsevier, 1974.

Zelitch, I., Ann. Rev. Biochem. **44**, 123–145, 1975. "Pathways of Carbon Fixation in Green Plants."

CHAPTER TEN

The Nature of the Gene

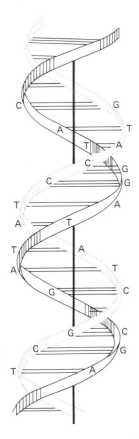

We will begin this chapter with a review of the major findings of the early geneticists and cytogeneticists with the intent of providing a background for the analysis of the molecular nature of the gene and gene expression. Much of this material will be presented in a historical framework in the hope of placing these studies into better perspective with respect to the information available at the time the studies were conducted. It is assumed that the reader is familiar with the basic findings from studies on classical genetic systems. In the following discussion, those sections dealing with genetic breeding studies are set in the smaller typeface and represent background material for the accompanying cytological topics.

The second half of the nineteenth century was an important period in biology for it marked an era when many of the most basic questions were posed and the first important insights were made. In no field was this more true than in the study of inheritance. During the period between 1850 and 1865, the findings and thoughts of two men, Charles Darwin and Gregor Mendel, were to lead to two of the most important statements made during the history of our search to understand ourselves and our universe. In essence, both studies were concerned with aspects of the same overall question, yet the response that their theories generated at the time could not have been more different. Darwin's theory of natural selection as the guiding principle behind organic evolution caused an immediate furor and became the central topic of conversation in both scientific and nonscientific circles. In contrast, the minutes of the meetings of the Natural History Society of Brunn in 1865, where Mendel reported the findings of his nearly ten years work, record no discussion of his presentation. Its publication in the society's journal generated no interest until 1900 when three different European botanists independently reached the same conclusions and independently rediscovered Mendel's paper which had been sitting on the shelves of numerous libraries throughout Europe for the preceding 35 years.

MENDELIAN CONCEPTS OF GENETICS

In the next few pages we will survey the findings of some of the early geneticists, not simply for their historical content, but as a means to review the basic foundation of information in this important area of cell biology. Mendel's laboratory was a small garden plot on the grounds of the Austrian monastery (now in Czechoslovakia) to which he belonged. His goal was to determine the principles, if such existed, governing the transmission of inherited characteristics. Mendel chose the garden pea for a number of practical reasons. His approach to the study of inheritance was to be via the analysis of hybrids. Having obtained 34 varieties of peas locally, he narrowed his studies down to 14 varieties, 2 each of 7 readily identifiable traits. The 2 distinguishing characteristics for each of these 7 traits are shown in Table 10-1. He began with a necessary control experiment: he made certain that each of the varieties he had obtained would breed true when self-fertilized. The next year he was ready to cross-fertilize his plants and examine the condition of the hybrids. Taking pollen from the stamens of one plant with a fine camel's hair brush, he dusted the stigma of a plant having the other characteristic for that trait. He did this same experiment for all seven traits, being certain to make the cross in both directions. Altogether he made 287 cross-fertilizations on 70 plants. For ease of discussion we will focus on the results of just one set of crosses, that of seed shape.

TABLE 10-1

Trait	Varieties
1. Seed color	Yellow, green
2. Seed shape	Round, wrinkled
3. Pod shape	Inflated, wrinkled
4. Flower position	Axial, terminal
5. Stem length	Long, short
6. Pod color	Green, yellow
7. Seed coat color	Colored, white

When plants having wrinkled seeds were crossed with plants having round seeds, Mendel found that all plants of the next generation (later termed the F_1 *generation*) had round seeds (Fig. 10-1a). Mendel referred to the characteristic that appeared in the hybrid as the dominant one and the characteristic that had seemingly been lost as the recessive one. The next experiment to perform, the following year, was to allow the hybrid plants to fertilize themselves and observe the nature of the subsequent offspring (the F_2 generation). Undoubtedly to his surprise, he found that the characteristic that had previously disappeared (wrinkled seeds) had now returned. Mendel then set about on a unique course—he counted the number of plants with wrinkled seeds and the number with round seeds and found that there were approximately three times as many plants having round seeds as wrinkled ones (Fig. 10-1a). This same 3 : 1 ratio was found for the other 6 traits as well; thus it was not an accidental occurrence, but had to be explained by some mathematical relationship. Even though each of the F_1 hybrid plants *seemed* to have the exact same properties as one of its original parents, it had to have been different since the original plant with round seeds, when self-fertilized, produced only plants with round seeds, while the F_1 produced a quarter of the plants with wrinkled seeds.

Mendel continued his breeding experiments by crossing the various F_2 plants with themselves and the original stocks. Some of the findings are shown in Fig. 10-1a. He found that the F_2 plants with wrinkled seeds, when allowed to self-fertilize, would produce only plants (F_3) with wrinkled seeds; for all practical purposes he seemed to have regained one of the stocks that he had started with. However, self-fertilization of the F_2 plants with round seeds indicated that these plants were of two types: one-third produced only plants with round seeds and the other two-thirds produced both types of plants, once again in the ratio of 3 : 1 in favor of round seeds. From these experiments Mendel drew a number of conclusions: The characteristics of the plants were governed by factors (or units) of inheritance. Each trait of the plant was determined by the presence of two independent factors, one derived from each parent. The two factors for each trait in a given plant could be of an identical or a nonidentical nature. For each of the seven traits, one of the two factors was dominant over the other. When both were present together as in the F_1, the existence of the recessive factor was masked by the dominant one.

Early in the twentieth century some new terms came into usage. The term *gene* was used to describe the units of inheritance and the term *allele* (or *allelomorph*) to refer to alternate forms for the same gene. The genes for smooth and wrinkled seeds were alleles. When two identical forms of a gene for a given trait were present, the individual was said to be *homozygous* for that trait and when two different alleles were present, the individual was said to be *heterozygous*. Mendel realized that, even though the F_1 hybrids resembled one of their parents, they showed different breeding properties—they had the same *phenotype* (appearance) but different *genotypes* (pair of genetic factors). If we assign the symbol R to the dominant gene for seed shape and r to the recessive gene, the genotype of each plant will be repre-

Figure 10-1. Principles underlying Mendel's crosses between different strains of pea plants in which the strains differed with respect to one trait. *(a)* Phenotypes of plants produced over several generations of crosses between strains having round as opposed to wrinkled seeds. *(b)* Genotypes of the plants shown in *a*. The initial F_1 generation is formed by crossing plants of the two strains while all subsequent generations are obtained by self-fertilization.

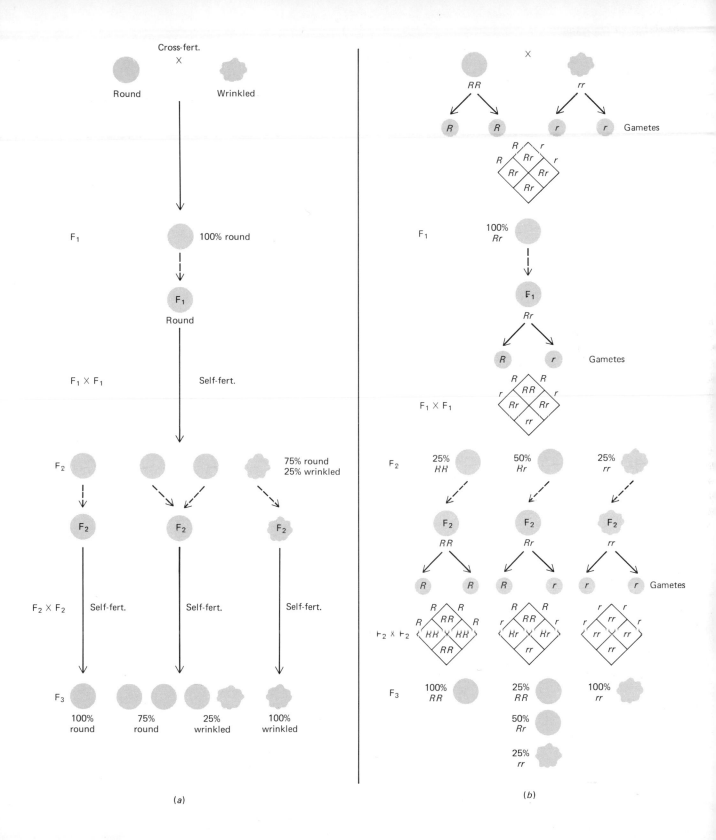

(a)

(b)

sented by some combination of these two symbols (Fig. 10-1*b*). The three possible genotypes are RR, Rr, and rr, i.e., the homozygous dominant, the heterozygous, and the homozygous recessive. Both RR and Rr will appear phenotypically dominant, i.e., will have round seeds, while rr will appear phenotypically recessive, i.e., will have wrinkled seeds. All F_1 plants were Rr. Among the F_2 plants all three possible genotypes were represented in the ratio 1RR : 2Rr : 1rr.

In addition, Mendel concluded that each of the gametes produced by a plant had only one factor for each trait. A particular gamete could have either the recessive (r) or the dominant (R) factor, but not both. Even though the pair of factors controlling a trait stayed together throughout the life of an individual plant, they had to separate (or segregate) from one another during the formation of the gametes (Fig. 10-1*b*). The factors in the gametes remained "pure"; they had not been altered by the presence of the other factor within the plant. Just as each plant produced gametes with but one factor, each plant arose by the union of two gametes, each carrying but a single factor. In other words, each individual was represented genetically by equal contributions from each parent in the form of a maternal and a paternal factor for each trait. In addition, based upon the numbers of various types of offspring, Mendel concluded that gametes bearing each of the two factors were present in equal numbers, and that the likelihood for a particular combination of factors arising in a given individual was strictly a chance affair.

Following these experiments, Mendel attempted crosses between plants that differed from one another in more than one trait (Fig. 10-2). For example, he crossed plants having round, yellow seeds (RRYY) with ones having wrinkled, green seeds (rryy). As expected, the seeds of the offspring showed only the dominant characteristics, i.e., seeds of a round, yellow variety (RrYy). However, when he allowed these hybrid plants to self-fertilize, four different types of seeds could be found among the next generation. Most of the seeds (approximately 9/16) were round and yellow, while approximately 3/16 were wrinkled and green. In addition, however, were two types of seeds that he had never seen before, round and green seeds (represented by approximately 3/16 of the population) and wrinkled and green seeds (represented by approximately 1/16 of the population). In order to account for these results, Mendel made a final conclusion. The segregation of the pair of factors for one trait had no effect upon the segregation of the pair of factors for another trait (Fig. 10-2). Just because an individual inherited a green factor and a round factor from the pollen cell of one of its parents, did not mean that, upon formation of its own gametes, those two factors must remain together. Rather, since the factors for seed color and seed shape segregated independently from one another, it would be just as likely that a particular *gamete* would have a green-round combination as a green-wrinkled combination or a yellow-round or a yellow-wrinkled combination. There would be as many types of gametes (RY, Ry, rY, ry) as possible combinations of characters, all in equal proportion. This law has become known as Mendel's *law of independent assortment* and its effect upon the phenotypes and genotypes of the next generation is illustrated in Fig. 10-2.

It is not surprising that in 1865 no one took Mendel very seriously. It would be quite a few years before the world would know about meiosis, fertilization, and chromosomes, yet here was a person proposing that invisible factors became shuffled together and later sorted themselves out. Mendel was obviously many years ahead of his time, which makes his work that much more remarkable. With very little background information at his disposal, he set out to learn about heredity. He carefully planned his experimental approach, he recorded his observations in the most meaningful way, he drew his conclusions for which no precedent existed, and then went about testing them with further experiments. There is one important feature of Mendel's work that has pervaded nearly all of the research in both classical and molecular genetics—irrefutable conclusions have been made concerning structures and processes that we cannot directly observe. Probably more than any other type of biologist, geneticists, both classical and molecular, have had

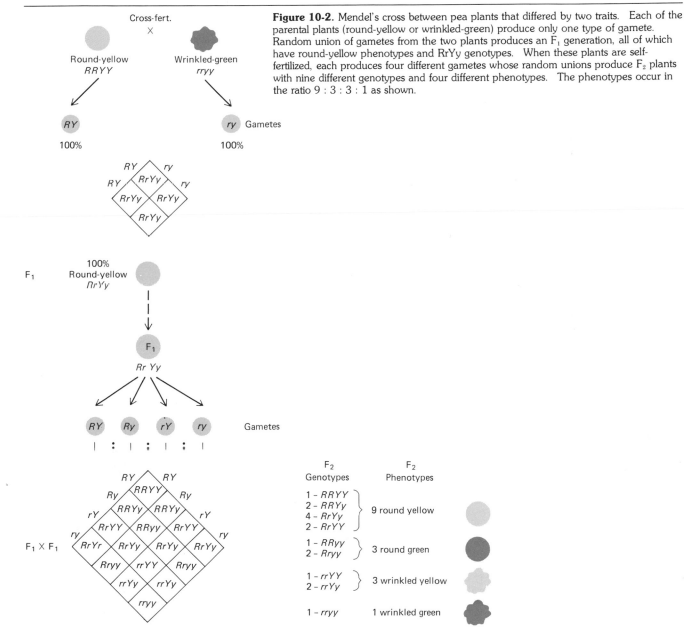

Figure 10-2. Mendel's cross between pea plants that differed by two traits. Each of the parental plants (round-yellow or wrinkled-green) produce only one type of gamete. Random union of gametes from the two plants produces an F₁ generation, all of which have round-yellow phenotypes and RrYy genotypes. When these plants are self-fertilized, each produces four different gametes whose random unions produce F₂ plants with nine different genotypes and four different phenotypes. The phenotypes occur in the ratio 9 : 3 : 3 : 1 as shown.

to rely upon their wits to design the appropriate experiment the results of which, one way or the other, would lead them to a conclusion based upon pure inductive reasoning. This will become apparent as we consider various types of experiments in this and the next few chapters and we will return to this point again.

CYTOGENETICS: EARLY OBSERVATIONS

Whereas Mendel provided convincing evidence of the behavior of theoretical factors responsible for determining the nature of inheritable traits, his research was totally unconcerned with the physical nature of these units or their location within the organism. Mendel was able to carry out his entire research project without ever having to observe anything in a microscope. During the time between Mendel's work and his rediscovery, a considerable number of biologists were concerned with this other aspect of heredity, i.e., its physical basis within the cell. This was a much more visual type of analysis, one involving the examination of cell behavior and of cell parts. Some of the most important observations of the 1880s and 1890s led a number of these early cytologists to propose that the chromosome was the structure in which the hereditary influences were kept. Some of the findings that led to this proposal will now be described.

Cell Division and Fertilization

A most important concept that emerged before the turn of the century was that of *genetic continuity*. It was realized that, whatever the hereditary structure might be, it would have to be passed on from cell to cell and from generation to generation. Examination of mitosis, i.e., cell division (particularly by Walther Flemming), revealed in a broad manner the way in which the cell contents were divided so as to provide each daughter cell with its rightful share. It was noted that most of the contents of the cell seemed merely to be shuttled into one daughter cell or the other as a matter of chance depending upon the plane through which the furrow happened to divide the cell. In contrast, it appeared that the cell went to great lengths to ensure that the nuclear contents became organized into threads which in 1888 were named chromosomes. At mitosis each of the chromosomes could be seen to be doubled. During the division, the members of each chromosome doublet split longitudinally from each other and were passed to each of the two cells (see Chapter 15 for details of mitosis). Once in the daughter cells, chromosomes would again disappear from view until the next division when they would reappear in a doubled form.

Careful observations indicated that a chromosome that disappeared after one division *seemed* to be the same structure that reappeared before the next. Even though the chromosomes were not visible in the "resting" nucleus, it was believed that their existence as an entity was maintained. In other words, there appeared to be a chromosomal continuity from mother to daughter cells, just as there was believed to be a genetic continuity. Just as cells only arise from preexisting cells, it appeared that chromosomes only arose from preexisting chromosomes. As would be expected, the chromosome number remained constant from cell to cell.

The process of fertilization was analyzed and the role of the two gametes, the sperm and the egg, was observed. Attention was turned to what might be present in these two cells that was responsible for endowing the offspring with its genetic inheritance. Even though the contribution from the male was a very tiny cell, it was known to be as important in a genetic sense as was that of the much larger egg. What was it that these two grossly different cells had in common? The only apparent feature was the nucleus and its chromosomes which were known to exist in both cell types. The processes occurring after fertilization were followed in the roundworm *Ascaris,* whose few chromosomes are large and were as readily observed in the nineteenth century as in introductory biology laboratories today. In 1883 it was noted by Edouard van Beneden that the body cells of this worm had four large chromosomes, but that the male and female nucleus present in the egg just after fertilization (before the two nuclei come together) had but two chromosomes apiece (Fig. 10-3). During the 1880s the process of meiosis was described, and in 1887 August Weismann proposed the existence of some type of reduction division that would occur in both the ovaries and testes to allow for the chromosome number to be divided in half during the process of gamete formation. If this type of reduction did not occur, the number of chromosomes would double from generation to generation, which obviously did not and

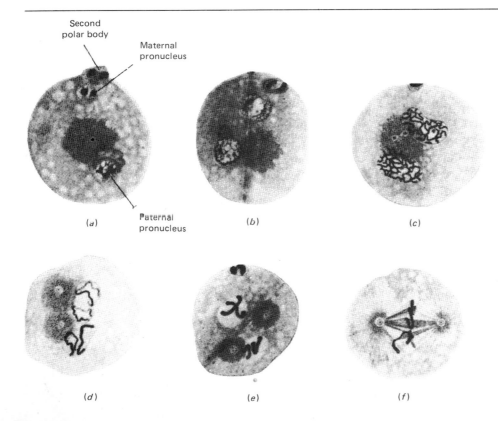

Second polar body

Maternal pronucleus

Paternal pronucleus

(a) (b) (c)

(d) (e) (f)

Figure 10-3. Events occurring in the roundworm *Ascaris* very soon after fertilization. Both the male and female gamete contain two chromosomes which, after fusion of the pronuclei in the egg cytoplasm (between e and f), leads to a chromosome number of four in the zygote. *(From T. Boveri,* Jenaische Zeit., *22: 685, 1888.)*

could not occur. However, these cytological observations bore no obvious relation to inheritable characteristics without the knowledge of the behavior of genetic markers gathered by Mendel, and, thus, a few speculations that were made on the role of the chromosomes in inheritance had only a very indirect basis.

The Chromosome as the Carrier of Genetic Information

The rediscovery of Mendel's work and its acceptance based upon confirming observations had an important influence on the cytological studies of the time. The results from the breeding studies demanded that, whatever the physical nature of the carrier of the hereditary units, it would have to behave in a manner consistent with Mendelian principles. In other words, the genetic studies set certain guidelines that cytological proposals would have to follow. At about the same time as the rediscovery of Mendel's work, an important finding was made by Theodor Boveri in his study of the eggs of sea urchins that had been fertilized by two sperm, rather than just one as is normally the case. This condition, referred to as *polyspermy,* is a very serious one and inevitably leads to the early death of the embryo. Why should the presence of one extra tiny sperm within the very large egg have such drastic consequences? The reason was discovered when the chromosomes of the cells of the early embryo were examined. As a result of the extra set of paternal chromosomes and an extra centriole (page 701), the machinery responsible for the early cell divisions is placed in an impossible situation. For certain reasons, the chromosomes are not passed to the various daughter cells in an equivalent manner, but rather the number of chromosomes present among the cells is quite variable. This finding suggested two important points: not all chromosomes are of the same nature, and a particular number (a complete *complement*) is needed to ensure the proper functioning of the cell. Up until Boveri's research there was no evidence of a *qualitative difference* among chromosomes, which would be expected for a set of carriers of genetic information. To the eyes of the early cytologists, the chromosomes all looked about the same. Now an explanation had been provided for the finding that the chromosomes number remained strictly constant from cell to cell; all of them were needed for the life of the cell.

In 1902 and 1903, Walter Sutton, a graduate student of the prominent American cytologist Edmund Wilson, published two papers which are generally accepted as the first definitive demonstration of the chromosomes as the physical basis of inheritance. Sutton's research centered on the nature of the chromosomal activities occurring during sperm formation (*spermatogenesis)* in the grasshopper, which, like *Ascaris,* has readily observable chromosomes and is also used in laboratory class exercises. There are two types of divisions occurring among germ cells within the male gonad (Fig. 10-4): the mitotic divisions whereby spermatogonia produce more spermatogonia, and the meiotic divisions whereby spermatogonia produce spermatids (which then are converted to mature spermatozoa without further division). Upon examination of the mitotic stages of the spermatogonia, Sutton was able to count 23 chromosomes. Most importantly, he found that, among 23 chromosomes, there were recognizable types, i.e., recognizable on the basis of size and shape. When all 23 chromosomes were considered, he found that the chromosomes were present in pairs. There were 11 pairs of distinguishable chromosomes and one additional chromosome, termed an accessory chromosome, which did not seem to have a mate.

Sutton realized that the presence of pairs of chromosomes (*homologous chromosomes*) correlated perfectly with the pairs of inheritable factors uncovered by Mendel. In addition, when he examined the chromosomes in cells just beginning meiosis (the primary spermatocytes), he found that the members of each pair of chromosomes were associated with one another forming what is termed a *bivalent*. Eleven bivalents were visible, each showing a transverse line of fusion (Fig. 10-5). The ensuing first meiotic division resulted in the separation of the two homologous chromosomes into separate cells (secondary spermatocytes). Sutton proposed that the two members of each pair of homologous chromosomes were the maternal and paternal chromosomes derived from those originally brought together by the fusion of a sperm and an egg at fertilization. One

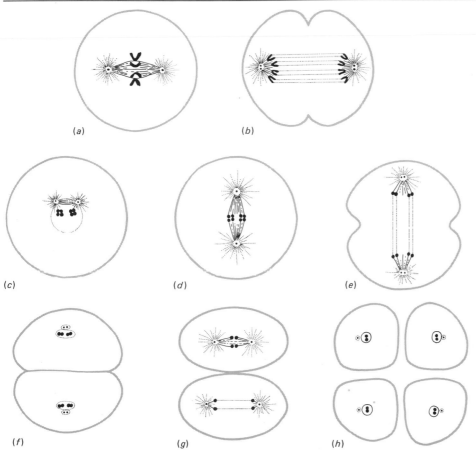

Figure 10-4. Diagrams showing the essential facts of meiosis in the male. The somatic number of chromosomes is four. The last mitotic division of this line of spermatogonia is shown in part *b*. The cell shown in *d* is termed a primary spermatocyte which undergoes the first meiotic division (the reduction division) in *e* to form two secondary spermatocytes which undergo a second meiotic division in *g* to form four spermatids. (This process is discussed at greater length in Chapter 15.) (*From E. B. Wilson, "The Cell in Heredity and Development," 3d ed., 1925. Reprinted with permission of Macmillan Publishing Co., Inc.*)

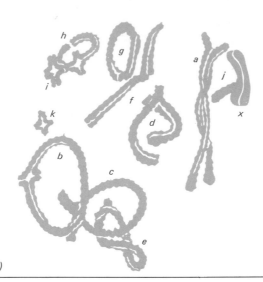

Figure 10-5. The appearance of the chromosomes from meiotic prophase in the male grasshopper. In addition to the X chromosome, 11 pairs of chromosomes *(a–k)* could be seen in association with one another. *(From W. S. Sutton,* Biol. Bull., **4**:24, 1902.)

complete set of chromosomes was present in each gamete. From that single cell, the fertilized egg, all of the cells of the body eventually form by a succession of mitotic divisions. Throughout all of these mitotic divisions (Fig. 10-6*a*), the maternal and paternal chromosomes remain together within the many nuclei, together controlling the physiological affairs of each cell. Only within the gonads and only during the formation of the gametes do these maternally and paternally derived chromosomes associate with each other and then undergo separation (Fig. 10-6*b*). This was the reduction division proposed many years earlier on theoretical grounds. Here also was the physical basis for Mendel's proposal that genes exist in pairs that remain together but independent throughout the life of the individual, and then separate from one another upon formation of the gametes. Since the maternal and paternal members of a homologous pair of chromosomes govern the same trait, the reduction division does not result in a loss of genetic information.

It would be expected that the two alleles governing the same trait would be located on the same site in each chromosome. When the individual was heterozygous for a particular trait, the homologous chromo-

Figure 10-6. The distribution of chromosomes into daughter cells following mitosis or meiosis. The two traits being considered (seed color and seed shape) are present on two different chromosomes, i.e., they are unlinked, and the individual plants are all heterozygous for both traits, i.e., RrYy. Chromosomes with a dotted edge are maternally derived, ones with a solid edge are paternal. *(a)* Mitosis. After duplication there is no association of homologous chromosomes and each lines up independently at the metaphase plate. Separation of the duplicates produces daughter cells with one of each of the four chromosomes; these are therefore RrYy. *(b)* Meiosis. After duplication, the homologous chromosomes associate to form two bivalents. Anaphase I separates one of the homologous chromosomes from the other, producing daughter cells with either one or the other allele for each trait. Among a population of cells undergoing meiosis, there will be two different types of division with respect to the genetics of the products. In approximately half of the cases, the maternal chromosomes go to the same pole and the paternal chromosomes to the other pole (right side). In the other half, each daughter cell from the first meiotic division inherits one paternal and one maternal chromosome (left side). After the second meiotic division each cell has produced four haploid gametes [either 2RY and 2ry (right side) or 2Ry and 2rY (left side)]. In the population as a whole the four types of gametes will be present in equal number.

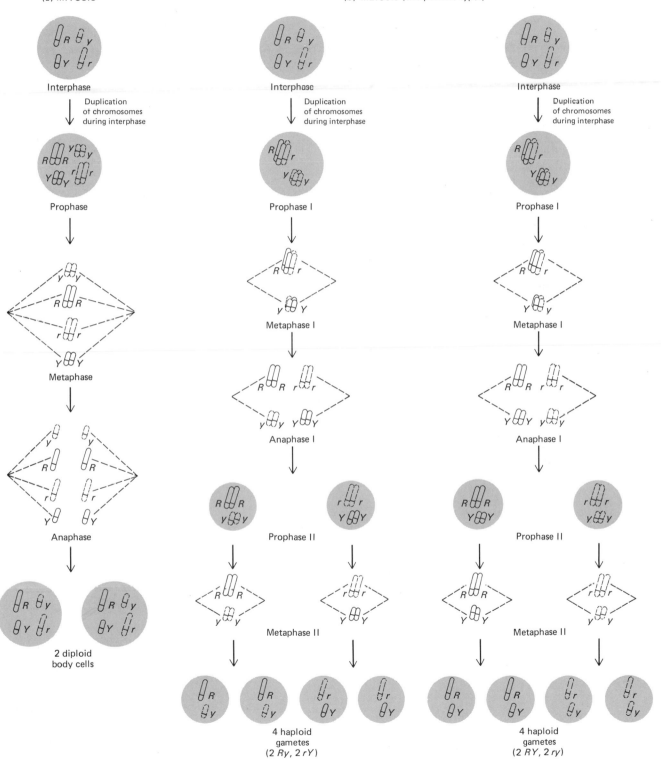

(a) MITOSIS

(b) MEIOSIS (two possible types)

Interphase

Duplication of chromosomes during interphase

Prophase

Metaphase

Anaphase

2 diploid body cells

Interphase

Duplication of chromosomes during interphase

Prophase I

Metaphase I

Anaphase I

Prophase II

Metaphase II

4 haploid gametes (2 Ry, 2 rY)

Interphase

Duplication of chromosomes during interphase

Prophase I

Metaphase I

Anaphase I

Prophase II

Metaphase II

4 haploid gametes (2 RY, 2 ry)

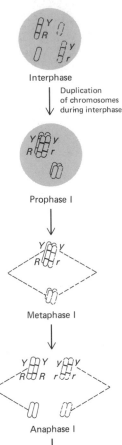

Interphase

Duplication
of chromosomes
during interphase

Prophase I

Metaphase I

Anaphase I

Prophase II

Anaphase II

4 haploid gametes
(2 RY, 2 ry)
no other possibilities

(a)

somes would contain different factors at the corresponding sites in the two homologous chromosomes (Fig. 10-6). The reduction division observed by cytologists explained several other findings of Mendel: the gametes would be pure for one characteristic or another (Fig. 10-6b), an equal number of each type of gamete would be formed (Fig. 10-6b), and fertilization would result in the reformation of an individual with two factors for each trait. The term *diploid* has been adopted in referring to a cell with two sets of chromosomes, while the term *haploid* refers to the condition in the sperm, in which only one set is present.

Linkage Groups

We can now reexamine Mendel's dihybrid cross between plants having smooth, yellow seeds with ones having wrinkled, green seeds and observe the basis for the distribution of traits in the offspring (Fig. 10-6b). Since the two traits, seed color and seed shape, are present on different chromosomes, they will be present within different bivalents. Since each bivalent, containing its pair of homologous chromosomes, is spatially distinct from every other bivalent, there is no way that the separation of homologous chromosomes of one pair can affect the separation of any other pair. As a result, the maternal and paternal members of each pair segregate in a random manner and gametes with all possible combinations will be formed.

As clearly as Sutton saw the relationship between chromosome behavior and Mendelian genetics, he also saw one very important basis for contention.

Figure 10-7. *(a)* The distribution of chromosomes into daughter cells following meiosis when the two traits being considered are on the same chromosome and the individual plants are all heterozygous for both traits, i.e., RrYy. If we assume that crossing-over does not occur, there will be only two types of gametes produced, RY and ry, in equal numbers. *(b)* A cross between two of these heterozygous individuals will produce offspring with only three different genotypes and two different phenotypes. The phenotypes will be in the ratio of 3 : 1.

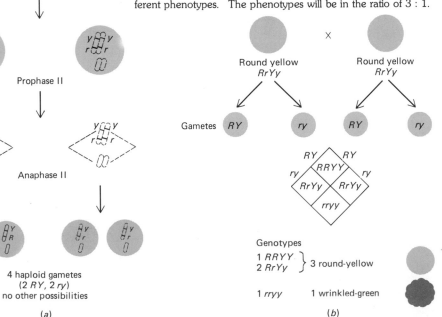

X

Round yellow
RrYy

Round yellow
RrYy

Gametes   RY   ry   RY   ry

Genotypes

1 *RRYY*
2 *RrYy* } 3 round-yellow

1 *rryy*   1 wrinkled-green

(b)

Mendel had examined the inheritance of seven traits and found that the manner in which each of these characteristics was inherited was independent of the inheritance of the others. This formed the basis of his law of independent assortment. We need to consider the situation once again. Since the number of different chromosomes of an organism, whether an insect or a mammal, is relatively small and the number of possible traits to be genetically determined is relatively large, it follows that each chromosome must bear numerous traits. Assuming that the genes for various traits are associated in groups which are physically equivalent to chromosomes, then intact groups of traits should be passed from each parent to their offspring, just as intact chromosomes are passed. Consider the effect of this conclusion on the F_2 plants had the two traits been on the same chromosome (Fig. 10-7) rather than different chromosomes. Under these conditions, the F_1 plants could produce only two types of gametes (RY and ry), and F_2 plants with only two phenotypes (round-yellow and wrinkled-green, ratio of 3 : 1) rather than four types of gametes and four different F_2 phenotypes as shown in Fig. 10-2. Traits present on the same chromosome should remain linked to one another, i.e., should occur in the same *linkage group*.

How is it that all of Mendel's seven traits demonstrated independent assortment? Were they all in different linkage groups, i.e., on different chromosomes? As it turned out, the garden pea has 7 different chromosomes (a diploid number of 14) and each of the 7 traits that Mendel reported on does occur on a different chromosome. Whether Mendel owed his findings to good fortune, divine intervention, or simply to the lack of interest in a trait or two which did not fit, will have to remain a topic of discussion. Sutton's prophecy of the existence of linkage groups was not long in the unfulfilled state. Within a couple of years the evidence for linkage of two traits (flower color and pollen shape) in sweet peas had appeared and other evidence would soon follow.

GENETIC ANALYSIS IN *DROSOPHILA*

Research in genetics soon came to focus on one organism, the fruit fly *Drosophila* (Fig. 10-8). Here was an animal with a generation time (from egg to maturity) of 14 days and the capability of producing up to 1000 eggs in a lifetime. In addition, it was very small so a large number could be kept on hand, they were easy to maintain and breed, and very inexpensive. In 1909, it seemed like the perfect organism to Thomas Hunt Morgan, and he began what was to be the start of a whole new era in genetic research. There was one large disadvantage in beginning work with this insect—there was only one type of fly available, the *wild type*. Whereas Mendel had simply purchased 34 varieties of seeds, Morgan had to generate his own variety of organisms. Morgan expected that variants or mutants from the wild type might appear if he bred sufficient numbers of flies. Within a year and thousands of flies, he found his first mutant, one having white eyes rather than the normal red-colored ones. By 1915 he and his students had found 85 different mutants, i.e., individuals having altered, inheritable characteristics from the wild type affecting a wide variety of structures of the fly. It was apparent that infrequently a spontaneous change or *mutation* occurred within a gene, altering it in some permanent fashion, so that it could be passed from generation to generation. The demonstration of a spontaneous, inheritable alteration in a gene had consequences far beyond the study of *Drosophila* genetics. Here was a demonstration of the origin of the variation that exists within popula-

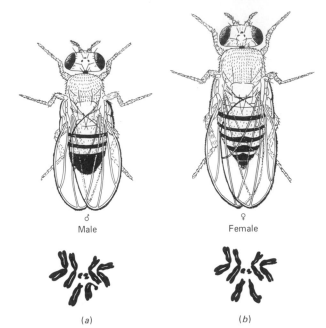

Male   Female

(a)   (b)

Figure 10-8. The appearance of the female and male *Drosophila melanogaster* and their complement of mitotic chromosomes. *(From T. H. Morgan, C. B. Bridges, and A. H. Sturtevant, "The Genetics of Drosophila," Martinus Nijhoff, 1925.)*

tions; evidence for a vital link in the theory of evolution. If new forms of genetic traits could arise in this manner, then natural selection could act upon these mutants, and new species could slowly emerge. Genetic mutation can be considered as the raw material needed for the eventual mutability of the species.

Whereas mutations are a necessary occurrence for evolution, they are a tool for geneticists, a marker providing contrast against the wild-type condition. As *Drosophila* mutants were isolated, they were bred and cross-bred and kept as stocks within the lab. As expected, all of these 85 different mutations did not independently assort; instead, Morgan found that they belonged to 4 different linkage groups, one of which contained very few mutant genes (only 2 in 1915). Perfectly correlated with this finding was the presence of 4 different chromosomes (Fig. 10-8) in the cells of *Drosophila*, one of which was very small.

Sex Determination and Sex-linked Inheritance

Meanwhile, during this period, a parallel story on the role of the chromosomes was unfolding. It was mentioned above that Sutton found 11 pairs of chromosomes in grasshopper spermatogonia and an accessory chromosome, later called the X chromosome. It was proposed as early as 1901 that accessory chromosomes were involved in the determination of the sex of the individual bearing them. During the next few years this proposal was confirmed and it was shown that the male of the species could be distinguished by its chromosomes from the female. In most species the male had a single X chromosome (together with a small Y chromosome) and the female had a pair of X chromosomes. It was soon shown that the X and Y chromosome associated with one another during meiosis in the testes, and the two markedly different chromosomes were separated from each other during the reduction division (Fig. 10-9). Whereas all female gametes will have an X chromosome, half of the sperm

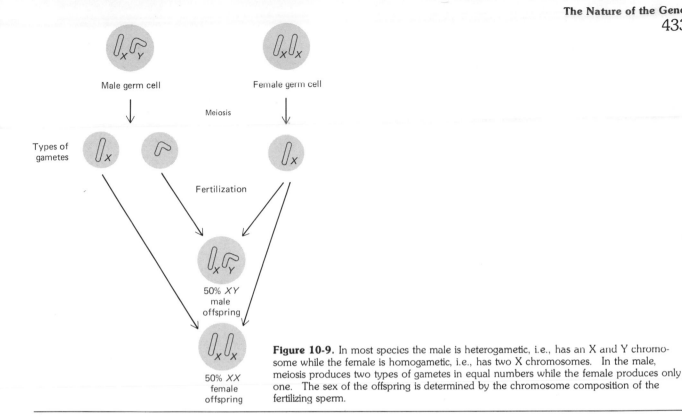

Male germ cell

Female germ cell

Meiosis

Types of gametes

Fertilization

50% *XY* male offspring

50% *XX* female offspring

Figure 10-9. In most species the male is heterogametic, i.e., has an X and Y chromosome while the female is homogametic, i.e., has two X chromosomes. In the male, meiosis produces two types of gametes in equal numbers while the female produces only one. The sex of the offspring is determined by the chromosome composition of the fertilizing sperm.

will have an X and the other half a Y. When an X-bearing sperm fertilizes an egg, the offspring will contain two X chromosomes, causing it to be a genetic female. When a Y-bearing sperm enters the egg, the XY offspring is a genetic male. Here was a clear-cut correlation between the presence of a particular chromosome and a genetically determined trait.

If we return briefly to Morgan's first mutant, the white-eyed fly, we can demonstrate another feature of genetic inheritance. Morgan's first mutant was a male, which, when crossed with a wild-type female, produced red-eyed offspring, both male and female. This is in keeping with Mendelian genetics; the white-eyed trait is simply recessive. Presumably, all of the offspring had one dominant factor and one recessive factor. However, when Morgan bred the F_1 flies with each other, he found about 75% red-eyed and 25% white-eyed flies, as expected, but all of the females of the F_2 were red-eyed and approximately half of the males were white-eyed. After an extensive series of crosses, Morgan concluded that the trait was sex-linked—it must be carried on the X chromosome (Fig. 10-10). Equally as important, there must be no corresponding site on the Y. Whereas Morgan showed the distribution of the white-eyed gene completely parallels the distribution of the X chromosome, a student of Morgan's, Calvin Bridges, provided convincing evidence; the identification of a particular gene on a particular chromosome had been made. Not only was the X chromosome involved in sex determination, it also carried genes whose function was unrelated to sexual characteristics. In fact, if the white-eyed characteristic had not been sex-linked, it would not have been detected since it is a recessive characteristic and its chance of appearing in a homozygous state in a population of wild, red-eyed flies is vanishingly small.

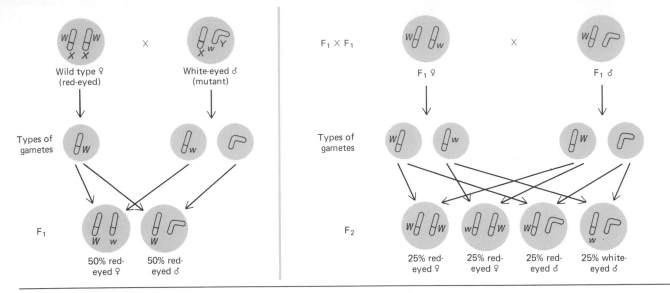

Figure 10-10. The inheritance of sex-linked characteristics. In most cases genes carried on the X chromosome have no corresponding allele on the Y chromosome leading to the distribution shown. The phenotype of the male is determined by a single allele inherited maternally.

Many examples of sex-linked characteristics are known in humans, the most familiar being color-blindness and hemophilia, both diseases which afflict primarily males.

Crossing-over and Recombination

Even though the association of genes into linkage groups was found to occur, it was usually the case that the linkage was incomplete, i.e., a certain percentage of new combinations of characteristics was found. For example, it was mentioned that the first demonstration of linkage was made on sweet peas, using genes for pollen grain shape and flower color. In this experiment, plants with purple flowers and long pollen grains were crossed with plants having red flowers and round pollen grains. All the F_1 had purple flowers and long pollen grains. The F_1 were then crossed with themselves. If the two traits were unlinked (as in Fig. 10-2), i.e., capable of independent assortment, one would expect a 9 : 3 : 3 : 1 distribution of the four possible phenotypes. If the two genes were in the same linkage group (as in Fig. 10-7b), one would expect a ratio of 3 : 1 distribution of two phenotypes (purple-long and red-round) and the total absence of the other two combinations. The actual ratios that were observed were approximately 3 : 1 : 0.15 : 0.15, the latter two representing the "forbidden" combinations, purple-round and red-long. The same breakdown in linkage was found to occur in *Drosphila* and it was shown that the "strength" of the linkage, i.e., the frequency with which the unexpected combinations occurred, varied depending upon the two genes being compared, but was constant for any given pair. For example, two genes located on the X chromosome, white eyes and yellow bodies, typically have approximately 1% new combinations while white eyes and miniature wings gave 34% new combinations (Fig. 10-11).

In 1911, Morgan offered an explanation for the breakdown in linkage as well as for the variation in its strength. Two years earlier, it had been

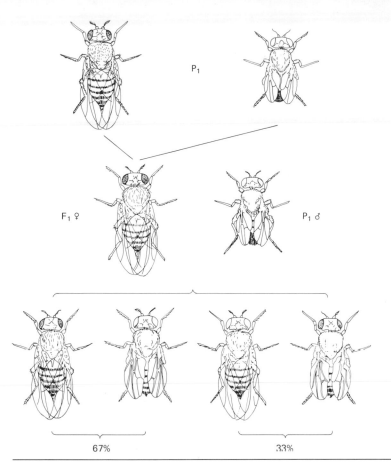

Figure 10-11. The genetic demonstration of crossing-over during the inheritance of two sex-linked traits. A female with red eyes and long wings (wild-type) is crossed with a male with white eyes and short wings. The F_1 females are back crossed with the white-eyed, short-winged parental male. Approximately one-third of the offspring have red eyes and short wings or white eyes and long wings. These individuals contain chromosomes having mixed maternal and paternal components. *(From T. H. Morgan, "The Theory of the Gene," Yale University Press, 1926.)*

observed by F. A. Janssens in a study of meiosis, that the homologous chromosomes of the bivalents actually became wrapped around each other (Fig. 10-12a) while they were pressed together. Janssens proposed that this interaction between maternal and paternal chromosomes might result in the breakage and exchange of pieces. Capitalizing on this observation and proposal, Morgan suggested that this phenomenon, termed *crossing-over,* could account for the new combinations present among the offspring. If exchange of chromosomal parts occurred during meiosis, then the potential existed for genes that previously had been present upon the same chromosome to become separated from one another and occupy positions on separate chromosomes (Fig. 10-12b). Offspring possessing chromosomes with one characteristic located on a maternal part of a chromosome and another characteristic located on a paternal part of the same chromosome were termed a *recombinant* (Fig. 10-11).

The fact that the percentages of recombinations were constant for a pair of genes was strong evidence that their position along the chromosome (their locus) was fixed regardless of which cell in an individual it happened to occur or which individual in the population. Crossing-over explained the breakdown in linkage, and the position of the genes ex-

(a)

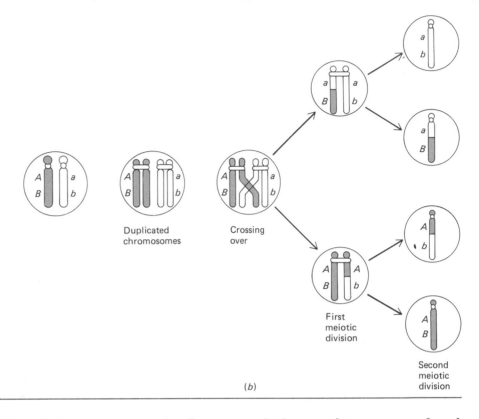

Duplicated chromosomes

Crossing over

First meiotic division

Second meiotic division

(b)

Figure 10-12. (a) A pair of homologous chromosomes from a late stage of meiosis in the grasshopper. The points at which the homologs are crossed over one another are termed chiasmata and are believed to represent sites at which crossing-over had occurred at an earlier stage. (b) The effect of crossing-over on the distribution of maternal and paternal alleles into the gametes. (From E. W. Sinnott, I. C. Dunn, and T. Dobzhansky, "Principles of Genetics," 5th ed., McGraw-Hill, 1958.)

plained the variation in the degree to which recombinants were found among the offspring population. Those genes which gave few recombinants, such as white eyes and yellow bodies, were located close to one another on the chromosome (Fig. 10-13) and were therefore unlikely to be separated by an intervening breakage point. However, genes located farther from one another would stand a greater chance of having a breakage occur between them and, therefore, as in the case of white eyes and miniature wings, should produce a much higher percentage of recombinant offspring. Once more a critically important correlation had been made between the physical activity of a cellular structure that could be observed microscopically and the behavior of the hereditary factor. These types of *cytogenetic* observations, i.e., ones linking cytological and genetic findings, were of the greatest importance in producing a solid foundation for our understanding of the cellular basis of information storage and transmission.

The understanding of variations in the strength of linkage clearly corroborated the concept of the chromosome as a linear array of independent genetic units. Remarkably, genes on chromosomes seemed like beads on a string. In 1911, Alfred Sturtevant realized, if there is a relationship between recombination frequency and distance of two sites on the chromosome, then one should be able to use these values for recombination to assign an order along the chromosome (as in Fig. 10-13) for any available gene. In other words, it should be possible, simply by breeding flies with the appropriate mutations located on the same chromosome, to obtain a

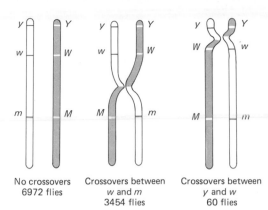

Figure 10-13. The assignment of genes to positions on chromosomes based upon the frequency of crossing-over. *(From "Heredity and Development, 2d ed. by J. A. Moore, Copyright © 1963, 1972 by Oxford University Press, Inc. Reprinted by permission.)*

detailed topographical map of the serial order of these genes along the chromosome. Using this concept, geneticists have mapped chromosomes from viruses to bacteria to a large number of eukaryotic species.

Mutagenesis and Giant Chromosomes

At the start of the century, genetics was an unknown science, whereas, by the end of the first quarter, the potential offered by classical breeding experiments had been essentially fully realized. Morgan, for example, had turned his attention to the analysis of the mechanisms of embryonic development, a frontier which has proved more resistant to exploration. The years following 1925 were primarily a period of filling in holes, compiling additional support for previous proposals, and looking toward the day when a whole new set of questions could be answered. There were at least two major findings in genetics between the early classical and molecular eras: the discovery of the mutagenic properties of x-rays by Hermann Müller and the discovery by Theophilus Painter of the usefulness of the giant chromosomes of the cells of the larval salivary gland of *Drosophila*.

During the early period of genetics, the search for mutants was a slow, tedious procedure dependent upon the *spontaneous* appearance of altered genes. Using a special strain of flies designed to reveal the presence of recessive alleles, Müller estimated the frequency of spontaneous mutations at about 1 in 1000. In contrast, he found that flies subjected to sublethal irradiation have 100 times the mutation frequency. This finding has had several important consequences. On the practical side, the use of mutagenic agents, such as x-rays, provided a great increase in the number of available mutants of many animals and microorganisms for research purposes. The finding also pointed out the hazard of the increasing use of radiation in the industrial and medical fields. In addition, it revealed an important new property of the genetic material: whatever its chemical nature, it must be sensitive to alteration by electromagnetic radiation.

The rediscovery of the giant chromosomes of certain insect cells in the 1930s illustrates a characteristic of biological systems: if you search long

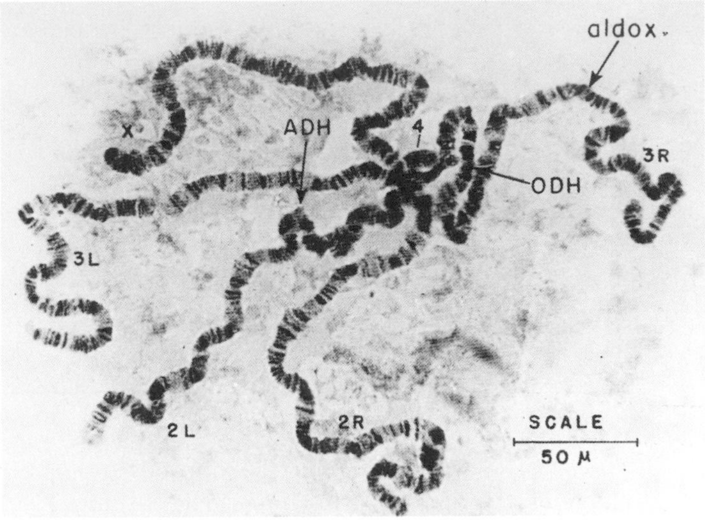

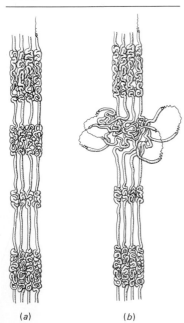

Figure 10-14. The giant polytene chromosomes of *Drosophila melanogaster*, prepared from a squash of a larval salivary gland. Selected gene loci are indicated on the photograph. *(From H. Ursprung, K. Smith, W. Sofer, and D. Sullivan, Science, 160: 1075, 1968. Copyright by the American Association for the Advancement of Science.)*

Figure 10-15. *(a)* A folded fiber model of band-interband structure in a giant polytene chromosome. *(b)* The same model, illustrating a possible mode of "puffing" in one of the bands. *(From E. J. DuPraw and P. M. M. Rae, Nature, 212:600, 1966.)*

enough you are likely to find just about anything. There is such a remarkable variability among plants and animals, not only at the obvious morphological levels, but also at the cellular and subcellular level, that often one particular type of cell may be much better suited for a particular type of research than all of the others. A micrograph of the giant chromosomes from one nucleus of a larval salivary gland is shown in Fig. 10-14. These chromosomes result from the duplication of most of the genetic material and the subsequent failure of the duplicates to separate. This process continues until there are over 1000 identical units attached to one another along most of their entire length. If we consider the chromosome as a type of long, linear, beaded strand, then the giant (or *polytenic*) chromosomes represent a large number of strands lying next to each other, side by side (Fig. 10-15a). Under the microscope, stained preparations of these chromosomes appear banded (Fig. 10-14), the bands reflecting the fact that the strands are in perfect "register." The banding pattern itself results from differences in the density of the components of the chromosome from one end to the other (Fig. 10-15a). Most importantly, the banding pattern is essentially constant from fly to fly and from tissue to tissue, though very great differences are observed between the chromosomes from flies of different species of the *Drosophila* genus.

It was realized by Painter that the giant chromosomes might provide the opportunity to obtain a visual portrait of the chromosomes of a species that had been mapped in the breeding experiments of the past 20 years. Could a correlation be made between the genetic maps obtained by inductive reasoning and the chromosomal banding patterns? Painter and others found that it could. Certain of the individual bands became correlated with specific genes and detailed examination of the patterns provided visual confirmation of the validity of the mapping procedure. It was estimated that one set of the giant chromosomes of *Drosophila* contained approximately 5000 bands and therefore the conclusion was reached that

(a)   (b)

there were approximately 5000 genes in the fruit fly. Though it does appear that only one genetic function can be assigned to a given band, the concept of "one band–one gene" is an oversimplification since each band contains a great deal more genetic material than would be expected for a single gene. The meaning of this situation (termed the *chromomere paradox*) remains to be unraveled.

Not surprisingly, the giant chromosomes of these insects have been useful in a number of ways. For example, they have been particularly valuable in furthering our understanding of chromosomal aberrations. Based upon ingenious efforts, the early geneticists had predicted that certain structural deformities arose in chromosomes on rare occasions. Mapping studies had suggested the presence of a variety of different types of chromosomal changes (Fig. 10-16) including duplications, in which one or more genes were present in an increased number; deletions, in which one or more genes were missing; inversions, in which a piece of the chromosome had become turned around within the body of the chromosome; and translocations, in which a part of a chromosome had broken off and had reattached to some other chromosome. Though translocations could often be detected by examination of normal mitotic chromosomes, the others could not. The existence of all of these deformities was confirmed by examination of the banding patterns of giant chromosomes.

Comparison of banding patterns among polytenic chromosomes of

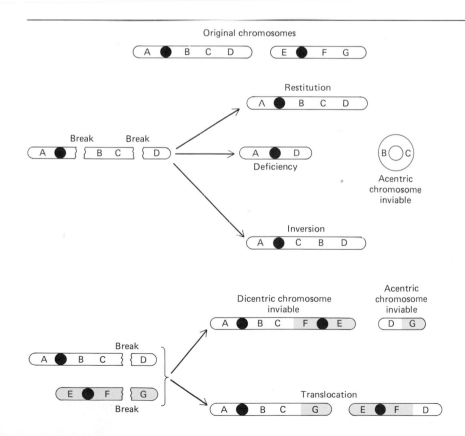

Figure 10-16. The origin of various types of chromosomal aberration through chromosome breakage. The centromeres are shown as black circles. (*From E. W. Sinnott, I. C. Dunn, and T. Dobzhansky, "Principles of Genetics," 5th ed., McGraw-Hill, 1958.*)

different species has provided an unparalleled opportunity for the investigation of evolutionary changes. In addition, it became apparent that these chromosomes were not simply a stable, unchanging reflection of the composition of the chromosomes, but underwent important physiological changes as well. Examination of the chromosomes revealed that specific regions were puffed out (Fig. 10-15b) and the locations of the puffs were characteristic for different tissues, or the same tissue at different stages of the life cycle. Further examination indicated the puffs were sites of gene activity, i.e., genes that were currently active in controlling specific cell functions. We will have more to say about this finding later (page 619).

MOLECULAR GENETICS
The Chemical Nature of the Gene

Classical genetics concerned itself with the rules governing the transmission of genetic characteristics and the relationship between genes and chromosomes. In his Nobel prize acceptance speech in 1934, Morgan said "At the level at which the genetic experiments lie it does not make the slightest difference whether the gene is a hypothetical unit or whether the gene is a material particle." In the 1940s, a new set of questions began to be considered. What is the chemical nature of the gene? What is the chemical nature of a mutation? How is the genetic material organized within the structure of the chromosome? What mechanism is employed to utilize the genetic information in the control of cell function? These will be the topics of this chapter and the following ones; there is a great deal of information to cover.

The Chemical Composition of Chromosomes

Knowing that the chromosomes are the carriers of the genetic material, one logical approach to the analysis of the chemical nature of the gene would be the analysis of the chemical composition of the chromosomes. However, the isolation of purified chromosomes is a sophisticated task which has only recently been accomplished. In contrast, the isolation of the material from nuclei and its analysis is much less technically demanding. The first analysis of the chemical composition of the nucleus was carried out by Friedrich Miescher in a brilliant and farsighted series of biochemical studies beginning in about 1868. Using pus cells and fish sperm, both of which had a large nuclear/cytoplasmic volume, Miescher made the first preparation of isolated DNA, a substance he called "nuclein." Later he isolated a protein associated with the nuclein from fish sperm which he called "protamine." It has since been shown that protamine is not a typical nuclear protein, but rather one found only in association with the DNA of certain sperm. Regardless of this latter fact, Miescher had succeeded in isolating the two major components of the chromosome, DNA and protein.

The next major step was to demonstrate the localization of these components in the chromosomes. Since chromosomes could not be isolated, an alternate procedure was necessary. As might be expected from the discussion in the previous chapters, a cytochemical method is ideally suited for the task. In 1924, R. Feulgen developed a cytochemical method which was specific for DNA, and he demonstrated the presence of this nucleic acid in mitotic chromosomes. Since proteins had already been shown to be present, the question became which of these two materials contained the hereditary information. There seemed little doubt. During the first half of the century the complexity, versatility, and importance of proteins became established. In contrast, little importance was attached to the nucleic acids, which were believed to have a structural role. The proteins with their large number of different amino acid building blocks would be a much more likely candidate for an information-storing molecule than nucleic acids with their four different building blocks. In addition, it was generally believed that the four nucleotides (page 446) comprising nucleic acid structure were simply organized into a strictly repetitive sequence much like the repeating disaccharide organization that is now known for the mucopolysaccharides (page 268). It was inconceivable that a molecule like DNA could be the gene; it might as well be starch or glycogen (see Fig. 4-23).

DNA as the Genetic Material

Up to this point in our discussion of genetics, we have considered plants and fruit flies. These types of organisms were suitable for classical genetic studies where the characteristics being observed were obvious morphological features such as seed color or wing shape. But these types of characteristics are biochemically complex and determined by a totally unknown battery of factors. By the late 1940s, a new era in biochemistry was opening and a new type of geneticist was on the scene, one with a background in the physical-chemical sciences. The morphological approach to genetics was suitable for the study of the principles governing inheritance, but a biochemical approach was needed if the physicochemical mechanisms were to be revealed. The emphasis turned from the complex higher organism to the much simpler viral and bacterial systems.

Though bacteria had been intensively studied during the first half of the century primarily due to their role in the cause of disease, their value in the study of genetics was not realized until about the midpoint in the century. Once it was accepted that bacteria had genetic systems much like those of higher organisms, they were obvious candidates. In addition to being much simpler, they were haploid, single-celled, able to grow on extremely simple media, and had incredibly short generation times. When grown under rich nutrient conditions, a culture of *E. coli* can double every 20 minutes. Billions of cells could be grown in small containers in a matter of hours.

Evidence from Transformation With this background behind us we can turn to the experiments that first demonstrated the chemical nature of the genetic material, those of Oswald Avery, Maclyn MacLeod, and Colin McCarty in 1944. In order to understand the nature of their research we have to go back to 1928 and the studies of Fred Griffith on the bacteria

Diplococcus pneumoniae (better known as pneumococcus), the agent responsible for pneumonia. Several infective strains of this bacteria had been isolated that differed from one another in the chemical nature of their polysaccharide capsule. The strains were termed type I, II, or III; all were virulent, i.e., capable of causing disease. In addition, mutants had been isolated from each of these strains that had metabolic defects making them unable to produce the capsule. Without the capsule, the bacteria were nonvirulent, i.e., unable to cause pneumonia when injected into a sensitive mouse. Colonies of capsuleless mutants growing on dishes could be distinguished from the virulent encapsulated form by their appearance. The mutants formed rough (R) colonies, whereas those with capsules grew into smooth (S) colonies.

Griffith made a surprising discovery when injecting various bacterial preparations into mice. Injections of either heat-killed S bacteria or living R bacteria, by themselves, were harmless to the mouse. However, if he injected both of these preparations into the same mouse, it contracted pneumonia and died. Virulent bacteria could be isolated from the mouse and cultured. To extend the findings (Fig. 10-17), he injected bacteria of different strains. When heat-killed type-I S bacteria were injected together with type-II R bacteria, he was able to isolate virulent type-I S bacteria from the infected mouse. Since there was no possibility that the heat-killed bacteria had been brought back to life, one had to conclude that the presence of the dead type I cells had *transformed* the noncapsulated type II mutants into a capsulated type I form. The transformed bacteria continued to produce type-I S cells when grown in culture, i.e., the change was a stable, permanent one. In other words, transformation was type-specific, predictable, and inheritable. During the next few years

Figure 10-17. The Griffith experiment on transformation as described in the text.

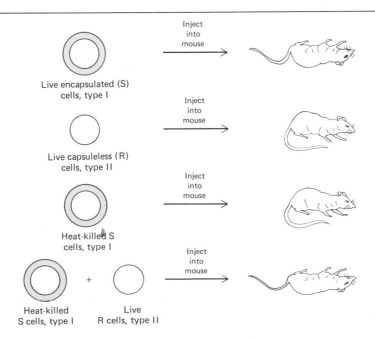

Live encapsulated (S) cells, type I

Inject into mouse

Live capsuleless (R) cells, type II

Inject into mouse

Heat-killed S cells, type I

Inject into mouse

Heat-killed S cells, type I   +   Live R cells, type II

Inject into mouse

it was found that transformation need not occur within a mouse. A crude extract of the dead virulent bacteria, when mixed with the mutant cells in vitro, was capable of converting them into the virulent form which was always of the type (I, II, or III) characterized by the dead cells.

In the early 1940s, Avery and his coworkers set out to determine the nature of the substance in the crude extract of the virulent cells that was capable of causing the transformation. Using various purification procedures, they succeeded in obtaining a substance which was active in causing transformation when present at only 1 part/100 million. Analysis of this material pointed clearly to DNA: (1) It had the properties of DNA, (2) no other type of material could be detected in the preparation, i.e., there was no evidence of protein, lipid, or polysaccharide, and (3) an enzyme that selectively destroyed DNA, i.e., DNase, was the only hydrolytic enzyme capable of destroying the transforming activity. Avery concluded that DNA was capable of causing an *inheritable* change in another type of substance, the polysaccharide of the capsule. Later studies were to show that the transformation event results from the uptake of a DNA fragment by the mutant bacterium and the replacement of its own corresponding DNA by the newly acquired piece.

Evidence from Bacteriophage Infection Relatively little attention was paid to this discovery during the next few years. It was not that the world doubted Avery's findings, but rather that the transforming substance was not believed to represent true genetic material. It would be eight years before an experiment by Alfred Hershey and Martha Chase was to confirm the claim that DNA was the material of the gene. By 1952 it was fairly well accepted that viruses, such as the type that infect bacteria (see Fig. 2-1), were in possession of a genetic program, just as were more complex organisms. The important question concerned the nature of the viral component that contained the information for the infection and the production of progeny from the infected cell. There were two possibilities, DNA or protein, since these were the only materials that the virus contained. Hershey and Chase assumed there were two properties that the genetic material should possess: first, it should pass into the infected cell if it were going to direct the events during infection, and, second, it should be passed onto the next generation. Electron microscopic observations had shown that, during the infection, the bulk of the phage remains outside the cell, attached to the cell surface by the phage tails (Fig. 10-18). Using bacteriophage with either ^{32}P-labeled DNA or ^{35}S-labeled protein, they determined the percentage of each type of radioactivity that actually enters the infected cell or is left behind in the attached viral coat (see Fig. 10-19 for a description of the experiment). They found that when protein-labeled phage were used, the bulk of the radioactivity remained outside the cell. In contrast, when DNA-labeled phage were used, the bulk of the radioactivity passed inside the host cell. When they monitored the radioactivity passed onto the next generation, they found that less than 1% of the labeled protein could be detected in the progeny, whereas approximately 30% of the labeled DNA could be accounted for in the next generation. Though not proving that DNA was the genetic material, the results

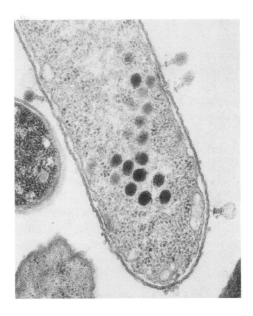

Figure 10-18. Electron micrograph of a section through an *E. coli* cell with bacteriophage particles attached to the cell surface by their tails. *(Courtesy of J. King and E. Hartwig.)*

were very suggestive. In later experiments, it was shown that purified bacteriophage DNA in the total absence of protein was capable of causing infection and the production of phage progeny.

Between the experiments of 1944 and 1952, the scientific climate had greatly changed toward the possibility of DNA as the genetic material. There were numerous reasons for this change of attitude. Most importantly, a large number of studies on bacteria and their viruses had indicated the validity of their use in genetic analysis. The concept was gradually being accepted that the same type of genetic activities were present in all organisms from viruses and bacteria to man. In addition, transformation experiments were extended to other genetically determined characteristics beside capsule formation. In other words, transformation was a more general phenomenon. Also, the concept of DNA as a monotonously repeating tetranucleotide was no longer held in much favor, thereby removing a restriction from DNA that had made it incompatible with an information-storing role. In addition, it was shown that the most highly mutagenic wavelengths of ultraviolet radiation were the ones most strongly absorbed by DNA.

Evidence from DNA Measurements Another piece of evidence became available that was concerned with the amount of DNA per nucleus in various types of cells. Just as Mendel's findings laid down specific requirements that had to be met by the chromosomes if they were to be the physical carriers of the genes, the behavior of the chromosomes makes predictions concerning the genetic material. If DNA is the genetic material, there should be a correlation between the number of sets of chromosomes in a nucleus (termed the *ploidy*) and the amount of DNA. For example, sperm that contain a haploid number of chromosomes should have half the DNA of most cells of the body. Cells just prior to mitosis should have twice the DNA as cells following mitosis, i.e., four times the haploid value.

How can one determine the amount of DNA in one particular nucleus, for example, within a cell just beginning mitosis? It obviously cannot be done

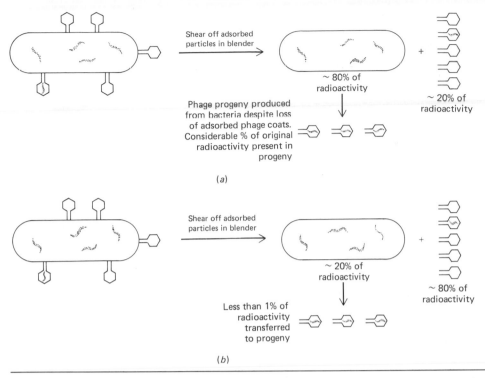

Shear off adsorbed
particles in blender

~ 80% of
radioactivity

+

~ 20% of
radioactivity

Phage progeny produced
from bacteria despite loss
of adsorbed phage coats.
Considerable % of original
radioactivity present in
progeny

(a)

Shear off adsorbed
particles in blender

~ 20% of
radioactivity

+

~ 80% of
radioactivity

Less than 1% of
radioactivity
transferred
to progeny

(b)

Figure 10-19. The Hershey-Chase experiment in which bacteria were infected with bacteriophage containing either ^{32}P-labeled DNA (a) or ^{35}S-labeled protein (b). At various times after the infection, the adsorbed bacteriophage particles were sheared from the surface of the bacterial cells and the radioactivity within the bacterium (as opposed to the sheared-off particles) was determined. It was found that, when DNA-labeled phage were used in the infection, approximately 80% of the radioactivity entered the cell, while in the case of protein-labeled phage only about 20% were found in the bacteria. In addition, it was found that a much higher percentage of the original radioactive DNA appeared in the progeny than did radioactive protein.

by isolating the DNA from that one cell, but rather must be done indirectly. Several labs in 1949–1950 made such measurements using a microspectrophotometer, an instrument that can measure the amount of light of specific wavelengths absorbed by a specimen being examined under a microscope. In order to make such a measurement, one observes a particular specimen, such as the nucleus of a dividing cell, and, knowing the amount of light used to illuminate the specimen and the amount of light transmitted by the specimen to a photocell, one can determine the amount of light absorbed by that single nuc'eus. All that is needed is to use a particular wavelength that is absorbed primarily by the substance being measured. In the case of DNA, the procedure utilizes the Feulgen stain, which is specific for DNA and absorbs strongly at 560 nm. When light of this wavelength is used as the illuminating source, one can determine the amount of light absorbed by the stained DNA. Since the amount of light absorbed is proportional to the amount of stain present, which is in turn proportional to the amount of DNA present in the field, the absorbance can be used for DNA determination in very small microscopic specimens. Using this procedure, a perfect correlation was found between the ploidy of a cell and its DNA content. The correlation even held for cells that were *polyploid,* i.e., cells with additional sets of chromosomes as occurs routinely, for example, in liver cells of vertebrates and endosperm of plants. No other substance in the cell was found to exist in such well-defined units or to parallel the chromosome content. All haploid cells of a species are said to have the 1C amount of DNA, all diploid cells the 2C amount, etc.

The Structure of DNA

Confident that DNA was the genetic material [except in a few viruses such as tobacco mosaic virus (TMV), where RNA (page 99) was shown to

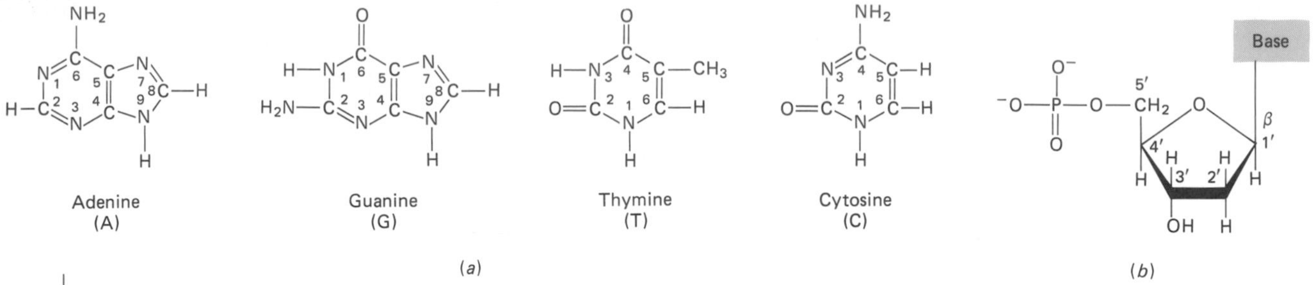

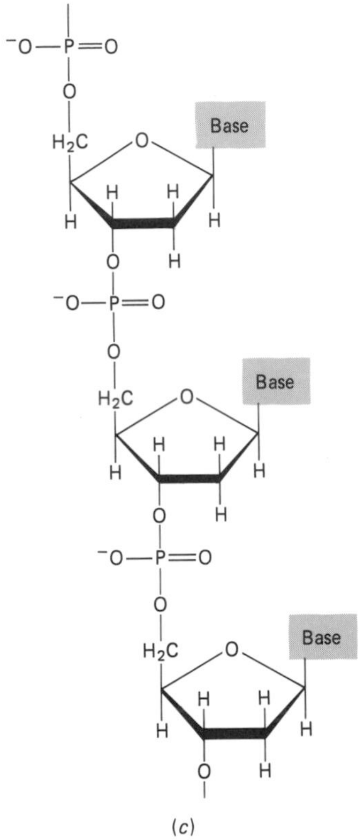

Adenine (A) Guanine (G) Thymine (T) Cytosine (C)

(a)

(b)

(c)

Figure 10-20. (a) The four nitrogenous bases of DNA. Adenine and guanine are purines, cytosine and thymine are pyrimidines. (b) Generalized structure of a deoxyribonucleotide (a deoxyribonucleoside 5'-monophosphate. (c) Generalized structure of a portion of a single strand of DNA.

contain the program], a tremendous new interest was generated in a molecule which had largely been ignored. The first question of importance was the nature of its structure, a problem which was pursued by several laboratories and solved by James Watson and Francis Crick in 1953. Before describing their proposed structure, we will consider the facts available at the time. Information on DNA structure had been derived from scattered biochemical studies first begun by Miescher, and x-ray diffraction analysis begun by William Astbury in the 1940s and followed by Linus Pauling, Rosalind Franklin, Maurice Wilkins, and others.

Base Composition The basic building block of DNA was the nucleotide (Fig. 10-20a) consisting of one molecule of deoxyribose to which one phosphate was esterified at the 5' position of the sugar and one nitrogenous base was attached at the 1' site. The bases were of two classes (Fig. 10-20b), the smaller pyrimidines and the larger purines. There were two pyrimidines, thymine and cytosine and two purines, adenine and guanine. The nucleotides were known to be covalently linked to one another to form a very large molecular weight, fibrous polymer whose backbone was composed of alternating sugar and phosphate groups joined by 3'5'-phosphodiester bonds (Fig. 10-20c). The bases attached to each sugar were believed to project from the backbone, forming a stacklike structure. Note that a structure such as that shown in Fig. 10-20c has direction or polarity; each unit is asymmetrical, having a 3' edge and a 5' edge and, therefore, the entire polymer is polarized having a 3' end and a 5' end. X-ray diffraction had indicated that the distance between the elements of the stack was 3.4 Å.

As stated above, the early concept of DNA as a simple repeating tetranucleotide (e.g., ATGCATGCATGC) inhibited its being considered as an informational macromolecule. In 1950, Erwin Chargaff made an important biochemical finding that provided the final blow to the tetranucleotide theory and provided vital information to Watson and Crick. Chargaff, believing that the sequence of nucleotides of the DNA molecule held the key to its importance, began a first step in any sequence analysis, the determination of the relative amounts of each base, i.e., the *base composition*. The analysis was accomplished by hydrolyzing the bases from their attached sugars, separating the bases in the hydrolysate on paper chromatography (Fig. 3-5), and determining the amount of material in each of the four spots to which the four bases migrated.

446

If the tetranucleotide theory were correct, each of the four bases should be present at about 25% of the total number. Chargaff found that the ratios of the four component bases were quite variable from one type of organism to another, often being much different from the 1 : 1 : 1 : 1 ratio predicted by the tetranucleotide theory. For example, the A:G ratio of the DNA of a tubercle bacillus was 0.4 while in humans the ratio was 1.56. It made no difference which tissue was used as the source of the DNA, the base composition remained constant for that species. Amidst this great variability in base composition of different DNAs, an important quantitative principle was discovered. The number of purines always equaled the number of pyrimidines in a given sample of DNA. More specifically, the number of adenines equaled the number of thymines and the number of guanines equaled the number of cytosines. In other words, Chargaff found the following rules of DNA base composition:

$$(A) = (T), \qquad (G) = (C), \qquad (A) + (T) \neq (G) + (C)$$

However, the significance of the equivalencies remained obscure.

The Watson-Crick Proposal   When protein structure was considered in Chapter 3, the importance of secondary and tertiary structure as the determinant of the protein's activity was stressed. A similar need for information about the three-dimensional organization of DNA was needed if its biological activity was to be understood. Using x-ray diffraction data and the construction of feasible models, Watson and Crick proposed the following structural properties of the DNA molecule (Fig. 10-21*a–d*):

1. The molecule is composed of two chains of nucleotides.

2. The two chains form a pair of right-handed helices that coil around the same axis.

3. The sugar-phosphate-sugar-phosphate backbone is located on the outside of the molecule with the bases projecting toward the center.

4. The bases occupy planes which are perpendicular to the long axis of the molecule and are therefore stacked one on top of another.

5. The two chains are held together by hydrogen bonds which occur between each base of one chain and an associated base on the other chain.

6. The distance from the phosphorus atom of the backbone to the center of the axis is 10 Å (thus the width of the double helix is 20 Å).

7. The 20-Å width of the fiber requires that a pyrimidine from one chain always be paired with a purine of the other chain since the association of two purines would extend beyond the specified width and the association of two pyrimidines would not be sufficiently wide.

8. The nitrogen atoms linked to carbon 4 of cytosine and carbon 6 of adenine are predominantly in the amino (NH_2) configuration (Fig. 10-20) rather than the imino (NH) state. Similarly, the oxygen atoms linked to carbon 6 of guanine and thymine are predominantly in the keto (C=O) configuration rather than the enol (COH) state. These structural restrictions on the configurations of the bases suggested that adenine was the only purine structurally capable of bonding to thymine, and guanine was the only purine capable of bonding to cytosine. Therefore, the only possible pairs were A-T and G-C (Fig. 10-21), which fit perfectly with the

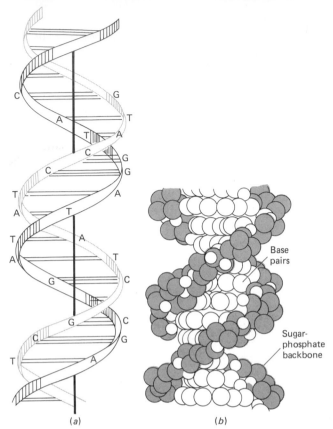

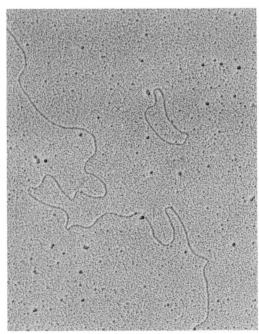

Figure 10-21. The structure of DNA. *(a)* A schematic representation of the double helix of DNA. The two ribbons represent the phosphate-sugar chains, and the horizontal rods represent the bonding between the pairs of bases. The vertical line indicates the fiber axis. *(From A. White, P. Handler, and E. L. Smith, "Biochemistry," 5th ed., McGraw-Hill, 1973.)* *(b)* Space-filling drawing of DNA. *(c)* Chemical structure of DNA indicating the angular and linear dimensions of the base pairs. *(From M. Wilkins.)* *(d)* Electron micrograph of a pair of metal-shadowed DNA molecules. *(Courtesy of C. Dykstra.)*

Base pairs

Sugar-phosphate backbone

(a)

(b)

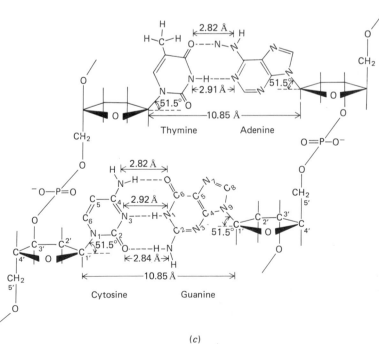

(c)

(d)

previous base composition analysis by Chargaff. A-T pairs were joined by two hydrogen bonds, G-C pairs by three hydrogen bonds.

9. The two chains comprising one double helix run in opposite directions, i.e., they are antiparallel. In other words, if one chain is aligned in a $5' \rightarrow 3'$ direction, its partner must be aligned in the $3' \rightarrow 5'$ direction.

10. The double helix makes one complete turn every 10 residues (34 Å) or 150 turns per million molecular weight.

11. There is no restriction on the sequence of bases in a given chain of the molecule. However, once a particular sequence is specified in one chain, the sequence of the other chain is automatically determined. The term *complementary* is used to express the relationship between the two chains of the double helix. For example, A is complementary to T, AGC is complementary to TCG, one entire chain is complementary to the other. As we shall see, the concept of complementarity is of overriding importance in nearly all of the activities and mechanisms in which nucleic acids are involved.

From the time biologists first began to consider the nature of the genetic material, there were three primary functions it was expected to fulfill: the storage of inheritable information, the means for self-duplication, and the capacity to direct the activities of the cell. The Watson-Crick model of DNA structure was of critical importance in revealing the manner in which two of these genetic functions was accomplished. The model strongly confirmed the belief that its information content resided in the linear sequence of the bases in a given molecule of DNA. A given segment of DNA would correspond to each gene. As for the second function, that of self-duplication, the model immediately suggested to its authors a mechanism previously unconsidered. Watson and Crick proposed that the two chains of the double helix were, in effect, a pair of templates that were available for use in duplication. Each chain has the information for the complementary sequence. If we denote one strand as the + strand and the other as the − strand, then, if the two chains were to separate (see Fig. 14-14), the + strand could serve as a template upon which the polymerization of a − strand could take place, and vice versa. The linking of the two chains by hydrogen bonds is ideally suited for its biological role. Each individual bond is relatively weak and readily broken when necessary, such as during the duplication process. On the other hand, the large number of hydrogen bonds makes the overall structure of the molecule extremely stable. In addition to the hydrogen bonds, the fact that the bases are flat and stacked closely on top of one another facilitates a degree of hydrophobic interaction and electronic or stacking interaction, which also adds to the stability of the structure.

Upon occasion, it might be expected that an inappropriate, i.e., non-complementary, nucleotide might be incorporated into one of the growing chains. If such a mistake occurs, the information content of the strand would be changed as would its future partners that were formed using it as a template. A mistake such as this would be detected as a stable mutation in the genetic material. The topic of mutation will be considered

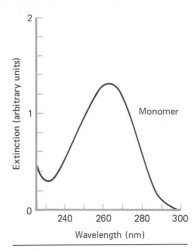

Figure 10-22. Absorption spectrum of thymine. All four nucleotides of DNA, and consequently DNA itself, absorb maximally near 260 nm. The absorbance at 260 nm is generally used to determine nucleic acid concentrations. The ratio of absorbance at 260 : 280 is one measure of the purity of a nucleic acid preparation since contaminating proteins generally absorb strongly at 280 nm. The 260 : 280 ratio of purified DNA is about 2. *(From R. K. Clayton, "Light and Living Matter," McGraw-Hill, 1972.)*

further in Chapter 11. Of the primary functions mentioned previously, only the mechanism by which DNA governs cellular activity remained a total mystery. Not only was the elucidation of DNA structure significant in its own right, but it provided the stimulus for investigation of all of the activities in which the genetic material must take part. Once the model for its structure was accepted, any theory of the genetic code, or of DNA synthesis, or information transfer had to be consistent with that structure.

Properties of DNA
Absorbance

As was mentioned for chlorophyll, molecules containing a conjugated system of double bands are strong absorbers of electromagnetic radiation, the specific wavelengths depending upon the electronic structure of the molecules. Purines and pyrimidines absorb most strongly in the ultraviolet region of the spectrum, having an absorbance maximum near 260 nm (Fig. 10-22). Even though there is some variation in the spectrum of the various bases, the absorbance at 260 nm is generally used as the basis for the determination of the amount of DNA. The use of the spectrophotometer and certain of the principles governing absorbance studies are considered in the appendix to this chapter.

Ionic Interactions

In its double helix condition, the outside surface of the DNA fiber is highly anionic due to the presence of the large numbers of phosphate groups, each of which is fully ionized at physiological pH. As a polyanion, DNA is capable of ionic interactions with a large number of positively charged molecules. In solution, DNA is generally complexed with small cations, but in the eukaryotic cell it is tightly associated with positively charged proteins whose role is being intensively studied (Chapter 13).

Denaturation

One of the first properties of the double helix to be investigated was its disruption. Since the two chains of DNA are held together by hydrogen bonds, it would be expected that agents which break these bonds might cause the strands to separate. The most commonly employed means to bring about the *denaturation* (strand separation) of DNA is to raise the temperature. Julius Marmar, Paul Doty, and coworkers using DNA from various sources found that, when the temperature of a DNA solution was slowly raised, a specific temperature was reached when strand separation began. Within a few degrees, the process was complete and the solution contained single strands of DNA that were completely separated from their original partners. In order to monitor the extent of denaturation at a particular temperature, it was necessary to distinguish between single-stranded DNA and the original double-stranded DNA termed *native DNA*. As will be seen in this section, single- and double-stranded DNA have markedly different properties. For example, DNA in these two states differ in their absorbance, density, viscosity, sensitivity to enzymatic digestion, and their biological activity (as measured, for example, by their transformation ability). The most commonly employed means to follow thermal denaturation, termed *melting,* is to measure the increase in absorbance of the single strands as compared to the duplexes from which they originated. This rise in absorbance, termed the *hyperchromic shift,* shown in Fig. 10-23, results from the change in the interactions among the bases. When present in the double-stranded state, there is very little freedom of individual bases because they are tightly stacked upon one another. Once the strands have separated, the stacking interactions are greatly decreased thereby changing the electronic nature of the bases and increasing their absorbance. The temperature at which the hyperchromic shift is half completed is termed the *melting temperature* (T_m).

When the melting profiles of DNA extracted from various sources were compared, it became apparent that the T_m for a given DNA preparation (at a given ionic strength) was a very sensitive indicator of the base composition of the DNA. The higher the *GC content* (%G + %C), the higher the T_m. This increased stability of GC-containing DNAs reflects the presence of the extra hydrogen bond between the bases as compared to A-T pairs. Figure 10-24*a* shows the relationship between base composition and T_m for a number of different DNAs. The electron micrograph of Fig. 10-24*b* reveals that, even within a single DNA molecule, AT-rich sections melt before GC-rich regions.

Viscosity

The viscosity of a solution is proportional to the resistance of the components to flow. Viscosity results from the frictional effects of the molecules sliding relative to one another. If one considers the shape of a DNA molecule, the basis for its extremely high viscosity is apparent. Each molecule is very long and thin and is stiff due to the nature of its double-helical structure. DNA molecules of only 20-Å diameter are believed to exist at lengths in excess of one inch, making them extremely elongated

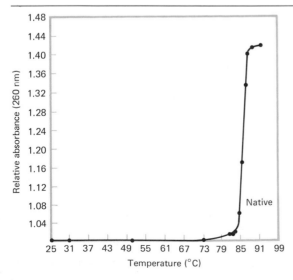

Figure 10-23. A thermal denaturation curve for native bacteriophage T_6 DNA in 0.3 *M* NaCl plus 0.03 *M* NaCitrate. The hyperchromic shift of purified DNA is relatively steep, particularly for DNA of small viral genomes. The temperature corresponding to half the increase in absorbance is termed the T_m. *(From J. Marmur and P. Doty,* J. Mol. Biol., **3:**593, 1961.)

in shape. As a result of their highly elongated nature, DNA molecules are readily broken into smaller and smaller fragments even under the gentlest conditions. However, even when quite fragmented, DNA solutions are highly viscous and the molecules can be removed from solution by winding them around a glass rod as if they were lengths of fine string. Molecules of single-standed DNA are much less viscous than double-stranded ones since they are no longer held in the rigid state, but rather tend to collapse upon themselves to form a more globular structure.

Molecular Weight

One of the early questions to be considered once DNA was shown to be the genetic material was the lengths of DNA molecules that existed in cells. The early studies on viral and bacterial DNAs indicated that the entire genetic program of these organisms was present as one DNA molecule, sometimes in a linear form and sometimes in a circular form. Table 10-2 provides the lengths, molecular weights, and nucleotide pairs of a number of viral and prokaryotic DNAs. Attempts to determine the molecular weight of DNAs present in eukaryotic cells proved much more dif-

Figure 10-24. (a) Dependence of the denaturation temperature, T_m, on the guanine-cytosine content of various samples of DNA. The greater the percentage of G-C bonds, the greater the stability of the duplex. (From J. Marmur and P. Doty, Nature, **183**:1428, 1959.) (b) An electron micrograph of a partially denatured DNA molecule from the phage lambda. In this case denaturation (as indicated by the single-stranded bubbles) is brought about by high pH, a treatment comparable to temperature elevation. (Courtesy of R. B. Inman.)

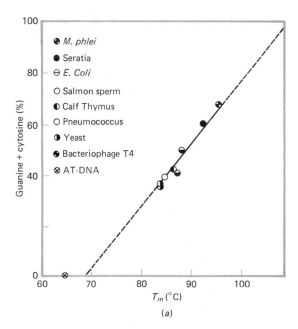

(a)

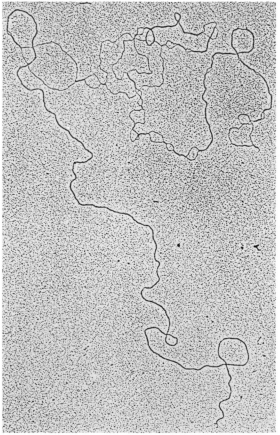

(b)

ficult due to their greater length and, consequently, their ease of being broken. The amount of DNA in a chromosome could be estimated and the primary question became whether that amount of DNA was present in one very long molecule that stretched from one end of the chromosome to the other, or whether a chromosome's DNA was present in pieces which might possibly be joined together by a non-DNA linking material. With improved isolation methods, very long DNA molecules have now been isolated from eukaryotic cells. For example, using viscoelasticity as a measure of DNA length, it has been found that molecules can be isolated from *Drosophila* cells that are as long as would be expected from the largest chromosome of the set, approximately 2.1 cm. In addition to the viscoelastic technique, extremely long DNA molecules have been detected by light microscopic autoradiography of radioactively labeled DNA molecules spread on a slide (Fig. 10-25). Based upon these types of findings, it is generally believed that chromosomes do contain just one long DNA molecule, though the results at present are preliminary.

In order to measure the molecular weight of smaller pieces of DNA or to determine the variety present in a mixture of DNAs of varying molecular weights, different types of procedures can be employed. Acrylamide gel electrophoresis, for example, can be used to separate DNA fragments according to length, i.e., molecular weight, on the same basis that proteins are separated in SDS-containing gels (page 115). In the case of nucleic acids, all molecules have a similar charge density, i.e., number of negative charges per unit length, and therefore have an equivalent potential for migration in an electric field. However, if the acrylamide gel is sufficiently concentrated, the larger the molecule, the slower it makes its way through the gel. Sephadex gel filtration (page 117) can also be used to fractionate small DNA molecules on the basis of their molecular weight. Another method, that of density gradient centrifugation, is described following.

Sedimentation Behavior

DNA (and RNA) molecules are extensively analyzed by techniques utilizing the ultracentrifuge. A brief introduction to some of the principles of ultracentrifugation can be found in the appendix to this chapter. For the present purpose, we will consider two of the most commonly employed centrifugation techniques for the study of nucleic acids, namely separation by sedimentation rate in a preformed density gradient and isopycnic separation in an equilibrium gradient. These procedures are illustrated

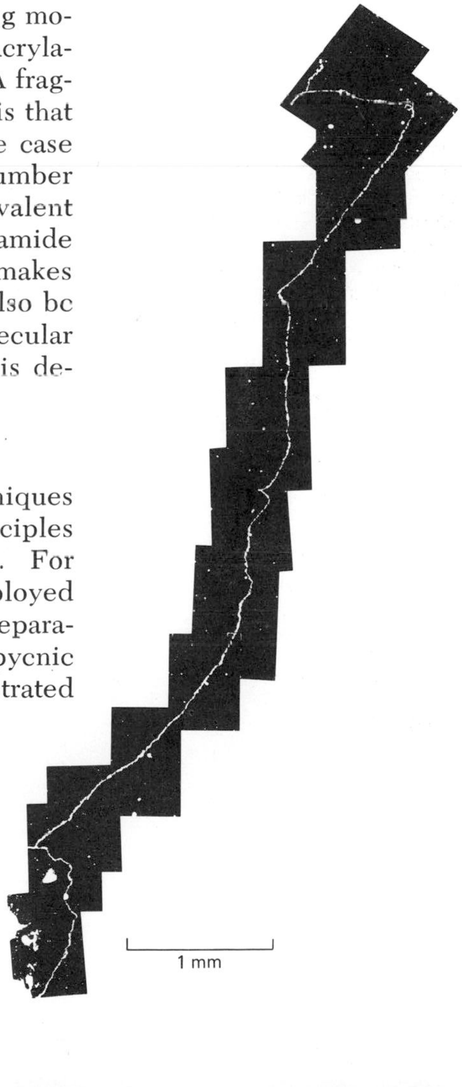

Figure 10-25. Autoradiograph of a *Drosophila* DNA molecule. Cultured cells were labeled with ³H-thymidine for 24 hours, lysed, pronase-treated and aliquots applied to coated slides for light microscopic autoradiography. The contour length for this molecule was 1.2 cm. *(From R. Kavenoff, L. C. Klotz, and B. H. Zimm,* Cold Spring Harbor Symp. Quant. Biol., **38:4, 1974.)**

1 mm

TABLE 10-2
Sizes of DNA Molecules and Genomes

Organism	Size of Genome		Shape
	Number of base pairs (thousands) (kb)	Total length° (mm)	
Viruses			
Polyoma, SV40	5.1	0.0017	Circular duplex
φX174	5.4	0.0018	Circular single strand; duplex replicative form
M$_{13}$ (fd, f1)	5.74	0.0019	Circular single strand; duplex replicative form
P$_4$	15.0	0.0051	Linear
T$_7$	35.4	0.0120	Linear
P$_2$, P$_{22}$	40.5	0.0138	Linear
λ	49	0.0166	Linear
T$_2$, T$_4$, T$_6$, P$_1$	180	0.061	Linear
Vaccinia	183	0.062	Linear
Bacteria			
Mycoplasma hominis	760	0.26	Circular
Escherichia coli	4000	1.36	Circular
Eukaryotes			
			No. chromosomes (haploid)
Yeast	13,500	4.6	17
Drosophila (fruit fly)	165,000	56	4
Humans	2,900,000	990	23
South American lungfish	102,000,000	34,700	19

° Length = (Kb)(3.4 × 10^{-4}) mm
SOURCE: DNA SYNTHESIS by A. Kornberg. W. H. Freeman and Company. Copyright © 1974.

in Fig. 10-26. In the former case, the sample containing the mixture of nucleic acid molecules is carefully layered over a solution containing an increasing concentration of sucrose (or other suitable substance). This preformed gradient increases in both density and viscosity from the top to the bottom. In the presence of large centrifugal forces, the molecules move through the gradient at a rate determined by their sedimentation coefficient and the nature of the medium. As a given molecule moves down the tube, the increasing distance from the center of the rotor tends to increase its velocity while the increasing sucrose concentration tends to slow its movement. Overall, the greater the sedimentation coefficient, the larger the molecule and the farther it will move in a given period of centrifugation. Since the density of the medium is less than that of the nucleic acid molecules, even at the bottom of the tube (approximately 1.2 g/cc for the sucrose solution and 1.7 g/cc for the nucleic acid), these molecules will continue to sediment as long as the tube is being centrifuged. In other words, the centrifugation never reaches equilibrium. After the desired period, the tube is removed from the centrifuge, its con-

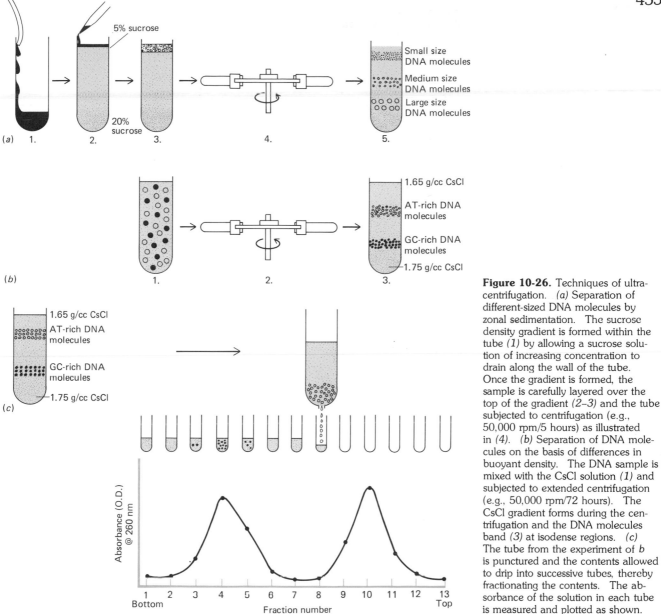

Figure 10-26. Techniques of ultra-centrifugation. *(a)* Separation of different-sized DNA molecules by zonal sedimentation. The sucrose density gradient is formed within the tube *(1)* by allowing a sucrose solution of increasing concentration to drain along the wall of the tube. Once the gradient is formed, the sample is carefully layered over the top of the gradient *(2–3)* and the tube subjected to centrifugation (e.g., 50,000 rpm/5 hours) as illustrated in *(4)*. *(b)* Separation of DNA molecules on the basis of differences in buoyant density. The DNA sample is mixed with the CsCl solution *(1)* and subjected to extended centrifugation (e.g., 50,000 rpm/72 hours). The CsCl gradient forms during the centrifugation and the DNA molecules band *(3)* at isodense regions. *(c)* The tube from the experiment of *b* is punctured and the contents allowed to drip into successive tubes, thereby fractionating the contents. The absorbance of the solution in each tube is measured and plotted as shown.

tents fractionated (as shown in Fig. 10-26*c*), and the relative positions of the various molecules determined. The presence of the sucrose in the tube blocks convection, thereby causing molecules of identical S value to remain in place in the form of a band. If molecules of known sedimentation coefficient, i.e., *markers*, are present, the S values of unknown components can be determined.

In the other type of technique, that of isopycnic sedimentation (or density gradient equilibrium sedimentation), nucleic acid molecules are separated on the basis of their *buoyant density*. In this procedure, one generally employs a highly concentrated solution of the salt of the heavy metal, cesium. The analysis is begun by mixing the DNA with the solution of cesium chloride or cesium sulfate in the centrifuge tube and then subjecting the tube to extended centrifugation (e.g., 2 to 3 days at high forces). During the centrifugation, the heavy cesium ions are very slowly driven toward the bottom of the tube, forming a continuous density gradient through the liquid column. After a time, the tendency for ions to be concentrated toward the bottom of the tube is counterbalanced by the opposing tendency for them to become redistributed by diffusion, and the gradient becomes stabilized. As the cesium gradient is forming, individual DNA molecules move either downward or upward in the tube until they reach a position which has a buoyant density equivalent to their own at which point they are no longer subject to further movement. Molecules of equivalent density form narrow bands within the tube.

As in the case of T_m, DNA molecules of varying base composition behave quite differently when subjected to isopycnic sedimentation. Guanine-cytosine pairs are held more closely together than adenine-thymine pairs due to the presence of the added hydrogen bond. As a result, the G-C pair is more dense, causing DNA molecules of higher G-C content to have a greater buoyant density (Fig. 10-27), and therefore capable of being separated from molecules of markedly lesser G-C content. Similarly, single-stranded DNA has a higher buoyant density than double-stranded DNA, RNA has a higher density than either forms of DNA, which are, in turn, more dense than proteins. If desired, the hydrogen bonding between nucleic acid strands can be broken during centrifugation by using solutions of sucrose or cesium that have an alkaline pH. Under these denaturing conditions, all molecules will remain single-stranded and extended.

Figure 10-27. Relationship of density to the guanine-cytosine content of various samples of DNA. *M. phlei* is *Mycobacterium phlei*. (From P. Doty, Harvey Lectures, **55**:103, 1960.)

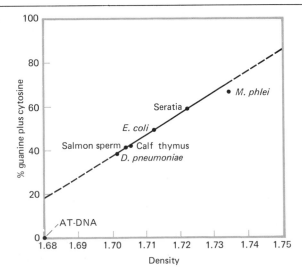

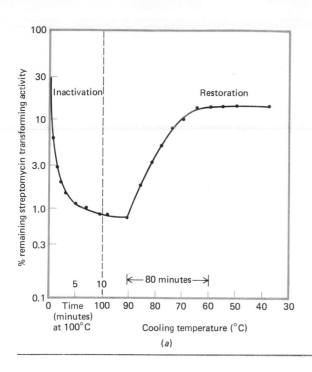

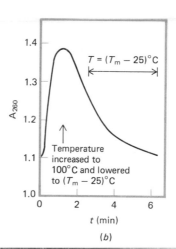

Figure 10-28. (a) Thermal inactivation and restoration of the transforming activity of DNA. Pneumococcal DNA was preheated for $1\frac{1}{2}$ minutes at 85.5° (no loss of transforming activity), then transferred to a boiling water bath at 0 minute. At the times shown during the first 10 minutes (left portion of figure), samples were removed and cooled quickly; these samples rapidly lost their biological activity indicating they were denatured. In contrast, if a sample were removed from the boiling water bath and slowly cooled as shown on the right, its biological activity was slowly restored indicating the renaturation of the DNA strands. [*From J. Marmur and D. Lane,* Proc. Natl. Acad. Sci. (U.S.), **46:**453, 1960.] (b) DNA was denatured by heating to 100°C, the temperature of the solution was then lowered to a value 25°C below the T_m, and the subsequent reannealing followed by observing the change in absorbance with time. (*From J. G. Wetmur and N. Davidson,* J. Mol. Biol., **31:**354, 1968.)

Renaturation

The separation of the two strands of the DNA duplex by heat is not an unexpected finding, but the *reassociation* of single strands into stable double-stranded molecules seems much less likely. However, in 1960 Julius Marmur and coworkers found that, if they *slowly* cooled a solution of bacterial DNA which had been thermally denatured, they regained a preparation of DNA that had properties of the double helix: it absorbed less ultraviolet light and was once again capable of transformation (Fig. 10-28a). Similar results were obtained by heating the DNA to 100°C to denature it, rapidly dropping the temperature of the solution to approximately 25°C below the T_m, and allowing the DNA to incubate at this temperature for a period of time (Fig. 10-28b). It became apparent that single-stranded DNA molecules, given enough time of incubation and an ionic strength and temperature (approximately 25°C below the T_m) that will promote hydrogen bond formation, were capable of reassociating with one another, an event termed *renaturation* or *reannealing*. When viral and bacterial DNA is allowed to reanneal, the duplexes formed are essentially indistinguishable from native DNA. In other words, the reannealing process is highly specific; only those strands that are truly complementary to one another are capable of association. The reannealing reaction demonstrates second-order kinetics, i.e., it is proportional to the product of the concentration of the two reactants. The greater the number of complementary molecules in a given volume, the greater will be their collision frequency, and the more rapidly they will form a stable duplex.

The finding that complementary, single-stranded nucleic acids are capable of association has proved to be one of the most valuable observations made in molecular biology. On one hand, the observation of reannealing has led to an investigation of the nature of nucleotide sequences within DNA that has provided a whole new insight into the nature and function of the genetic material. On the other hand, it has led to the development of a technique, termed *molecular hybridization* (see Appendix), which has provided a wealth of information about all types of nucleic acids. During the course of this part of the book we will have several occasions to describe examples of this powerful analytical method.

The Complexity of the Genome
Viral and Bacterial Genomes

There are several factors which determine the rate of renaturation of a given preparation of DNA. They are (1) the ionic strength, (2) the temperature of the solution, (3) the concentration of DNA, (4) the period of incubation, and (5) the size of the interacting molecules. With these in mind, consider what might happen if one were to compare the rate of renaturation of three different DNAs, that of a small virus, such as SV40 (5.4×10^3 nucleotide pairs), that of a larger virus, such as T_4 (1.8×10^5 nucleotide pairs), and *E. coli* (4.5×10^6 nucleotide pairs). The entire genetic material of either of these viruses or the bacterium exists as one DNA molecule. In each case, this one molecule constitutes its *genome*, i.e., its total genetic content. The primary difference between these DNAs is their length. In order to compare their renaturation, it is important that the reacting molecules are of equivalent length. DNA molecules can be fragmented into smaller pieces of uniform length by forcing them through a tiny orifice at high pressure or shearing them in a special type of homogenizer.

When these three types of DNA, all present at the same length and DNA concentration (mg/ml), are allowed to reanneal, they do so at distinctly different rates (see Fig. 10-30). The smaller the genome, the faster the renaturation. The reason for this becomes apparent when one considers the concentration of complementary sequences in the three preparations. Since all of these preparations have the same amount of DNA in a given volume of solution, it follows that the smaller the genome size, the greater number of genomes present in a given weight of DNA, and the greater the number of complementary fragments. This is illustrated by the hypothetical case shown in Fig. 10-29. The same reannealing profiles are observed, regardless of whether the three DNAs reanneal in separate tubes (Fig. 10-29b) or together in the same solution (Fig. 10-29c). This illustrates an important principle of hybridization reactions: as long as there are no long stretches of nucleotide sequences in common in these three genomes, there is no reason why the presence of any one genome should interfere with the reactions between DNA sequences of any other type of genome.

It is important to keep in mind that the renaturation shown in Fig. 10-29 is plotted against time on a log scale in order to condense several orders of magnitude into one graph. The impression received from a quick examination of this type of plot is that each DNA waits for a certain period of time before it reanneals and then rapidly forms double-stranded molecules. This is an

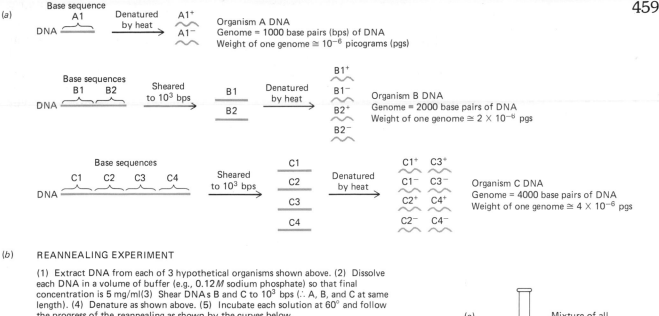

(b) REANNEALING EXPERIMENT

(1) Extract DNA from each of 3 hypothetical organisms shown above. (2) Dissolve each DNA in a volume of buffer (e.g., 0.12 M sodium phosphate) so that final concentration is 5 mg/ml (3) Shear DNAs B and C to 10^3 bps (∴ A, B, and C at same length). (4) Denature as shown above. (5) Incubate each solution at 60° and follow the progress of the reannealing as shown by the curves below.

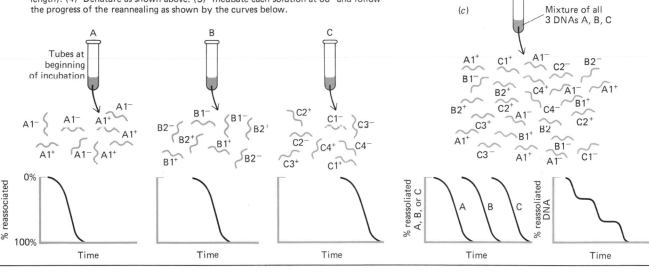

Figure 10-29. A hypothetical experiment involving the renaturation of DNA from three different organisms. (a) The nature of the genome of each organism. Each DNA preparation is sheared to 1000 base pairs and denatured into single strands by heat. Each genome from A yields one pair of complementary sequences, each from B yields two pairs of complementary sequences, and each from C yields four pairs. (b) A comparison of the rates of reannealing of the three different DNAs when all are present at the same DNA concentration. (c) The reannealing of DNA when all three preparations are present in the same mixture. The plot on the left indicates the reannealing curve of the DNA from each of the three organisms. The plot on the right follows the reannealing of total DNA. In all of the hypothetical curves of this figure, the time variable of the abscissa is plotted on a log scale.

illusion of the semilog nature of the plot. If the concentration of single-stranded molecules remained constant, the likelihood of a given molecule undergoing reannealing in the first minute of the incubation would be as great as any other minute. In fact, the overall rate of renaturation decreases with time due to the decreasing concentration of available molecules.

In the hypothetical experiment of Fig. 10-29 all three DNAs were present at the same concentration for the sake of comparison. In most actual cases, attempts to monitor DNA reannealing with a single concentration of DNA would prove infeasible. Consider the experiment described in Fig. 10-29c. If a single concentration of DNA was employed, the sequences of genome A would reanneal at too fast a rate to follow while those of genome C would require too great a time period. Reannealing of genome A would best be followed at low DNA concentration while that for C would best require high DNA concentration. Fortunately, the kinetics of reannealing are such that comparisons can be made between solutions of DNA at different concentrations. Incubation time and DNA concentration both affect reannealing in the same way, i.e., the greater the time *or* concentration, the greater the percentage of DNA molecules that will be reannealed. The formation of double-stranded molecules is generally plotted against a measure of both of these variables, rather than just time. The term C_0t (initial DNA concentration × time of incubation) used in Fig. 10-30 combines these two variables and allows one to compare experiments using DNAs at different concentrations. A solution containing a high concentration of DNA incubated for a short time will have the same C_0t as one of low concentration incubated for a correspondingly longer time; both will have the same percentage reannealed. If all of the other variables are kept constant, the C_0t for half renaturation ($C_0t \frac{1}{2}$) is a sensitive measure of the molecular weight of these simple types of genomes. For example, the $C_0t \frac{1}{2}$ of T_4 is approximately 35 times that of SV40, exactly proportional to their molecular weights.

The Eukaryotic Genome

Up to this point we have purposely restricted the discussion of renaturation to viral and bacterial DNAs. The curve for renaturation of these DNAs is symmetrical and approximately 80% complete over about a 2-log interval of C_0t (Fig. 10-30). This is due to the fact that, for a particular DNA, any given nucleotide sequence in the population is as likely to find

Figure 10-30. Reassociation of sheared strands of nucleic acids from various sources. The genome size (where applicable) is indicated by the arrows near the upper nomographic scales. The shape of the various renaturation curves is similar, although the C_0t (discussed in the text) over which it occurs is very different and depends upon the concentration of complementary fragments. In this particular figure, the renaturation of a small viral genome is exemplified by MS-2 (an RNA-containing virus) rather than SV40 as mentioned in the text. These two viral genomes would reanneal with roughly similar kinetics. Fig. 10-29 and the text provide explanatory information concerning these curves. *(From R. J. Britten and D. E. Kohne, Science, **161**:530, 1968. Copyright by the American Association for the Advancement of Science.)*

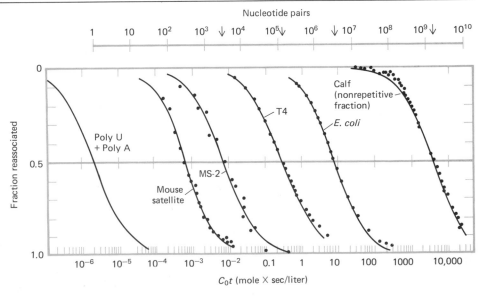

Sequences in DNA

DNA | 1 | 2 | 1 | 3 | 4 | 1 | 5 | 3 | 6 | 1 | 2 | 7 |

DNA of organism having genome of 7000 bps (weight of one genome $\cong 7 \times 10^{-6}$ pgs)

1. Extract DNA and dissolve at 5 mg/ml
2. Shear to 10^3 bps
3. Denature by heat
4. Incubate at $60°C$ and follow progress of reannealing

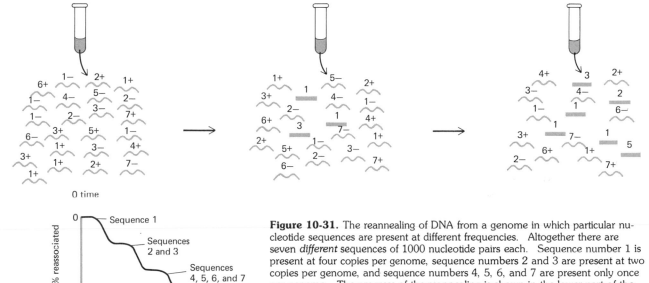

Figure 10-31. The reannealing of DNA from a genome in which particular nucleotide sequences are present at different frequencies. Altogether there are seven *different* sequences of 1000 nucleotide pairs each. Sequence number 1 is present at four copies per genome, sequence numbers 2 and 3 are present at two copies per genome, and sequence numbers 4, 5, 6, and 7 are present only once per genome. The progress of the reannealing is shown in the lower part of the figure. The greater the frequency, the faster the tendency to reanneal. Time (or C_0t) is plotted on a log scale.

a partner in a given time period as any other; they are all present (with the exception of a very few sequences in bacterial DNA) at the same concentration. As was shown by Roy Britten and David Kohne in 1966, the curve for eukaryotic DNA is much more complex. This added complexity reflects the fact that the various nucleotide sequences in a solution of DNA of higher organisms are not all present at the same concentration. In other words, unlike the DNA of viruses and bacteria, a particular nucleotide sequence in eukaryotic DNA may not be present in only one copy per genome, i.e., *per haploid amount of DNA*. Rather, Britten and Kohne found that different sequences could be present at different concentrations. There are generally three recognizable classes of fragments when eukaryotic DNA is broken into pieces of several hundred nucleotide length. The three classes are distinguished by their rate of reannealing, which reflects the number of times their nucleotide sequence is repeated within the population; the greater the number of copies in the genome, the greater their concentration, and the faster they reanneal. This is illustrated in the hypothetical experiment of Fig. 10-31 and the actual case of Fig. 10-32a. The classes are termed the *highly repetitive* (or *satellite*) fraction,

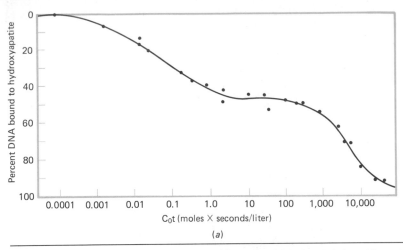
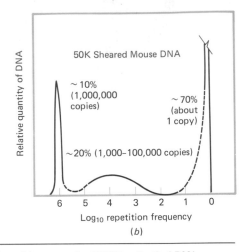

Figure 10-32. *(a)* A curve showing the reannealing of fragmented calf DNA. Purified DNA was sheared to 400 nucleotide lengths and incubated to different C_0t values in 0.12 M phosphate buffer at 60°C. The percentage of DNA that had reannealed in each sample was determined by hydroxyapatite chromatography. The reannealing of purified nonrepeated DNA is shown in Fig. 10-30. *(From R. J. Britten and J. Smith, Yearbook of the Carnegie Inst. Wash., **68**:368, 1968.)* *(b)* Spectrogram of the frequency of repetition of nucleotide sequences in the DNA of the mouse. Relative quantity of DNA plotted against the logarithm of the repetition frequency. Ten percent of the sequences are highly repeated, 20% moderately repeated, and 70% apparently nonrepeated. The dashed segments of the curve represent regions of considerable uncertainty. *(From R. J. Britten and D. E. Kohne, Science, **161**:536, 1968. Copyright by the American Association for the Advancement of Science.)*

the *moderately repetitive* fraction, and the *nonrepetitive* (or *unique*) fraction (Fig. 10-32).

The highly repeated fraction generally constitutes about 10% of the genome and reanneals so rapidly that its progress can only be followed in dilute solutions where the concentration of the reactants is very low. This fraction consists of a very short nucleotide sequence, repeated over and over again. In some species, the specific sequence being repeated may be as short as six nucleotides, while in other species the sequence may be a few hundred nucleotides in length. In the calf, one of the best-studied DNAs, the C_0t $\frac{1}{2}$ for this fraction is approximately 10^{-3} molar-seconds, which indicates the presence of over 1 million copies per genome. In many species, satellite DNA is sufficiently different in base composition from the remainder of the DNA that it can be separated by centrifugation to equilibrium in a density gradient (Fig. 10-33). In some animals, more than one band of satellite DNA can be distinguished indicating the presence of more than one highly repeated base sequence. For example, in *Drosophila*, there are three different sequences, each seven nucleotides in length, and all three quite similar, indicating a common evolutionary origin. As discussed at length in the appendix to this chapter, satellite DNAs are localized in the centromere of the chromosomes.

The moderately repeated fraction of the DNA, which can vary from about 20% to about 80% of the total depending upon the type of organism, includes sequences that are repeated to a *varying* degree within the genome. Sequences that reanneal in this fraction can range from several

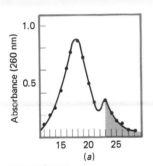 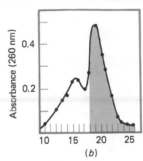 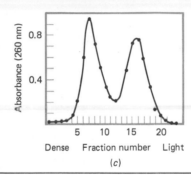

(a) (b) (c)

hundred thousand copies down to approximately ten. Within this fraction, the more highly repeated sequences will constitute that part of the curve (Fig. 10-32a) closer to the origin. One particular characteristic of a large percentage of the moderately repeated DNA is that the copies of a sequence are not all identical to one another, but rather comprise a *family* of closely related sequences. The moderately repeated fraction of the DNA consists of a variable number of families whose members are capable of complexing with other sequences within the family but not outside of it. Since the repeated sequences undergoing duplex formation are not necessarily of a precisely complementary nature, there is generally a considerable degree of mismatching of bases within the reannealed molecules formed. This mismatching is detectable by determination of the T_m of the reannealed duplexes since mismatched base pairs lower the temperature at which the duplex melts.

The remaining fraction of the DNA is composed of nucleotide sequences which reanneal with kinetics indicating they are present in one copy per genome. For example, the $C_0t \frac{1}{2}$ for the nonrepeated fraction of calf DNA is 3×10^3 molar-seconds, over six orders of magnitude greater than that for calf satellite DNA. Considering the variety of different sequences present, the nonrepeated fraction contains by far the greatest amount of genetic information.

Analysis of the role of these various DNA fractions is the subject of a large amount of research at present, though much of the significance of the complexity remains unknown. We will consider the satellite DNA in further detail in the appendix to this chapter and defer discussion of the other two classes until we have covered some of the basic aspects of transcription. Suffice it to state here that the repeated DNA sequences may be primarily regulative in function, though the genes responsible for the production of the transfer RNAs, ribosomal RNAs, and histone mRNAs are among this class. The nonrepeated fraction is believed to contain the DNA sequences that code for nearly all the messenger RNAs. Satellite DNA is not transcribed and its function is unknown.

Size and Organization of the Genome
As would be expected, if one measures the DNA content of the genome of various organisms, there is an overall increase as one progresses up the evolutionary ladder (Fig. 10-34). Viruses have the least amount of DNA, a

Figure 10-33. Banding pattern of mouse DNA in CsCl density gradients. *(a)* Profile obtained by sedimentation of whole DNA. The smaller peak of lower density DNA represents satellite. *(b)* The DNA from the portion of the *a* gradient that is shaded was collected and recentrifuged in a second density gradient. The peak on the left indicates the extent to which this fraction is now contaminated by nonsatellite DNA. *(c)* Satellite DNA centrifuged in an alkaline gradient which causes the denaturation of the duplex. The two strands that make up the duplex can be separated and each used for further experiments. [*From W. G. Flamm, M. McCallum, and P. M. B. Walker,* Proc. Natl. Acad. Sci. (U.S.), **57**:1730, 1967.]

Figure 10-34. The increase in the DNA content of the genome (haploid value) with increasing phlyogenetic complexity. Values used in this plot are minimal sizes for a genome of each class of organism listed; the genomes of other members of the class may have considerably greater DNA content. The ordinate is not a numerical scale, but simply a reflection of general phylogenetic order, and the exact shape of the curve has little significance. *(From R. J. Britten and E. H. Davidson,* Science, **165:** *352, 1969. Copyright by the American Association for the Advancement of Science.)*

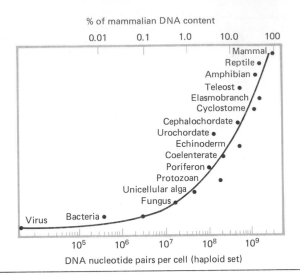

few of them enough for only a handful of genes. Bacterial genomes contain several thousand genes, while many vertebrates have sufficient DNA to account for over a million genes. However, there are animals, particularly a variety of amphibians, that have a totally disproportionate amount of DNA for their evolutionary position. For example, *Amphiuma,* a salamander, has about 25 times the amount of DNA per haploid set of chromosomes than humans. The chromosomes of some of the amphibians are huge which has made them excellent but puzzling research subjects. One possible explanation for these abnormally large genomes might be simply a disproportionately high amount of repeated sequences with no increase in the number of the nonrepeated variety. If this were the case, the genome would simply be larger but possess no more information. Analysis of these sequences, however, indicates a large increase in all frequency classes, including the nonrepeated, which leaves the question of the role of all of this DNA unanswered.

Even if we ignore these abnormally large genomes, the question of the role of such a large amount of DNA remains uncertain. The fact that there are a few thousand genes in a bacterium correlates well with the number of expected functions, approximately 20% of which have already been defined to some degree. However, if we were to assume that the primary function of DNA is to code for the amino acid sequence of different proteins, we would expect to find within our body nearly 10 million different proteins, a seemingly unimaginable number. From what we know about the relationship of amino acid sequence to the structure and function of proteins, it seems unlikely that so many different-shaped proteins could have any meaningful function. In other words, it is likely that a large percentage of the DNA sequences do not code for proteins, but have some as yet unknown role. We will have more to say about this aspect when we consider RNA synthesis and the extent of the genome that seems to be transcribed. It is also possible that a large amount of DNA is involved in the functioning of the particularly complex and less well understood tissues of the immune and nervous systems.

We have described at some length the presence within the genome of nucleotide sequences of widely ranging copy frequency. How are these various types of sequences organized within the huge DNA molecules present in the chromosomes? If we are to understand the basis for the complexity of the genome, it is imperative that the organization of these sequences be determined. Studies carried out by Eric Davidson, Roy Britten, and coworkers have provided information on the sequence organization of a number of organisms including an echinoderm, an amphibian, and a mammal (the calf). The primary observations are that the organization is (1) similar in all of these animals and (2) highly ordered. Based on a variety of hybridization techniques, it was found that approximately half of the genome of these animals contains closely interspersed repeated and unique sequences. The repeated sequences average about 300 nucleotides in length while the unique are about 800 nucleotides. A significant fraction, about 20% in the sea urchin, is made of essentially uninterrupted unique sequences, while about 6% apparently contains relatively long regions of repeated DNA. The remaining DNA is made of repeated sequences interspersed with unique sequences of considerable length (4000 nucleotides). The *effect* of some of these arrangements will become apparent when we consider the RNA molecules made from these DNA templates; however, the *basis* for the arrangement remains poorly understood.

APPENDIX

Molecular Hybridization

It was mentioned previously that renaturation studies have led to the development of a technique termed *molecular hybridization,* by which the presence of particular nucleotide sequences can be sought. Though there are many ways in which the technique can be used, all of them depend upon the formation of a duplex, or *hybrid,* between two types of single-stranded molecules. The hybrids formed can be of a DNA-DNA, DNA-RNA, or even an RNA-RNA nature. In almost all cases, one of the members is radioactive and therefore easily followed. The extreme importance of this technique in all of its ramifications is its ability to distinguish between nucleic acid molecules that differ from one another by their nucleotide sequence. The greatest difficulty in attempting to discriminate among nucleic acid molecules is that their properties are so similar. Consider two fragments whose length and overall base composition are identical, but whose nucleotide sequence is quite different. How can one go about distinguishing between these two molecules? They cannot be separated by sedimentation, electrophoresis, gel filtration, or any other bulk biochemical technique that one might use to separate two proteins. Differences in base sequence of a molecule hundreds of nucleotides long is a very subtle distinction. The way to distinguish between molecules of different sequence is to use *complementary* molecules as probes. The only characteristic that a labeled nucleic acid probe uses in its search for a complement is base sequence. These points will be clearer after the presentation of an example of the technique's great discriminatory powers.

The Localization of Satellite DNA

Where is satellite DNA located in the chromosomes? This question has been answered using a type of molecular hybridization, termed *in situ hybridization* developed by Mary Lou Pardue and Joseph Gall. All hybridization reactions involve the interaction between two populations of nucleic acids. In this case, one of the reactants is a labeled population of satellite DNA molecules and the other reactant is the DNA of the chromosomes. The term *in situ* means "in place." In this experiment the term refers to the

DNA of the chromosomes which is kept *in place* in the chromosomes during the incubation. Before describing the results we need to briefly describe how the two reactants are obtained.

---

1. *Preparation of labeled molecules that contain only the sequence for satellite DNA.* First, satellite DNA must be separated from all others of the genome. Pardue and Gall took advantage of the difference in base composition between satellite DNA and the remainder of the DNA and separated the two fractions by equilibrium sedimentation (by a modification of that shown in Fig. 10-33) and collected the DNA of the satellite band. In one type of experiment, the DNA they prepared was radioactively labeled by virtue of growing the cells in ^{3}H-thymidine before extracting the DNA. In another type of experiment, they isolated unlabeled satellite DNA and used it as a template to make radioactive RNA in vitro using purified RNA polymerase and radioactive RNA precursors. Since the RNA formed from the satellite DNA will be complementary to it, it can be used to search for satellite DNA in the chromosomes as readily as labeled satellite DNA itself. In this case, the only difference between the use of labeled DNA and labeled RNA is that the latter, by virtue of being made in vitro, will be much more radioactive. The greater the specific activity (cpm/μg) of the probe, the easier its detection.

2. *A preparation of mitotic chromosomes.* Since the question to answer concerns the location of satellite DNA in the chromosomes, chromosomal DNA must be one of the reactants. The procedure to obtain mitotic chromosomes from blood cells is shown in Fig. 15-14. Once the chromosomes are dried onto a slide, the slide can be gently immersed in various solutions without the release of the chromosomes. The next step is to denature the DNA within the chromosomes so that it will be rendered capable of hybridization. Treating the chromosomes with an alkaline solution causes the strands to separate and remain unjoined, even though the two strands remain close together within the chromosomes.

The hybridization step simply involves the incubation of the two reactants together to allow the radioactive satellite DNA (or the labeled RNA copy, termed *cRNA*) to attach to complementary sequences within the chromosomes. Once the incubation period for the hybridization has ended, the soluble

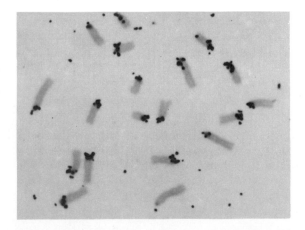

Figure 10-35. Autoradiograph of mitotic chromosomes from a cultured mouse cell after in situ hybridization with radioactive RNA copied in vitro from mouse satellite DNA. The RNA has bound to its complementary sequences in the DNA of the chromosomes, showing that these DNA sequences are localized in the centromeric region of the chromosomes. *(From M. L. Pardue and J. G. Gall, in C. D. Darlington and K. R. Lewis, eds., "Chromosomes Today," vol. 3, Oliver and Boyd Ltd., 1972.)*

labeled nucleic acid can be washed away or digested and the slide prepared for autoradiography (Fig. 6-7) in order to observe the location of the bound radioactive DNA (or cRNA) within the chromosomes. Figure 10-35 shows an autoradiograph prepared in this manner. It is apparent from this photograph that the DNA sequences complementary to the labeled satellite DNA (or cRNA) are localized in very specific sites within the mitotic chromosomes. These sites are the regions of the centromeres of each of the chromosomes, the places at which the mitotic spindle fibers (Fig. 15-17) normally make their attachment. The precise function of these short, repeating sequences present within the centromere region remains unknown, though the evidence suggests it is not transcribed into RNA. When this same type of in situ hybridization experiment is carried out with total DNA, radioactivity is found distributed rather uniformly over the entire chromosome.

Spectrophotometry

It was pointed out on page 388 that, in order for a molecule to absorb light of a particular wavelength, it must absorb an entire quantum of light, the energy of which boosts an electron from a lower to a higher orbital. In order for this to occur, the energy of light must exactly balance the energy difference between the electrons in the two positions. A spectrophotometer is an instrument which is capable of measuring the amount of light of a specific wavelength that is absorbed by a given solution. If one knows the absorbance characteristics of a particular type of molecule, then the amount of light of the appropriate wavelength absorbed by a solution of that molecule is a measure of its concentration. The greater the number of absorbing molecules present in the light path, the greater will be its absorbance.

In order to make this type of measurement, the solution is placed in a special flat-sided quartz container (quartz is used because unlike glass it does not absorb ultraviolet light), termed a *cuvette*, which is then placed in the light beam. The amount of light which passes through the solution unabsorbed, i.e., the transmitted light, is measured by photocells on the other side of the cuvette. The path length, i.e., the distance through the solution, is equal to the internal width of the cuvette. The greater the path length (l), the greater the number of molecules that have a chance to absorb the light and the greater the absorbance. In order to standardize this factor, cuvettes are built so that their path length is equal to one centimeter. The *absorbance* (A) of the solution is defined as the log of the ratio of the intensity of the incident light (I_0) to the intensity of the transmitted light (I) as stated by the Beers-Lambert law.

$$A = \log \frac{I_0}{I}$$

which is equal to

$$a \cdot c \cdot l$$

where c is the concentration, l the path length, and a (or ϵ) the absorptivity, a measure of the likelihood that a given photon will be absorbed by a given type of molecule. The absorptivity of a 1 molar solution of a particular solute is given the term *molar extinction coefficient* (ϵ_m). Given the molar extinction coefficient, which can be looked up, one can calculate the concentration (c) of that solute from the absorbance measurement using the equation

$$c = \frac{A}{\epsilon_m} \cdot l$$

Ultracentrifugation

Common experience indicates that the stability of a solution (or suspension) depends upon the nature of the components. Cream floats to the top of raw milk, a fine precipitate gradually settles to the bottom of its container, while a solution of sodium chloride remains stable indefinitely. Numerous factors determine whether or not a given component will settle through a liquid medium, including the size, shape, and density of the substance and the density and frictional resistance, i.e., viscosity, of the medium. In order for a substance to sediment toward the bottom of a tube, it must have a greater density than the surrounding medium. Even if the substance meets the density requirement, the sedimentation process, which tends to concentrate the molecules, will be counteracted by their tendency to be redistributed uniformly as a result of diffusion. Other factors remaining constant, the sedimentation of a particular population of molecules will depend upon its rate of diffusion as opposed to the centrifugal force being applied. Although the rate of diffusion of a molecule can be markedly dependent upon its shape, among a similar group of molecules, the diffusion coefficient varies inversely with molecular weight. Macromolecules of the type found within cells have sufficiently large diffusion coefficients so that they do not settle in an aqueous medium at room temperature under the force of gravity. However, with the development of ultracentrifuges capable of generating tremendous centrifugal forces, it has become feasible to analyze the sedimentation properties of macromolecules.

The velocity with which a given particle moves through a liquid medium during centrifugation is directly proportional to the square of the angular velocity (ω) and the distance of the particle from the center of the rotor (r).

$$v = s\ \omega^2 r$$

In this equation, $\omega^2 r$ is termed the *radial acceleration* and is given in cm/sec². The term s, which is given in seconds, is referred to as the *sedimentation coefficient*. The sedimentation coefficient is equivalent to the average velocity per unit of acceleration. Centrifugal accelerations are generally expressed relative to the earth's gravitational acceleration which has a value of 980 cm/sec². For example, a value for radial acceleration of 4.9×10^7 cm/sec² is said to be equivalent to 50,000 times the force of gravity, i.e., 50,000 g.

Throughout this book we have referred to various macromolecules and their complexes as having a particular S value. The unit S (or Svedberg, after the inventor of the ultracentrifuge) is equivalent to a sedimentation coefficient of 10^{-13} seconds. Since the velocity by which a particle moves through a liquid column is dependent upon a number of factors, including shape, determination of the sedimentation coefficient does not, by itself, provide the molecular weight. However, molecular weight can be determined by centrifugal techniques whose discussion is beyond the scope of this book (see any of the major biochemistry texts). Even though sedimentation coefficients are not generally proportional to molecular weight, as long as one is dealing with the same type of molecule, such as double-stranded linear DNA, the S value provides a good measure of the molecular weight. For example, in the case of the three ribosomal RNAs of *E. coli*, the 5S, 16S, and 23S molecules have nucleotide lengths of 120, 1600, and 3200, respectively.

There are basically two types of ultracentrifuges, analytical models and preparative models. The types of experiments outlined in Fig. 10-26 are performed using the simpler preparative ultracentrifuge. Centrifuges of this type are designed simply to generate large centrifugal forces for a set period of time. The centrifugation proceeds in a near vacuum so as to minimize frictional resistance. In some cases, the rotors (centrifuge heads) allow the tubes to swing out (as illustrated in Fig. 10-26) so that the particles move in a direction parallel to the walls of the tube. This type of rotor is said to contain *swinging buckets.* In the other type of rotor, the *fixed-angle* variety, the tubes are maintained at a specific oblique angle. In the analytical ultracentrifuge, the rotors and tubes (termed *cells*) are very different and the centrifuge contains instrumentation that allows one to follow the progress of the substances in the cell during the centrifugation. In this manner various estimates of sedimentation velocity can be made during the centrifugation and measurements of molecular weight, purity, etc., can be determined. In preparative ultracentrifugation, all determinations are made after the tube is removed from the centrifuge. The term *preparative* refers to the use of this type of centrifuge to purify components for further use.

REFERENCES

Avery, O. T., MacLeod, C. M., and McCarty, M., J. Exptl. Med. **79,** 137–158, 1944. "Studies on the Chemical Nature of the Substance Inducing Transformation of Pneumococcal Types. Induction of Transformation by a Desoxyribonucleic Acid Fraction Isolated from Pneumococcus Type III."

Benzer, S., "The Elementary Units of Heredity," in *The Chemical Basis of Heredity,* W. D. McElroy and B. Glass, eds., pages 70–93, Johns Hopkins Press, 1957.

Britten, R. J., and Kohne, D. E., Science **161,** 529–540, 1968. "Repeated Sequences in DNA."

———— and Kohne, D. E., Sci. Am. **222,** 24–31, April 1970. "Repeated Segments of DNA."

Carlson, E. A., *The Gene: A Critical History.* W. B. Saunders, 1966.

Chargaff, E., Experientia **6,** 201–209, 1950. "Chemical Specificity of Nucleic Acids and Mechanism of Their Enzymatic Degradation."

————, Science **172,** 637–642, 1971. "Preface to a Grammar of Biology."

Chromosome Structure and Function. Cold Spring Harbor Symp. Quant. Biol. **38,** 1974.

Ciba Foundation Symposium Number 28, "The Structure and Function of Chromosomes." Elsevier, 1975.

Crick, F., Pauling, L., Gurdon, J. B., Chargaff, E., Olby, R., Brenner, S., and Klug, A., Nature **248,** 765–788, 1974. "Molecular Biology Comes of Age."

Davidson, J. N., *The Biochemistry of Nucleic Acids,* 7th ed. Academic, 1975.

Edström, J.-E., and Lambert, B., Prog. Biophys. Mol. Biol. **30,** 57–82, 1975. "Gene and Information Diversity in Eukaryotes."

Flamm, W. G., Int. Rev. Cytol. **32,** 1–51, 1972. "Highly Repetitive Sequences of DNA in Chromosomes."

Fraenkel-Conrat, H., and Singer, B., Biochim. Biophys. Acta **24,** 540–548, 1957. "Virus Reconstitution II. Combination of Protein and Nucleic Acid from Different Strains."

Goodenough, U., and Levine, R. P., *Genetics*. Holt, Rinehart and Winston, 1974.

Hennig, W., Int. Rev. Cytol. **36**, 1–44, 1973. "Molecular Hybridization of DNA and RNA *in situ*."

Hershey, A. D., Science **168**, 1425–1427, 1970. "Idiosyncrasies of DNA Structure."

——— and Chase, M., J. Gen. Physiol. **36**, 39–56, 1952. "Independent Functions of Viral Protein and Nucleic Acid in Growth of Bacteriophage."

Kolata, G. B., Science **182**, 1009, 1973. Research News. "Repeated DNA: Molecular Genetics of Higher Organisms."

Mendel, G., de Vries, H., Correns, C., and Tschermak, E., Genetics **35**, Supp., 1–47, 1950. "The Birth of Genetics." Papers in English translation.

Mirsky, A. E., Sci. Am. **218**, 78–88, June 1968. "The Discovery of DNA."

Moore, J. A., *Heredity and Development*, 2d ed. Oxford University Press, 1977.

Novitski, E., and Blixt, S., Bioscience **28**, 34–35, 1978. "Mendel, Linkage, and Synteny."

Pardue, M. L., and Gall, J. G., Sci. **168**, 1356–1358, 1970. "Chromosomal Localization of Mouse Satellite DNA."

Rosenberg, H., Singer, M., and Rosenberg, M., Science **200**, 394–402, 1978. "Highly Reiterated Sequences of SIMIANSIMIANSIMIANSIMIANSIMIAN."

Schroedinger, E., *What is Life?* Anchor Books, Doubleday, 1956.

Skinner, D. M., Bioscience **27**, 790–795, 1977. "Satellite DNAs."

Stent, G., *Molecular Genetics*. W. H. Freeman, 1971.

Strickberger, M. W., *Genetics*, 2d ed. Macmillan, 1976.

Sturtevant, A. H., J. Exptl. Zool. **14**, 43–59, 1913. "The Linear Arrangement of Six Sex-Linked Factors in Drosophila, as Shown by Their Mode of Association."

———, *A History of Genetics*. Harper & Row, 1965.

Suzuki, D. T., and Griffiths, A. J. F., *An Introduction to Genetic Analysis*. W. H. Freeman, 1976.

Swift, H., Proc. Natl. Acad. Sci. (U.S.) **36**, 643–654, 1950. "The Constancy of Desoxyribose Nucleic Acid in Plant Nuclei."

Taylor, J. H., *Selected Papers on Molecular Genetics*. Academic, 1965.

Tomasz, A., Sci. Am. **220**, 38–44, Jan. 1969. "Cellular Factors in Genetic Transformation."

Watson, J. D., *Molecular Biology of the Gene*, 3d ed. W. A. Benjamin, 1976.

——— and Crick, F. H. C., Cold Spring Harbor Symp. Quant. Biol. **18**, 123–131, 1953. "The Structure of DNA."

——— and Crick, F. H. C., Nature **171**, 737–738, 1953. "Molecular Structure of Nucleic Acids. A Structure of Deoxyribose Nucleic Acid."

Wilkins, M. H. F., Stokes, A. R., and Wilson, H. R., Nature **171**, 738–740, 1953. "Molecular Structure of Deoxypentose Nucleic Acids."

Wilson, E. B., *The Cell in Development and Heredity*, 3d ed. Macmillan, 1925.

CHAPTER
ELEVEN

The Flow of Information
Through the Cell

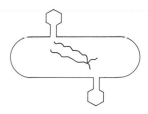

Now that we have established the identity of the genetic material we can turn to the next logical question, that of its utilization. The DNA molecules that are passed from cell to cell and organism to organism contain the totality of the information which can be inherited. The questions of physiological importance concern the means by which the cell can put the information stored in the genetic blueprint to work in governing the cell's many activities. The first point which must be established is the manner in which specific inheritable traits can be determined by the presence and activity of single genes.

THE RELATIONSHIP BETWEEN GENES AND PROTEINS
The One Gene–One Enzyme Concept

The first meaningful insight into gene function was made by Archibald Garrod during the first quarter of this century, the same general period when the principles of classical genetics were being formulated. Garrod was interested in certain congenital diseases which seemed to affect a person's metabolism. One of the best studied of these diseases was *alcaptonuria,* a condition readily detected as a result of the blackening of the urine upon exposure to air. The cause for this curious effect was the oxidation by air of a substance, homogentisic acid, which is present in the urine of persons with this condition and absent from the urine of normal individuals. Since the condition was transmitted recessively in a strict Mendalian fashion, it seemed clearly to have a genetic basis. Garrod seized on an opportunity to correlate a specific genetic defect with a specific metabolic deficiency. The point to be ascertained was the nature of the deficiency—why was homogentisic acid present in the urine of these persons? Homogentisic acid was shown to be a product of the breakdown of the aromatic amino acids, phenylalanine and tyrosine (Fig. 11-1). If these amino acids were ingested by persons with this disease, large amounts of this breakdown product appeared in their urine. It was then shown that persons with alcaptonuria lacked an enzyme in their blood that normally altered this compound, thereby causing it to accumulate. Garrod concluded that the normal gene is responsible for the production of this enzyme. The relationship between a specific gene, a specific enzyme, and a specific chemical reaction had been made. As seems to have happened with other observations of basic importance in genetics, this finding was essentially ignored for many years.

It was obvious that genetic studies of human traits were drastically limited, so during the 1930s George Beadle and Boris Ephrussi turned to *Drosophila*, the organism whose genetics were best understood, and to a trait, eye color, for which a large number of mutants were known. It was hoped that an analysis of the biochemical deficiencies of the pigments of these various mutants might provide an insight into the nature of the genetic functions that were defective in these strains. After considerable effort and limited success, the concept that genes and enzymes were closely interrelated had gained some support. However, it was realized that if convincing evidence were to be obtained, a much simpler biochemical system would have to be examined.

Beginning in about 1940, Beadle, along with Edward Tatum, decided that *Neurospora*, a tropical bread mold, would be an ideal organism for a combined biochemical and genetic attack on this problem. *Neurospora*

Figure 11-1. Steps in the pathway leading to the breakdown of the aromatic amino acids.

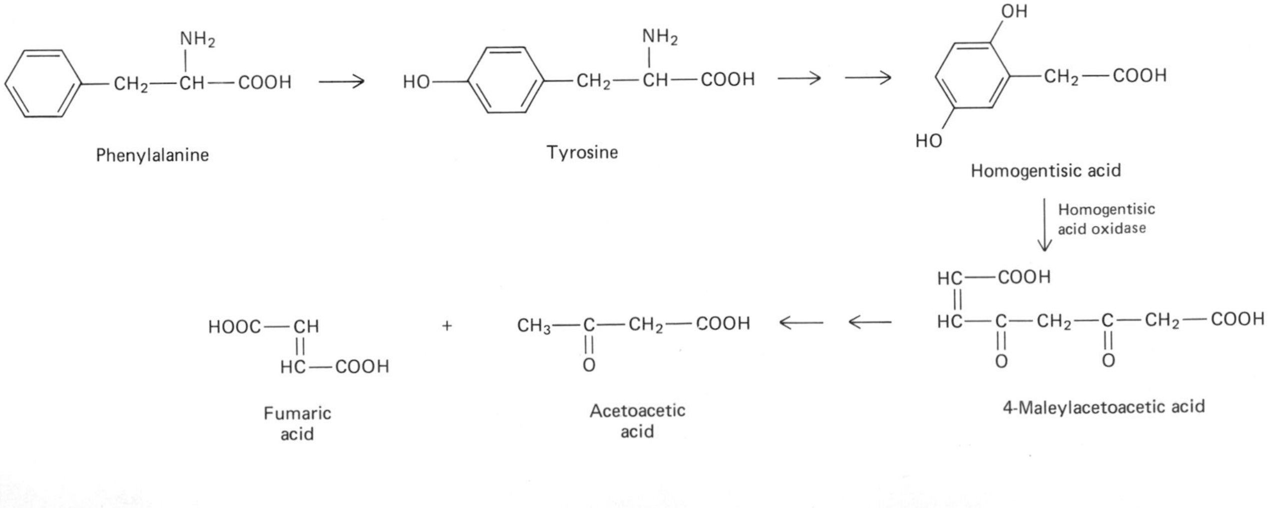

had a number of advantages. In the first place, its genetics were understood as a result of extensive analysis. Secondly, it had both a sexual and an asexual stage. When individuals of the opposite sex were brought together, sexual reproduction resulted in the formation of fruiting bodies which contained haploid spores that would grow into haploid individuals. The system was ideal; the haploid nature of the organism eliminated the worry about recessive mutations being masked by dominant alleles, while the sexual stage allowed them to determine whether or not a particular metabolic deficiency was due to a mutation in a single gene. In addition, its growth requirements were ideal; it needed a single organic carbon source (e.g., some sugar), inorganic salts, and biotin (a B vitamin). Together, these ingredients form what is termed a *minimal medium* for *Neurospora,* i.e., the simplest medium that will support its growth. Since it needed so little to live on, it had the capability of synthesizing all of its required metabolites; an organism with such a wide synthetic capacity should be very sensitive to enzymatic deficiencies which should be easily detected.

In order to fully discuss Beadle and Tatum's experimental procedure, we would need to describe the life cycle of *Neurospora,* which is unwarranted. Basically, they irradiated spores and allowed each of them to produce a genetically identical population of haploid cells which they could test for specific metabolic deficiencies (Fig. 11-2). They did this by growing the cells on media supplemented with a great variety of organic compounds and then tested representatives of each colony on the minimal medium. If they found a strain of cells that could grow on the supplemented but not on the nonsupplemented medium, they had induced a mutation suitable for study. Once mutant strains had been identified, there were two tasks to accomplish. One was to prove that the defect was due to a mutation in a single gene, which was determined by crossing the mutant with the wild-type and examining the progeny. The second was to determine the nature of the substance present in the supplemented medium that the mutant cells could no longer synthesize. This was accomplished by determining if cells could grow in various types of supplemented media, each lacking different ingredients, and identifying the necessary substance by a process of elimination.

Fearing that the radiation-induced mutation rate might be very low, Beadle and Tatum began by culturing over a thousand different irradiated cells on supplemented medium before they did any of their tests of cells on minimal medium. Of the first group, two cultures of cells (the 299th and the 1085th) proved unable to grow on the simple medium. It was found that one needed pyridoxin (B_6) and the other thiamine (B_1) in order to survive. Eventually, about 100,000 irradiated spores were tested and dozens of different types of mutants were isolated whose gene defects were manifested in their inability to catalyze a great variety of reactions. The results were clear-cut: genes acted to specify, in some unknown way, particular enzymes which could then catalyze specific reactions. The finding became known as the "one gene–one enzyme" hypothesis. Once it was realized that enzymes are often composed of more than one polypeptide

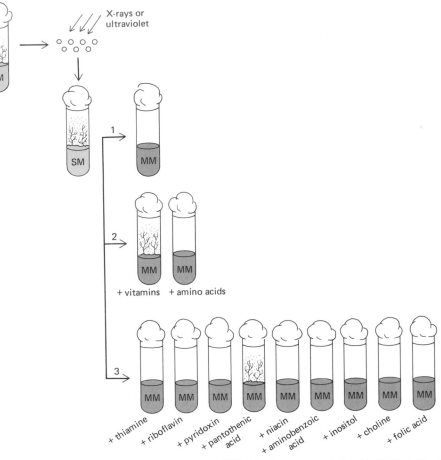

Figure 11-2. The procedure of the Beadle-Tatum experiment for the isolation of genetic mutants in *Neurospora*. Spores were irradiated to induce mutations and then allowed to grow in a tube containing supplemented medium (SM). Cells of each were then tested for their ability to grow on the minimal medium (MM). Those that failed to grow (as in step 1 of the figure) were mutants and the task was to determine the nature of the mutant gene. In step 2, cells were found to grow in the minimal medium supplemented with "vitamins" but not that supplemented with amino acids. This step indicates the deficiency is one leading to the formation of a vitamin. In step 3, growth of the cells in minimal medium supplemented with one or another of the vitamins indicates the deficiency resides in a gene involved in the formation of pantothenic acid (part of coenzyme A). (*From A. Roller, "Discovering the Basis of Life," McGraw-Hill, 1974.*)

chain (page 99), each of which is determined by a different gene, the con-concept became modified to the presently accepted form of "one gene–one polypeptide chain."*

In order to better appreciate the experimental rationale behind the development of the "one gene–one enzyme–one reaction" concept, we will briefly consider a group of *Neurospora* mutants studied by Adrian Srb and Norman Horowitz. All of the mutants in this group were unable to synthesize the amino acid arginine, i.e., they were arginine *auxotrophs*. Seven strains of arginine auxotrophs had been isolated whose genetic defects mapped out by genetic recombination in seven different genes. Even though none of these seven strains could grow on minimal medium, and all of them could grow on medium supplemented with arginine, they were not identical in their growth requirements. Of the seven, four could

---

* It should be noted that special cases have been uncovered where this relationship does not hold. As discussed on page 795, polypeptides of antibody molecules arise, in a sense, from two genes. Conversely, a few genes in certain viruses are known to code for two different polypeptides (see page 568).

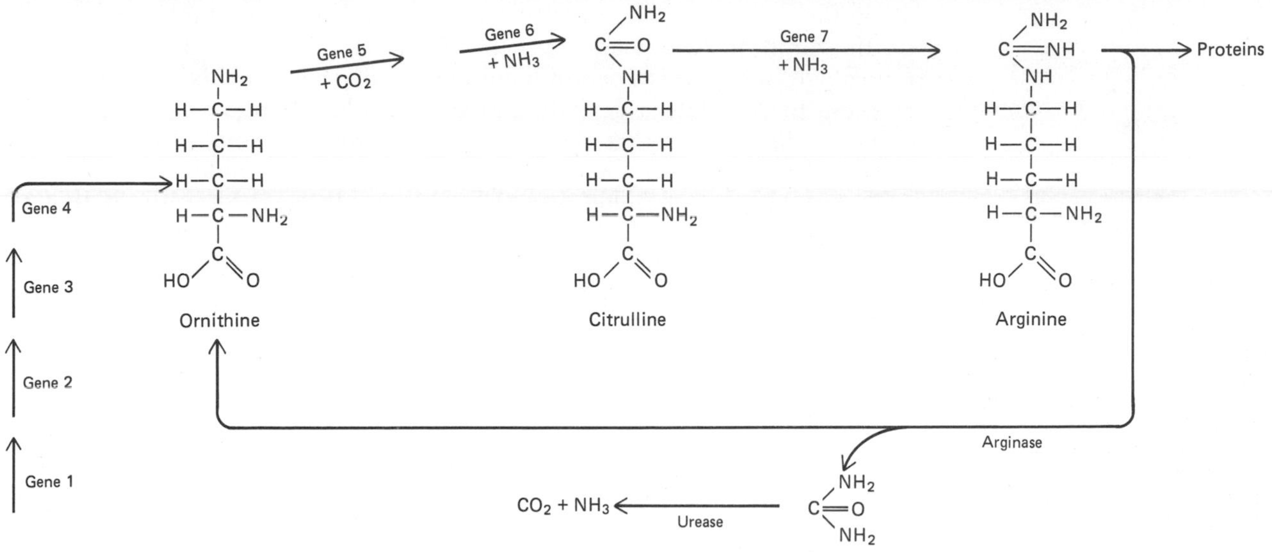

Figure 11-3. The ornithine cycle in *Neurospora*, showing the steps leading to the biosynthesis of arginine. (*From* American Scientist, *34:31, 1946, based on the work of A. Srb and N. Horowitz.*)

grow if given minimal medium plus either ornithine or citrulline, two could grow if given citrulline but not ornithine, while the other had to have arginine in order to grow. Figure 11-3 shows the metabolic pathway leading to the biosynthesis of arginine and provides a clear explanation for the growth requirements of these strains on the basis of the one gene–one enzyme proposal. The strains that would grow on ornithine, citrulline, or arginine were deficient in various of the enzymes in the pathway prior to ornithine. The two mutant strains that would grow only on citrulline or arginine were ones with deficiencies in the pathway between ornithine and citrulline. There are two enzyme-catalyzed reactions involved in the conversion of ornithine to citrulline, each being deficient in one of the two strains. The mutant with the strict requirement for arginine had its mutation in the gene responsible for the last enzyme of the arginine pathway. By these types of experiments, each mutant was eventually correlated with a specific enzymatic deficiency. In addition to their genetic analysis, these types of mutants have been invaluable in the elucidation of a number of metabolic pathways as well as other types of sequential physiological processes.

Mutations as Changes in Nucleotide Sequence

The findings of Beadle and Tatum provided a whole new perspective in the embryonic field of biochemical genetics and research branched out in several directions. One of the paths that was taken was to establish the nature of the enzymatic defect for which a genetic mutation was responsible. It should be kept in mind that these studies on *Neurospora* were performed a full decade before Sanger had worked out the primary sequence of insulin or Watson and Crick the structure of DNA. Very little was known about the primary, secondary, and tertiary structure of proteins and it was not at all obvious how the gene and the protein were related.

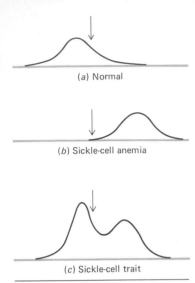

(a) Normal

(b) Sickle-cell anemia

(c) Sickle-cell trait

Figure 11-4. Differences in the electrophoretic mobility of hemoglobin extracted from cells of normal persons, or persons heterozygous or homozygous for the sickle cell allele. Heterozygotes which have both normal and mutant hemoglobin are said to have "sickle cell trait" while homozygotes have sickle cell anemia. *(After L. Pauling, H. A. Itano, S. J. Singer, and I. C. Wells,* Science, *110:543, 1949.)*

Therefore, it came as a surprise when in 1949 an important breakthrough on this problem was made by Linus Pauling. Like Garrod, Pauling took advantage of the existence of a congenital disease, in this case sickle cell anemia, to correlate a genetic and a molecular change. Pauling and co-workers discovered that the hemoglobin from persons with sickle cell anemia had a different electrophoretic mobility and isoelectric point (page 115) from normal hemoglobin (Fig. 11-4). It was determined that the nature of the molecular change in the sickle cell hemoglobin caused it to have extra positive (or less negative) charge. Persons that were heterozygous for the gene were found to have both types of hemoglobin within their blood cells. Presumably, each allele in the heterozygote was responsible for the production of a particular species of the protein.

Within the next few years the techniques of amino acid sequencing were developed and the primary structure of insulin was described. Hemoglobin, however, was very large in comparison to insulin and it was not feasible to attempt the determination of its amino acid sequence in order to discover the molecular differences between the normal and the sickle cell proteins. However, a shortcut to the answer was devised by Vernon Ingram in 1956. Using the rationale developed for amino acid sequencing, Ingram cleaved the two types of hemoglobin into a number of pieces using the proteolytic enzyme, trypsin. He then subjected these peptide fragments to paper chromatography to determine if any of the products of trypsin digestion could be distinguished between the two hemoglobin species. The pattern of the spots on the paper formed what is termed a *fingerprint*, reflecting the fact that each protein has a unique primary sequence. Of the 30 or so peptides in the mixture, one peptide migrated in a different manner in the two preparations (see Fig. 3-6); this one difference was responsible for all of the symptoms of the anemia. Once the peptides had been separated, Ingram had only one small piece to analyze rather than an entire protein. The difference proved to be the presence of a valine in the sickle cell hemoglobin as compared to a glutamic acid in the normal molecule. Ingram had demonstrated that a mutation in a single gene had caused a single substitution in the amino acid sequence.

Colinearity of Nucleotide and Amino Acid Sequences

The research of the early 1950s established that DNA consisted of a linear sequence of nucleotides in a double-stranded molecule and that proteins consisted of a linear sequence of amino acids. Once it was established that one gene determined the production of one protein, and that an alteration in that gene caused an alteration in the amino acid sequence of the corresponding protein, a logical conclusion could be drawn. The conclusion was that the linear sequence of nucleotides in a segment of DNA, i.e., a gene, *determined* the linear sequence of amino acids in the corresponding protein. Making this assumption and proving it were entirely different propositions. In its strictest sense, the proof of *colinearity,* i.e., the point-to-point correspondence between nucleotide and amino acid sequence, would require the determination of the nucleotide sequence of a gene and its correlation to the primary structure of its protein. By the late 1960s this feat had been essentially accomplished, but in the 1950s some alternate type of evidence would have to suffice. Before we can describe this evidence, we need to digress briefly and cover some of the most basic research in molecular genetics that we will need to draw upon in several places in this and the next chapter.

Though the genetic analysis of auxotrophs (mutants unable to synthesize a particular substance) began with *Neurospora*, its focus soon switched to bacteria, particularly *E. coli*, and the horizons of biochemical genetics rapidly expanded. The first important step in the use of bacteria as genetic subjects was to prove that they underwent spontaneous genetic mutation. This was unequivocally demonstrated by Max Delbruck and Salvador Luria in 1943 for the genetic trait controlling the sensitivity of bacteria to infection by bacteriophage. As a result of a rare spontaneous event, it was found that one in 10^7 to 10^8 cells were no longer killed by these viruses, but would instead continue to multiply in their presence. Later studies showed the insensitivity to the virus to be the result of a surface change in the bacteria making the virus unable to attach. This finding set the stage for Joshua Lederberg to begin an analysis of a number of metabolic mutants and demonstrate the phenomenon of bacterial recombination.

To avoid any complications that might arise from the spontaneous reversion of a mutation to the wild type, Lederberg and Tatum began by attempting to "cross" bacteria that had mutations in more than one gene, i.e., they were multiple auxotrophs. For example, one of the "crosses" involved two strains of bacteria, $B^-M^-P^+T^+$ and $B^+M^+P^-T^-$, where $^-$ indicates a mutation in a gene responsible for the production of that metabolite (biotin, methionine, proline, and threonine) and $^+$ indicates the wild-type condition for the gene. Their approach was to grow the two strains of multiple auxotrophs together in a medium containing all of these metabolites, then wash the medium away and replace it with the minimal medium in which neither of the original strains could grow. The result was the appearance of a few colonies (about 1 in 10^7) of bacteria. Since, in order to grow in the minimal medium, a bacterium had to have a genotype, $B^+M^+P^+T^+$, this finding indicated that the cells were some type of genetic recombinant. They concluded that, in a few cases, two bacterial cells had undergone fusion followed by some type of meiosis complete with crossing-over. To make a long and interesting story very short, it was later found that the recombination in bacteria was not the result of the fusion of two cells, but rather involved the transfer of a piece of genetic material from one individual to another (Fig. 11-5). The situation is analogous to the transformation experiment of Avery (page 443) in which a piece of DNA is taken up by a recipient bacterium and then finds its corresponding site on the host chromosome and undergoes a genetic exchange, i.e., a recombination. The event by which the transfer of a chromosome occurs from one cell to another is termed *conjugation*. Once recombination had been demonstrated to occur between bacterial chromosomes, it was shown it could also occur between two bacteriophage chromosomes that happened to infect the same bacterial cell. The demon-

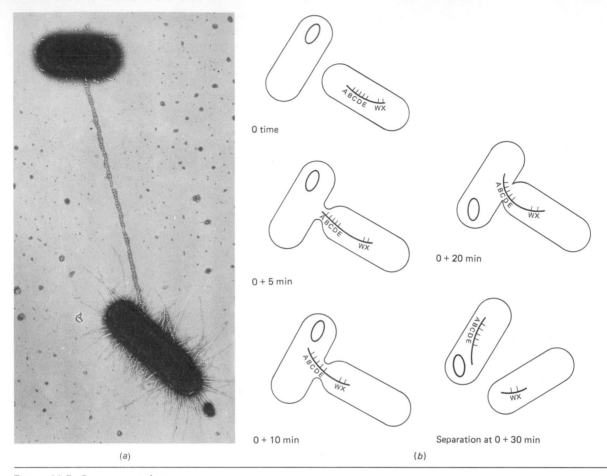

0 time

0 + 5 min

0 + 10 min

0 + 20 min

Separation at 0 + 30 min

(a)

(b)

Figure 11-5. Conjugation in bacteria. (*a*) Electron micrograph showing a male (Hfr) and female (F⁻) cell joined by a structure of the male cell termed the *F pilus*. (*Courtesy of C. C. Brinton.*) (*b*) Schematic representation of the time course of the conjugation process. With increasing time, more and more of the chromosome is transferred until a point is reached where the cells separate and transfer is stopped. Chromsome transfer always begins at the same site, and the percentages of the genes transferred can provide information on their linear order on the chromosome.

stration of recombination in bacterial and viral systems was to prove invaluable in later research in molecular genetics.

In the early work on recombination in higher organisms, little attention was paid to where the actual breakage in the chromosome was occurring since it was assumed to happen between genes. If one reconsiders the phenomenon in terms of a chromosome made of one long DNA molecule, there is no reason to suppose that the breakage cannot occur at some place within a gene as opposed to some special type of site that might exist between adjacent genes. In 1953, Seymour Benzer began experiments designed to demonstrate that recombination could occur between two sites in the same gene, a phenomenon termed *intragenic recombination*. To carry out his initial experiments, Benzer infected bacterial cells with pairs of phage strains carrying mutations in the same gene and searched for wild-type progeny that would indicate recombination had occurred.

Consider what might happen if two phage DNA molecules were present in the same cell, each carrying a mutation in a different part of the same gene (Fig. 11-6). If the breakage of the DNA molecules occurred between the two sites of mutation, then crossing-over would be expected

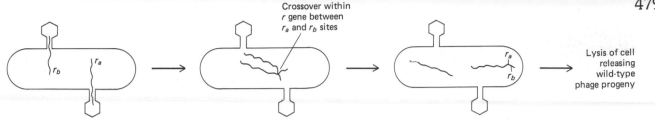

Crossover within
r gene between
r_a and r_b sites

Lysis of cell
releasing
wild-type
phage progeny

Figure 11-6. Recombination between two phage chromosomes carrying different mutations within the same gene. If crossing-over occurs between the two sites, wild-type progeny can be produced.

to produce one chromosome of the wild type and one chromosome containing a double mutation. How often can one expect crossing-over to occur between two sites *in a given gene?* One would expect recombination within a *particular* gene to be a very rare event, which it is. However, one of the advantages of working with bacterial and viral systems is the tremendous number of progeny that can be screened in an attempt to observe a "rare" event. Within a relatively short period of time, Benzer had not only demonstrated that intragenic recombination could occur, but he had obtained recombination between two sites separated from one another by only one or a very few nucleotides.

In the brief discussion of recombination in the previous chapter (page 434), it was shown that the recombination frequency could provide a measure of the distance between two genes in the same chromosome. In principle, the same rules should apply to the distance between two sites of mutation within the same gene: the farther apart they are, the greater should be their frequency of recombination. The only difference between intergenic recombination and intragenic recombination is the percentages of the original population that undergo recombination. Using these types of procedures, Benzer constructed a detailed genetic map of several hundred sites within one small portion of the T_4 genome.

We can now return to our discussion of the relationship between the nucleotide sequence of a gene and the amino acid sequence of the corresponding protein. Even though the technique of intragenic mapping provides no information as to the specific nucleotide sequence of a gene, it does allow the relative sites of mutations within a gene to be ordered. Using several techniques of intragenic mapping, Charles Yanofsky and coworkers determined the relative positions of a number of mutations within the A gene of the enzyme tryptophan synthetase of *E. coli*. Simultaneous with the genetic analysis, the entire amino acid sequence of the normal protein was determined and the analysis of mutant proteins began.

In lieu of being able to obtain the nucleotide sequence of the tryptophan synthetase gene, Yanofsky set out to determine if the relative positions of the sites of mutation within the gene correlated with the relative positions of the amino acid substitutions within the protein. Using the same approach taken by Ingram, tryptic digests of mutant proteins were subjected to chromatography and the search for the altered peptide was undertaken. Analysis of the amino acid sequence of the abnormal peptide provided an assignment for the position of each of the amino acid substitutions. In every case, the relative distances between altered nucleotides

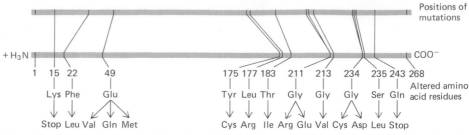

Figure 11-7. The demonstration of colinearity between the nucleotide sequence in the DNA and the amino acid sequence in the corresponding polypeptide chain (the alpha chain of tryptophan synthetase). Genetic mutants were isolated, the site of the alteration was determined by genetic fine-structure mapping, and the amino acid substitution was determined by peptide analysis. The upper bar in the figure represents the gene with the location of the mutations indicated. The lower bar represents the polypeptide chain with the number and nature of the corresponding substitution indicated below. (*From "Gene Structure and Protein Structure" by C. Yanofsky. Copyright © May, 1967 by Scientific American, Inc. All rights reserved.*)

correlated with relative distances between altered amino acids (Fig. 11-7). A similar study carried out by Sidney Brenner on mutations within the gene for the head protein of T_4, arrived at exactly the same results. Together these studies demonstrated the *colinearity* between the DNA molecule which carries the genetic code and the protein molecule for which the DNA is coding.

THE FLOW OF INFORMATION: DNA TO RNA TO PROTEIN

Up to this point we have established the relationship between genetic information and amino acid sequence, but this knowledge by itself provides no clue as to the mechanism by which the specific polypeptide chain is generated. As we now know there is an intermediate between the genetic material and the polypeptide, a molecule of RNA in which the genetic material is carried. RNA which contains information for a particular polypeptide chain is termed *messenger RNA (mRNA)*. The use of a messenger-type molecule allows the cell to separate its information storage function from its information utilization function (Fig. 11-8). Whereas DNA remains "locked" in the nuclei within the chromosomes, some of its information can be imparted to a mRNA which is a more mobile nucleic acid. The mRNA is much smaller than the chromosome-sized DNA molecules and can pass into the cytoplasm. Once in the cytoplasm, the mRNA can provide the template to direct the incorporation of amino acids in a particular order.

The use of a messenger RNA molecule interspersed between the DNA and its corresponding protein offers another great advantage to the cell —

Figure 11-8. Schematic representation of the flow of information in a cell. The chromosomes within the nucleus contain all of the information. Selected sites on the DNA, in this case the A gene, are transcribed into messenger RNAs (the A-mRNAs in this example), which leave the nucleus and enter the cytoplasm where they are translated into proteins. The translation process, which in this case produces A-proteins, occurs in combination with ribosomes.

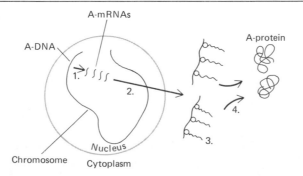

the opportunity for the information to be amplified. One DNA molecule can serve as the template to form a large number of RNA molecules, each containing the same information as the original template (Fig. 11-8). Each one of these mRNAs can then serve as a template for the formation of many polypeptide chains. If the DNA were to serve directly as the template for protein synthesis without further amplification, the cell would possess a much reduced potential for protein synthesis.

In addition to messenger RNA, two other major types of RNA molecules, *ribosomal RNA (rRNA)* and *transfer RNA (tRNA)*, are found in cells, but their role in protein synthesis is quite different. In this section we will briefly describe some of the early research that established the existence of the three main classes of RNA and discuss the role of these RNAs in protein synthesis. We will return for an in-depth analysis of the synthesis of all types of RNA in the next chapter, after a description of the genetic code and the mechanism of protein synthesis has been presented. RNA, in general, differs from DNA in three ways (Fig. 11-9): (1) the sugar is ribose rather than deoxyribose, (2) thymine is not found in RNA but is replaced by another pyrimidine, uracil, which lacks the methyl group, and (3) RNA is generally a single-stranded molecule, though base-pairing is still possible and many RNAs have double-stranded regions. In fact, in certain viruses, totally double-stranded RNA molecules are formed.

Even before Avery, MacLeod, and McCarty had discovered the nature of the transforming principle, some type of relationship had been established between RNA and protein. Primarily as a result of the work of

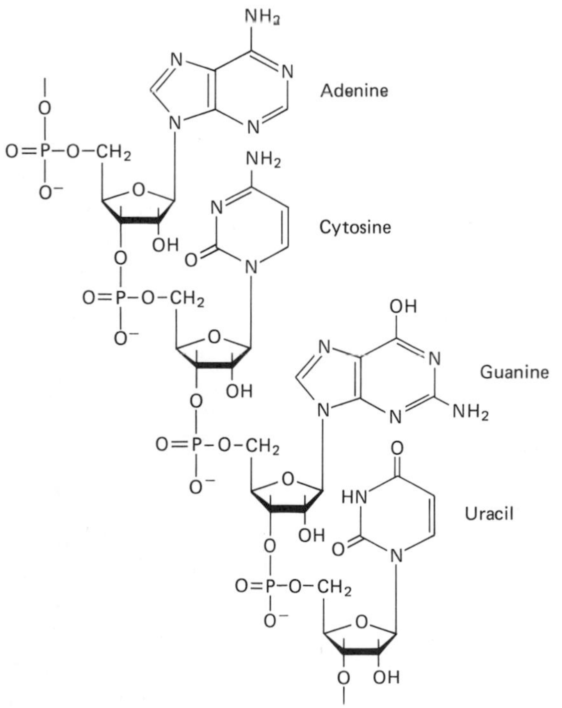

Figure 11-9. The chemical structure of a small section of a polynucleotide chain of RNA. The backbone consists of alternating sugar and phosphates linked by 3'5'-phosphodiester bonds with the four bases attached at the C1 of the ribose.

Torbjörn Caspersson and Jean Brachet, a strong correlation had been found between cells actively engaged in protein synthesis and the presence of high concentrations of RNA. In addition, when the incorporation of ^{32}P into DNA and RNA was compared, it was found that cells active in protein synthesis could be synthesizing RNA at very rapid rates without any detectable synthesis of DNA. During the 1950s, as the role of DNA became accepted, a concentrated research effort on RNA metabolism began to provide an insight into the role of this second nucleic acid.

One of the most important of the early observations was the discovery that RNA synthesis occurred in the nucleus, the organelle known to contain the genetic material. From the nucleus, RNA was found to migrate to the cytoplasm, that part of the cell in which the synthesis of protein occurs. This feature of RNA metabolism was demonstrated autoradiographically in a variety of different cells. The experimental procedure generally called for the administration of a radioactively labeled RNA precursor, such as a ^{3}H-pyrimidine, for a brief period of time, and the determination of the site of incorporation autoradiographically. The experiment is completely analogous to that described in connection with the synthesis of secretory proteins (Fig. 7-8). In this case, however, virtually all of the silver grains are localized over the nucleus. Again, as in the case of the secretory proteins, the fate of the newly synthesized RNA was followed by transferring the labeled cells to fresh medium containing unlabeled RNA precursors. When cells were chased in this manner, radioactivity in ribonuclease-sensitive material, i.e., RNA, was found to migrate into the cytoplasm. An example of this type of experiment is shown in

Figure 11-10. The migration of RNA from its site of synthesis in the nucleus to the cytoplasm as revealed autoradiographically. (*a*) A portion of a rabbit embryo that had been incubated in ^{3}H-uridine for 10 minutes. Virtually all grains are localized over the nucleus. (*b*) When the incubation period is extended to 60 minutes, silver grains begin to appear over the cytoplasm indicating the movement of newly synthesized RNA out of the nucleus. (*c*) When an embryo labeled for 1 hour is chased in unlabeled medium for an additional 8 hours, radioactive RNA becomes distributed in a nearly uniform manner throughout the cells.

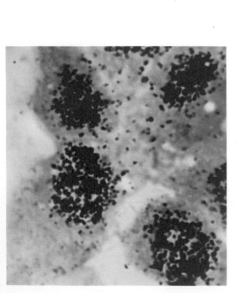

(*a*)

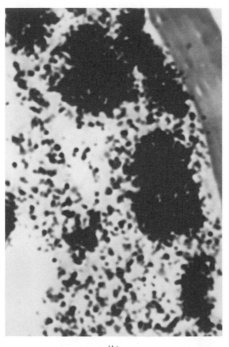

(*b*)

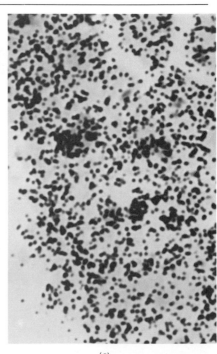

(*c*)

Fig. 11-10. We will return to the early studies of RNA metabolism after consideration of the mechanism of RNA synthesis.

Transcription

If RNA is to serve as a functional intermediate in the flow of information between DNA and proteins, it is imperative that its nucleotide sequence be a precise reflection of that of the genetic material from which it was synthesized. The process of RNA synthesis, i.e., transcription, has been intensively studied and is relatively well understood, particularly in prokaryotic cells. It is the complex series of events that occurs between the transcription of RNA and its involvement in protein synthesis that has proven to be the most difficult to elucidate. Unlike the process of protein synthesis, the formation of RNA molecules is accomplished by a particular enzyme, albeit one whose activities are more complex than most enzymes involved in metabolic pathways. The enzymes responsible for transcription in prokaryotic and eukaryotic cells are termed *DNA-dependent RNA polymerases*. They catalyze the reaction

$$\text{RNA}_n + \text{NPPP} \rightarrow \text{RNA}_{n+1} + \text{PP}_i \ (\rightarrow 2 \ \text{P}_i) \qquad \text{(Fig. 11-11a)}$$

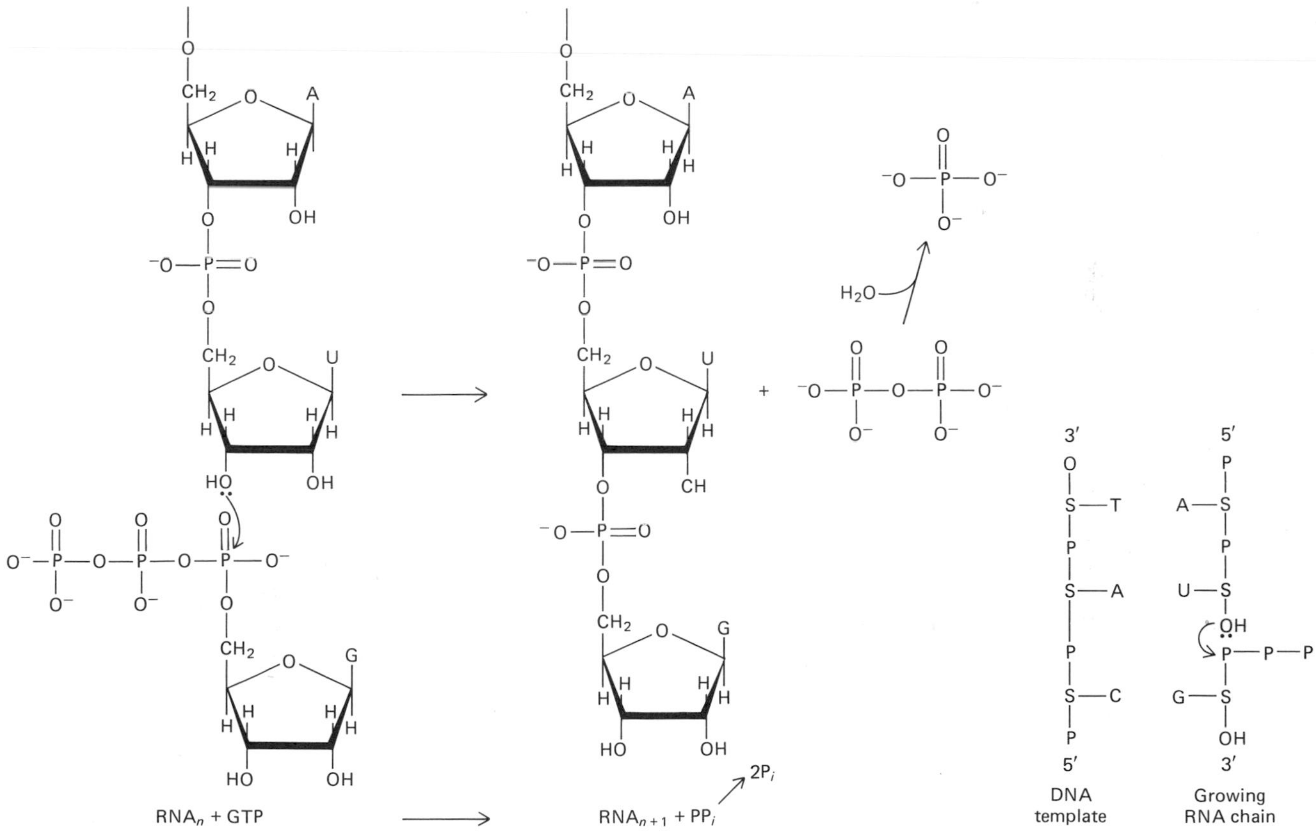

Figure 11-11. The synthesis of RNA. (*a*) Chain elongation occurs as a result of an attack by the free 3′ hydroxyl group on the alpha phosphate of the 5′ position of the incoming nucleotide. The pyrophosphate released is subsequently cleaved which further drives the reaction to polymerization. (*b*) The relationship between the template and the growing RNA chain.

RNA$_n$ + GTP

RNA$_{n+1}$ + PP$_i$

2P$_i$

(a)

3′   5′

DNA template   Growing RNA chain

5′   3′

(b)

in which ribonucleoside triphosphates are polymerized into a chain of nucleotides with the simultaneous release of pyrophosphate (PP$_i$) which is then hydrolyzed. In addition to ribonucleoside triphosphates and Mg^{2+}, the reaction requires the presence of DNA. The polyribonucleotide chain produced by the enzyme has a base sequence which is complementary to one of the DNA strands of the duplex (Fig. 11-11b), that which served as the *template*. In prokaryotic cells it appears that one type of RNA polymerase is responsible for the synthesis of all types of cellular RNAs, whereas in eukaryotic cells at least three distinct types of RNA polymerases have been characterized, each required for the production of particular types of RNA. The transcription process in prokaryotic and eukaryotic cells is shown in Figs. 11-36 and 12-14, respectively. We will begin this discussion with a brief description of transcription in prokaryotic cells.

The first step in the synthesis of an RNA molecule is the association of the polymerase with the DNA template. This brings up a question of more general interest, namely the specific interactions of proteins and nucleic acids, macromolecules of a very different nature. We mention examples of specific RNA-protein interactions in several places in this book. Although these types of interactions are not very well understood, it appears that particular proteins can recognize specific nucleotide sequences and/or specific types of secondary structures in nucleic acid molecules. If one considers the *E. coli* chromosome as essentially a naked piece of circular DNA, how does the RNA polymerase molecule "know" where to attach to the DNA template in order to produce an RNA molecule with the appropriate information? There is no doubt that specific RNA molecules have a defined sequence from one end to another, and therefore it is evident that RNA polymerase molecules begin and terminate transcription at defined sites on the template. Analysis of the enzyme's binding to sites in the bacterial chromosome has led to the finding that the bacterial polymerase exists in two states. Under certain conditions of extraction and purification, an RNA polymerase molecule is obtained which has the ability to bind to DNA in vitro and transcribe from it, but the RNA molecules produced are not the same as those found within the cell. The populations of RNA molecules produced in vitro and in vivo are different because in the test tube the enzyme has attached to places it would have normally ignored and begun its transcription at inappropriate sites. However, if another purified polypeptide termed the *sigma factor* (σ) is added to the RNA polymerase before it attaches to the DNA, transcription begins at the proper location. It appears that the attachment of the sigma factor to the "core enzyme" increases its affinity for the proper DNA binding site, termed the *promoter* region, and decreases its affinity for DNA in general. Once the *complete enzyme* (core enzyme plus sigma factor) has attached to the DNA at the promoter (Fig. 11-12), the next step is to begin to synthesize a ribonucleotide chain. Once the polymerization process starts, the sigma factor is no longer needed and it dissociates from the complex, leaving the core enzyme attached to the DNA to catalyze the elongation of the nascent RNA chain.

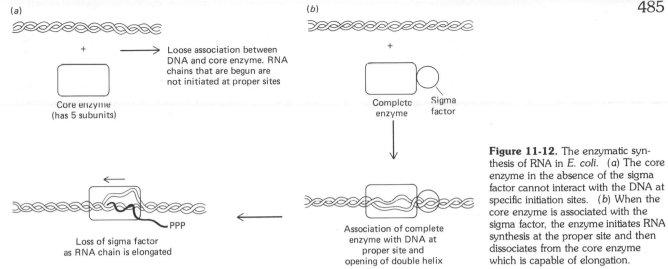

(a)

+

Core enzyme
(has 5 subunits)

→ Loose association between
DNA and core enzyme. RNA
chains that are begun are
not initiated at proper sites

Loss of sigma factor
as RNA chain is elongated

PPP

(b)

+

Complete
enzyme

Sigma
factor

Association of complete
enzyme with DNA at
proper site and
opening of double helix

Figure 11-12. The enzymatic synthesis of RNA in *E. coli*. (*a*) The core enzyme in the absence of the sigma factor cannot interact with the DNA at specific initiation sites. (*b*) When the core enzyme is associated with the sigma factor, the enzyme initiates RNA synthesis at the proper site and then dissociates from the core enzyme which is capable of elongation.

An early question to be raised was whether or not both strands of the DNA duplex were transcribed. If one considers that amino acid sequences are specified by a code contained in the linear sequence of nucleotides of a single polynucleotide chain, it is very difficult to imagine how two complementary sequences of a DNA molecule could both contain information for a meaningful polypeptide chain. The mutual complementarity of the duplex provides an elegant means for duplication—each strand acts as a template for the production of its partner. Information content, however, is an entirely different matter and is most logically conceived as being present in only one of the two strands *within a given gene*. Analysis of this question by molecular hybridization has indicated that in both prokaryotic and eukaryotic systems, RNA molecules are present that are complementary to only one strand of the DNA, therefore only one strand is transcribed. Even though only one strand is transcribed within a given gene, the other strand may hold the information-containing sequence in other genes at different sites within the same chromosome.

RNA polymerase is a remarkable enzyme. In the first step it not only has the ability to recognize the proper place on the DNA to attach but also the proper strand to transcribe. Proper strand selection is also a consequence of the presence of the sigma factor. In the process of binding and initiation of transcription, the enzyme appears to be able to break the hydrogen bonds holding the two strands of the template together and cause a local unwinding of the duplex in that region at the time (Fig. 11-12). The transcription process requires that the polymerase move along the DNA strand in the 3′ to 5′ direction, separating the strands of the DNA as it travels and fabricating a complementary antiparallel RNA strand which grows from its 5′ terminus in a 3′ direction (Fig. 11-11). As indicated in Fig. 11-11*a*, the precursor of each nucleotide residue in RNA is a nucleoside triphosphate. Incorporation occurs as a result of a nucleophilic attack by the hydroxyl group of the 3′ end of the growing strand on the innermost phosphate (the alpha phosphate) of the triphosphate group. The hydrolysis of this phosphate ester bond is a highly exergonic reaction with its equilibrium far toward the direction of incorporation. The reaction is further driven in this direction by the hydrolysis of the pyrophosphate by a pyrophosphatase to form inorganic phosphate ions. Reactions leading to the formation of nucleic acids (and proteins) are inherently different from those of intermediary metabolism discussed in Chapter 4 in that it is imperative that the reaction proceed in only one direction, i.e., that the reaction be essentially irreversible. Whereas some of the reactions leading to the formation of small molecules, such as amino acids, may be close enough to equilibrium so that a considerable reverse reaction can be measured, those reactions leading to the synthesis of nucleic acids and proteins must occur under conditions where there is virtually no back reaction. This condition is met by coupling these reactions to the highly exergonic hydrolysis of molecules such as pyrophosphate.

The 5' terminal of bacterial RNAs contains the intact triphosphate of the first nucleotide (Fig. 11-12) and in all cases it appears that the first nucleotide of the chain is a purine, either pppG · · · or pppA · · · . As the polymerase moves along it must insert the proper nucleotide into the growing chain at each site. Presumably it is capable of selecting the proper ribonucleoside triphosphate for incorporation by its ability to form complementary hydrogen bonds with the nucleotide in the DNA strand being transcribed (Fig. 11-11*b*). Once the polymerase has moved past a particular site, the double helix is reformed; the RNA chain does not remain associated with its template as a DNA-RNA hybrid (Fig. 11-12). This can be shown, for example, by the sensitivity of the nascent RNA to ribonucleases which would not be capable of digesting the nascent RNA molecules if they were present in a DNA-RNA hybrid. It has been estimated that bacterial RNA polymerase is capable of incorporating approximately 50 nucleotides into a growing RNA molecule per second. Just as transcription is initiated at specific points in the chromosome, it is also terminated when a specific nucleotide sequence is reached. In some cases, a protein, the *rho factor,* is required for termination to occur whereas in other cases it appears that the polymerase can stop transcription and carry out the release of the RNA chain without additional factors.

As stated above, eukaryotic cells have at least three distinct transcribing enzymes, each specific for a particular type of RNA. RNA polymerase I (or A) is responsible for the synthesis of 18S and 28S ribosomal RNA, RNA polymerase II (or B) synthesizes messenger RNA, and RNA polymerase III (or C) synthesizes small molecular weight RNAs including the transfer RNAs and the 5S ribosomal RNA. As in the case of many other properties, the single versus multiple RNA polymerase seems to represent a sharp distinction between the prokaryotic and eukaryotic condition: no prokaryote has been found with these multiple enzymes while the lowliest eukaryotes, yeast, have the same three types that are present in mammalian cells. All three of the eukaryotic polymerases are complex enzymes containing several polypeptides. Some of the subunits appear to be present in only one of the types of polymerase while others may be shared among the group.

One of the most important questions under present investigation is the basis for the specificity of these enzymes with respect to the type of DNA template. Does RNA polymerase I transcribe only rRNA because it recognizes a particular nucleotide sequence in the ribosomal RNA gene to which it can bind, or does it do so because of some property of the chromosome, such as the presence of a particular protein? Given the tremendous size of the eukaryotic genome and the great diversity of sequences, one might expect that some factor in addition to simply nucleotide sequence would have to be involved in this recognition process, but this issue remains a topic of research. Consideration of the activities of RNA polymerase II raises entirely different questions, ones concerning the basis for the transcription of particular messenger RNAs in one type of cell and not in another. Questions of this sort direct one to the very heart of the regulation of gene expression in eukaryotic cells and extend far beyond the specificities of a particular enzyme. The basis for the selection process will be considered at length in the next two chapters. In the remaining sections of the present chapter we will proceed to the consideration of each of the three major types of RNA and then turn to the role of these molecules in the synthesis of proteins. In addition to the RNAs discussed below, a variety of small molecular weight RNAs have been discovered which do not appear to be involved in protein synthesis but rather may act as structural components of various organelles. Since relatively little is known about these RNAs, they will not be discussed further.

Messenger RNA

Early Studies on Bacteriophage

The first experiment to provide evidence of the existence of an RNA molecule that carried the same information as was present in the DNA was carried out by Elliot Volkin and L. Astrachan in 1956. These investigators were studying a type of RNA that was made soon after the infection of *E. coli* cells with the bacteriophage T_2. They found that when this RNA was labeled with ^{32}P and the radioactivity in its four component nucleotides was determined, its $A + U/G + C$ ratio of 1.7 closely resembled that of the phage DNA (ratio of 1.8), but bore no resemblance to that of either the bacterial DNA (1.0) or bacterial RNA (0.85). A few years later it was shown that the formation of this particular type of RNA was required in order for bacteriophage production to result. It appeared that one of the consequences of the introduction of viral DNA into the bacterial cell was the synthesis of an RNA copy which was somehow involved in the process of infection. However, the fact that the base composition of a particular RNA molecule is similar to one type of DNA as opposed to another is not convincing evidence of the origin of that molecule. It was important to know something more about the RNA produced after phage infection than simply its base composition.

In order to obtain further information about the phage RNA, Benjamin Hall and Sol Spiegelman devised an important new technique in 1961, that of DNA-RNA hybridization (Fig. 11-13), which has already been alluded to in the previous chapter (page 465). Hall and Spiegelman

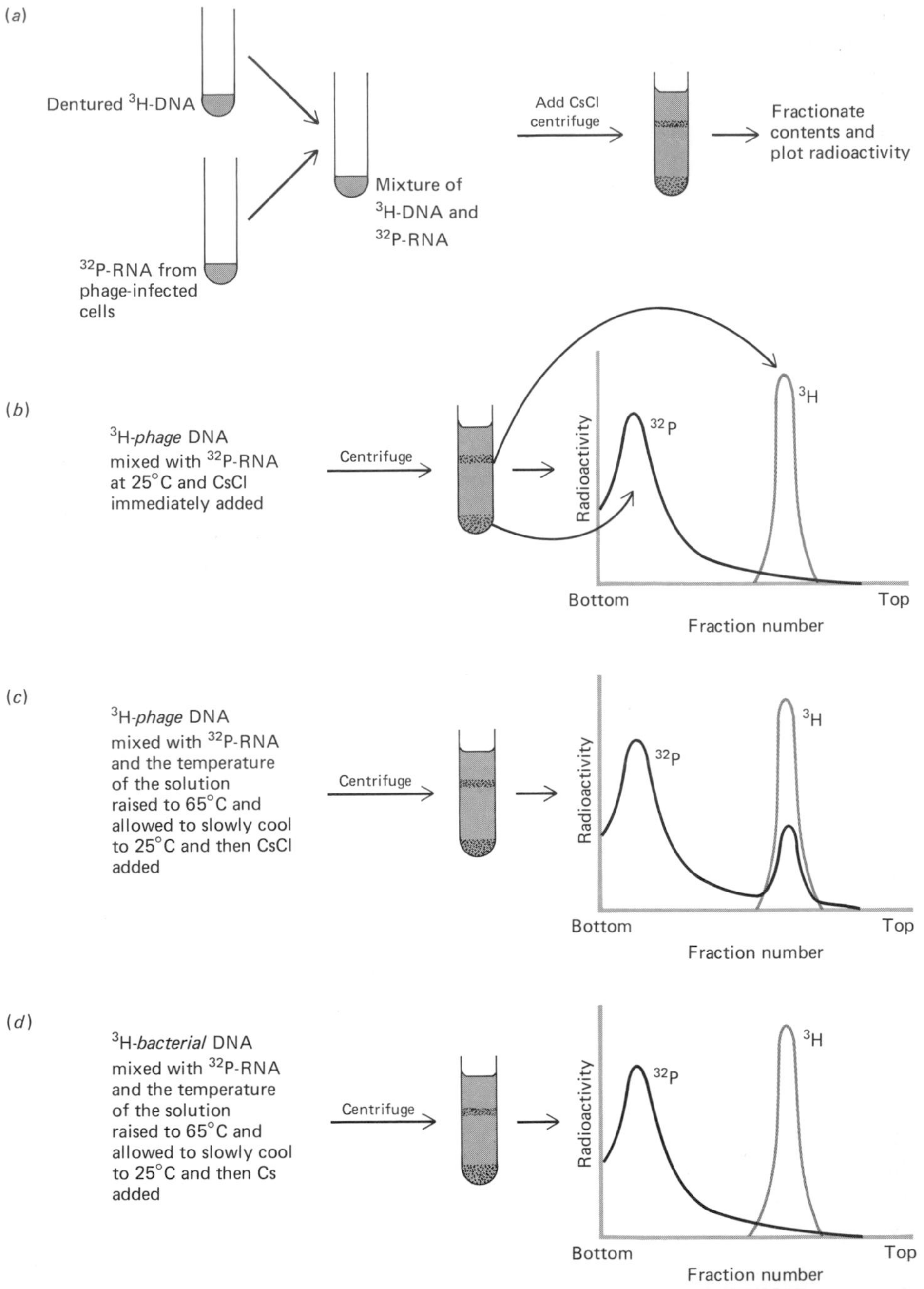

(a)

Dentured ^{3}H-DNA

^{32}P-RNA from
phage-infected
cells

Mixture of
^{3}H-DNA and
^{32}P-RNA

Add CsCl
centrifuge

Fractionate
contents and
plot radioactivity

(b)

^{3}H-*phage* DNA
mixed with ^{32}P-RNA
at 25°C and CsCl
immediately added

Centrifuge

Radioactivity

^{32}P

^{3}H

Bottom   Top

Fraction number

(c)

^{3}H-*phage* DNA
mixed with ^{32}P-RNA
and the temperature
of the solution
raised to 65°C and
allowed to slowly cool
to 25°C and then CsCl
added

Centrifuge

Radioactivity

^{32}P

^{3}H

Bottom   Top

Fraction number

(d)

^{3}H-*bacterial* DNA
mixed with ^{32}P-RNA
and the temperature
of the solution
raised to 65°C and
allowed to slowly cool
to 25°C and then Cs
added

Centrifuge

Radioactivity

^{32}P

^{3}H

Bottom   Top

Fraction number

Figure 11-13. The experiment by
Hall and Spiegelman which demonstrated the synthesis during bacteriophage infection of RNA molecules
that were complementary to the phage
DNA.

reasoned that, if two complementary strands of DNA can come together
in a specific manner, why should not one strand of DNA and a complementary strand of RNA? In order to carry out this experiment they prepared
^{32}P-labeled RNA from phage-infected cells and DNA from purified phage
that had been labeled with ^{3}H.   The ^{3}H-DNA was denatured by heat and

then allowed to slowly cool in the presence of the ^{32}P-RNA. Once the solution had cooled, the nucleic acids were centrifuged to equilibrium in a cesium chloride density gradient capable of separating the various types of molecules which might be present. When the radioactivity in the centrifuge tube was determined after the sedimentation, it was found that a significant percentage of the added ^{32}P-RNA was present in the band containing ^{3}H-DNA of the phage (Fig. 11-13c), completely separated from the bulk of the unhybridized ^{32}P-RNA which had moved toward the bottom of the tube. In other words, they had demonstrated the presence in the tube of a DNA-RNA hybrid. When ^{3}H-bacterial DNA was substituted in the experiment for ^{3}H-phage DNA, none of the radioactive RNA was found associated with it after sedimentation (Fig. 11-13d). These results showed the true complementary nature between the *base sequence* of the RNA synthesized after phage infection and the phage DNA itself.

While the work on RNA synthesized during phage infection was proceeding, another line of research was coming to similar conclusions based on RNA synthesized from a bacterial genome. In 1961, a landmark paper was published by Francois Jacob and Jacques Monod on the concept of a "messenger" RNA as an intermediate in the synthesis of the bacterial enzyme, β-galactosidase. The advantage of the bacteriophage system just described in the study of the possible existence of a messenger RNA is the sudden transition that occurs in the activities of the cell as a result of the virus infection. Before the phage attaches to the bacterium it is carrying out one set of metabolic activities geared to the growth of the cell and the production of more bacteria. Suddenly after infection, the phage genome has taken over and the activities in the cell are geared toward the production of phage proteins and phage progeny. This is an ideal system to search for the production of a new type of informational RNA molecule carrying information for phage proteins that were not previously being synthesized.

Early Studies on Bacteria

Jacob and Monod, in their search for an informational RNA produced from the bacterial genome, also took advantage of a system in which a sudden activation occurred. In designing the type of experiments that would be best suited for the demonstration of a messenger RNA, they utilized bacteria that up until a specific point in time were unable to manufacture β-galactosidase and then suddenly gained the ability. The gain in ability to synthesize β-galactosidase was brought about in two different ways. In one set of experiments it was *induced* as a result of the addition to the medium of a substrate for the enzyme (page 586). In the other type of experiment, which will be discussed presently, the transition was brought about by the introduction of a gene (Fig. 11-14a). Jacob and Monod had isolated strains of bacteria that contained mutations in this gene (z^-) and were unable to produce active enzyme. When DNA containing the wild-type gene (z^+) was introduced into these mutant cells by conjugation, they found that within two minutes of the transfer of the normal gene, active β-galactosidase could be measured. It was proposed

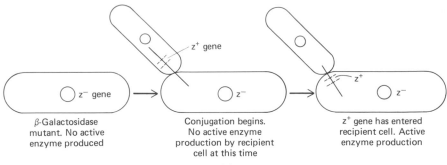

z⁺ gene

z⁻ gene

z⁻

z⁺

z⁻

β-Galactosidase mutant. No active enzyme produced

Conjugation begins. No active enzyme production by recipient cell at this time

z⁺ gene has entered recipient cell. Active enzyme production

(a)

Figure 11-14. (a) Diagrammatic representation of the transfer of a portion of the *E. coli* chromosome during conjugation. (b) The kinetics of gene transfer and enzyme synthesis in z⁻ cells that have received the z⁺ gene during conjugation. Cells of this type are referred to as *merozygotes*. Curve 1 shows the kinetics of transfer of the z⁺ gene to recipient cells. The z⁺ locus begins to be transferred at about 15 minutes after contact between cells is made. The formation of z⁺/z⁻ merozygotes follows a linear course reflecting the formation of new mating pairs with time. Curve 2 follows the production of β-galactosidase in the mixture. The genetics of the two parental strains are such that only the z⁺/z⁻ merozygotes can produce active enzyme. Curve 3 follows the synthesis of β-galactosidase in subsamples of a sample in which mating had been interrupted, and further mating prevented, at 5 minutes after the z⁺ gene had begun to be transferred. The time of interruption is indicated by the arrow. The rate of enzyme synthesis after interruption of mating is constant and equal to the rate achieved at the time of interruption. This latter finding is consistent with the existence of an unstable intermediate, i.e., an mRNA. (*From W. Hayes, "The Genetics of Bacteria and Their Viruses," 2d ed., Wiley, 1969, after the work of M. Riley et al., J. Mol. Biol., 2:216, 1960.*)

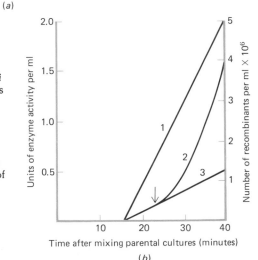

(b)

that during this brief period, "copies" of the gene in the form of an RNA molecule had been made, and these "messenger" RNA molecules were being rapidly translated into protein.

Another observation was soon made: once the z⁺ gene had entered the cell, the rate of production of active enzyme molecules quickly reached a constant and maximum level (Fig. 11-14b). Since the *rate* of formation did not increase with time, this suggested that the proposed messenger RNA was not accumulating within the cells in greater and greater amounts. There are two explanations to account for a relatively constant mRNA concentration.

1. The mRNA is synthesized for a period of time and then its further production is halted so that its concentration remains constant (or increases just enough to keep pace with the growth of the cells).

2. Messenger RNA molecules continue to be synthesized, but the rate of destruction of existing molecules is equal to the rate of their production, so that a steady-state concentration is maintained.

It may seem that the first alternative makes the most sense; why waste energy simply forming and destroying the same species of RNA molecule?

The second alternative is rather costly, but it gives the cell a great deal of flexibility. When the enzyme is no longer needed, i.e., when its substrate is no longer present in the medium, mRNA production can be halted (Fig. 13-4) and the continuing destruction will remove the existing RNA messages so that no further enzyme will be produced. Where the cell may lose energy by producing extra mRNA molecules in one situation, it saves energy and vital storage space by not having to produce unnecessary protein molecules in another case.

On the basis of a number of experiments, Jacob, Monod and colleagues accumulated evidence in favor of the second alternative. In one experiment, for example, they took advantage of the tendency for DNA containing radioactive phosphorus atoms (^{32}P) to have its backbone broken as a result of being struck by particles emitted from the phosphorus atoms as they decay. Normally the time of an experiment is short relative to the half-life of the radioactive atoms being used and there is very little danger to the molecules from this type of "autodestruction" by these disintegrations. However, if the time period is extended by keeping the labeled cells frozen for a period of time, the damage becomes significant, its degree being proportional to the time the cells are kept before they are thawed. In their experiment, Monica Riley, Arthur Pardee, Jacob, and Monod allowed ^{32}P-containing DNA with the z$^+$ gene to be transferred to recipient z$^-$ cells, then allowed about 25 minutes so that maximal enzyme production was well underway, and then froze the cells for varying lengths of time. After the cells were thawed, the level of production of β-galactosidase was determined. It was found that the longer the cells were frozen, the less likely the enzyme would be produced after the cells were thawed (Fig. 11-15). Since the freezing is only destructive to the *labeled* DNA and has no effect upon mRNA molecules present in the cell at the time they were frozen, they concluded that the enzyme production was halted as a result of the normal breakdown of existing unlabeled mRNAs and the inability to produce new ones due to damage to the DNA template.

In a second experiment they obtained corroborating evidence using a compound, 5-fluorouracil, which is incorporated into RNA chains and alters

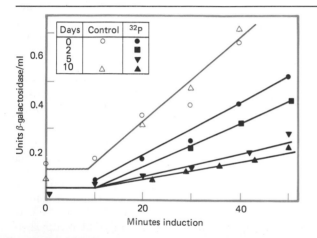

Figure 11-15. Kinetics of enzyme formation as influenced by the decay of ^{32}P (half-life of 14 days). *E. coli* cells were grown on ^{32}P and mixed with nonradioactive F$^-$ (female) bacteria. Mating was arrested at 35 minutes and cells were allowed to develop further for 25 minutes before storage in liquid nitrogen. When the cells were thawed at various time intervals and checked for their ability to produce active enzyme, it was found that, the longer the freezing period, the less enzyme was produced by the culture. These results indicated that the mRNA molecules present in the cell at the time of freezing were quickly lost upon thawing the cells as a result of normal turnover. Since the cells contained ^{32}P-labeled DNA, the template properties of the DNA were progressively destroyed during freezing, leading to the reduced ability to produce new mRNA molecules. (*From M. Riley, A. B. Pardee, F. Jacob, and J. Monod,* J. Mol. Biol., **2**:223, 1960.)

their template activity, so that they direct the formation of abnormal proteins. When 5-fluorouracil is added to cells producing β-galactosidase, there is an almost immediate production of abnormal enzyme. Since this compound can only be incorporated into mRNAs being manufactured in its presence and can have no effect upon preexisting RNA templates, one can conclude that mRNA production normally continues despite the fact that enzyme production remains constant.

Based on these experiments, Jacob and Monod suggested that messenger RNA should have the following properties.

1. It should have a very high rate of turnover, i.e., any given molecule should be rapidly broken down. If mRNA is necessary but unstable, it should account for a large percentage of the RNAs *being made at any given moment*, yet at the same time it should account for a relatively small percentage of the total RNA present in the cell. This is exactly what has been found in nearly every case examined. If a radioactive RNA precursor is administered to bacterial cells for a brief period of time, a large percentage of the radioactivity is found to be incorporated into the mRNA fraction. However, if the total amount of mRNA present in these same cells is determined, it is found to amount to only about 3% of the total RNA population. Despite the fact that the other two types of RNA, the tRNA and the rRNA, constitute approximately 97% of the RNA of the cell, relatively little radioactivity is incorporated into these molecules in short labeling periods. The reason for the preponderance of tRNA and rRNA in the cell, despite their relatively low rates of synthesis, is their very great stability. Once an rRNA or tRNA molecule is produced, its chance of being destroyed in the near future is very remote.

Though the situation is somewhat more complicated in eukaryotic cells (page 606), the same basic principle applies. The major difference is the greater stability of the mRNAs. Half-lives of eukaryotic mRNAs are generally measured in hours (Fig. 11-16) rather than minutes, and in some cases specific messengers are very stable (page 607). It appears that the differences reflect the different types of conditions faced by prokaryotic and eukaryotic cells. Whereas the former are subject to drastic environmental changes, to which they must rapidly adjust their metabolism, the cellular environment within a higher plant or animal is much more constant and predictable.

2. Even though one particular species of mRNA would be expected to have a definite, predictable molecular weight, mRNA, *as a class* would be expected to have a heterogeneous molecular weight, reflecting the heterogeneity in the size of the various proteins that have to be coded for. Once again this has been amply confirmed in this case by the analysis of mRNA on sucrose gradients (Fig. 12-20) or acrylamide gels. It is generally found that a great range of molecular weights exists among the mRNA group. However, in some cases, when the synthesis of a particular protein greatly dominates a cell's activities, as in the case of hemoglobin production in the red blood cell line, then a distinct peak can be seen in the sucrose gradient reflecting the presence of large amounts of this one mRNA species.

3. Since, *as a class*, mRNAs would be expected to be synthesized from a wide variety of different genes one would expect the A + U/G + C ratio of the total mRNA fraction to be similar to the T + A/C + G ratio of the DNA itself. This is also generally found to be the case as was previously illustrated in the example

Figure 11-16. HeLa cells were labeled for 3 hours in ³H-uridine, washed, and allowed to grow in unlabeled medium. At various periods of time, cells were removed and RNA extracted. The ratios of mRNA to 18S rRNA labeling is plotted on a log scale against time. The results indicate the existence of more than one population of mRNA: one with a short half-life (approximately 6 to 7 hours) and one with a long half-life (approximately 24 hours). (*From R. H. Singer and S. Penman,* J. Mol. Biol., **78:**329, 1973.)

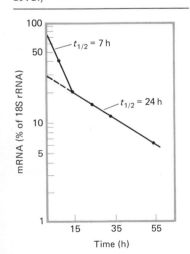

of the RNA synthesized by the phage genome following infection of the bacterial cell. Not only should its base composition reflect the overall composition of the DNA, it should have the ability to hybridize with a considerable fraction of the DNA since it is formed by the transcription of a large number of genes. This latter property, i.e., its hybridization capacity, was not predicted by Jacob and Monod since the technique was developed in the same year as their report. However, it follows directly from their other predictions and has been found to be true for mRNA preparations derived from many sources.

4. Messenger RNAs should be found associated with the ribosomes during protein synthesis. As will be described later in this chapter, when the translation of an mRNA template into protein is occurring, ribosomes form clusters, termed *polyribosomes*, which are held together by an attachment to an mRNA strand. When a relatively pure mRNA fraction is being sought, one generally begins by isolating the polyribosomes from which the mRNAs can be released (page 522).

Transfer RNA *

When the concept of colinear sequences of nucleotides and amino acids first came to be accepted, it was proposed that there must be some type of direct structural relationship between the nucleotide sequence in the DNA and the amino acid sequence in the corresponding protein. In other words, nucleic acids were believed capable of serving as some form of direct template by which particular amino acids could be ordered in a polypeptide. Later, Francis Crick formulated an entirely different concept which radically altered the role of nucleic acids in protein synthesis. Crick believed that the information-containing nucleic acids (the DNA and mRNA) were not structural templates for amino acid sequences, but codes which had to be *translated*. In a sense, nucleic acids and proteins were like two languages with two different types of letters. The question then became how information coded in one type of molecule, the nucleic acid, could dictate the sequence of a totally unrelated class of molecules, the amino acids. Based on his belief that the predominant manner in which nucleic acid specificity could be expressed was via the formation of hydrogen bonds to complementary nucleic acids, Crick proposed the existence of a nucleic acid "adaptor" molecule. The adaptor was originally conceived as a very small oligonucleotide, possibly a trinucleotide which was complementary to a particular nucleotide sequence of the template on one hand, while on the other hand it would be attached to a particular amino acid. In this way, a specific base sequence could be translated into a defined polypeptide chain. With the discovery of the transfer RNA molecules, the verification of this type of adaptor molecule was made.

Since each tRNA is able to recognize two different types of molecules (the mRNA and an amino acid), we must consider the manner in which both of these recognition events take place. Before we can describe the nature of these specific interactions, we need to discuss the structure of

---

* The discussion in this section presupposes some knowledge of the genetic code. If the reader is unfamiliar with this topic, it may be best to read pages 499 to 512 before beginning this section.

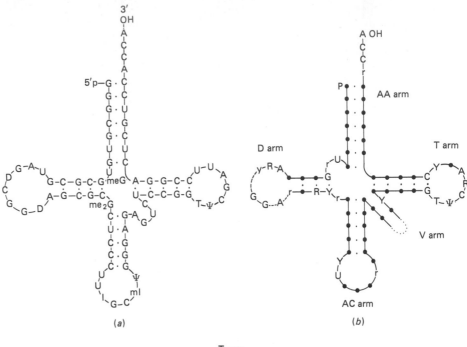

(a)

(b)

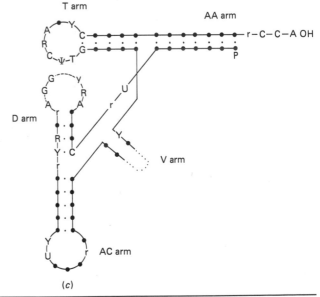

(c)

Figure 11-17. (a) The nucleotide sequence of the cloverleaf form of a yeast tRNA$_{ala}$. The amino acid becomes linked to the 3' end of the tRNA, while the opposite end of the molecule bears the anticodon, in this case IGC. In addition to the four bases A, U, G, and C, this tRNA contains ψ, pseudouridine; T, ribothymidine; mI, methyl inosine; I, inosine; me$_2$G, dimethyl guanosine; D, dihydrouridine; meG, methyl guanosine. (b) and (c) Generalized representation of tRNA in the cloverleaf form and in the L form, respectively. The bases common to all tRNAs which participate in peptide elongation are indicated. Other symbols are R for a purine nucleoside and Y for a pyrimidine nucleoside in all tRNAs, r for a purine nucleoside and y for a pyrimidine nucleoside in most tRNAs. The dotted lines are the regions in the chain where the number of nucleotides varies among tRNAs. (From S-H. Kim, Prog. Nucleic Acid Res., **17:**185, 1976.)

tRNA molecules, which allows them to perform these functions. In 1965, after seven years of work, Robert Holley reported the first base sequence analysis of an RNA molecule, that of a transfer RNA specific for the amino acid alanine (Fig. 11-17a). This tRNA molecule was composed of 77 nucleotides, 10 of which were found to be somewhat different from the standard 4 nucleotides, A, G, C, and U of RNA. Over the following years, a number of other tRNA species were purified, their nucleotide sequences determined, and a number of distinct similarities present in all of the different tRNAs became evident. All of them consisted of roughly the same number of nucleotides, between 75 and 100, giving them a sedimentation coefficient of 4S. All of them had a significant percentage of unusual bases which were found to result from enzymatic modifications which occurred upon one of the four common bases *after* it had been incorporated into the RNA chain. In other words, the conversion of the standard bases to the unusual ones was a *posttranscriptional modification* of a highly predictable nature. When all of the tRNA nucleotide *sequences* were compared, it became apparent that they had a number of common features, the most important of which was the potential for these molecules to become folded in a similar manner (Fig. 11-17*b*, *c*). Scattered along the primary sequence of nucleotides were stretches which were complementary to other stretches located elsewhere on the molecule (Fig. 11-17). The locations of these various complementary sections were present in every tRNA species and were compatible with the formation of a cloverleaf-type structure. The stems and loops of the cloverleaf-shaped tRNAs are shown in Fig. 11-17. The amino acid becomes attached to the A at the 3′ end of the molecule. The unusual bases, which are concentrated in the loops, act to disrupt hydrogen bond formation in those regions.

As is apparent from the preceding account, transfer RNAs have an unusual structure in comparison with DNA or other types of RNA. Up to this point we have simply considered the secondary or two-dimensional structure of these adaptor molecules. Unlike the case of DNA, which simply forms an extended double helix, transfer RNA molecules are flexible and the various parts of the cloverleaf are capable of folding into a unique and defined tertiary structure. Recent x-ray-diffraction analysis on crystallized tRNA molecules has revealed the overall three-dimensional structure of tRNA molecules: they exist in the shape of an L (Fig. 11-18), i.e., a molecule with two axes of symmetry. As in the case of the tertiary

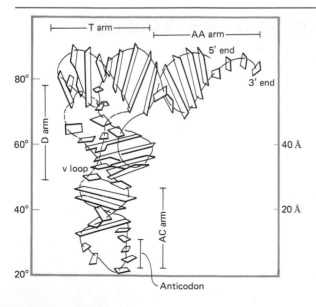

Figure 11-18. Schematic drawing of the three-dimensional structure of yeast phenylalanine tRNA. The amino acid acceptor arm (AA) and the TψC (T) arm form a continuous double helix and the anticodon (AC) arm and the DHU (D) arm form the other partially continuous double helix. These two helical columns form an L. The phosphate-ribose backbone is shown as a continuous wire except for the three dotted segments in which the nucleotide number varies among different tRNAs. The molecule is rather flat with the approximate thickness of 20 Å. (*From S-H Kim, Nature, **256**:680, 1975.*)

structure of proteins, the folding of the tRNA molecule results from the formation of weak bonds between otherwise distant parts of the molecule. All of the models of tRNA conformation seem to agree that those bases which tend to be at the same sites in all tRNAs (the invariant ones) are particularly important in generating the common L-shaped tertiary structure. The questions of greatest current interest in tRNA analysis concern the relationship between the conformation of the molecules and their function. Transfer RNAs are unusual nucleic acids in that they are less of informational importance than structural importance. In a sense, they perform functions more commonly suited to proteins than nucleic acids and as such are characterized by a more unusual secondary and tertiary structure. As will be described below, tRNAs must be capable of being distinguished from one another by various enzymes so that the proper amino acid can be attached. In addition, tRNAs must be capable of specific interaction with the ribosomes as well as certain soluble proteins required in polypeptide formation. These interactions between tRNAs and all of these various proteins are presumably dependent upon the shape and electron densities of the tRNA molecules, as well as their base sequence, per se.

As stated previously, transfer RNA molecules bear the responsibility of reading the words in the nucleotide alphabet of the mRNA and ensuring that they become translated into the proper words in the amino acid alphabet of the protein. Transfer RNAs, therefore, must be specific for words in both types of language. The recognition between a given tRNA molecule and a given mRNA triplet of nucleotides, i.e., a codon which specifies a particular amino acid (page 499), is accomplished by the formation of hydrogen bonds between a complementary base sequence on the transfer and messenger RNAs (see Fig. 11-28). The part of the tRNA which participates in this specific interaction with the codon of the mRNA is a stretch of three nucleotides, termed the *anticodon*, which is located in the middle loop of the tRNA molecule. In all tRNAs, this loop is composed of seven bases, the middle three of which form the anticodon. The specificity of this interaction ensures that only the proper tRNA will be brought into contact with the mRNA during the process of polypeptide formation on the ribosome. Tertiary structure analysis indicates that the anticodon is located at one end of the L-shaped tRNA molecule and the amino acid at its opposite end (Fig. 11-19).

Figure 11-19. As described in the text, the alteration of a cysteine to an alanine after it has been attached to a tRNA$_{cys}$ does not affect its incorporation into the polypeptide chain, i.e., it is still incorporated into sites normally occupied by cysteine residues.

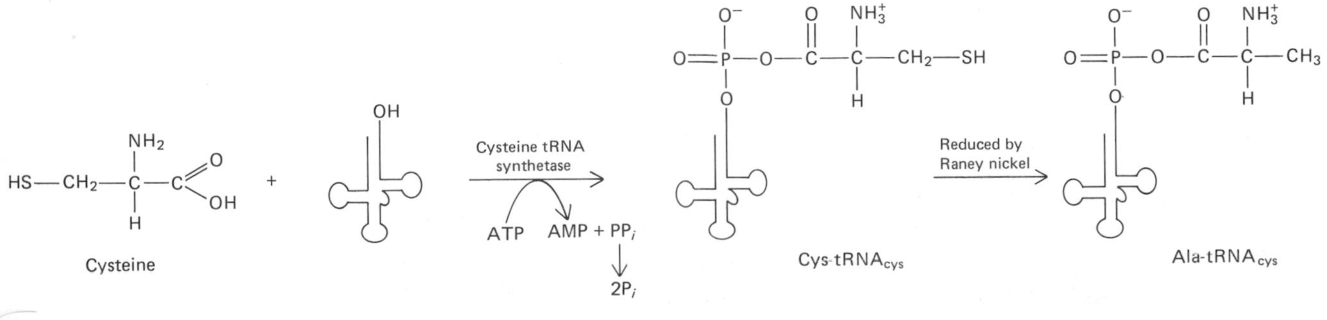

Amino Acid Activation

It is of critical importance for the subsequent process of protein synthesis that each transfer RNA molecule becomes attached only with the amino acid that is appropriate for its anticodon. The specificity in the linkage between the amino acid and the tRNA is accomplished by the enzymes that catalyze the reaction whereby the two molecules are linked. As a group, these enzymes, termed *aminoacyl-tRNA synthetases* or *amino acid activating enzymes,* accomplish the attachment by catalyzing the following two-step reaction:

$$\text{ATP} + \text{amino acid} + \text{enzyme} \rightarrow \text{enzyme-amino acid-AMP} + \text{PP}_i \, (\rightarrow 2 \, \text{P}_i)$$

$$\text{Enzyme-amino acid-AMP} + \text{tRNA} \rightarrow \text{aminoacyl-tRNA} + \text{enzyme} + \text{AMP}$$

It appears that each amino acid has one specific aminoacyl-tRNA synthetase, which is capable of "charging" all of the tRNAs that are specific for that amino acid (page 510). For example, in the case of leucine and serine, there are at least five different tRNAs specific for each of these amino acids, yet all of them are recognized by the same aminoacyl-tRNA synthetase. The enzyme-tRNA interactions form a remarkable example of the specificity of macromolecular interactions. Some common features must exist among all of the tRNA species of a given amino acid to allow the one appropriate enzyme to recognize all of them but at the same time discriminate against all of the tRNAs of the other amino acids. Presumably, certain of the subtle differences in the tertiary structure of the tRNA molecules cause them to be selected or rejected by the various enzymes. Similarly, these enzymes are capable of discriminating among some very closely related amino acids, for example, valine and isoleucine, which differ from one another by just one methyl group.

In the first step of the two-step charging reaction, the amino acid is said to be "activated" by the formation of an adenylated amino acid-enzyme complex. This is the energy-requiring step and, from this point, the transfer of the amino acid to the tRNA molecule, and eventually to the growing polypeptide chain, is an exergonic event. In other words, the energy required to convert amino acids into polymers via the formation of peptide bonds is actually expended in the formation of the aminoacyl-tRNA complex. The second step in the formation of the charged tRNA molecule occurs when the enzyme with its bound, adenylated amino acid happens to collide with the appropriate tRNA, causing the displacement of the enzyme and AMP. The PP_i produced in the first step is subsequently hydrolyzed to P_i, driving the overall reaction further to the right. We are now in a position to briefly summarize the chain of events that is responsible for the flow of information from a single gene to a single polypeptide chain. The specificity of the nucleotide sequence of the DNA is passed on in the formation of a complementary molecule of mRNA, whose specific information is brought to bear on the successive incorporation of amino acids via the attachment of specific tRNA molecules.

From the preceding description, it is apparent that the amino acid itself plays no direct role in determining where in the amino acid sequence it will be placed, rather the sole basis for its being incorporated correctly resides in the recognition abilities of the aminoacyl-tRNA synthetases. This was demonstrated in an early experiment by Francois Chapeville, Fritz Lipmann, and coworkers, when they chemically altered an amino acid *after* it had already been attached to its specific tRNA (Fig. 11-19). In this case they prepared cysteine tRNAs charged with this amino acid, and once the attachment had been made they converted the cysteine to alanine and observed the effect this would have upon its incorporation. It had no effect. The recognition group of the tRNA, i.e., its anticodon, remained unaltered, and therefore continued to recognize the same nucleotides of the mRNA molecule that it would if the chemical alteration of the amino acid had not been made. Consequently, wherever a cysteine was coded for in the message, an alanine was incorporated into the polypeptide chain.

Ribosomal RNA

As described on page 100, ribosomal RNA is an integral component of each of the two ribosomal subunits. Ribosomes can be purified from cell homogenates by differential centrifugation (Fig. 3-29). The larger subcellular particles are first removed by lower speed centrifugation and the ribosomes (after release from the membranes by detergents) are then pelleted from the supernatant by greatly increasing the centrifugal force. The RNA can then be extracted from ribosomes and subjected to sucrose density gradient centrifugation; three distinct peaks are visible. The precise location of each of the peaks depends upon the source of the ribosomes from which the RNA was taken. Ribosomal RNA of *E. coli* sediments as peaks of 23S, 16S, and 5S (Fig. 11-20), while RNA from mammalian ribosomes sediment at 28S, 18S, and 5S. The 5S peak from eukaryotic cells actually contains two different rRNAs, the 5S RNA and the 5.8S RNA, the latter representing a species which has only recently been discovered.

The manner in which these various ribosomal RNAs are synthesized in eukaryotic cells will be covered in the following chapter. The RNAs of the ribosomal subunits are believed to function in various ways. Their role in the assembly and maintenance of ribosomal structure was briefly discussed in Chapter 3. In addition, rRNAs are believed to have impor-

Figure 11-20. Sucrose density gradient profiles of RNA extracted from *E. coli* cells. The two larger peaks contain the 23S and 16S ribosomal RNA while the smaller peak toward the top of the tube contains a mixture of 5S ribosomal RNA and 4S transfer RNA molecules.

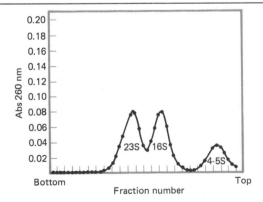

tant specific roles to play during the synthesis of the polypeptide chain. For example, the ribosomal RNA of the small subunit has been implicated in the initiation of polypeptide formation and the 5S RNA of the large subunit is believed to interact with one of the loops of the transfer RNA as it is positioned within the ribosome. Like tRNA molecules, rRNAs are also subject to posttranscriptional modification, which in this case involves primarily the addition of methyl groups to the bases (page 537).

The Genetic Code

Theoretical Aspects

Once it was realized that amino acids were specified by some type of nucleotide *code*, as opposed to some type of direct structural template, the focus of attention in molecular genetics turned to speculations on the properties of that code. One of the first formal models for the nature of the code was presented by the physicist George Gamow who proposed that the code for each amino acid was contained in the sequence of three adjacent nucleotides, i.e., the *codons* (code words) were nucleotide *triplets*. Gamow arrived at this conclusion by a bit of armchair logic: he reasoned that it would *require* at least three nucleotides for each amino acid to have its own codon. Consider the number of words that can be spelled using an alphabet containing four different letters corresponding to the four possible bases that can be present at a particular site in the DNA (or mRNA). There are 4 possible one-letter words, 16 possible two-letter words, and 64 possible three-letter words that can be made. Since there are 20 different amino acids (words) that have to be specified, they must contain at least 3 successive nucleotides (letters). The triplet nature of the code has been verified in a number of ways.

In addition to proposing that the code was triplet in nature, Gamow also proposed that it was *overlapping*. Consider the following sequence of nucleotides.

$$\cdots \text{AGCATCGCATCGA} \cdots$$

If the code is overlapping, then the ribosome would move along the mRNA one nucleotide at a time, recognizing a new codon with each move. In the above sequence, AGC would specify an amino acid, GCA would specify the next amino acid, CAT the next, and so on. However, if the code is *nonoverlapping*, each nucleotide along the mRNA is part of one, and only one, codon. There are two general types of nonoverlapping codes which could easily be imagined. In one type, AGC might specify an amino acid, ATC the next, GCA the next, and so on. A nonoverlapping code of this type is said to have no *punctuation*, i.e., there are no nucleotides to act as spacers between codons. In this type of code, the ribosome would be expected to "jump" three nucleotides, from one codon to the next as it moved along the mRNA tape. The other type of nonoverlapping code that might exist is a punctuated one in which successive codons are set off by some type of spacer nucleotide(s). For example, in the above sequence, AGC might specify an amino acid, TCG, the next, ATC the next, and so on.

The question then becomes how can one distinguish between an overlapping and a nonoverlapping code, and, if the code happens to be nonoverlapping, how does one determine if it is punctuated? The simplest means to distinguish between an overlapping and a nonoverlapping code is to consider the effect of a simple mutation on the resulting amino acid sequence of the corresponding protein. A change in the nature of one base in the DNA of an overlapping code would be expected to affect the nature of three successive codons (Fig. 11-21), and, therefore, three successive amino acids in the corresponding polypeptide. If, however, the code is nonoverlapping, and each nucleotide is part of only one codon, then only one amino acid replacement might be expected (Fig. 11-21). If one examines the nature of the changes in the primary sequence between various proteins and their mutant forms, the results almost invariably indicate a substitution of only one amino acid, as was described in the case of hemoglobin prepared from persons with sickle cell anemia. This is very strong evidence against the possibility of an overlapping code. Even without this direct evidence, the use of an overlapping code is unlikely since each codon would place restrictions on the next two codons to be read, i.e., the first two letters of the next codon would have to be compatible with the last two letters of the preceding codon, etc. This restraint in the freedom of successive codons would result in a restriction on the possible sequence of amino acids, a condition which is not found and probably would not be tolerated.

Although the use of a nonoverlapping code removes restraints on possible amino acid sequences, it adds a technical difficulty that has to be solved. Assuming that the codons are triplets, then the nature of the amino acids being specified is precisely dependent upon the proper triplets of the mRNA being read. If a ribosome should become attached to the mRNA tape at an incorrect position, the three nucleotides it would read would be expected to specify an incorrect amino acid. In order to ensure that the proper triplets are read, the ribosome attaches to the mRNA at a precise site, termed the *initiation codon*, AUG, which automatically puts it in the proper "frame" for reading the entire message. Once the ribosome attaches to this codon, it always moves along in blocks of three nucleotides so it never becomes associated with three successive nucleotides that do not compose a proper codon. In the following case:

$$\cdots \text{CUAGUUAC} \underline{\text{AUG}} \underline{\text{CUC}} \underline{\text{CAG}} \underline{\text{UCC}} \underline{\text{GU}} \cdots$$

$$\quad\quad\quad\quad\quad 1 \quad 2 \quad 3 \quad 4 \quad 5$$

Figure 11-21. The distinction between an overlapping and nonoverlapping code and the effect on each of a single base substitution.

Base sequence	Codons	
Original sequence ...AGCATCG...	Overlapping code ..., AGC, GCA, CAT, ATC, TCG, ...	Nonoverlapping code ..., AGC, ATC, ...
Sequence after single-base substitution ...AGAATCG...	Overlapping code ..., AGA, GAA, AAT, ATC, TCG, ...	Nonoverlapping code ..., AGA, ATC, ...

the ribosome moves from the first three nucleotides, AUG, to the next three, CUC, and so on along the entire page. There is no punctuation between the codons; all nucleotides of the message are involved in the specification of amino acids.

Given the use by organisms of a triplet code which has the capability of specifying 64 different amino acids, and the reality that there are only 20 amino acids to be specified, the question arises as to the function of the extra 44 triplets. There are three possibilities: either all of them, some of them, or none of them code for amino acids. If either of the first two possibilities are correct, then there will be at least some amino acids that will have more than one codon, a condition termed *degeneracy*. As it turns out, the code is highly degenerate since nearly all of the codons are involved in the specification of particular amino acids. Those that are not (3 of the 64) have a special function—they are recognized by the ribosome as termination codons and cause the reading of the message to stop.

The degeneracy of the code was originally predicted by Crick on theoretical grounds when he considered the great range in the base composition among the DNAs of various bacteria. It had been found, for example, that the $G + C$ content of the DNA of various organisms could range from 20% to 74% while at the same time there was little overall variation among the amino acid composition of their proteins. This suggested that the same amino acids were being coded for by different base sequences indicating the degeneracy of the code.

Genetic Studies

Up to this point our discussion of the genetic code has been restricted to theoretical points as was much of the early literature on this subject. However, certain types of genetic studies were also being performed that provided some evidence for the code's properties. One particularly important series of experiments was carried out by Crick, Brenner, and co-workers, from which several important conclusions were drawn. Before these experiments, which utilized a particular chemical *mutagen,* can be described, we need to consider briefly the types of mutations that can be induced in a nucleic acid molecule.

After Müller found that x-irradiation was an effective means of causing changes in the genetic material, it was found that mutations could also be induced by a number of chemicals. When the effect of these chemicals on DNA was analyzed, it was found that they were causing changes in the base sequence. The changes, i.e., the base substitutions, were found to be of two general types, *transitions* or *transversions,* depending upon the nature of the mutagen. Transitions are changes from one pyrimidine to the other pyrimidine or one purine to the other purine, while transversions arise from the replacement of a purine with a pyrimidine or vice versa. Nitrous acid, for example, which causes the deamination of certain bases (Fig. 11-22) results in transitions. In one case, the nitrous acid-induced transition (from cytosine to uracil) is brought about directly by the deamination, while in the other case the transition (from adenine to guanine) comes about after a round of replication (page 656). Transver-

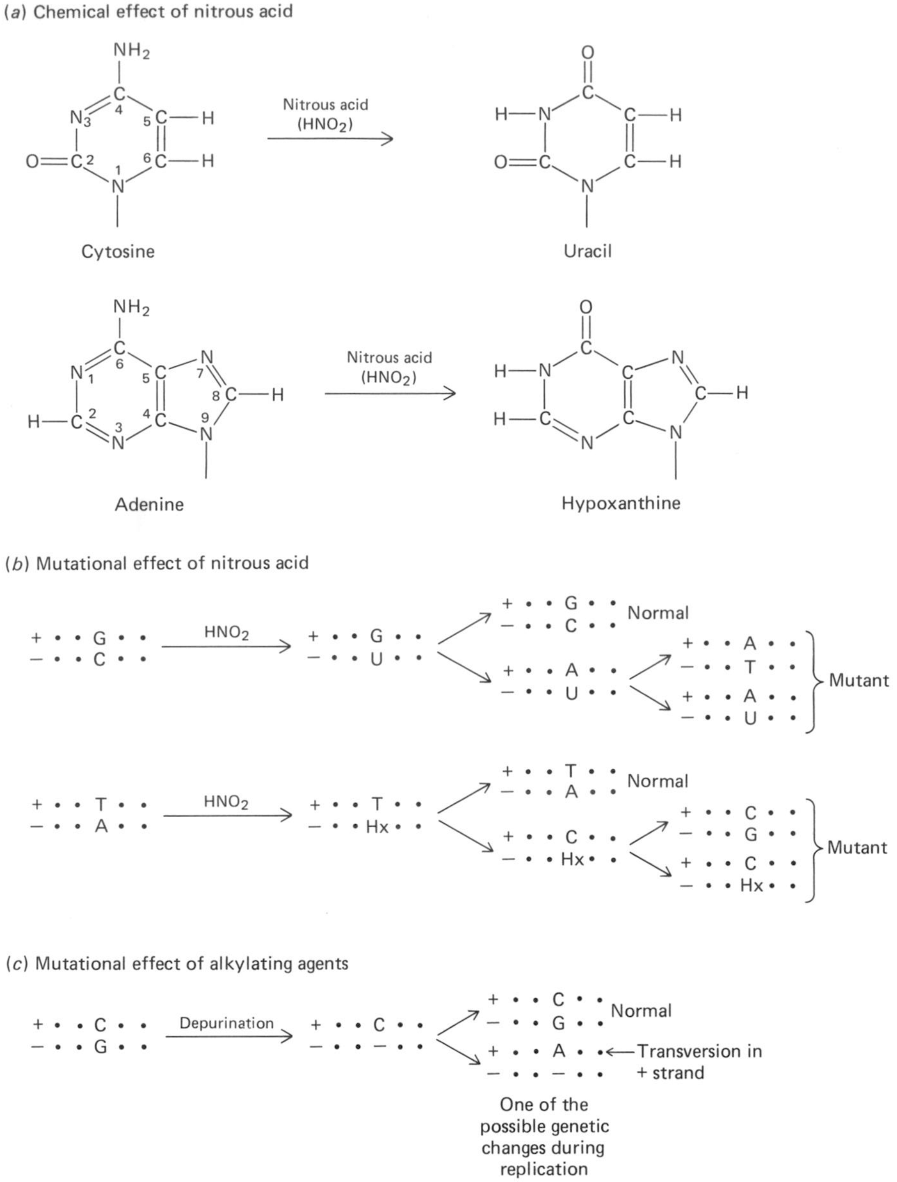

(a) Chemical effect of nitrous acid

Cytosine → Uracil

Adenine → Hypoxanthine

(b) Mutational effect of nitrous acid

Figure 11-22. (a) The chemical effect of nitrous acid on DNA. Cytosine residues can be altered to uracil, and adenine can be altered to hypoxanthine, both as a result of deamination. (b) The chemical changes induced by nitrous acid lead to transitions in the DNA because the bases formed in the reactions of a have different pairing properties from the original bases. (c) Alkylating agents, which remove the entire base from the sugar group, can lead to transversions since the site of the missing base does not restrict the selection of the opposing base in the complementary DNA strand.

(c) Mutational effect of alkylating agents

sions are generally produced by agents which cause the hydrolysis of a base from its attachment site on the sugar. For example, alkylating agents lead to the removal of purines. Once the base is gone, so is its template activity at the next round of replication when the complementary strand is produced. If the complementary strand happens to incorporate a purine across from the missing base, then a transversion has occurred.

Crick and Brenner, in their studies on bacteriophage T_4, used a chemical, *proflavin*, which causes an entirely different type of mutation than those just described (Fig. 11-23). Mutagens, such as nitrous acid, which cause "point" mutations, i.e., those of a single base *substitution*, generally lead to the production of a protein product with some activity. Even though the protein is abnormal, and therefore its alteration is readily detectable, it is not so distorted that it does not have residual activity. Mutations in genes whose products retain a measure of activity are said to be "leaky." In contrast, proteins produced by genes that have undergone proflavin alteration tend to be totally nonfunctional and even unrecognizable by antibodies capable of reacting with the normal protein. In other

---

(a)

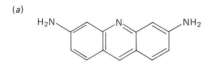

(b) Original base sequence

(c) Addition of one base

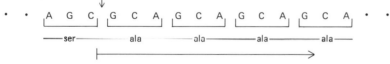

(d) Deletion of one base

(e) Addition of one base and deletion of another base

(f) Addition of two bases

(g) Addition of three bases

Figure 11-23. (a) The chemical structure of proflavin. (b) The base sequence which would code for polyserine. (c)–(e) The effects on the polypeptide by various types of additions or deletions induced by proflavin. The addition (or deletion) of one or two bases totally disrupts the template properties of the polynucleotide. However, if there are compensating changes, such as one addition and one deletion, or if there are three additions (or three deletions), the original coding property of the polynucleotide is restored after the altered section has been passed. The length of the altered amino acid sequence is indicated by the line drawn beneath each polypeptide.

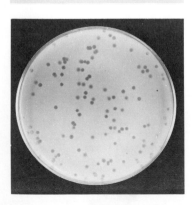

words, proflavin brings about a drastic alteration in amino acid sequence. It does this by causing the affected genes to contain either one extra nucleotide (an addition) or one less nucleotide (a deletion). The mechanism by which proflavin accomplishes this effect will not concern us here. In order to understand the reason for the production of such altered proteins, we need to consider the effect of the insertion (or deletion) of an extra nucleotide into a specific base sequence (Fig. 11-23c, d). Since the code is nonoverlapping, once the reading of the sequence begins, it automatically moves along in groups of three nucleotides. If the sequence is altered by the addition of a nucleotide (or the deletion of a nucleotide), then from that point throughout the remainder of the gene, the sequence of codons will be changed from that of the normal sequence of triplets. Proflavin is said to induce "frameshift" mutations, i.e., ones which cause the ribosome to shift in position relative to each successive codon along the message. Consequently, frameshift mutations would be expected to cause an entirely abnormal sequence of amino acids from the point of the deletion or addition in the DNA and subsequent mRNA.

The proposal that proflavin mutagenesis was occurring via the addition and deletion of single bases was derived by an extensive series of genetic experiments on bacteriophage mutants of the rII genes, the same ones studied by Benzer in his fine structure analysis employing intragenic recombination. The rII locus is a useful one in genetic analysis because the absence of its gene product, a protein of undetermined function, does not prevent the phage from reproducing when grown on a particular strain of *E. coli*, strain B. In contrast to the wild type, nonleaky rII mutants cannot grow on strain K, and even when growing on strain B, the colonies they produce are recognizably mutant in nature (Fig. 11-24). The first mutant induced by proflavin was termed *FC0* and was eventually shown to be a result of the addition of an extra nucleotide in the gene. Using this mutant strain to infect additional B-type cells in the presence of proflavin, they isolated what appeared to be a revertant of this mutation, i.e., a strain of phage that would once again grow on *E. coli* strain K. When a gene containing a single base substitution reverts to the wild type, it invariably does so by having the abnormal codon change to a codon which once again specifies the proper amino acid. When the proflavin revertant was tested by genetic analysis, rather than finding that the original codon had been altered once again, it was found that the phage had instead accumulated a second site of mutation within the gene close to that of the original alteration in base sequence. The second mutation was acting to *suppress* the effect of the first mutation. The two mutations were of opposite sign; whereas the original effect was caused by an addition of a nucleotide, the suppressor mutation was caused by the deletion of a nucleotide. The

Figure 11-24. A petri dish (bottom photograph) whose surface is covered with a "lawn" of growing bacterial cells. The blotches on the plate (upper photograph) are sites where bacteriophage infections are lysing the host bacterial cells. Infections caused by wild-type phage can be distinguished from those caused by r mutants. Whereas the plaques formed by wild-type phage are small and rough edged, those of the mutant are large and smooth edged. Rapid lysis is signified by r and results in large plaques. (*Courtesy of D. P. Snustad, from D. O. Woodward and V. Woodward,* Concepts of Molecular Genetics, *McGraw-Hill, 1977.*)

ability of two mutations of opposite sign to cancel one another out and result in a wild-type (or more properly, a *pseudowild*-type) phenotype can be readily understood by considering the effect of these mutations upon the reading frame of the message (Fig. 11-23e). Whereas the addition of a nucleotide retards the reading frame, the deletion once again advances it back to its proper position. The degree to which the double mutant will return to the wild-type condition will depend roughly upon the distance between the two mutations: the greater the distance, the the greater the number of altered amino acids in the resulting polypeptide. From these types of results, the following conclusions were drawn:

1. The code showed no evidence of punctuation (page 499). Punctuating nucleotides would have been expected to have reoriented the ribosome back into the proper reading frame without the need for a compensating second mutation.

2. Since the code was known to be nonoverlapping, and since there was no evidence of punctuation, there must be some recognizable site on the message to which the ribosome initially attaches (page 500). Once the ribosome has become attached to the message in the proper reading frame, it would continue to read the tape three nucleotides at a time from that fixed point.

3. The code is degenerate. Double mutants are capable of growing on strain K, because the polypeptide chain reverts to the normal amino acid sequence despite the fact that there must be some stretch in the protein, between the two mutations, which is abnormal (Fig. 11-23e). If only 20 of the possible 64 triplets actually coded for amino acids, then one would expect "nonsense" codons to be generated in the region between the two alterations where the reading frame has been shifted. If a nonsense codon were present, no amino acid could be specified at the corresponding site in the polypeptide chain, and the growing polypeptide would be terminated rather than continue through this altered region to regain its normal amino acid sequence at the other side. In fact, there was one region where double mutants of opposite sign did not cause the phage to revert to the wild-type and this provided genetic evidence for the existence of at least one codon that specified termination of the reading message (page 521).

All of these conclusions were later verified by the *amino acid sequence analysis* of lysozyme from phage-bearing proflavin-induced mutations in the lysozyme gene.

In additional experiments on proflavin-induced mutations in the rII locus, Crick and coworkers provided the strongest evidence up to that point for the triplet nature of the code. They found that, just as two mutations of *opposite* sign located close together (Fig. 11-23e) would cause the phage to regain its ability to grow on *E. coli* strain K, so too would *three* mutations of the *same* sign (Fig. 11-23g). In other words, triple mutants, ones with either three additions or three deletions, all relatively close together within the gene, also caused the phage to revert to the wild-type condition. As before, this result indicates that once the disturbed region of the mRNA has been passed, the ribosome is once again reading the message in the proper frame.

This study is of interest beyond its confirmation of the nature of the genetic code. It serves as an excellent example of the kinds of conclusions that can be reached about biochemical activities using nonbiochemical analysis and inductive reasoning. This study was entirely of a genetic nature. In fact, the protein for which this region of the phage genome codes had not been isolated, nor was its function even known. All that was known in this regard was that it was required for the phage to grow on one strain of bacteria and not another. This system was chosen strictly for its advantages in genetic analysis, yet the conclusions that were reached concerned, to a large degree, a biochemical mechanism of great complexity.

The Identification of the Codons

By 1961 the general properties of the code were known, but not one of the coding assignments of the specific triplets had been discovered. The theoretical attack on the nature of the code had essentially ended and the biochemical phase was ready to begin. It would be a surprisingly short period of time before all of the codon assignments would be made. The most straightforward approach would have been to compare nucleotide sequences of a gene and the corresponding amino acid sequences of the protein. However, since there was little hope at the time of isolating and sequencing a polynucleotide, some alternate procedure would have to be designed. The breakthrough came in 1961, much sooner than expected, when Marshall Nirenberg devised a means whereby a number of codons could be determined. In a sense, Nirenberg reversed the approach that had been used with such success up to that time. Whereas all of the studies we have described in this chapter have involved the analysis of polypeptide chains made in the cell under different conditions, Nirenberg set out to make, in essence, his own "genetic message" and then determine what kind of a "protein" it might code for. In order to carry out this type of experiment, he had two prerequisites: he had to be able to make his own polynucleotide and he had to be able to use it to make a polypeptide.

A few years before Nirenberg began his studies on the code, a bacterial enzyme had been discovered, *polynucleotide phosphorylase*, which was capable of polymerizing RNA precursors in vitro into a polyribonucleotide. The enzyme has two major properties that set it apart from *RNA polymerase*, the enzyme that normally produces RNA in the cell: it uses ribonucleoside diphosphates (GDP, UDP, CDP, and ADP) as precursors rather than triphosphates, and it polymerizes them without directions from a template. This enzyme has these unexpected properties because in the cell it presumably acts to hydrolyze RNAs, one nucleotide at a time, rather than synthesizing them. Nirenberg used the enzyme to catalyze the reverse reaction in vitro and manufacture synthetic messengers (Fig. 11-25) which could then be tested with respect to the directions they contained for amino acid incorporation.

The other necessary component of Nirenberg's experiment was the system for testing the template activity of the synthetic mRNA. As described in the chapters on membranes and membranous organelles, one of the most important areas of research in cell biology has been concerned

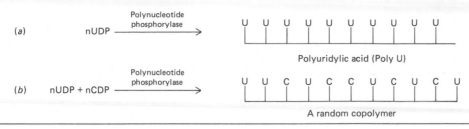

Figure 11-25. The synthesis of poly-ribonucleotides using the enzyme polynucleotide phosphorylase. (a) The formation of a homopolymer when incubated in the presence of a single ribonucleoside diphosphate. (b) The formation of a random copolymer when incubated with two different nucleoside diphosphate precursors.

with the development of in vitro systems in which various activities can be examined in a much simpler and defined chemical environment. Using these various reconstituted systems, experiments can be carried out in vitro that cannot be accomplished within the cell. In the field of molecular biology, one of the most important types of in vitro systems is one capable of the synthesis of DNA, RNA, or protein. The in vitro synthesis of DNA and RNA requires fewer components, since it is essentially a straightforward enzymatic polymerization. On the other hand, as determined primarily by Paul Zamecnik, Malon Hoagland, Fritz Lipmann, and co-workers, the in vitro synthesis of proteins requires a complex mixture including ribosomes, charged transfer RNAs, many protein factors, and an ATP-generating system.

Even though our knowledge of the components required for protein synthesis has greatly increased, cell-free systems of amino acid incorporation are still basically undefined, i.e., are composed of a mixture of materials of unspecified nature. The mixture of materials (Fig. 11-26) is obtained by using the soluble cytoplasm of the cell as part of the reaction mixture. If cells are homogenized and the particulate material is centrifuged out of suspension, then the supernatant contains all of the tRNAs, enzymes, and protein factors needed to incorporate amino acids. If the high-speed supernatant is combined with ribosomes, messenger RNA, and amino acids, one or more of which are radioactively labeled, then this cell-free system will incorporate these amino acids into a labeled polypeptide.

The first synthetic polyribonucleotide that Nirenberg and J. Heinrich Matthei tested was a polymer of uracil, poly U (Fig. 11-25). When poly U was added to the cell-free protein-synthesizing system, they found that a polypeptide containing phenylalanine was produced. The synthesis of polyphenylalanine in the presence of poly U indicated that the codon for this amino acid must be UUU. The first codon assignment had been made. The accepted terminology is to refer to the codon as the nucleotides of the mRNA, the molecule which is actually involved in protein synthesis. The triplet in the strand of DNA that codes for phenylalanine is AAA. It was soon found that poly A coded for lysine and poly C for proline. Poly G could not be tested due to structural difficulties in the polymer. Once the homopolymers, i.e., ones containing only one base, were tested, copolymers containing two or more bases were synthesized (primarily in the laboratories of Nirenberg and Severo Ochoa) and tested for their template activity (Fig. 11-25). The problem with the synthesis of copolymers using the enzyme polynucleotide phosphorylase, is that the

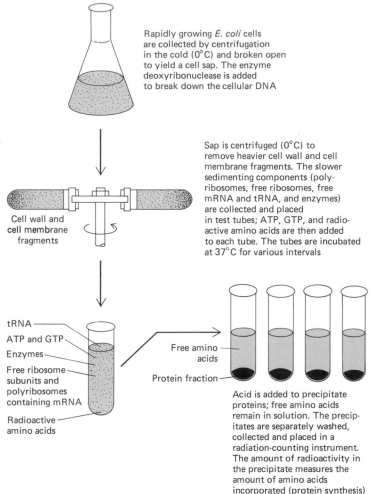

Rapidly growing *E. coli* cells are collected by centrifugation in the cold (0°C) and broken open to yield a cell sap. The enzyme deoxyribonuclease is added to break down the cellular DNA

Sap is centrifuged (0°C) to remove heavier cell wall and cell membrane fragments. The slower sedimenting components (poly-ribosomes, free ribosomes, free mRNA and tRNA, and enzymes) are collected and placed in test tubes; ATP, GTP, and radio-active amino acids are then added to each tube. The tubes are incubated at 37°C for various intervals

Cell wall and cell membrane fragments

tRNA

ATP and GTP

Enzymes

Free ribosome subunits and polyribosomes containing mRNA

Radioactive amino acids

Free amino acids

Protein fraction

Acid is added to precipitate proteins; free amino acids remain in solution. The precip-itates are separately washed, collected and placed in a radiation-counting instrument. The amount of radioactivity in the precipitate measures the amount of amino acids incorporated (protein synthesis)

Figure 11-26. Experimental details of the in vitro studies of protein synthesis. *(From J. D. Watson, "The Molecular Biology of the Gene," 3d ed., W. A. Benjamin, 1976.)*

the incorporation of the specific nucleotides was random, thereby generating an unpredictable nucleotide sequence.

Consider the nature of a synthetic polymer produced by the random incorporation of 2 different precursors, UDP and CDP. There will be 8 different codons within the polymer (UUU, UUC, UCU, CUU, CCU, CUC, UCC, CCC). These 8 codons can be divided into 4 classes on the basis of their base composition. Among the group, there are codons with 0, 1, 2, or 3 of each of the bases. Next consider what will happen if the relative concentration of the 2 precursors is varied. For example, the greater the concentration of UDP in the reaction mixture relative to CDP, then the greater will be the presence within the polymer of U-containing codons. In other words, the proportions of the various types of codons reflects the concentrations of the precursors. If the concentration of UDP is 3 times that of CDP, then on the average for every 64 codons in the polymer, 27 of them will have 3 U's, 27 of them will have 2 U's (9 of each of the 3 possible codons, 9 will have 1 U (3 of each of the 3 possible codons) and

1 will have no U's. By determining the amino acid compositions of the various polypeptides produced by this and the other copolymers, it was possible to determine the base composition of the codons that coded for the various amino acids. For example, it was determined that codons with 2 U's and 1 C coded for phenylalanine, leucine, and serine, but it was impossible to decide which of the codons (UUC, UCU, or CUU) correspond to each of the 3 amino acids.

Information leading to the assignment of particular codons came in two different types of experiments, one in Nirenberg's laboratory and the other in H. Gobind Khorana's. The two approaches were totally different and their results complemented one another perfectly. The method of Philip Leder and Nirenberg stemmed from a report that, in the presence of poly U, the specific transfer RNA for phenylalanine would bind to ribosomes. In other words, the attachment of a particular tRNA to a complementary codon on an mRNA-ribosome complex could occur in the absence of protein synthesis. Leder and Nirenberg set out to determine how small a message could attach to a specific tRNA (in the presence of a ribosome), and found that "minimessengers" of only three nucleotides, i.e., one codon, would suffice. These investigators synthesized the 64 possible codons and determined which aminoacyl-tRNA would attach to each. Although for technical reasons a number of codon assignments could not be made by this technique, the amino acid specifications of most of the codons were established.

As in the first technique of Nirenberg's, the third technique for codon assignment also involved the use of synthetic ribonucleotide polymers. The polymers made by Khorana differed in one very important respect from the previous polymers—their sequence was defined. Using complex organic and enzymatic procedures, Khorana was able to make mRNAs with specific repeating sequences. For example, he could synthesize a polymer of the sequence · · · UCUCUCUCUC · · · ·. A polymer of this nature is composed of two alternating codons, UCU and CUC. When this polymer was added to a cell-free protein-synthesizing system, only one type of polypeptide was produced—one made of alternating amino acids, serine and leucine. If he added a polymer with a repeating trinucleotide sequence, such as UCUUCUUCU · · ·, he found that three different types of polypeptides would be produced: polyserine (codon of UCU), polyleucine (codon of CUU), and polyphenylalanine (codon of UUC). If the synthetic polymer happened to contain one of the three termination codons, very small peptides of two or three amino acids were generated.

It was clear from these results that a ribosome could attach at any frame on the message, but, once it had attached, it continued to read the message in groups of three nucleotides. In contrast, in the cell, the attachment of the ribosome to the messenger RNA is much more specific (page 513), recognizing only the initiation codon, AUG. In the test tube, the abnormally high Mg^{2+} concentrations seem to be the most responsible for allowing the ribosome to attach initially in a random manner. Although this method by itself does not allow precise codon assignments to be made, when taken together with information obtained from the other methods,

Second position

		U		C		A		G		
U		UUU	Phe	UCU	Ser	UAU	Tyr	UGU	Cys	U
		UUC	Phe	UCC	Ser	UAC	Tyr	UGC	Cys	C
		UUA	Leu	UCA	Ser	UAA	Stop	UGA	Stop	A
		UUG	Leu	UCG	Ser	UAG	Stop	UGG	Trp	G
C		CUU	Leu	CCU	Pro	CAU	His	CGU	Arg	U
		CUC	Leu	CCC	Pro	CAC	His	CGC	Arg	C
		CUA	Leu	CCA	Pro	CAA	Gln	CGA	Arg	A
		CUG	Leu	CCG	Pro	CAG	Gln	CGG	Arg	G
A		AUU	Ile	ACU	Thr	AAU	Asn	AGU	Ser	U
		AUC	Ile	ACC	Thr	AAC	Asn	AGC	Ser	C
		AUA	Ile	ACA	Thr	AAA	Lys	AGA	Arg	A
		AUG	Met	ACG	Thr	AAG	Lys	AGG	Arg	G
G		GUU	Val	GCU	Ala	GAU	Asp	GGU	Gly	U
		GUC	Val	GCC	Ala	GAC	Asp	GGC	Gly	C
		GUA	Val	GCA	Ala	GAA	Glu	GGA	Gly	A
		GUG	Val	GCG	Ala	GAG	Glu	GGG	Gly	G

First position (5' end) — Third position (3' end)

Figure 11-27. The genetic code. Of the 64 possible codon assignments, 61 specify amino acids while the other 3 specify termination. The systematic organization of the various codons is discussed in the text.

all of the codon assignments were determined. The results of Khorana's experiments are shown in Table 11-1 and all of the code words in Fig. 11-27.

Nature and Universality of the Code

If one examines the codons of Figure 11-27, it is apparent that the amino acid assignments are distinctly nonrandom; they demonstrate a

TABLE 11-1
The Translation of Repeating-Sequence Copolymeric RNAs

Repeating-sequence copolymer	Codons involved	Polypeptides into which translated
(AAG)$_n$	AAG, AGA, GAA	poly Lys, poly Arg, poly Glu
(UUC)$_n$	UUC, UCU, CUU	poly Phe, poly Ser, poly Leu
(UUG)$_n$	UUG, UGU, GUU	poly Leu, poly Cys, poly Val
(AAC)$_n$	AAC, ACA, CAA	poly Asn, poly Thr, poly Gln
(UAC)$_n$	UAC, ACU, CUA	poly Tyr, poly Thr, poly Leu
(AUC)$_n$	AUC, UCA, CAU	poly Ile, poly Ser, poly His
(GAU)$_n$	GAU, AUG, UGA	poly Asp, poly Met, °
(GUA)$_n$	GUA, UAG, AGU	poly Val, † poly Ser
(AG)$_n$	AGA, GAG	(Arg · Glu)$_n$
(AC)$_n$	ACA, CAC	(Thr · His)$_n$
(UC)$_n$	UCU, CUC	(Ser · Leu)$_n$
(UG)$_n$	UGU, GUG	(Cys · Val)$_n$
(UAUC)$_n$	UAU, CUA, UCU, AUC	(Tyr · Leu · Ser · Ile)$_n$
(UUAC)$_n$	UUA, CUU, ACU, UAC	(Leu · Leu · Thr · Tyr)$_n$
(GAUA)$_n$	GAU, AGA, UAG, AUA	None‡
(GUAA)$_n$	GUA, AGU, AAG, UAA	None‡

° No polypeptide formed corresponding to the UGA codon; see discussion of peptide chain termination punctuation.

† No polymer formed corresponding to the UAG codon; UAG is implicated in peptide chain termination punctuation (page 521).

‡ The exact reason for failure to observe peptide synthesis in these cases is not clear. Each one does contain one of the codons implicated in peptide chain termination punctuation, but it is also possible that secondary structure of the polymer may prevent translation.

SOURCE: Woese, C. R., *The Genetic Code,* Harper and Row, 1967, from work of H. G. Khorana.

great deal of systematic order. There is an obvious similarity in codons which specify the same amino acid. As a result of this pattern, spontaneous mutations causing single base changes in a gene will be less likely to produce a change in the amino acid sequence of the corresponding protein. Comparison of nucleotide sequences for the same protein among different species has indicated that many more nucleotide substitutions accumulate which do not change the amino acid sequence of the protein than substitutions which do make a change. The "safeguard" aspect of the code appears to go beyond its degeneracy. The nature of the codon assignments is such that *similar* amino acids tend to be specified by similar codons. For example, the codons of all of the hydrophobic amino acids are related so that a base substitution that does cause a change in an amino acid is most likely to substitute one hydrophobic residue for another.

The greatest similarities between amino acid-related codons occur in the first two nucleotides of the triplet, while the greatest variability occurs in the third nucleotide. Consider those 16 codons ending in U. In every case, if the U is changed to a C, the same amino acid is specified (first two columns of each box in Fig. 11-27). Similarly, in most cases, a switch between an A and a G at the third site is also without effect upon amino acid determination. The interchangeability of the base of the third position led Crick to propose that the same transfer RNA species may be able to recognize more than one codon. His proposal, termed the *wobble* hypothesis, suggested that, whereas the steric requirement between the anticodon of the tRNA and the codon of the mRNA may be very strict for the first two positions, it may be more flexible at the third position, allowing two codons that specify the same amino acid and differ only at the third position to use the same tRNA in their protein synthesis. The rules governing the wobble at the third position of the codon (Fig. 11-28) are as follows: U of the anticodon can pair with A or G; G of the anticodon can pair with U or C; and I (inosine, which is derived from guanine in the original tRNA molecule) of the anticodon can pair with U, C, or A. The wobble hypothesis has been verified by showing that certain purified tRNA molecules will bind to more than one codon for that amino acid. As a result of the wobble, the six codons for leucine, for example, require only three tRNAs.

Since the code was determined totally as a result of the analysis of in vitro systems, it was important that some verification of the assignments be made upon polypeptides synthesized within cells. A number of different approaches have been taken along this line. For example, amino acid sequences of mutant proteins have been determined and checked

Figure 11-28. Schematic illustration of the wobble hypothesis. The first nucleotide of the tRNA anticodon, i.e., the nucleotide on the 5' end, is capable in some cases of pairing with more than one nucleotide at the third position (3' end) of the mRNA codon. Consequently, more than one codon can use the same tRNA. The rules for pairing in the wobble scheme are indicated in the figure as well as in the text.

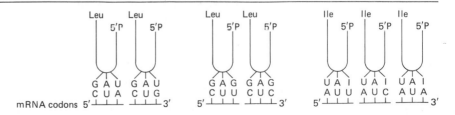

mRNA codons

against the code to see if the substitution can be explained by a single base change. Studies of this type on a variety of proteins, including a large number of hemoglobin mutations, have checked out in complete accordance with the code. In fact, the one amino acid substitution in a bacterial protein that was not consistent with a single nucleotide change was also the only one that would not spontaneously revert to the wild type. This correlation indicated that this mutant protein arose by a rare double mutation occurring in the same codon. In recent years, the ability to obtain nucleotide sequences of both prokaryotic and eukaryotic genes has provided the final touch to the entire study of the code and the co-linearity of the gene and its protein. All of the results agree to form one undisputed body of knowledge of great scientific importance. An additional feature of the gene-sequencing work is the finding that all three of the termination codons are present at the ends of genes, verifying their role in the translation process as well.

The last point to mention concerning the code is its universality. Of all of the organisms studied, from viruses to humans, there is no evidence of any evolutionary change in the assignments of the 64 possible codons. For example, messenger RNAs can be obtained from one source and combined with the protein-synthesizing machinery from another source, and the protein produced is always the same one that the message would have produced with its own machinery. For example, one of the means by which the template activity of a messenger RNA can be tested is to inject it into amphibian oocytes which, due to their very large size, are good targets for a hypodermic needle and a tiny volume of RNA solution. When rabbit globin mRNA is injected into an amphibian oocyte, globin polypeptide chains are produced in the oocyte which are indistinguishable from those made within a rabbit. Attempts to translate eukaryotic mRNAs in prokaryotic cells and vice versa are generally less successful. The problems arise from differences in regions of the mRNA that are not directly involved in amino acid specification (page 514) rather than differences in codon assignments, which are universal. More than any other single fact, the universality of the code provides the strongest evidence for genetic continuity among all organisms present today on earth. It is virtually inconceivable to explain this universality by having the present forms of life originate from more than one source.

The Mechanism of Protein Synthesis

Protein synthesis, or translation, is certainly the most complex of any synthetic activity occurring in the cell. Whereas the other molecules of the cell are manufactured as a result of relatively straightforward enzymatic reactions, the assembly of a protein requires all of the various transfer RNAs with their attached amino acids, ribosomes, messenger RNA, a number of proteins having a variety of functions, cations, and GTP. The complexity is not surprising when one considers that protein synthesis requires the incorporation of each of 20 different amino acids in precisely the proper sequences as dictated by a coded message written in a language which uses different symbols. In the description to follow, we will draw most heavily upon translation mechanisms as they operate in bacterial cells where they were first studied and best understood. The process is remarkably similar in eukaryotic cells, the primary differences reflecting the number of soluble (nonribosomal) protein factors and their precise functions. The synthesis of a polypeptide chain can be divided into three rather distinct activities: the *initiation* of the chain, the *elongation* of the chain, and its *termination*. We will consider each of these activities separately.

Initiation

One of the points that was stressed in a previous section was the danger inherent in a nonpunctuated, nonoverlapping code of having the ribosome attach to the message in the wrong reading frame. It is essential that the ribosome attach at a fixed point in the message, specifically at the initiation codon AUG. If one examines the codon assignment (Fig. 11-27), it can be seen that AUG is more than just an attachment codon, it is the only codon for methionine. The fact that AUG specifies methionine was of prime importance in the recognition of its role as an initiation codon.

An early analysis of the N-terminal amino acids of the total protein of *E. coli* turned up a surprising finding: approximately 45% of the residues at the amino end of the polypeptide chains were methionine, a relatively uncommon amino acid. Later, an unusual aminoacyl-tRNA, one with *N*-formylmethionine, was discovered among the substances in bacterial homogenates and the possibility was raised that this particular tRNA was involved in the initiation of polypeptide chains. Additional work using natural phage mRNAs to direct amino acid incorporation in vitro revealed that *N*-formylmethionine was indeed the first amino acid incorporated in the chain. In the cell, the formyl group is removed enzymatically leaving the newly forming chain with a methionine at its N-terminus. In approximately 45% of the polypeptides, the methionine is left in that position while in the remainder it too is removed. Initiation of polypeptide chains in eukaryotic cells also begins with the AUG codon, but in higher cells the methionine is incorporated as such rather than being formylated. As in the prokaryotic chains, the methionine may or may not be removed. Interestingly, the mitochondria and chloroplasts of eukaryotic cells use *N*-formylmethionine as their first amino acid rather than methionine, a finding which provides one of the strongest pieces of evidence in the argument for the prokaryotic origin of these organelles.

The use of *N*-formylmethionine in the initiation process raises several questions. The most troublesome aspect is that methionine has only one codon, AUG. How does AUG at the beginning of the message lead to the incorporation of *N*-formylmethionine while AUG elsewhere in the message results in the incorporation of methionine? Are there different tRNAs for *N*-formylmethionine and methionine? If there are, they presumably have the same anticodon (CAU). It has been found that there are two different tRNAs whose anticodons recognize AUG in the mRNA (Fig. 11-29a). One of these tRNAs ($tRNA_m$) is responsible for the incorporation of methionine in the internal positions within the polypeptide, while the other

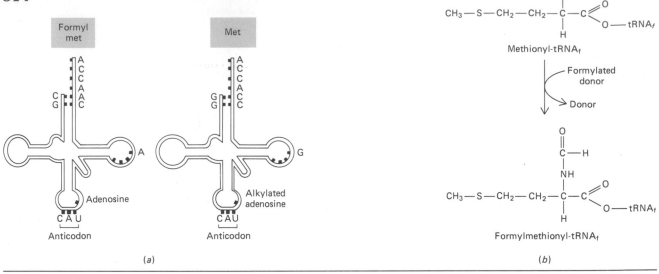

Figure 11-29. (a) The differences between the tRNA$_{met}$ and the tRNA$_{fmet}$. (*From J. D. Watson, "The Molecular Biology of the Gene," 3d ed., W. A. Benjamin, 1976.*) (b) The formylation of methionine.

one (tRNA$_f$) is responsible for bringing N-formylmethionine in at the start. Both tRNAs are recognized by the same aminoacyl synthetase which places a methionine on the two tRNAs. The formylation of the methionine on tRNA$_f$ occurs after the amino acid is attached to the tRNA (Fig. 11-29b); the formylating enzyme recognizes tRNA$_f$ as opposed to tRNA$_m$, and adds the formyl group to the amino acid. Even though eukaryotic cells begin their polypeptide synthesis with methionine, the tRNA that brings this first methionine into position is different from the methionyl-tRNA that is used for internal AUG codons. The situation, other than the presence of the formyl group itself, is therefore quite similar in the two types of organisms.

The first step in the chain of events leading to the formation of a specific polypeptide occurs as a result of the interaction of a 30S ribosomal subunit with the first AUG codon of the message. How does the 30S subunit select the initial AUG as opposed to some internal one? It is now believed that the recognition of this particular AUG of the message results from the interaction between complementary nucleotides of the 3' end of the 16S *ribosomal* RNA of the 30S subunit with a region of the message just before the initiation codon. As will be described below, the mRNA is read in a 5' to a 3' direction. The analysis of several viral and bacterial mRNAs have indicated that just to the 5' side of the initial AUG codon is a stretch of nucleotides very rich in purine, both A and G. For example, the phages that infect *E. coli* have the sequence 5'-AGGAGGU-3' in a similar position just to the 5' side of the initiation codon. The 3' terminus of the 16S rRNA of *E. coli* has the complementary sequence of 5'-ACCUCCU-3' near its terminus. This and other data suggests that the 30S subunit attaches at the initial AUG, as opposed to some internal one, as a result of the interaction between these two complementary nucleotide sequences. The binding of the 30S subunit to the mRNA requires a soluble protein factor, IF3 (or f3). The process of initiation, which is depicted in Fig. 11-30, ensures that the proper reading frame will be used by the ribosome.

Once the 30S subunit attaches, the N-formylmethionyl-tRNA becomes associated with the mRNA, and the 50S ribosomal subunit joins the complex by attaching to the 30S subunit to form the complete ribosome. Each ribosome has two sites for association with transfer RNA molecules. These two sites, termed the A (aminoacyl) site and the P (peptidyl) site, have very different roles in relation to the process of amino acid incorporation as will be described below. The A site is generally the one in which aminoacyl-tRNAs enter the ribosome-mRNA complex while the P site is the one from which the tRNA leaves after having donated its amino acid. In the case of the first aminoacyl-tRNA, that of N-formylmethionyl-tRNA$_f$, the entry appears to be directly into the P site of the 30S subunit. Another of the soluble protein factors, IF2 (or f2) is involved in the specific binding of the N-formylmethionyl-tRNA$_f$ to the 30S subunit. This initiation factor specifically recognizes this form of the methionyl-tRNA and binds to it so as to allow it to act during initiation. The role of the third factor, IF1 (or f1) is less well defined.

In addition to the protein factors, GTP is also required for successful initiation. It appears that the GTP binds to the IF2-N-formylmethionyl-

Figure 11-30. Steps in the initiation of protein synthesis as described in the text. (*1*) The 30S subunit forms a complex with IF3 allowing it to bind to the mRNA at the AUG initiation codon. (*2*) The N-formylmethionine tRNA with its associated IF2 and GTP binds to the 30S subunit-mRNA complex. (*3*) The attachment of the 50S subunit and release of the various factors. The precise order in which these various events occur has not been determined with certainty.

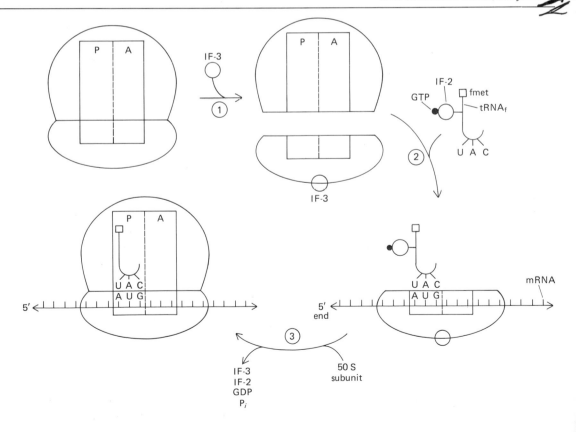

tRNA$_f$ complex before it attaches to the mRNA-30S subunit complex. Once the initial tRNA becomes attached, the 50S subunit enters into the association, and it is at this or a later point that the GTP appears to be hydrolyzed, and the GDP and IF2 released.

Figure 11-31. (*a*) Steps in the elongation of the polypeptide chain as described in the text. (*1*) The entrance of the aminoacyl-tRNA into the empty A site of the ribosome. (*2*) The binding of the tRNA is accompanied by the release of GDP-Tu which is recycled as shown. (*3*) Peptide bond formation is accomplished by the transfer of the nascent polypeptide chain from the tRNA at the P site to the aminoacyl-tRNA of the A site. (*4*) The binding of factor G and the hydrolysis of GTP results in the release of the tRNA from the P site and the translocation of the ribosome relative to the mRNA. (*b*) Peptide bond formation and the subsequent displacement of the deacylated tRNA of the P site.

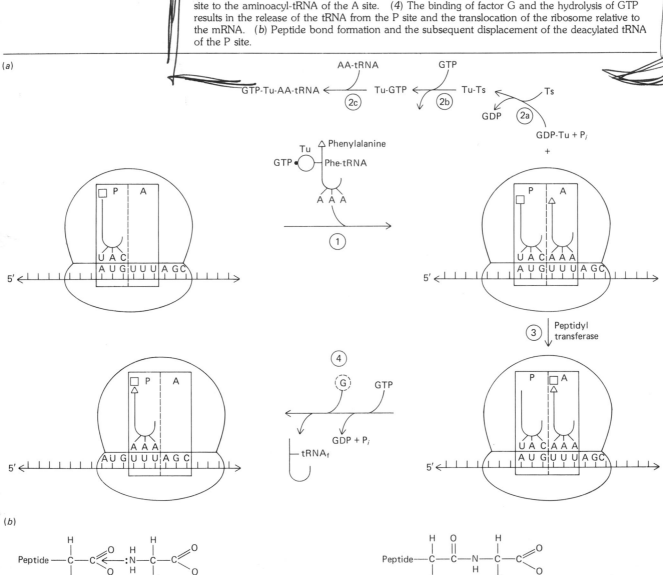

Elongation

As in the case of initiation, elongation (Fig. 11-31) in bacterial cells also requires three known soluble protein factors, termed Tu, Ts, and G. The first step in the elongation of the polypeptide involves the entry of the second aminoacyl-tRNA to the A site of the ribosome. As in the case of the binding of the initial aminoacyl-tRNA, the binding of the subsequent ones is also a relatively complex event. Before the second aminoacyl-tRNA or any of the following ones can effectively bind to the mRNA in the A site, it must combine with one of the elongation factors (Tu) and GTP (Fig. 11-31a). Though the precise function of Tu is not known, it is believed to be involved in codon recognition and/or orientation of the incoming aminoacyl-tRNA. Once the binding of Tu-GTP-AA-tRNA to the ribosome occurs, the GTP is hydrolyzed and a Tu-GDP complex is released. A second soluble elongation factor (Ts) is required to rejuvenate the Tu-GTP in preparation for its binding to another aminoacyl-tRNA to be used at another site.

The selection of the appropriate tRNA by the mRNA-ribosome complex raises some very important questions. Although any of the possible charged tRNAs can presumably enter the open A site, only one with the suitable complementary anticodon can actually combine in a specific manner with the mRNA codon being read at the time. The entire basis for specificity in the selection of particular aminoacyl-tRNAs is believed to reside in the nature of the triplet codons of the mRNA. However, the evidence suggests that the interaction between the triplets of the mRNA and tRNA does not occur with sufficient strength to account by itself for the known accuracy of protein synthesis (only about one in 10,000 amino acids incorporated incorrectly). Results from various types of analyses indicate that the binding of the tRNA to the mRNA is stabilized by a subsequent interaction between the tRNA and the ribosome. The interaction is believed to occur between the T and/or D loops of the tRNA (see Fig. 11-17) and certain sections of the ribosomal RNAs. It is believed that the interaction between the tRNA and the ribosome occurs only upon the previous association of the tRNA anticodon with the complementary mRNA codon. In other words, stabilization would occur only upon entrance of the appropriate aminoacyl-tRNA. Examination of the tertiary structure of the isolated tRNA molecule (see Fig. 11-18) indicates that the D and T loops are folded in toward the center of the molecule, a position that would make them essentially unavailable for interaction with the ribosome. It has been proposed that the association between complementary codon and anticodon triplets leads to a conformational change in the tRNA causing the loops of the molecule to shift in such a way as to promote their interaction with the ribosome.

Once the new aminoacyl-tRNA is bound to the ribosome-mRNA complex, the second step in the elongation cycle, i.e., the formation of the peptide bond between the first and second amino acids, can occur. The reaction is accomplished by the transfer of the N-formylmethionine on the tRNA of the P site to the amino acid attached to the tRNA that has just entered the ribosome. The reaction is catalyzed by one of the proteins of the large ribosomal subunit, termed *peptidyl transferase*. The formation of the first peptide bond leaves one end of the tRNA molecule of the A site still attached to its complementary codon on the mRNA and the other end of the molecule attached to a dipeptide (Fig. 11-31*b*). The tRNA of the P site is now devoid of any linked amino acid. The third step in the elongation cycle involves (1) the loss of the uncharged tRNA from the P site and (2) the movement of the ribosome one codon along the mRNA in the 3′ direction. This last step, termed *translocation*, is accompanied by the movement of the tRNA-dipeptide to the P site of the ribosome, still hydrogen-bonded to the second codon of the mRNA (Fig. 11-31*a*). The translocation requires the other elongation factor (G, also termed the *translocase*) and the hydrolysis of GTP. Since translocation involves the movement of a particle approximately 10 Å along the route, it is believed that this step is accompanied by marked conformational changes in the structure of the ribosome. Once the translocation has moved the tRNA with its attached peptide to the P site, the A site is once again open to the entry of another aminoacyl-tRNA, in this case one whose anticodon is complementary to the third codon. Once the third charged tRNA is associated with the mRNA in the A site, the dipeptide from the tRNA of the P site is transferred to the amino acid on the tRNA of the A site forming the second peptide bond and a tripeptide attached to the tRNA of the A site. Peptide bond formation is followed by translocation and the cycle is ready to begin again.

Each round of elongation requires the hydrolysis of two molecules of GTP, one during aminoacyl-tRNA selection and the other during mRNA-ribosome translocation. The GTPase responsible for this hydrolysis (and that which occurs during initiation and termination) may be one of the ribosomal proteins of the 50S subunit. The hydrolysis of GTP during elongation does not appear to be directly involved in peptide bond formation. The transfer of the peptide from the tRNA of the P site to the aminoacyl-tRNA of the A site would be expected to be, in itself, an exergonic reaction, one that need not be directly coupled to the hydrolysis of GTP in the sense that the formation of glutamine is coupled to the hydrolysis of ATP (page 133). It has been proposed that the hydrolysis of GTP is coupled to the generation of conformational changes in the protein-synthesizing machinery. In this way, its function may be analogous to that of ATP in muscle contractility (page 743). More recently it has been proposed that the steps of protein synthesis are coupled to GTP hydrolysis to provide a means whereby these steps become essentially irreversible. In this proposal by Charles Kurland, the hydrolysis of GTP during protein synthesis is roughly analogous to the hydrolysis of pyrophosphate which occurs during the synthesis of nucleic acids (page 486).

Several inhibitors acting at various steps of the elongation process have been discovered (Table 11-2). For example, two important inhibitors of the peptidyl transferase reaction, chloramphenicol and cycloheximide, are known. The former inhibitor is effective against peptide formation by prokaryotic 50S subunits, while the latter by eukaryotic 60S subunits.

Other inhibitors which distinguish between prokaryotic and eukaryotic translation are the commonly prescribed antibiotics streptomycin and tetracycline. Both selectively bind to the 30S ribosomal subunit of prokaryotes; streptomycin causes certain codons of the mRNA to be misread while tetracycline inhibits the binding of aminoacyl-tRNAs to the ribosome-mRNA complex. The streptomycin-ribosome interaction has been extensively studied and has led to interesting conclusions concerning the role of certain of the 30S ribosomal proteins. Streptomycin appears to exert its effect by binding to the ribosomal protein termed S12 (page 101), which is somehow involved in controlling the accuracy of the interaction between specific tRNAs and the ribosome-mRNA complex. Resistance by bacteria to streptomycin can be traced to changes in ribosomal proteins, particularly S12. In addition to its effect on misreading, streptomycin also interferes

TABLE 11-2
Antibiotics Inhibiting Protein Synthesis

| Antibiotic | Site of Action | | | | Specific effects |
	Cell type	Subunit	Step	Altered in resistant mutants	
Puromycin	Eu, Pro	L	R, P	———	Releases peptidyl-puromycin
Tetracyclines	Pro	S	R	———	Blocks binding in A site
Streptomycin, other aminoglycosides	Pro	S	R	Protein S12	Irreversible; blocks recognition, causing ribosome release; distorts recognition, causing misreading
Chloramphenicol, lincomycin	Pro	L	P	———	Blocks fragment binding to P site, puromycin reaction with pp-tRNA
Sparsomycin	Eu, Pro	L	P	———	Promotes fragment binding in unreactive position; blocks puromycin reaction
Erythromycin	Pro	L	P/T	23S RNA	Blocks P site, inhibiting both peptidyl transfer and translocation
Siomycin, thiostrepton	Pro	L	T, R	———	Irreversible; blocks binding of EFG + GTP, and of EFT · aa − tRNA · GTP
Fusidic acid	Eu, Pro	L	T	EFG	Blocks release of EFG and GDP
Rifamycins	Pro	Transcription		β-subunit	Blocks initiation
Streptolydigin	Pro	Transcription		Polymerase	Inhibits extension
Actinomycin D	Eu, Pro	Transcription		———	Binds to DNA

Eu, Pro = eukaryotic, prokaryotic cells; L, S = large, small subunit; R = recognition; P = peptidyl transfer; T = translocation.
SOURCE: Davis, B. D., et al., *Microbiology*, 2d ed., Harper and Row, 1973.

with the initiation process when the 30S subunit attaches to the mRNA. Once again this drug-induced deficiency is mediated by the reaction of the antibiotic with the S12 protein, indicating this protein plays some role in initiation as well. Use of a number of inhibitors has provided one of the most important tools for understanding the mechanism of protein synthesis, particularly the role of various ribosomal proteins.

As the ribosome continues to move along the message, the growing polypeptide chain, termed the *nascent* polypeptide, increases in length. The direction of elongation, i.e., from the amino end of the polypeptide to the carboxyl end, was first shown in early experiments by Howard Dintzis on cells actively engaged in the synthesis of hemoglobin. In essence, the experiment consisted of providing these cells with a radioactively labeled amino acid for a very brief period, extracting the *completed* globin chains, digesting them with trypsin, and subjecting the peptide fragments to two-dimensional chromatography to obtain their fingerprints. Since the amino acid sequences of the globin chains were known, the various peptides could be ordered according to their relative positions from the amino to the carboxyl end. When the radioactivity of the various peptides was determined, it was found that the closer the peptide was to the carboxyl end of the polypeptide, the greater was its radioactivity (Fig. 11-32). In very brief exposures to radioactivity, none of the peptides of the far amino end of *completed* chains were labeled. The amino ends of these polypeptides were unlabeled because they had been formed before the radioactive amino acids had been added. These results clearly indicate that the carboxyl end, which contains the greatest radioactivity, must be the last part of the chain to be synthesized.

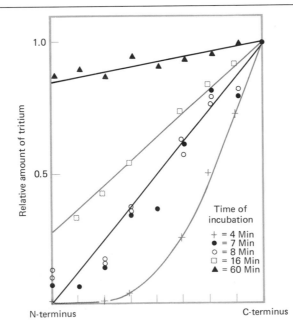

Figure 11-32. Distribution of ^{3}H-leucine among tryptic peptides of the α chain of rabbit hemoglobin after various times of incubation in the radioactive amino acid. After a four-minute incubation, the peptides, that are closest to the C-terminus of *completed* chains have the greatest radioactivity. There is no radioactivity at the N-terminus of the completed chains in the four-minute incubation. With longer periods of exposure to the label before killing the cells, radioactivity appears in the N-terminal portion, but the radioactivity of the peptides increases as they are situated closer to the C-terminus. [*From H. M. Dintzis, Proc. Natl. Acad. Sci. (U.S.), **47**:255, 1961.*]

Relative amount of tritium

Time of incubation

+ = 4 Min
● = 7 Min
○ = 8 Min
□ = 16 Min
▲ = 60 Min

N-terminus   C-terminus

Termination

As mentioned previously, 3 of the 64 possible trinucleotide codons serve in the process whereby polypeptide chains are terminated rather than in the specification of amino acids. There are no tRNAs with anticodons complementary to these triplets. When the ribosome reaches one of these codons, either UAA, UAG, or UGA, the signal is received to stop further elongation and terminate the polypeptide. Though the manner in which termination is accomplished is poorly understood, it is known to involve soluble proteins, termed *release factors*, as well as GTP. Whereas two of the release factors, R1 and R2, are involved in the recognition of the termination codons, a third release factor, termed S, is somehow involved in the release of the completed polypeptide from its attachment at its C-terminal end to the last tRNA that had entered. The hydrolysis of the protein from the tRNA is believed to be catalyzed by the same enzyme responsible for peptide bond formation. Once the termination process is complete and the completed polypeptide has been severed from its tRNA, the ribosome separates from the message and dissociates once again into its subunits in preparation for additional translation activity.

Since the three termination codons can be readily formed by single base changes from many other codons, one might expect mutations to arise that produced termination codons *within* a gene and resulted in the premature termination of the growing polypeptide chain. Mutations of this type (Table 11-3), termed *nonsense* mutations, have long been known and studied. Another way in which premature termination of chain growth can be brought about is by the addition of the antibiotic puromycin. This molecule (Fig. 11-33) resembles the 3′ end of a charged tRNA and is capable of entering the A site of the ribosome. Once in place in the site, the peptide of the P site is actually transferred to puromycin of the A site and a covalent bond is formed. The transfer of the nascent polypeptide chain to puromycin places a cap on the carboxyl end of the chain so no further amino acids can be attached. Instead, the peptidyl-puromycin complex falls off the ribosome. Puromycin has been the most important of the inhibitors in the analysis of the mechanism of protein synthesis.

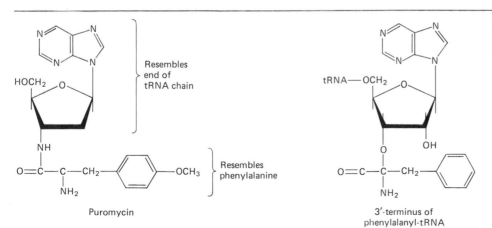

Figure 11-33. The similarity between puromycin and the 3′ terminus of phenylalanyl-tRNA.

TABLE 11-3

Original codon	Original amino acid	Mutant codon	Position changed
AAG	LYS	UAG	1st
CAG	GLN	UAG	1st
GAG	GLU	UAG	1st
UGG	TRP	UAG	2nd
UCG	SER	UAG	2nd
UUG	LEU	UAG	2nd
UAU	TYR	UAG	3rd
UAC	TYR	UAG	3rd

Polyribosomes

In the previous sections on protein synthesis, we followed the various activities that occur as a single ribosome attaches to a messenger RNA, moves down a part of its length, and finally becomes detached at the termination site. Investigations by several laboratories into protein synthesis during the early 1960s revealed that more than one ribosome is involved at one time in the translation of a single message. When a messenger RNA in the process of being translated is examined with the electron microscope (Fig. 11-34a, b), one sees a number of ribosomes attached along the length of the thread. This complex of ribosomes and mRNA is termed a *polyribosome* or *polysome*. Each one of the ribosomes initially attaches to the mRNA at the initiation codon and then moves from that point toward the 3′ end. As soon as a recently attached ribosome moves a sufficient distance along the tape, the next ribosome becomes associated with the mRNA and begins to translate the message simultaneously with its translation by previously attached ribosomes (Fig. 11-34c). The simultaneous translation of the same mRNA by several different ribosomes greatly increases the potential for protein synthesis within the cell. This is particularly true in the bacterial cell where the message often begins to be degraded even before its synthesis is complete. The more ribosomes that have a chance to attach at the initiation codon before that end of the molecule is attacked by a nuclease, the more polypeptides will be formed.

The presence and activity of polyribosomes can be determined by techniques other than ones which directly visualize the complex. The most commonly employed means for polysome analysis is the sucrose density gradient. This type of analysis can be illustrated using some of the early profiles from Alexander Rich's laboratory, where the polysomes from reticulocytes were first examined. The reticulocyte is a commonly employed cell type in studies of mammalian protein synthesis since its synthetic picture is so dominated by the production of the α and β chains of hemoglobin. If reticulocytes are gently homogenized and the cell supernatant is sedimented through a sucrose gradient, two distinct peaks are resolved (Fig. 11-35a). The peak toward the top of the gradient contains single ribosomes (*monosomes*) while the peak closer to the bottom of the tube contains polysomes. As long as the preparation is handled gently enough, those ribosomes that were present in the cell as polysomes will

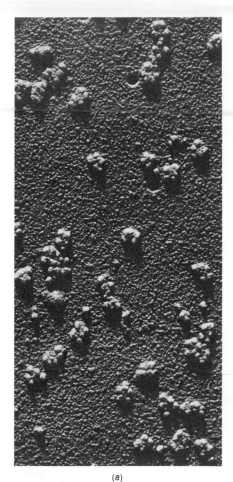

(a)

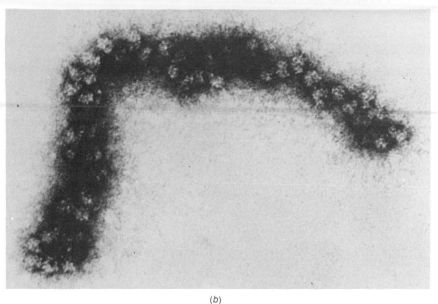

(b)

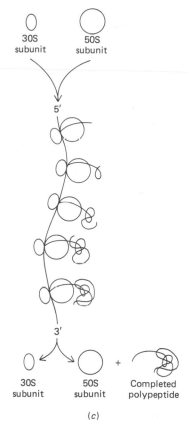

Figure 11-34. (a) Electron micrograph of metal-shadowed polyribosomes isolated from reticulocytes engaged in hemoglobin synthesis. (*Courtesy of A. Rich.*) (b) Electron micrograph of a negatively-stained polyribosome engaged in the synthesis of a polypeptide of the giant muscle protein myosin. [*From S. M. Heywood, R. M. Dowben, and A. Rich,* Proc. Natl. Acad. Sci. (U.S.), **57**:1007 (1967).] (c) Schematic representation of a polyribosome.

remain together as polysomes as they sediment through the gradient. Examination of the material of the lower peak in the electron microscope reveals its polysome nature. In reticulocytes, nearly all of the polysomes contain from 4 to 6 ribosomes (primarily 5) which reflects the fact that they are nearly all involved in the translation of only two particular mRNAs of nearly identical molecular weight. When the polysomes of another type of cell, such as a HeLa cell, is examined (Fig. 11-35a), material of a much more heterogeneous nature is found. Rather than forming a single peak as in the case of the reticulocyte, the polysomes of the HeLa cell are spread throughout the gradient. The varied distribution of the polysomes of the HeLa cell reflects the great range in molecular weights of the mRNAs. The larger the polysomes, i.e., the greater the number of attached ribosomes (generally a reflection of the length of the mRNA to which they are attached), the greater the sedimentation velocity and the farther it will move toward the bottom of the tube after a given period of time in the centrifuge. Polysomes containing as many as 40 ribosomes have been observed.

The involvement of polysomes in protein synthesis can be demon-

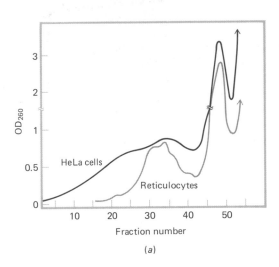

(a)

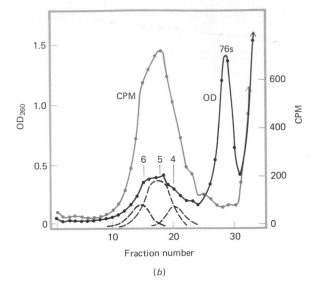

(b)

Figure 11-35. Polyribsome analysis by sucrose density gradient centrifugation. (a) A composite diagram of the absorbance profile of two gradients: one showing the sedimentation of material from reticulocytes, the other from HeLa cells. In both cases there is a large peak near the top of the tube (about fraction 48) that contains single ribosomes. Material below fraction 40 in the gradient represents polyribosomes which are much more heterogeneous in the HeLa cell than the reticulocyte. (b) When reticulocytes are incubated in ^{14}C-amino acids for 45 seconds before lysing the cells, the location of the radioactivity in the gradient indicates the role of the various polyribosome fractions in protein synthesis. The arrows labeled 4, 5, and 6 indicate the positions of polyribosomes with 4, 5, and 6 associated ribosomes. The three dashed peaks indicate the calculated amounts of polysomes containing 4, 5, and 6 ribosomes that would give the composite peak of the absorbance profile. (c) Treatment of the reticulocyte lysate with RNase for one hour before centrifugation destroys the integrity of the polyribosomes by degrading the mRNA thread. (*From A. Rich, J. B. Warner, and H. M. Goodman,* Cold Spring Harbor Symp. Quant. Biol., **28:**269, 1963.)

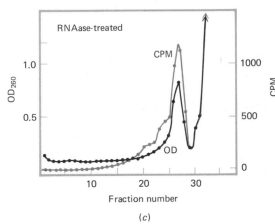

(c)

strated by adding a radioactive amino acid to the cells for a brief period just prior to homogenization. This label will be taken up and incorporated into the growing polypeptide chains that remain attached to the polysomes (Fig. 11-35b). The distribution of radioactivity in the gradient becomes an excellent measure of the size of the polysomes engaged in protein synthesis. In order to verify that material seen in a sucrose gradient does indeed represent polysomes, one of the samples is generally treated with a very low concentration of ribonuclease before centrifugation. When this is done, the material in the enzyme-treated sample sediments entirely as a single peak near the top of the gradient (Fig. 11-35c). The ribonuclease has broken the mRNA between the ribosomes, causing the integrity of the polysome to be destroyed. When this preparation is then sedimented through the gradient, all of the particles are now present as single ribosomes and thus sediment together in a single peak. If radioactive amino acids had been added prior to homogenization, the radioactivity would also have shifted from the heavy regions of the gradient into the monosome peak.

A Comparison between Eukaryotes and Prokaryotes

The analysis of the mechanism of protein synthesis, though revealing many significant features, remains incomplete. The process is so complex that it has been particularly difficult to obtain precise information as to the roles of the particular components. The least understood component in the process is the ribosome. It is clear that the rRNAs as well as numerous of the 50 or so ribosomal proteins are involved in one way or another in protein synthesis, but, at the present time, only a few poorly understood, tentative roles have been assigned. Nothing is known about the possible conformational changes that the RNA and proteins of the ribosome must undergo. Not only is the process of protein synthesis complex, but it is very rapid. It is estimated that the polypeptide chains of *E. coli* are produced at the rate of about 20 amino acids per second. Though the actual mechanism of protein synthesis is strikingly similar between prokaryotic and eukaryotic cells, there are basic differences in the nature of the ribosomes (page 100) as well as the nature of the mRNAs being translated. Whereas the messenger RNAs of eukaryotic cells contain only the information for one polypeptide chain, those of bacterial mRNAs (or viral mRNAs being translated in bacterial cells) often include the information for more than one chain. For example, one of the bacterial mRNAs we will discuss at some length in Chapter 13 contains the nucleotide sequence for β-galactosidase as well as two other proteins. Messages of this type are said to be polycistronic. The term *cistron* is one to emerge from the fine structural analsysis of the rII locus by Benzer (page 478) and for our purposes is synonomous with the term *gene*. Polycistronic messages are ones transcribed from *contiguous*, i.e., adjacent genes. The RNA polymerase responsible for the production of the mRNA simply moves along from one gene to the next without stopping to interrupt the message (Fig. 13-4). When one analyzes the nature of the synthesis of the proteins made from polycistronic mRNAs, one finds that each of the "parts" of the message has its own initiation codons to which ribosomes can attach and its own termination codons from which they can detach. In other words, ribosomes can attach to any of the initiation codons, independently of one another, and carry out the translation activity of the particular polypeptide to which that initiation codon belongs.

A particularly unusual case related to the difference between eukaryotic and prokaryotic messages can be seen in the mRNAs produced by certain of the viruses which infect eukaryotic cells. Polio virus, for example, produces one long mRNA containing the information for more than one polypeptide, but apparently because it is operating in a eukaryotic cell, the entire message is translated into one long *polypeptide* chain, which only later becomes cleaved into its separate component proteins. Other important differences between prokaryotic and eukaryotic mRNAs result from special processing steps which occur in eukaryotic cells, steps which appear to chop large mRNA precursors into smaller molecules and

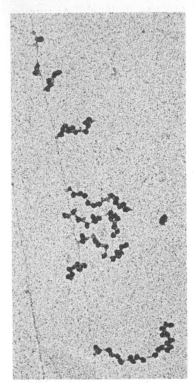

modify the 5′ and 3′ ends. These events are discussed at length in the next chapter.

Another basic difference between prokaryotic and eukaryotic protein synthesis concerns the temporal relation between the transcriptional and translational processes. In eukaryotic cells, as will be described in detail in the next chapter, there are a considerable number of steps interposed between the synthesis of the RNA in the nucleus and its translation into protein in the cytoplasm. In contrast, protein synthesis begins on mRNA templates of bacteria well before the mRNA has even been completed. RNA synthesis proceeds in the same direction as RNA translation, i.e., from the 5′ to the 3′ end. Consequently, as soon as an RNA molecule has begun to be produced, the 5′ end is available upon which ribosomes can attach. The simultaneous activities of transcription and translation of bacterial messages is elegantly revealed in the electron micrographs of Oscar Miller and his coworkers (Fig. 11-36). Examination of the micrograph of Fig. 11-36, reveals the presence of the various components used in the storage and transfer of information within the cell. One can resolve the DNA thread upon which transcription is occurring, the mRNAs of increasing length from one end of the DNA to the other, and the ribosomes attached to each nascent mRNA forming polyribosomes. The nascent protein chains are not visible in the micrograph. Fig. 11-37 shows a micrograph of protein synthesis from a eukaryotic cell (a cell from the silk gland of the silk moth). Since protein synthesis in eukaryotic cells occurs in the

Figure 11-36. Electron micrograph of portions of *E. coli* chromosomes engaged in transcription. The DNA is seen as faint lines running the length of the photos, while the nascent mRNA chains are seen to be attached at one of their ends, presumably by an RNA polymerase molecule. The particles associated with the nascent RNAs are ribosomes in the act of translation — in bacteria, transcription and translation occur simultaneously. The RNA molecules increase in length reflecting the distance from the initiation site. Structures on the DNA near the point of initiation are believed to be RNA polymerases. The strand of DNA on the left which lacks nascent RNA represents inactive, i.e., repressed, DNA. (*From O. L. Miller, Jr., B. A. Hamkalo, and C. A. Thomas, Jr., Science, 169:392, 1970. Copyright by the American Association for the Advancement of Science.*)

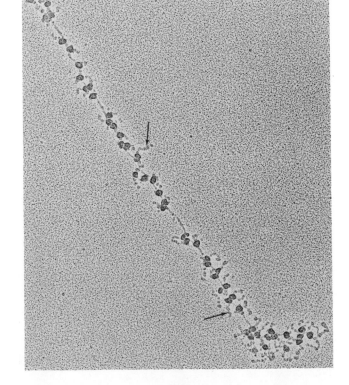

Figure 11-37. Electron micrograph of a polyribosome isolated from cells of the silk gland of the silk worm, cells that produce large quantities of the silk protein fibroin. This protein is of large molecular weight and therefore visible in the electron micrographs (arrows point to nascent polypeptide chains). (*Courtesy of S. L. McKnight and O. L. Miller, Jr., copyright reserved.*)

526

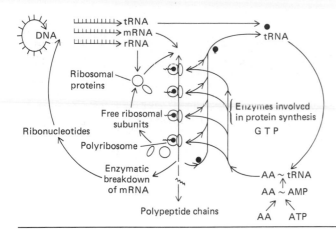

Figure 11-38. Schematic view of the role of RNA in protein synthesis. (*From J. D. Watson, "The Molecular Biology of the Gene," 3d ed., W. A. Benjamin, 1976.*)

cytoplasm, totally removed from the chromosomes, no DNA is visible as was present in Fig. 11-36. However, since the silk protein being synthesized is of particularly large molecular weight and fibrous in nature it is clearly visible extending from each of the ribosomes of the polyribosome. The development by Oscar Miller of the techniques to visualize transcription and translation, have provided a fitting visual demonstration of processes that had been discovered and described over the previous couple of decades. The photographs of "genes in action" represent precisely the type of portrait expected on the basis of the bulk of the nonvisual evidence. An overall summary of the basic information presented in this chapter is shown in Fig. 11-38.

REFERENCES

Benzer, S., Sci. Am. **206**, 70–84, Jan. 1962. "The Fine Structure of the Gene."

Bosch, L., ed., *The Mechanism of Protein Synthesis and Its Regulation.* North Holland, 1972.

Brenner, S., Jacob, F., and Meselson, M., Nature, **190**, 576–581, 1961. "An Unstable Intermediate Carrying Information from Genes to Ribosomes for Protein Synthesis."

Brimacombe, R., Nierhaus, K. H., Garrett, R. A., and Wittman, H. G., Prog. Nucleic Acid. Res. and Mol. Biol. **18**, 1–44, 1976. "The Ribosome of *Escherichia coli.*"

Brown, D. D., chairman, symposium on "Genome Organization in Higher Organisms." Federation Proc. **35**, 11–35, 1976.

Caskey, C. T., Adv. Protein Chem. **27**, 243–276, 1973. "Peptide Chain Termination."

Chambon, P., Ann. Rev. Biochem. **44**, 613–638, 1975. "Eukaryotic Nuclear RNA Polymerases."

Crick, F. H., Sci. Am. **207**, 66–74, Oct. 1962 "The Genetic Code."

———, Sci. Am. **215**, 55–63, Oct. 1966. "The Genetic Code III."

Dunn, J. J., ed., "Processing of RNA," Brookhaven Symposium in Biology, Number 26, 1974.

Fraenkel-Conrat, H., Sci. Am. **211**, 47–54, Oct. 1964. "The Genetic Code of a Virus."

The Genetic Code. Cold Spring Harbor Symp. Quant. Biol. **31**, 1966.

Hall, B. D., and Spiegelman, S., Proc. Natl. Acad. Sci. (U.S.) **47**, 137–146, 1961. "Sequence Complementarity of T2-DNA and T2-Specific RNA."

Hoagland, M. B., Stephenson, M. L., Scott, J. F., Hecht, L. I., and Zamecnik, P. C., J. Biol. Chem. **231**, 241–257, 1958. "A Soluble Ribonucleic Acid Intermediate in Protein Synthesis."

Holley, R. W., Sci. Am. **214**, 30–39, Feb. 1966. "The Nucleotide Sequence of a Nucleic Acid."

Jacob, F., and Wollman, E. L., Sci. Am. **204**, 92–107, June 1961. "Viruses and Genes."

Khorana, H. G., Harvey Lectures **62**, 79–106, 1966. "Polynucleotide Synthesis and the Genetic Code."

Kim, S.-H., et al., Science **185**, 435–440, 1974. "Three-Dimensional Tertiary Structure of Yeast Phenylalanine Transfer RNA."

Kurland, C. B., Ann. Rev. Biochem. **46**, 173–200, 1977. "Structure and Function of the Bacterial Ribosome."

Lane, C., Sci. Am. **235**, 60–71, Aug. 1976. "Rabbit Hemoglobin from Frog Eggs."

Leder, P., Adv. Protein Chem. **27**, 213–242, 1973. "The Elongation Reactions in Protein Synthesis."

Lederberg, J., and Tatum, E. L., Cold Spring Harbor Symp. Quant. Biol. **11**, 113–114, 1946. "Novel Genotypes in Mixed Cultures of Biochemical Mutants of Bacteria."

Lewin, B., *Gene Expression,* vol. 1. J. Wiley, 1974.

Losick, R., and Chamberlin, M., eds., *RNA Polymerase,* Cold Spring Harbor, 1976.

McElroy, W. D., and Glass, B. H., eds., *The Chemical Basis of Heredity.* Johns Hopkins University Press, 1956.

The Mechanism of Protein Synthesis. Cold Spring Harbor Symp. Quant. Biol. **34**, 1969.

Miller, O. L., Jr., Sci. Am. **228**, 34–42, March 1973. "The Visualization of Genes in Action."

Nirenberg, M., Sci. Am. **208**, 80–94, March 1963. "The Genetic Code."

Nobel Lectures in Physiology-Medicine, 1901–1970, 4 vols. Elsevier, 1970.

Nomura, M., Tissieres, A., and Lengyel, P., eds., *Ribosomes,* Cold Spring Harbor, 1974.

Rich, A., Sci. Am. **209**, 44–53, Dec. 1963. "Polyribosomes."

—— and RajBhandry, U. L., Ann. Rev. Biochem. **45**, 805–860, 1976. "Transfer RNA: Molecular Structure, Sequence and Properties."

——, and Kim, S.-H., Sci. Am. **238**, 52–62, 1978. "The Three-Dimensional Structure of Transfer RNA."

Shine, J., and Dalgarno, L., Nature **254**, 34–38, 1975. "Determinant of Cistron Specificity in Bacterial Ribosomes."

Sigler, P. B., Ann. Rev. Biophys, and Bioeng. **4**, 477–528, 1975. "An Analysis of the Structure of tRNA."

Spiegelman, S., Sci. Am. **210**, 48–56, May 1964. "Hybrid Nucleic Acids."

Stent, G. S., *Molecular Genetics*. W. H. Freeman, 1971.

Synthesis and Structure of Macromolecules. Cold Spring Harbor Symp. Quant. Biol. **28**, 1963.

Taylor, J. H., *Selected Papers on Molecular Genetics*. Academic, 1965.

———, ed., *Molecular Genetics*. Academic, 1967.

Transcription of Genetic Material. Cold Spring Harbor Symp. Quant. Biol. **35**, 1970.

Volkin, E., and Astrachan, L., Virology **2**, 433–437, 1956. "Intracellular Distribution of Labeled Ribonucleic Acid after Phage Infection of *Escherichia coli*."

Watson, J. D., *Molecular Biology of the Gene*, 3d ed. W. A. Benjamin, 1976.

Weissbach, H., and Ochoa, S., Ann. Rev. Biochem. **45**, 191–216, 1976. "Soluble Factors Required for Eukaryotic Protein Synthesis."

Yanofsky, C., Harvey Lectures **61**, 145–168, 1965. "Gene Structure and Protein Structure."

———, Sci. Am. **216**, 80–94, May 1967. "Gene Structure and Protein Structure."

Zamecnik, P. C., Harvey Lectures **54**, 256–281, 1960. "Historical and Current Aspects of the Problem of Protein Synthesis."

CHAPTER TWELVE
Gene Expression in Eukaryotic Cells

RIBOSOMAL RNA

The study of the synthesis of ribosomal RNA has progressed gradually over a considerable number of years until at the present time it is the best understood of any gene transcript. The choice made by investigators in concentrating on rRNA synthesis is a logical one considering the tremendous number of rRNA-containing ribosomes present in a cell and their importance. There are several basic topics to be covered in this section; these include the nature of the DNA that serves as the template, the steps leading from the transcription of the rRNA genes to the incorporation of the rRNAs into mature ribosomes, and the nature of the cell organelles involved in these activities.

The Genes for Ribosomal RNA

One of the earliest approaches to the study of RNA metabolism was to cytochemically demonstrate its presence in various regions of the cell. The two regions of an active cell that generally stain most intensely for RNA are the cytoplasm and the *nucleolus* (plural *nucleoli*), a nuclear organelle whose structure will be described shortly. Once the various types of RNA were described, it became apparent that the RNA of the

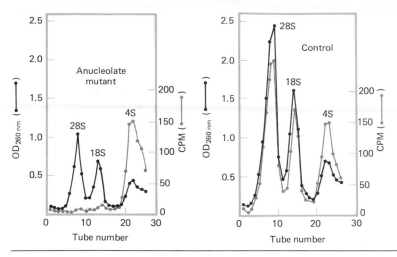

Figure 12-1. Sucrose density gradients of RNA extracted from anucleolate (left) and control (right) embryos of *Xenopus* after incubation for a long period in a radioactive precursor. The long incubation period ensures that radioactivity will accumulate in the stable ribosomal RNA species. The absorbance (OD) reflects the relative quantities of rRNA and tRNA in these two types of embryos while the radioactivity provides a measure of the relative synthesis by the embryos of these stable RNA species. It is apparent that the anucleolate mutants do not synthesize ribosomal RNA. The ribosomes that are present in these mutants (measured by absorbance) were produced in the oocyte within the female prior to fertilization. [*From D. D. Brown and J. B. Gurdon,* Proc. Natl. Acad. Sci. (U.S., **51:**141, 1964.]

nucleolus was of the ribosomal variety, but it was less clear whether ribosomal RNA was actually produced in the nucleolus or was simply transported there from its site of synthesis elsewhere in the nucleus.

During the 1960s an important series of observations were made on nucleolar function and rRNA synthesis in an amphibian, *Xenopus*, the clawed toad. As will be obvious from the remainder of the discussion, this organism has continued to be of great importance in this area of research. In 1958, a mutation in *Xenopus* was described which affected the nucleolus. Whereas the cells of normal animals had two nucleoli, those of heterozygotes had only one nucleolus. When two heterozygotes were mated, one quarter of the offspring were homozygous for the mutation; these individuals had cells totally lacking a nucleolus and did not survive past a relatively early embryonic stage. Analysis of these *anucleolate* embryos by Donald Brown and John Gurdon indicated that, unlike normal embryos, they were not engaged in the synthesis of ribosomal RNA (Fig. 12-1). Messenger RNA and transfer RNA synthesis did not seem to be affected by the mutation; the effect was specific for rRNA. Interest in the anucleolate mutants then focused on the genetic basis for the absence of nucleoli and rRNA synthesis.

The analysis of genetic function through the use of mutation can be a difficult undertaking. Mutations in single genes are often *pleiotropic*, i.e., cause a wide variety of phenotypic disturbances, and it can be very difficult to decide which of these alterations is the primary consequence of the mutation and which are secondary manifestations. Using molecular hybridization experiments with purified rRNA, Hugh Wallace and Max Birnstiel were able to determine the genetic nature of the anucleolate condition. They found that approximately 0.1% of the DNA extracted from wild-type individuals was complementary to rRNA while there was essentially no hybridization to this RNA when DNA from anucleolate embryos was tested (Fig. 12-2). The DNA of the wild type that is responsible for rRNA synthesis is termed *rDNA*. Calculations indicated that there are approximately 450 copies of the genes for rRNA in a normal

Figure 12-2. Hybridization of radioactive 28S ribosomal RNA to DNA extracted from normal embryos (2n), heterozygotes (1n), and *anucleolate* homozygotes (0n). The amount of radioactive RNA bound to the DNA is a relative measure of the amount of ribosomal DNA present in the genome. It appears that the homozygous embryos, i.e., the anucleolate mutants, have virtually no ribosomal DNA while those with one nucleolus have half the amount of controls. (*From H. Wallace and M. L. Birnstiel,* Biochim. Biophys. Acta, **114:**306, 1966.)

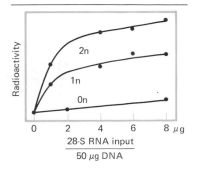

Figure 12-3. (a) Light micrograph of a section through liver tissue showing portions of several parenchymal cells and their nuclei. Unlike most cells of the body, these liver cells are often characterized by having more than one nucleus (they are often multi-nucleate) as well as being polyploid (more than two sets of chromosomes per nucleus). Each of the nuclei shown in this micrograph contains one or more distinct nucleoli as well as scattered clumps of chromatin. (*Courtesy of E. Leduc.*) (b) Electron micrograph of an oocyte nucleolus from the amphibian *Triturus*. The compact fibrous core is surrounded by the granular cortex. A somewhat different nucleolar organization is seen in the micrograph of Fig. 13-11. (*Courtesy of O. L. Miller, Jr. and B. R. Beatty.*)

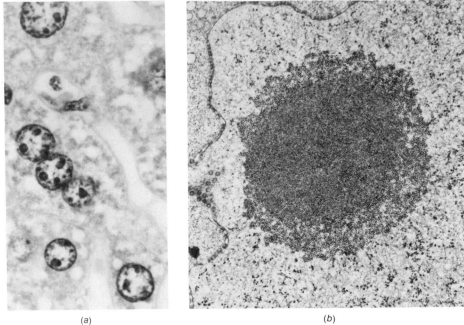

(a)   (b)

haploid genome of *Xenopus* whereas the genome of the anucleolate mutants contained a deletion for all of these sequences. The primary effect of the deletion was the inability of these embryos to manufacture rRNA due to the lack of its template. To understand the relationship of this

Figure 12-4. The nucleolar organizer region of *Drosophila* is that section of one of the chromosomes with which the nucleolus is associated. Mutant strains were isolated that had varying numbers of copies of this region. The deletion or duplication could be monitored by examination of the giant larval chromosomes where the number of bands in the NO region can be directly observed. DNA was extracted from these various strains of flies and its content of ribosomal DNA was determined by hybridization with ribosomal RNA. As shown in the figure, the degree to which rRNA bound to DNA under saturating conditions was proportional to the dosage of the NO (given in parentheses). The presence of a particular DNA sequence could be correlated with the appearance of the banding pattern of the giant chromosomes. [*From F. M. Ritossa and S. Spiegelman*, Proc. Natl. Acad. Sci. (U.S.), **53**:742, 1965.]

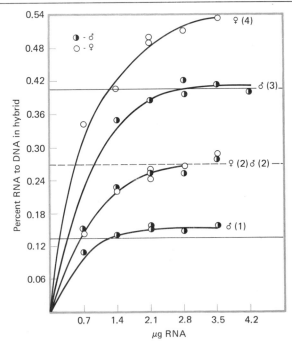

metabolic defect to the absence of a nucleolus, some background information on this organelle must be provided.

Nucleoli are large organelles that can be either spherical or irregular in shape. Since nucleoli (Fig. 12-3) are clearly visible in the light microscope, their behavior has been followed over a long period of time. In many species, such as *Xenopus*, typical cells contain two nucleoli. In other species either one nucleolus (e.g., humans) or more than two nucleoli (e.g., *Chironomus*) can be seen. The nucleolus is an example of an organelle whose presence is usually related to the cell's division cycle. Except in the case of various algae and fungi, nucleoli typically become dispersed during prophase of mitosis and reappear again during the following telophase. One of the basic observations in cytology is the association of nucleoli with specific chromosomes. In those species which contain only two nucleoli per diploid set of chromosomes, only one chromosome of the haploid set bears the specific site where association with the nucleolus occurs. The identification of the chromosome(s) involved in a relationship with the nucleolus is best seen during prophase when the chromosomes become distinctly visible before the nucleolus has disappeared. In many cases, a constriction (termed the *secondary constriction* to distinguish it from the constriction at the centromere) can be seen on the chromosome(s) at the site where the nucleolus is in contact (actually surrounding) the chromosome. This region of the chromosome has been known as the *nucleolar organizer region* in the cytological literature. Based on the type of molecular hybridization experiments described below, and another one illustrated in Fig. 12-4, the nucleolar organizer regions of chromosomes have been identified as the sites which contain the ribosomal DNA.

Since there are several hundred genes coding for rRNA in *Xenopus*, and since all of them can be removed by a single deletion, one would expect that the rDNA of *Xenopus* exists in a large cluster. This can be shown in numerous ways, most graphically by the technique of in situ hybridization. When radioactive 18S rRNA or 28S rRNA (or both) is incubated with a set of *Xenopus* chromosomes whose DNA has been denatured (page 466), the radioactivity becomes bound to one region on both members of one of the 18 pairs of homologous chromosomes (Fig. 12-5). That region is the nucleolar organizing region.

Since both 18S and 28S rRNA hybridize with the same region of the chromosome, the question was raised about the arrangement of the genes for these two rRNAs in the DNA of that chromosome: Do they alternate with each other or are they organized into separate blocks of sequences? The answer to this question was first clearly obtained by molecular hybridization experiments which examined the size of a DNA fragment to which both species of rRNA can still attach. Consider the two possible organizations. If genes for 18S (*or* 28S) rRNA exist together within a large DNA molecule, then rDNA-containing molecules could be fragmented into relatively large pieces (several genes in length) and they would still be capable of hybridizing only with the 18S or 28S rRNA molecules, but not both. However, if the DNA that coded for the two rRNA species alternated, then these same sized fragments would be capable of binding both species of RNA. Analysis of the hybridization ability of various sized fragments showed conclusively that the two sequences

Figure 12-5. The localization of ribosomal DNA sequences by in situ hybridization. Ribosomal DNA was isolated and used as a template in vitro to synthesize ³H-cRNA (complementary RNA). This labeled RNA was then used to hybridize with cytological preparations of metaphase chromosomes of *Xenopus*. Radioactivity bound to the chromosomes (arrow) is localized over the terminal secondary constriction, i.e., the constriction which marks the nucleolus organizer region. Since the NO regions of the two members of a homologous pair of chromosomes are associated with one another, the radioactivity appears localized in one site on the set of chromosomes. (*From M. L. Pardue,* Cold Spring Harbor Symp. Quant. Biol., ***38**:476, 1973.*)

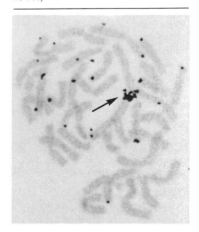

alternated with one another. It was found that, in order to obtain fragments capable of binding only one RNA species, the rDNA had to be broken into pieces of the same approximate size as one RNA molecule itself. The alternation of the two sequences has important consequences for the synthesis of ribosomal RNA as will be described in the next section.

When radioactivity labeled 5S rRNA is used as a probe in in situ hybridization experiments, an entirely different localization pattern of complementary DNA within the chromosomes is revealed. Rather than becoming localized in the nucleolar organizing region of the *Xenopus* chromosomes, it becomes localized on nearly all of the chromosomes particularly at their tips (the *telomeres*). Hybridization analysis has indicated that there are approximately 24,000 copies of the 5S rRNA gene in a haploid set of *Xenopus* chromosomes and the synthesis of this RNA is not related to the synthesis of the 18S and 28S species, even though all ribosomal RNAs must come together during the formation of mature ribosomes.

One of the properties that sets rRNA apart from other cellular RNAs is its particularly high G + C content, approximately 62%. Since the rRNA has a high G + C content, the same would be expected from rDNA, upon

Figure 12-6. Buoyant density gradient centrifugation involving *Xenopus* ribosomal DNA. (*a*) Cesium chloride gradient showing the position of the bulk of the DNA at 1.698 g/cc and a minor peak at 1.723 g/cc. The dotted peak is marker DNA from a different organism. (*b*) Demonstration that the minor peak at 1.723 g/cc contains rDNA as indicated by the preferential hybridization of ^{14}C-labeled 28S rRNA to DNA in the gradient at this position. The minor peak containing the rDNA does not appear in the absorbance profile unless the scale is expanded as in *a*. [(*a*) and (*b*) from M. L. Birnstiel, H. Wallace, J. L. Sirlin, and M. Fischberg, Natl. Cancer Inst. Mono., **23**:438,440, 1966.] (*c*) Comparison of the profiles obtained with DNA from the germinal vesicle of the oocyte (upper trace) and that obtained from somatic DNA (lower trace). The band at 1.679 is a dAT polymer used as a marker, the peak at 1.699 contains the non-rDNA, and that at 1.729 is the rDNA which has been selectively amplified in the oocyte. (*From D. D. Brown and I. B. Dawid, Science,* **160**:272, 1968. *Copyright by the American Association for the Advancement of Science.*)

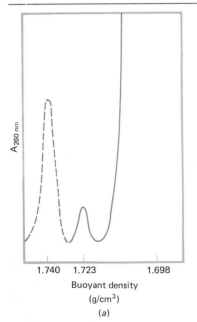

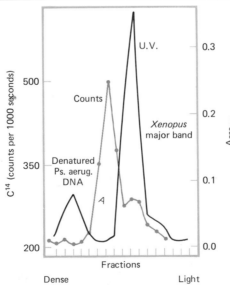

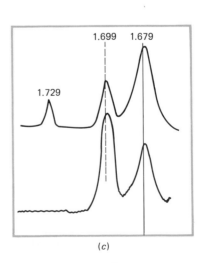

(*a*)

(*b*)

(*c*)

which it is transcribed. This fact has proved of particular importance in rRNA research and led directly to the first isolation of a specific eukaryotic gene. If rDNA is particularly unusual in its high G + C content, then it would be expected to separate from the bulk of the DNA when subjected to cesium chloride density gradient centrifugation (page 455). This has proved to be the case. When DNA extracted from *Xenopus* is centrifuged in this manner, two distinct bands of DNA are detected, a main band having a density of about 1.699 g/cc and a very minor band whose density corresponds to 1.723 g/cc (Fig. 12-6*a*). The presence of rDNA within this minor band can be verified by its ability to hydridize to rRNA (Fig. 12-6*b*); the main band DNA lacks this ability.

The isolation of rDNA and its subsequent analysis received a great boost when it was found that oocytes of *Xenopus* (as well as various other organisms) had an unusually large amount of rDNA. This is apparent in the marked increase in the amount of DNA present in the heavier band (Fig. 12-6*c*) when oocyte DNA is subjected to density centrifugation as compared to DNA extracted from other cell types. Before one can understand the basis for this situation, we need to consider the nature of the unfertilized amphibian egg and the process of oogenesis by which it is produced.

Amphibian eggs are giant cells by most standards, measuring approximately 1.5 mm in diameter at the time of fertilization. If one examines the cytoplasm of the egg it is found to contain large numbers of ribosomes which are needed to carry out the protein synthesis required by the early embryo as it develops. One *Xenopus* egg is estimated to contain approximately 10^{12} ribosomes. If one examines the ovary of these animals with the intention of finding the cells from which these giant eggs initially develop, it is found that they begin as oogonia approximately 50 μm in diameter. The conversion from an oogonia to a mature oocyte is accomplished with an increase in volume of approximately 27,000 fold (Fig. 12-7*a*), a process accomplished under the direction of a single nucleus. Even though one would expect an oocyte (which has a tetraploid number of chromosomes) to contain approximately 1900 copies (4×450 copies) of the genes for rRNA, this number is still drastically below that needed to produce 10^{12} ribosomes in the allotted period of time (a couple of years). In fact, based on normal rates of RNA synthesis, it would take 1900 copies of rDNA approximately 500 years to produce the required number of ribosomes. The evolutionary solution to this problem has been to increase the amount of rDNA by a selective replication process—only rDNA is amplified. One of the consequences of the additional rDNA is the formation of a greatly increased number of nucleoli within the large oocyte nucleus. Whereas the average *Xenopus* cell has two nucleoli, the oocytes contain over a thousand of these organelles (Fig. 12-7*b*), each one actively producing rRNA.

The reason for the relatively extensive discussion of the increased amount of rDNA of oocytes is that it brings up one of the most basic questions in molecular biology, that of the constancy of the DNA among various cells. Is the selective amplification of a particular DNA sequence of wide-

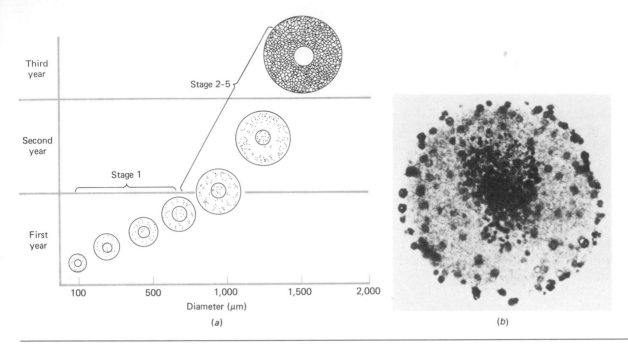

(a)

(b)

Figure 12-7. (*a*) Diagram showing the tremendous growth of an amphibian oocyte over a period of three years. (*From S. Wischnitzer.*) (*b*) Photomicrograph of an isolated nucleus from a *Xenopus* oocyte stained to reveal the hundreds of nucleoli. (*From D. D. Brown and I. B. Dawid, Science, **160**:272, 1968. Copyright by the American Association for the Advancement of Science.*)

spread occurrence in different cells? Is this the manner by which different cells are able to synthesize large amounts of different proteins, i.e., they increase the number of DNA templates for those RNAs that are most needed? As we will see in a later section (page 579), selective gene amplification is not of general occurrence, but rather seems to be restricted to rDNA in oocytes and a peculiar type of insect chromosome which we will not discuss.

Ribosomal RNA Synthesis and Processing

Whereas the nature and organization of the rDNA has been most intensively studied in *Xenopus*, the synthesis of ribosomal RNA and the formation of ribosomes is best understood as the result of studies on mammalian cells (particularly human HeLa cells by Robert Perry, by Sheldon Penman, and by James Darnell and their coworkers) growing in culture. Human ribosomes, as with all eukaryotic ribosomes examined, have four species of rRNA: the large subunit contains a 28S, 5.8S, and 5S molecule while the small subunit contains an 18S molecule. As in the case of *Xenopus*, the 5S rRNA is synthesized from entirely different sites in the chromosomes from the other rRNA species, and will not concern us for the present. As will be described shortly, the other three rRNA molecules of the human ribosome are synthesized in what is originally one large precursor molecule. This precursor, termed the 45S RNA, contains approximately 14,000 nucleotides, nearly twice the number required for its three final products. The nature of the *processing* steps by which this conversion is accomplished has been one of the central problems in the molecular biology of eukaryotic cells and is fairly well understood.

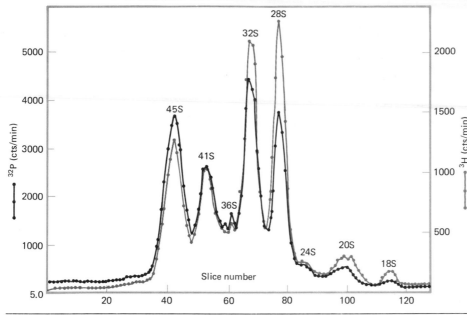

Figure 12-8. A profile of the various species of RNA found in the HeLa cell nucleolus as revealed by acrylamide gel electrophoresis. One of the lines traces the radioactivity present within methyl groups which are added to sections of the rRNA precursor after transcription. The other line indicates the radioactivity in ^{32}P that had been incorporated into the nucleotide precursors. Cells were incubated in both isotopes for three hours prior to extraction. (*From R. A. Weinberg and S. Penman, J. Mol. Biol.,* **47***:169, 1970.*)

Since nucleoli are the sites of rRNA synthesis and ribosome formation, they provide an excellent source for the isolation of purified rRNA in various stages of its processing. When purified nucleoli are prepared, and the RNA extracted and fractionated, a variety of distinct RNA species are detected (Fig. 12-8). This collection of molecules includes species of 45S, 41S, 32S, 28S, 20S, 18S, and 5.8S. Since ribosomal RNAs are heavily methylated molecules, one of the best ways to label them is to incubate cells in the presence of radioactive methionine (labeled in methyl position). Since methionine acts as a methyl donor, i.e., a compound from which methyl groups can be transferred to the appropriate receptor, ribosomal RNA and its precursors become particularly radioactive (Fig. 12-8). If HeLa cells are incubated in ^{14}C-methionine, the first species to become labeled is the 45S precursor. Since the 45S species is the one actually produced from the DNA template, it is referred to as the *primary transcript,* while the segment of DNA on which it is transcribed is termed a *transcriptional unit.* After a matter of minutes, radioactivity appears in various of the other species (Fig. 12-9) indicating the breakup of the large precursor into products that will lead to the formation of the mature rRNA species. The information from Fig. 12-9 is summarized in Fig. 12-10.

There are several questions to consider. How are the three final rRNA products positioned within the much larger precursor? What are the steps that occur during the processing of the precursor molecules and how are they controlled? How is the processing of the RNA related to the formation of mature ribosomes? What is the role of the nucleolus in this process?

The disposition of the final rRNA molecules within the 45S precursor has been determined using various techniques including fingerprinting of enzymatic digests, molecular hybridization, and electron microscopy. We will consider only the last of these methods. When rRNA molecules are purified

Figure 12-9. Sedimentation analysis of the RNA of various cellular fractions at various times following a ten-minute incubation with ^{14}C-methionine, a compound which donates labeled methyl groups to the rRNA precursor. The continuous line represents the absorbance of each cellular fraction (which does not change with time), the dotted line gives the radioactivity at various times during the chase. The graphs of nucleolar RNA (upper profiles) show the synthesis of the rRNA precursor and its subsequent conversion to a 32S species, a precursor of the 28S molecule. The other major product of the 45S precursor leaves the nucleolus very rapidly and, therefore, does not appear prominently in the profile. The lower profile shows the time course of the appearance of the mature rRNA molecules in the cytoplasm. The 18S rRNA appears in the cytoplasm well in advance of the larger species which correlates with its rapid exodus from the nucleolus. (*From H. Greenberg and S. Penman, J. Mol. Biol.,* **21:**531, 1966.)

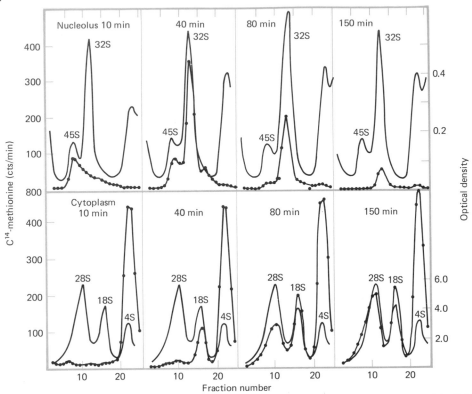

and present in solution, they assume a relatively complex shape as a result of interactions among nucleotides located at various places within the molecule (Fig. 12-11). A considerable portion of the molecule exists in double-

Figure 12-10. A summary of the data of Fig. 12-9. The total radioactivity in the various RNA species was measured and is plotted as a function of time. (*From H. Greenberg and S. Penman,* J. Mol, Biol., **21:** 532, 1966.)

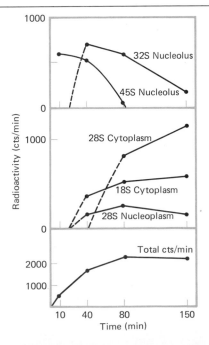

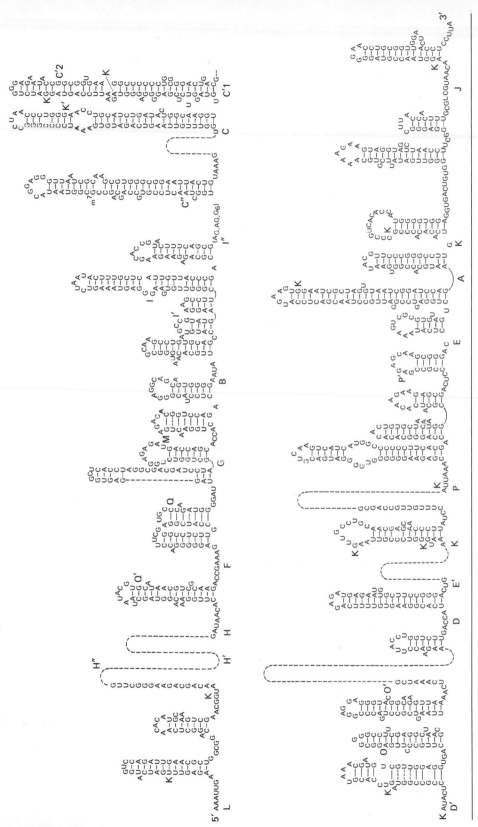

Figure 12-11. Primary and hypothetical secondary structure of *E. coli* 16S rRNA. This figure is included to illustrate the extensive base pairing potential that exists between sequences within ribosomal RNA molecules. (*From H. G. Wittmann, Europ. J. Biochem.,* **61**:2, *1976.*)

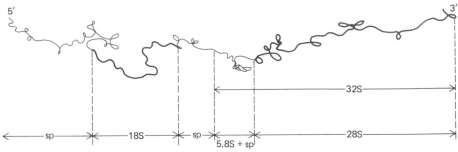

Figure 12-12. Topographical map of HeLa cell 45S RNA. The 28S sequence is shown at the 3′ end. The 5.8S sequence is in the 5′ direction from the 28S sequence, but its precise location in the overall 32S precursor fragment is not known. Transcribed spacers are indicated by sp. (*From B. E. H. Maden, Trends in Biochem. Sci.,* **1**:*150, 1976*.)

stranded sections. A comparison between the structure of the 45S precursor and the 18S and 28S molecules allows one to locate the positions of the latter species within the precursor. The determination of the 5′ and 3′ end of the molecule has been controversial in the past, but is now known to occur as shown in Fig. 12-12. The 5.8S molecule, whose presence has more recently been discovered, has not yet been visualized in the electron microscope.

Those regions of the 45S precursor molecule which do not become parts of the final products are referred to as the *transcribed spacer* sequences. The transcribed spacer sequences account for approximately half of the molecule and they are rapidly destroyed during the processing reactions. The presently accepted processing scheme is illustrated in Fig. 12-13. The first reaction appears to be the cleavage of the primary transcript toward its 5′ end to remove the outside spacer on that side of the molecule. The product formed is the 41S intermediate (see Fig. 12-8) which can be detected in labeled nucleolar preparations after a relatively short chase period. The next cleavage step is believed to be the one that separates the 28S and 18S sections by the formation of 32S and 20S intermediates. The final steps involve the separation of the 28S and 5.8S sections from the 32S molecule as well as the trimming away of smaller spacer sequences from these and the 18S RNA to produce the final products. The 5.8S and the 28S molecules remain noncovalently associated with each other in the ribosome.

At the present time very little is known about the underlying mechanisms involved in the complex, precision-requiring steps in the processing of the primary rRNA transcript. The breakage of the RNA strands is accomplished by

Figure 12-13. The steps occurring during the maturation of the primary rRNA transcript. The arrows a, b, c′, and c″ indicate the sites and the temporal order of endonuclease attack during maturation as described in the text. (*From A. A. Hadjiolov, Trends in Biochem. Sci.,* **2**:*85, 1977*.)

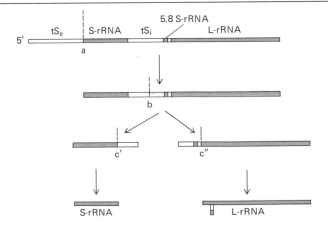

one or more enzymes that hydrolyze specific phosphodiester bonds of the RNA backbone. In order to better understand the nature of the substrate upon which these enzymes must act, we need to consider some of the other types of steps that occur during rRNA processing in addition to the breakage of larger RNA molecules into smaller ones. Analysis of the 45S precursor that is extracted from nucleoli reveals that an extensive number of modifications are made on the nucleotides after they are incorporated into the growing RNA chain. At the time the 45S precursor is first cleaved, over 100 of the uridine residues have been converted to pseudouridine (uracil is attached to ribose at 5′ C of U not 3′ N) and over 100 methyl groups have been added to the molecule. Nearly all of the methylations occur on the ribose, though a few are placed on bases. The nucleotides bearing the modifications are located at specified positions, which tend to be very similar among rRNAs of different species. They are believed to be very important in the recognition process whereby the enzymatic cleavages occur.

A comparison of the numbers of methylated nucleotides in the 45S precursor to those in the mature rRNA molecules indicates that the methyl groups are completely conserved, i.e., those nucleotides that are methylated remain as part of the final products while the unmethylated sections are the ones destroyed during processing. It would appear that the presence of the methylations serves to protect the regions containing them from enzymatic cleavage either directly or as a result of changes they impose on the secondary structure of the molecule. It is evident that the secondary structure of the rRNA precursor is of particular importance in determining the cleavage sites. In fact, the bacterial enzymes (RNAase III) most responsible for processing of rRNA in bacterial cells is specific for double-stranded helical regions in RNA molecules and there is evidence that a similar type of enzyme is active in nucleolar preparations. In addition to the methyls added to the 45S RNA, a few additional ones are added to the 18S rRNA after its separation from the remainder of the precursor.

Another important feature of the processing reaction is that the sequential changes do not occur on isolated nucleolar RNA, but rather on RNA molecules tightly associated with protein. The association with protein begins even before the 45S precursor has been completely transcribed. These proteins are synthesized on cytoplasmic polyribosomes and subsequently migrate into the nucleus and then the nucleolus. There are two types of proteins that are associated with the RNA during the processing steps: those proteins that will eventually remain in the ribosomal subunits and those proteins of the nucleolus which have only a transient interaction with the rRNA intermediates. Analysis of the ribonucleoprotein (RNP) of the nucleolus indicates that the 45S RNA precursor exists together with the 5S RNA and a considerable variety of proteins in a particle which sediments at about 80S. The proteins of the RNP particles are believed to be of great importance in determining the sensitivity of the RNA components to the processing enzymes. When the proteins are stripped from the precursor RNAs, nucleolar enzyme preparations appear to be much more destructive of the RNA molecules.

Once the separation of the two major rRNAs occurs, the 28S intermediate is detected in a 55S nucleolar RNP particle while the smaller segment containing the 18S rRNA and its associated proteins seems to rapidly find its way out of the nucleolus and into the cytoplasm (Figs. 12-9 and 12-10). In other words, the RNA and proteins of the larger subunit remain in the nucleus for an additional period of time after the 18S species has already departed. The conversion of the 80S RNP particle into the 40S and 60S ribosomal subunits appears to be a gradual process requiring interaction of specific proteins at specific times. Certain of the proteins of the precursor particles are lost as the maturation process continues while others are added. Some of the proteins probably serve to protect sites from cleavage while others probably cause sites to be more accessible. Presumably, one of the major reasons for the presence of the spacer sequences is to bring about the proper interactions with proteins that lead to the eventual formation of ribosomes. One point is clear; there is more remaining to be discovered in the area of ribosome formation than is already known.

The Structure and Function of the Nucleolus

Electron microscopic examination of nucleoli has revealed three distinct components, a granular region, a fibrillar region, and filamentous material (see Figs. 12-3 and 13-11). In addition, all of these components may be embedded in an amorphous proteinaceous matrix. In order to understand the nature of these components, we will consider some observations made on the nucleoli present in the large nucleus (termed a *germinal vesicle*) present in the *Xenopus* oocyte. The nucleoli of these oocytes contain a well-defined central fibrillar *core* and a surrounding granular *cortex* (see Fig. 12-3b). Using the appropriate techniques, the fibrillar cores of these nucleoli can be gently dispersed to reveal the presence of a circular filament whose composition is indicated by enzymatic digestion. While treatment with proteolytic enzymes and ribonuclease greatly reduce the diameter of the filament, they do not break it. However, treatment with the DNA-digesting enzyme DNase destroys the integrity of the filament. We can conclude that within the core of the nucleoli there exist

Figure 12-14. (*a*) A section of an rDNA molecule isolated from a nucleolar core of a nucleolus of a *Triturus* oocyte. The rDNA, caught in the act of transcription, is seen to have nascent RNA molecules (associated with protein) attached. The lengths of the nascent rRNA primary transcript increase with increasing distance from the point of initiation. The RNA polymerase molecules at the base of each fibril can be seen as dots. (*Courtesy of O. L. Miller, Jr., and B. R. Beatty.*) (*b*) Diagrammatic representation of the type of electron micrograph shown in *a*. The rRNA genes are arranged tandemly with untranscribed spacer DNA (S) in between. (*From E. Stubblefield*, Int. Rev. Cytol., **35**:26, 1973, based on the work of O. L. Miller, Jr., and B. R. Beatty.)

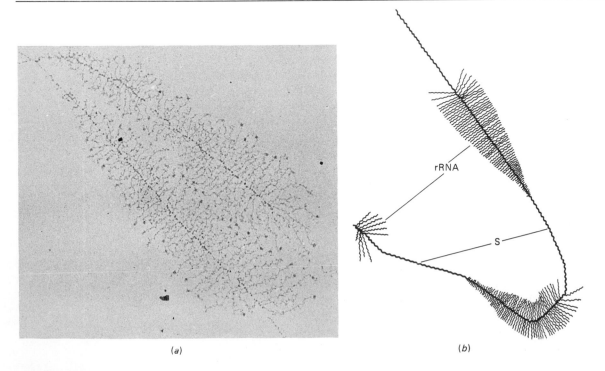

(*a*) (*b*)

circular ribosomal DNA molecules containing associated RNA and protein materials.

When the circular filaments are examined in the electron microscope, the activity of the nucleolar DNA is dramatically revealed. At low magnification certain of the regions of the filament appear "bushlike" while other regions appear naked and devoid of additional material. Examination of a "bush" at higher magnification reveals the presence of approximately 100 fibrils which are attached to the axial filaments at one end and project outward from that point of attachment (Fig. 12-14). There is a distinct polarity with respect to the length of the fibrils along the axial filament; they are shortest at one end and then become longer as they progress along the length of the filament until their presence is abruptly terminated.

As in the previous discussion of this type of electron micrograph (page 526), these portraits of gene expression are readily understood in terms of increasing lengths of nascent RNA chains. Those fibrils of shorter length represent nascent RNA molecules of fewer nucleotides, i.e., closer to the site of initiation of transcription. At their longest length, the fibrils represent nearly completed ribosomal RNA precursor molecules, which in the *Xenopus* case are approximately 40S. If one examines the site of attachment of each of the fibrils with the axial filament, a small particle can be seen which most likely represents the RNA polymerase molecules responsible for transcription. In the case of the ribosomal genes of these oocytes, approximately one-third of the DNA appear to be covered by these particles, indicating the extreme density with which the DNA is bound, a reflection of the intensity of rRNA production in these cells. Another feature which is apparent from these types of studies is the essentially immediate association with protein of the incorporated ribonucleotides of the nascent chain. All of the fibrils seen extending from the axial filament are in the form of ribonucleoprotein, presumably on its way to becoming the 80S ribonucleoprotein particles previously discussed.

The Existence of Spacer DNA

An additional point which is readily discerned from the electron micrographs of the type shown in Fig. 12-14 is the alternation of inactive and active regions of the DNA. There are two possible explanations: either these inactive regions represent ribosomal RNA genes that are not being transcribed, or they represent some type of nonribosomal DNA. Various types of analyses have indicated that the DNA between the active stretches represents spacer DNA that is not transcribed. In other words, in addition to the spacer regions of the rDNA which are transcribed as part of the 40S rRNA precursor of *Xenopus* (45S in humans) and later destroyed, these DNA sequences do not serve as templates for RNA synthesis but rather they serve to separate transcriptional units (Fig. 12-15). As more types of gene transcripts become studied, the presence of both transcribed and nontranscribed spacers among repeated, clustered genes appears to be widespread. These include the tRNAs, the 5S rRNA, the 18S and 28S rRNAs, and the histone genes. In all of these cases, the genes exist in

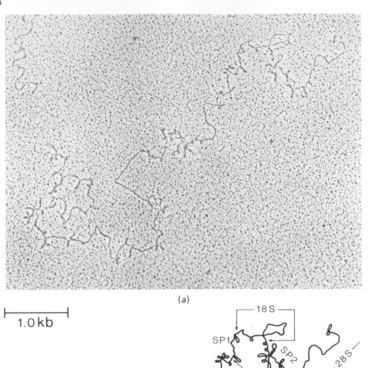

(a)

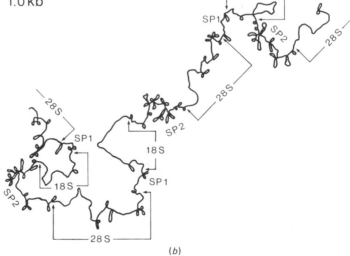

1.0 kb

(b)

Figure 12-15. (*a*) Electron micrograph and (*b*) accompanying diagram of a single-stranded molecule of *Xenopus* rDNA displaying a highly regular secondary structure pattern. This particular rDNA strand displays two complete and two partial repeating units. The genes coding for 28S and 18S rRNA (28S, 18S) are separated by a transcribed spacer (Sp1) and alternate with spacer regions (Sp2) which are not transcribed. (*From P. K. Wellauer and I. B. Dawid,* J. Mol. Biol., **89:**379, 1974.)

clusters within specified chromosomal regions. The DNA sequences exist as repeating units present one after the other, i.e., *in tandem,* along the DNA molecule. There appears to be a section of DNA within the repeating unit that is not transcribed, i.e., represents nontranscribed spacer, and sections within the transcribed RNA that do not end up in the final product.

The ubiquitous presence of spacer sequences suggests that they play some role of basic importance in the organization and/or transcription of these genes. However, it appears that, whatever their role, the precise nucleotide sequence is not of great importance. The basis for this statement comes from the analysis of nucleotide sequences of the same molecules among closely and distantly related organisms. One generally accepted indication of the importance of a specific sequence, whether it be an amino acid sequence or a nucleotide

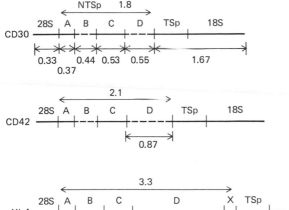

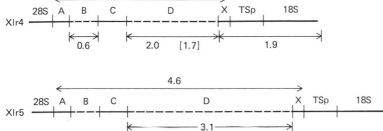

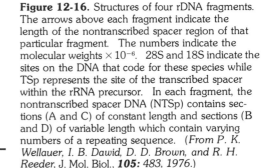

Figure 12-16. Structures of four rDNA fragments. The arrows above each fragment indicate the length of the nontranscribed spacer region of that particular fragment. The numbers indicate the molecular weights $\times 10^{-6}$. 28S and 18S indicate the sites on the DNA that code for these species while TSp represents the site of the transcribed spacer within the rRNA precursor. In each fragment, the nontranscribed spacer DNA (NTSp) contains sections (A and C) of constant length and sections (B and D) of variable length which contain varying numbers of a repeating sequence. (*From P. K. Wellauer, I. B. Dawid, D. D. Brown, and R. H. Reeder,* J. Mol. Biol., **105**: 483, 1976.)

sequence, is the extent to which that sequence is preserved over evolutionary time. If two sequences are very similar among distantly related organisms, one can conclude that mutations that arose in that sequence must have been deleterious ones and were strongly selected against. Therefore, the specific sequence must be of particular importance. When one examines the nucleotide sequences of the mature rRNAs of different organisms, the conservative nature of the sequence is quite evident. In contrast, comparisons of nucleotide sequences of spacer regions indicate rapid evolutionary divergence. For example, the nucleotide sequences of rRNAs are closer between *Xenopus* and a plant species than the spacer sequences are between two different species of the *Xenopus* genus.

Examination of electron micrographs of nucleolar DNA indicates that the distances between actively transcribing sections of the filament are quite variable in length. Molecular analysis of rDNA has indicated that there is considerable variation in the nontranscribed spacer between transcriptional units even within the same nucleolar organizer. Further analysis by Peter Wellauer, Igor Dawid, Donald Brown, and Ronald Reeder has indicated the basis for the heterogeneity in the lengths of the nontranscribed spacers. There exist within the nontranscribed spacer, repeating sequences of relatively short length (e.g., 50 nucleotides in length). The variation in length of the spacers reflects variations in the number of these short repeating sequences (Fig. 12-16). The situation is analogous to having a variable number of satellite DNA sequences present within the nucleolar organizing regions. Though the precise function of these repeating sequences and the variable number is not understood, they are believed to be involved in a recognition process between DNA molecules which facilitates crossing-over and recombination (page 723). The finding of these short repeating sequences is believed to relate to a very important question concerning repeated genes which we will consider briefly.

The presence of repeated identical DNA sequences within the genome that code for rRNAs, tRNAs, and histone raises an interesting question. How can these sequences be maintained as exact copies of one another over periods of evolutionary time? Ribosomal RNAs, for example, while evolving in a conservative manner, are different among various organisms; therefore, one can presume that mutations must occur within these DNA sequences. However, within a particular individual or among a particular species, there is very little evidence of sequence heterogeneity in ribosomal RNA. Considering that particular mutations arise as a result of some alteration in one nucleotide at one site in the genome, there must be some mechanism to ensure that a particular mutation is either eliminated from the population of identical repeated sequences or is spread throughout all of them. Otherwise all of the sequences could not evolve in parallel. At the present time there is considerable speculation and very little evidence as to the mechanism by which these identical sequences are maintained. Generally the theories suggest some means whereby mutated DNA can be corrected. One of them suggests that mutations are either accepted or eliminated on the basis of a process of unequal crossing-over. If this theory proves to be correct, then the short repeating sequences within the nontranscribed spacer regions of rDNA may play a key role in the unequal crossing-over process.

We began this discussion on ribosomal RNA synthesis with a few points concerning the nucleolus, which was the center of some of the early studies in this area of research. More recent studies have provided a better basis for the understanding of nucleolar structure in terms of its molecular functions. It was stated previously that there are three recognizable structural elements in nucleoli: deoxyribonucleoprotein filaments, ribonucleoprotein fibrils, and ribonucleoprotein granules. The filaments represent the rDNA-containing chromatin of the nucleolar organizer region of the chromosomes, the fibrils represent the nascent rRNA precursor molecules with their associated protein, and the granules represent the various types of ribonucleoprotein particles in various stages of maturation.

In some cells, the fibrillar and granular regions are organized into two distinct regions as in the case of *Xenopus* oocyte nucleoli, whereas in other cells the two regions are more loosely intermixed (see Fig. 13-11). Regardless

Figure 12-17. Autoradiographs of isolated nucleoli from oocytes that had been incubated with ³H-uridine. On the left oocytes were labeled for 80 minutes and immediately fixed. Only the core contains incorporated radioactivity indicating this is the site of synthesis of ribosomal RNA. On the right oocytes were labeled for 90 minutes and chased for 5 hours; the cortex is primarily labeled, indicating the movement of newly synthesized rRNA from the core to the cortex. [*Courtesy of N. K. Das, J. Micou-Eastwood, G. Ramamurthy, and M. Alfert,* Proc. Natl. Acad. Sci. (U.S.), **67**:970, 1970.]

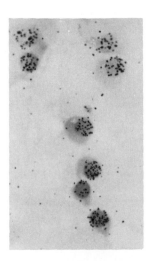
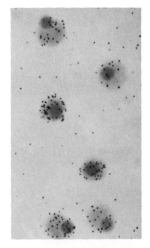

of the organization of the nucleolus, the chromatin is associated with the fibrillar parts, as would be expected if the fibrils represented nascent rRNA precursor-protein complexes. Similarly, autoradiographic studies indicate that RNA synthesis occurs in the fibrillar regions and later moves into the granular regions (Fig. 12-17). Analysis of the granules of the nucleolus indicates that, as expected, they contain preribosomal subunits in various stages of maturation. The nucleolus being of simple construction and lacking any membranous components, has become one of the best understood of the cell organelles in terms of the relationship between structure and molecular function.

TRANSFER RNA

Many of the general properties of rDNA and its transcription are shared by the DNA coding for tRNA. The DNA sequences that code for a particular tRNA species are also repeated and clustered within the genome as revealed by in situ hybridization. However, the DNA for different species of tRNA molecules are found in different regions of the genome. For example, studies on *Drosophila* indicate that there are approximately 60 different tRNA sequences within the genome, each sequence being present in approximately 10 copies. Within a particular gene cluster, the genes are arranged in tandem with relatively long nontranscribed spacers separating the short transcribed portions. The primary transcript is somewhat longer than the final product which must arise by undefined processing steps. In addition to the shortening of the tRNA precursor, a large percentage of the nucleotides are chemically modified (see Fig. 11-17). In some cases there is evidence for tissue-specific differences with respect to posttranscriptional modifications, resulting in tissue-specific differences in tRNA populations. Presumably, these differences have important regulatory roles to play in eukaryotic cellular metabolism.

One of the most important recent accomplishments in molecular biology has been the successful synthesis of a complete bacterial tyrosine tRNA gene, including the nontranscribed promoter region (Fig. 12-18), by H. Gobind Khorana and many of his associates. The gene, totalling 126 nucleotide pairs was put together from over 20 segments each individually synthesized and later joined enzymatically. This artificial gene has been introduced into bacterial cells carrying mutations for this tRNA and the synthetic DNA has been shown to successfully replace the pre-

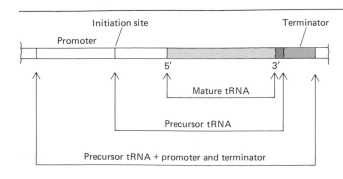

Figure 12-18. A diagram of the tRNA gene indicating transcriptional control elements on either side: promoter region to the left and terminator region to the right. (*From A. Kornberg,* "DNA Synthesis," *Freeman, 1974, courtesy of H. G. Khorana.*)

viously deficient function. The advantage in producing a complete functional DNA sequence from its component pieces, as was done in this case, as opposed to simply allowing a DNA polymerase to copy an isolated DNA or RNA template, is that specific alterations or "mutations" can be made in the gene by modifying one or more of the segments before they are put together. In this manner the effect of specific changes in the nucleotide sequence upon such activities as polymerase binding, etc., can be studied.

MESSENGER RNA

Messenger RNAs are those RNAs which contain the information for the synthesis of polypeptide chains. As described in the previous chapter, bacterial mRNAs often contain the information for more than one polypeptide chain, i.e., they are polycistronic, while those of eukaryotic cells are monocistronic, capable of being translated into only one polypeptide. In bacteria the formation and utilization of mRNA molecules seems quite straightforward—they are translated as they are synthesized (see Fig. 11-36) and then rapidly degraded by nucleases. In the remainder of this chapter we will be concerned exclusively with mRNA formation in eukaryotic cells, though we will return again to prokaryotic systems when we discuss the control of gene expression in the next chapter.

In eukaryotic cells, the manner of mRNA formation is complex and not well understood. For our purposes we will define a messenger RNA as one which (a) codes for a polypeptide, (b) is found in the cytoplasm, and (c) is either attached to ribosomes or capable of such an attachment in order to be translated. Even though eukaryotic messenger RNAs contain information for only one polypeptide chain, there are several parts of the molecule that are not directly involved in the coding process. For example, in the case of the mRNA for the polypeptides of hemoglobin, approximately 25% of the mRNA is not involved in coding for amino acids in the protein. Nucleotide sequencing analysis has indicated that noncoding sections are present on both the 5' and the 3' sides of the coding section; the termination codon (UAA) in globin mRNA is approximately 100 nucleotides from the 3' end of the message itself.

In addition to noncoding nucleotides, eukaryotic mRNAs have special modifications at their 5' and 3' termini that are not found on prokaryotic messages. The 5' end of essentially all eukaryotic mRNAs has a "cap" on it, which is added after the original 5' end has been synthesized. Both prokaryotic and eukaryotic RNAs are synthesized with a triphosphate at their 5' terminus which is the first part of the RNA molecule to be manufactured. Once the 5' end has been formed, several enzymes present in the eukaryotic cell convert the tip into one having a radically different structure (Fig. 12-19). The first step appears to be the removal of the last of the three phosphates, converting the 5' terminus to a diphosphate. The next step is to add a GMP to the end in an *inverted* direction so that the 5' end of the added guanine is facing the 5' end of the RNA chain. In between these last two nucleosides and joining them is a triphosphate. The last steps in the process involve the methylation of this last pair of

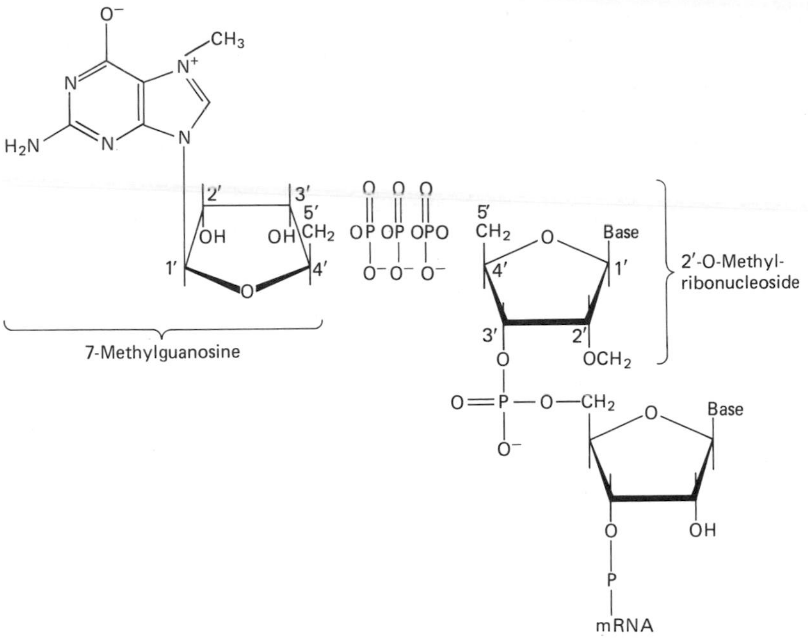

Figure 12-19. The structure of the 5′ end of eukaryotic mRNAs.

nucleotides. The terminal, "inverted" guanine is methylated at the 7′ position on its base while the nucleotide on the internal side of the triphosphate bridge is methylated on the 2′ position of the ribose. In addition to methylation of the terminal nucleotides of the mRNA, various internal nucleotide residues may or may not be methylated as well. For a long time it was believed that the tRNAs and rRNAs of the cell were the only ones to be methylated, but with the development of techniques for greater mRNA purification, the presence of methyl groups were also found in various locations in mRNA.

Though uncertainties still exist, there is good evidence that the 7′-methylguanine cap at the 5′ end is important in translation. These results are based primarily on the study of animal and plant viral mRNAs which are also capped. In situations where the 5′ methylated cap is either prevented from being added to these viral mRNAs or removed after it is already present, the ability of these mRNAs to be translated in vitro can be greatly reduced. However, since several eukaryotic viruses (including polio virus) are known to have mRNAs that are not capped, it appears that the presence of the 7′ methyl guanine is not an absolute requirement for protein synthesis. It appears that, at least in some cases, the presence of the "cap" is important in the ability of the 40S subunit of the eukaryotic ribosome to bind to the mRNA. The presence of the cap may also be involved in protecting the mRNA from digestion by ribonucleases as well as playing some type of role in mRNA processing reactions (page 553).

The base sequence of the 3′ end of most mRNAs is very unusual; it is composed exclusively of adenine. These adenosine residues which make up the last 50 to 200 nucleotides are referred to as *poly A*. Like the 5′ cap, the poly A is also added to the RNA molecule after it has been synthesized, i.e., it is a posttranscriptional modification. Even though the presence of the 3′ poly A has been known for a number of years, evidence of its function has been very difficult to obtain. Results of some experiments have suggested that it is involved in the transport of the mRNA from the nucleus to the cytoplasm since in some cases, if the addition of the adenines is inhibited, the mRNA does not appear on the polysomes. However, as in the case of the 5′ methylated cap, not all mRNAs have poly A, yet those normally lacking it seem to have no difficulty in escaping from the nucleus. At the present time, the histone mRNAs are the only specific mRNAs known to lack poly A tails. Another function that has been attributed to poly A is a role in the stability of the molecule in the cytoplasm. It has been found, for example, if the poly A is removed from mRNA, it appears to be degraded much more rapidly in the cytoplasm. This finding has been made by injecting normal mRNA and mRNA whose poly A has been removed, into large amphibian oocytes (page 512) and comparing their lifetimes.

Heterogeneous Nuclear RNA

In the previous section a number of properties of mRNA populations were described but little was said about the origin of these template active molecules. In this section we will consider some of the data and speculations on the synthesis of mRNAs in eukaryotic cells. The conclusions are primarily speculative since, despite extensive research in this area, we are still not sure about the formation of these most important RNA molecules. As just described, a messenger RNA molecule consists of more than simply a polyribonucleotide which codes for the amino acid sequence of a polypeptide chain. In addition, there are substantial stretches of nucleotides on both sides of the coding section of the molecule as well as a methylguanine cap at the 5′ terminus and a stretch of AMPs (the poly A) at the 3′ terminus. All of these parts of the mRNA must be taken into account.

In the 1960s a large number of studies were carried out on the nature of the RNAs made by various types of cells. The experiments generally consisted of supplying the cells with a radioactive RNA precursor for some period of time and then extracting the RNA and analyzing it by one or more fractionation techniques, such as sucrose density gradient centrifugation or acrylamide gel electrophoresis. During the mid-1960s a surprising finding was made by several laboratories. When cells were given very short exposures to labeled RNA precursors and the RNA immediately extracted, RNAs having huge molecular weights could be isolated. These RNAs were found to have sedimentation coefficients up to 80S with lengths estimated to 50,000 nucleotides. Most importantly, these RNAs were restricted to the nucleus. A great deal of effort went into verifying the existence of these giant molecules and determining their properties. These *rapidly labeled, nucleus-restricted* RNAs were found to:

1. Represent a large percentage of the RNAs being synthesized by the cell at a particular instant

2. Represent a large variety of different size classes

3. Hybridize to a large number of different DNA sequences

4. Have a very short half-life

5. Contain a 5′ methylguanine cap, as well as, in many cases, 3′ poly A tails

These properties are essentially the same ones that were previously described for messenger RNA (page 492). The differences between mRNAs and this class of RNA (termed *heterogeneous nuclear RNA* or *hnRNA*) are seen most clearly when the following *quantitative* differences are taken into account:

1. The hnRNAs, as a class, are much larger than the mRNAs present in the same cell (Fig. 12-20). In a mammalian cell the average hnRNA is approximately

Figure 12-20. (*a-b*) Electron micrographs of shadowed preparations of poly (A)-mRNA (*a*) and poly (A)-hnRNA (*b*) molecules. Representative size classes of each type are shown. Reference molecule is φX-174 single-strand DNA. [*From J. A. Bantle and W. E. Hahn, Cell, 8:145 (1976).*] (*c*) Size distribution of hnRNA and mRNA from mouse L cells as determined by density gradient sedimentation. Dark lines represent rapidly labeled hnRNA while the lighter lines represent poly A-containing mRNA isolated from polyribosomes after a four hour labeling period. The upper curves show the actual sedimentation profiles of the RNAs. The abcissa has been converted from fraction number (indicated by the points) to molecular size by calibration of the gradients. In the lower curves the mass distribution of the RNA is converted to molecular distributions by dividing each ordinate value by its abcissa value. The data were plotted cumulatively as percent molecules greater than a particular size. (*From R. P. Perry et al., Prog. N. A. Res. Mol. Biol., 19:278, 1976.*)

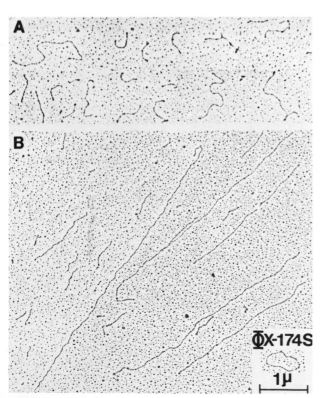

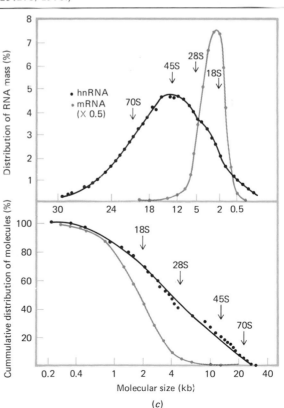

(*c*)

13,000 nucleotides in length while the average mRNA is approximately 1500 nucleotides. However, there is considerable overlap in the size of these two types of RNA. It is estimated, for example, that the average mRNA may be as large or larger than about 20% of the hnRNAs.

2. The sequence complexity of the hnRNAs is much greater than that of the mRNAs. As discussed at length in the appendix to this chapter, the mRNA population of a cell may correspond to approximately 1 to 2% of the DNA sequences (2 to 4% of the duplex DNA), whereas the population of hnRNAs typically correspond to 5 to 10% of the DNA sequences (10 to 20% of the duplex DNA). A recent analysis of hnRNA from the most complex of all tissues, the brain, indicates that as much as 40% of the DNA duplex may be transcribed in this organ (see Fig. 12-36b).

3. Although mRNAs are not considered highly stable molecules, when they are compared to hnRNAs they are *relatively* long-lived. Messenger RNA molecules generally have half-lives in a range from a few hours to a day or so (see Fig. 11-16). In contrast, most hnRNAs have half-lives measured more in minutes than hours. If one labels a cell for a few minutes and then chases it for over an hour before extracting the RNA, most of the radioactivity has been chased from the large molecular weight hnRNAs (Fig. 12-21).

4. The hnRNA molecules are restricted to the nucleus while the mRNAs are prepared from cytoplasmic fractions of the cell. This does not mean that mRNAs as such cannot exist in the nucleus before arriving in the cytoplasm; however, there is little evidence for large concentrations of mRNAs in the nucleus.

The Relationship between hnRNA and mRNA

It is clear from the above points that hnRNAs and mRNAs represent two distinct classes of molecules. However, these data are perfectly consistent with there being a "precursor-product" relationship between the two RNAs. It is generally believed that mRNAs are not synthesized in the same form as that found when they are associated with cytoplasmic polyribosomes. Rather, they are synthesized as parts of much longer, more complex hnRNA molecules. The situation is analogous in principle to the formation of the 18S and 28S rRNA species from the 45S precursor. A *primary transcript* is first produced as a large precursor molecule which

Figure 12-21. (a) Curves showing the sedimentation pattern of total RNA extracted from duck blood cells after exposure to $^{32}P_i$ for 30 minutes. (b) Curves illustrate the degradation of the large molecular weight hnRNAs when cells pulse-labeled as above are chased for three hours in the presence of actinomycin D, which prevents the synthesis of additional RNA. (*From G. Attardi, H. Parnas, M-I. H. Hwang, and B. Attardi, J. Mol. Biol. 20:160, 1966.*)

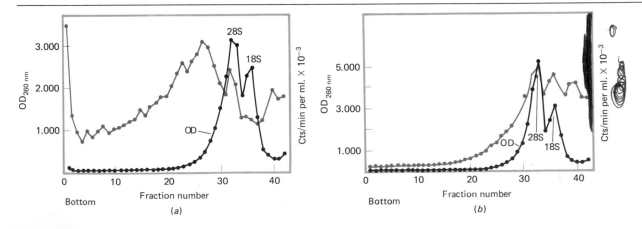

then undergoes some type of *processing* in the nucleus to produce one (or more) smaller mRNA(s) for release to the cytoplasm. The evidence that hnRNAs represent precursors to cytoplasmic mRNAs is largely circumstantial, but nevertheless convincing. The main points in favor of the relationship are as follows:

1. *Evidence from kinetic studies.* If, in fact, hnRNAs serve as precursors to cytoplasmic mRNA products, one would expect radioactive precursors to be incorporated initially into hnRNAs, whose radioactivity would then decrease with time as radioactivity in the mRNA fraction increased. Though there are numerous problems in the interpretation of studies of labeling kinetics, this course of events is generally found. Short exposures to labeled RNA precursor produce highly labeled populations of hnRNA having a wide range of sedimentation constants and virtually no radioactivity in cytoplasmic messengers. If one chases these cells for a period of an hour or more, the radioactivity in the large nuclear molecules has decreased and radioactivity can now be found in much smaller cytoplasmic molecules that are also heterogeneous in their sedimentation.

The evidence from the kinetic studies is strengthened by results utilizing actinomycin D, a potent inhibitor of RNA synthesis which acts by binding to guanine residues in DNA. Whereas actinomycin D is effective in blocking rRNA synthesis at relatively low concentrations (due to the greater G content of rDNA), higher concentrations are required to effectively block hnRNA and mRNA synthesis. If hnRNA served as a precursor to mRNA, one would expect both of these RNAs to have the same actinomycin D sensitivity, which they do.

2. *Posttranscriptional modifications.* The finding that both hnRNAs and mRNAs contain 5' methylguanine caps and 3' poly A tails has provided additional indirect evidence for a relationship between the two types of molecules. Since the discovery of poly A preceded the discovery of the 5' caps by a number of years, the fact that both hnRNAs and mRNAs contained poly A was interpreted as indicating that the mRNA was derived from the 3' end of the original hnRNA transcript. However, this conclusion is no longer valid based on this evidence since it appears likely that a poly A tail can be added to a nuclear RNA, followed by the cleavage of the 3' end and the addition of another poly A to the new 3' end. Analysis of the 5' ends of hnRNAs and mRNAs has led to the suggestion that at least some of the mRNA may be derived from the 5' end of the initial precursor. Although a considerable amount of data could be discussed, it is probably best at the present time to leave the question of the site of the mRNA within hnRNAs, since it is a topic of considerable controversy. This same comment applies to the role of the two types of terminal modifications, which have variously been suggested as being involved in mRNA processing, transport, translation, and/or stability (page 550).

3. *Sequence analysis.* The most direct evidence for a relationship between the hnRNAs and mRNAs comes from molecular hybridization studies. It has been generally found that the population of hnRNAs of a given cell contains all of the sequences present in the mRNA population, a fact which provides strong evidence for their being a precursor to the cytoplasmic mRNAs. In contrast, the hnRNAs contain much more information (4 to 10 times) than is present in the mRNA population, a fact consistent with the belief that as the messenger portions are carved from the hnRNA, the remainder of the precursor is degraded.

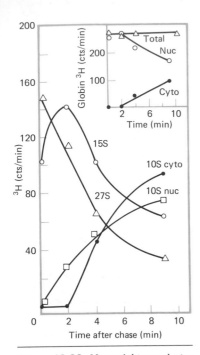

Figure 12-22. Hemoglobin-producing cultured cells were pulse-labeled with ³H-uridine for five minutes and then transferred to unlabeled medium for the chase. Samples were taken at the indicated times and nuclear and cytoplasmic RNA was prepared and analyzed on sucrose gradients. Globin RNA sequences in the various sucrose gradient fractions were determined at increasing chase times by hybridization to excess globin cDNA cellulose. Open triangles represent radioactivity in 27S globin precursor nuclear RNA, open circles in 15S globin precursor nuclear RNA, open squares in nuclear globin RNA of mRNA size, and closed circles in cytoplasmic globin mRNA. Insert shows labeled globin RNA sequences in nuclei versus cytoplasm. (*From R. N. Bastos and H. Aviv, Cell,* **11**: 647, 1977.)

The type of sequence analysis just described makes use of total hnRNA and mRNA populations. Another series of molecular hybridization studies have set out to locate *specific* mRNA sequences in hnRNA precursors. The question asked in these studies is whether or not RNA molecules of molecular weight greater than the particular mRNA are found to contain the mRNA sequence. The results of probes of hnRNAs with labeled cDNAs (page 578) made from various messengers (globin, ovalbumin, fibroin) have proven somewhat controversial. Results employing globin cDNA probes have indicated that the code for the globin polypeptides is contained in considerably larger nuclear RNA precursors; however, there is disagreement whether the precursors are very large or rather only 2 to 3 times as large as the final product.° One study in which a 27S and 15S RNA precursor of the globin message was reported is shown in Fig. 12-22. In contrast the analysis of hnRNAs from cells that are not producing hemoglobin (such as brain) indicates the absence of the globin sequence in these molecules.

If the hnRNAs are not the precursors to mRNA it leaves us in an embarrassing situation. If the mRNAs are synthesized at the same length they are found in the cytoplasm, then all of the larger nuclear transcripts, representing over 90% of the total number of nucleotides being incorporated at any given time, would have a totally unknown function.

Even if we assume that hnRNAs are the precursors to mRNAs, there are more unanswered questions than answered ones. The most important question in the hnRNA to mRNA story is why the cell should synthesize molecules much larger than what appears to be needed. It is generally believed that the extra sequences, those that are destroyed during processing are somehow involved in the regulation of transcription and/or processing. The nature of this regulation is one of the central unanswered questions in molecular biology. Another question concerns how the cell knows which part or parts of the molecule to keep and ship to the cytoplasm and which part to degrade by the action of nucleases. As in the case of ribosomal RNA processing, the secondary structure of the larger precursor molecules may be of particular importance. Large hnRNA molecules have been shown to contain regions of double-stranded hairpin loops, which may be particularly important in determining the sites of cleavage within the molecule.

Another important unanswered question is whether or not every hnRNA molecule produced serves as a precursor to mRNA and whether it is possible for more than one mRNA to exist within one primary transcript. Questions also relate to the various size classes of the hnRNAs. Are the smaller hnRNAs derived from even larger molecules, or are they synthesized in these widely ranging lengths? If they are gradually chopped into smaller and smaller lengths, do any of these processing steps

---

° The recent finding of gene inserts, i.e., noncoding sections of DNA within a gene itself (discussed in the next section), makes the interpretation of these studies even more difficult. For example, it becomes unclear as to whether or not the extra portions of the precursor RNA molecule are present within the coding portion, external to the coding portion, or in both places. Recent success in RNA purification and sequence analysis leads one to believe that many of the steps in mRNA processing will soon be unraveled.

occur while the RNA is still being synthesized or only after transcription is complete?

Gene Inserts: An Unexpected Finding

Throughout the discussion of transcription in this and the preceding chapter, it has been tacitly assumed that a continuous linear sequence of nucleotides in a messenger RNA is complementary to a continuous sequence of nucleotides on one strand of a DNA template. This had also been the assumption made by investigators studying the molecular biology of gene expression until 1977, when a remarkable new finding was made. It was found that an RNA molecule seemed to be transcribed from DNA sequences separated from one another along the DNA template strand. The first indication of this occurrence appeared in studies of ribosomal DNA in *Drosophila* and was then extended by a comprehensive analysis of transcription in adenovirus, a virus which infects various mammalian cells (see Fig. 2-5). It was found for example that a number of different adenovirus messenger RNAs were composed of the same 150 to 200 nucleotide 5' terminus. One might expect that this sequence represented a repeated stretch of nucleotides located near the promoter region of each of the genes for these mRNAs. However, further analysis revealed that the leader sequence itself was transcribed from three distinct and separated segments of DNA within the adenovirus genome (represented by blocks x, y, and z on the first part of Fig. 12-23). Furthermore, these segments were not located even near many of the genes whose mRNA transcripts bore this 5' leader sequence. In other words, the mRNAs of adenovirus were

Figure 12-23. A model for processing adenovirus mRNAs based on the removal of RNA loops and subsequent splicing. The model presents a mechanism by which different segments of an RNA molecule encoded at widely spaced sites on the DNA are brought together to form a contiguous leader sequence (XYZ), which is then attached to various structural RNAs encoded downstream. The three segments of the leader are brought into close proximity by looping out the intervening sequences (S_1 and S_2) and are then covalently joined by intramolecular ligation. This leader is coupled to various structural RNAs (G_1 to G_4) encoded downstream by another looping out of any intervening sequences (S_3, $S_3 + G_1$, or $S_3 + G_1 + G_2$ and so on) followed by intramolecular ligation. (*After D. F. Klessig*, Cell, **12**:19, 1977.)

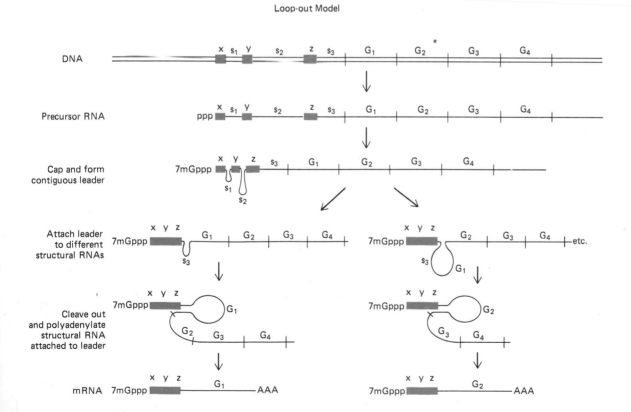

composed of nucleotides transcribed from discontinuous and remote regions in the genome.

There are several possible ways to account for the structure of adenovirus mRNAs. For example, the RNA polymerase may transcribe along one stretch of DNA template and then jump over a DNA loop or move across a spacer sequence (while holding on to its nascent RNA molecule) and begin to transcribe again on the other side. A second possibility is that the mRNA is transcribed as completely separate pieces that are joined together to form a continuous covalent strand. A third possibility is the formation of a large RNA precursor containing the entire DNA complement followed by the removal of sections of the RNA molecule and subsequent fusion of the fragments. The evidence, some of which will be presented below, strongly suggests that the latter possibility is the correct one. The sequence of events which might be expected to occur by this mechanism leading to the production of various adenovirus mRNAs are shown in Fig. 12-23. Note the difference between the type of processing discussed in the previous sections, which results in the removal of terminal segments of the RNA precursor leaving smaller and smaller fragments (shown in the last step of Fig. 12-13), from the processing discussed in this section which results in the removal of internal sections of a molecule and the splicing together of the newly formed ends. More recent studies have indicated that a surprisingly large number of splicing steps occur during adenovirus RNA processing and the situation has become very complex and perplexing.

At the end of 1977 and beginning of 1978, several laboratories reported that the DNA of certain genes in mammalian cells were longer than the mRNAs transcribed from them. For example, it was found that the β globin gene of the rabbit and mouse genomes contained a stretch of approximately 600 nucleotides *within the coding region of the gene* which did not appear in the globin mRNA isolated from polysomes. In other words, the coding sequence of the globin gene appeared to be interrupted

Figure 12-24. Drawings based on electron micrographs of R-loop structures between (a) 15S globin precursor RNA and a DNA fragment containing the globin gene and (b) 10S globin mRNA and the same DNA fragment. The procedure for the formation of R loops is described briefly in the text and in detail in the paper by *L. T. Chow et al.*, Cell, **11:**819, 1977. The dotted lines in the drawings indicate the positions of the RNA molecules. When the 10S globin mRNA is used in the hybridization procedure, a double-stranded segment of DNA is formed in the middle of the gene. This double-stranded portion represents a section within the gene for which there is no complementary region in the mRNA. [*From S. Tilghman et al.*, Proc. Natl. Acad. Sci. (U.S.), **75:**312, 1978.]

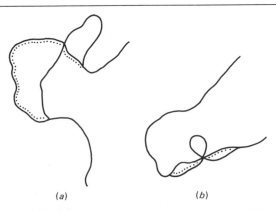

(a)　　　　　(b)

by a large noncoding section. This noncoding section was present in all of the DNA samples tested, regardless of the tissue from which it was extracted. Furthermore, this noncoding sequence was not one that was present throughout the genome, but rather was repeated, at most, approximately 15 times. Further studies on the globin gene insert have indicated that mRNA formation occurs by the removal of an internal sequence of ribonucleotides from a larger precursor as proposed in Fig. 12-23 for adenovirus. One line of evidence is shown in Fig. 12-24, which represents drawings made of DNA-RNA hybrids prepared as described in the following paragraph.

The interpretation of Fig. 12-24 requires a brief description of the technique of forming the R loops shown in the diagram (and in the micrograph of Fig. 12-27). Double-stranded DNA containing the globin sequence is incubated with globin mRNA in solutions containing 70% formamide. Under these conditions, the DNA remains double-stranded only along its G-C rich sites. Consequently, most of the DNA in these partially denatured molecules is single-stranded and available to hybridize to complementary mRNA sequences. (DNA-RNA hybrids remain stable under these conditions while DNA-DNA duplexes are primarily denatured.) Once the mRNA has hybridized to the single-stranded DNA portions, the conditions are changed to allow the single-stranded portions to reanneal, which can occur in all regions except those containing an RNA-DNA hybrid. Single-stranded DNA portions appear visible in these regions. Figure 12-24b shows the types of complexes visualized when the mature globin mRNA is used in the hybridization analysis. The double-stranded DNA loop is formed in the middle of the globin gene due to the lack of mRNA bound to this region. Consequently, one can conclude that this section represents the internal coding portion of the gene. Figure 12-24a reveals the type of molecular complex formed when globin precursor is used in the analysis. The double-stranded loop is no longer seen since the internal noncoding DNA sequence is now complementary to a nucleotide sequence in the precursor RNA.[*]

The findings just described raise several extremely important questions. For example, how is the noncoding section of the precursor RNA removed? This question is complicated by the fact that it is essential that the removal of the segment and joining of the new ends be done precisely to the nucleotide. Any errors in this process of *RNA splicing* will result in additional or missing nucleotides in the polypeptide template. One clue to the mechanism involved in RNA splicing has come from an analysis of the DNA sequence of a mouse immunoglobulin (antibody) gene and its corresponding mRNA. In this case, there are at least two regions within the gene that are absent from the mature mRNA. Sequence analysis of the DNA indicates that the codons for the first 15 amino acids of the N-terminus (beginning with the initiating methionine) are separated from the codons for the adjoining amino acids by a stretch of 93 nucleotides. (The expected sequence of this region of the RNA transcript is shown in Fig. 12-25.) The sequence of nucleotides in the pre-

Figure 12-25. A hypothetical hairpin structure that could form in a precursor RNA to the immunoglobulin message (based on known sequence data). The nontranslated sequences (above the dashed line) would be excised and the coding sequences subsequently joined. The base-paired sequences are from the ends of the 93-base noncoding segment. Proper splicing of this precursor RNA would require a symmetric pair of cuts in the antiparallel sequences boxed in the stem, followed by ligation of the strands below to preserve one copy of the CAGG sequence. The strands to be joined, however, are not favorably oriented for ligation. [*From S. Tonegawa et al.,* Proc. Natl. Acad. Sci. (U.S.), **75:**1489, 1978.]

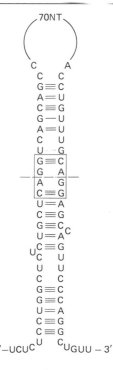

[*] It should be noted that the 15S globin mRNA precursor of Fig. 12-22 corresponds roughly in size to the mature mRNA plus the 600 base insert. It is possible, therefore, that the processing of the hnRNA containing the globin sequence involves simply the removal of the insert as opposed to the removal of sections at the 5' and/or 3' end of the precursor. More recent studies on the ovalbumin gene indicate the presence of seven distinct noncoding inserts. The evidence suggests that the ovalbumin precursor RNA is carved into successively smaller fragments by the stepwise removal of the extra RNA sequences.

cursor RNA is such that it can form a hypothetical hairpin loop containing the ends of the noncoding insert. In order for this precursor to be converted to the mature mRNA, both strands of the double-stranded stem would have to be cut at the point shown by the dotted line and the ends sealed to form the continuous mRNA. Enzymes capable of recognizing particular nucleotide sequences have been found in bacterial cells (discussed later in this chapter on page 562), but the presence of enzymes of this type has not been demonstrated to occur in eukaryotic cells. Needless to say, there is an intensive search proceeding at the time of this writing for enzymes involved in RNA splicing.

An even more basic question concerning RNA splicing than how it occurs is why it occurs. The demonstration of internal noncoding DNA sequences came as a totally unexpected finding. No obvious reason for their existence or evolution can be formulated at this time, though some speculation has appeared. Since noncoding gene inserts have not been found in prokaryotic cells, they probably represent one more example of an important difference in the mechanism of gene expression between prokaryotes and eukaryotes (along with their viruses).

THE NEED FOR RNA AND PROTEIN SYNTHESIS

Throughout these chapters on gene expression and RNA metabolism we have been concentrating on molecules without much consideration to their ultimate role in cell function. If we assume that the overall function of RNA metabolism is to lead to the production of specific proteins, which in turn are needed to carry out the cell's many different activities, we should consider these molecular functions, at least briefly, in terms of their effects on higher levels of organization. To what degree does a cell depend upon the minute-to-minute synthesis of RNA and protein molecules to support its activities and its life? As one might expect, the answer obtained varies greatly with the type of activity and the type of cell being studied. The experimental approach to this question generally involves the addition of inhibitors and the subsequent observation of the cells for varying lengths of time. The most commonly employed inhibitors of RNA and protein synthesis are actinomycin D and puromycin (or cycloheximide), respectively. The most rapid and profound effects of treatment with these inhibitors are seen when cells are in the process of acquiring some new phenotype. In other words, cells are most sensitive to the inhibition of RNA and protein synthesis when changing their physiological activities. Most cells are capable of altering their state as a consequence of changes in their environment. The agents responsible include drugs, hormones, ions, cyclic AMP, carcinogenic agents, mitogens (mitosis-producing chemicals), injury, etc. The effect of many of these agents is mediated via the synthesis of new mRNAs and protein. The situation is analogous to the appearance of beta galactosidase in bacterial cells in the presence of lactose or the appearance of phage progeny in these same cells after viral infection; all these events are very sensitive to the presence of inhibitors of transcription and translation.

A particularly well-studied series of predictable changes in the nature of cell function occurs during embryonic development when cells are continually gaining new phenotypic states and losing old ones. It is generally found that the ability of a cell to shift its activities depends on the synthesis of new species of RNA and subsequently the synthesis of new species of protein; when these are blocked the modifications cannot occur. Whereas the effect of the inhibition of protein synthesis is usually seen immediately, the effect of blocking RNA synthesis is usually delayed. For example, if a newly fertilized sea urchin or frog egg is treated with puromycin, development is blocked before the first division (Fig. 12-26), while the same eggs treated with actinomycin D continue to develop normally for several hours and generally are capable of reaching a blastula stage (Fig. 12-26) composed of several hundred cells. Similarly, if these same types of embryos are treated with puromycin at the beginning of *gastrulation*, development comes to an almost immediate halt whereas, if the embryos had been given actinomycin D at that same stage, gastrulation would have occurred before development ceased.

It appears that embryonic development is continually dependent upon a supply of newly synthesized proteins. Since there is a period of time between the synthesis of an RNA molecule and its translation into the corresponding protein, there is a delay in the effect of the blockage of new RNA molecules by actinomycin D. However, once the proteins coded for by the blocked RNAs are needed, then the effect of actinomycin D also becomes felt and development slows down and then stops. A similar effect is seen, for example, after the administration of a hormone to a susceptible tissue. If actinomycin D or puromycin is given at the same time as the hormone, the effect of the hormone on the cells is generally not seen. In other words, the change in phenotype induced by the added hormone requires the synthesis of new RNA and proteins. However, if one waits until just before the cell's response becomes visible and then adds the inhibitors, it is generally found that the phenotypic change can occur in the presence of actinomycin D but is prevented by the addition of puromycin. The cell has already synthesized

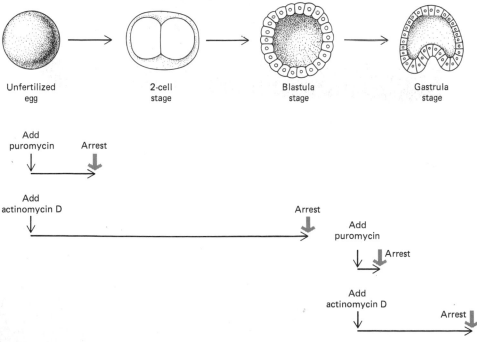

Figure 12-26. The effects of actinomycin D and puromycin on the early development of echinoderms. A few of the major stages of development are shown in the diagram. While treatment with puromycin brings about a very rapid cessation of development at any stage it is applied, the effects of actinomycin are considerably delayed.

the necessary RNAs in response to the presence of the hormone and can carry out its new activities in spite of its inability to produce additional RNAs. In contrast, the change in phenotype is closely accompanied in time by the synthesis of new proteins, and if this synthetic capacity is blocked by the addition of puromycin, so too is the hormone-induced response.

DNA CLONING AND SEQUENCING
DNA Cloning

During the past few years two types of techniques, DNA cloning and DNA sequencing, have become available which have been heralded by many molecular biologists as the key to a new era in our understanding of DNA organization and function. The first of these techniques involves the use of recombinant DNA molecules, i.e., DNA molecules made of nucleotide sequences derived from more than one species (Fig. 12-27b) and has caused more controversy than any other single development in the biological sciences. Some of the controversy has centered on the principle of combining DNAs from different species, but most of the discussion has been concerned with the dangers of combining the "wrong" kinds of DNAs. Numerous meetings have been held on the subject and the types of risks have been discussed and various guidelines have been formulated. In many cases, depending upon the estimated risk factor, this type of research is performed using laboratories and laboratory equipment which takes great precaution in ensuring that the organisms being studied are totally contained. As a secondary precaution, the bacteria used as carriers of the recombinant molecules are ones that have been genetically altered to be certain that they are incapable of growth in any environment other than the complex medium provided by the investigator. The types of recombinant molecules that are believed to pose the greatest risks are those in which the DNA from *E. coli* (a human gut bacterium) is combined with the DNA from mammalian tumor viruses. Since there is evidence that these viruses may be relatively widespread among mammalian genetic material, this potential threat exists in any experiment utilizing extracted mammalian DNA. For this reason, and others, most of the research that has been performed in the early period of cloning experimentation has concentrated on nonmammalian DNAs such as those of the frog, sea urchin, fruit fly, and yeast. Research with recombinant DNA has been performed with an eye toward the analysis of basic problems in molecular biology as well as toward the development of a new, highly applied technology. In this latter regard, many scientists envision the development of new strains of bacteria with highly unusual properties; for example, bacteria that produce large quantities of human hormones, or animal protein as a food source, or substances capable of serving as an energy source. In the following pages we will restrict the discussion to research with recombinant DNA molecules that serves to further our basic understanding of cellular and molecular phenomena. We will survey a few of the techniques involved in this type of research and consider the results obtained in a few studies.

There have been two types of vehicles most commonly employed to introduce a fragment of eukaryotic DNA into a bacterial population; the chromosome of the bacteriophage lambda and bacterial *plasmids*. The use of plasmids requires a brief introduction. The term *plasmid* refers to extrachromosomal DNA molecules present in many bacteria. Like the main bacterial chromosome, they are double-stranded circles but much smaller in molecular weight and of variable genetic constitution (Fig. 12-27*a* and the photograph of Fig. 10-21*d*, which shows a small bacterial plasmid along with a fragment of the much larger bacterial chromosome). Since the plasmid and chromosomal DNAs have very different properties, they can be readily separated from one another (Fig. 12-27*a*). A wide variety of different types of plasmids have been isolated, some of which carry genes for resistance to antibiotics; these are of particular value in recombinant research because they allow one to select for bacteria that have taken up a plasmid (Fig. 12-27*c*). If the plasmids are added to a bacterial culture and the bacteria are then treated with a particular antibiotic such as tetracycline, then only those bacteria that have taken up the plasmid, which in this case has the tetracycline-resistant gene along with the fragment of eukaryotic DNA, will be able to survive. Once the plasmid is taken up from the medium, which happens with considerable frequency, it replicates in the bacterium and is passed to its progeny.

Figure 12-27. (*a*) Schematic representation of the separation of plasmid DNA from chromosomal DNA. (*b*) This electron micrograph shows a plasmid to which a rabbit globin cDNA has been attached (see Fig. 12-32 for details). The position of the globin sequence is indicated by the location of the bubble, termed an R loop. As described on page 567, the R loop represents a section of the molecule in which one of the DNA strands is hybridized to an RNA probe (in this case the globin mRNA) and the other DNA strand (the thinner limb) remains single stranded. In this case, the circular recombinant DNA (plasmid + globin cDNA) has been made linear by cleavage with a restriction endonuclease. (*From the work of L. T. Chow, T. R. Broker, and T. Maniatis, in a paper by T. Maniatis et al., in W. A. Scott and R. Warner, eds., "Molecular Cloning of Recombinant DNA," Academic, 1977.*) (*c*) Schematic illustration of the use of an antibiotic resistance marker on the plasmid to select for bacteria that have taken up the plasmid and with it the eukaryotic DNA fragment to be cloned.

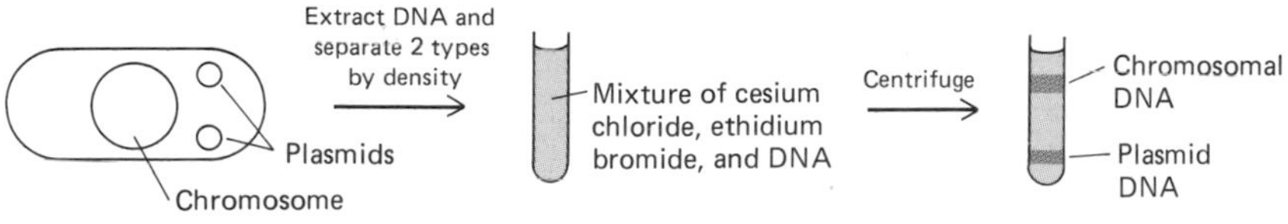

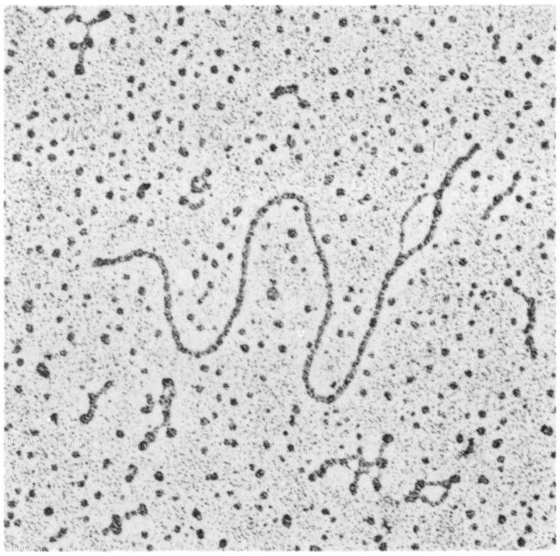

(b)

(a)

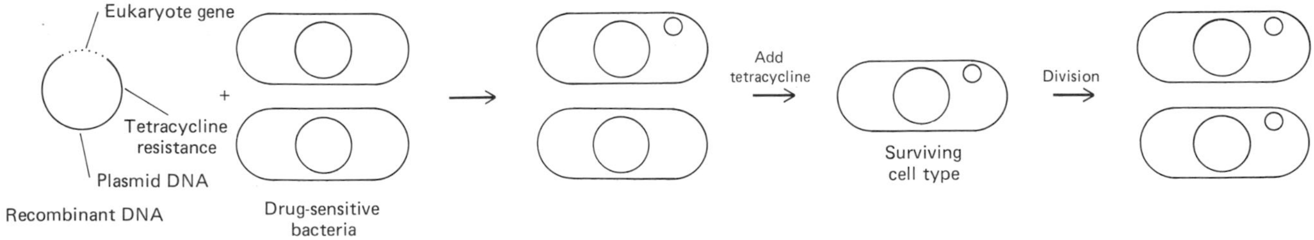

(c)

Cloning DNA Fragments

There are basically two types of eukaryotic DNA sequences that have been cloned: those isolated from extracted DNA, and those synthesized in the lab as cDNA from purified RNA molecules. The first type of cloning experiment to be described involves the preparation of the DNA from the eukaryotic cells, the cleavage of the DNA into fragments, and the joining of the fragments to the DNA vehicle for subsequent cloning. One of the most important steps in this technique is the fragmentation of the DNA, a process accomplished by a special type of enzyme, a *restriction endonuclease*. The term endonuclease refers to any enzyme capable of cleaving internal phosphodiester bonds of a nucleic acid backbone. The term restriction, in this case, refers to the ability of these enzymes to recognize particular nucleotide sequences (generally 4 to 6 nucleotides long) at which the cleavage occurs. As more bacteria are examined, a growing number of these enzymes have been isolated, each specific for a particular nucleotide sequence (Fig. 12-28). The bacteria uses these enzymes to destroy the DNA of invading bacteriophage while biologists use them to fragment DNA at specific sites in a DNA molecule. The availability of restriction enzymes has proved invaluable in the study of DNA sequences. For example, the finding that genes contain internal noncoding DNA sequences was discovered largely by the use of these enzymes (together with cloning and sequencing techniques). The most important restriction endonucleases being used in the cloning experiments are those that recognize sequences in DNA that have internal symmetry. Consider the particular sequence recognized by the enzyme *Eco R1*.

$$\downarrow$$
$$3'\text{-CTTAAG-}5'$$
$$5'\text{-GAATTC-}3'$$
$$\uparrow$$

A sequence with this type of symmetry is termed a *palindrome*. If one starts reading from one end of the sequence to the center and then switches to the other strand and continues reading, the two halves are symmetrical.

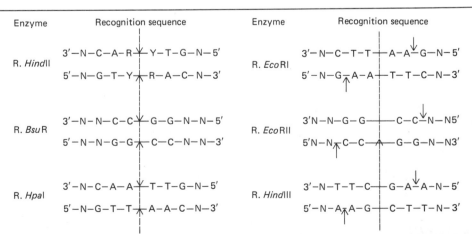

Figure 12-28. The cleavage of double-stranded DNA molecules by restriction endonucleases occurs at sequences of 4 to 6 specific nucleotides. The sequences recognized by several of these enzymes are illustrated along with the site on each strand where the cut is made. Those enzymes on the left break both strands at the same point while those on the right produce a staggered pair of strand breaks. N is any nucleotide (opposite nucleotide must be complementary), R is a purine, and Y is a pyrimidine. (*From K. Murray, Endeavour, **35**:130, 1976.*)

In addition, if one reads the sequence in the same direction (3′ to 5′ or 5′ to 3′) on both strands, the same sequence is observed. When the enzyme *Eco R1* attacks this palindrome, it breaks each strand at the same site *in the sequence* (Fig. 12-28). As a consequence, when parts of the DNA molecule on either side of the breaks separate from one another they are left with "sticky" complementary ends. Since a particular sequence of six base pairs would be expected to occur simply by chance approximately every 4000 nucleotides, any type of DNA is susceptible to this enzyme. The manner in which the recombinant DNA is formed in these experiments is shown in Fig. 12-29. A plasmid DNA containing one susceptible site for the restriction endonuclease is utilized; the action of the enzyme converts this plasmid to a linear molecule having the same ends as that in the eukaryotic DNA after its fragmentation by the enzyme. The two types of DNA are mixed and circular recombinant molecules are produced, which are capable of being taken up by bacterial hosts.

There have been two types of experiments involving recombinant DNA molecules in which the eukaryotic fragment is obtained from extracted DNA. In one type, a particular DNA sequence is sought, such as that for histones or for ovalbumin, while in the other case random fragments are cloned. The cloning of a particular sequence allows one to prepare large amounts of the desired sequence while the cloning of random sequences allows one to examine the various types of DNA fragments that

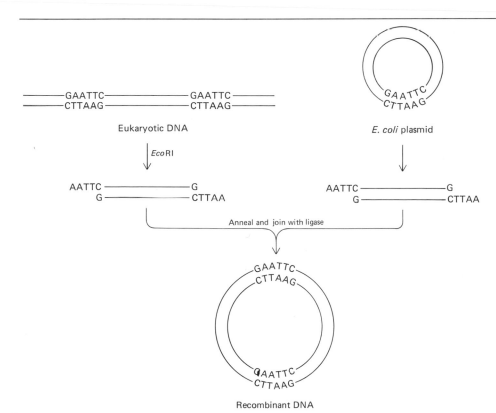

Figure 12-29. The procedure for the preparation of a recombinant DNA molecule containing a eukaryotic DNA fragment. Both the eukaryotic DNA and the bacterial plasmids are treated with a restriction endonuclease which makes staggered breaks in each double helix. The enzyme breaks the plasmid at one site converting it to a linear molecule. The enzyme cleaves the eukaryotic DNA into fragments whose number depend upon the frequency of the sequence recognized by the enzyme. The ends of the plasmid and the eukaryotic fragments are complementary and can be joined and sealed to form recombinant circular DNA molecules.

can be generated. Most importantly, the preparation of fragments from isolated DNA (as in the case of the sea urchin histone genes discussed in this chapter) allows one to obtain DNA samples which contain the regulatory sequences of the genes that are not transcribed. When one studies the cDNAs obtained from mRNAs with the reverse transcriptase procedure, nontranscribed portions of the gene will not be represented. It is currently believed that the cloning of DNA sequences containing intact regulatory elements will be of critical importance in unraveling the regulatory mechanisms involved in eukaryotic gene expression.

The cloning of a DNA fragment containing a particular nucleotide sequence involves several steps. Once the extracted DNA is fragmented enzymatically, some attempt is made to partially purify the sequence being sought. Purification in the case of histone DNA has been carried out by

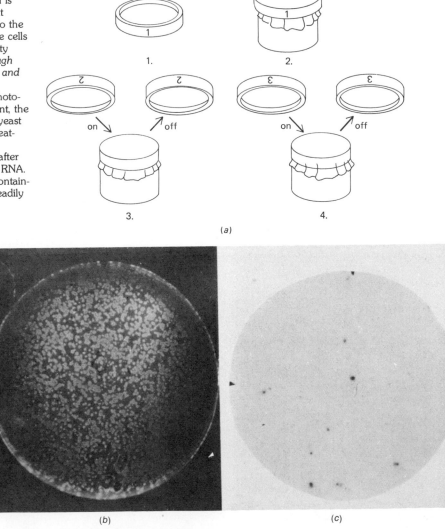

Figure 12-30. (a) The technique of replica plating. Colonies are established on a petri dish, which is then inverted over mounted filter paper so that some cells become transferred by adsorption to the paper. Replica plates are then made using the cells adsorbed to the filter paper to innoculate empty culture dishes as shown. (*After U. Goodenough and R. P. Levine, "Genetics," Holt, Rinehart, and Winston, 1974.*) (b) Plate containing phage plaques stained with ethidium bromide and photographed with ultraviolet light. In this experiment, the infecting phage contain random fragments of yeast DNA generated by restriction endonuclease treatment. (c) Autoradiograph of a nitrocellulose replica of the phage plaque plate shown in b after hybridization with ^{32}P-labeled yeast ribosomal RNA. Those plaques produced by phage progeny containing the yeast ribosomal DNA sequences are readily located. [(b)–(c) *Courtesy of R. W. Davis.*]

(a)

(b)

(c)

buoyant density separation, while in the case of ovalbumin DNA it has been accomplished via affinity chromatography by passing the DNA fragments through a column in which ovalbumin mRNA is adsorbed. Once the particular DNA sequence is partially purified, the recombinant DNA molecules can be prepared and provided to bacteria. Those bacteria that have picked up the recombinant DNAs are selected for by antibiotic resistance and are plated out on dishes by a technique of *replica plating* (Fig. 12-30a) which ensures that duplicate plates are set up in which members of the same clone are located in the same position in the duplicate plates. Once the bacteria begin to grow and form various colonies, each containing a different original DNA fragment, the task is to determine which of these colonies contains the DNA sequence being sought. The selection of the appropriate colonies is made by taking one of the duplicate plates, killing the cells, preparing the DNA for in situ hybridization (page 465) and identifying those cells that contain the particular sequence. The identification is made (Fig. 12-30b) by determining which of them contains DNA that hybridizes to a particular labeled probe (e.g., histone or ribosomal cDNA). Once this has been accomplished, the corresponding colonies can be isolated from the duplicate plates and the particular DNA sequence is therefore purified as well as increased greatly in amount. With this technique, any DNA sequence can be cloned as long as the corresponding RNA molecule can be prepared to identify the DNA fragment among a large number of colonies.

In a different type of experiment from those just described, DNA is extracted, fragmented enzymatically, and cloned without concern for the particular function of the fragment. In one set of experiments in David Hogness's lab (Fig. 12-31a), *Drosophila* DNA has been fragmented, random fragments have been cloned in bacterial hosts, and the locations of a few of the DNA sequences in the chromosome have been determined. In order to make this determination, the eukaryotic DNA is recovered from a bacterial colony, a radioactive complement of the DNA is synthesized (a labeled cRNA) and this labeled probe is then used for in situ hybridization to the DNA present in giant salivary gland chromosomes. It has been found that in most cases a randomly selected cloned fragment will hybridize to only one band (Fig. 12-31b) in the entire set of chromosomes. Analysis of the kinetics of hybridization suggests that the complementary sequence within the chromosomes is present in only one copy per genome. In a few cases, it has been found that a cloned fragment is complementary to DNA present in numerous chromosomal bands. For example, one of the fragments appeared to hybridize to 15 distinct locations scattered throughout the genome. It is proposed that fragments of this type contain nucleotide sequences with a regulatory function not unlike those proposed by Britten and Davidson in their model of the regulation of gene expression (page 624).

Cloning cDNAs

In this type of experiment (Fig. 12-27), the investigator begins with a purified mRNA molecule and synthesizes a cDNA copy using the reverse transcriptase. The single-stranded cDNA molecule is then converted to a double-stranded state by the addition of a DNA polymerase. The next step is to modify the DNA fragment and the DNA vehicle in such a way as to make them combine with one another. In this type of experiment, this is accomplished by enzymatically adding a string of nucleotides to one strand of the fragment and a complementary string of nucleotides

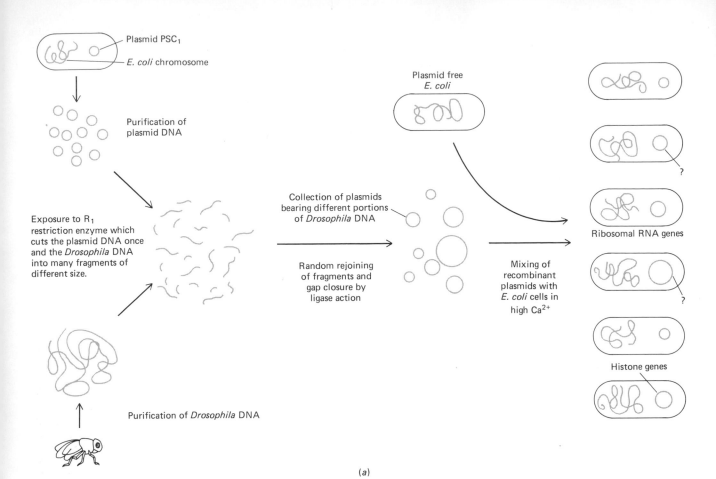

(a)

Figure 12-31. (a) Use of specific staggered nuclease (Eco Rl) cuts to insert fragments of *Drosophila* DNA into *E. coli* plasmid DNA. (*From J. D. Watson, "The Molecular Biology of the Gene," 3d ed., W. A. Benjamin, 1976.*) (b) In situ hybridization of a particular ^{3}H-labeled RNA sequence to polytene chromosomes of *Drosophila.* The labeled RNA was synthesized in vitro using a DNA template that had been cloned in *E. coli.* The ^{3}H-cRNA is found to hybridize at only one site in the genome, the 62E region in the left arm of chromosome 3. (*From P. C. Wensink, D. J. Finnegan, J. E. Donelson, and D. S. Hogness, Cell, 3:315, 1974.*)

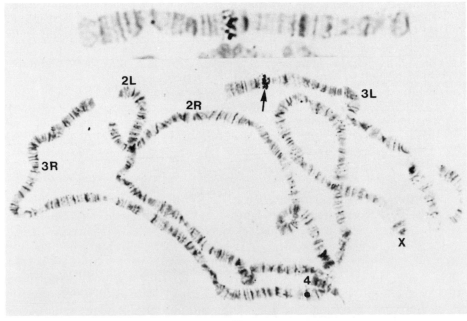

(b)

566

to the opposite strand of the vehicle (Fig. 12-32). When these two pieces of DNA are mixed they bind to one another by their "sticky ends" and can later be covalently linked by the action of a bacterial enzyme (a *ligase*). Bacteria are then incubated with the recombinant DNA and those cells which take up the DNA are selected for with the use of antibiotics. In this manner large amounts of the eukaryotic "gene" can be isolated for further study (Fig. 12-32). An example of one of the best-studied recombinant DNA molecules (one containing the globin sequence) is shown in Fig. 12-27*b*).

DNA Sequencing

The other recent advance in nucleic acid technology has come in the area of nucleotide sequencing. The first nucleic acid to be sequenced was that of a transfer RNA by Robert Holley in 1965 (page 494) after considerable work. Over the next decade or so, several advances were made so that a number of nucleotide sequences were determined. The procedure was similar in principle to that developed by Frederick Sanger for insulin (page 82); they involved the use of various nucleases to gen-

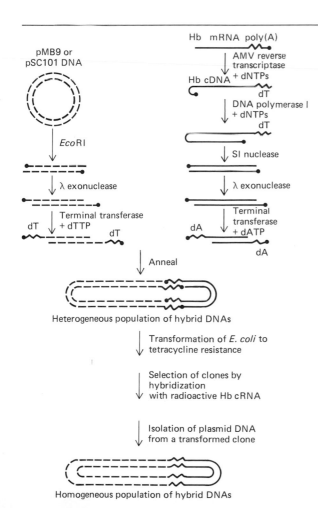

Figure 12-32. (*a*) The steps taken in the preparation of a plasmid containing the nucleotide sequence of the hemoglobin messenger RNA. The circular plasmid is cut once in a staggered manner by the restriction endonuclease *Eco* RI, the ends are extended by the addition of a small stretch of poly dT. The Hb mRNA is copied into DNA with the reverse transcriptase, made double-stranded, and given a small stretch of poly dA. These two types of DNA fragments now have complementary ends and will combine to form a circular molecule, i.e., a plasmid containing the eukaryotic mRNA sequence. These plasmids are then used to transform bacteria to tetracycline resistance which can be used as a basis for the selection of bacteria that had taken up the plasmid. If the mRNA that one starts with is not homogeneous, the sequence can be purified by this type of technique, as indicated. One would produce a heterogeneous population of hybrids. Those clones of bacteria containing the message being sought can be identified by hybridization to the radioactive mRNA. [*From R. Higuchi, G. V. Paddock, R. Wall, and W. Salser, Proc. Natl. Acad. Sci (U.S.), **73**:3147, 1976.*]

erate different types of overlapping fragments, the chromotography of the fragments, and their sequencing. For a variety of reasons, RNA sequencing was in a much more advanced state than that for DNA, until an entirely new approach to the problem of DNA nucleotide sequences was worked out in the laboratories of Sanger and of Walter Gilbert. Surprisingly, the techniques do not require direct sequencing of the DNA fragments. In this section we will briefly consider the technique developed by Allan Maxam and Walter Gilbert.

One begins with a homogeneous population of DNA fragments of, for example, 100 nucleotides in length. All of the copies are radioactively labeled at one end (the marked end). From this polynucleotide, an entire series of smaller fragments is generated, in this case ranging from 1 to 100 nucleotides. Most importantly, the fragments are produced by techniques which allow one to identify the terminal nucleotide, i.e., that nucleotide at the other end from the radioactive one. The fragments are then separated from one another by gel electrophoresis under conditions where each of the 100 different length fragments migrates to a specific location in the gel. Since the fragments are of increasing length, and their terminal nucleotide is known, the sequence of the entire molecule is automatically determine

The analysis begins with a homogeneous preparation of some DNA fragment (such as the one in Fig. 12-33a), each having the "marked end" (which can be the 3' or 5' end) labeled with ^{32}P The preparation is divided into several samples which are treated in different ways. One sample is subjected to reagents which break the polynucleotide by preferentially removing a guanine residue. Other treatments on other samples break the chain by preferentially removing an adenine or a cytosine residue. There is no procedure specific for thymine, therefore a pyrimidine-specific treatment is used instead. The conditions of the cleavage reactions are such that on the average, only one base is attacked in a given molecule. Figure 12-33a illustrates the products that one would expect from the cytosine-cleaving reaction. Five different fragments — °AT, GACTAG, °ATCGA, TAG, and GA — would be expected in addition to the uncleaved molecule, °ATCGACTAG, where the asterisk signifies the radioactively labeled initial nucleotide. Since the eventual identification of the fragments in the gel depends upon their radioactivity, we can ignore fragments which lack the marked radioactive end. When this particular preparation is subjected to electrophoresis, we find that the site in the gel to which a dinucleotide and a pentanucleotide should migrate is radioactive. This result indicates that the third and sixth positions from the radioactive marker must be occupied by a cytosine residue which had been selectively removed prior to electrophoresis. An example of an actual set of gels is shown in Fig. 12-33b.

Whereas a few years ago it might have taken two years or longer to sequence a DNA fragment of only 20 nucleotides, a piece this length can now be sequenced in a single day. In 1977, the nucleotide sequence of an entire viral genome was worked out by Frederick Sanger and his coworkers. In this case the DNA consisted of 5375 nucleotides which make up the genome of the single-stranded bacterial virus ϕX174. The sequencing data confirmed an earlier belief that the genome of this virus contained examples of overlapping genes, i.e., stretches of DNA which code for two different proteins. For example, the complete nucleotide sequence coding for the E protein is contained within the D gene which codes for an entirely different protein. The two different proteins are produced as a result of the movement of the ribosome along the RNA transcript in two different reading frames. This remarkable situation in which genetic information for different proteins is com-

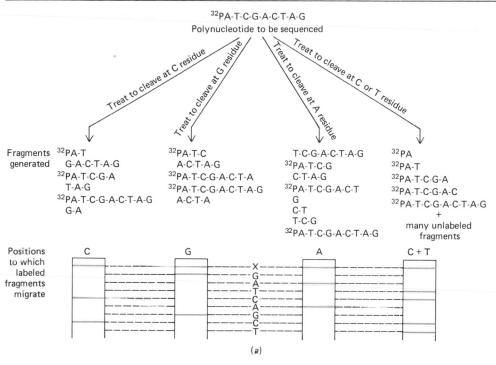

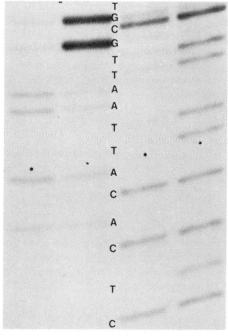

Figure 12-33. DNA sequencing by the technique of Maxam and Gilbert. (a) Sequencing of a small hypothetical fragment as described in the text. (b) Photograph of four gels used to sequence a particular DNA fragment. [(b) *From A. Maxam and W. Gilbert*, Proc. Natl. Acad. Sci. (U.S.), **74**:564, 1977.]

pressed into the same region of the chromosome has been uncovered in other viral genomes as well and may be widespread. The cases of overlapping genes provide investigators with the opportunity for interesting evolutionary speculation. It is hoped that the new techniques of DNA sequencing, which are certain to generate a tremendous number of gene sequences, will lead to a new period in our understanding of eukaryotic gene function.

Histone Genes: An Example of the Use of Molecular Cloning

The histones are a group of proteins which are tightly associated with DNA within the chromosomes. There are five different classes of these small basic proteins, and their structure and interactions with DNA will be described in detail in the next chapter. As discussed in the first part of this chapter, investigators had early success in the analysis of the genes that code for ribosomal RNA for a number of reasons: rDNA was found to be repeated, clustered in one section of the genome, unusual in base composition, and its complementary RNA was readily available in a purified form. These same properties exist in the case of the genes for the histones and have provided the basis for a similar, although less extensive, analysis.

Histone DNA has been best studied in the sea urchin primarily due to an opportune biochemical condition that exists during the early development of the embryos of these animals. The principal activity of early development is cell division, an event which occurs with unparalleled frequency in small, nonyolky embryos like that of the sea urchin. Since a rapidly increasing number of cells requires a corresponding increase in the number of chromosomes, a large percentage of the mRNA and proteins made by these embryos are histones. As a result, it has been possible to isolate the histone mRNAs from these embryos. The first step is to isolate the entire histone mRNA population which sediments at about 9S in a sucrose gradient, and then subsequently purify each of the five component mRNAs by additional fractionation techniques. These individual mRNAs can then be used as probes for various types of experiments with histone DNA.

Since histone DNA, like rDNA, is clustered, it lends itself to isolation. Fragments containing histone DNA are first partially purified by buoyant density centrifugation and are then treated with nucleases that cleave the DNA only within specific nucleotide sequences (see page 562). It has been found, for example, that treatment of sea urchin histone DNA with certain of these nucleases produces fragments of histone DNA of approximately 6000 nucleotide pairs which contain the information for all five types of histone. In other words, like rDNA, histone DNA is organized into repeating units arranged in tandem; in this case each unit contains the code for five different molecules. These 6000 base-pair fragments of double helix have been amplified greatly in amount by molecular-cloning techniques. As described above, DNA cloning involves the introduction of a fragment of foreign DNA, such as sea urchin histone DNA, into a bacterium where it replicates along with the bacterium's own DNA and is passed on to its progeny. All of the offspring of that bacterial cell will contain copies of the eukaryotic DNA fragment thereby producing a tremendous increase in the amount of this particular DNA sequence. Once the histone DNA is amplified in this way it can be recovered and subjected to further investigation such as nucleotide sequencing. The analysis of cloned sea urchin DNA has been carried out primarily by two groups of investigators, that of Max Birnstiel and coworkers and that of Norman Davidson, Eric Davidson, Laurence Kedes and coworkers. Although the details of their approaches have differed, both groups have found essentially the same results. They have found, for example, that there is one copy of each of the five histones in each repeating unit and that there is a specific order of these five genes. In addition, they have found that the information for all five histones is on the same strand of the duplex in the order 5' H4-H2B-H3-H2A-H1 3'. The fact that all five codes are present on the same strand can be proven by actually separating the two strands from each other and demonstrating that all five mRNAs bind to the same isolated strand. The order of the five genes (Fig. 12-34a) has been obtained by fragmenting the 6000 nucleotide DNA into five pieces by a collection of specific nucleases in such a way that each of the fragments is complementary to a different histone mRNA. Once the identity of each of the fragments was determined, their location within the entire repeating unit could be assigned.

In between the code for each of the five histones there exists a spacer whose existence can be visualized in the electron microscope by a technique termed *denaturation mapping*. The technique is based on the fact that the temperature at which a particular stretch of DNA melts is dependent upon its base composition. In this case, the DNA that actually codes for the histone proteins contains a significantly higher GC content than the DNA of the spacers (52% vs. 36%). Consequently, it is possible to incubate the DNA at a temperature which causes the lower melting spacer regions to become single

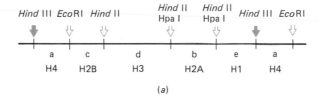

(a)

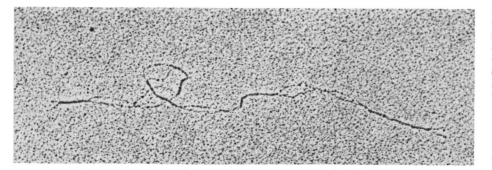

Figure 12-34. (a) The location of the five histone mRNA-coding sequences within the repeating histone unit of the sea urchin. The sites at which restriction endonuclease cleavage occurs is indicated. (b) A denaturation map of histone DNA cloned as a hybrid bacteriophage molecule. Those regions which undergo strand-separation at 61°C under the conditions employed are indicated in this electron micrograph of shadowed DNA. Those regions undergoing denaturation under these conditions represent the AT-rich spacers between the DNA sequences that code for the histones. [(a) From W. Schaffner, K. Gross, J. Telford, and M. L. Birnstiel, Cell, **8**:475, 1976. (b) Courtesy of R. Portmann and M. L. Birnstiel, copyright reserved.]

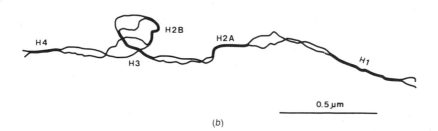

(b)

stranded while the GC-rich histone genes remain in the duplex state. Since single-stranded regions can be distinguished from double-stranded regions in the electron microscope (Fig. 12-34b), one is left with a striking map of the location of the coding and noncoding regions.

APPENDIX ON PROPERTIES OF MESSENGER RNA AS DETERMINED BY MOLECULAR HYBRIDIZATION

The study of any particular type of molecule generally depends upon the availability of a method for its purification. Since mRNAs are involved in protein synthesis, the best way to obtain them is to extract them from the cells' polyribosomes which are readily prepared from a cell homogenate (page 522). Messenger RNAs are released from polyribosomes by the removal of Mg^{2+} which is accomplished by the addition of the chelating agent EDTA. However, when one examines the material released from the polyribosomes, one does not find soluble mRNA molecules, but rather mRNAs that are tightly complexed with proteins. Studies of these complexes, termed *mRNP particles* (messenger ribonucleoprotein particles), from a number of different types of cells have revealed that the proteins are very restricted in number (2 to 3 species) and are quite similar in all mammalian cells examined. It appears that mRNAs are not translated in a naked state but rather are present in the polyribosomes in conjunction with specific proteins of unknown function. Since the mRNP particles are similar in some respects to the ribonucleoprotein subunits of the ribosome, it has been difficult to separate these two types of particles. Consequently, preliminary extractions of mRNA are generally highly contaminated with rRNA.

With the discovery that a large percentage of the mRNAs of the cytoplasm contained poly A at their 3' end (page 549), the opportunity became available for the simple and rapid purification of all poly A-containing messengers by affinity chromatography (page 118). The isolation technique consists of simply passing a solution of RNA, regardless of how impure, through a column containing either poly U or poly dT linked to some type of column material such as cellulose. Either of these synthetic nucleic acid polymers are complementary to the poly A portion of the mRNA and under the appropriate conditions will adsorb the poly A-containing mRNAs and remove them from solution. The adsorbed mRNAs can then be eluted in a purified state by changing the ionic strength of the solution. Unfortunately, mRNAs which do not contain poly A (a smaller percentage, though a considerable number) cannot be as highly purified and are less well characterized. At the time of this writing, the relationship between poly A-containing and poly A-lacking mRNAs remains uncertain. In some studies, these two classes of RNAs have been reported to represent different populations of species, while in other studies they are essentially similar. At the present, the only known mRNAs lacking poly A are those that code for histones, and even in this case not all of the histone messengers necessarily lack this added stretch of nucleotides.

Basically, studies with purified mRNA can be divided into two types; those that utilize a particular mRNA species and those that utilize an entire population of mRNA from a cell or tissue. The studies employing particular mRNAs are limited to those species that can be purified, i.e., those which are present in one kind of cell or another in very large amounts. Examples of this type include the mRNAs for globin in reticulocytes (red blood cell precursors), fibroin in silk gland cells, crystallins in lens cells, ovalbumin in chick oviduct cells, immunoglobulins in antibody-producing cells, and histones in sea urchin embryonic cells. In each of these cases the production of this one type of protein so dominates the synthetic activities of the cell that its mRNA can be obtained in sufficient amounts for further studies. Regardless of whether the study employs a particular mRNA species or a population of diverse mRNAs, once the RNA has been purified, a number of important questions concerning gene expression can be answered. In the next few pages we will consider several separate, but related questions, whose analysis requires different types of molecular hybridization experiments (page 465). The questions to be discussed are the following:

1. How great a variety of mRNAs are there in a particular cell?

2. How similar are the populations of mRNA in different types of cells?

3. Are mRNAs synthesized from repeated sequences, nonrepeated sequences, or both? Is there any evidence of selective gene amplification, i.e., an increase in the number of selected DNA sequences?

4. Are mRNAs present in cells in different numbers of copies?

1. *How great a variety of mRNAs are there in a particular cell?* The typical human cell contains approximately 5.6 picograms (5.6×10^{-12} grams) of DNA. In terms of information content this tiny amount of DNA has sufficient information to code for over one million different polypeptide chains of average molecular weight. A very important question concerns the percentage of this potential information that is actually put to use in coding for different proteins. When we consider the amount of genetic material present in a virus or bacterium, we find sufficient DNA (or RNA in the case of many viruses) in the genome to code for a reasonable number of polypeptides, from a very few in certain viruses to a few thousand in bacteria. However, in the case of mammals or other higher organisms with their much larger genomes, it becomes very difficult to imagine that so many different types of proteins could serve any meaningful function. It is generally believed that most of the DNA of higher eukaryotes does not code for proteins, but rather has some undefined role. A first step toward understanding that part of the genome which does code for proteins is to determine the fraction of the DNA sequences that are involved in the production of messenger RNAs.

Questions concerning the variety of DNA sequences that are involved in a particular template activity are generally answered by measuring the percentage of the DNA that is found to be complementary to the population of RNA molecules under study. Because experiments with repeated and nonrepeated DNA sequences are handled quite differently, the question becomes: How great is the variety of mRNA molecules that are complementary to each of the two types of DNA sequences? In the present discussion we will examine only those mRNAs complementary to nonrepeated DNA sequences. The overall experimental procedure is shown in Fig. 12-35. Experiments utilizing repeated DNA sequences are much more complex, much more difficult to interpret, and therefore fewer such experiments have been performed.

The first step in the experiment is to prepare radioactive DNA, fragment it, denature it, and separate the nonrepeated sequences from the repeated ones. The separation is accomplished by allowing the total denatured DNA fragments to reanneal long enough for all of the repeated sequences to hy-

I. PREPARE RADIOACTIVE DNA (e.g., ^{3}H-DNA)

Incubate cells with labeled DNA precursor such as ^{3}H-thymidine, extract and purify DNA

II. ISOLATE LABELED NONREPEATED SEQUENCES

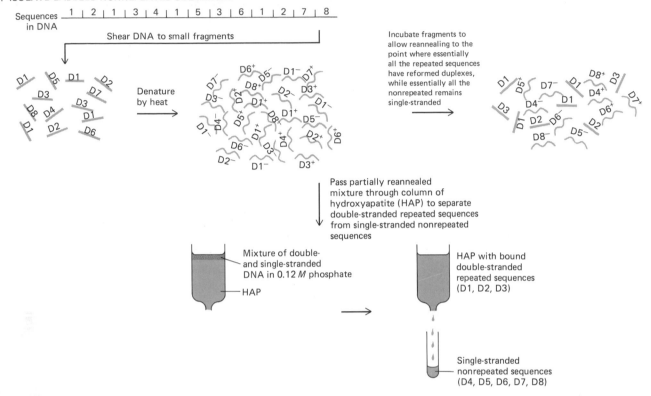

Pass partially reannealed mixture through column of hydroxyapatite (HAP) to separate double-stranded repeated sequences from single-stranded nonrepeated sequences

III. DETERMINE PERCENTAGES OF NONREPEATED SEQUENCES THAT ARE COMPLEMENTARY TO POPULATION OF RNA

Incubate nonrepeated labeled DNA with large excess of unlabeled RNA and determine percentage of DNA sequences that hybridize with the RNA molecules

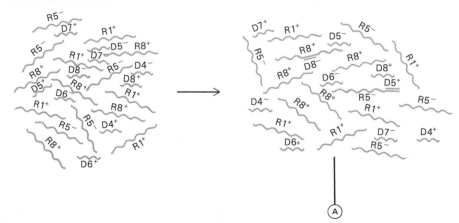

Figure 12-35. Step-by-step procedure for determining the percentage of nonrepeated DNA that is complementary to a given population of RNA.

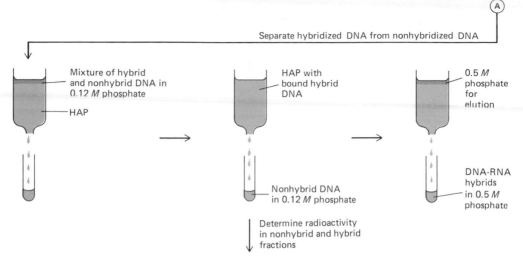

Separate hybridized DNA from nonhybridized DNA

Mixture of hybrid and nonhybrid DNA in 0.12 M phosphate

HAP

HAP with bound hybrid DNA

0.5 M phosphate for elution

Nonhybrid DNA in 0.12 M phosphate

DNA-RNA hybrids in 0.5 M phosphate

Determine radioactivity in nonhybrid and hybrid fractions

Results: 2% of nonrepeated sequences are complementary to RNA

or   4% of nonrepeated genome is complementary to RNA

bridize to one another leaving the nonrepeated ones present in solution as single strands. When the reannealing has proceeded to the desired point, the double-stranded DNA fragments are separated from the single-stranded ones, and the latter population, representing the nonrepeated sequences, is used in the following experiment. The labeled, nonrepeated sequences are combined with a great excess of the purified, unlabeled mRNA population to be tested and the two nucleic acids are allowed to hybridize. In a sense, the labeled DNA sequences can be divided into two groups: those to which the RNAs can bind and those to which they cannot. Presumably, those DNA sequences that become hybridized were active as templates for the synthesis of the complementary mRNA molecules. Even though there is only one copy of each DNA sequence per haploid amount of DNA, the RNA is in such great excess that there are sufficient collisions and the reaction is driven to completion, i.e., all DNA sequences that have RNA complements in the preparation will hybridize. One of the main difficulties with this type of experiment is one cannot always be sure that there are not some very rare mRNAs in the population that are not at sufficient concentration to ensure hybridization of their complementary DNA sequences.

In order to determine the percentage of the DNA sequences that has hybridized with RNA molecules at various times of incubation, reaction mixtures are separated by hydroxyapatite (Fig. 12-35) into a DNA-RNA hybrid fraction and a nonhybrid fraction, and the percentage of the radioactivity in each fraction determined. Figure 12-36a shows the profile of this type of hybridization experiment at increasing times of incubation. When the percentage of radioactivity in the hybrid fraction reaches a plateau value, we can assume that all of the available DNA sequences have been *saturated* and we can estimate the fraction of the nonrepeated genome involved in mRNA production in this case. Studies of this type using mRNAs extracted from mammalian cells growing in culture or particular tissues removed from a chick or mammal generally reveal that approximately 1 to 2% of the nonrepeated DNA sequences become saturated by mRNA. Since only one of the two

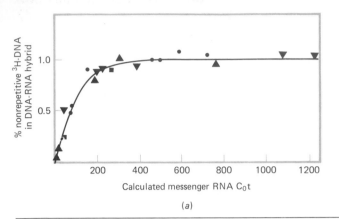

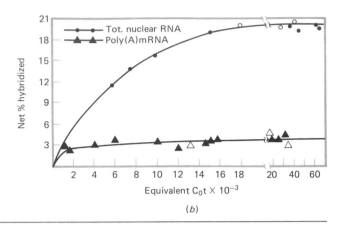

(a)

(b)

Figure 12-36. The results of hybridization experiments which determine the percentage of nonrepeated DNA that is complementary to (a) the population of messenger RNA that is present in sea urchin embryos and (b) the population of messenger RNA and hnRNA present in mouse brain tissue. Whereas the saturation values using mRNA are relatively low, in the order of 1 to 4%, that obtained using the hnRNA of brain nuclei is relatively high suggesting that a large percentage of the nonrepeated DNA sequences are being transcribed, at least within the brain. [(a) *From G. A. Galau, R. J. Britten, and E. H. Davidson, Cell, 2:13, 1974;* (b) *From J. A. Bantle and W. E. Hahn, Cell, 8:140; 1976.*]

strands of the DNA duplex is being transcribed, the value can be doubled to give approximately 2 to 4% of the nonrepeated genes being involved in mRNA synthesis. When these percentages are translated into numbers of different average-sized polypeptides that can be synthesized, values in the order of 2500 are obtained for *Drosophila* cells, 10,000 for sea urchin embryos, and 30 to 40,000 for cultured mammalian cells (including human cells). It is important to keep in mind that these types of studies do not prove that so many thousands of different polypeptides are actually synthesized; the function of these cytoplasmic, poly A-containing sequences as templates for protein synthesis is only assumed.

One tissue, that of brain, appears to have considerably greater complexity than any of the others. One recent study finds that 7.6% of the nonrepeated genes (3.8% of the nonrepeated DNA sequences) are complementary to poly A-containing mRNA of mouse brain (Fig. 12-36b). This level of saturation corresponds to approximately 100,000 different active genes, approximately 25 times the amount of information present in the entire genome of *E. coli*. In the same study, preparations of hnRNA from brain tissue were found to saturate approximately 21% of the nonrepeated DNA sequences corresponding to approximately 40% of the nonrepeated genome. When one considers the fraction of DNA that appears to be involved in hnRNA synthesis, it no longer seems out of the question that a very large percentage of the DNA, possibly all but satellite DNA, may be transcribed at one time or another during an organism's life.

2. *How similar are the populations of mRNA in different types of cells?* At the present time too few experiments have been performed with purified polysomal mRNA populations to be able to fully discuss the extent to which cell- and tissue-specific differences exist; however, it is worthwhile to at least speculate on their significance. *If* the 2 to 4% of the genes producing mRNAs in one type of cell were different from the genes producing mRNAs in other types of cells, and *if* we could add up the efforts of all the different cell types, then we might expect to arrive at a total percentage utilization for mRNA production roughly equivalent to the total genetic material of the organism. Another way it might be possible to arrive at a value for total genetic utilization in terms of mRNA synthesis would be to take all of the mRNAs present in an entire organism and attempt to determine the percentage of the DNA to which this total population is complementary. At the present time this latter type of experiment has not been performed on purified mRNA, but some idea

of the similarity of mRNA populations present in different cells is available. Similarities between two populations of RNA (complementary to nonrepeated sequences) are compared by carrying out what is termed an *additive* experiment. First consider a hybridization experiment using two populations of RNA containing totally different sequences. Assume each population is complementary to, i.e., capable of saturating, 1% of the nonrepeated DNA sequences. If these two populations of RNA were combined they would be expected to hybridize with 2% of the available DNA sequences. In contrast, if the two populations had identical RNA sequences, then the combined mixture would still be capable of saturating only 1% of the DNA. In an actual case, the saturating value obtained (somewhere between 1 and 2%) provides an indication of the similarity of the two populations. Figure 12-37 illustrates an experiment of this type in which RNA populations of several mouse tissues are compared. The results of the experiment shown in Fig. 12-37*b* suggest that the populations of RNA present in the liver and spleen or liver and kidney are partially, but not completely, overlapping. In other words, some of the RNAs represent tissue-specific gene products, while others may be common to many cell types. Although the information is not yet completely available, it is generally believed that, if one could add up all of the mRNAs present in all cells of the body at all stages in an organism's life cycle, it would still represent only a fraction (*perhaps* a maximum of 30%) of the total information content of the genome. Presumably, most of the remaining DNA sequences are responsible for producing regions of hnRNA molecules which are not translated.

3. *Are mRNAs synthesized from repeated sequences, nonrepeated sequences, or both? Is there any evidence of selective gene amplification, i.e., an increase in the number of selected DNA sequences?* The best way to discuss these two related questions is to select a particular protein and examine the possibilities in light of the particular mRNA that codes for its polypeptide chain(s). The best-studied protein with respect to this question is that of hemoglobin, therefore we will consider the mRNAs that code for the globin chains. Before we discuss the molecular hybridization studies, consider for a moment the genetic situation. The entire Mendelian concept rests on the presence of two genetic factors, one on each of the two homologous chromosomes, being responsible for each genetically determined trait. In molecular terms this becomes translated into one DNA sequence per haploid genome. Similarly, the analysis of hemoglobin mutants, such as that found in sickle cell anemia, indicates that all of the hemoglobin molecules in the sickle cell anemia victim are identical—they all possess the single amino acid substitu-

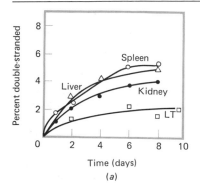

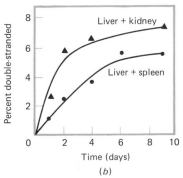

Figure 12-37. (*a*) Hybridization of RNA populations extracted from four different mouse organs to mouse unique DNA. Each preparation of RNA is complementary to a specific percentage of the total DNA sequences under saturating conditions. (*b*) Hybridization of mixtures of liver and kidney or liver and spleen RNA populations to unique DNA sequences. In both cases, an apparent saturation plateau was obtained which exceeded that obtained for RNA from either organ alone. (*Reprinted with permission from L. Grouse, M. D. Chilton, and B. J. McCarthy, Biochemistry, **11**:803, 1972. Copyright by the American Chemical Society.*)

tion. We can conclude from this type of evidence that there is either (1) a single copy of the DNA sequence for the globin polypeptides per haploid set of chromosomes or (2) there is more than one copy of the sequence, but the copies are always identical. In other words, if the globin polypeptides are coded by repeated sequences, then when a mutation occurs in one copy all of the copies somehow become altered. We will now turn to the molecular hybridization studies which bear on these questions.

In principle, the hybridization procedure to answer this question is somewhat the reverse of that used to answer the previous questions. Instead of a reaction mixture containing an excess of unlabeled RNA (RNA-driven) and a low concentration of radioactive DNA, in this case one uses a high concentration of fragmented unlabeled DNA (DNA-driven) and a low concentration of radioactive RNA. The unlabeled DNA of the reaction mixture consists of the total unfractionated DNA (both repeated and nonrepeated sequences) and a very low concentration of the particular radioactive mRNA to be tested. The reaction mixture is heated to convert the DNA to single strands and then incubated to allow complementary sequences to reanneal. Two types of hybrid molecules will be formed during the incubation. The formation of DNA-DNA hybrids represents a straightforward reannealing reaction and will follow the same type of course as that shown in Fig. 10-32a, the more repeated the sequence the more rapidly it will reanneal. The formation of DNA-DNA hybrids, i.e., the reannealing process, is monitored by measuring the absorbance of the unlabeled DNA in the double-stranded versus the single-stranded fraction at various times of incubation (page 460). The formation of the second type of hybrid, i.e., the DNA-RNA hybrid, is monitored by measuring the radioactivity in the same double-stranded versus single-stranded fractions in which the absorbance is measured. Since the concentration of the particular labeled mRNA is negligible in comparison to any given sequence (even nonrepeated ones) in the DNA, it will not affect the course of the reannealing reaction. Therefore, if the labeled sequence is complementary to a sequence in the repeated DNA, it should be found in the rapidly reannealing fraction (it should hybridize at a relatively low $C_0t \frac{1}{2}$). If it is complementary with a sequence in the nonrepeated fraction, it should take as long to find a partner as all of the nonrepeated DNA sequences; radioactivity should appear in the hybrid fraction only after very long times of incubation (high $C_0t \frac{1}{2}$).

Before describing the results of this experiment, an important modification must be mentioned. In actual fact, it is virtually impossible to obtain a preparation of mRNA with sufficient radioactivity to be able to use at such low concentrations and still be detectable. To get around this problem, as well as for other reasons, labeled mRNA is not used, instead a highly radioactive complementary DNA copy (a cDNA) is employed. The labeled cDNA is synthesized in vitro by an enzyme, reverse transcriptase, which uses RNA as a template for the synthesis of a complementary copy of DNA. The enzyme is prepared from certain RNA-containing tumor viruses (Chapter 19). The procedure is analogous to the formation of the cRNA of satellite DNA described on page 466. In this case the cDNA is made using the globin mRNA as a template. We can now consider the results. When a very low concentration of highly radioactive globin cDNA is present during the reannealing of total unlabeled DNA fragments, the radioactivity appears in the hybrid fraction with the same kinetics (same $C_0t \frac{1}{2}$) as the reannealing of the unlabeled *nonrepeated* DNA sequences (Fig. 12-38). In other words, the globin gene is represented, *as closely as can be determined,* by a nonrepeated DNA sequence. At most there could be only five or so copies.

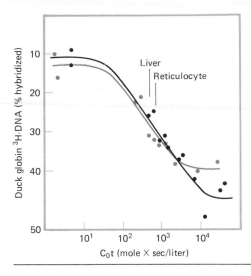

Figure 12-38. Reassociation of single-stranded, ^{3}H- duck globin DNA in the presence of a large excess of unlabeled duck reticulocyte or liver DNA. In both cases, the labeled DNA sequences hybridize with kinetics that suggest they are present at one or, at most, a few copies. (*From S. Packman, H. Aviv, J. Ross and P. Leder,* Biochem. Biophys. Res. Comm., **49**:816, 1972.)

Most importantly, the same results are obtained regardless of the tissue from which the DNA is extracted. DNA extracted from cells that are actively engaged in hemoglobin synthesis appear to have the same number of globin genes as cells from a tissue that is not involved in the synthesis of this particular protein (Fig. 12-38). Even though synthesis of hemoglobin can account for 90% of the protein being produced by certain cells of the erythrocyte line, there is no evidence of any increase in the number of globin genes to allow for increased transcription. It would appear that two genes per cell can account for the very large concentration of hemoglobin present within a red blood cell. The ability of a cell to produce such a large amount of protein coded for by such few genes provides an indication of the complex regulatory mechanisms that exist within a cell.

The only established case of selective gene amplification is that for ribosomal DNA in the oocyte, which is necessary as a result of the tremendous volume of the egg being packaged. Another very important reason for the need for amplification of rDNA and the lack of such amplification of DNA that codes for globin mRNA can be appreciated by considering their activities. Whereas one rRNA molecule can only be involved in the formation of one ribosome, one mRNA molecule can direct the formation of thousands of polypeptide chains.

The analysis of several other specific messenger RNAs has also indicated that they are transcribed from a DNA sequence present within the non-repeated fraction of the genome. An important exception to this condition is found in the case of the histone mRNAs which are transcribed from genes that are repeated to the extent of a few hundred copies per genome. Since histone synthesis must keep pace with DNA synthesis if chromosomes are to be produced, the larger number of copies must be required to meet the demands in rapidly dividing cells. It is believed that the same number of histone genes are present in all cells, i.e., there is no evidence of amplification of histone DNA in one or another cell type.

Up to this point in the discussion of this type of experiment, we have followed the hybridization of a labeled cDNA made from a specific mRNA. The experiment can also be performed using labeled cDNA preparations made from an entire mixed population of mRNA molecules. When the hybridization of this type of cDNA population is followed in the presence of an excess population of unlabeled total DNA, it is generally found that some of the cDNA sequences hybridize to repeated DNA fragments while others hybridize to nonrepeated ones (Fig. 12-39). In most cases less than 15% of the cDNA molecules hybridize to repeated sequences and in some cases these may represent molecules with both repeated and nonrepeated sections.

4. *Are mRNAs present in cells in different numbers of copies?* It has been known for a long time that certain cells engaged in the synthesis of a large amount of one or a few proteins have an extremely large number of mRNAs for these polypeptides. For example, globin mRNA can be present to the extent of about 2% of the total RNA in a cell whose protein-synthesizing activity is dominated by hemoglobin production. In addition to this type of determination, molecular hybridization has provided a means for the analysis of the abundance of mRNAs in a large population of different species. The procedure involves the kinetic analysis of the hybridization of mRNA molecules with a population of complementary cDNA molecules made from them.

The first step in this type of experiment is to purify the poly A-containing mRNA population, the second step is to use this entire population of mRNAs as templates for the production of complementary radioactive cDNA molecules with the viral reverse transcriptase, and the third step is to allow these two populations (the mRNAs and the cDNAs) to hybridize. Consider the effect of mRNAs present in the population at different frequencies. Specifically, consider two different species of mRNA, one present at one copy per cell and the other at 1000 copies per cell. When the mRNA population from these cells is used as a template for cDNA production, there will be 1000 times as many cDNAs having the sequence of the common mRNA as the rare one. When these two types of nucleic acid molecules (the mRNA and the cDNA) are allowed to hybridize, under conditions of RNA excess, the cDNA having the common sequence will hybridize to the RNA approximately one thousand times faster than the cDNA containing the rare sequence. The experiment is similar to that described in the analysis of the first question on page 573, the main differences being that: (1) Since mRNA was used as a template, the DNA population contains only those sequences that are present

Figure 12-39. Kinetics of the hybridization reaction between a small amount of pulse-labeled mouse L-cell RNA and a large excess of unlabeled mouse DNA. Light circles indicate the course of DNA-DNA reannealing as discussed in reference to Fig. 10-32. Dark circles indicate the percentage of the labeled RNA present appearing in a DNA-RNA hybrid. Labeled RNA molecules are found to be complementary to both repeated and nonrepeated DNA sequences. The greatest hybridization occurs between RNA and nonrepeated DNA which, however, constitutes the majority of the sequences of the genome. (*Reprinted with permission from L. Grouse, M. D. Chilton, and B. J. McCarthy,* Biochemistry, ***11:804, 1972.** Copyright by the American Chemical Society.*)

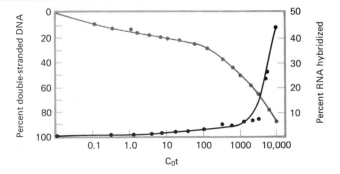

in the mRNA population and (2) the relative proportion of the various DNA sequences reflects the relative proportions of those sequences in the mRNA population rather than the proportions that happen to be present in the genome.

The experiment just described was originally performed by J. O. Bishop and coworkers in 1974 using mRNA extracted from human HeLa cells and has since been carried out on several other mRNA populations. The results from HeLa cells indicated that there were roughly three different classes of mRNA based on their average abundance in the population. Approximately 22% of the mRNAs were present in about 10^4 copies per cell and represented transcripts from an estimated 17 different genes. In other words, it appeared that a handful of genes accounted for nearly one-quarter of the mRNAs in the cell. Another fraction comprising approximately 28% of the mRNA were present in about 450 copies per cell and represented the transcripts of approximately 370 different genes. The remaining 50% of the mRNA sequences were present in about 10 copies per cell and represented the activities of an estimated 33,000 different genes. It would appear that the vast majority of the genes involved in mRNA production are responsible for producing rather few copies. Analysis of this same question in other types of cells provides similar results: a few genes are responsible for a large percentage of the mRNA molecules though a large number of genes are actively producing small numbers of transcripts. This type of analysis once again points out the regulation of gene expression and mRNA levels in eukaryotic cells.

REFERENCES

Arber, W., Prog. Nuc. Acid Res. Mol. Biol. **14,** 1–38, 1974. "DNA Modification and Restriction."

Barrell, B. G., Air, G. M., and Hutchison, C. A., III, Nature **264,** 34–41, 1976. "Overlapping Genes in Bacteriophage $\phi \times 174$."

Berg, P. B., et al., Proc. Natl. Acad. Sci. (U.S.) **72,** 1981–1984, 1975. "Summary Statement of the Asilomar Conference on Recombinant DNA Molecules."

———, Science **188,** 991–994, 1975. "Asilomar Conference on Recombinant DNA Molecules."

Birnstiel, M. L., Schaffner, W., and Smith, H. O., Nature **266,** 603–607, 1977. "DNA Sequences Coding for the H2B Histone of *Psammechinus miliaris.*"

Bishop, J. O., Cell **2,** 81–86, 1974. "The Gene Numbers Game."

———, Morton, J. G., Rosbash, M., and Richardson, M., Nature **250,** 199–204, 1974. "Three Abundance Classes in HeLa Cell Messenger RNA."

Brawerman, G., Ann. Rev. Biochem. **43,** 621–642, 1974. "Eukaryotic Messenger RNA."

Brown, D. D., Sci. Am. **229,** 20–29, 1973. "The Isolation of Genes."

——— and Dawid, I. B., Science **160,** 272–280, 1968. "Specific Gene Amplification in Oocytes."

——— and Stern, R., Ann. Rev. Biochem. **43,** 667–694, 1974. "Methods of Gene Isolation."

Cohen, S. N., Sci. Am. **233**, 24–33, July 1975. "The Manipulation of Genes."

———, Nature **263**, 731–738, 1976. "Transposable Genetic Elements and Plasmid Evolution."

———, Science **195**, 654–657, 1977. "Recombinant DNA: Fact and Fiction."

Daneholt, B., Cell **4**, 1–9, 1975. "Transcription in Polytene Chromosomes."

Darnell, J. E., Jelinek, W. R., and Molloy, G. R., Science **181**, 1215–1221, 1973. "Biogenesis of Messenger RNA: Genetic Regulation in Mammalian Cells."

Fiddes, J. C., Sci. Am. **237**, 54–67, Dec. 1977. "The Nucleotide Sequence of a Viral DNA."

Furuichi, Y., LaFiandra, A., and Shatkin, A. J., Nature **266**, 235–239, 1977. "5′-Terminal Structure and mRNA Stability."

Ghosh, S., Int. Rev. Cytol. **44**, 1–28, 1976. "The Nucleolar Structure."

Greenberg, J. R., J. Cell Biol. **64**, 269–288, 1975. "Messenger RNA Metabolism of Animal Cells."

Grobstein, C. S., Sci. Am. **237**, 22–33, July 1977. "The Recombinant DNA Debate."

Helinski, D. R., chairman, symposium on Plasmids, Federation Proc. **35**, 2024–2043, 1976.

———, Trends Biochem. Sci. **3**, 10–14, 1978. "Plasmids as Vehicles for Gene Cloning: Impact on Basic and Applied Research."

Herman, R. C., Williams, J. G., and Penman, S., Cell **7**, 429–437, 1976. "Message and Non-Message Sequences Adjacent to Poly (A) in Steady State Heterogeneous Nuclear RNA of HeLa Cells."

Hood, L. E., Wilson, J. H., and Wood, W. B., *The Molecular Biology of Eukaryotic Cells.* W. A. Benjamin, 1975.

Jeffreys, A. J., and Flavell, R. A., Cell **12**, 1097–1108, 1977. "The Rabbit β-Globin Gene Contains a Large Insert in the Coding Sequence."

Kedes, L. H., Cell **8**, 321–331, 1976. "Histone Messengers and Histone Genes."

Kolata, G. B., Science **192**, 645, 1976. Research News. "DNA Sequencing: A New Era in Molecular Biology."

———, Science **196**, 1187, 1977. Research News. "Overlapping Genes: More Than Anomalies."

Lappe, M., and Morrison, R. S., eds., symposium on "Ethical and Scientific Issues Posed by Human Uses of Molecular Genetics," Ann. N.Y. Acad. Sci. **265**, 1976.

Lewin, B., *Gene Expression*, vol. 2. Wiley, 1974.

———, Cell **4**, 77–93, 1975. "Units of Transcription and Translation: Sequence Components of Heterogeneous Nuclear RNA and Messenger RNA."

Macgregor, H. C., Biol. Revs. **47**, 177–210, 1972. "The Nucleolus and Its Genes in Amphibian Oogenesis."

Marx, J. L., Science **191**, 1160, 1976. Research News. "Molecular Cloning: Powerful Tool for Studying Genes."

———, Science **197**, 853, 1977. Research News. "Viral Messenger Structure: Some Surprising New Developments."

Maugh, T. H., Science **194**, 44, 1976. Research News. "The Artificial Gene: It's Synthesized and It Works in Cells."

Miller, O. L., Jr., and Hamkalo, B. A., Int. Rev. Cytol. **33**, 1–25, 1972. "Visualization of RNA Synthesis on Chromosomes."

Molloy, G., and Puckett, L., Prog. Biophys. Molec. Biol. **31**, 1–38, 1976. "The Metabolism of Heterogeneous Nuclear RNA and the Formation of Cytoplasmic Messenger RNA in Animal Cells."

Nathans, D., and Smith, H. O., Ann. Rev. Biochem. **44**, 273–293, 1975. "Restriction Endonucleases in the Analysis and Restructuring of DNA Molecules."

Nobel Lectures, Physiology and Medicine, 1942–1962. Elsevier, 1964.

Perry, R. P., Ann. Rev. Biochem. **45**, 605–630, 1976. "Processing of RNA."

Portmann, R., Schaffner, W., and Birnstiel, M., Nature **264**, 31–33, 1976. "Partial Denaturation Mapping of Cloned Histone DNA from the Sea Urchin *Psammechinus miliaris.*"

Proudfoot, N. J., and Brownlee, G. G., Nature **263**, 211–214, 1976. "3′ Non-Coding Region Sequences in Eukaryotic Messenger RNA."

Rosbash, M., Campo, M. S., and Gummerson, K. S., Nature **258**, 682–686, 1975. "Conservation of Cytoplasmic Poly (A)-Containing RNA in Mouse and Rat."

Sanger, F., et al., Nature **265**, 687–695, 1977. "Nucleotide Sequence of Bacteriophage $\phi \times 174$ DNA."

———— and Coulson, A. R., J. Mol. Biol. **94**, 441–448, 1975. "A Rapid Method for Determining Sequences in DNA by Primed Synthesis with DNA Polymerase."

Scott, W. A., and Werner, R., *Molecular Cloning of Recombinant DNA.* Academic, 1977.

Shafritz, D. A., et al., Nature **261**, 291–294, 1976. "Evidence for Role of $m^7G^{5'}$-Phosphate Group in Recognition of Eukaryotic mRNA by Initiation Factor IF-M_3."

Shanmugam, K. T., and Valentine, R. C., Science **187**, 919–924, 1975. "Molecular Biology of Nitrogen Fixation."

Sinsheimer, R. L., Ann. Rev. Biochem. **46**, 415–438, 1977. "Recombinant DNA."

Smith, J. D., Prog. Nuc. Acid. Res. Mol. Biol. **16**, 25–73, 1976. "Transcription and Processing of Transfer RNA Precursors."

Tartof, K. D., and Dawid, I. B., Nature **263**, 27–30, 1976. "Similarities and Differences in the Structure of X and Y Chromosome rRNA Genes of Drosophila."

Tilghman, et al., Proc. Natl. Acad. Sci. (U.S.) **75**, 1309–1313, 1978. "The Intervening Sequence of a Mouse β-Globin Gene Is Transcribed within the 15S mRNA Precursor."

Watson, J. D., *Molecular Biology of the Gene*, 3d ed. W. A. Benjamin, 1976.

Yankofsky, S. A., and Spiegelman, S., Proc. Natl. Acad. Sci. (U.S.) **48**, 1466–1472, 1962. "The Identification of the Ribosomal RNA Cistron by Sequence Complementarity. II. Saturation of and Competitive Interaction at the RNA Cistron."

CHAPTER THIRTEEN
The Control of Gene Expression

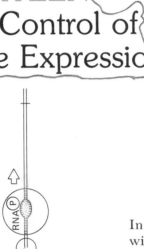

In the last three chapters we have covered a great variety of phenomena within the general topic of gene expression. The ultimate goal in the study of any biological process is the attainment of a full understanding of the means by which the process is controlled. In the area of gene expression, the central activity to be regulated is the synthesis of particular proteins. Each individual organism from a bacterium to a human inherits a full complement of genetic material with which to run its cellular affairs. If one considers the genome as a type of blueprint for survival, then it is clear that different measures are required to meet different types of challenges. Every bacterial cell lives in direct contact with its environment, a constantly changing chemical milieu. At certain times a given type of molecule may become available for use while at other times that compound is absent. It is of obvious selective advantage for these cells to utilize their available resources in the most efficient way, and mechanisms have evolved to allow them to respond to specific environmental changes by selective gene expression. Multicellular organisms, on the other hand, face a different type of requirement for the selective utilization of their genetic inheritance. From its evolutionary onset, the appearance of the multicellular state has been accompanied by the cellular division of labor. The opportunity for cell specialization has resulted in the evolution of a large number of different types of cells within complex eukaryotic organisms, each drawing upon a different but overlapping subset of the total number of DNA sequences. In this chapter we will explore some of the concepts that have been formulated to explain the manner in which the phenotypes of cells are generated, maintained, and/or altered by the control of specific protein syntheses. We will begin the discussion with genetic control mechanisms that exist within prokaryotic cells, where at least some types of control are well understood. With these types of systems as background we will discuss genetic regulation in eukaryotic cells in which the problems are much more complex and the cells more difficult to study.

PROKARYOTES: THE CONCEPT OF THE OPERON

In Chapter 4 metabolic pathways were divided into two broad groups, catabolic types and anabolic types, depending upon whether the enzymes were responsible for the breakdown of molecules into less complex compounds, or the reverse. Consider the consequences of transferring a culture of bacteria from a minimal medium to one containing either (1) lactose or (2) tryptophan.

1. Lactose is a disaccharide (Fig. 13-1) of glucose and galactose whose oxidative breakdown can provide the cell with metabolic intermediates and energy. The first step in the breakdown of this molecule is the hydrolysis of the bond (a β-galactoside linkage) that joins the two sugars, a reaction catalyzed by the enzyme β-galactosidase. When these bacterial cells were growing under the

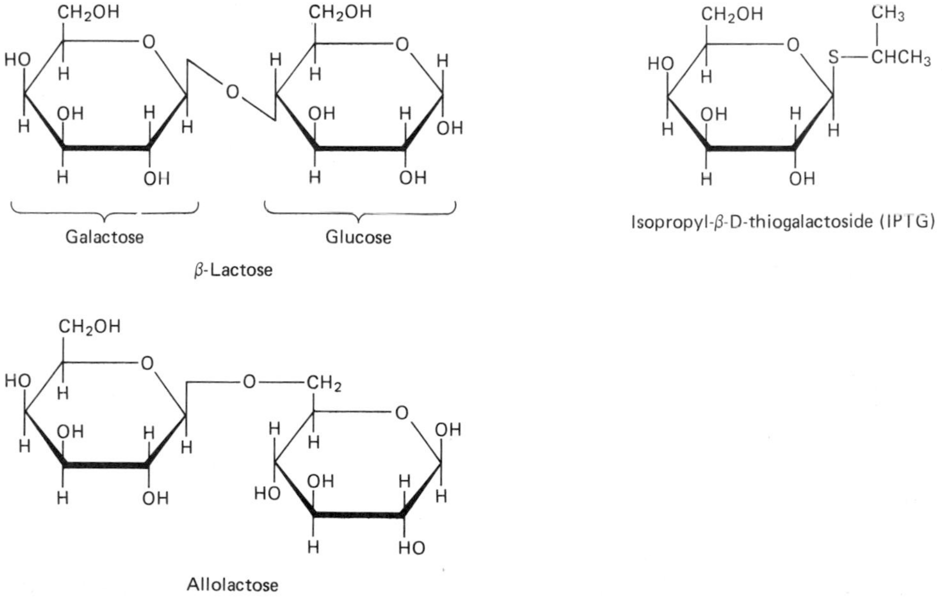

Figure 13-1. The structure of lactose, a disaccharide which is commonly used to induce the synthesis of β-galactosidase in E. coli; allolactose, which is formed from lactose and serves as the actual inducer within the cell; and IPTG, an artificial inducer of the enzyme and one which is not altered by the enzyme.

minimal conditions prior to being transferred, there was no need for the production of β-galactosidase and, had an analysis been made at that time, one would have found fewer than approximately five copies of the enzyme and one copy of its mRNA in the average cell. The introduction of lactose into the culture medium leads to the production of approximately 1000 times the number of β-galactosidase molecules in these cells. The presence of lactose has resulted in the *induction* of the synthesis of this enzyme (Fig. 13-2).

2. Tryptophan is an essential amino acid and, in the absence of this compound in the medium, the bacterium must expend a certain amount of energy for its synthesis to ensure its availability for incorporation into protein. Cells growing in the absence of tryptophan contain the enzymes, and their corresponding mRNAs, needed for tryptophan manufacture. If, however, this substance should suddenly become available, there is no reason for the cell to continue to produce the enzymes of this anabolic pathway when their presence is no longer needed. It is found that, under these conditions, the production of the mRNAs for these proteins stops; the genes responsible are *repressed*.

The *Lac* Operon

The genetic and biochemical basis for the phenomena of induction and repression was formulated by Jacob and Monod in 1961 on the basis of an extensive series of investigations carried out primarily in their laboratory. This paper, published in the Journal of Molecular Biology (3:318, 1961), set forth the principal elements in the proposal for two of the most important concepts in the field of molecular biology, that of the messenger RNA and of the *operon*. We have already considered the section of this paper dealing with the messenger RNA (page 489), we will now turn to their proposal of the operon. The question that had to be answered was how the presence of an inducer, such as lactose [the actual inducer

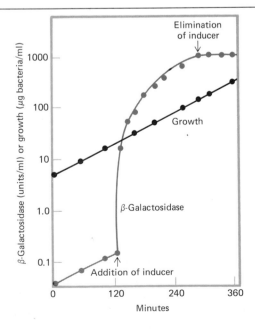

Figure 13-2. The kinetics of β-galactosidase induction in *E. coli*. The inducer is IPTG. *(From D. S. Hogness, in J. L. Oncley, ed., "Biophysical Science; A Study Program," Wiley, 1958.)*

is allolactose (Fig. 13-1), a derivative of lactose] could evoke the synthesis of an enzyme responsible for its hydrolysis. There were several important pieces of evidence they had to consider.

One of the clues to the puzzle was the finding that there was no precise structural relationship between the ability of a molecule to act as a substrate of β-galactosidase and its ability to induce the synthesis of the enzyme. It was found that all active inducers were β-galactosides, but not all of them were capable of being hydrolyzed by the enzyme. For example, one of the most potent inducers (Fig. 13-1), IPTG (isopropyl thiogalactoside), is not hydrolyzed by the enzyme. Therefore, the ability of a compound to call forth the transcription of β-galactosidase did not appear to be related to its ability to directly bind to the enzyme. Jacob and Monod proposed that the action of the inducer was mediated via its binding to some unknown component (or "receptor") in the cell.

Another clue to the mechanism of induction was the finding that the induction of β-galactosidase was always accompanied by the induction of two other proteins: galactoside permease, a membrane protein which is responsible for allowing the galactoside into the cell, and thiogalactoside acetyltransferase, an enzyme which transfers an acetyl group from AcCoA to the galactoside. The genes, termed z, y, and a, for these three functions, were found to map together in the same region of the chromosome (see Fig. 13-4) and were presumably adjacent to one another. The relative amounts of these three enzymatic activities remained constant regardless of the nature and concentration of the inducer. It was proposed that the expression of the three genes responsible for these proteins was coordinately controlled via the action of some common element. On the basis of their data, Jacob and Monod reasoned that the common element governing the rate of synthesis of these three proteins was not represented by the *structural genes* (z, y, or a) that coded for these proteins, but rather some other genetic element they termed a *regulator gene* (denoted as i). If, indeed, there were two different classes of genes, structural and regulator genes, then mutations should be detectable which affect the regulatory abilities of the cell without affecting the nature of the enzymes themselves. Various mutations of this sort had been discovered. Mutations of the i gene (termed i$^-$) resulted in the production of the proteins coded for by the structural genes regardless of whether inducer was present or not. Enzyme synthesis under these conditions is termed *constitutive,* signifying its continual production regardless of environmental conditions. The availability of mutations in the regulator gene provided the opportunity to map their sites on the chromosome. It was found that the i$^-$ mutations were located in the same region of the chromosome as the three structural genes, but separated from them by a definite space (see Fig. 13-4).

As the result of work by various bacterial geneticists during the 1950s, conjugation techniques were developed whereby pieces of one bacterial chromosome could be introduced into a recipient cell in a way that caused the recipient to be diploid for the particular genes that were transferred. Jacob, Monod, and coworkers set out to test the properties of cells that were

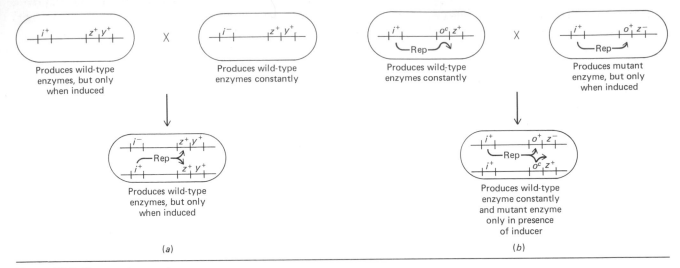

(a)

(b)

Figure 13-3. The genotypes and phenotypes of bacterial cells before and after conjugation as discussed in the text. (The o gene is not shown in part *a* simply because the operator is discussed at a later point in the text.)

made diploid for the various lactose genes. They found, for example, if a piece of one chromosome containing the genes i⁺z⁺y⁺ was introduced into a cell whose own chromosome was i⁻z⁺y⁺, the resulting diploid individuals (i⁻z⁺y⁺/i⁺z⁺y⁺) were wild-type in nature, i.e., they were inducible (Fig. 13-3a). In other words, a cell i⁻z⁺y⁺ that formerly had been making β-galactosidase in a constant and unregulated manner, i.e., constitutively, was once again placed under control by the inducer as a result of receiving a piece of DNA that contained an i⁺ allele. This finding indicated that a wild-type i⁺ gene on one piece of DNA could control the expression of structural genes on an entirely separate piece of DNA. The i⁺ gene was said to be *trans dominant*. They reasoned that, in order to accomplish this feat, the i gene must be responsible for the production of an actual regulatory product, one which could diffuse through the space of a cell and interact with another piece of DNA. They termed the product of the i gene a *repressor*, because it acted to repress the expression of the structural genes in the absence of an inducer. Although Jacob and Monod had no indication of the nature of the repressor, and in fact believed it to be of a nonprotein nature, they proposed that the small molecular weight inducers acted as antagonists of the repressor by forming a stereospecific complex with it (see Fig. 13-6).

Having established that the expression of the structural genes z, y, and a is controlled by a product of the i gene, the next point is to consider the manner in which the repressor accomplishes the inhibition. Jacob and Monod assumed that the repressor must act by combining with some type of control element involved in the flow of information from gene to protein. They termed this control element the *operator(o)*. Since the nature of the operator must be genetically determined, one should be able to find mutants in which the specific affinity of the operator for the repressor is altered or abolished. The search for operator mutants turned up a class of mutations that mapped within the *lac* region of the *E. coli* chromosome between the i gene and the z gene, next to the latter. Bacteria carrying these mutations (termed oᶜ) produced the three *lac* enzymes constitutively, presumably because the repressor was no longer capable of complexing to the operator to inhibit the transcription of the structural genes. When a piece of DNA containing $i^+o^cz^+$ was introduced into a cell $i^+o^+z^-$ (Fig. 13-3b), the diploid recipient began producing β-galactosidase constitutively. The o^c allele is said to be *dominant* over the wild type because it cannot be shut off by the repressor produced by the i^+ genes. The site of the o^c mutations identified the genetic position of the operator gene, but not the nature of the operator itself.

Analysis of the products of the structural genes in various types of diploid cells provided valuable information on the nature of the operator. Consider the diploid just described, $i^+o^+z^-/i^+o^cz^+$. It was found in this particular case that in the *presence* of an inducer, both the z^+ and z^- genes were active. The z^+ activity was demonstrated by the presence of β-galactosidase activity and the z^- by the presence of an altered protein that had no enzymatic activity but was still able to react with antibodies against the enzyme. However, in the absence of the inducer, only the active form of the enzyme, i.e., the product of the z^+ gene, is detected. Conversely, in the diploid $i^+o^+z^+/i^+o^cz^-$, in the absence of the inducer, only the z^- gene is expressed. These results indicated that the o^c allele only allows the constitutive expression of the structural genes in the same DNA molecules as the operator mutation itself. The o^+ allele on the other DNA molecule remains fully repressor-sensitive, even though the o^c allele is present in the cell. Unlike the i^- mutants, the o^c mutants are *cis dominant*. This type of data, along with certain other types of i gene mutations, led Jacob and Monod to conclude that the operator did not act by producing a cytoplasmic product but rather was a site on the chromosome to which the repressor combined.

Jacob and Monod concluded that the operator was a site on the DNA adjacent to the structural genes which determined whether or not the structural genes would be transcribed. In the absence of lactose, the

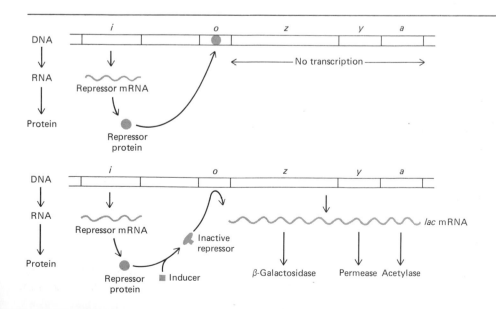

Figure 13-4. The mechanism of repression and induction in the *lac* operon. In the upper diagram it is seen that, in the absence of inducer, the product of the i gene acts to repress transcription of the structural genes as a consequence of its binding to the operator site. In the lower diagram, the presence of the inducer leads to the formation of an inactive repressor complex, thereby allowing transcription of the structural genes to proceed.

product of the i gene was combined with the operator and no transcription of the three structural genes could occur (Fig. 13-4). However, when lactose was present in the medium, it entered the cells, combined with the repressor and altered the binding properties of the repressor such that it no longer attached to the operator site. In the absence of a repressor-operator complex, transcription of the structural genes was maintained. The i⁻ and oᶜ mutants were constitutive due to their inability to form an effective operator-repressor complex. The i⁻ cells produced a faulty repressor while the oᶜ cells had a faulty binding site. They coined the term operon to refer to a genetic unit consisting of an operator gene and the adjacent structural genes, which were coordinately expressed. Operons were envisioned as the means by which functionally related enzymes could be simultaneously controlled with a minimum of regulatory machinery. It was soon demonstrated that the basis for the coordinate expression of the three structural genes resided in the transcription of a single messenger RNA molecule containing the information for all three polypeptides (Fig. 13-4). The means by which repressor-DNA interaction blocks transcription is discussed later in this chapter.

The Repressor and Its Interaction with DNA

The concept of the operon introduced in 1961 was a farsighted proposal and one which has stimulated a tremendous amount of research. Jacob and Monod had completely described the genetic basis for the operon concept; the most pressing questions concerned the nature of the repressor and its mode of action with both the inducer and the operator. The greatest difficulty in attempting to purify the repressor, which was shown to be a protein, was its very low concentration (approximately 0.002% of the protein or about ten copies per cell) in wild-type bacteria. Despite the obstacles, Walter Gilbert and Benno Müller-Hill began an attempt to purify the repressor. Their approach depended on the expected property of the repressor to bind inducer molecules in a highly specific manner. Their first goal was to demonstrate that cells contained a specific inducer-binding protein. They did this by the technique of *equilibrium dialysis* (Fig. 13-5). They placed a very concentrated extract from bacterial cells within a closed cellophane bag (a dialysis bag) and placed the bag into a solution containing radioactive IPTG, the synthetic inducer of the *lac* operon. Whereas macromolecules in the cell extract were trapped in the bag, the small molecular weight inducer molecules were capable of passing freely in and out through the cellophane so that the concentra-

Figure 13-5. Schematic representation of the assay for the purification of the repressor protein by equilibrium dialysis. The presence of the repressor protein within the dialysis bag leads to the increased concentration of radioactivity in that compartment.

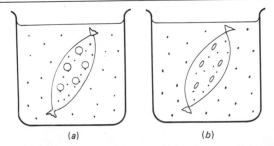

(a)   (b)

tion of the IPTG molecules *in solution* on both sides of the cellophane rapidly became equalized. The labeled inducer molecules bound to repressor within the bag was then reflected in an increase in the radioactivity of the bag's contents as opposed to a comparable volume of the surrounding fluid (Fig. 13-5). When extracts from mutant bacterial cells that were unable to produce the repressor were tested, there was no evidence of excess radioactivity within the bag at equilibrium; the binding of IPTG was specific for the repressor. Once they had demonstrated the presence of the repressor protein in the extract from wild-type cells, they had the necessary assay (page 109) to allow it to be purified. The purification of the repressor was greatly aided by the isolation of mutants that produced repressor in very large quantities. The repressor, a protein of approximately 160,000 molecular weight made of four identical subunits, was soon purified in sufficient quantities to allow its amino acid sequence to be determined.

The properties of the purified repressor were found to be exactly those predicted by Jacob and Monod. The repressor was capable of tightly binding to DNA of the *lac* regions but only if prepared from o⁺ cells; DNA-containing mutant operator sites did not effectively bind the repressor. The maintenance of the repressor-operator complex was sensitive to the presence of inducer. If IPTG was added to a preparation of DNA to which repressor molecules were complexed, the interaction of the repressor and inducer changed the affinity of the repressor such that the DNA was released. One model of the repressor-operator interaction is shown in Fig. 13-6.

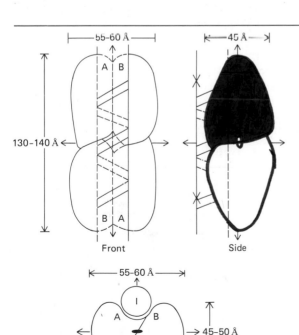

Front   Side

Top

Figure 13-6. A schematic drawing showing approximately to scale the major features of a proposal for the mode of repressor-operator DNA interaction. The tetrameric repressor is visualized in the electron microscope as an elongated, dumbbell-shaped molecule. Each subunit would have two different surfaces, A and B, which can interact with the DNA, resulting in the possibility of two operator-binding sites (I and II) per repressor. In this model, it is proposed that the repressor binds to the DNA with its long axis aligned with the long axis of the DNA. The two-fold axis of the operator sequence would be coincident with the two-fold axis of the repressor protein. As indicated in the diagram, the DNA would fit between the four subunits, perhaps within a groove much like a hotdog fits in a hotdog bun. [*From T. A. Steitz, T. J. Richmond, D. Wise, and D. E. Engelman*, Proc. Natl. Acad. Sci. (U.S.), **71**:596, 1974.]

The most recent approach to the analysis of the interactions that govern the regulation of the *lac* operon has centered on the determination of the nucleotide sequence of the control region (Fig. 13-7). The identification of the particular site to which the repressor binds has been determined by taking advantage of the fact that the interaction of a nucleic acid with a protein protects the nucleic acid from nuclease digestion. In this case, the DNA-repressor complexes are incubated with DNase, after which the enzyme can be removed, the repressor-DNA complex dissociated, and the protected DNA fragment sequenced. When this is carried out it is found that 27 base pairs of DNA (shown in Fig. 13-7) are present in the protected fragment and represent the basic part of the operator site. An important aspect of the operator sequence is its extensive two-fold symmetry (Fig. 12-28). Sixteen of the 27 base pairs are involved in the symmetry (a palindrome, page 562). It is believed that base sequence symmetry is an important property in the binding of proteins, such as the symmetrical repressor molecule to the DNA.

When a similar type of nuclease protection experiment is carried out with *lac* DNA to which RNA polymerase molecules are bound at the point where initiation occurs, the basis for the action of the repressor protein becomes apparent. The site where transcription of the *lac* operon begins overlaps to a considerable extent with the site to which the repressor binds (Fig. 13-7). In other words, the promoter and operator regions of the *lac* operon share the same piece of DNA and, therefore, the presence of the repressor attached to the DNA would physically block the polymerase from reaching its "start site." It has been proposed that the site where the RNA polymerase first enters the *lac* operon is not the same site at which transcription is initiated, but rather is approximately 35 bases "upstream" in a region of the DNA that is AT-rich (Fig. 13-7, 13-8). As a result of its high AT content, this section of the DNA would be expected to undergo strand separation more readily than other regions and may represent the point where the polymerase first causes the duplex to open. From this point, the polymerase would drift "downstream" to the initiation site where, in the absence of a bound repressor, RNA synthesis would begin. Analysis of both the DNA and the mRNA indicate that the AUG codon of the β-galactosidase gene is 38 bases from the 5' end of the polycistronic mRNA.

Positive Control by cAMP

The control exerted over the *lac* operon by the product of the i gene can be considered as *negative control* since the interaction of the DNA with this protein inhibits the ability of the *lac* operon to be expressed. Although it is more difficult to conceive of a molecular mechanism, one might imagine that transcription of certain genes might be under positive control in which the product of some regulatory gene actually promotes

Figure 13-7. The DNA sequence of the regulatory region of the *lac* operon. Indicated above the sequence are the proposed locations of the i gene, the promoter (CAP and RNA polymerase sites), the o gene, and the first part of the z gene. The regions of symmetry in the CAP site and o are shown as is the mRNA start site. *(From R. C. Dickson, J. A. Belson, W. M. Barnes, and W. S. Reznikoff,* Science, **187**:32, 1975.)

(a)

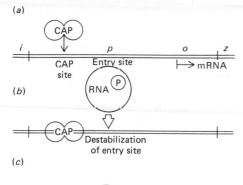

(b)

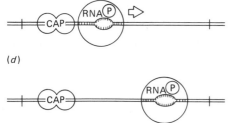

(c)

(d)

Figure 13-8. A model for the initiation of *lac* transcription. The diagram is drawn to scale with RNA polymerase [RNA (P)] shown as a sphere 31 nucleotide pairs in diameter and CAP shown as two spheres each 11 nucleotide pairs in diameter. The CAP site is located at the region of symmetry noted in the previous figure. The entry site is positioned at an AT-rich block in the sequence and the start site is at the 5′ end of the *lac* mRNA. (a) CAP binds to the CAP site. (b) CAP destabilizes the entry site, thereby facilitating RNA polymerase entry. (c) RNA polymerase in the entry site "drifts" to the start site. (d) RNA polymerase at the start site. *(From R. C. Dickson, J. A. Belson, W. M. Barnes, and W. S. Reznikoff, Science, **187**:34, 1975.)*

transcription rather than repressing it. Surprisingly, further investigation of the *lac* operon revealed that it was also under positive control.

The study of this second type of control circuit for the *lac* operon dates back to an early observation concerning bacterial metabolism. It was found that, if bacterial cells were supplied with glucose, as well as a variety of other types of substrates such as lactose, arabinose, or galactose, the cells would metabolize the glucose and ignore the other compounds. The presence of glucose, which is a particularly rich source of cellular energy and intermediates, served to repress the production of various enzymes such as β-galactosidase, which were needed to degrade these other substrates. The phenomenon was termed *catabolite repression* and was studied primarily in the laboratory of Boris Magasanik. In 1965 an unusual finding was made: cyclic AMP, previously believed to be involved only in eukaryotic metabolism, was detected in cells of *E. coli*. Most importantly, it was found that the concentration of cAMP in the cells was related to the presence of glucose in the medium; the higher the glucose concentration, the less the cAMP. The relationship between cAMP and catabolite repression was soon discovered when Ira Pastan and coworkers found that, despite the presence of glucose in the medium, the addition of cAMP, and only cAMP, served to stimulate the synthesis of all of the enzymes normally subject to catabolite repression.

Although the means by which glucose lowers the concentration of cAMP remains obscure, the mechanism by which cAMP overcomes the effect of glucose has been elucidated. As might be expected, a small molecule such as cAMP (Fig. 5-46) could not serve to directly stimulate the expression of a variety of specific genetic loci. As in the case of its role as a second messenger in the response of cells to hormones, cAMP acts to change the conformation of a particular protein, in this case the *catabolite gene-activator protein* (CAP). Like the repressor, CAP has been shown to interact with a specific symmetrical segment of the *lac* control region (Figs. 13-7, 13-8), but only when complexed with cAMP. The binding of the cAMP-CAP complex to DNA somehow acts to promote the transcription of the *lac* operon. In the absence of glucose, i.e., under conditions when the *lac* operon is inducible, the cAMP-CAP complex is always associated with the chromosome, thereby providing the potential for the structural genes to be transcribed. Whether transcription actually occurs or not depends on the presence or absence of the bound repressor. The means by which CAP exerts its positive control over transcription remains unclear, though it is believed to be mediated by its effect on the RNA polymerase-DNA interaction.

The *Trp* Operon

In the beginning of this section it was indicated that the presence of a compound such as tryptophan serves to repress the synthesis of the enzymes responsible for its manufacture. The original model for the operon proposed by Jacob and Monod was designed to apply to both inducible enzymes such as β-galactosidase and repressible enzymes such as those of the tryptophan biosynthetic pathway. The primary difference between the two types of regulatory mechanisms is the relationship between the operator site, the i gene product, and the small molecular weight ligand, lactose (actually allolactose) or tryptophan. In the tryptophan *(trp)* operon (Fig. 13-9*a*), which includes the structural genes for five coordinately produced proteins, the product of the i gene by itself is not capable of attaching to the operator DNA. It is the complex formed by the association of the repressor protein with tryptophan itself (Fig. 13-9*b*) that serves as the active repressor. In the absence of tryptophan, the operator site is open to the binding of an RNA polymerase, and the transcription of the long polycistronic mRNA can occur. When tryptophan becomes available, the enzymes are no longer needed and the tryptophan-repressor protein complex blocks transcription.

As in the case of the analysis of the *lac* operon, subsequent studies beyond the initial proposal of Jacob and Monod, have uncovered a second means by which the expression of the *trp* operon is controlled. As a result of various types of experiments by Charles Yanofsky and coworkers, it has been found that, in the presence of excess tryptophan, those mRNAs which are initiated despite the presence of the active repressor are terminated prematurely after approximately 130 to 140 nucleotides have been incorporated. Fragments of this length, representing the 5′ end of the natural polycistronic message of the *trp* operon, can be detected in extracts of cells growing in the presence of tryptophan. When tryptophan is absent, the termination of transcription at this early site, termed the *attenuator site,* does not occur and the complete message is produced. Although the mechanism by which tryptophan brings about this premature termination has not been completely unraveled, it appears to be mediated by the charged tryptophanyl-tRNA molecule. The reason why the *trp* operon should be subject to both repression and attenuation in the presence of tryptophan is unclear, but it serves to once again indicate the variety of ways in which gene expression can be regulated.

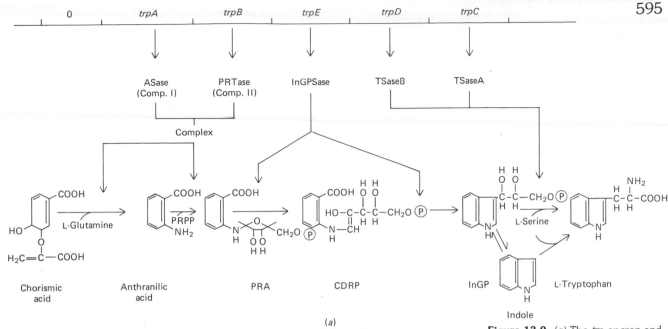

(a)

Figure 13-9. (a) The *trp* operon and the gene-enzyme relationships in the tryptophan biosynthetic pathway in *Salmonella typhimurium*. (*From G. Wuesthoff and R. H. Bauerle*, J. Mol. Biol., **49**:172, 1970.) (b) The mechanism of repression in the *trp* operon. In the absence of tryptophan in the medium, the product of the i gene is unable to bind to the repressor, thereby allowing the transcription of the structural genes to proceed. In the presence of tryptophan, the formation of repressor-tryptophan complex converts the protein into an active state and transcription is blocked.

(b)

EUKARYOTES
The Interphase Chromosome

The analysis of genetic control mechanisms in prokaryotic cells has been greatly aided by two important properties of their chromosomes: the amount of DNA is limited and that which is present is not permanently associated with other macromolecules. The greatly increased complexity of both the genome *and* the chromosome of higher organisms suggests a priori that different and more complicated regulatory mechanisms are to be expected. With the acceptance of the operon concept in bacteria, an extensive search for similar types of regulatory systems in eukaryotic cells was begun. Although recent data has pointed to the possibility of operons in a fungus, *Aspergillus*, very little evidence has been obtained for the existence of these types of control circuits in higher cells. Consequently, different types of models have been proposed to account for the control of

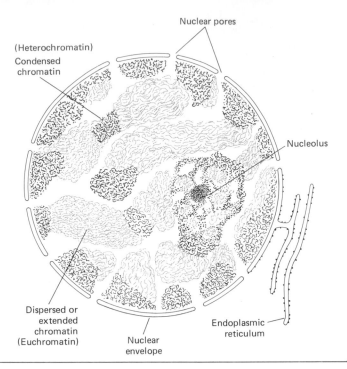

Nuclear pores

(Heterochromatin)
Condensed
chromatin

Nucleolus

Dispersed or
extended
chromatin
(Euchromatin)

Nuclear
envelope

Endoplasmic
reticulum

Figure 13-10. Schematic interpretation of the state of chromatin in the interphase nucleus. The condensed portions of the chromosomes (heterochromatin) are relatively inactive. The extended or uncoiled segments (euchromatin) are found to be the sites of active transcription. *(From W. Bloom and D. W. Fawcett, "A Textbook of Histology," 10th ed., Saunders, 1975.)*

gene expression in higher organisms and later in this chapter we will explore one of these proposals in some detail. However, the first topic to consider is the organization of the eukaryotic chromosome, paying particular attention to any clues it might provide as to the molecular mechanism involved in its expression.

When viewed in the electron microscope most subcellular organelles reveal a structure that provides considerable insight into the organization and function of the organelle. Unfortunately, the analysis of chromosome structure has proven to be a particularly difficult undertaking. Chromosomes seem to appear out of nowhere during prophase and disappear once again during telophase, thereby providing cytologists with one of the most important, yet challenging questions, namely, what is the nature of the chromosome in the nonmitotic, i.e., interphase, cell? Electron micrographs of thin sections cut through the nucleus generally reveal little of the nature of the interphase chromosome besides an indication that it is fibrous and can exist in either a condensed, compact configuration or in a more dispersed, diffuse state (Figs. 13-10 and 13-11). The diameters of the fibers present in the nucleus are variously reported to range between 100 and 300 Å, and numerous models have been formulated to account for DNA-containing fibers having a diameter in this range.

Whereas the organization of the chromosome has proven difficult to resolve, its composition has long been known to consist of DNA, associated protein, and RNA. The term *chromatin* is used to denote this complex of macromolecules which makes up the chromosomal material of the nonmitotic cell. The proteins of chromatin are generally divided into

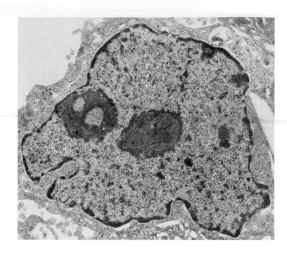

Figure 13-11. Electron micrograph of an interphase HeLa nucleus showing a pair of nucleoli and scattered heterochromatin. Heterochromatin is particularly prevalent around the entire inner surface of the nuclear envelope. Similarly, there is an intimate association between the perinucleolar chromatin and the nuclear envelope. *(From W. W. Franke,* Int. Rev. Cytol. Supp., **4:**130, 1974.)

two major groups, the histones and the nonhistones. The histones, as previously mentioned (page 570), represent a collection of well-defined basic proteins while the nonhistone chromosomal proteins include a large number of uncharacterized species. Both of these groups of proteins will be discussed in detail. The RNA of chromatin is a relatively minor component (approximately 3%) and is composed primarily of nascent RNA chains at various stages in the process of being elongated.

The Interaction between Histones and DNA

The five classes of histone (Fig. 13-12) are most readily distinguished from each other on the basis of their content of the two basic amino acids, arginine and lysine (Table 13-1). The "arginine-rich" classes are termed H3 and H4, the "slightly lysine-rich" classes are termed H2A and H2B, and the "lysine-rich" class is termed H1. Of these five classes, the only one that generally shows heterogeneity, i.e., variants with respect to amino acid sequence, is the H1 class. The other four histones are typically represented by only one species per class. Amino acid sequence analysis of the various histones among different organisms has revealed several im-

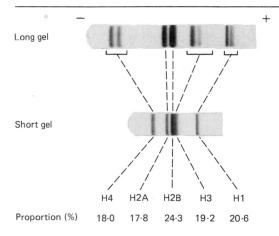

Long gel

Short gel

	H4	H2A	H2B	H3	H1
Proportion (%)	18·0	17·8	24·3	19·2	20·6

Figure 13-12. Acrylamide gel electrophoresis of histone proteins. Fractionation of a mixture of histones on short gels produces five bands representing the five major classes of histones—H1, H2A, H2B, H3, and H4. If the mixture is run on longer gels to obtain greater separation, each of the five classes is seen to be heterogeneous. In the case of the H1 group, the various subfractions are believed to represent different gene products. In other words, the H1 histones represent a family of related genes. The subfractions of the other histone classes are believed to represent different posttranslational modifications of the same initial polypeptide. *(Photographs by R. Chalkley, in B. Lewin,* "Gene Expression," *vol. 1,* "Bacterial Genomes," *Wiley-Interscience, 1974.)*

TABLE 13-1
Characterization of Calf Thymus Histones[a]

Class	Fraction	Minimum number of subfractions	Ratio of lysine to arginine	Total residues	Molecular weight
Very lysine-rich	I (f1)	3–4	20	~215	~21,000
Lysine-rich	IIb1 (f2a2)	2	1.25	129	~14,500
	IIb2 (f2b)	0–2	2.5	125	13,774
Arginine-rich	III (f3)	3	0.72	135	15,324
	IV (f2a1)	4	0.79	102	11,282

[a] SOURCE: R. J. DeLange and E. L. Smith, *Accts. of Chem. Res.*, **5**:368 (1972); *ARB*, **40**:279 (1971). From DNA SYNTHESIS by A. Kornberg. W. H. Freeman and Company. Copyright © 1974.

portant characteristics. First of all it has been shown that the evolution of these proteins has been extremely conservative, particularly that of the two arginine-rich classes. For example, it has been found that there are exactly 102 amino acids that make up the H4 histone in both the pea plant and the calf and that between these advanced members of the plant and animal kingdom, there have been only two amino acid changes, both of which are physiologically minor. The extreme conservatism among the histones suggests that each and every amino acid is of importance in their function. Another important characteristic of histone structure is that in each type of molecule the basic amino acids tend to be clustered into one-half of the molecule, leaving the remaining part with a relatively hydrophobic, uncharged character. When the relative amounts of histone and DNA in chromatin are measured, a basic constancy in the relative proportions of each type of macromolecule is found: the DNA/histone weight ratio is approximately 1 : 1. If the relative amounts of the individual histone classes are determined, these too generally reveal a distinct stoichiometry; the H2A, H2B, H3, and H4 classes are present in approximately equal numbers of molecules, while that of H1 is generally present at a more variable level.

One of the overriding questions in the analysis of the molecular organization of chromatin concerns the nature of the interaction between the DNA and the histones, a topic which has been approached using a wide variety of techniques. An important observation in this area was made in the early 1970s when chromatin was subjected to digestion by certain types of nucleases. It was found that, when chromatin was incubated with staphylococcal nuclease under mild conditions, most of the DNA was converted to fragments of approximately 200 base pairs in length or some multiple of that size. This finding suggests that uniform pieces of DNA are being protected from enzymatic attack, presumably by their tight association with protein. In 1974, using the evidence obtained from nuclease digestion together with various types of biochemical data and the results from x-ray diffraction analysis, Roger Kornberg proposed a new type of structure for chromatin. He proposed that the basic structure of chromatin consisted of a repeating subunit composed of DNA and histones. He proposed that each subunit was composed of a stretch of DNA approxi-

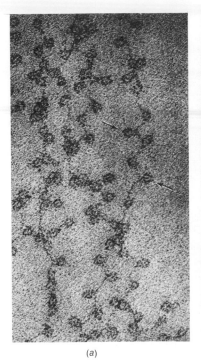
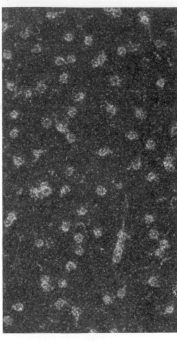

(a)   (b)

Figure 13-13. *(a)* Electron micrograph of a negatively stained preparation of chromatin fibers streaming out of a chicken erythrocyte nucleus. The v bodies (nucleosomes) are approximately 100 Å in diameter and the connecting strand is approximately 140 Å in length. *(From D. E. Olins and A. L. Olins in P. O. P. Ts'o, ed., "The Molecular Biology of the Mammalian Genetic Apparatus," Elsevier/ North Holland Biomedical, 1977.)* *(b)* A dark-field electron micrograph of a negatively stained preparation of poly (dI-dC)·poly (dI-dC) reconstructed with inner histones. *(Courtesy of D. E. Olins and A. L. Olins.)*

mately 200 base pairs in length together with eight molecules of histone, two each of H2A, H2B, H3, and H4. It was suggested that the histone-DNA subunits were connected to each other by stretches of DNA so that the overall structure was essentially that of a beaded string. The DNA filaments between the subunits act like structural joints allowing the chromatin fiber to fold back on itself or to coil into a more compact superstructure. The histone-DNA subunits have been termed *nu bodies*, or *nucleosomes*. At about the same time as the Kornberg proposal was made and soon after, a number of different reports were published which provided considerable substantiating evidence for this model and, like the fluid mosaic model for membrane structure, it has become rapidly accepted. Electron microscopic observations of chromatin that had been gently prepared from lysed nuclei provided a clear portrait of chromatin fibers (Fig. 13-13a) that corresponded perfectly with the subunit model. These electron micrographs depicted the chromatin strands as being made of beads of approximately 100 Å diameter, connected to one another by strands of DNA 15 to 25 Å in diameter. This portrait is particularly apparent after the H1 histone molecules are removed, an extraction accomplished by subjecting the chromatin preparation to $0.7M$ NaCl before observation. Whereas the H2A, H2B, H3, and H4 species are present within the nucleosome particles themselves, the H1 protein appears to be associated with the links of DNA connecting the nucleosomes (Fig. 13-14).

The core of the nucleosome is composed of the eight histone molecules while the DNA is wrapped around the outside of the histone aggre-

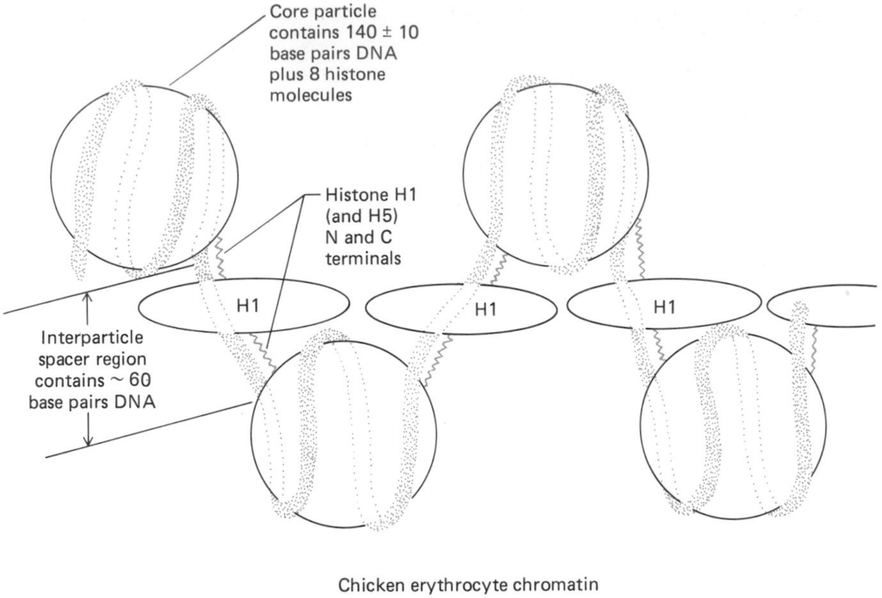

Figure 13-14. A model of the nucleosome organization of inter-phase chromatin, as discussed fully in the text. The H5 histone noted in the drawing is a species found in the nucleated erythrocytes of fish, am-phibians, and birds which replaces the H1 class usually found in more active chromatin. (From a drawing by B. R. Shaw, presented in 1976.)

gate (Fig. 13-14). Current evidence suggests that approximately 140 base pairs of DNA make two turns around a disk-shaped histone octamer. The separation of the amino acid residues of the histone molecule into two re-gions is ideally suited for the organization of the nucleosome. The more hydrophobic, uncharged regions of the histones are directed toward the center of the nucleosome, thereby promoting their aggregation. In con-trast, the positively charged regions of the histones are directed toward the outside of the particle where the positively charged, basic amino acid residues can form ionic interactions with the negatively charged phosphate groups of the DNA backbone. Since all DNA molecules, regardless of the source and nucleotide sequence, contain the identical sugar-phosphate backbone with which the histone interacts, it is not surprising that there would be little basis for evolutionary change in these molecules. Even satellite DNA with its highly unusual nucleotide sequences exists in the same type of nucleosome beads as does the remainder of the chromatin. The interaction of the DNA and histone of the nucleosomes serves to greatly compact the DNA. If the nucleosomes are isolated after digesting the interconnecting DNA fragments and the DNA of the isolated particles is extracted and examined in the electron microscope, it is found that there is sufficient DNA present in the approximately 100-Å particle to stretch approximately 700 Å in length. In other words, the *packing ratio* of the DNA in the nucleosomes is approximately 7 : 1.

The interaction between histone and DNA is an unusual protein-nucleic acid interaction since it is structural in nature and not related to nucleotide sequence. This can be readily demonstrated in the following type of experiment. Whereas $0.7M$ NaCl is of sufficient ionic strength to remove the lysine-rich H1 molecules, the remainder of the histones can be removed by further raising the salt concentration to about $2.0M$. If one carries out this dissociation and then reverses the procedure and lowers the salt concentration back to the point where the DNA and histones can reassociate, the organization of chromatin into nucleosomes and interconnecting strands is regained. The rapid reconstitution of the native type of chromatin structure illustrates the nonspecific nature of the interaction and the capacity of the structure for self assembly. These points are made even clearer by the demonstration that DNAs that are normally not associated with histones, such as those of bacteriophage or synthetic double-stranded polynucleotides (Fig. 13-13b) can be converted into the nucleosomal configuration by incubation with vertebrate-derived histones.

The "beads on a string" nature of chromatin is generally considered to represent its lowest level of organization; however, chromatin does not appear to exist in this relatively extended state within the cell. For example, if one examines electron micrographs of sections cut through nuclei, the chromatin fibers typically appear approximately 250 Å in diameter. Similarly, when chromatin is prepared from nuclei in the presence of divalent ions, a fiber of 250 to 300 Å in thickness is observed. It seems that the 100-Å nucleosomes are organized into some type of thicker filament in the interphase cell. Furthermore, this higher-order structure, which is believed to involve a type of helical coiling of the thinner 100-Å filament, is dependent upon the presence of the H1 internucleosomal histone molecules and divalent ions. Even though the overall character of the interphase and mitotic chromosomes are very different, the 250-Å chromatin fiber is found in the chromosomes of both phases of the cell cycle. Figure 13-15 shows a whole mount of a highly condensed mitotic chro-

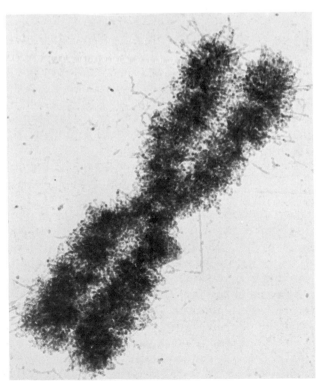

Figure 13-15. Electron micrograph of a moderately condensed chromosome No. 1 from a human lymphocyte culture. This unfixed chromosome was prepared by a surface-spreading technique. *(Reprinted from G. F. Bahr,* Federation Proceedings, *34:2211, 1975)*

mosome. A close look at this electron micrograph clearly reveals the fibrous nature of the mitotic chromosome. These looping, intertwining, knobby fibers are believed to represent the same 250-Å fibers observed in the interphase nucleus. The nature of mitotic chromosomes will be considered in further detail in Chapter 15.

The ability to isolate the material of the interphase chromosomes in an essentially intact state has provided one of the most important opportunities for the analysis of gene expression in eukaryotic cells. We will consider a few experiments that use isolated chromatin as a template for RNA synthesis in vitro in a later section, but first it is necessary to discuss in a broad sense the various levels at which gene expression can be regulated.

Levels of Control of Gene Expression

As described in the previous three chapters, the chain of events leading to the synthesis of particular proteins is made up of a number of separate and very different steps. One of the most basic goals of molecular biology is to understand the means by which various types of eukaryotic cells become capable of synthesizing different species of proteins and thereby attain recognizably different phenotypes. If this goal is to be achieved, it is important to examine all of the various steps leading to protein synthesis in an attempt to determine which ones might be sites for important regulatory influences. In the following discussion, the various *possible* types of regulation that one might conceive are briefly explored. Those that appear to hold the most promise are discussed further in the following sections.

1. *Loss of DNA templates.* One of the earliest proposals to explain cellular *differentiation,* i.e., the process whereby cells become histologically different from one another, was made by August Weismann, before the turn of the century. In essence, Weismann proposed that the genetic determinants, later shown to be the chromosomes, were parceled out in some manner that determined the ultimate path of differentiation that a given cell might take. In other words, cells would achieve a particular specialized state by retaining those parts of the chromosome that are required for that condition and eliminating those that are not needed. Putting this theory into more molecular terms, it would predict that different DNA sequences should be present in different types of cells. The loss of genetic information as an explanation for the direction of cells along one or another pathway of differentiation is a reasonable theory, but one which has had virtually no experimental support. Throughout this book we have assumed that all of the cells have all of the DNA; it is time to provide some evidence that this is actually the case.

One could argue on the basis of the constancy of DNA, i.e., the fact that all diploid cells have the same amount of DNA, that there is no loss of genetic information accompanying differentiation. However, this evidence is only suggestive until there is some reason to believe that the type of genetic information is the same in all cells, rather than just the amount. One convincing piece of evidence was presented in the previous chapter (page 579) without being discussed in this regard. Figure 12-38 illustrates the nature of the hybridization occurring between a labeled cDNA probe made from globin mRNA and total DNA prepared from two different types of cells, one of which normally produces hemoglobin while the other does not. Just as this curve indicates the absence of amplification of the globin gene in the reticulocyte, it reveals the presence of the globin gene in the liver. Clearly the path of differentiation leading to the formation of liver is not accompanied by the loss of genes, in this case for hemoglobin, that will not be needed for that cell.

A more elegant demonstration of this critically important conclusion has been achieved by the technique of *nuclear transplantation*. The first successful demonstration of the removal of a nucleus from one cell of a multicellular organism and its transfer into the cytoplasm of another cell was developed in the early 1950s by Robert Briggs and Tom King using embryonic cells of the frog *Rana pipiens*. The experimental approach taken by Briggs and King was to remove the nucleus from a donor cell of a developing embryo and transplant that nucleus into the cytoplasm of a recipient egg whose own nucleus was subsequently removed (Fig. 13-16). The basic question to be asked in this experiment was whether or not the nucleus taken from the embryonic cell was capable of supporting the normal development of the egg into which it had been transplanted. It was reasoned that if a nucleus could be removed from a cell of an advanced embryonic stage, transplanted into an enucleated egg, and the egg develop into a normal embryo, then this transplanted nucleus must have retained the full complement of genetic information. In other words, despite the

Figure 13-16. Diagram illustrating the method for transplanting blastula cell nuclei into enucleated eggs. The donor nucleus is sucked out of a cell with a surrounding layer of protective cytoplasm. The nucleus is then injected into an egg whose own nucleus is subsequently removed. This latter enucleation step is accomplished by causing the egg's own chromosomes to flow into an exudate which is removed with fine needles. *(From R. Briggs and T. J. King, J. Exp. Zool.,* **122:**488, *1953.)*

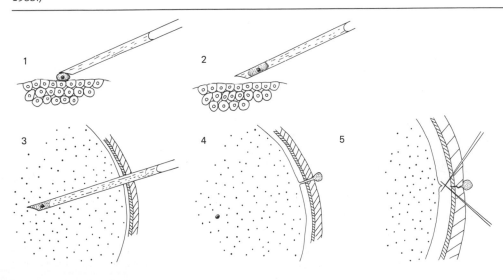

fact that the various nuclei of the donor embryo had undergone many divisions and had come to reside in different types of cytoplasm, they would have remained equivalent. Using nuclei from early embryonic stages (gastrula and before), Briggs and King found that the eggs receiving the nucleus were capable of normal development. However, for what is believed to be technical reasons, they found that nuclei from later embryonic stages were not capable of supporting normal development and the recipients of these later-stage nuclei died quickly. The reasons for these results are complex and beyond the scope of this book.

It was not until a new series of nuclear transplant experiments were performed, this time by John Gurdon and coworkers using *Xenopus* rather than *Rana,* that the equivalence of nuclei was demonstrated. Carrying out a similar type of experiment, Gurdon found that at least a small percentage of *Xenopus* nuclei could be removed from a fully differentiated intestinal epithelial cell of a tadpole larva, transplanted back into an enucleated egg (Fig. 13-17), and normal development could ensue using the genetic information in the transplanted nucleus as the directing force. In subsequent experiments, Gurdon and coworkers found that they could remove nuclei from adult skin cells, transplant them back into an egg, and normal development of that egg would result. The conclusion from these experiments seems irrefutable: nuclei of specialized cells contain genetic information required for the differentiation of many different types of cells. There is no evidence of the loss of genetic material from vertebrate cells as they undergo differentiation.

2. *The number of DNA templates.* As discussed in the previous chapter, with the exception of the DNA coding for ribosomal RNA, there is no evidence of any alteration during the life cycle in the number of DNA sequences per genome that are present to code for the polypeptides of any protein. This is clearly shown in the curve of Fig. 12-38, as well as

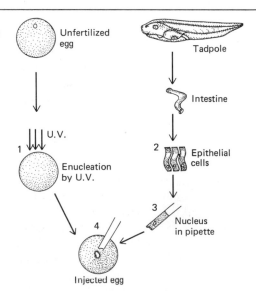

Figure 13-17. The experiment of John Gurdon demonstrating the ability of a nucleus from a differentiated intestinal epithelial cell to support the normal development of *Xenopus*. In this procedure the recipient egg is enucleated by ultraviolet irradiation rather than surgically. *(From "Readings in Genetics and Evolution," chapter 4 by J. B. Gurdon, Oxford Press, 1973.)*

in various other studies with different types of mRNA. The only known alteration in the nature of the genome accompanying differentiation in vertebrates occurs in the cells of the developing immune system (discussed in Chapter 18). In this case, there is good evidence to suggest that new DNA sequences are generated during embryonic development and that subsequent rearrangements of certain DNA sequences occur. However, there is no evidence to support the existence of selective gene amplification of the type discussed in the previous chapter.

3. *The accessibility of DNA templates.* The analysis of the control of protein synthesis in prokaryotes has revealed that the focus of the cell's regulatory effort is centered on the ability or inability of RNA polymerase molecules to reach their template. In both the *lac* operon and the *trp* operon, transcription is blocked when the repressor protein is attached to the operator site, thereby blocking the polymerase from reaching the site at which transcription is initiated. Regulation at the attenuator site of the *trp* operon similarly centers on the ability of the polymerase to move past this premature termination site and reach the DNA coding for the structural genes themselves. Regulation of this type is termed *transcriptional level control* since it determines whether or not a particular part of the chromosome will be transcribed. As in prokaryotic cells, it appears that transcriptional level control is the single most important means by which eukaryotic cells govern the nature of their protein synthesis. We will consider this topic in considerable detail in the following section.

4. *The rate of transcription.* Molecular hybridization studies (page 580) have clearly indicated that different mRNAs exist in the eukaryotic cell at greatly different frequencies. Whereas most species of mRNA are present in very few copies per cell, a smaller number of species are present in much greater abundance. There are several possible ways in which the number of copies of a particular mRNA might be regulated, all of which may be operating simultaneously in a given cell. For example, the more abundant mRNAs might reflect transcripts synthesized over a greater period of time or transcripts that are not destroyed as rapidly. It is also very possible that different genes may be transcribed at varying rates. This does not mean that the actual rate of movement of the RNA polymerase along the DNA template has to be controlled, but rather the number of polymerases operating per unit length of template may be quite variable. Examination of chromatin in the electron microscope has provided direct visual support for the existence of variable rates of transcription. For example, when the chromatin of cells of the silk gland are examined, it is found that there is one length of DNA, clearly nonribosomal in nature, that is very densely covered by nascent RNA chains. All of the other stretches of nonribosomal DNA are either inactive or contain a much lower nascent RNA density. It is believed that this particularly active segment of the chromosome represents the single copy of the fibroin gene which is responsible for the large amount of this protein produced by this gland. In this and other cases, there is good evidence that the rate of *initiation* of transcription need not be the same for all DNA templates.

5. *The processing and transport of newly synthesized RNAs.* Considering the extensive changes that seem to be required before a given mRNA is able to leave the nucleus for the cytoplasm, it is apparent that there is a great deal of room at this posttranscriptional level in which regulatory mechanisms might be operating. The problem in this case is that we know so little about these activities per se that we are essentially unable to determine whether such regulation even exists. The strongest indication of the existence of widespread posttranscriptional control can be found in a recent paper by B. J. Wold et al., 1978.

6. *The translation of mRNA templates.* There is no reason to assume that the transport of the mRNA into the cytoplasm is a guarantee that the RNA will be immediately translated. The process of translation, with all of its required factors, is complex and there are many places at which control mechanisms might conceivably operate. Regulation at this level is termed *translational control* and its existence is well documented. The primary criterion for translational level control is the disproportionate relationship between the amount of cytoplasmic mRNA and the synthesis of the corresponding protein. In other words, if one finds the presence of significant amounts of a particular mRNA or a population of mRNA in a cell and a corresponding deficiency in the synthesis of the corresponding protein, then there is a strong likelihood that translational control mecha-

Figure 13-18. *(a)* The cumulative incorporation of ^{14}C-leucine by unfertilized (dots) and fertilized (circles) eggs of the sea urchin. The 0 time marks the point of fertilization which is followed, after a brief lag, by a marked elevation in the rate of protein synthesis. *(b)* Rates of incorporation of ^{14}C-valine into fertilized eggs of the sea urchin, with and without actinomycin D. The initial increase in protein synthesis which follows fertilization, *a*, is not inhibited by actinomycin D. These results indicate that protein synthesis in the preblastula is not dependent upon newly synthesized mRNA templates but rather is taking place on mRNAs present in the egg at the time of fertilization. In contrast, the second rise in protein synthesis beginning at about ten hours postfertilization does require new mRNA templates because it is inhibited by the drug. [*(a) From D. Epel,* Proc. Natl. Acad. Sci. (U.S.), **57**:*901, 1967; (b) from P. R. Gross, L. I. Malkin, and W. A. Moyer,* Proc. Natl. Acad. Sci. (U.S.), **51**:*409, 1964.*]

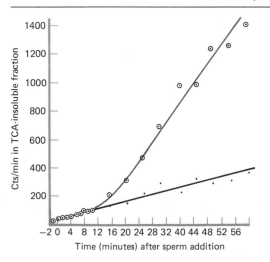

(a)

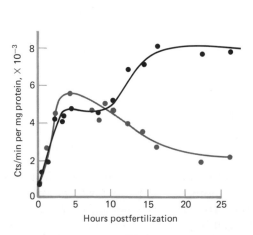

(b)

nisms are operating. Other than transcriptional level control, regulation at the translational level is the best studied and the most important.

One of the best-studied cases of translational control occurs during the early embryonic development of the sea urchin. The unfertilized sea urchin egg is almost totally inactive with respect to protein synthesis. However, within a very few minutes after fertilization, the incorporation of labeled amino acids becomes detectable (Fig. 13-18a) and the rate of incorporation slowly rises over a period of a few hours until it levels off at a relatively high rate about the time when the blastula stage is reached. Since the change from the inactive to the active state is so rapid and the synthetic rate becomes so much greater, it would be difficult to understand how the synthesis of mRNAs could be rate-limiting under these circumstances. As can be demonstrated in several ways, the messenger RNAs being translated after fertilization are primarily those that were present in the egg before its contact with the sperm, but were not being translated. For example, if sea urchin eggs are fertilized and raised in the presence of actinomycin D, the activation of protein synthesis after fertilization occurs in much the same manner as in control cultures (Fig. 13-18b). Since new RNAs cannot be produced in the presence of the drug, the proteins must be synthesized on preformed mRNA templates.

Extraction of RNA from unfertilized sea urchin eggs provides convincing evidence of the presence of the messengers, but the basis for their inactivity is not well understood. One point is clear, namely that the mRNAs of the unfertilized egg are not associated with ribosomes; there are virtually no polyribosomes present at that stage. Once fertilization occurs, more and more of the mRNAs of the egg become recruited into polyribosomes, particularly those responsible for the synthesis of the histones. It appears that the inhibited state of the unfertilized egg is somehow related to the inability of mRNA and ribosomes to associate. The inhibition appears to be a property of the mRNA, not the ribosomes, and is believed to be imposed by the presence of certain proteins associated with the mRNA. Regardless of the molecular mechanism, early development in the sea urchin (and many other embryos) is a time of rapid biochemical activity and morphological change and appears to be best carried out without having the protein-synthesizing machinery completely dependent upon newly synthesized mRNAs. Another example of translational control which has come under intensive recent examination concerns the regulation of globin synthesis by hemin, the iron-containing porphyrin group of the hemoglobin molecule. In this case regulation is accomplished via the phosphorylation of the eukaryotic elongation factor eIF-2. A final example of translational control is seen in the marked drop in protein synthesis that occurs during mitosis, a phenomenon which is not accompanied by a loss of mRNA content.

7. *Messenger RNA stability.* As was mentioned previously (page 492), eukaryotic mRNAs are considerably more stable than their prokaryotic counterparts. Not only are eukaryotic mRNAs relatively stable, but there appears to be considerable variability as to how long a particular mRNA is likely to exist before it is degraded. This suggests that

mRNA lifetime may be an important aspect in the control of protein synthesis since the longer a given mRNA is present in the cell, the greater is its template potential. Unfortunately, as in the case of mRNA processing, the underlying factors involved in mRNA stability are not understood and therefore its importance cannot be measured. One point seems well established—those mRNAs responsible for the production of the dominant proteins of a cell, such as hemoglobin in the reticulocyte, tend to be very long-lived and therefore somehow indentifiable to the cell.

8. *The modification of polypeptides.* Many polypeptides must be modified after they are synthesized before they are active. A particularly widespread example of posttranslational modification is the formation of disulfide bridges. Numerous examples are known where a polypeptide is retained until needed in its inactive premodified state; however, this type of regulation applies to a relatively small number of proteins and is not of basic importance in the transfer of information within the cell.

Transcriptional Level Control

One of the questions discussed in the previous chapter was the degree to which mRNA populations varied among different cell types (page 576). The presence of tissue-specific mRNAs is strong evidence for transcriptional level control, i.e., differential transcription among different cell types. The reason it cannot be considered conclusive evidence stems from the various steps that might occur between the formation of the primary transcript on the chromatin and the appearance of the mRNA in the cytoplasm. It is not inconceivable that the reason the globin message, for example, is present in a reticulocyte and absent in a brain cell is that it is synthesized and rapidly destroyed in the brain cell and retained in the reticulocyte. In other words, the analysis of mRNAs present in various cells cannot rule out the possibility that much of the control of gene expression occurs at some posttranscriptional level, such as that of mRNA processing. What is needed is some means to assay the template activity of the chromosomes under conditions where posttranscriptional activities cannot occur. If we could demonstrate, for example, that under these conditions, the globin mRNA sequences are made by reticulocyte chromatin and not by brain chromatin, then we could attest to the existence and importance of transcriptional level control with much greater assurance.

Template Activity of Isolated Chromatin

It has been found that chromatin can be removed from cells and used, in an isolated state, as a template for gene expression. In these experiments (Fig. 13-19*a*), the chromatin is incubated with an RNA polymerase and the four nucleoside triphosphates (generally radioactive). The polymerase attaches to the DNA of the chromatin and transcribes it and the nature of the RNA synthesized is determined. The question of greatest importance in this type of experiment is whether or not the RNAs being made using chromatin as a template bear any relationship to the RNAs made in vitro. On the basis of various types of molecular hybridization experiments, it is generally believed that the isolation procedure does not

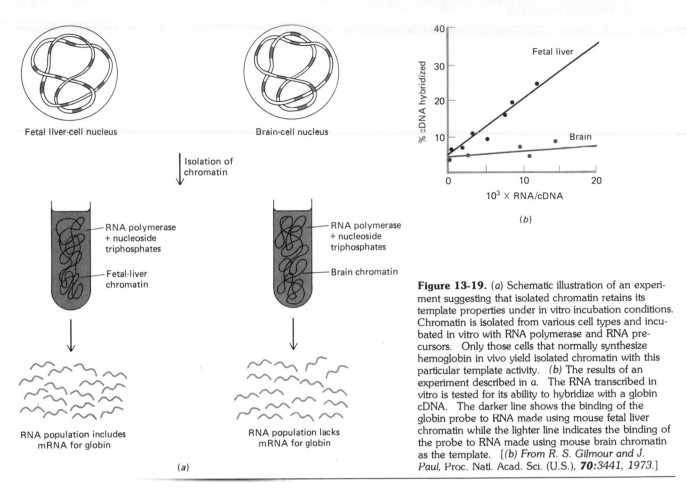

Figure 13-19. (a) Schematic illustration of an experiment suggesting that isolated chromatin retains its template properties under in vitro incubation conditions. Chromatin is isolated from various cell types and incubated in vitro with RNA polymerase and RNA precursors. Only those cells that normally synthesize hemoglobin in vivo yield isolated chromatin with this particular template activity. (b) The results of an experiment described in a. The RNA transcribed in vitro is tested for its ability to hybridize with a globin cDNA. The darker line shows the binding of the globin probe to RNA made using mouse fetal liver chromatin while the lighter line indicates the binding of the probe to RNA made using mouse brain chromatin as the template. [(b) From R. S. Gilmour and J. Paul, Proc. Natl. Acad. Sci. (U.S.), **70**:3441, 1973.]

greatly disrupt the template activity of the chromatin. For example, chromatin isolated from hemoglobin-producing tissues (reticulocytes or fetal liver) produces the globin mRNA sequences in vitro, while chromatin isolated from other tissues does not (Fig. 13-19b). The same type of conclusion is drawn from experiments where the entire population of mRNAs made in vitro and in vivo is compared. Since tissue-specific transcriptional patterns are retained in isolated chromatin, we can conclude that the state of the chromatin in one tissue is different from its state in another tissue and, therefore, that transcriptional level control is of basic importance in the formation and maintenance of the differentiated state.

The demonstration of tissue-specific template activity by preparations of isolated chromatin has led to a large-scale analysis of the basis of its template specificity. What is it about the nature of fetal liver or reticulocyte chromatin that allows it to serve as a template for globin synthesis while the chromatin from brain does not? How is the structure of the chromatin at a site where transcription is occurring different from an inactive site? The analysis of these questions is still in the preliminary stages, but considerable information is available.

Some insight to the structure of transcriptionally active chromatin as opposed to inactive chromatin can be gained simply by considering the nature of the experiments with isolated chromatin. Most of these experiments employ the *E. coli* RNA polymerase, which has been shown to transcribe essentially the same sequences from eukaryotic chromatin as the organisms own RNA polymerase. However, it is evident that the bacterial polymerase is not capable of recognizing specific eukaryotic initation sequences. This can be illustrated by the fact that both strands of the DNA duplex are transcribed, an occurrence which does not happen with the homologous polymerase. It seems somewhat paradoxical that an enzyme is capable of synthesizing essentially the "correct" RNA sequences without being able to recognize the actual promoter sites. The explanation for this type of occurrence must reside in the physical structure of the chromatin. Presumably, those regions of the chromosome that are transcriptionally active in the cell are of sufficiently different configuration, that a *nonspecific* "probe" such as the bacterial polymerase, can selectively attach to them in vitro, leading to their transcription.

There is a substantial body of evidence to support the concept that transcriptionally active and inactive sites have different structural properties. For example, there are techniques to shear chromatin into pieces and to at least partially separate active from inactive fractions. Agents that bind to DNA such as actinomycin D and ethidium bromide are found to complex to active segments of the chromosome to a greater degree than inactive segments. Similarly, pancreatic DNase is much more effective in digesting the DNA of active regions of the chromatin. This can be shown by examining the effect of the nuclease on a particular DNA sequence. Treatment of reticulocyte chromatin, for example, with pancreatic DNase results in the destruction of the globin DNA sequences while treatment of fibroblast chromatin under the same conditions has very little destructive effect on the globin genes.

Having established that active genetic sites can be distinguished from inactive ones, we can begin to consider the components of chromatin which might be responsible for these differences. For our purposes, we will assume that chromatin is composed of three basic components: DNA, histones, and nonhistone proteins. The first step in this analysis is to compare the template activity of DNA when present in chromatin to DNA from which all of the associated protein has been removed. When protein-free DNA is prepared from chromatin and used as a template in an in vitro RNA-synthesizing system, one finds that a much greater variety of RNA sequences are produced than were found when the corresponding intact chromatin served as the template (Fig. 13-20*a*). For example, in one study it was found that, when purified DNA was used as a template, the bacterial polymerase initiated transcription on the average of once every 700 base pairs of DNA. In contrast, when chick oviduct chromatin was the template, initiation occurred only once in 40,000 base pairs. It is generally found that a chromatin template yields only 1 to 10% of the variety of RNA sequences as the DNA itself. We can conclude that the proteins associated with DNA are in some manner responsible for restricting its template activity, not altogether an unexpected finding. The great value of the in vitro system, as in all of the previous topics covered in this book, is the opportunity for experimental manipulation. What happens to the template activity of DNA when some of these proteins are added back under conditions where they reassociate? During the 1960s these types of experi-

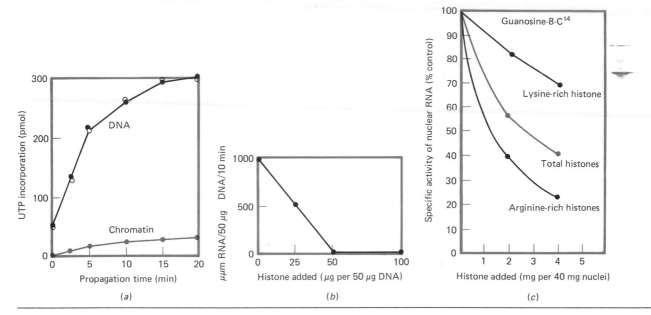

Figure 13-20. (a) Comparison of the template activity of DNA versus chromatin. The conditions are such that the incorporated UTP provides a measure of the propagation of RNA chains that had been initiated prior to the incubation. *(From H. Cedar and G. Felsenfeld, J. Mol. Biol., 77:241, 1973.)* (b) The influence of increasing amounts of added histone on the level of RNA synthesis by purified DNA. *[From R-C. Huang and J. Bonner, Proc. Natl. Acad. Sci. (U.S.), 48:1221, 1962.]* (c) The effects of adding different histones on RNA synthesis in isolated thymus nuclei. RNA synthesis is greatest in the absence of added histones and then drops upon histone addition. *[From V. G. Allfrey, V. C. Littau, and A. E. Mirsky, Proc. Natl. Acad. Sci. (U.S.), 49:416, 1963.]*

ments were carried out, primarily in the laboratories of Ru-Chih Huang and James Bonner (Fig. 13-20b) and of Vincent Allfrey and Alfred Mirsky (Fig. 13-20c), and it was observed that the histones, in contrast to the nonhistones, were powerful inhibitors of RNA synthesis (Fig. 13-20b, c). Although there was a major disagreement over which of the histones were responsible for the inhibition of template activity, it was generally agreed that, as a group, the histones were responsible for the repression of gene expression. The difficulties arose when the nature of the histones were considered. Analysis of histones from a great variety of different tissues revealed their limited number and lack of tissue-specificity. The same handful of histone species are found in most tissues. The most important properties one would expect from a class of genetic repressors is variety and tissue-specificity. This type of analysis suggested that there was much more to the story of the control of gene expression than simply the presence or absence of histones associated with the DNA.

Reconstitution of Chromatin

Greater insights into the nature of eukaryotic genetic regulation were obtained when more complete types of reconstitution experiments were performed. As a result of the initial studies by R. Stewart Gilmour and John Paul, attention turned to the other major class of chromosomal proteins, the nonhistones.   These investigators began a series of experiments where chromatins from different tissues were separately dissociated into their components, then mixed together in various combinations and allowed to reassociate.   Once the reconstituted chromatin had formed, its template activity could be assayed and conclusions could be made as to the role of the various ingredients in determining the specificity.   The results of this type of experiment have indicated that the source from which the nonhistone proteins are derived determines the template activity of the reconstituted chromatin (Fig. 13-21).   For example, if the DNA and histones from thymus are allowed to reassociate in the presence of nonhistone proteins isolated from reticulocyte chromatin, then the resulting complex serves as a template for the synthesis of globin mRNA sequences. However, if the DNA and histone from reticulocyte chromatin is allowed to reassociate with nonhistone proteins from thymus, the reconstituted chromatin does not produce the globin mRNA sequences when incubated with RNA polymerase.

The Regulation of Chromatin Activity

The general picture that has emerged from a variety of experiments of this type indicates that the histone and nonhistone proteins somehow work together in the control of gene expression.   The addition of histones to DNA in the absence of nonhistone proteins serves to repress the tem-

Figure 13-21. Experiments on the dissociation and reconstitution of chromatin suggest that the nonhistone chromosomal protein fraction includes molecules involved in the regulation of gene expression.   In this experiment thymus chromatin and reticulocyte chromatin were dissociated into their three primary components (a) which were then mixed in various combinations.   The reconstituted chromatin was then assayed for its ability to serve as a template for globin mRNA (b)   Those preparations which synthesize this messenger contain reticulocyte nonhistones.   Those containing thymus nonhistones do not produce the message.

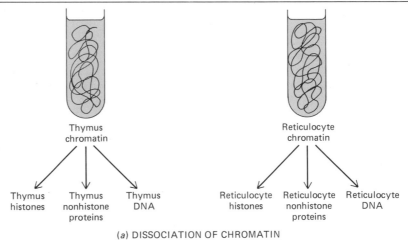

(a) DISSOCIATION OF CHROMATIN

Source of			Presence of globin mRNA among transcription products
Histones	Nonhistones	DNA	
Thymus	Thymus	Thymus	—
Thymus	Thymus	Reticulocyte	—
Thymus	Reticulocyte	Thymus	+
Reticulocyte	Thymus	Thymus	—
Thymus	Reticulocyte	Reticulocyte	+
Reticulocyte	Thymus	Reticulocyte	—
Reticulocyte	Reticulocyte	Thymus	+
Reticulocyte	Retlculocyte	Reticulocyte	+

(b) RECONSTITUTION OF CHROMATIN

plate activity of the DNA in a totally nonspecific manner. Histones from reticulocyte chromatin, for example, are as likely to repress the activity of the globin DNA sequences as are histones from nonerythroid (non-hemoglobin-producing) tissues. In contrast, nonhistone proteins added to DNA in the absence of histones are generally without much effect on its template activity even though a fraction of the nonhistone proteins bind to DNA. This binding of the nonhistone proteins to DNA is sequence-specific. For example, nonhistone proteins isolated from rat chromatin preparations bind to rat DNA but not to *E. coli* DNA; they are capable of recognizing specific nucleotide sequences in their own DNA. When this type of data is put together, it suggests that the role of the nonhistone proteins is specifically to interfere with the repressive action of the histones. It is envisioned that the nonhistone proteins, presumably as a result of their affinity to particular nucleotide sequences in the DNA, somehow attach themselves to the chromatin to bring about a derepression, i.e., an activation. Although it is important to keep in mind that this is a very simplified scheme for a complex and obscure process, this type of theory allows investigators to design various types of experiments which might provide support or contradictions to the proposal.

A number of important questions are generated by the preceding scheme. What is the nature of the nonhistone protein population and which of these molecules are acting as "derepressors"? Do the nonhistone proteins show the tissue-specificity one would expect from genetic activators? Are the histones displaced from active DNA regions? In contrast to the histones, the non-histone chromosomal proteins represent a very diverse group of molecules with a great variety of functions. Included among these proteins are a number of enzymes, including polymerases, methylases, kinases, phosphatases, etc., as well as proteins that may have some structural role. When nonhistone chromosomal proteins are prepared from various tissues and fractionated, certain tissue-specific differences are detected. Altogether, as many as 500 different proteins (Fig. 13-22) can be detected by two-dimensional electro-

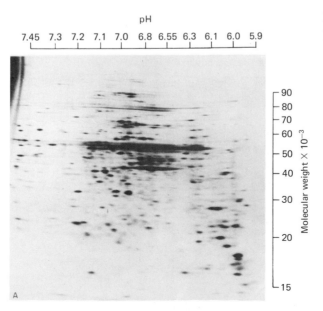

Figure 13-22. A two-dimensional acrylamide gel of HeLa cell non-histone chromosomal proteins labeled with ^{35}S-methionine. *(From J. L. Peterson and E. H. McConkey, J. Biol. Chem., **251**:550, 1976.)*

phoresis (page 117), most of which are present in a relatively low number (approximately 500 to 2000 copies per haploid amount of DNA). Since it is possible that certain regulatory proteins may be at too low a concentration to detect in these procedures, the total number of proteins might even be greater.

A particularly active area of current research is centered on fractionating the nonhistone proteins so as to purify those species that contain specific regulatory activity. The overall scheme of these experiments is to generate various fractions of nonhistone proteins and then test each of them in combination with DNA and histones in an attempt to restore a particular type of template activity. One important aspect that has come to light concerning the metabolism of the nonhistone proteins is their phosphorylation by particular kinases. In the case of the nonhistone chromosomal proteins, the presence of the phosphate groups (attached primarily to serine residues) may be important in allowing them to activate previously inactive DNA sequences. In support of this contention, it has been found that phosphorylated nonhistone chromosomal proteins are particularly active in stimulating RNA synthesis by isolated DNA templates (Fig. 13-23a). Similarly, when these phosphorylated nonhistone proteins are dephosphorylated in vitro by passing them through a column to which a bound phosphatase is present, these nonhistone proteins lose most of their ability to stimulate the synthesis of RNA by reconstituted chromatin (Fig. 13-23b).

Since the addition of phosphate groups to nonhistone proteins would be expected to cause them to compete with the phosphate groups of DNA for binding to histones, one might expect phosphorylation to have important regulatory consequences. In addition the control of gene expression via protein phosphorylation provides a direct means by which variations in the level of cyclic AMP might bear on the cell's genetic activity. Since the only clearly established activity of cAMP is to modify the activity of protein kinases (page 248), changes in the level of this cyclic nucleotide would indirectly involve this second messenger in the control of gene expression via the phosphorylation of chromosomal proteins. The extent to which phosphorylation and dephosphorylation act as determinants in regulation remains to be determined.

Regardless of the effect of the nonhistone proteins, we are still left with the question of how the histone-mediated repression is alleviated in activated

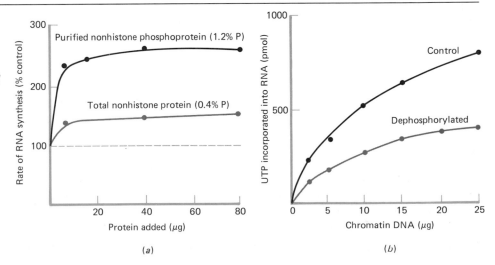

Figure 13-23. (a) A comparison between the effect of the total nonhistone protein fraction versus a purified nonhistone phosphoprotein fraction on the rate of RNA synthesis by DNA in vitro. (From M. Shea and L. J. Kleinsmith, Biochim. Biophys. Res. Comm., **50**:475, 1973.) (b) The template activity of chromatin reconstituted with control versus dephosphorylated histone. [From L. J. Kleinsmith, J. Stein and G. Stein, Proc. Natl. Acad. Sci. (U.S.), **73**:1176, 1976.]

regions of chromatin. Is the histone displaced entirely from the DNA or is its interaction with the DNA simply disrupted? Although the question as to whether or not active regions of chromatin remain in the nucleosomal configuration remains controversial, the evidence does suggest that DNA being actively transcribed is associated with histones. Since the nonhistone proteins are believed to be responsible for converting an inactive genetic locus to an active one, it is likely that they are involved in changing the nature of the DNA-histone interaction so as to relieve the repression of the histones. In this last regard, it has been noted that histones undergo a variety of enzymatic modification—they are methylated, acetylated, and phosphorylated. If one considers the extremely conservative evolution of the histones which attests to the importance of the amino acid sequence of these molecules, then it becomes apparent that these types of modifications are likely to have significant effects on their function. The sites of modification appear to be very specific. For example, the activation of a cell by glucagon results in the cAMP-mediated phosphorylation of one particular serine residue of H1 histone molecules with 15 minutes after the hormone is administered. It has been noted in several different systems that the acetylation of histones precedes a stimulation in genetic activity. Although the significance of these types of modifications are not understood, they are believed to have an influence on the nature of the association of the histones with DNA. Whatever the basis, the apparent presence of the histones associated with the DNA of active genes indicates that the RNA polymerase must be capable of moving along a DNA strand regardless of the close proximity of histone molecules.

Heterochromatin and Euchromatin

In the previous section on transcriptional level control we discussed the means by which a cell might select specific sequences to transcribe and others to repress. We can consider this type of regulation as "fine tuning" in comparison to a different type of inactivation also employed by cells. The examination of sections of cells in the light microscope has long revealed the presence of clumps of chromosomal material in the nonmitotic nucleus (see Fig. 13-3a). It was noted by cytologists that, unlike most of the material of the chromosomes which became dispersed following mitosis, a portion of the chromosomes remained in their condensed, compacted form. The term *heterochromatin* was coined to refer to this condensed interphase chromatin (see Fig. 13-11), and the term *euchromatin* to refer to the majority of the chromatin which becomes dispersed after mitosis. When a radioactively labeled RNA precursor such as ^{3}H-uridine is given to cells which are subsequently fixed, sectioned, and autoradiographed, one finds that the heterochromatin is particularly unlabeled, indicating its general lack of RNA synthetic capacity.

What is the nature of the DNA sequences present in heterochromatin as opposed to the euchromatin? Heterochromatic sections of the chromosome can be divided into two categories, *constitutive heterochromatin* and *facultative heterochromatin,* depending on whether that chromatin is always condensed or only under certain conditions. The DNA of constitutive heterochromatin is permanently inactivated and remains in the condensed state at all times. In the case of constitutive heterochromatin, the same regions on *both* members of the pair of homologous chromo-

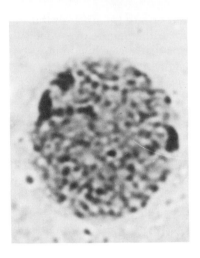

Figure 13-24. The interphase nucleus of a person having an abnormal chromosome complement containing four X chromosomes. In this XXXX individual, three of the X chromosomes become heterochromatized within each nucleus. The three sex chromatin bodies, i.e., Barr bodies, are visible at the periphery of the nucleus. In a normal female, only one Barr body is present. *(Courtesy of J. de Grouchy.)*

somes are permanently inactivated. The best-studied example of constitutive heterochromatin is the centromere regions of the chromosomes, which contain the satellite DNA (page 466) and remain in a permanently compacted organization. Other scattered sections of chromosomes, particularly their tips (the telomeres), are often permanently heterochromatic.

Facultative heterochromatin, on the other hand, is a physiological state to which part of the chromatin is subjected. The best-known example of facultative heterochromatin can be seen by comparing the appearance of cells taken from a female as opposed to a male. Although females have a chromosome complement which includes two X chromosomes, in many cells only one of the pair is transcriptionally active, the other remains condensed to form the heterochromatic *Barr body* (Fig. 13-24). The same genetic information is present in the two X chromosomes, but some mechanism acts to cause one to remain condensed while the other passes through the typical condensation-dispersion cycle of dividing cells. The inactivation of one of the two X chromosomes of cells occurs at some time during early development after a considerable number of cells have been formed. Heterochromatization is a random process in the sense that in any given cell both the paternally derived X chromosome and the maternally derived X chromosome stand an equal chance of inactivation. In other words, the paternal X can become the Barr body in one cell at the time of inactivation and the maternal X the Barr body in another cell. From that point on, all of the progeny of each of those cells will retain the same condition. Since maternally and paternally derived X chromosomes may contain different alleles for the same trait, adult females are in a sense *genetic mosaics* since different alleles will be functioning in different cells. This is most apparent when one examines the coats of mammals which are heterozygous for X-linked genes that determine fur color. Individuals with this type of genetic constitution show patches of different colors (Fig. 13-25) reflecting the particular X chromosome which happened to be inactivated in the cell from which that patch was subsequently derived.

One of the most interesting and best-studied cases of facultative heterochromatin is found in the mealy bug. In the male of this species, the entire set of chromosomes it receives from its father is inactivated by heterochromatization. Cells of male mealy bugs, therefore, are essentially haploid with respect to transcriptionally active genetic material. In contrast, both sets of chromosomes present in females remain euchromatic.

Steroid Hormones as Genetic Activators

When one considers that a mammalian genome contains sufficient DNA to code for over a million different RNA sequences of average polypeptide length and that, among this massive number, a given type of cell is found to synthesize approximately 30 to 40,000 different mRNAs (page 576), the magnitude of the regulation problem becomes apparent. At this point we have very little information about the nature of the intracellular regulatory proteins and even less about the nature of the DNA sequences that might be involved in transcriptional control. The analysis of eukaryotic control mechanisms is made even more difficult in comparison to prokaryotic systems by the relative inability to obtain mutants with alterations in the regulatory sequences. As mentioned at the end of the last chapter, one of the most promising techniques in the study of regulatory DNA sequences is that of DNA cloning, which allows one to amplify particular DNA fragments containing both coding and noncoding (regulatory?) nucleotides. Once these purified segments can be obtained, it becomes possible to add chromosomal proteins in an attempt to reconstitute the region of the chromosome from which the DNA fragment was derived. Hopefully this type of experiment may allow investigators to dissect out the roles of the various constituents. One of the best-studied systems in which transcriptional control mechanisms are known to operate, and one which may lend itself to analyses of the type just discussed, is the response by cells to steroid hormones.

The administration of steroid hormones to previously deprived organisms has profound effects on the metabolism of the various types of target cells. To a large degree the response by target cells to the sudden presence of the particular hormone is mediated at the transcriptional level, i.e., results in the activation of certain regions of the chromatin. One of the greatest values in the use of hormones in the study of genetic activation is the opportunity they provide to compare the system before and after hormone administration. In this way the hormone response is analogous to the response by bacterial cells to the sudden presence of lactose or tryptophan in their culture medium. In all of these cases, a specific outside agent enters the cells and brings about a well-defined response — one that would not have occurred had the hormone or inducer not been supplied. The best-studied systems are those involving the response by vertebrate tissues to steroids of the gonads such as estrogen, progesterone, or testosterone, or the steroids of the adrenal cortex, such as the glucocorticoids. Before turning to the vertebrate steroid hormones, it is worthwhile to briefly consider the response by insect cells to the steroid hormone, ecdysone.

Figure 13-25. The female mouse shown in this photograph is heterozygous for an X-linked gene that affects coat color. All cells of the animal contain the same pair of alleles; however a different member of the pair of alleles may be expressed in different cells. In this highly unusual, i.e. statistically rare, case, the pigment cells of three-quarters of the animal are derived from cells in which the same allele remained active. In the other quarter of the animal, a mixture of the two alleles are expressed among the cell population *(From S. Ohno, L. N. Geller, and J. Kan, Cell, **1**:182, 1974.)*

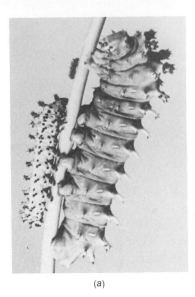
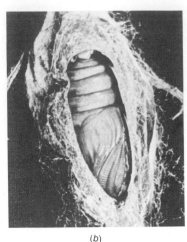

(a)   (b)

Figure 13-26. Stages in the life history of the *Cecropia* silk worm. *(a)* First, third, and fifth instar larvae. *(b)* Pupa within cocoon. *(Courtesy of L. I. Gilbert.)*

Insect Systems

The larval period in the life cycle of insects is characterized by a series of molts followed by the transformation of the larva into a pupa (Fig. 13-26). If one examines the giant chromosomes of the salivary glands (page 438) of a *Drosophila* larva at about 15 hours before pupation, approximately 10 prominent bulges can be seen. These bulges, termed "puffs" (Fig. 13-27a), represent sites on the chromosome where the DNA that had previously been compacted into a band is now in a dispersed state and being transcribed. In a few cases, the puffs become very large and are referred to as a *Balbiani ring*. The transcriptional activity of the DNA of the puffs can be demonstrated by providing the cells with radioactive RNA precursors and determining the sites of incorporation by autoradiography (Fig. 13-27b, c). Since we can equate the presence of a puff with a major site of gene expression, the observation of the giant chromosomes provides a direct visual display of the genetic activity of the salivary gland chromosomes.

Approximately ten hours before pupation occurs, a dramatic change in the nature of the pattern of chromosomal puffs is seen. From this point to about two hours after pupation, approximately 125 different bands undergo puffing, each appearing and disappearing at specified times in the process (Fig. 13-28). The appearance of these many puffs during the period preceding pupation is an indication of a sweeping genetic activation occurring in these cells. This activation, which was originally studied in detail by Wolfgang Beermann and Ulrich Clever, was found to result from the secretion of the hormone ecdysone by one of the endocrine glands of the fly. Remarkably, the entire series of puffs can be induced to appear in *isolated* salivary glands incubated in vitro with ecdysone. The two puffs which appear first in vivo are the same two puffs to appear first after the administration of ecdysone in vitro. These two puffs are seen within five minutes after the addition of the hormone and represent

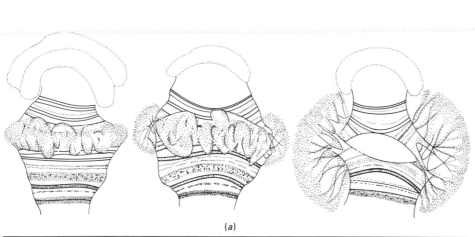

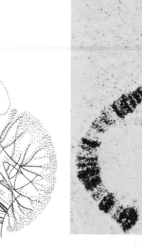

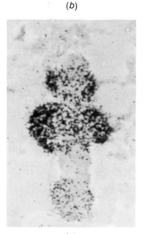

Figure 13-27. Puffing of specific sites in giant salivary gland chromosomes. (a) One of the three large puffs (Balbiani's rings) of a salivary gland chromosome of *Chironomus*. Three different stages of puffing are shown. (From W. Beerman, Am. Zool., **3**:24, 1963.) Autoradiograph of a *Chironomus* chromosome showing that puffs are the sites of the major incorporation of ³H-uridine into RNA. (c) Autoradiograph of a *Chironomus* chromosome showing extensive incorporation of ³H-uridine by the large Balbiani ring. [(b)-(c) From C. Pelling, Chromosoma, **15**:92, 98, 1964.]

(b)

(c)

one of the very best examples of the activation of a genetic function as a consequence of a change in a cell's environment.

This conclusion is supported by more recent experiments employing the technique of in situ hybridization, in this case using polytene chromosomes. As described in previous sections, the existence of polytene chromosomes provides an unparalleled opportunity in the study of gene expression since virtually every genetic locus of the organism is available for direct observation. In the in situ hybridization studies, the flies (or their cells) can be incubated briefly with radioactively labeled RNA precursors, the RNA extracted and hybridized to the DNA of the polytene salivary gland chromosomes. The location of sequences in the DNA complementary to the newly synthesized RNA can then be determined autoradiographically as described before (page 566). It has been found that labeled RNA synthesized after ecdysone treatment hybridizes to a different set of genetic sites from RNA synthesized prior to hormone treatment. Furthermore, it has been shown that different tissues of the larva synthesize different species of RNA in response to ecdysone. This last finding would be expected from the fact that different tissues respond very differently to the presence of the same hormone, ecdysone. For example, some of the tissues of the larva (such as the salivary glands) undergo degeneration in the ecdysone-induced larval-to-pupal molt while other tissues (such as the imaginal disks) undergo differentiation into adult-type structures.

There is little doubt that ecdysone acts either directly or indirectly at the transcriptional level to activate, i.e., derepress, a particular set of DNA

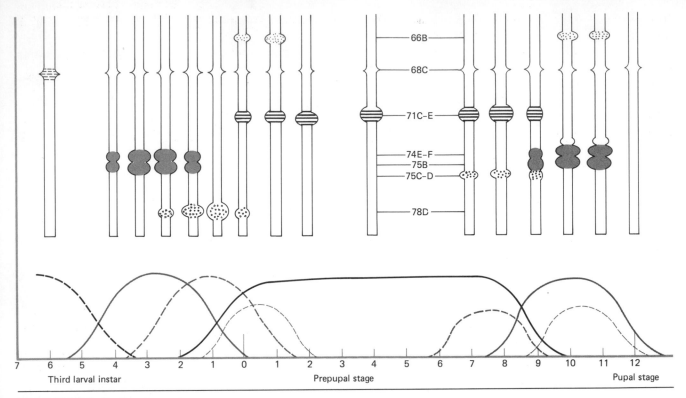

Figure 13-28. Puffs appearing and disappearing during the third larval instar and the prepupal stage at the base of chromosome 111 L of *Drosophila melanogaster* salivary glands. Numbers indicate hours before or after puparium formation. (*After H. J. Becker,* Chromosoma, **10**:670, 1959.)

sequences. Although ecdysone is of the utmost importance as a stimulant of transcription in the insect, a very different stimulus has proven even more important in the study of gene expression in the insect system. It was found a number of years ago that the elevation of the temperature at which insect larvae were kept resulted in a dramatic activation of a handful of genes (as seen by the appearance of new puffs in polytene chromosomes) and the repression of many genes that were previously active. Although the reason for the activation of these "heat shock loci" is not understood, this system has been widely exploited (see the papers by L. M. Silver et al., Cell, **11**:971, 1977, and S. Menikoff and M. Meselson, Cell, **12**:441, 1977, for information and recent references).

Vertebrate Systems

The first major clue concerning the mechanism of steroid hormone action in vertebrates was obtained by Elwood Jensen in 1961. Jensen found that injected radioactive estradiol (an estrogen) was retained only in those cells that were known to be targets for the action of the hormone (Fig. 13-29*a*). Further analysis revealed that a soluble protein was present in the cytoplasm of target cells to which labeled estradiol bound with very high affinity. This protein was an estrogen receptor, and, like the glucagon or insulin receptors (which are located in the plasma membrane rather than the cytoplasm), identifies the cell as a target for hormone action. There are approximately 8000 estrogen receptors in an average uterine cell (Fig. 13-29*b*). A great deal of research has centered on the nature of this protein, as well as receptor proteins for the other steroid hormones, and the nature of the interaction between receptor and hormone is relatively well understood. In the absence of circulating estrogen (such as occurs after the removal of the ovaries) the receptors are located in the cytoplasm (Fig. 13-30). If labeled estrogen is injected into an animal whose ovaries have been removed, there is a rapid migration of the labeled hormone-receptor complexes into the nucleus (Fig. 13-29*b*, 13-30) where they become attached to the chromatin. It appears that the interaction of the receptor protein with the steroid changes the conformation of the receptor, causing it to become translocated into the nucleus and capable of attaching to the chromatin thereby leading to the synthesis of new species of RNA. The nature of the interaction between the chromatin and the steroid-receptor complex has been the center of considerable controversy. In the following account we will describe results obtained primarily in the laboratory of Bert O'Malley.

The best-studied effects of steroid hormones are those of estrogen and progesterone in the oviduct of the hen. As a female bird matures, the rising levels of these ovarian hormones are responsible for converting the oviduct from a rudimentary, undifferentiated state to a tissue having a very active secretory role. The differentiation of the oviduct is accompanied by the onset of the synthesis of a number of specific proteins, molecules which are added to the fertilized eggs as they pass through the oviduct on their way to being laid. The proteins secreted by the oviduct make up the materials present in the white of the hen's egg. Two of the best-studied proteins secreted by the oviduct are ovalbumin and avidin. Ovalbumin synthesis is induced by estrogens and avidin synthesis is induced by progesterone. It is estimated that the cells of the oviduct of an active laying hen have almost 100,000 molecules of ovalbumin mRNA, which account for approximately 60% of the protein synthesis of the cell. It is for these reasons that the ovalbumin messenger has been purified and used so extensively. Avidin, in contrast, is a relatively minor protein (constituting approximately 0.1% of the egg white) and its mRNA has not been highly purified.

Since the type of system most amenable to analysis is one that can be

Figure 13-29. (a) Radioactivity present in various tissues of the rat after a single subcutaneous injection of ³H-estradiol. *(From E. V. Jensen et al.,* Gynecol. Invest., **3:**110, 1972.) (b) Autoradiograph of a section through the uterus of an ovariectomized mouse one hour after the injection of ³H-estradiol. Radioactivity becomes localized in the nuclei of several different uterine cell types. *(From W. E. Stumpf and M. Sar,* in J. Pasqualini, ed., ''Receptors and Mechanism of Action of Hormones,'' part I, *Marcel Dekker, 1976.)*

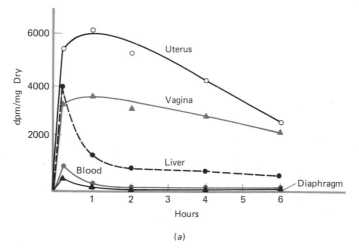

(a)

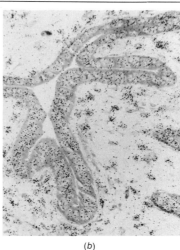

(b)

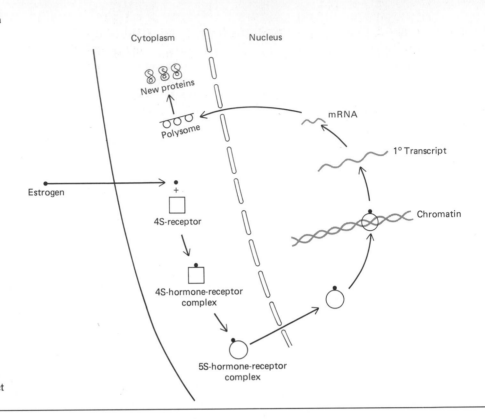

Figure 13-30. Steps in the mechanism of estrogen action on a target cell.

activated by the administration of the "inducer," the immature chick oviduct has received the most attention as opposed to the oviduct of the mature hen. Analysis of the RNA extracted from the immature chick oviduct reveals the total absence of the ovalbumin mRNA sequence. Even though there is no evidence of the transcription of mRNA for this protein, the presence of the estrogen receptor is readily verified by the administration of estrogen to the chick. The animal responds to the injection of estrogen by the subsequent synthesis of ovalbumin mRNA and its translation into the corresponding protein. The appearance of the ovalbumin mRNA is quite rapid and throughout the entire period of estrogen stimulation there is a close correlation between the amount of ovalbumin mRNA in the cell and the level of ovalbumin synthesis (Fig. 13-31). In other words, it appears that the synthesis of additional ovalbumin is dependent on synthesis of additional ovalbumin mRNA; the mRNA is rate-limiting. In addition, the administration of actinomycin D at the same time as the hormone blocks the appearance of the ovalbumin message. These three characteristics (rapidity of response, correlation between mRNA concentration and protein synthesis, and actinomycin D sensitivity) of the estrogen response suggest that the hormone is acting at the transcriptional level to activate the synthesis of new mRNAs. If hormone treatment is maintained for periods of 18 days, the immature chick oviduct is converted into a highly differentiated tissue whose cells contain over 30,000 copies of the ovalbumin mRNA. If the hormone injections are stopped, the active state of the tissue is not maintained and in 12 days the number of mRNAs falls to essentially the prestimulated level.

The analysis of the template activity of chromatin isolated from the chick oviduct has provided strong support for the proposal that steroid hormones

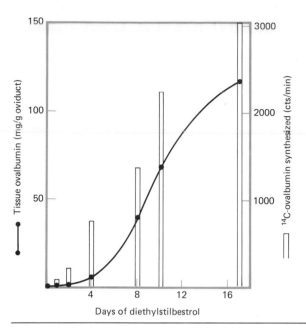

Figure 13-31. Effect of estrogen on the level of ovalbumin mRNA and the tissue concentration of ovalbumin of the immature chick oviduct. Ovalbumin mRNA was determined by its ability to direct the incorporation of ^{14}C-valine into ovalbumin in a cell-free protein-synthesizing system. The concentration of ovalbumin was determined immunochemically. The parallel relationship between the levels of ovalbumin mRNA and ovalbumin within the cells suggests that mRNA concentrations are rate-limiting and, therefore, estrogen is acting at the transcriptional level in stimulating ovalbumin synthesis. [*From J. P. Comstock, G. C. Rosenfeld, B. W. O'Malley, and A. R. Means*, Proc. Natl. Acad. Sci. (U.S.), **69**:2379, 1972.]

act at the transcriptional level. Chromatin isolated from unstimulated immature chick oviduct is totally unable to promote the synthesis of ovalbumin mRNA sequences in the presence of RNA polymerase and RNA precursors (Fig. 13-32). In contrast, chromatin isolated from chick oviduct after estrogen treatment produces a much greater variety of mRNAs including the mRNA for ovalbumin. Approximately 0.014% of the RNA synthesized in vitro from chromatin isolated from the estrogenized oviduct represents ovalbumin mRNA. Once estrogen is withdrawn from the chick, the chromatin isolated from oviduct cells once again drops in its capacity to serve as a template for ovalbumin mRNA synthesis. Less than 0.001% of the RNA synthesized from chromatin isolated from chicks 12 days after estrogen treatments have ceased is ovalbumin mRNA.

These experiments indicate that the synthesis of ovalbumin under different physiological conditions (unstimulated, stimulated, and withdrawn) is closely paralleled by the template activity of the chromatin of these cells. In addition to providing evidence that estrogen acts at the transcriptional level, these in vitro experiments illustrate the effectiveness of the cells' regulatory machinery in allowing the cells of the oviduct to become dominated by the secretion of ovalbumin. In the test tube, the ovalbumin DNA sequence is just one of the many sequences (0.014% of them) available to the polymerase for transcription. In the oviduct cell, however, this one DNA sequence is responsible for producing approximately half of the total mRNA in the cell which in turn accounts for the synthesis of approximately 60% of the protein.

As in the case of the synthesis of the globin mRNA sequence by reticulocyte chromatin, it appears that it is the nonhistone fraction of the chromosomal protein that is responsible for the "opening" of the ovalbumin gene for transcription. Evidence obtained by O'Malley and coworkers indicates that there is a particular nonhistone protein present in the chromatin of oviduct cells (and absent from the chromatin of lung or spleen) which acts as an "acceptor" for the binding of the hormone-receptor complex to the chromatin. In other words, not only do target cells respond to steroid administration be-

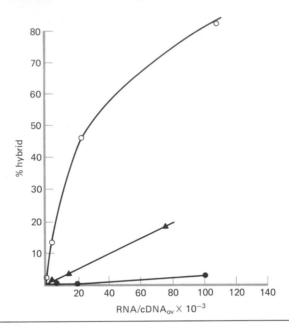

Figure 13-32. Measurement of oval-bumin mRNA transcribed from ovi-duct chromatin isolated from the chick during different states of estrogen treatment. Closed circles, unstimulated chick oviduct chromatin; open circles, 14-day estrogen-stimulated oviduct chromatin; closed triangles, 12-day estrogen-withdrawn oviduct chromatin. *(From S. E. Harris, R. J. Schwartz, M-J. Tsai, B. W. O'Malley, and A. K. Roy, J. Biol. Chem., **251:** 527, 1975.)*

cause they have the appropriate receptor, but also because they have the necessary chromosomal proteins. The situation is analogous to that for the hormones discussed in Chapter 5. The presence of the receptor determines whether a cell is a target cell or not, while the nature of certain components within the target cell are presumably responsible for the nature of the cell's response.

The Britten-Davidson Model

Several models have been presented in attempts to explain, in a general sense, the mechanism by which selective gene activation can be accomplished. The most extensive proposal has been that of Roy Britten and Eric Davidson made in 1969 and reexamined by these authors in 1973. Although a detailed discussion of their model is beyond the scope of this book, a glimpse at its most salient features is warranted. The model attempts to take into account the following observations concerning the organization of the genome and its expression.

1. The genome is composed of both repeated and unique DNA sequences which, to a large degree, are interspersed (page 465).

2. The primary gene transcripts in eukaryotic cells are hnRNAs which include a large amount of nucleus-restricted RNA sequences in addition to the relatively small fraction of the transcripts which are mRNA.

3. Genes coding for functionally related proteins, such as the α- and β-polypeptides of globin, are not located close to one another in the chromosomes. Therefore, the activation of a particular process, such as hemoglobin production in differentiating red blood cells, would require the simultaneous activation of widely scattered DNA sequences.

4. Cells respond to the presence of activating agents, such as estrogen, by the transcription of a particular set (or *battery*) of genes. It is expected that dif-

ferent agents can activate overlapping batteries of genes. For example, the administration of one hormone to an animal might activate genes A, B, and C while the injection of a different hormone might activate genes C, D, and E. If this were the case, some provision must be made so that gene C is responsive to two different cytoplasmic signals.

The Britten-Davidson model attempts to divide the DNA sequences of the genome into several categories depending upon their function. The nature of the various types of sequences is best understood by considering the effect on a cell of the uptake of a particular activating agent such as a steroid hormone. The first stage in the genetic response occurs as the result of the association of the hormone-receptor complex with specific DNA sequences termed *sensor genes* (Fig. 13-33a). The sensor genes, as the name implies, serve to monitor the presence of cellular effector molecules (activating agents). The interaction between the sensor sequence and the effector molecule, in this case the hormone-receptor complex, promotes the transcription of an adjacent region of the DNA, termed an *integrator gene*. The function of the integrator gene is to code for an *activator* molecule, which might be the RNA itself or its corresponding protein. The activators would serve as diffusible gene products capable of recognizing and binding to specific DNA sequences; the binding sites are termed *receptor sites*. The function of the activator is analogous to that of the cAMP-CAP complex in *E. coli* — it would bind to receptor sites and stimulate the transcription of adjacent *structural genes*.

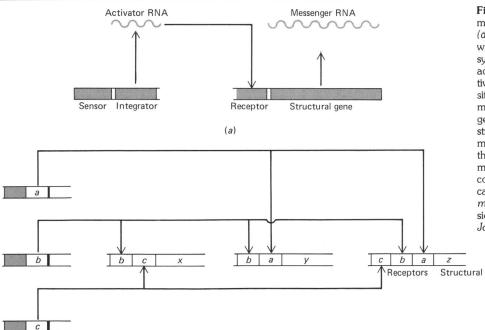

Figure 13-33. The Britten-Davidson model for eukaryotic gene control. *(a)* Interaction of an effector molecule with the sensor sequence leads to the synthesis of an activator RNA at the adjacent integrator gene. The activator binds specifically to a receptor site leading to the synthesis of a mRNA from the adjacent structural gene. *(b)* By proposing that a single structural gene can be activated by more than one activator RNA, and that one activator RNA can activate more than one structural gene, very complex responses to specific stimuli can occur. *(Reproduced with permission from Vol. 1. "Gene Expression," B. Lewin. Copyright © 1974. John Wiley & Sons Ltd.)*

In the Britten-Davidson model, the receptor sites for the binding of the activators are represented by repeated DNA sequences adjacent to structural genes which in turn are represented by nonrepeated sequences (such as those coding for the globin polypeptides). The value in having repeated sequences as receptors is that the same sequence can be located throughout the genome. Those nonrepeated sequences adjacent to a particular repeated sequence would be subject to coordinated control by the same activator molecule. As a result, a set or battery of structural genes is capable of becoming activated by the same stimulus, such as the presence of estrogen. The interspersion of repeated and unique sequences provides an ideal organization for this type of control circuit. In addition, if one assumes that a given structural gene is subject to transcriptional control by the nearby presence of more than one receptor sequence, then it can be understood how two sets of overlapping genes can be activated by two distinct effector molecules (Fig. 13-33b). The extent to which the Britten-Davidson model can explain the control of eukaryotic gene expression remains to be seen.

NUCLEOCYTOPLASMIC INTERACTIONS

The discussion of induction and repression in bacterial cells and the action of steroid hormones in eukaryotic cells brings out an extremely important point in the topic of gene expression: the nature of chromosomal activity depends upon the composition of its environment. In bacterial cells the chromosome is folded up into a centrally located region of the cell termed the *nucleoid* or *nuclear body* (see Fig. 2-6). Since there is no membrane surrounding the bacterial chromosome, there is presumably no barrier to the exchange of materials with the remainder of the cell. In eukaryotic cells, a membranous *nuclear envelope* surrounds the nucleus and must be taken into account when considering nuclear-cytoplasmic interactions. There is no doubt that a great deal of communication exists between the two major compartments of the cell across the nuclear envelope. We have mentioned several important macromolecules that pass from the cytoplasm into the nucleus, including the steroid hormone receptors, ribosomal proteins, and chromosomal proteins. Similarly, the movement in the opposite direction of ribosomal subunits and messenger ribonucleoprotein particles has been pointed out. We can conceive of the nucleus and cytoplasm as two interdependent compartments. The nucleus transports information in the form of RNA, and synthesizing machinery in the form of ribosomes, to the cytoplasm. Some of the proteins produced by the cytoplasm, such as the steroid receptors and nonhistone proteins, feed back to the nucleus to determine its activity. This type of reciprocal interaction is nicely illustrated by the response of the salivary gland to ecdysone (see Fig. 13-28). The first response to ecdysone, which occurs in a few minutes after hormone administration, is the activation of a handful of new genes as reflected in the formation of the "early" puffs. Approximately three hours after the early puffs have formed, an extensive series of "late"

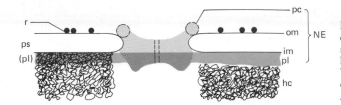

Figure 13-34. Schematic diagram of the nuclear periphery. Nuclear pore complex, pc; outer nuclear membrane, om; perinuclear space, ps; inner nuclear membrane, im; amorphous (peripheral) lamina, pl; nuclear envelope, NE; heterochromatin, hc; ribosome, r. The left half of the drawing indicates the appearance when the chromatin obscures the lamina. [*From R. P. Aaronson and G. Blöbel.* Proc. Natl. Acad Sci. (U.S.), **72**:1010, 1975.]

puffs develop. If inhibitors of protein synthesis are added to the cells at the same time as the hormones, the early puffs form despite the absence of newly synthesized proteins but the late ones do not—presumably certain cytoplasmic products of the early puffs are directly or indirectly responsible for the induction of the later puffs. A number of different types of systems have indicated that, just as there is nuclear control over cytoplasmic activity, there is also cytoplasmic control over nuclear activity. Before discussing some of these investigations, we will consider the nature of the nuclear envelope which separates the two major cellular compartments.

The Nuclear Envelope

The term nuclear *envelope* as opposed to nuclear *membrane* denotes the complex nature of the structure at the boundary between the nucleus and cytoplasm. The nuclear envelope consists of several components (Fig. 13-34): a double membrane, each approximately 70 to 80 Å thick, separated by a perinuclear space of 100 to 500 Å; nuclear pores of complex organization; and a continuous layer or lamina attached to the inside of the inner membrane. Most of the attention paid to the nuclear envelope has centered on the structure and function of the nuclear pores (Fig. 13-35), for they appear to be the most likely sites of exchange between the two compartments. The pores are sites in the envelope formed by the fusion of the two membranes (Figs. 13-34 and 13-36). If we define the pores as that opening circumscribed by the fused membranes, then their diameter can range from about 400 to about 1000 Å. However, electron micrographs clearly indicate that actual openings in the envelope of this diameter do not exist as such. Instead, nuclear pores contain material, termed *annular material*, which fills the majority of the space of the pore. The annular material occupies roughly the shape of an hourglass. It projects into both the cytoplasmic and nucleoplasmic space extending behind the rim of the pore so that the outer diameter of the annular material is greater than that of the pore itself (Figs. 13-34 and 13-36). Located toward the periphery of the annular material on both the cytoplasmic and nucleoplasmic sides are eight granules (Fig. 13-36) which give the annular material an octagonal outer border. Running through the center of the annular material, there appears an indication of an "open" channel connecting the nuclear and cytoplasmic spaces. A technique has now been developed for the isolation of the pore complexes free of the membranous elements of the envelope and studies are underway on the nature of the proteins of which the annular material is composed. When pore complexes are isolated, it is found that they are attached to the lamina which abuts against the inner surface of the inner membrane (Fig. 13-34) and probably acts as a structural support for the envelope. Another feature commonly encountered in electron micrographs of the nucleus is the presence of heterochromatin associated with the inner surface of the inner membrane (see Fig. 13-11), presumably attached to the lamina and annular material. The number of pores present per unit space is variable and can undergo change during a physiological or developmental process. For example, there is a reduction in nuclear pore density that accompanies the development of the spermatozoa. The pores of nuclear membranes of most cells cover from 5 to 25% of the envelope's surface area.

The most important physiological question concerning the nuclear envelope is the restraints, if any, that it places on communication between the nucleus and cytoplasm. In one approach to the analysis of nuclear envelope permeability, Werner Loewenstein and coworkers have sought to determine whether or not a membrane potential could be detected across the envelope. The presence of a potential difference across the membrane would be indicative of the maintained separation of ionic charge, which in turn indicates the presence of a barrier to the free diffusion of ions. Utilizing electrodes placed within the nucleus and cytoplasm of the large insect salivary gland cells, they found that a considerable voltage existed across the envelope (approximately 15 millivolts, which

Figure 13-35. (a) Electron micrograph of a freeze-etch replica of the nuclear envelope of a *Xenopus* oocyte. (*Courtesy of W. W. Franke.*) (b) Electron micrograph of a negatively stained preparation of a newt oocyte nuclear envelope. Each pore is delimited by a thin white line, which is interpreted as the edge-on view of a unit membrane. (*From J. G. Gall*, J. Cell Biol., **32**:392, 1967.)

(a)

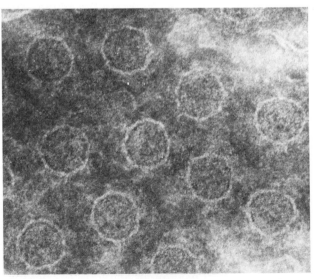

(b)

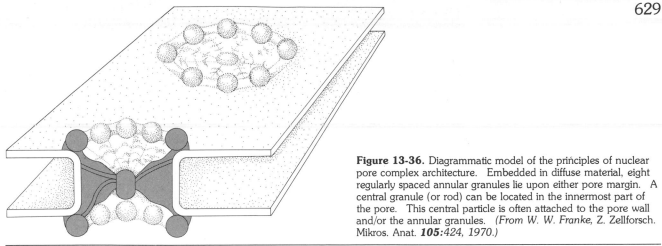

Figure 13-36. Diagrammatic model of the principles of nuclear pore complex architecture. Embedded in diffuse material, eight regularly spaced annular granules lie upon either pore margin. A central granule (or rod) can be located in the innermost part of the pore. This central particle is often attached to the pore wall and/or the annular granules. *(From W. W. Franke, Z. Zellforsch. Mikros. Anat. 105:424, 1970.)*

translates into a resistance of about 1.5 ohm-cm^2). On this basis, one would conclude that, despite the presence of nuclear pores, some feature of the envelope is blocking the movement of ions across the boundary. In contrast, when these same techniques were applied to oocyte nuclei, very little potential was detected indicating the free diffusion of ions across the envelope in these cells. Since there was no obvious difference in the electron microscope between the nuclear envelopes of these two types of cells, the basis for the physiological difference is unclear.

In another approach to the question of permeability, the penetration of various types of substances have been followed. When cells are immersed into various types of labeled substances, including ions and amino acids, it is found that they rapidly enter the nucleus and, in some cases, even become concentrated there. In another series of experiments, those of Carl Feldherr, colloidal gold particles (coated with an inert material) were injected into the cytoplasm of cells (amoebae and oocytes) and their ability to penetrate into the nuclear space was determined in the electron microscope. Electron micrographs of amoebae fixed within minutes after injection of the gold particles clearly revealed that particles could enter the nucleus and they did so by passing through the central region of the nuclear pores (Fig. 13-37). Particles greater than 125 to 145 Å were not found in the nucleus suggesting that this is the approximate upper limit for the penetration of the nuclear envelope in these protozoa. Electron micrographs of various types of cells fixed in the normal course of their activities have revealed the presence of particulate material squeezing through the center of the nuclear pores (Fig. 13-38) indicating that this is the preferred route for passage between nucleus and cytoplasm. Other possible routes have also been described (Fig. 13-39).

An interesting aspect of the physiology of the nuclear envelope is its disappearance and reappearance with every mitotic cycle. If one follows the breakdown of the nuclear envelope at the end of prophase, it appears to undergo a fragmentation process to produce small elements that become indistinguishable from the formed elements of the cytoplasmic reticular membranes.

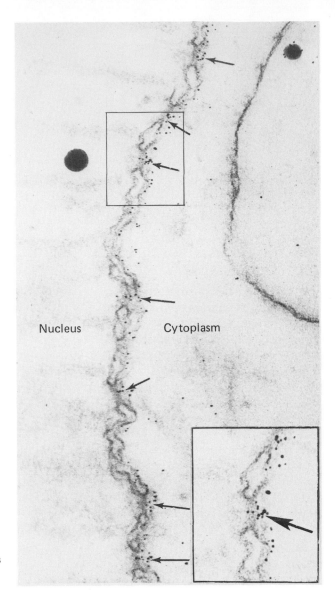

Nucleus   Cytoplasm

Figure 13-37. Electron micrograph of the nuclear-cytoplasmic border of an amoeba after injection with coated gold particles. These particles are seen to penetrate through the nuclear envelope via the center of the pores. Insert shows a portion of the figure at higher magnification. Arrows indicate nuclear pores. *(Courtesy of C. M. Feldherr.)*

Similarly, the reformation of the nuclear membrane at the end of anaphase appears to arise by the coalescence of membranous elements of the endoplasmic reticulum. The first indications of the reformation of the nuclear envelope is the coalescence of membrane around the condensed mitotic chromosomes that have moved to opposite poles of the cell. Chromosomes at this stage can often be seen to be completely surrounded by double membrane in which pores can sometimes be discerned. As the chromosomes disperse, their surrounding membrane fuses to form the definitive nuclear envelope. Although the formation of the nuclear envelope from endoplasmic reticulum is only inferred on the basis of electron micrographs, evidence of a basic relationship between these two membranous organelles is given added support by the existence of connections between them (see Fig. 7-19) in a

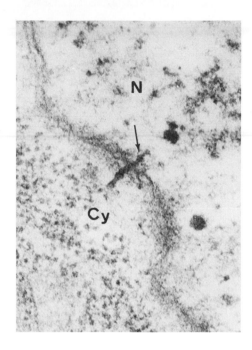

Figure 13-38. The movement of granular material through the center of a nuclear pore into the cytoplasm of a *Chironomus* cell. *(From B. J. Stevens and H. Swift,* J. Cell Biol., **31:72,** *1966.)*

wide variety of cells and by the presence of ribosomes on the outer membrane of the envelope.

Is the membrane involved in the reformation of the nuclear envelope the same membrane that had undergone fragmentation minutes earlier? This is a difficult question to answer, but in one organism at least, it does appear to be the same membrane. In this study, the investigators took advantage of the ability to transplant nuclei, in this case radioactively labeled ones, from one amoeba to another. In order to obtain a culture of amoebae with radioactively labeled nuclear envelopes, they incubated the organisms in ^{3}H-choline, a precursor to phosphatidylcholine of membrane lipids (page 320). Once the membranes were labeled, they transplanted a labeled nucleus into an unlabeled cell and followed the fate of the radioactivity during and after mitosis. They found, as expected, that the radioactivity became distributed throughout the cytoplasm upon the breakdown of the nucleus at the end of prophase. However, when the daughter cells were examined by autoradiography, virtually all of the radioactivity of the cell was once again part of nuclear membrane (Fig. 13-40), in this case evenly distributed between the envelopes of the nuclei of the two daughter cells. It appears, at least in amoebae, that the

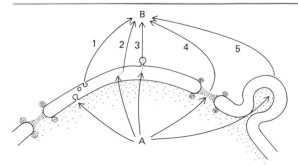

Figure 13-39. Scheme of the possible pathways in which material is translocated from the nucleoplasm (A) to the cytoplasm (B): *(1)* transport by vesicles; *(2)* transmembranous migration or transport; *(3)* a combination of *2* and a subsequent vesicle pinching-off process; *(4)* translocation through pores; *(5)* extrusion of nuclear material by delamination of sacs of both nuclear membranes. *(From W. W. Franke,* Int. Rev. Cytol. Supp., **4:204, 1974.)*

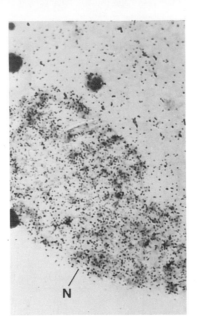

Figure 13-40. Autoradiograph of an amoeba fixed three days after implantation of a ³H-choline-labeled nucleus. The recipient of the nucleus has undergone mitosis and the radioactivity originally present in the donor nuclear envelope is now concentrated in the nuclear membrane of the daughter cells, one of which is shown here. *(From H. Maruta and L. Goldstein,* J. Cell Biol., **65**:638, 1975.)

membrane of the nuclear envelope does not simply become part of the cytoplasmic membrane system but instead retains its identity.

Cytoplasmic Control of Nuclear Activity

The most informative approach in the study of the influence of the cytoplasm over nuclear activity has involved the transfer of nuclei from one type of cytoplasm into an entirely different cytoplasmic environment. There are two general types of experimental procedures where this is most readily accomplished; nuclear transplantation (page 603) and cell fusion (page 222). We will begin with a discussion of studies utilizing cell fusion.

In the chapter on membranes, cell fusion proved valuable in allowing investigators to follow the location of membrane proteins of one species as opposed to another. In the present experiments it is the nuclei of the fused cell which we will examine. Cells that have undergone fusion to form hybrids pass through several different stages. For a considerable period following the fusion event, the nuclei of the fused cells retain their separate identities even though they reside together in a common cytoplasm (see Fig. 5-28). It is during this period when the effects of the cytoplasm on the nuclei can be examined. In this section we will consider only one example of cell fusion, a case that clearly reveals the potential for nuclear reprogramming. In this example, the cells being fused are the hen erythrocyte and the human HeLa cell. The hen erythrocyte is a circulating red blood cell that, unlike its mammalian counterpart, has retained its nucleus. The nucleus of this cell is synthetically inactive; it makes no DNA and no RNA and its chromatin is highly condensed. In contrast, the HeLa cell is a rapidly dividing, active cell that makes RNA continuously and DNA during its periodic S phases. One of the peculiar features of using a hen erythrocyte in these fusion experiments is that the virus added to induce the fusion process causes the loss of the cytoplasm of the cell, producing, in essence, a nucleated erythrocyte ghost which then fuses with the HeLa cell. As a result of the loss of the erythrocyte cytoplasm, the bird nucleus will be totally immersed in human cytoplasm and its effect on the activity of the foreign nucleus can be determined. When these two cells are fused there is a rapid enlargement of the erythrocyte nucleus (a twenty- to thirty-fold volume increase), a loosening of the condensed chromatin into a more "synthetic" state, the accumulation of cytoplasmic materials, and the reactivation of both DNA and RNA synthesis. The chicken erythrocyte nucleus is quite dormant, yet it can be rapidly reprogrammed under cytoplasmic influence to a full synthetic state. Since the hen nucleus is immersed in human cytoplasm, it must be responding to activation signals of a very different species. The results of a variety of fusion experiments suggest that the cytoplasmic influences operating in eukaryotic cells are primarily positive, i.e., stimulatory in nature, rather than inhibitory as in the case of the i gene products of bacterial cells.

A very important point, and one not answered by this particular experiment, is the nature of the RNAs being synthesized by the activated bird nuclei. Are the RNA species produced in response to the human cytoplasm the result of the activation of a meaningful battery of genes? The results of one cell fusion experiment indicate that influences from foreign cytoplasm do seem capable of activating genes that are appropriate to the new cytoplasm of the hybrid (Fig. 13-41). In this case the fusion occurred between human white blood cells and mouse liver tumor cells. Liver cells synthesize albumin, while white blood cells do not. Human and mouse albumins are sufficiently different in molecular structure to be separable by electrophoresis. Therefore, if the synthesis of human albumin is induced by the presence of mouse liver cytoplasm, then the specific human protein should be detectable. The results of this analysis have indicated the presence of the human form of this protein in these cells, which indicates that the human nucleus is responding to the presence of specific induction signals in the mouse liver cytoplasm by the synthesis of specific mRNAs. Not only does this experiment indicate the presence of specific cytoplasmic signals but reveals that the albumin gene, which has long been repressed in the human white blood cell line, is still capable of activation leading to protein synthesis.

The other means by which a nucleus can be brought into direct contact with cytoplasm of a different nature is by utilization of the nuclear transplantation technique (Fig. 13-17). In a series of experiments on *Xenopus*, John Gurdon and coworkers have determined the effect on the synthetic activity of various types of nuclei suddenly immersed into very different types of cytoplasm. Two related types of *Xenopus* cells have generally been used as recipients for the transplanted nuclei in these experiments, the oocyte and the activated egg. Although there are many similarities between these two cells (one is derived from the other), their synthetic activities are very different. The oocytes used in these experiments are characterized by the synthesis of large amounts of RNA, both ribosomal and messenger. However, these oocytes

Figure 13-41. The activation of a specific human gene by factors in mouse cell cytoplasm.

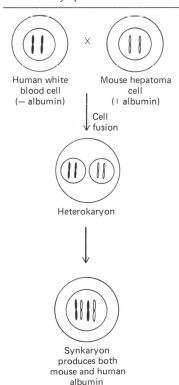

Human white blood cell (− albumin) ✕ Mouse hepatoma cell (⊦ albumin)

Cell fusion

Heterokaryon

Synkaryon produces both mouse and human albumin

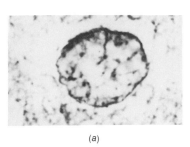

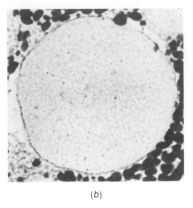

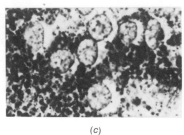

(a)

(b)

(c)

Figure 13-42. Blastula nuclei swell following their injection into (a) an egg or (b) an oocyte. The nucleus in the oocyte enlarges 200 times. (c) Blastula nuclei of normal dimensions. (Courtesy of J. B. Gurdon.)

do not synthesize DNA, an activity which occurred at a much earlier stage of differentiation within the ovary. In contrast, the newly fertilized (or artificially activated) egg is an active synthesizer of DNA but highly inactive with respect to the synthesis of RNA.

It is generally found that nuclei transplanted into one or the other of these two cells begin to take on the synthetic character in keeping with their new environment, regardless of their activity at the time of transplantation. For example, if a blastula nucleus, which is characterized by a low rate of RNA synthesis and a high rate of DNA synthesis, is transplanted into an oocyte, the nucleus stops making DNA and markedly increases its synthesis of RNA. The nucleus has responded to the state of cytoplasm in which it finds itself and becomes reprogrammed. Not only does the synthetic activity of the transplanted nucleus become modified, so too does its morphological appearance. The normal nucleus of the oocyte, which is termed a *germinal vesicle,* is huge (see Fig. 12-7) in comparison to the relatively normal-sized blastula nucleus (Fig. 13-42c). Transplantation of the blastula nucleus into the oocyte results in a rapid swelling of the transplant (Fig. 13-42b) so as to mimic the morphology of the oocyte's own large germinal vesicle. If the blastula nucleus had been transplanted into an activated egg, rather than an

Figure 13-43. The results following the injection of frog brain nuclei into an oocyte (a) and an egg (b). Frog brain nuclei synthesize RNA but rarely DNA. In the oocyte, a, where RNA synthesis is progressing, autoradiographs indicate that the transplanted nucleus is not synthesizing DNA. In contrast, when the nuclei are injected into the activated egg, b, the brain nuclei switch from RNA to DNA synthesis in response to conditions in the host-cell cytoplasm. (Courtesy of J. B. Gurdon.)

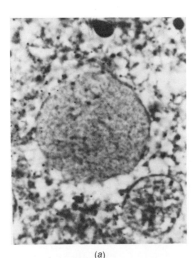

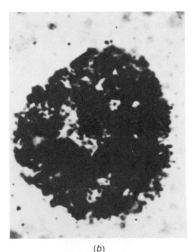

(a)

(b)

oocyte, a less marked swelling would have occurred (Fig. 13-42a) causing the nucleus to become similar in size to that typical of the newly fertilized egg.

What is there in egg cytoplasm that produces these rapid changes in the injected nuclei? The answer is not known, but it has been shown that proteins from the egg cytoplasm rapidly enter the nucleus and become concentrated there as it swells. This can be shown by transplanting an unlabeled nucleus into an egg whose cytoplasm contains labeled protein. It is believed that entering proteins become concentrated within the nucleus as a result of their specific attachment to the chromosomes.

Although most of the transplantation studies with *Xenopus* have utilized embryonic nuclei, similar results are found when adult nuclei are transplanted into oocyte or activated egg cytoplasm. For example, if nuclei taken from a brain cell homogenate are injected into an egg, the synthesis of DNA begins in these nuclei (Fig. 13-43). Since nerve cells are incapable of cell division, the synthesis of DNA in these transplanted nuclei represents the activation of a function that had been turned off years earlier at the approximate time of the neuron's differentiation.

REFERENCES

Adler, K., et al., Nature **237**, 322–327, 1972. "How *lac* Repressor Binds to DNA."

Back, F., Int. Rev. Cytol. **45**, 25–64, 1976. "The Variable Condition of Heterochromatin and Euchromatin."

Barr, M. L., Int. Rev. Cytol. **19**, 35–95, 1966. "The Significance of the Sex Chromatin."

Baserga, R., and Nicolini, C., Biochim. Biophys. Acta **458**, 109–134, 1976. "Chromatin Structure and Function in Proliferating Cells."

Beckwith, J., and Rossow, P., Ann. Rev. Genetics **8**, 1–14, 1974. "Analysis of Genetic Regulatory Mechanisms."

Beerman, W., and Clever, U., Sci. Am. **210**, 50–58, April 1974. "Chromosomal Puffs."

Berendes, H. D., Int. Rev. Cytol. **35**, 61–116, 1973. "Synthetic Activity of Polytene Chromosome."

Bertrand, K., et al., Science **189**, 22–26, 1975. "New Features of the Regulation of the Tryptophan Operon."

Bonner, J. J., and Pardue, M. L., Cell **12**, 227–234, 1977. Polytene Chromosome Puffing and In Situ Hybridization Measure Different Aspects of RNA Metabolism."

Bourgeois, S., and Pfahl, M., Adv. Prot. Chem. **30**, 1–99, 1976. "Repressors."

Brinkley, B. R., and Porter, K. R., eds., symposia on "Cytoplasmic Control of Nuclear Expression *and* Chromatin Structure and Function." *International Cell Biology 1976–1977,* 439–486, Rockefeller University Press, 1977.

Britten, R. J., and Davidson, E. H., Science **165**, 349–357, 1969. "Gene Regulation for Higher Cells: A Theory."

—— and ——, Quart. Rev. Biol. **46**, 111–138, 1971. "Repetitive and Non-Repetitive DNA Sequences and A Speculation of the Origins of Evolutionary Novelty."

Busch, H. B., ed., *The Cell Nucleus*, 3 vol. Academic, 1974.

Carpenter, G., and Sells, B. H., Int. Rev. Cytol. **41**, 29–58, 1975. "Regulation of the Lactose Operon in *Escherichia coli* by cAMP."

Cattanbach, B. M., Ann. Rev. Gen. **9**, 1–18, 1975. "Control of Chromosome Inactivation."

Chamberlin, M. J., Ann. Rev. Biochem. **43**, 721–776, 1974. "The Selectivity of Transcription."

Datta, A., de Haro, C., Sierra, J. M., and Ochoa, S., Proc. Natl. Acad. Sci. (U. S.) **74**, 3326–3329, 1977. "Mechanism of Translational Control By Hemin in Reticulocytes."

Davidson, E. H., *Gene Activity in Early Development*, 2d ed. Academic, 1976.

Davidson, R. L., Ann. Rev. Genetics **8**, 195, 1974. "Gene Expression in Somatic Cell Hybrids."

Dickson, R. C., Abelson, J., Barnes, W. M., and Reznikoff, W. S., Science **187**, 27–35, 1975. "Genetic Regulation: The Lac Control Region."

Elgin, S. C. R., and Weintraub, H., Ann. Rev. Biochem. **44**, 725–774, 1975. "Chromosomal Proteins and Chromatin Structure."

Englesberg, E., and Wilcox, G., Ann. Rev. Genetics **8**, 219–242, 1974. "Regulation: Positive Control."

Feldherr, C. M., Adv. Cell Mol. Biol. **2**, 273–307, 1972. "Structure and Function of the Nuclear Envelope."

Fitzsimons, D. W., and Wolstenholme, G. E. W., eds., Ciba Foundation Symposium 28 (new series). "The Structure and Function of Chromatin," Elsevier, 1975.

Franke, W. W., Int. Rev. Cytol. Supp. **4**, 72–236, 1974. "Structure, Biochemistry and Functions of the Nuclear Envelope."

———, Phil. Trans. Roy. Soc. (London **B268**, 67–93, 1974. "Nuclear Envelopes. Structure and Biochemistry of the Nuclear Envelope."

Gilbert, W., and Müller-Hill, B., Proc. Natl. Acad. Sci. (U.S.) **56**, 1891–1898, 1966. "Isolation of the Lac Repressor."

Gorski, J., and Gannon, F., Ann. Rev. Phys. **38**, 425–450, 1976. "Current Models of Steroid Hormone Action: A Critique."

Gurdon, J. B., *The Control of Gene Expression In Animal Development*. Harvard, 1974.

Gurpide, E., ed., Ann. N.Y. Acad. Sci. **286**, 1977. Symposium on "Biochemical Actions of Progesterone and Progestins."

Hamkalo, B. A., and Miller, O. L., Jr., Ann. Rev. Biochem. **42**, 379–396, 1973. "Electron Microscopy of Genetic Activity."

Harris, H., *Nucleus and Cytoplasm*, 3d ed. Oxford University Press, 1974.

Jacob, F., Science **152**, 1470–1478, 1966. "Genetics of the Bacterial Cell."

——— and Monod, J., J. Mol. Biol. **3**, 318–356, 1961. "Genetic Regulatory Mechanisms in the Synthesis of Proteins."

Jensen, E. V., and De Sombre, E. R., Science **182**, 126–134, 1973. "Estrogen-Receptor Interaction."

Jones, K. R., and Brandham, P. E., eds., *Current Chromosome Research*. North Holland, 1976.

Kasai, T., Nature **249**, 523–527, 1974. "Regulation of the Expression of the Histidine Operon in *Salmonella typhimurium*."

Kassell, B., and Kay, J., Science **180**, 1022–1027, 1973. "Zymogens of Proteolytic Enzymes."

Kornberg, R. D., Science **184**, 868–871, 1974. "Chromatin Structure: A Repeating Unit of Histones and DNA."

————, Ann. Rev. Biochem. **46**, 931–954, 1977. "Structure of Chromatin."

Lewin, B., Cell **2**, 1–7, 1974. "Interaction of Regulator Proteins with Recognition Sequences of DNA."

————, *Gene Expression*, vols. 1 and 2. Wiley, 1974.

Li, H. J., and Eckhart, R., eds., *Chromatin and Chromosome Structure*. Academic, 1977.

Lodish, H. F., Ann. Rev. Biochem. **45**, 39–72, 1976. "Translational Control of Protein Synthesis."

Lyon, M., Biol. Revs. **47**, 1–36, 1974. "X-Chromosome Inactivation and Developmental Patterns in Mammals."

MacLean, N., and Hilder, V. A., Int. Rev. Cytol. **48**, 1–54, 1977. "Mechanisms of Chromatin Activation and Repression."

Maniatis, T., and Ptashne, M., Sci. Am. **234**, 64–76, Jan. 1976. "A DNA Operator-Repressor System."

Maurer, R., Maniatis, T., and Ptashne, M., Nature **249**, 221–223, 1974. "Promoters Are in the Operators in Phage Lambda."

Müller-Hill, B., Prog. Biophys. Molec. Biol. **30**, 227–252, 1975. "Lac Repressor and Lac Operator."

Nagl, W., Ann. Rev. Plant Phys. **27**, 39–69, 1976. "Nuclear Organization."

Nierlich, D. P., Rutter, W. J., and Fox, C. F., *Molecular Mechanisms in the Control of Gene Expression*. Academic, 1976.

Olins, A. L., and Olins, D. E., Science **183**, 330–332, 1974. "Spheroid Chromatin Units."

O'Malley, B. W., and Means, A. R., Science **183**, 610–620, 1974. "Female Steroid Hormones and Target Cell Nuclei."

———— and Schrader, W. T., Sci. Am. **234**, 32–43, Feb. 1976. "The Receptors of Steroid Hormones."

————, Towle, H. C., and Schwartz, R. J., Ann. Rev. Genetics **11**, 239–275, 1977. "Regulation of Gene Expression in Eucaryotes."

Oudet, P., Gross-Bellard, M., and Chamdon, P., Cell **4**, 281–300, 1975. "Electron Microscope and Biochemical Evidence That Chromatin Structure Is a Repeating Unit."

Pastan, I., Adhya, S., Bacteriol. Rev. **40**, 527–551, 1976. "Cyclic Adenosine 5'-Monophosphate in *Escherichia coli*."

Peacock, W. J., and Brock, R. D., *The Eukaryotic Chromosome*. Australian National University Press, 1975.

Ptashne, M., and Gilbert, W., Sci. Am. **222,** 36–44, June 1970. "Genetic Repressors."

———— et al., Science **194,** 156–161, 1976. "Autoregulation and Function of a Repressor in Bacteriophage Lambda."

Ruiz-Carrillo, A., Wangh, L. J. and Allfrey, V. G., Science **190,** 117–128, 1975. "Processing of Newly Synthesized Histone Molecules."

Schimke, R. T., et al., Rec. Prog. Hormone Res. **31,** 175–211, 1975. "Hormonal Regulation of Ovalbumin Synthesis in Chick Oviduct."

Stein, G. S., Spelsberg, T. C., and Kleinsmith, L. J., Science **183,** 817–824, 1974. "Nonhistone Chromosomal Protein and Gene Regulation."

Stein, G. S., Stein, J. S., and Kleinsmith, L. J., Sci. Am. **232,** 46–57, Feb. 1975. "Chromosomal Proteins and Gene Regulation."

Travers, A., Nature **263,** 641–646, 1976. "RNA Polymerase Specificity and the Control of Growth."

Van Holde, R. E., and Isenberg, I., Accts. Chem. Res. **8,** 327–341, 1975. "Histone Interactions and Chromatin Structure."

Watson, J. D., *Molecular Biology of the Gene,* 3d ed. W. A. Benjamin, 1976.

Weintraub, H., and Groudine, M., Science **193,** 848–856, 1976. "Chromosomal Subunits in Active Genes Have an Altered Conformation."

Wischnitzer, S., Endeavour **35,** 27–31, 1976. "The Lampbrush Chromosomes: Their Morphology and Physiological Importance."

Wold, B. J., et al., Cell **14,** 941–950, 1978. "Sea Urchin Embryo mRNA Sequences Expressed in the Nuclear RNA of Adult Tissues."

Yamamoto, K. R., and Alberts, B. M., Ann. Rev. Biochem. **45,** 721–746, 1976. "Steroid Receptors: Elements for Modulation of Eukaryotic Transcription."

CHAPTER FOURTEEN
Cell Growth and Replication

Throughout this book we have stressed the approach in cell biology that attempts to understand particular processes by their analysis in a simplified, controlled in vitro system. It has generally been found that, if the components involved in a particular activity can be removed from the cell and reassembled in vitro, then the mechanisms involved in their function can be studied without interference from all of the other influences normally present within the cell. This same approach can be applied to the study of the entire cell since it too can be removed from the influences it is normally subject to within a complex multicellular organism. As discussed in the first chapter, cells are units of living systems capable of independent life. It has long been felt that the study of cells growing and dividing in culture might reveal many of the basic properties of cells, whose study is more difficult within the organism. The many references to the analysis of the synthetic activities of cultured cells in the chapters on gene expression support the validity of this belief. Even though there are great advantages in working with subcellular and cellular systems under in vitro conditions, there is also an overriding disadvantage. Cells and their components do not exist in isolation but are dependent upon other cells and other components in order to carry out their activities.

The removal of a part from the whole allows one to focus on the part but often at the cost of losing sight of the whole.

CELL GROWTH °

The life of a mammal begins with the formation of a fertilized egg of approximately 100 μm in diameter. From this inauspicious beginning, animals containing trillions of cells and weighing over a ton can develop. The transition from the single cell to the multicellular adult, i.e., the growth of the organism, is accomplished by controlled cell growth and cell division, processes which in many tissues continue throughout the life of the individual. Each organ and tissue of a newborn mammal has a certain genetically determined shape and size. Somehow, the information is present within each developing individual to dictate an *orderly* course of cell growth, cell division, and cell differentiation for each specific tissue of the body. From the time of infancy to that of maturity, the growth processes continue in an orderly fashion and the number of cells that make up the various tissues increase in a regulated manner. In the adult, cell growth and division continue, but only so far as is needed to replace cells that are normally lost. Certain tissues, such as those that form the blood cells, maintain relatively high levels of mitotic activity, while others, such as nervous and muscle tissue, are mitotically inactive.

The means by which the growth rates of various types of cells are controlled is one of the least understood aspects of cell biology, although its presence can be dramatically revealed. For example, the mammalian liver is normally an organ with a very low rate of mitotic activity; its cells are characterized by a relatively long life. However, if one were to surgically remove a large piece of the liver (a partial hepatectomy), a whole new program of activities would be initiated. Cells would begin to grow and divide in an extremely rapid manner and, in a short period of time, the original mass of the liver would be regained. As the cell number of the organ approached its proper value, its growth rate would greatly diminish until it had once again reached the low rate characteristic of the undisturbed tissue. Similarly, under normal conditions the cells of the epidermis of the body grow, divide, and differentiate at a rate which provides the organism with a continual supply of skin cells to replace those that normally die and are lost; the rate of growth is tightly regulated. If the epidermis should be wounded, the tissue shifts into a higher gear. The process of wound healing initiates a greatly increased level of mitotic activity and the damage is soon repaired after which the normal growth rate is resumed.

We know very little about the mechanisms involved in the regulation of cell growth and division, but the existence of such regulatory mechanisms is very much in evidence. It is widely believed that the analysis of cell proliferation in culture can provide valuable information concern-

Figure 14-1. Diagrammatic representation of a very small piece of tissue being cultured in vitro by the hanging-drop method.

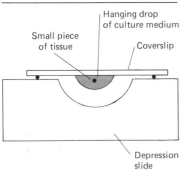

Hanging drop of culture medium

Small piece of tissue

Coverslip

Depression slide

° The term *growth* is generally used to denote the growth of populations of cells as well as the growth of individual cells. In the latter case, growth leads to an increase in cell volume, while in the former case it leads to an increased number of cells.

ing its regulation in vivo. The analysis of the regulation of cell growth is important in its own right but even more so in the study of the malignant cell and the formation of tumors. Cancer cells are somehow released from the body's normal control mechanisms and they grow and divide in an unrestrained manner. Fortunately, the aberrant behavior of cancer cells is readily observed in cell culture, providing investigators with the opportunity to study the basis of the malignant state in vitro. Our presently limited understanding of the cancer cell, as opposed to its normal counterpart, is almost entirely due to its study in cell culture. Some of these findings will be discussed in Chapter 19.

CELL CULTURE

The first successful attempt to culture living vertebrate cells outside of the body was made in 1907 by Ross Harrison, who at the time was interested in the differentiation of neurons. Harrison removed a very small piece of tissue from the neural tube (prespinal cord) of an early amphibian embryo and cultured it in a drop of nutrient-rich lymph. The drop was inverted over a depression slide (Fig. 14-1) and the behavior of the tissue was followed over the next few days. Harrison found that under these conditions the cells not only remained alive, but they differentiated into neurons whose axons grew out into the culture medium, i.e., the lymph fluid. Harrison had observed the process of nerve outgrowth just as it occurs in the embryo when the nerve cells of the brain and spinal cord grow out to innervate all of the tissues of the body. Whereas Harrison was primarily interested in a problem, namely nerve outgrowth, another investigator, Alexis Carrel, soon became interested in cell culture itself. Carrel began growing cells under various conditions and found that, if the proper environment was provided for the cells and the medium was changed on a regular basis, cells could grow in culture for long periods of time, longer in fact than the animal from which they were taken. By 1923, Carrel was growing cells aseptically in flasks, much as they are grown today. By the late 1940s, methods were developed (primarily in the laboratories of Wilton Earle and Renato Dulbecco) whereby pieces of tissue could be dissociated into suspensions of single cells, which could then be used to initiate cell cultures.

Once the cells of a tissue have been separated, there are basically two ways in which these cells can be cultured. In a *mass culture* a large number of cells are added to a culture dish; they settle and attach to the bottom and form a relatively uniform, scattered layer of cells. These cells grow and divide and, after a number of generations, form a monolayer of cells that covers the bottom of the dish (Fig. 14-2a). In the other type of culture procedure, a *clonal culture,* a relatively small number of cells are added to a dish, each of which, after settling and attaching to the surface, will be at some distance from its neighbors. In this case, the proliferation of the cells results in the formation of individual colonies (Fig. 14-2b) or *clones* of cells whose members are all derived from the same original cell. These two types of cell cultures tend to have different growth re-

Figure 14-2. *(a)* Light micrograph showing a portion of a monolayer of mouse L cells growing over the surface of a culture dish in a chemically defined medium. *(Courtesy of C. Waymouth.) (b)* Low-power photograph of colonies scattered over the surface of a culture dish. In this clonal culture, each colony contains a large number of HeLa cells, all descendants of a single original cell. This culture was begun by the addition of only 100 cells to the dish. *(From T. T. Puck, "The Mammalian Cell as a Microorganism," Holden-Day, 1972.)*

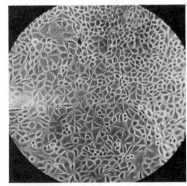

(a)

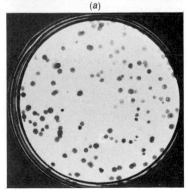

(b)

quirements, reveal different characteristics as they grow, and have different advantages in the study of cell activities.

The initial attempts at growing many types of cells, particularly normal ones (as opposed to malignant cells) were met with a great number of difficulties. It was found to be particularly difficult to grow cells at low densities in attempts to obtain cell clones. Attention turned to the nature of the culture media and the ingredients which would best promote the growth of various types of cells. The early tissue culture studies employed media containing a great variety of unknown substances. Cell growth was accomplished by adding fluids obtained from living systems such as lymph, blood serum, or embryo homogenates. What was it in these complex, undefined mixtures that was needed? Not surprisingly, it was found that cells required a considerable variety of nutrients, i.e., small molecular weight substrates and cofactors, in order to grow. However, it was also found that nutrients alone were not sufficient to support growth of cells in culture; larger molecular weight growth factors of an unknown nature were also required. Most culture media included large amounts (10 to 20%) of serum (generally dialyzed to remove small molecular weight substances). The need for a fluid such as blood serum in order for cells to grow raises an interesting question. Are the required substances present in serum the same ones that are involved in the regulation of cell growth in the organism? The serum from an animal would be expected to contain a large variety of molecules secreted into the blood by various tissues, many of which might have important regulatory influences. With this in mind, a large-scale search has gone on in the last few years to identify the nature of various macromolecular substances that have effects upon cell growth in culture. Before we can discuss these studies, it is necessary to provide additional background on the growth of cultured cells.

When a piece of tissue is dissociated into its single cells and a suspension of these cells is added to a dish so as to start a mass culture, those cells that survive begin to grow and divide. Observation of living cultured cells in the phase-contrast microscope generally reveals them to be much less differentiated than when they were present in the intact tissue. Cultured cells are usually rather elongated (spindle-shaped) and have an appearance that resembles the cells of connective tissue, the fibroblasts (Fig. 14-3). If the culture conditions are optimal in nature, the cells will grow as rapidly as they can. Mammalian cells typically grow in culture with generation times from 12 to 24 hours. Cells growing under these conditions are said to be in an *exponential* (or *log*) phase of growth (Fig. 14-4). If one follows the culture over time, cell density increases exponentially for a period but then gradually slows and finally reaches a point where cell growth and division has virtually stopped, even if sufficient *nutrients* are present. The cells are said to be in a quiescent or *stationary* phase. The reason for the cessation of growth at higher cell density is not completely understood.

Density-Dependent Regulation of Growth

If one examines a culture of normal cells in stationary phase, the cells are seen to form a monolayer in which each cell is completely surrounded

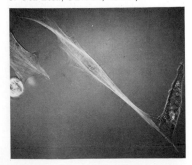

Figure 14-3. A fully spread fibroblastlike cell observed with polarized light optics. The birefringence apparent in this micrograph reflects the longitudinal orientation of the cell's cytoskeletal elements (discussed in Chapter 16). *(From R. D. Goldman, J. Cell Biol., **51**:754, 1971.)*

by its neighbors. It was generally believed that the cessation of growth in each cell was the result of that cell's being surrounded on all sides by other cells. The term *contact inhibition of growth* was used to denote this condition. It was believed that contact with other cells produced a response by the plasma membrane which in turn passed that signal on to the metabolic machinery of the cell signaling it not to initiate DNA synthesis or other growth-related activities. Contact inhibition may be an important factor in the control of cell growth; however, more recently investigators have proposed that it is not the contact per se that is responsible for the cessation of growth, but rather that cells at higher density seem to require a higher concentration of growth factors such as those present in serum. For example, it has been found that, in many cases, the density to which cells grow is dependent upon the concentration of serum in the medium; the greater the percentage of serum (up to a point), the more cells one finds on the plate before the stationary phase is reached. Furthermore, if the cells are grown to a stationary phase at one serum concentration and then additional serum is added (Fig. 14-5), the cells will continue to grow and reach another stationary phase at a higher cell density. Instead of the term contact inhibition of growth, a more neutral term, *density-dependent regulation of growth*, is now more commonly employed to indicate the fact that, as cell densities increase, their growth potential decreases; no mechanism is implied with this term.

The phenomenon of density-dependent regulation of growth, regardless of its basis, is an extremely important property since it is believed to reflect in tissue culture the same properties that maintain normal cells under growth control in the body. The most dramatic evidence of this relationship is seen when the growth behavior of normal cells is compared to those that are malignant (page 819). Cancer cells within the body have somehow escaped the body's control mechanisms and they continue to grow in an unchecked manner. Similarly, these same cancer cells grown

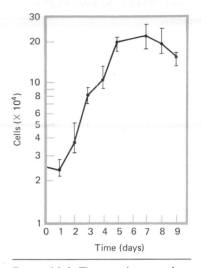

Figure 14-4. The growth curve of human WI-1 cells. An initial lag of 24 hours is found after which the cells enter the log phase for 96 hours. After approximately 9 days, the cell numbers remained constant with viability remaining at 90% for one month. *(From L. Hayflick and P. S. Moorhead,* Exp. Cell Res., **25:**602, 1961.)

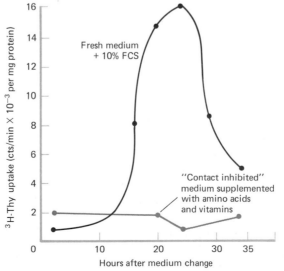

Figure 14-5. Release of cells from growth inhibition by the addition of fresh serum. *(From J. E. Froelich and M. Rachmeler,* J. Cell Biol., **55:**22, 1972.)

in culture generally lack the density-dependent inhibition of growth and they continue to divide, piling up on top of each other in clumps (see Chapter 19).

Growth Factors

Whether normal cells stop growing at high density because of contacts they make with other cells or because of an insufficient concentration of growth factors in their medium or for some other reason, it does appear that the phenomenon is a reaction by the cells to the environment, either cellular or chemical, in which they find themselves. It is generally believed that the response that cells make to these environmental stimuli is mediated by their cell membrane. In order to consider the nature of this response we need to return to the question of the types of substances that seem to promote cell growth. In the following discussion we will concentrate on factors that are believed to have a positive effect on growth processes. It should be kept in mind that many claims have been made for the existence of factors that inhibit cell growth, and, ultimately, it may be shown that growth levels are dependent upon some type of balance between stimulatory and inhibitory agents.

Recent analyses have uncovered a large and growing number of factors that appear to be able to promote growth, at least under certain conditions with certain cells. Factors have been isolated from serum, for example, which have the ability to cause serum-starved stationary phase cells to initiate DNA synthesis and undergo division. Other factors have been isolated from various tissues; for example, the fibroblast growth factor (FGF) from brain or epidermal growth factor from the submaxillary gland. Certain hormones, for example, insulin or somatomedin, can promote growth. A few types of cells seem to be able to grow in a completely defined medium in the absence of added serum or tissue extracts. It appears that, in this latter case, the cells produce their own growth factors and secrete them into the medium where they can act on other cells to promote their growth. The medium in which these cells have been growing is said to be "conditioned" by the cells, and substances can be isolated from conditioned medium which will promote the growth of other cell types. Treatment of density-inhibited cells with low concentrations of proteolytic enzymes such as trypsin also has a general growth-promoting activity. In some cases, such as with lymphocytes, the presence of a lectin (page 225) such as concanavalin A can have dramatic growth-stimulating effects. All of these agents are said to be *mitogenic*, i.e., mitosis-stimulating. Furthermore, in most, if not all of these cases, it is believed that the active agent brings about the response as a result of the interaction with the cell's external surface, presumably causing some type of organizational change within the lipid-protein assembly.

If we assume that the signals that control the growth rate of cells are mediated by the plasma membrane, then the question of greatest importance concerns the means by which these changes at the cell surface bring about the ensuing internal changes. Although a considerable amount of speculation has been offered, the answer to this question is far from clear.

Most of the speculation on this subject centers around two related groups of theories. In one group, the potential role of the cyclic nucleotides is stressed, while in the other group the role of cytoskeletal elements, primarily the microtubules and microfilaments, is emphasized. However, since cyclic nucleotides are believed to be one of the major regulatory substances involved in microtubule-microfilament formation, the two types of theories are actually related. We will begin with a consideration of the potential role of cyclic AMP and cyclic GMP in growth control.

In many ways the phenomenon of growth control seems analogous to that of the second messenger concept which was discussed in Chapter 5. As in the case of the hormone-stimulated changes in glucose metabolism (page 247), the greatest attention regarding growth control has centered on the role of the cyclic nucleotides. In many cases, but not necessarily all cases, a correlation is found between the levels of cyclic AMP and cyclic GMP and the growth rate. Rapidly dividing cells tend to have lowered concentrations of cAMP and elevated concentrations of cGMP. The reverse relationship is found in stationary phase cells. When additional serum is supplied to density-inhibited stationary cells, one of the very first responses to the added serum is a dramatic rise in the cGMP levels and a decrease in the cAMP levels (Fig. 14-6). These changes occur rapidly, they are transient, and they precede the initation of DNA synthesis by many hours. The question of importance is whether or not these changes are causally responsible for the variety of metabolic changes that occur when stationary cells are stimulated to divide or growing cells are inhibited by increased cell density.

A considerable body of evidence suggests that cyclic nucleotides are involved either directly or indirectly (for example, via changes in permeability) in growth control. For example, it has been reported in a number of cases that the addition of cyclic AMP to growing cells markedly reduces their growth rate. Similarly, the addition of cAMP can block

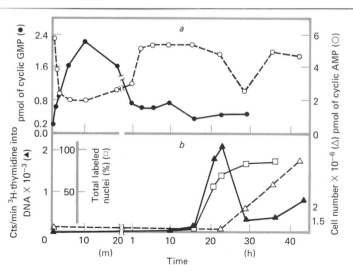

Figure 14-6. The effect of the addition of fresh serum on the levels of cyclic AMP and cyclic GMP (upper curves) and the incorporation of ³H-thymidine into DNA (lower curves). The most dramatic changes in nucleotide levels occur very soon after the addition of serum, long before the synthesis of DNA. *(From W. E. Seifert and P. S. Rudland,* Nature, ***248***:138, 1974.)*

the mitogenic effect of serum on density-inhibited cells. There are numerous attractive features of a cyclic nucleotide-mediated control of growth. They are known to be synthesized by membrane-bound enzymes in response to the interaction of the membrane with external influences. In addition, they are known to have an influence on a diverse variety of seemingly unrelated processes within the cell (Fig. 14-7). It is not unreasonable to suppose that changes in the levels of cyclic nucleotides could be responsible for many of the differences that have been noted between growing and nongrowing cells, and furthermore between normal and malignant cells. Other investigators raise the question of how many different responses can one or two compounds be expected to control. Different external agents seem to produce seemingly opposite types of physiological response in different types of cells using the same nucleotide.

The proposal that cytoskeletal elements are involved in growth control has emerged from the demonstration of the relationship between plasma membrane receptors and underlying microfilament-microtubule components (page 229). It was shown in Chapter 5 that the interaction between externally disposed membrane receptors and specific ligands, such as lectins or antibodies, resulted in a physical linkage between the receptor and actin-containing elements on the inner surface of the mem-

Figure 14-7. Schematic illustration of the variety of types of processes affected by cAMP. All of these effects may be mediated by the activation of one or a variety of protein kinases. *(Drawing by B. Tagawa, from Cyclic Nucleotides, by N. D. Goldberg, Hospital Practice, Vol. 9, No. 5 and "Cell Membranes: Biochemistry, Cell Biology & Pathology," G. Weissmann and R. Claiborne, Eds., HP Publishing Co., Inc., New York, NY 1975. Reprinted with permission.)*

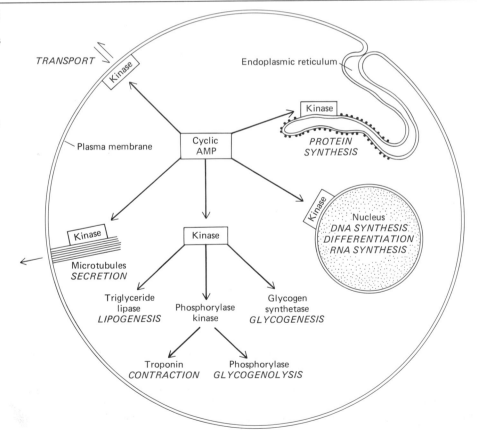

brane. Since numerous growth-promoting substances are believed to act by forming complexes with plasma membrane receptors, the linked cytoplasmic microfilaments could serve to transmit the signal to the interior of the cell. The proposal is analogous in many ways to that described for cyclic nucleotides. In both cases there is an opportunity for the involvement of the cyclic nucleotide or cytoskeletal element in growth control via its relationship to surface receptors as well as a correlation between the cytoplasmic agent (nucleotide or microfilament) and the growth state of the cell. In the case of the cytoskeletal hypothesis, the best correlations are made on malignant cells which are considered to be out of growth control. In various studies (page 808), it has been found that the transformation from the normal to the malignant state is accompanied by a marked disorganization of the microfilament-microtubule system. Whether this change in cytoplasmic organization is an underlying cause of the loss of growth control or an unrelated consequence of transformation remains to be determined.

SOMATIC CELL GENETICS

The study of the genetics of higher organisms by classical genetic techniques has been, for all practical purposes, limited to those organisms such as the fruit fly or maize which have a short generation time and produce a large number of offspring. Although many mutants have been isolated among various mammals and these have proved invaluable in the analysis of an innumerable variety of physiological processes which these genetic alterations affect, very little is known about the chromosomal locations of these mutations. Although this may not represent a serious deficiency in our understanding of genetic activities, this type of information may be very important in the case of human genetic disorders.

The value of microorganisms in genetic analysis was stressed in earlier chapters. These cells can be grown in large numbers in culture, mutations can be induced by the use of various mutagens, the mutants can be selected by the use of various types of culture media, and the location of the genetic alteration can be mapped by recombination. During the period in which cell culture methods were being developed, the concept arose that some of these same genetic techniques might be applicable to cultured mammalian cells. Like bacteria, mammalian cells contain genetic information which is expressed by the cells as they grow. It should, therefore, be possible to induce detectable mutations in cultured cells and isolate mutant strains.

Treatment of cultured mammalian cells with physical or chemical mutagens was found to result in detectable mutations in a small percentage of cells. How does one select for these genetically deficient cells? The extensive research on bacterial mutants had provided certain guidelines for selection techniques for mammalian cells. The general approach (developed primarily by Theodore Puck and coworkers) that has been used in the isolation of the mutant cell is to place the culture (after treatment with mutagen) into a medium which is capable of supporting the growth of the wild-type cells but not the particular mutant being sought, and then add some substance which will kill all of the cells that are growing. The technique is best described by the use of an example. One of the best-studied groups of cell mutants contains alterations in the gene that codes for thymidine kinase (TK), an enzyme which catalyzes phosphorylation of thymidine (forming dTMP) (Fig. 14-8a). In a normal medium, TK⁻ cells can do without this enzyme since they can form dTMP from dUMP (Fig.

Figure 14-8. (a) Reactions involved in a technique for cell selection as described in the text. (b) The procedure for selecting against growing sufficient cells. [(b) From F-T. Kao and T. T. Puck, Proc. Natl. Acad. Sci. (U.S.), **60**:1275, 1968.]

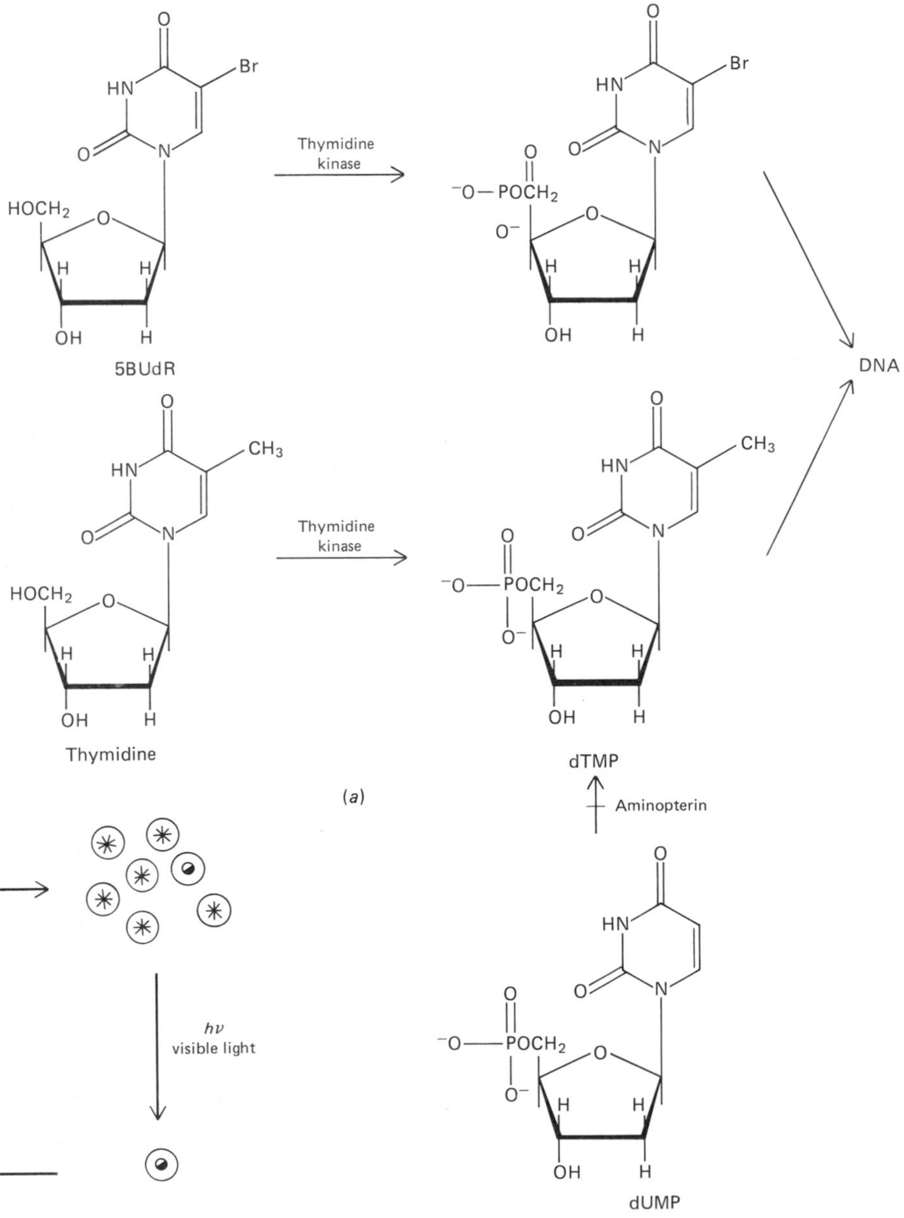

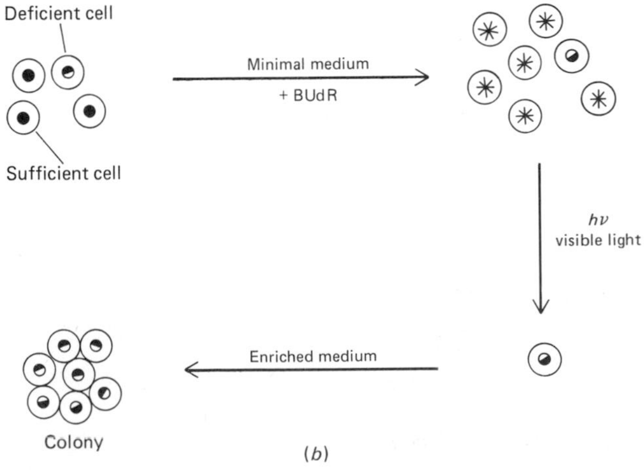

(a)

(b)

8a). In contrast to wild-type cells, TK⁻ cells cannot use thymidine even when it is supplied to them. Similarly, TK⁻ cells cannot use bromodeoxyuridine (BUdR), an analog of thymidine, while wild-type cells incorporate BUdR into DNA. Since BUdR, once incorporated, is lethal in the presence of light, the wild-type cells can be selectively eliminated while the TK⁻ cells, which have not been able to incorporate BUdR, will survive and can be grown into colonies after transfer to an enriched medium (Fig. 14-8b).

How are the chromosomal sites of these mutations determined? The classical technique for gene mapping, whether in viruses, bacteria, or fruit flies, is to determine the frequency with which recombination occurs between two mutations on the same chromosome. Recombination between genetic markers requires the exchange of genetic material between two chromosomes present in the same cell. In viruses this can occur if two particles infect the same host cell, in bacteria it normally occurs during conjugation, and in higher organisms it occurs during meiosis. Unfortunately, genetic recombination does not occur in mammalian cells growing in culture, so an alternate means to carry out genetic analyses had to be devised. The technique that has proved amenable for genetic mapping in cultured cells involves cell fusion.

During the mid-1960s, it was noted that, when certain types of cells were fused with one another, there was a tendency for chromosomes of one of the species to be lost as the hybrid cells proliferated. A particularly important finding in the field of human genetics was made when human cells were fused with various types of rodent cells; the hybrids lost human chromosomes. While the entire rodent chromosomal complement was usually retained, most of the human chromosomes were lost (at first rapidly and then more slowly) in a completely *random* manner. Consequently, if one were to produce hybrids between mouse and human cells, allow these hybrids to divide many times, add colchicine to these cells (page 693) to stop mitosis in metaphase, and then make chromosome preparations from these cells (Fig. 15-14a), one would find that most of the human chromosomes were missing among the chromosomes of the hybrids and that different cells would have different chromosomes that remained. Since human chromosomes can be identified (Fig. 15-14c), correlations can be made between a particular human gene function in the hybrid and a particular chromosome that is retained.

To illustrate how this technique can be used in gene mapping, we will briefly consider its first successful application by Mary Weiss and Howard Green in 1967 on the thymidine kinase locus. We have already described the way in which one selects *for* TK⁻ cells. There is also a procedure by which one can select *against* these mutants. It was stated that TK⁻ cells survive by forming dTMP from dUMP, a reaction which is indirectly blocked by the inhibitor aminopterin (Fig. 14-8a). Normal cells can grow in the presence of aminopterin as long as they are supplied with thymidine, while TK⁻ cells cannot grow in aminopterin under any conditions. Selection against TK⁻ cells is accomplished by growing cells in a medium which contains both aminopterin and thymidine (the HAT medium).

Weiss and Green mixed TK⁻ mouse cells with wild-type human cells in normal medium and allowed cell fusion to occur. After four days in culture, they transferred the cells to the HAT medium; unfused mouse cells were killed and unfused human cells survived. The survival of hybrid cells depends upon their chromosome complement. Those hybrid cells that retain the human chromosome which codes for thymidine kinase will proliferate while those that lose it anywhere along the line will die. Growth of the cells in HAT medium left two types of cells, unfused human cells and thymidine kinase-containing hybrids. Mouse cells happen to grow much more rapidly in culture than do human cells and the more human chromosomes a hybrid loses the faster it generally grows. This situation provided a built-in system to select against the unfused human cells which were soon outgrown by the thymidine kinase-containing hybrids, upon which chromosome analysis could be made. It was soon determined that only human chromosome number 17 was always present in hybrid cells that survived the HAT medium. Chromosome 17 was therefore the one carrying the thymidine kinase gene. Various modifications and advances in these techniques have been made (including new means to introduce chromosomes into cells) and by 1974 over 100 different human genes had been localized on nearly all of the 23 different human chromosomes. Included among these genes are ones responsible for certain congenital metabolic disorders including Tay-Sachs disease (the deficient gene codes for hexosaminidase A) and for galactosemia (the deficient gene codes for galactose-phosphate uridylyl transferase).

The technique of mammalian cell culture, the isolation and identification of cell mutants, and the genetic mapping of somatic cells (nongerm cells) have already made important contributions in clinical areas. For example, the technique of amniocentesis (Fig. 14-9), whereby cells are taken from the amniotic fluid of a human fetus, provides fetal cells upon which tests for chromosomal abnormalities and metabolic disorders can be made. The early identification of congenital defects provides the parents with the opportunity to abort the fetus or apply the proper corrective actions upon the newborn baby. In addition, there has been a great deal of speculation about the possibility of cor-

Figure 14-9. The technique of amniocentesis by which cells of the fetus can be removed from the amniotic fluid and cultured in preparation for chromosome anlaysis. *(From "Prenatal Diagnosis of Genetic Disease" by T. Friedmann. Copyright © Nov. 1971 by Scientific American, Inc. All rights reserved.)*

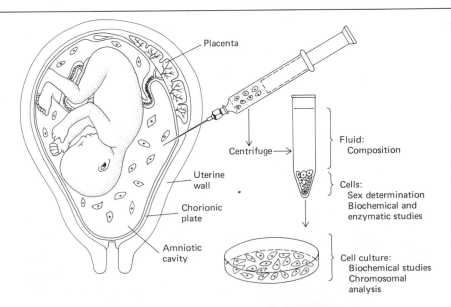

recting genetic defects in individuals. The types of procedures discussed in Chapter 18 concerning gene synthesis and cloning provide the potential for many unforeseen genetic manipulations; the controversies in this area have just begun. Regardless of the course that these events may take, one can be certain that cultured human cells will be of central importance in both the research and its applications.

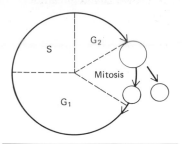

Figure 14-10. The cell cycle. *(From E. D. P. de Robertis, F. A. Saez, and E. M. F. de Robertis, "Cell Biology," 6th ed., Saunders, 1975.)*

THE CELL CYCLE

We are used to thinking of biological processes in terms of cycles, including life cycles, metabolic cycles, and physiological cycles. The life process is a continuum in time and there must be a continual renewal or return to an initial state so that the process can occur once again. Any given cell can be said to have a life, just as any given organism. In the case of a particular cell, life begins with its formation by cell division of a parental cell and ends either with the formation of daughter cells or with its death. Since the same process, that of cell division, is responsible for both the gain and loss of identity of cells, we can speak of the cell's life cycle or simply the *cell cycle* (Fig. 14-10). There are two readily observed phases of the cell cycle, that of mitosis and that of interphase. Based on the relatively small percentage of cells in an average tissue or culture dish that exist in mitosis at any given time, one can conclude that the vast majority of the cell cycle is spent in interphase while the mitotic event is a relatively fleeting process. This can be readily verified by actually following a particular cell from the time of its formation to the time it undergoes division itself.

Whereas the alternation of interphase and mitosis forms a morphologically visible (cytological) cycle, the analysis of various biochemical events, primarily DNA synthesis, reveals a very different cycle. The most important biochemical activity that a cell undertakes in preparation for cell division is the synthesis of its DNA, i.e., *replication*. When does the replication of chromosomal DNA occur in the cell cycle? The definitive answer to this question was supplied soon after the development of the technique of autoradiography and the availability of radioactive DNA precursors, first ^{32}P and then ^3H-thymidine. The initial experiments were performed in 1953 by Alma Howard and Stephen Pelc and later by other investigators using more refined techniques.

If one adds ^3H-thymidine for a relatively brief period (e.g., 30 minutes) to a culture of growing cells, one finds that only a fraction (generally about 40%) of the cells are actively engaged in the synthesis of DNA at a particular time. If we assume that the percentage of cells engaged in a particular activity is an approximate measure of the percentage of time that this activity occupies in the lives of cells, then it becomes apparent that DNA synthesis is not a continual interphase activity. When in interphase does DNA synthesis occur? The answer to this question can be determined by adding ^3H-thymidine to a culture of cells (growing asynchronously, page 654), washing the label from the culture, removing samples at subsequent periods, and determining autoradiographically the percentage of *cells in mitosis that are labeled* (such as in Fig. 14-32). Any cell that is in mitosis which contains labeled chromosomes must have been synthesizing DNA during the period when the ^{3}H-thymidine was available. The experiment is most easily understood by considering the results.

If ^{3}H-thymidine is given to a culture of cells for 30 minutes and an aliquot of cells are fixed, dried onto a slide, and autoradiographed, one finds that there are no labeled mitoses. A standard percentage of the cells are in mitosis, but their chromosomes are not labeled because these cells were not engaged in replication during the labeling period. If one waits an hour or two or three before removing an aliquot of cells, there still are generally no cells with labeled mitoses (Fig. 14-11). It appears from these results that there is a definite period (termed G_2 or Gap_2) between the end of DNA synthesis and the beginning of mitosis (M). The duration of G_2 is revealed by continuing to take aliquots from the culture until one finds labeled mitotic figures. The first labeled mitoses to appear represent cells that were at the last stages of DNA synthesis at the time the ^{3}H-thymidine was present. The length of time between the start of the labeling period and the appearance of cells with labeled mitotic figures represents the duration of G_2. The phase of the cell cycle in which replication occurs is termed *S phase*, the duration of which can be determined by continuing to follow the percentage of labeled mitoses. When this is done, one finds that the percentage increases, plateaus, and then falls off (Fig. 14-11). The duration of time over which labeled mitoses continue marks the duration of the S phase. When one adds up the periods of $G_2 + S + M$, it is apparent that there is an additional period in the cell cycle to be accounted for. This fourth phase, termed G_1 (*Gap$_1$*), is sandwiched between M and S and represents a period preceding the initiation of DNA synthesis.

In a previous section of this chapter it was pointed out that cells growing in culture reach a density at which they stop further growth and division. Analysis of cells in the stationary growth phase indicates that they possess a diploid amount of DNA. Furthermore, if stationary phase cells are given ^{3}H-thymidine, there is very little incorporation of radioactivity into nuclear DNA. We can conclude from these two observations that each cell had passed through its last mitosis and then stopped in G_1. Although there are exceptions,

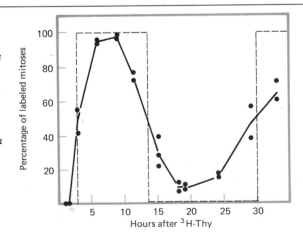

Figure 14-11. Percentage of labeled mitoses in monolayers of HeLa cells as a function of time after a 30-minute pulse with ^{3}H-thymidine. The broken line represents the curve that would be obtained if there were no individual variation among cells in the lengths of the cell cycle and its phases. The solid line represents the curve obtained experimentally as described in the text. *(From R. Baserga and F. Wiebel,* Int. Rev. Exp. Path., **7**:6, 1969.)

one generally finds that cells that have stopped dividing, whether they occur in the body or in culture, exist in a stage prior to the initiation of DNA synthesis. Cells that are "locked" in this state are usually said to be in the G_0 state to distinguish them from the typical G_1 cell that will soon enter S phase. There is evidence that the G_0 and G_1 have different physiological properties. It appears that the most important decision concerning whether or not a cell will continue to divide is made in the G_1 phase, and that decision is whether or not to initiate DNA synthesis. Once the cell begins to replicate its DNA, it almost invariably completes that round of synthesis and goes on to mitosis. The evidence suggests that a cell in G_1 must receive a positive signal in order to proceed into S phase. If the signal is not received at the proper time, the cell goes into the quiescent state (G_0) from which it can be retrieved. In culture, the G_0 cell can be reprogrammed to begin preparation for DNA synthesis and mitosis by the various factors (fresh serum, proteases, etc.) previously discussed. It is as if there is an internal cell cycle which is open at one point in G_1 to influences from the outside. These influences act via the plasma membrane, probably via cyclic nucleotide intermediates, to generate a positive environment in which the cycle can continue. The evidence for the existence of positive signals involved in the initiation of DNA synthesis can be directly revealed by cell fusion experiments described in a following section.

Cell Cycles In Vivo

One of the properties that distinguishes various types of cells within an organism is their capacity to grow and divide. We can broadly recognize three categories of cells.

1. Cells with extreme structural specializations, such as nerve cells, muscle cells, or red blood cells, have lost the ability to undergo division. Once the differentiation process has occurred, they will remain in that state until they die.

2. Cells that normally do not divide, but can be induced to begin DNA synthesis when faced with the appropriate stimulus. Cells of this type are analogous to the stationary phase cultured cell; both can be thought of as G_0 cells. Included in this group are liver cells that can be induced to undergo proliferation by the surgical removal of part of the liver, or lymphocytes that can be induced to proliferate by interaction with the appropriate antigen (page 796).

3. Cells that normally possess a certain level of mitotic activity. Certain tissues of the body are subject to continual destruction and must be replaced on an ongoing basis. Included in this category are the blood cells, the epithelial cells that line the many body cavities, spermatogonia, and the cells of the skin. In each of these cases there exist large populations of reserve cells whose responsibility lies in the replacement of cells that are destroyed under normal physiological conditions. These reserve cells, or *stem cells*, can be thought of as being partially differentiated. Although they appear morphologically unspecialized and are still capable of division to form additional reserve cells, they have reached a state of differentiation where the choices before them are very limited. Melanoblasts, for example, are cells that give rise to *melanocytes* (pigment cells) of the skin or other part of the body. Melanoblasts are found in considerable numbers in the inner layers of the skin and are responsible for producing more melanoblasts by mitosis and for differentiating into pigment cells to replace those that are sloughed from the body's surface.

The analysis of the cell cycles of a large variety of cells has revealed a great deal of variability among various types present in the same organism. Cell cycles can range from approximately 30 minutes in the very rapidly dividing cells of cleaving embryos (other than mammals which cleave very slowly) to cycles lasting several days in certain slowly growing tissues. Of the three stages of interphase, G_1 is the most variable, although major differences also exist for S and G_2. The cell cycles of early embryonic development are unusual for reasons other than their rapidity; they occur without an intervening period of growth. The fertilized egg is a huge cell by most standards, one with a great amount of cytoplasm and only one nucleus. The mitotic divisions that occur during cleavage (Fig. 14-12) serve to divide the cytoplasm of the egg into smaller and smaller cellular compartments, each with its own nucleus, until the cells produced are approximately equivalent to those of the adult organism. The dissociation of the growth process from the division process as it occurs during early development, although highly unorthodox, illustrates that the two events are not necessarily causally related. On the other hand, cases are known where growth occurs in the absence of cell division. The formation of the oocyte during oogenesis is a prominent example. In numerous other cases where excessive cell enlargement occurs, as in the case of the salivary gland cells of *Drosophila*, the chromosomal material continues to increase despite the lack of separation of the chromosomes by mitosis, resulting in the formation of polytene chromosomes.

Cell Cycle Activities

Having established that DNA synthesis is confined to a certain period within the cell cycle, the question arises as to whether other activities are also restricted to one or another of the various stages. The discovery of DNA synthesis during the S phase could be made upon cells that were growing asynchronously, i.e., present at all stages in the cell cycle, because of the unique ability of ^{3}H-thymidine to label one specific type of macromolecule which could be subsequently identified in single cells at a specific stage (mitosis) by autoradiography. If we were to ask whether a particular enzyme was synthesized at one particular period during the cycle, we could not generally answer this question on a population of cells growing in an *asynchronous* manner. For a variety of types of studies, it has been necessary to develop techniques whereby all cells in the population are present at the same stage.

Figure 14-12. A diagram of cleavage in the echinoderm egg. Although the total cellular mass does not change, the cell number increases from 1 to over 1000 cells.

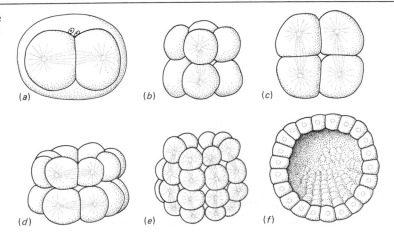

Measurements made on *synchronized* cells are theoretically, at least, equivalent to making measurements on single cells. In physiological terms, the entire culture can be thought of as one giant cell.

There are generally two broad types of synchronization procedures. In one type, cells of a particular stage are *selected* from among the entire population and this subpopulation can then be cultured in a somewhat synchronous manner. The most common selective technique takes advantage of the fact that cells in mitosis tend to round up and lose their attachments to the plastic or glass substratum of the culture dish (see Fig. 16-21). Consequently, mitotic cells are readily washed free from nonmitotic cells. The other type of synchronization approach is to *induce* synchrony by reversibly inhibiting some basic process of the cell cycle such as mitosis (with colchicine, page 693) or DNA synthesis (by various agents) and allow all of the cells to gradually accumulate at this one stage at which arrest has occurred. Once all of the cells of the culture have reached this same point, the inhibition can be relieved and the cells continue from that point in an approximately synchronized condition. Each approach has its advantages and disadvantages and both have been extremely useful in uncovering the extent to which various activities are restricted within the cell cycle.

When the overall rates of RNA or protein synthesis are measured in synchronized populations of cells, it is generally found that the rates of these synthetic activities are relatively constant throughout the interphase period, followed by a marked drop in protein synthesis and virtual cessation of RNA synthesis during mitosis. In contrast, many *specific* proteins do appear to be synthesized primarily at one stage or another of the cell cycle. The most dramatic stage-specific synthesis is seen in the case of the histones. Histone synthesis is found to occur exclusively during S phase (Fig. 14-13). Not only is the synthesis of the histones restricted to the S phase, but so too is the synthesis of the histone mRNAs. When chromatin from various stages of the cell cycle is isolated and its template activity determined by in vitro incubation with RNA polymerase (page 608), chromatin isolated from S-phase cells is found to produce histone mRNAs, while chromatin isolated from G_1 or G_2 is not

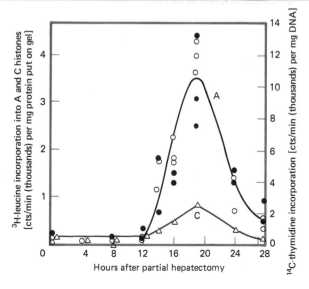

Figure 14-13. The coordinated synthesis of DNA and histone. Rats were partially hepatectomized (liver removal) to stimulate division of liver cells. At various times after surgery, animals were injected with ³H-leucine and ¹⁴C-thymidine and the livers removed one hour later to determine the radioactivity in DNA and histones (two electrophoretically separable histone classes, designated A and C in the figure). Open circles indicate radioactivity in DNA while the solid circles and open triangles indicate radioactivity in the two histone fractions. *(From S. Takai, T. W. Borun, J. Muchmore, and I. Lieberman, Nature, **219**:861, 1968.)*

active in this regard. Experiments of this type strongly argue that stage-specific synthesis of histones is under transcriptional control. As in the case of the steroid hormones (page 617), the template activity appears to be mediated by specific nonhistone chromosomal proteins.

When synchronized populations of cells are examined for the presence of histone mRNAs, it has been found that these specific messages are present only within S-phase cells. In other words, not only is the synthesis of the histone mRNAs restricted to S phase, but, once replication has been completed, the messages appear to be selectively destroyed. The synthesis of the DNA and histones of the chromosomes during S phase is more than a reflection of two independetly controlled activities. The tight manner in which DNA and histone synthesis is coupled is revealed by the effects on the synthesis of one, by blocking the synthesis of the other. If DNA synthesis is inhibited by various agents, histone synthesis soon comes to a halt. In fact, the histone mRNAs disappear prematurely from the cytoplasm. It would appear that the presence of the histone templates requires the continued synthesis of the DNA with which the histones produced will associate. Once the synthesis of the DNA stops, whether at the end of S phase or prematurely by the action of inhibitors, some type of cytoplasmic state is established that leads to the destruction of the histone mRNAs.

Control of the Cell Cycle

The nuclear-transplantation and cell-fusion experiments described in the previous chapter indicated that the cytoplasm from a cell actively engaged in DNA synthesis (an amphibian egg or a HeLa cell) was capable of activating DNA synthesis in a nucleus that was no longer synthesizing DNA. These experiments suggested that the activation of DNA synthesis results from the positive action of some type of cytoplasmic factor or factors. Cell-fusion experiments carried out by Potu Rao and Robert Johnson on cells at different stages of the cell cycle have provided further support for this concept. In these experiments it was found that, if cells in G_1 were fused with cells in S, the nucleus of the G_1 cell was activated to begin replication in advance of the time it would have if fusion had not occurred. In contrast, when cells of G_2 were fused with S-phase cells, each nucleus continued with its previous activities. The G_2 nucleus, which had already replicated its DNA prior to fusion, was not induced to initiate replication again, nor was the S-phase nucleus subjected to influences to suppress its DNA synthetic activity. Other aspects of these cell-fusion experiments which focus on the transition of a cell from G_2 to M will be considered in the following chapter.

REPLICATION

One of the most basic characteristics of living systems, regardless of how primitive, is their ability to reproduce. The self-duplication process can be observed at several levels: organisms produce offspring, cells undergo division, and the genetic material is replicated. In this section we will concentrate on the latter activity. The formulation of the structure of DNA by Watson and Crick in 1953 was accompanied by a proposal for its self-duplication. The proposal centered on the hydrogen bonds which hold

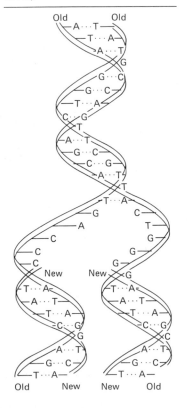

Figure 14-14. The initial proposal of the structure of DNA by Watson and Crick was accompanied by a suggestion on the means of replication. In this concept, illustrated here, the two strands separated and each served as a template for the synthesis of a new strand. *(From J. D. Watson, "The Molecular Biology of the Gene,"* W. A. Benjamin, 3d ed., *1976.)*

the duplex together. Individually these hydrogen bonds are weak and readily broken nonenzymatically. Watson and Crick envisioned that replication occurred by the gradual separation of the strands of the double helix (Fig. 14-14), which resulted from the progressive breakage of hydrogen bonds much like two halves of a zipper become separated. Since each strand of the duplex contains the information required for the construction of its complement, the unzipping provides the cell with two single-stranded templates, each capable of bringing about the restoration of the double-stranded state. This proposal makes certain predictions concerning the behavior of DNA during replication, foremost among these is the physical separation of the parental strands during the process. Not only are the two strands of the original duplex separated during replication, but they are permanently displaced to different cells following division.

Semiconservative Nature

Before considering some of the evidence either for or against this scheme, it is important to consider other possibilities and the means by which they can be distinguished. The Watson-Crick proposal for replication leads to the production of daughter duplexes each of which contains one complete strand from the parental duplex and one complete strand that has been newly synthesized. Replication of this type (Fig. 14-15, scheme 3) is termed *semiconservative* since one-half of the parent structure is conserved in passage to each of the daughter cells. Without having information on the mechanism responsible for replication, two other types of replication had to be considered. In one case, that of *conservative* replication (Fig. 14-15, scheme 1), the two original strands remain together (after serving as templates) as do the two newly synthesized strands. As a result, one of the daughter cells contains only the fully conserved duplex, while the other daughter cell contains only newly synthesized DNA. In the third case, that of *dispersive* replication (Fig. 14-15, scheme 2), the integrity of each of the parental strands becomes disrupted. As a result, the daughter cells contain duplexes in which each strand is a composite of the two types of DNA, i.e., neither the strands or the duplex itself is conserved.

Figure 14-15. Diagram illustrating the distribution of labeled atoms present in a parental DNA duplex among daughter and granddaughter duplexes, according to three different schemes of DNA replication. Scheme 1 illustrates conservative replication, scheme 2 illustrates dispersive replication, and scheme 3 illustrates semiconservative replication. *(From W. Hayes, "The Genetics of Bacteria and Their Viruses," 2nd ed., Wiley, 1968.)*

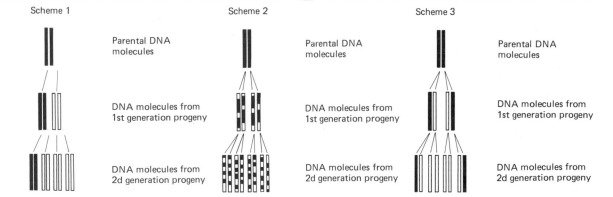

In order to distinguish among these three possibilities, it is necessary to distinguish between those atoms present in the original DNA molecules and those incorporated into the newly synthesized DNA so that the distribution of each can be followed. About 1958, two sets of experiments were performed, one employing bacteria and the other cultured plant cells, that clearly established the semiconservative nature of the replication process. In the experiments with bacteria performed by Matthew Meselson and Franklin Stahl, original and newly synthesized DNA were distinguished by the use of heavy versus light isotopes, while, in the experiments with eukaryotic cells performed by J. Herbert Taylor, the DNAs were distinguished by the use of nonradioactive versus radioactive isotopes. Meselson and Stahl used density gradient centrifugation to assay their results; Taylor used autoradiography. We will begin with the experiments of Meselson and Stahl.

These investigators began their experiments with bacteria that had been grown for many generations in medium containing ^{15}N-ammonium chloride as the nitrogen source. Consequently, all of the nitrogen-containing bases of the DNA of these cells were composed of the heavy nitrogen isotope. Cultures of "heavy" bacteria were then abruptly exposed to a medium with ^{14}N-containing substances, and samples were re-

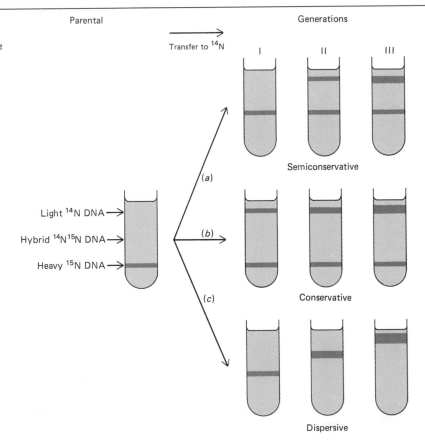

Figure 14-16. Results expected by isopycnic sedimentation for each of the three schemes of replication in the Meselson-Stahl experiment. The single tube on the left indicates the position of the parental DNA and the positions at which totally light or hybrid DNA fragments would band.

moved at increasing intervals over a period of several generations. DNA was extracted from the bacterial samples and subjected to cesium chloride centrifugation (page 455). The position at which a particular DNA molecule is found at equilibrium in a cesium chloride density gradient is directly related to its buoyant density which, in turn, is directly related to the percentage of ^{15}N atoms as opposed to ^{14}N atoms. If replication is semiconservative, one would expect that the density of the DNA would gradually decrease up to the point where one generation time had been reached (Fig. 14-16a). This decrease in density would occur as light strands were synthesized in association with heavy strands. After one generation, all of the DNA molecules should be hybrids with respect to the presence of ^{15}N and ^{14}N, and their buoyant density should be halfway between that expected for totally heavy and totally light DNA (Fig. 14-16a). As replication continues beyond the first generation, the newly synthesized strands continue to contain only light isotopes and two types of duplexes appear in the gradients: those containing DNA of a hybrid nature and those containing only light strands. As the time of growth in the light medium continues, a greater and greater percentage of the DNA molecules present will be light. However, as long as replication continues semiconservatively, the original heavy parental strands should remain intact and present in hybrid DNA molecules which come to occupy a smaller and smaller percentage of the total DNA (Fig. 14-16a). The results of the density gradient experiments obtained by Meselson and Stahl are shown in Fig. 14-17 and they conclusively indicate that each of the strands of the double helix remains intact during replication and each is passed on to one of the two daughter molecules. Replication must be semiconservative. The results that would have been obtained had replication been conservative or dispersive are diagrammed in Fig. 14-16b and c.

The experiments by Taylor and coworkers (Fig. 14-18) that established the semiconservative nature of replication in eukaryotic cells were carried out on cultured root cells of the bean *Bellevalia*. In these experiments, cells were labeled during replication with ^3H-thymidine and then treated with colchicine prior to the next division. Preparations of condensed chromosomes were prepared and the location of the silver grains was determined autoradiographically. Chromosomes that were prepared from cells labeled during the preceding S phase were found to contain radioactivity in both members of the pair of duplicated chromatids (Fig. 14-18) which remained attached to each other in the metaphase chromosome. As in the case of the Meselson-Stahl experiment, it was important to determine the distribution of the labeled atoms after a second round of replication. In order to carry out further examination, Taylor allowed cells to pass through one S phase in medium containing ^3H-thymidine and then a second S phase in medium lacking radioactivity before preparing the colchicine-blocked metaphase chromosomes. Examination of the chromosomes indicated that one of the chromatids of each associated pair was radioactive while the other was unlabeled. Each chromatid must be made of two subunits, i.e., DNA strands. During the formation of daughter chro-

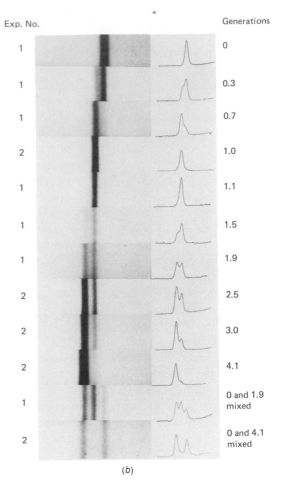

Exp. No.		Generations
1		0
1		0.3
1		0.7
2		1.0
1		1.1
1		1.5
1		1.9
2		2.5
2		3.0
2		4.1
1		0 and 1.9 mixed
2		0 and 4.1 mixed

(b)

Figure 14-17. Experimental results obtained by Meselson and Stahl. *(a)* The photograph on the left shows the bands obtained by simply mixing samples of heavy and light DNA prior to centrifugation. Tracing on the right indicates the absorbance of the peaks from which calculations as to amount of DNA can be made. *(b)* The effects on the density of DNA (fragments of the bacterial chromosome) with time after transferring cells from the heavy medium to the light medium. The appearance of a hybrid band and the disappearance of the heavy band by one generation eliminates conservative replication. The subsequent appearance of two bands, one light and one hybrid, eliminates the dispersive scheme. [*From M. Meselson and F. Stahl,* Proc. Natl. Acad. Sci. (U.S.), **44:**671, 1958.]

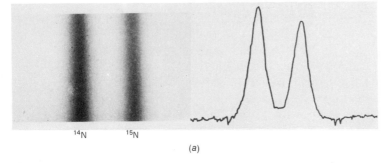

^{14}N ^{15}N

(a)

mosomes, one of the strands is conserved and passed onto the daughter chromatids, while the other strand is newly synthesized.

Replication in Bacteria

According to the semiconservative scheme of replication, the double helix would slowly unzip, leading to the synthesis of a pair of comple-

Figure 14-18. The experiment by Taylor and colleagues as described in the text. Cells were allowed to undergo one round of replication in medium containing ^{3}H-thymidine and the distribution of radioactivity followed autoradiographically. Dotted line indicates a labeled DNA strand.

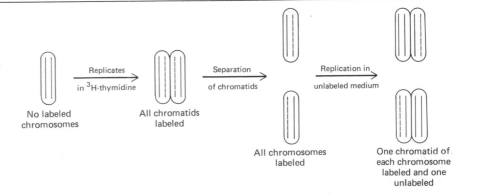

No labeled chromosomes

Replicates in ^{3}H-thymidine

All chromatids labeled

Separation of chromatids

Replication in unlabeled medium

All chromosomes labeled

One chromatid of each chromosome labeled and one unlabeled

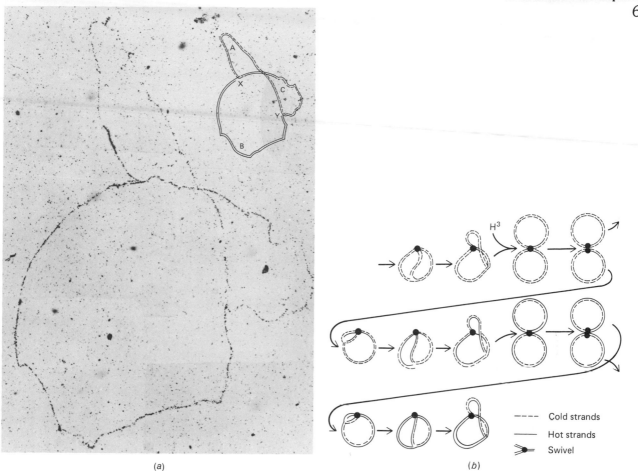

Figure 14-19. *(a)* Autoradiograph of a chromosome of *E. coli* that had undergone approximately 1.8 rounds of replication in a medium containing ³H-thymidine. The density of the silver grains provides a measure of the label in that portion of the chromosome. The dashed line in the explanatory insert represents an unlabeled strand, the solid line a labeled strand. *(From J. Cairns, Cold Spring Harbor Symp. Quant. Biol., **28**:44, 1963.)* *(b)* Diagrammatic illustration of the events that would lead to the autoradiograph shown in *a*, on the basis of a single growing fork. Later research revealed that bacterial chromosome replication occurred by the movement of a pair of replication forks from a single origin. *(From J. Cairns, Cold Spring Harbor Symp. Quant. Biol., **28**:43, 1963.)*

mentary strands in its wake. In the early 1960s, an autoradiographic technique developed by John Cairns provided a striking visual portrayal of this replication process. Cairns found that bacteria could be lysed very gently and their chromosomes spread out on a surface without further manipulation. If the bacteria had been growing on ³H-thymidine prior to the preparation of the chromosomes, then light microscopic autoradiography could reveal the outline of the labeled DNA strands, even though the molecules themselves were far below the resolution of the microscope. Autoradiographs of this type (Fig. 14-19*a*) clearly revealed that the bacterial chromosome was circular in nature, a property that had been inferred from genetic mapping studies. Secondly, these autoradiographs revealed the overall pattern by which replication occurred. In

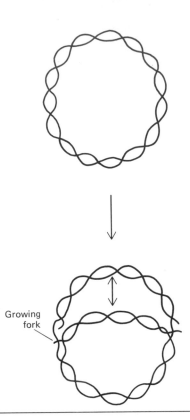

Growing
fork

Figure 14-20. Model of a circular chromosome undergoing bidirectional semiconservative replication. Two replication forks spread in opposite directions from a single origin (arrows).

many cases, outlines of the silver grains formed what are termed *theta* (Θ) structures. This profile represents a circular duplex caught in the act of replication as illustrated in Fig. 14-19*b*. They reveal the structure one would expect if one double-stranded circle were unraveling in the process of becoming two double-stranded circles. Each theta structure (Fig. 14-20) is composed of three distinct lengths of DNA, one of which represents the unreplicated portion of the chromosome and the other two the two daughter molecules in the process of formation. The points at which the pair of replicated stretches come together and join the nonreplicated section are termed the *growing points* or *replication forks*. The autoradiographs indicate that the replication forks correspond to sites where (a) the parental double helix is undergoing strand separation and (b) nucleotides are being incorporated into the newly synthesized strands.

On the basis of these early autoradiographs and many subsequent studies, it has been shown that the replication of the bacterial chromosome is characterized by the following properties. There is a specific point on the chromosome, the *origin*, at which replication always begins. Replication proceeds from the origin in both directions *(bidirectionally)*. The two replication forks move in opposite directions until they meet at a point approximately across from that at which the process begins. Following the termination of replication, the two newly replicated duplexes detach from one another and are ultimately directed into two different cells. It would appear that, if we are to understand the process of replication, we must determine the activities that occur at the replication fork.

The double helix is a spiraling molecule, one that makes a complete turn every 10 residues. The separation of the two strands of a circular, double-helical molecule poses certain difficulties. In order to picture the difficulties we will briefly consider an analogy between a DNA duplex and a two-stranded helical rope. Consider what would happen if you placed a linear (noncircular) piece of this rope on the ground, took hold of the two strands at one end, and then began to pull the strands apart just as DNA is pulled apart during replication. It is apparent that the separation of the halves of a double helix is also a process of *unwinding* the structure. In the case of the rope, which is free to rotate, the separation of strands at one end would be accompanied by the rotational movement of the entire fiber as a means to resist the development of tension in the structure. Then consider the result of pulling the strands apart if the other end of the rope is attached to a hook on a wall. Under these circumstances, separation of the two strands at one end generates increasing torque in the rope and causes the unseparated portion to become more tightly wound. When the number of turns per unit length of rope (or DNA) increases, as shown in Fig. 14-21, the structure is said to be *supercoiled*. Separation of the two strands of a circular DNA molecule is analogous to having one end of a linear molecule attached to a wall; in both cases the tension developed in the molecule cannot be relieved by rotation of the entire molecule. When one considers that the complete circular chromosome of *E. coli* contains approximately 400,000 turns and is replicated by 2 forks within 40 minutes, the magnitude of the problem becomes apparent. There are two ways in which the supercoiled situation might be avoided. In the case of the rope attached to the wall, if the ends had been turned (in the opposite direction of the turn of the helix) as they separated, the rope would not have become supercoiled. Alternatively, one could conceivably allow the rope to become somewhat supercoiled, then cut one of the strands of the pair, allowing it to rotate freely relative to the other one until the tension had been

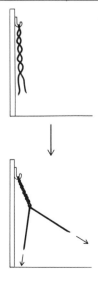

Figure 14-21. Diagram illustrating the effect of unwinding a two-stranded rope that has one end attached to a hook. The same problems arise when attempting to unwind a double-stranded circular molecule.

relieved. The cut strand could then be mended and the process of separation could continue and the release of the supercoil repeated.

Although the first of these two alternatives appears to be the simplest, in order for it to accomplish the unwinding of a circular DNA molecule, an added feature must be incorporated into the model. The added feature is the presence of some type of swivel as shown in Fig. 14-22a. If the strands of the bacterial chromosome were turning at the replication fork as they were separating, and there was no swivel, this would only cause the stretch of chromosome that had already been replicated to become supercoiled. Although it was originally believed that a swivel of some type did exist within the replicating chromosome, more recent evidence has suggested that the second alternative (Fig. 14-22b) is the correct one. As the DNA unwinds, a supercoiled condition is generated in the rear of the replication fork. A type of protein has been isolated from various types of cells which is capable of relieving the supercoiled state. If a preparation of this protein, termed the ω *protein* or the *relaxing enzyme* or the *swivelase*, is added to a supercoiled DNA molecule, the DNA returns to its normal helical state. The protein is believed to function as described above. It makes a nick in one of the two strands in the region under tension thereby allowing that strand to rotate around the intact strand until the strain has been dissipated. The same protein then seals the gap by catalyzing the formation of the phosphodiester bond that it had previously broken, and the replication fork can continue to move along the circular DNA molecule.

Figure 14-22. Two alternative models by which a circular molecule could unwind without meeting unsurmountable difficulties. *(a)* The introduction of a swivel so that the two ends of the replicating chromosome could turn independently of one another. *(From "The Bacterial Chromosome" by J. Cairns. Copyright © Jan. 1966 by Scientific American, Inc. All rights reserved.)* *(b)* In this second model, an enzyme acts to nick one of the strands, which is then capable of rotation around the intact strand until the duplex is no longer supercoiled. The free ends are enzymatically rejoined to reform the intact strand.

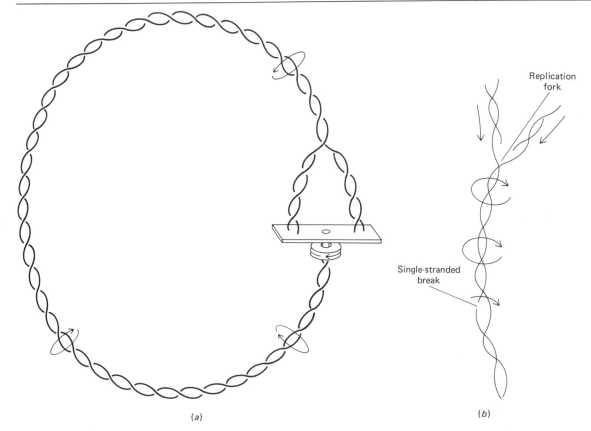

Replication fork

Single-stranded break

(a)

(b)

The rapid unwinding of the double helix is accomplished via the attachment to the DNA of "unwinding proteins." These proteins have been isolated from a wide variety of cells by taking advantage of their preferential binding to single-stranded DNA. When cellular extracts are passed through columns containing single-stranded DNA that has been immobilized by attachment to cellulose, the unwinding proteins are selectively removed from solution. Purified preparations of these proteins are capable of causing the complete denaturation of added DNA molecules at temperatures 40°C below that normally required to separate the strands. All double-stranded DNA molecules are characterized by a certain level of opening and closing of the double helix at various locations within the molecule. In other words, individual hydrogen bonds are continually being broken and reformed as the DNA molecule "breathes." It is believed that the initial interaction between unwinding proteins and DNA occurs as a result of the attachment of the proteins to AT-rich sites that have been transiently converted to the single-stranded state. Once one of the proteins attaches, there is a greatly increased likelihood that additional proteins will attach alongside the first ones; the proteins interact cooperatively with one another (page 155). Unwinding proteins attach to the DNA in large numbers in advance of the replication and bring about the separation of the strands.

The Mechanism of Replication

In order to understand the steps that occur during the process of replication, we have to turn from the studies on intact bacterial cells and consider a large body of evidence that has accumulated on the in vitro analysis of DNA synthesis. Very soon after the proposal of the double-helical structure of DNA, investigations began on the nature of the enzymatic machinery responsible for its synthesis. Despite the extensive research that had been carried out on the enzymatic reactions involving small molecular weight metabolites, virtually nothing was known about the types of reactions responsible for the polymerization of the precursors of the macromolecules. The first polymerization of a nucleic acid by an enzyme acting in vitro was reported by Marianne Grunberg-Manago and Severo Ochoa in 1955. They had discovered the enzyme polynucleotide phosphorylase (later used by Nirenberg and Ochoa in their studies on the analysis of the genetic code, page 507) which was capable of polymerizing ribonucleoside diphosphates into polyribonucleotides. However, as described in Chapter 11, this enzyme does not operate on instructions of a template and does not synthesize nucleic acid within the cell. In this same year, Arthur Kornberg began a search for the enzymes responsible for the synthesis of DNA.

In their initial experiments, Kornberg and his coworkers sought to find activity in bacterial extracts that was capable of incorporating radioactively labeled precursors (ones they had synthesized) into an acid-insoluble polymer. Very slight incorporation activities were discovered, which proved to be all that was needed to provide the necessary assay (page 109) for the purification of the responsible enzymes. Within a short period of time they had greatly purified the activity, they had determined that the enzyme required the *presence of DNA* before incorporation would begin, and they had found that four deoxyribonucleoside triphosphates

Figure 14-23. *(a)* Examples of DNA structures that do not stimulate the synthesis of DNA in vitro by DNA polymerase prepared from *E. coli.* *(b)* Examples of DNA structures which can be used as template primers by the enzyme. In all cases the molecules in *b* contain a strand to copy and a 3′ OH on which to add nucleotides.

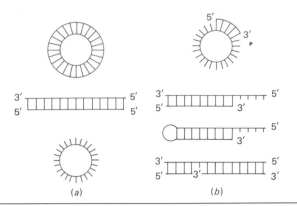

(a)   (b)

(dTTP, dATP, dCTP, and dGTP) were needed. An important question concerned the relationship between the added DNA and that which was synthesized. Analysis of the radioactive product indicated that the DNA being synthesized in the experiment always had the same base composition of the unlabeled DNA that had been originally added. As expected, the added DNA was serving as a *template* for the polymerization reaction. However, as more of the properties of this enzymatic reaction were uncovered, the situation became more complex.

When various types of template DNAs were tested, it was found that the added DNA had certain structural requirements if it was to promote the incorporation of labeled precursors (Fig. 14-23). An intact double-stranded DNA molecule, for example, was not capable of stimulating incorporation. This does not seem surprising considering the requirement that the strands of the helix must be separated for replication to occur. However, it is less obvious why a single-stranded circular molecule is also devoid of activity; one might expect this to be an ideal template to direct the manufacture of a complementary strand. In contrast, if a partially double-stranded DNA molecule is added to the reaction mixture containing the polymerizing enzyme and labeled precursors, an immediate incorporation of nucleotides is noted. The reason that the single-stranded circle cannot serve in this capacity is because the enzyme is not able to *initiate* the formation of a strand. Rather, it is only capable of adding nucleotides to the 3′ hydroxyl terminus of an already existing strand.

A DNA strand that provides the enzyme with the appropriate 3′ OH terminus is termed a *primer*. The nature of the condensation reaction between the incoming nucleoside triphosphate and the end of the primer is essentially the same as shown in Fig. 11-11 for RNA formation.

It appeared that the DNA polymerase purified by Kornberg (termed DNA polymerase I) has two requirements: a template DNA strand to "copy" and a primer DNA strand to which nucleotides can be added. An intact linear double helix provides the 3′ hydroxyl terminus but lacks template activity. In contrast, a circular single strand provides a template but lacks primer activity. The partially double-stranded molecule satisfies both requirements and promotes incorporation. The finding that the Kornberg enzyme is not capable of initiating the synthesis of a DNA strand leaves us with a distinct problem: How is the synthesis of a new strand initiated in the cell? We will return to this question below.

There was another property of the Kornberg enzyme, i.e., DNA polymerase I, that was difficult to understand in terms of its role as a replicating enzyme—it was only capable of polymerizing DNA molecules in a 5' to 3' direction. The diagram first presented by Watson and Crick (see Fig. 14-14) embodies the essence of what would be *expected* to occur at the replication fork. The diagram suggests that one of the newly synthesized strands is polymerized in a 5' to 3' direction while the other strand is polymerized in a 3' to 5' direction. The autoradiographs of Cairns, which clearly showed that label was incorporated into newly synthesized strands at both limbs of a replication fork, seemed to confirm this concept of synthesis of strands in both directions. Is there some other enzyme responsible for the construction of the 3' to 5' strand? Does the enzyme work differently in the cell than under in vitro conditions? We will also return to this question below.

Another difficulty arose when mutants were isolated that retained only about 0.5 to 1.0% of the normal DNA polymerase I activity but were still capable of normal growth and development. Was this residual activity sufficient for the enzyme's role in replication or was DNA polymerase I an enzyme that was not used in replication or at least not absolutely required for replication? The situation in the 1960s with regard to the three questions raised in this discussion (the inability of the enzyme to initiate, its inability to polymerize in the 3' to 5' direction, and its apparent dispensability) was uncertain. On the brighter side, one very important goal had been accomplished. In 1968, Kornberg and coworkers had finally been successful in the synthesis in the test tube of a complete, functional DNA molecule. Using a single-stranded viral chromosome as a template, they had synthesized a fully active complementary molecule. Most importantly, they had shown that DNA polymerase I was capable of replicating DNA and that its product was constructed with great fidelity with respect to the template being copied.

During the late 1960s and early 1970s, several important findings were made that shed a great deal of light on the questions raised above. Two other DNA polymerases were found in bacterial extracts and were named DNA polymerase II and III. In normal bacteria there are approximately 300 to 400 molecules of DNA polymerase I per cell and only about 40 copies of DNA polymerase II and 10 of DNA polymerase III. The presence of polymerases II and III had been masked by the much greater amounts of DNA polymerase I in the cell, even though DNA polymerase III is a much more active molecule than either of the other species. Analysis of the properties of the newly discovered DNA polymerases revealed certain basic similarities among all three enzymes as well as certain differences. None of the three enzymes were able to initiate DNA chains nor were any of them capable of constructing strands in a 3' to 5' direction.

The lack of polymerization activity in the 3' to 5' direction has a straightforward explanation—there is no synthesis of strands in that direction. It turns out that both of the newly synthesized strands are synthesized in a 5′ to 3′ direction. The enzymes involved in the construction of the two new antiparallel strands move oppositely along their respective templates, both proceeding in a 3′ to 5′ direction *along the template* and

Figure 14-24. Discontinuous synthesis of Okazaki fragments on both template strands. Each *fragment* is synthesized from its 5′ to 3′ end, even though one strand is synthesized in a 3′ to 5′ direction overall. The fragments are subsequently joined to previously synthesized DNA by the action of a ligase. Note that the region just in front of the fork has one double-stranded limb and one single-stranded limb. This condition arises due to the fact that all fragments are initiated at their 5′ end. Consequently, the initiation of this fragment on the strand growing from 3′ to 5′ overall awaits the duplex being unwound to a sufficient length to expose the template. *(Reproduced with permission from Vol. 1. "Gene Expression," B. Lewin. Copyright © 1974 John Wiley & Sons Ltd.)* *(b)* An electron micrograph of a replicating T₇ DNA molecule. The Y-shaped structure corresponds to the replicated (left two limbs) and unreplicated (right end) portions. Note the single-stranded portion of one limb (to the arrow) in the region adjoining the fork. *(Photograph by J. Wolfson and D. Dressler, from D. Dressler,* Ann. Rev. Microbiol., **29**:540, 1975.)

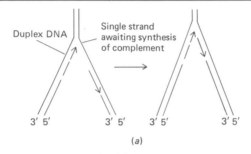

(a)

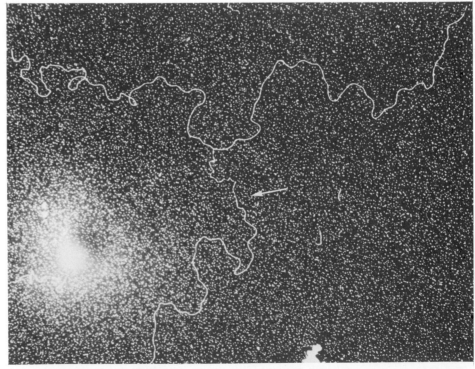

(b)

constructing a chain that grows from its 3′ hydroxyl terminus (Fig. 14-24a). Consequently, one of the newly synthesized strands should be growing toward the replication fork, while the other strand should be growing away from it. The reason that both of the newly synthesized strands of Fig. 14-19 and 14-31 as viewed by autoradiography appear to emerge from the replication fork is that these autoradiographs lack the resolution to actually show that incorporation is not occurring on both strands precisely at the site of unwinding. Whereas one new strand can be extended immediately behind the replication fork, the synthesis of the other new strand is delayed until some length of its template has been exposed (Fig. 14-24a). Consequently, when replicating bacterial chromosomes are examined in the electron microscope (Fig. 14-24b), one of the limbs of unwound DNA emerging from the replication fork is seen to be single-stranded and has yet to serve as a template. The other limb is double-stranded, revealing the presence of both the parental strand and its newly synthesized complement.

As stated previously, one of the newly synthesized strands should be growing toward the point of unwinding and the other away from it. It is apparent that the strand that grows away from the replication fork cannot be constructed in a continuous manner. Rather, this strand must *necessarily* be synthesized as fragments, each growing away from the replication fork toward the 5′ end of a previously formed fragment to which it would be subsequently linked. The other strand, i.e., the one that is constructed by the addition of nucleotides in a direction toward the replication fork, has the potential to be synthesized in a continuous manner. Although a considerable amount of research has been performed on this subject, at the time of this writing it is still uncertain as to whether the "potentially continuous" strand is actually synthesized in a continuous manner or whether it is synthesized discontinuously as its counterpart (see Cell, **12**:1029, 1977, for a discussion of this subject). In the following discussion, it will be assumed that both strands are synthesized discontinuously as shown in Figs. 14-24a and 14-26.

The synthesis of DNA strands in small pieces that are subsequently joined together was discovered by Reiji Okazaki in various types of labeling experiments. Okazaki found that, if bacteria were given very short pulses of ^{3}H-thymidine (e.g., 5 seconds), all of the incorporated radioactivity could be found as part of small fragments approximately 1000 to 2000 nucleotides in length (Fig. 14-25). In contrast, if this short pulse was followed by a brief chase before the cells were killed, all of the radioactivity was found associated with much greater molecular weight molecules (Fig. 14-25). Okazaki concluded that both of the newly synthesized strands were constructed in small fragments which were rapidly linked to the much longer piece that had previously been synthesized. The presence of an enzyme capable of joining adjacent pieces of DNA had been discovered in 1967 and termed the *DNA ligase*. Presumably this was the enzyme responsible for the attachment process.

How are DNA strands initiated? With the discovery of the synthesis of DNA strands as replication fragments (i.e., Okazaki fragments), the dif-

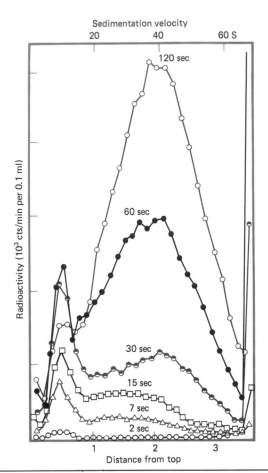

Figure 14-25. Sucrose density gradient profile of DNA from a culture of T_4-infected *E. coli*. The cells were labeled for increasing amounts of time and the sedimentation velocity of the labeled DNA determined. When DNA was prepared after very short pulses, radioactivity appeared in very short pieces of DNA. After longer labeling periods, the labeled DNA fragments had become joined to large molecular weight molecules. *(From R. Okazaki et al., Cold Spring Harbor Symp. Quant. Biol., **33**:130, 1968.)*

ficulties for initiation become even greater. How does the formation of each of these fragments begin when none of the DNA polymerases are capable of strand initiation? On the basis of various types of experiments, it became apparent that initiation was not accomplished by a DNA polymerase, but rather by an RNA polymerase. It was found, for example, that the 5′ terminus of the replication fragments consisted of RNA of approximately 50 to 100 ribonucleotides. It appears that, for each of the fragments, the first nucleotides are inserted by an RNA polymerase and the process is continued by a DNA polymerase. Once the fragment is completed, the RNA stretch at the 5′ end is removed, the gap in the strand is filled in with DNA, and then sealed. This arrangement involving temporary pieces of RNA appears to be a curious one and its basis is not understood. There is certainly no reason to believe that a DNA polymerase could not have evolved that was capable of chain initiation so one must conclude that it makes more biological sense to do it this way. It is possible that the likelihood of mistakes occurring is greater during initiation than elongation and the use of a removable RNA stretch avoids the presence of errors.

We are still left with questions concerning the need for three DNA polymerases. Based primarily upon the analysis of different types of mutant strains, the following assignments have been made (Fig. 14-26). The primary enzyme involved in strand formation during replication in *E. coli* is DNA polymerase III. This enzyme, though present in few copies (only a few are needed), is capable of incorporating approximately 2000 nucleotides per second, a rate 15 times greater than that found for DNA polymerase I. The function of DNA polymerase II is as yet uncertain. Mutants lacking this polymerase have been isolated and have no evident deficiency. DNA polymerase I is believed to be involved in both replication and DNA repair (page 678), a process by which damaged sections of DNA are corrected. Its function in replication centers on the replacement of the RNA with DNA.

In order to understand the proposed role of DNA polymerase I in replication, it is necessary to consider another property of the DNA polymerases: their ability to destroy a nucleic acid polymer as well as to synthesize it. DNA polymerase I, for example, is a potent exonuclease as well as a polymerase. An exonuclease is an enzyme that degrades nucleic acids by removal of the terminal nucleotides one at a time. There are $5' \rightarrow 3'$ exonucleases and $3' \rightarrow 5'$ exonucleases depending upon the direction in which the strand is degraded. DNA polymerase I has both $3' \rightarrow 5'$ and $5' \rightarrow 3'$ exonuclease activity, but the two functions are located in entirely different parts of the molecule. The $3' \rightarrow 5'$ activity involves the same active site region used in polymerase function, while the $5' \rightarrow 3'$ exonuclease activity is located in another part of the molecule. Remarkably, the two exonucleases can be separated by splitting the protein; the protein is essentially two different enzymes in one. The two exonuclease activities are believed to have entirely separate roles in replication. We will consider the $5' \rightarrow 3'$ exonuclease activity first.

Figure 14-26. Hypothetical sequence of steps in the replication of DNA. *(From A. L. Lehninger, "Biochemistry," 2d ed., Worth, 1975.)*

Recognition of the origin by RNA polymerase

Unwinding of DNA strands

Formation of RNA primers for leading and following strands

Formation of DNA on RNA primers by pol III*

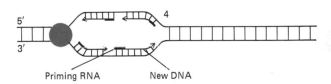

Excision of RNA primers by endonucleases to yield Okazaki fragments

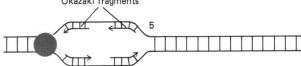

Filling of gaps by DNA polymerase I and sealing by DNA ligase

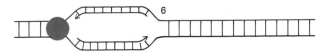

Most exonucleases are specific for either DNA or RNA, but the $5' \rightarrow 3'$ exonuclease of DNA polymerase I can degrade both types of nucleic acids. The initiation of replication fragments by RNA polymerase leaves a stretch of RNA at the 5' end of each fragment that must be removed. It is believed that the $5' \rightarrow 3'$ exonuclease activity of DNA polymerase I is responsible for the removal of the RNA. It is further believed that the enzyme, as it moves along degrading the RNA, uses its polymerase activity to fill the resulting gap with DNA. Once the enzyme has completed its task of RNA removal and DNA patching, there is only one function that remains—the formation of a phosphodiester bond (Fig. 14-27) between the 3' OH of the DNA patch and the 5' phosphate of the adjacent DNA, i.e., that segment to which the RNA primer had been covalently linked. This is the reaction catalyzed by the ligase.

Up to this point we have considered replication as a series of steps that begins with the opening of the double helix and ends with the sealing of a newly synthesized fragment to the main body of the forming DNA strand. It is appropriate at this point to move in and take a closer look at the activity of a DNA polymerase molecule as it is carrying out the incorporation of nucleotides. RNA and DNA polymerases appear to have similar functions, i.e., the construction of a complementary nucleic acid strand, although they are structurally very different types of enzymes. The most critical aspect of the process of replication is accuracy. Whereas a mistake made in a messenger RNA molecule results in the synthesis of defective proteins, one mRNA molecule is only one short-lived template among a large population of such molecules, therefore, little lasting damage results from the mistake. In contrast, a mistake occurring during replication results in a permanent mutation and very likely the elimination of the progeny of that cell. It is estimated that the chance that a given nucleotide will be copied incorrectly during replication in *E. coli* may be as low as 10^{-9}. In other words, only one out of one billion nucleotides incorporated are not complementary to the corresponding nucleotide on the template strand. Since there are approximately 4×10^6 nucleotides in the *E. coli* genome, there may be less than 1 nucleotide change for every 100 replication cycles.

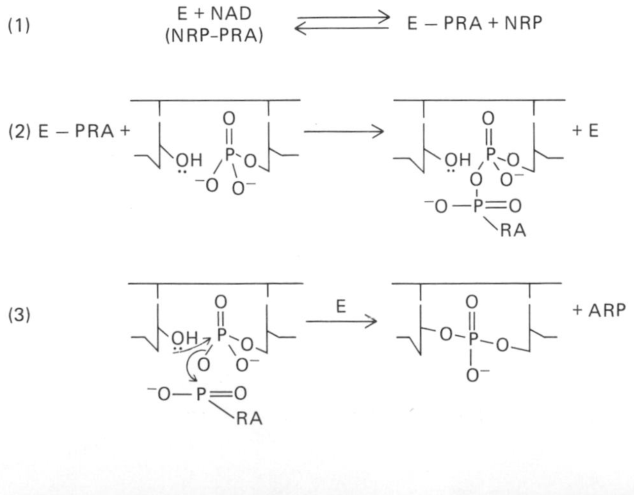

Figure 14-27. Postulated mechanism of the reaction catalyzed by the *E. coli* ligase. NAD is written as NRP-PRA to emphasize the pyrophosphate bond linking the nicotinamide mononucleotide (NRP) and adenylic acid (PRA) moieties of the NAD molecule. [*From B. M. Olivera, Z. W. Hall, and I. R. Lehman,* Proc. Natl. Acad. Sci. (U.S.), **61:**237, 1968.]

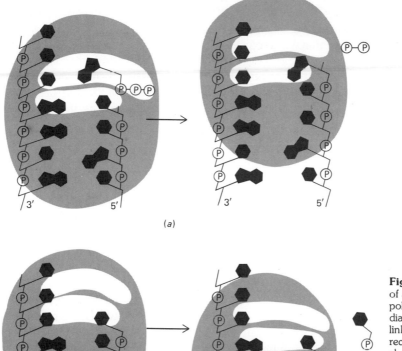

Figure 14-28. Model for the control of accuracy of DNA synthesis by DNA polymerase I of *E. coli.* In the upper diagram the incoming triphosphate is linked to the growing strand and is recognized by some type of stereo-chemical criteria as a proper fit, i.e., as part of an A-T or G-C pair. In the lower diagram, the newly linked nucleotide is recognized as having been a mistake and is removed by the 3' → 5' exonuclease activity of the enzyme. *(Reprinted with permission from Vol. 1. "Gene Expression," B. Lewin. Copyright © 1974 John Wiley & Sons Ltd.)*

It is believed that the basis for the incorporation of a particular nucleotide into the growing strand at a particular site is dependent upon the ability of that nucleoside triphosphate to form an acceptable base pair with the nucleotide of the template DNA strand. The enzyme, which has one site to which all four nucleoside triphosphates attach, is somehow capable of measuring the geometry of the base pairs between potential precursors and the template as the precursors move in and out of the active site. When an AT or GC pair is formed, the enzyme catalyzes the incorporation of the nucleotide into the growing strand (Fig. 14-28*a*). The incorporation of each nucleotide leaves an elongated primer with a new 3' OH to which the next nucleotide can be added. With each nucleotide incorporated, the polymerase moves one step along the template.

On occasion it can be expected that, as a result of a shift in the electronic state of the nitrogenous base of an incoming nucleotide or of the template nucleotide, an incorrect base pair (an A with a C or a G with a T) might form. For example, guanine and thymine exist predominantly in the keto (C—O) configuration, although, at a given instant, there is a certain chance that they might exist in the enol (C—OH) state. If

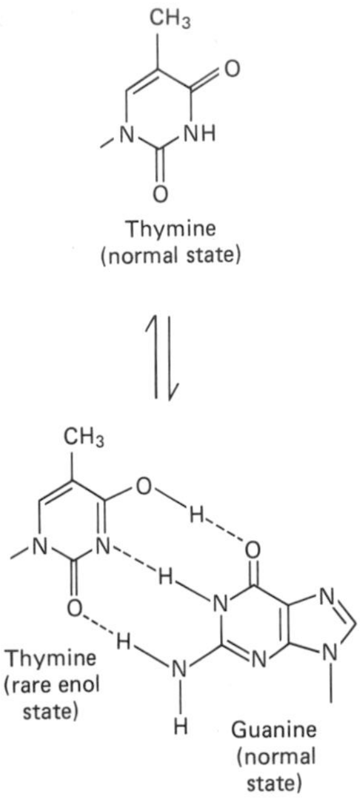

Figure 14-29. When thymine exists in its rare enol state it possesses a structure capable of forming an acceptable interaction with guanine, thereby leading to a transition-type mutation.

either of these bases were to exist in the enol form at the instant the base pairing took place (Fig. 14-29), then there is a strong likelihood of the formation of a G—T bond, thereby causing a mutation. It has been estimated that an incorrect pairing of this sort would be expected to occur about once every 10^4 nucleotides incorporated, a frequency much greater than the spontaneous rate. How is the normal mutation rate kept so low? The answer seems to lie in the other of the two exonuclease activities mentioned above, the $3' \rightarrow 5'$ exonuclease, which is present in all three polymerases. It appears that the responsibility of this exonuclease activity is to remove an incorrectly incorporated nucleotide before the next nucleotide is added to it (Fig. 14-29). The incorporation of an incorrect nucleotide produces a certain distortion of the double helix which is detected by the polymerase before it has moved from the spot. Recognizing its mistake, the polymerase immediately corrects it by the removal of the last nucleotide incorporated, and its replacement with the proper nucleotide. This job of "proofreading" is one of the most remarkable of all enzymatic activities and illustrates the sophistication to which molecular machinery has evolved.

It is apparent from the above discussion that the process of replication is very complex. We have described the role of unwinding proteins, a relaxing enzyme, an RNA polymerase, two DNA polymerases (I and III), and a ligase. Their specific roles are illustrated in Fig. 14-26. Each of these proteins has a specific responsibility, and it is believed that various of these proteins (in addition to ones as yet undescribed) are grouped together into a type of multienzyme complex, i.e., a replicase, which together brings about replication. The reason for believing that additional proteins are involved in replication comes from the analysis of mutants with defects in genes that do not code for any of the known proteins yet are required for either the initiation or continuation of replication. The isolation of mutants unable to replicate their chromosome seems somewhat paradoxical—how can cells with this defect be cultured? Unlike the case of metabolic auxotrophs (page 474), which can be raised simply by supplying the required metabolite, there is nothing one can simply provide mutants deficient in DNA polymerase or DNA ligase or some other protein to enable them to overcome their deficiency. The solution to this type of experimental difficulty is solved by the isolation of *temperature-sensitive* (ts) mutants. Temperature-sensitive mutants are ones whose deficiency is such that it only reveals itself at an elevated temperature termed the *restrictive* temperature. When grown at the lower temperature, the *permissive* temperature, the mutant protein can hold itself together sufficiently to carry out its required activity and the cells can continue to grow and divide. Temperature-sensitive mutants have been isolated in virtually every type of physiological activity and have been particularly important in the study of DNA synthesis as it occurs in replication, repair, and recombination.

It was pointed out in an earlier chapter (page 27) that bacteria, although lacking true cytoplasmic membranes, do have infoldings of the plasma membrane within which certain activities may be carried out.

Although the subject remains controversial, there is a substantial body of evidence that indicates this membrane may be important in the replication process. It appears that the origin of the chromosome where replication always begins is attached to the membrane suggesting that at least part of the machinery for the start of replication is somehow membrane-associated. There is also some evidence that the replication forks are also attached to membranous components. If this latter point is convincingly demonstrated, then the role of membrane in the entire process of bacterial replication will have to be considered. Even if membrane components are not directly involved in the enzymatic aspects of replication, the attachment of the origin of the chromosome to the membrane provides a straightforward means by which the daughter chromosomes are caused to be distributed into the daughter cells (Fig. 14-30).

One of the most remarkable features of bacterial replication is the rate at which it occurs. The replication of an entire bacterial chromosome in approximately 40 minutes requires that each replication fork move about 850 nucleotides per second. If one considers that the length of an Okazaki fragment is in this same range, this indicates that the entire process including the formation of an RNA primer, its elongation and simultaneous proofreading by the DNA polymerase, the excision of the RNA, its replacement with DNA, and its being joined to the remainder of the strand must occur within about one second. The accomplishment becomes more striking when one compares the rates of replication with that of transcription, a seemingly simpler process, which occurs at approximately 50 nucleotides per second.

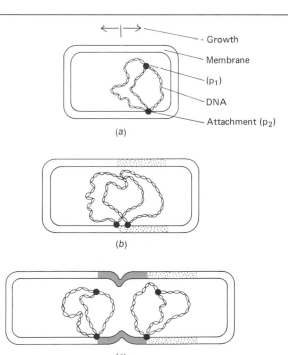

(a)

(b)

(c)

Figure 14-30. Mechanism of distribution of the bacterial chromosome. *(a)* A bacterial cell has just been produced. (Since cell division is not synchronized with replication, the cell contains a partially replicated chromosome.) *(b)* Replication is completed when the cell is halfway through its division cycle. The stippled portion represents newly formed membrane. *(c)* Additional growth of the membrane (gray portion) has resulted in separation of the chromosome. *(From MOLECULAR GENETICS: AN INTRODUCTORY NARRATIVE by Gunther S. Stent. W. H. Freeman and Company. Copyright © 1971.)*

Replication in Eukaryotic Cells

As might be expected, our understanding of replication in eukaryotic cells is at a far more primitive state than that previously outlined for prokaryotic systems. Taken at an overall view, replication seems quite similar in most viruses, bacteria, and eukaryotes. All of these utilize RNA primers, and the DNA strands are synthesized discontinuously as small replication fragments. There is evidence for the involvement of RNA and DNA polymerases, ligases, unwinding proteins, and relaxing enzymes in all of these cases. Similarly, eukaryotic cells appear to have several distinct DNA polymerases (termed α, β, γ, mitochondrial) although their role in various types of DNA synthetic activities is poorly understood. As in the case of the prokaryotic enzymes, all of the eukaryotic DNA polymerases elongate in the 5′ to 3′ direction by the addition of nucleotides to a 3′ hydroxyl group. Unlike the prokaryotic enzymes, the presence of exonuclease activity in the eukaryotic polymerases (outside of the mitochondria and chloroplast) has not been demonstrated. In fact, if the eukaryotic enzymes are supplied with a template-primer in which the 3′ terminal nucleotide of the primer is mismatched from the template, the enzyme will add nucleotides to the incorrect 3′ terminus of the primer causing the mismatched nucleotide to become internalized. Since there is no evidence of a higher mutation rate in eukaryotic cells as opposed to bacteria, it seems that some as yet unidentified process must occur in eukaryotic cells to ensure replication fidelity.

Whereas the early autoradiographic experiments on replicating bacteria indicated that replication of their chromosome began at only one site, similar types of experiments with eukaryotic cells suggested that there were many sites (Fig. 14-31a) in a chromosome at which replication was occurring simultaneously. Cells of higher organisms have much more DNA than bacteria and their polymerizing enzymes incorporate nucleotides into DNA at greatly reduced rates. In order to accomplish the replication of their genome in a reasonable amount of time, eukaryotic cells have evolved a means whereby their chromosomes are divided into units, termed *replicons,* which replicate independently. Replicons in eukaryotic chromosomes are generally between 15 and 100 μm in length and each has its own origin from which replication forks proceed in both directions (Fig. 14-31b). The presence of bidirectional replicated units of DNA can be observed in electron micrographs of DNA gently isolated from replicating chromosomes (Fig. 14-31c).

If one assumes that the average replicon includes approximately 50 μm of DNA, then a typical diploid mammalian cell would contain up to 35,000 such units. It is estimated that replication forks travel at the speed of 0.2 to 2.0 μm/minute. If we assume an average rate of fork movement of 1μm/minute, then we would expect the average replicon to be replicated using two growing points in approximately 25 minutes. Since the average S phase lasts approximately 6 hours, it is apparent that only a fraction of the replicons of a cell are active at any given time within the S phase. The time at which a particular replicon is involved in the replication process is not a random event but at least somewhat controlled. Those replicons active early in one S phase appear to be active at a com-

(a)

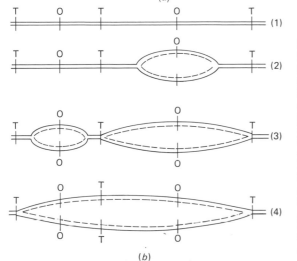

(b)

Figure 14-31. (a) Replication in cultured mouse cells. Cells were incubated for 10 minutes in 50 µC/ml ³H-thymidine (hot pulse) and then transferred to 5 µC/ml for two hours (warm pulse) before preparation of DNA fibers for autoradiography. Autoradiograph shows two adjacent replication units. Each has a small stretch of high grain density with lateral regions of lesser grain density. This is interpreted to mean that synthesis of DNA in these units began (at arrows) at the center of the stretch of dense grains *after* the hot pulse had begun. When the cells were transferred, the bidirectional replication forks began to incorporate less radioactive precursors, hence the lowered grain density. *(From R. Hand, J. Cell Biol., **67:766, 1975.**) (b)* Summary of the bidirectional model for DNA replication of two adjacent replication units. Solid lines represent parental strands, dashed lines represent newly synthesized strands, O and T indicate the positions of the origins and termini of the replication units. *(From J. A. Huberman and A. D. Riggs, J. Mol. Biol., **32:340, 1968.**) (c)* Electron micrograph of *Drosophila* DNA in the act of replication. Numerous sites of replication (as indicated by the single-stranded bubbles) are apparent along the length of the molecule. [*From H. J. Kriegstein and D. S. Hogness, Proc. Natl. Acad. Sci. (U.S.), **71:139, 1974.**]

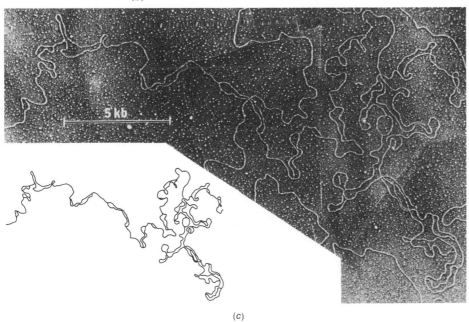

(c)

parable stage in succeeding S phases. In addition, it is generally observed that replicons located in one part of a chromosome tend to be active at the same time causing each part of a particular chromosome to have a characteristic time within the overall S phase at which it undergoes replication. Heterochromatic regions (both facultative and constitutive) of chromosomes tend to be replicated later in S phase than euchromatic regions. For example, in cells of a mammal in which one of the X chromosomes has been heterochromatized (page 616), the euchromatic X chromosome undergoes replication before the heterochromatic one (Fig. 14-32).

Although in bacteria and eukaryotes replication occurs by the movement of replication forks in opposite directions from some point or points of origin, there does exist another type of replication pattern with a much more restricted occurrence. This other scheme, termed the *rolling-circle* mechanism (Fig. 14-33), is the means by which a variety of viral genomes are replicated and appears also to be the basis for the selective amplification of rDNA in amphibian oocytes (page 535). In the rolling-circle model a nick is introduced in one of the strands of a circular duplex DNA molecule and the newly formed 3' OH serves as a primer for the addition of nucleotides. The 5' end of the nicked strand is most likely anchored in some manner to a cell structure. With the 5' end of the nicked strand immobilized, the rotation of the intact strand as shown in Fig. 14-33 would continually open new template surface on the intact strand which could be copied. The replication of mitochondrial DNA appears to occur by a different means than either of the ones previously described in this chapter.

DNA REPAIR

Of all the differences that set living machines apart from manufactured machines, probably the most basic and important is the ability of the former to regulate their own activity. When a computer or an automobile is operating improperly, it has to be serviced by outside help. In contrast, if the living machinery finds itself in need of repair, it has to make the

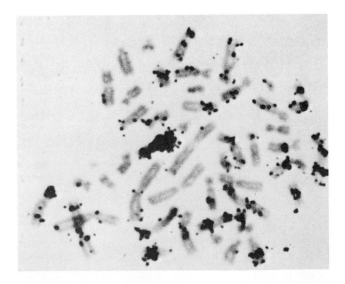

Figure 14-32. Mitotic figure of an erythroblast that had been exposed to ^{3}H-thymidine only at the very end of its S phase. The chromosome with the greatest label is the heterochromatized X, which is very late replicating. *(From F. Gavosto, L. Pegoraro, P. Masera, and G. Rovera, Exp. Cell Res., **49:**348, 1968.)*

necessary adjustments on its own. The first level of defense of a cell or tissue or the entire organism resides in the battery of regulatory mechanisms that exist to maintain the homeostatic conditions within the system. At least, in certain cases, a second level of defense exists which allows the system to repair damage should it occur, despite preventive measures. For example, in the case of DNA synthesis, errors are kept to a minimum as a result of an intricate proofreading mechanism present within the polymerase. Errors in base sequence that occur during replication despite the proofreading cannot be repaired at a subsequent stage and therefore represent permanent genetic changes. However, this type of error is not the only type of change that can occur in DNA; damage can also occur within preformed DNA as a result of environmental influences. Fortunately, most of this latter type of damage can be repaired and the mechanisms responsible provide an elegant example of the means by which evolution has provided for the maintenance of the living state.

There are several types of damage that can occur to DNA. In one case, the absorption of the energy of ultraviolet light results in an interaction between two adjacent pyrimidines on a DNA strand with the con-

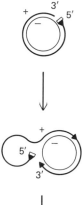

Figure 14-33. (a) The replication of the phage φX174 by the rolling-circle mechanism. This virus utilizes a single-stranded, circular DNA molecule as its genome. During infection, the single-stranded genome (+ strand) serves as the template for the synthesis of an inner minus strand, thereby forming a double-stranded replication intermediate. The outer strand is nicked and replication proceeds by the addition of nucleotides to its 3′ OH end (at a rate of about 200 nucleotides per second) using the rolling inner circle as a template. The closed triangles mark the origin-terminus of replication of the positive strand and the presumed site at which a second nicking event occurs during the final packaging steps. (b) Electron micrograph showing the rolling-circle intermediate found during replication of φX174 within the bacterial cell. A thinner, less rigid, single-stranded tail is seen emerging from the thicker, double-stranded circle. [From K. Koths and D. Dressler, Proc. Natl. Acad. Sci. (U.S.), **75**:605, 1978.]

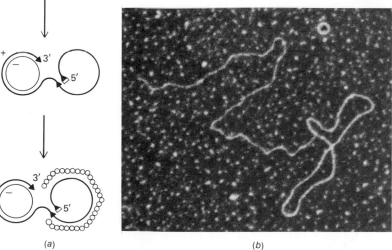

(a)   (b)

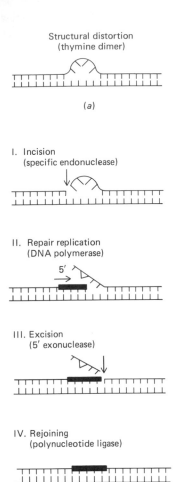

Structural distortion
(thymine dimer)

(a)

I. Incision
(specific endonuclease)

II. Repair replication
(DNA polymerase)

5'

III. Excision
(5' exonuclease)

IV. Rejoining
(polynucleotide ligase)

(b)

Figure 14-34. The postulated steps in excision repair. The repair patch is shown as a heavy line. The vertical arrows indicate the locations of nuclease cuts in the damaged parental strand and the horizontal arrow indicates the direction of repair replication, beginning at a 3' OH end of the parental strand. *(From P. C. Hanawalt, Endeavour, **31**:84, 1972.)*

sequent formation of a covalent complex or dimer. In another case, the loss of a purine from the backbone of the strand results from an acid-catalyzed hydrolytic reaction. If either of these types of damage were to remain in the DNA, serious consequences would result. To counteract the effects of this type of damage, DNA repair processes are capable of correcting the situation. In the best-studied repair mechanism, the defective region is removed and replaced by a new stretch of nucleotides of the proper sequence. This type of repair, termed *excision repair* (Fig. 14-34), is widespread in both prokaryotic and eukaryotic cells and involves the coordinated activity of a number of gene products. The process begins by the action of an endonuclease, i.e., an enzyme capable of hydrolyzing a phosphodiester bond between two nucleotides lying *within* a strand. It would appear that cells contain endonucleases which exist to "patrol" the DNA in search of sites in which the double helix is distorted in a particular way. When one of these sites is found, the nuclease apparently breaks a phosphodiester bond on the 5' side of the distortion on the strand where the dimer or missing base is found. If the cell in question is *E. coli,* once the nick in the strand is made, it appears that DNA polymerase I is responsible for the next series of events. Mutants lacking this enzyme, although capable of replication, are highly sensitive to ultraviolet light as a result of their repair inabilities. It is believed that DNA polymerase I, which can attach at a nick between two adjacent bases, removes the damaged region by use of the 5' → 3' exonuclease activity. As the exonuclease activity of the enzyme clears a path on the strand, the polymerase activity of the enzyme presumably lays down a path of correctly paired nucleotides using the undamaged strand as the template. The final step in the process is accomplished by the DNA ligase, which seals the break by the formation of a phosphodiester bond between an adjacent 3' OH and 5' phosphate group.

The importance of the DNA repair process can be appreciated by examining the effects in its absence. There is a very rare human disease condition, termed *xeroderma pigmentosum,* in which pyrimidine dimer removal is highly deficient. The molecular basis for the disease has been revealed by determining the effects of radiation on cultured cells from these persons. Persons with this condition are extremely sensitive to sunlight and run a high risk of the development of skin cancers. The experimental analysis of cultured cells from these patients illustrates another very useful application of the technique of cell fusion. Consider the question of how many genes are involved in the repair of pyrimidine dimers in man. The congenital disease xeroderma pigmentosum describes a particular syndrome characterized by the inability to remove ultraviolet-damaged DNA sequences. There is no a priori reason to expect that all patients with this condition have the same defective gene if more than one gene product is required for the repair process. If there is more than one gene involved in DNA repair, and if different persons with this disease have mutants in different gene loci, than one might expect some hybrids resulting from the fusion of cells of different patients to regain their repair abilities. This is exactly what has been found in cell fusion experiments. As a result of the fusion of cells in various combinations from many different persons, it has been found that the mutant loci fall into five different groups. Hybrids formed between cells in different groups complement one another since the two genomes together are capable of coding for all of the gene products necessary to remove the dimers and replace them with undamaged DNA.

REFERENCES

Baserga, R., ed., *Multiplication and Division in Mammalian Cells*. Dekker, 1976.

———— and Kisieleski, W. E., Sci. Am. **209**, 103–110, Aug. 1963. "Autobiographies of Cells."

Bernhard, H. P., Int. Rev. Cytol. **47**, 289–325, 1976. "The Control of Gene Expression in Somatic Cell Hybrids."

Bollum, E. J., Prog. Nuc. Acid Res. Mol. Biol. **15**, 109–144, 1975. "Mammalian DNA Polymerases."

Bradbury, E. M., Inglis, R. J., and Matthews, H. R., Nature **247**, 257–261, 1974. "Control of Cell Division by Very Lysine Rich Histone (F1) Phosphorylation."

Brinkley, B. R., and Porter, K. R., eds., symposium on "The Eukaryotic Cell Cycle," *International Cell Biology 1976–1977*, 409–435, Rockefeller University Press, 1977.

Bullough, W. S., Biol. Revs. **50**, 99–128, 1975. "Mitotic Control in Adult Mammalian Tissues."

Cairns, J., J. Mol. Biol. **6**, 208–213, 1963. "The Bacterial Chromosome and Its Manner of Replication as Seen by Autoradiography."

————, Sci. Am. **214**, 36–44, Jan. 1966. "The Bacterial Chromosome."

Chargaff, E., Prog. Nuc. Acid Res. Mol. Biol. **16**, 1–24, 1976. "Initiation of Enzymatic Synthesis of Deoxyribonucleic by Ribonucleic Acid Primers."

Clarkson, B., and Baserga, R., eds., *Control of Proliferation in Animal Cells*. Cold Spring Harbor, 1974.

Cleaver, J. E., and Bootsma, D., Ann. Rev. Genetics **9**, 19–38, 1975. "Xeroderma Pigmentosum; Biochemical and Genetic Characteristics."

Deering, R. A., Sci. Am. **207**, 135–144, Dec. 1962. "Ultraviolet Radiation and Nucleic Acid."

Dressler, D., Ann. Rev. Microbiol. **29**, 525–559, 1975. "The Recent Excitement in the DNA Growing Point Problem."

Edelman, G. M., Science **192**, 218–226, 1976. "Surface Modulation in Cell Recognition and Cell Growth."

Edenberg, H. J., and Huberman, J. A., Ann. Rev. Genetics **9**, 245–284, 1975. "Eukaryotic Chromosome Replication."

Ephrussi, B., and Weiss, M. C., Sci. Am. **220**, 26–35, April 1969. "Hybrid Somatic Cells."

Friedmann, T., Sci. Am. **225**, 34–42, Nov. 1971. "Prenatal Diagnosis of Genetic Disease."

Goldberg, N. D., Ann. Rev. Biochem. **46**, 823–896, 1977. "Cyclic GMP Metabolism and Involvement in Biological Regulation."

Gospodarowicz, D., and Moran, J. S., Ann. Rev. Biochem. **45**, 531–558, 1976. "Growth Factors in Mammalian Cell Culture."

Grossman, L., Braun, A., Feldberg, R., and Mahler, I., Ann Rev. Biochem. **44**, 19–44, 1975. "DNA Replication."

Hanawalt, P. C., and Setlow, R. B., eds., *Molecular Mechanisms for Repair of DNA*. Plenum, 1975.

Holley, R. W., Nature **258**, 487–490, 1975. "Control of Growth of Mammalian Cells in Cell Culture."

Howard, A., and Pelc, S., Heredity **6**, 261–273, 1953 (Supp.). "Synthesis of Deoxyribonucleic Acid in Normal and Irradiated Cells and Its Relation to Chromosome Breakage."

Jacob, F., Ryter, A., and Cuzin, F., Proc. Royal Soc. (London) **B164**, 267–278, 1966. "On the Association between DNA and Membrane in Bacteria."

Jovin, T. M., Ann. Rev. Biochem. **45**, 889–920, 1976. "Recognition Mechanisms of DNA-specific Enzymes."

Kornberg, A., Science **131**, 1503–1508, 1960. "Biologic Synthesis of Deoxyribonucleic Acid."

———, *DNA Synthesis*. W. H. Freeman, 1974.

Lehman, I. R., Science **186**, 790–797, 1974. "DNA Ligase: Structure, Mechanism, and Function."

——— and Uyemura, D. G., Science **193**, 963–969, 1976. "DNA Polymerase I: Essential Replication Enzyme."

Lewin, B., *Gene Expression*, vol. 1. J. Wiley, 1974.

Lieberman, M. W., Int. Rev. Cytol. **45**, 1–23, 1976. "Approaches to the Analysis of Fidelity of DNA Repair in Mammalian Cells."

Martz, E., and Steinberg, M. S., J. Cell Phys. **79**, 189–210, 1972. "The Role of Cell-Cell Contact in 'Contact' Inhibition of Cell Division: A Review and New Evidence."

Marx, J. L., Science **192**, 455, 1976. Research News. "Cell Biology: Cell Surfaces and the Regulation of Mitosis."

Mazia, D., Sci. Am. **230**, 54–64, Jan. 1974. "The Cell Cycle."

McKusick, V. A., and Ruddle, F. H., Science **196**, 390–405, 1977. "The Status of the Gene Map of the Human Chromosomes."

Meselson, M., and Stahl, F. W., Proc. Natl. Acad. Sci. (U.S.) **44**, 671–682, 1958. "The Replication of DNA in *Escherichia coli*."

Mitchison, J. M., *The Biology of the Cell Cycle*. Cambridge University Press, 1971.

Pastan, I. H., Adv. Metab. Disorders **8**, 7–16, 1975. "Regulation of Cellular Growth."

———, Johnson, G. S., and Anderson, W. B., Ann. Rev. Biochem. **44**, 491–522, 1975. "Role of Cyclic Nucleotides in Growth Control."

Paul, J., *Cell and Tissue and Culture*, 5th ed. Churchill Livingstone, 1975.

Pitot, H. C., Federation Proc. **34**, 2207–2232, 1975. Symposium on "Genes, Chromosome Loci, and Disease."

Pollack, R., *Readings in Mammalian Cell Culture*, 2d ed., Cold Spring Harbor, 1975.

Pollard, E. C., Am. Sci. **57**, 206–236, 1969. "The Biological Action of Ionizing Radiation."

——, "The Control of Cell Growth," in *Cell Biology in Medicine,* E. E. Bittar, ed., J. Wiley, 1973.

Prescott, D. M., Adv. Genetics **18**, 99–177, 1976. "The Cell Cycle and the Control of Cellular Reproduction."

——, *Reproduction of Eukaryotic Cells.* Academic, 1976.

Puck, T. T., *The Mammalian Cell as a Microorganism.* Holden-Day, 1972.

Rao, P. M., and Johnson, R. T., Adv. Cell Mol. Biol. **3**, 135–189, 1974. "Induction of Chromosome Condensation in Interphase Cells."

Rebhun, L. I., Int. Rev. Cytol. **49**, 1–54, 1977. "Cyclic Nucleotides, Calcium and Cell Division."

Reich, E., Rifkin, D. B., and Shaw, E., eds., *Proteases and Biological Control.* Cold Spring Harbor, 1975.

"Replication of DNA in Microorganisms," Cold Spring Harbor Symp., Quant. Biol. **33**, 1968.

Ruddle, F. H., and Kucherlapati, R. S., Sci. Am. **231**, 36–44, July 1974. "Hybrid Cells and Human Genes."

—— and Creagan, R. P., Ann. Rev. Genetics **9**, 407–486, 1975. "Parasexual Approaches to the Genetics of Man."

Schekman, R., Weiner, A., and Kornberg, A., Science **186**, 987–993, 1974. "Multienzyme System of DNA Replication."

Shall, S., *The Cell Cycle.* Halsted, 1977.

Taylor, J. H., Proc. 10th Intern. Congr. Genet. Montreal, 1958 **1**, 63–78, 1959. "The Organization and Duplication of Genetic Material."

——, Sci. Am. **198**, 37–42, June 1958. "The Duplication of Chromosomes."

——, Int. Rev. Cytol. **37**, 1–20, 1974. "Units of DNA Replication in Chromosomes of Eukaryotes."

Weiss, M. C., and Green, H., Proc. Natl. Acad. Sci. (U.S.) **58**, 1104–1111, 1967. "Human-Mouse Hybrid Cell Lines Containing Partial Complements of Human Chromosomes and Functional Human Genes."

Weissbach, A., Cell **5**, 101–108, 1975. "Vertebrate DNA Polymerases."

——, Ann. Rev. Biochem. **46**, 25–48, 1977. "Eukaryotic DNA Polymerases."

Wickner, R. B., *DNA Replication and Biosynthesis.* Dekker, 1974.

Willingham, M. C., Int. Rev. Cytol. **44**, 319–363, 1976. "Cyclic AMP and Cell Behavior in Cultured Cells."

CHAPTER FIFTEEN
Microtubules, Microfilaments, and Cell Division

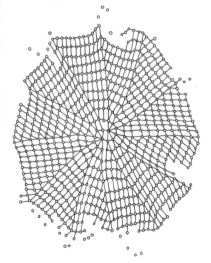

Throughout this book we have concentrated on biochemical aspects of cell function. Most of these processes involve various types of proteins, which in general owe their activity to small changes in conformation. Conformation changes are shifts in the positions of atoms relative to one another and hence constitute movement and with it the potential to do mechanical work. Consider the nature of DNA, RNA, or protein synthesis. In each case it appears that the elaboration of the macromolecule is accomplished by the gradual movement of an enzyme or a ribonucleo-protein complex along the length of the appropriate template. Larger-scale movements of materials within cells or movements of cells themselves are also believed to involve conformational changes in proteins, changes which are coordinated in such a way that the individual molecular movements become additive. In order for this to occur there must be an organized, structurally oriented machinery within the cell. In order to understand the nature of mechanical work, such as that involved in cell division or cell movement, we will need to consider the nature of the machinery utilized, the manner in which the various components function relative to one another, the means by which cellular energy is made to run the machines, and the means by which the force brings about directed movements.

Research on a variety of activities involving the movement of cells or subcellular organelles has consistently implicated two types of structures, microtubules and microfilaments, as being of central importance in the process. In this chapter we will consider these two types of organelles, concentrating on their structure, particularly as it relates to cell division. In the following chapter we will discuss the role of these organelles in contractility and movement.

MICROTUBULES

Structure

As the name implies, microtubules are small, hollow, cylindrical structures, rigid in nature, which have been found to occur in nearly every eukaryotic cell that has been scrutinized with the electron microscope. The tubule typically has an outer diameter of approximately 250 Å, a wall diameter of approximately 50 Å, leaving an internal (or lumen) diameter of 150 Å. The structure of microtubules (Fig. 15-1) is strikingly similar in a very wide variety of organisms. Careful ultrastructural analysis of negatively stained microtubules has revealed that the wall is a polymer composed of globular subunits. Examination of a cross section of the wall of a microtubule (Fig. 15-1*a, c*) nearly always reveals 13 subunits making up the complete circumference of the wall. When the surface of isolated microtubules is examined, the subunits are seen to be arranged in longitudinal rows, termed *protofilaments,* that are aligned parallel to the long

Figure 15-1. The structure of microtubules. *(a)* Electron micrograph of a cross section through a microtubule of a *Juniperus* root tip cell revealing the 13 subunits arranged within the wall of the organelle. The microtubules of these plant cells are most abundant in a cortical zone about 1000-Å thick just beneath the plasma membrane (shown in the micrograph). *(Reprinted with permission of M. C. Ledbetter, J. Agr. Food Chem.,* **13:**406, 1965. *Copyright by the American Chemical Society.) (b)* Electron micrograph of negatively stained microtubules from brain. Microtubule-associated proteins can be seen as regularly spaced projections on the surface of the microtubule. *(From L. A. Amos,* J. Cell Biol., **72:**645, 1977.) *(c)* and *(d)* Diagrams of a cross section and longitudinal section of a microtubule model. The model has 13 protofilaments composed of heterodimers. Each half of the dimer is assumed to be a 35 × 40 Å globular subunit (either α- or β-tubulin). The 80-Å longitudinal repeat is indicated. *(Reprinted from J. Bryan,* Federation Proceedings, **33:**156, 1974.)

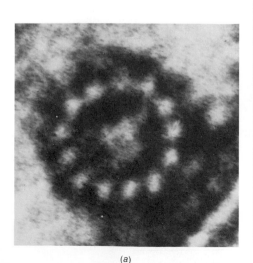

(a)

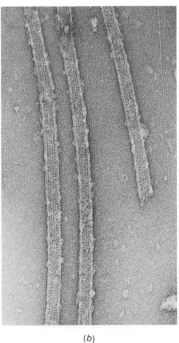

(b)

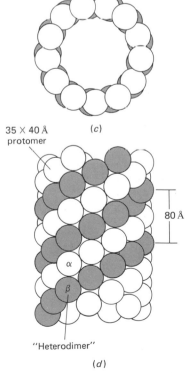

35 × 40 Å protomer

80 Å

α

β

"Heterodimer"

(c)

(d)

axis of the tubule (Fig. 15-1b, d). When microtubules are caused to split open and flatten out, the 13 protofilaments making up the wall of the tubule can be seen, indicating the association of the subunits with one another in these rows. If one traces the subunits around the wall of the tubule, they are seen to spiral in a helical pattern (Fig. 15-1d).

When microtubules are isolated (a procedure most readily accomplished using mammalian brain tissue) and their chemical nature analyzed, they are found to be composed totally of protein. Each subunit is believed to consist of two molecules of protein, one of α-tubulin and one of β-tubulin. These two proteins have nearly identical molecular weights (approximately 54,000 daltons), have related amino acid sequences, and are presumably derived from a common ancestral protein in an earlier evolutionary period. In addition to the tubulins, which account for 80 to 95% of the protein of the microtubule, a number of other proteins, termed *microtubule-associated proteins* (MAPs) are also present in the organelle (Fig. 15-1b), and are currently being intensively investigated.

In some cases microtubules are found within organized cellular structures such as flagella, cilia, and the mitotic spindle, whereas, in other cases, microtubules are seen either singly or in clusters in otherwise unstructured cytoplasm. The overall distribution of microtubules within a cell is most strikingly revealed by examination of the cell under the fluorescence microscope after treatment of the cell with fluorescent antitubulin antibodies (Fig. 15-2). Although microtubules appear morphologically quite similar regardless of their location within the cell, there are marked differences in microtubule stability. Microtubules of the mitotic spindle, for example, are extremely *labile*, i.e., sensitive to "destruction." Before the introduction of glutaraldehyde as a fixative around 1960, these microtubules disappeared during the fixation process and their existence went undiscovered. In fact, for a while, the morphology of the mitotic spindle was more apparent in examinations of living cells in the light microscope than in electron microscope studies. Even though an

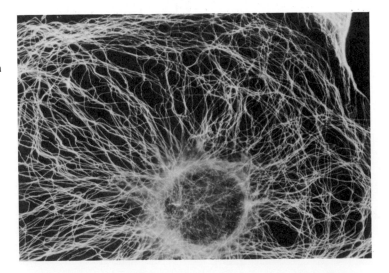

Figure 15-2. The microtubules of a large flat cultured mouse cell as revealed by fluorescent antitubulin antibodies. Individual microtubules can be followed over long distances within the cell and the localization of microtubules around the nucleus and beneath the plasma membrane is apparent. *(From M. Osborn and K. Weber,* Cell, **12:**563, 1977.)

individual microtubule has a diameter far below the limit of resolution of the light microscope, when present in large numbers having a parallel orientation, their presence becomes strikingly apparent when observed using polarization optics (see Fig. 15-7). In contrast to the microtubules of the mitotic spindle, those of cilia and flagella are much sturdier structures and relatively resistant to a variety of treatments (discussed below) that brings about the disappearance of labile microtubules.

Function

Microtubules are believed to function in two interrelated activities: acting as a sort of cellular skeleton by providing structural support, and providing part of the machinery required in certain types of movements. It is important to keep in mind in the following discussion that the assignment of microtubules as the basis for a particular function is usually indirect. There are two commonly employed criteria for deciding that microtubules are responsible for a particular process: the presence of microtubules in the area of activity, and the loss of the function after treatments which cause the disappearance of microtubules. We will begin by discussing the role of microtubules in supportive functions, i.e., as a "cytoskeleton."

The natural shape of a free body with liquid properties is spherical, as seen in the suspended drop of water in oil or in a soap bubble. When free bodies are clustered together, again as seen most readily in a mass of soap bubbles, the exposed surfaces remain curved, but adjoining surfaces become more or less flat. Surface tension and mutually adhesive forces are responsible for these configurations. When other shapes appear, as in most cells, then other agencies must exist that either produce or maintain the distortions. Microtubules seem to be the most likely agent. Microtubules are commonly seen to be lined up parallel with and close to cell surfaces and in general conform to whatever particular shape a cell may exhibit (Fig. 15-3). This arrangement is most strikingly seen in heliozoan protozoa which extend long processes (Fig. 15-4) out from a body cell or in cells such as neurons (see Fig. 15-5). The axons of nerve cells are filled with microtubules (termed *neurotubules* in these cells) which are oriented parallel to the long axis of the axon. Neurotubules appear to be involved, at least during the initial period of axon elongation, in maintaining the elongated structure. The addition of microtubule-disrupting agents to nerve cells in the process of axon elongation results in the withdrawal of the process and the rounding up of the nerve cell. Once tracts of nerve fibers have become established, the cells are also held in their shape by attachments at their external surface and are much less sensitive to disruption. In plant cells, the bulk of the cell's microtubules are typically located very close to the cell membrane (as in Fig. 15-1a) forming a distinct cortical zone of approximately 1000 Å in thickness. Furthermore, the orientation of the microtubules within the cell is closely mirrored by the orientation of the cellulose microfibrils in the adjacent cell wall. This geometrical relationship suggests that the microtubules may be an important influence on the deposition of the oriented cellulose microfibrils.

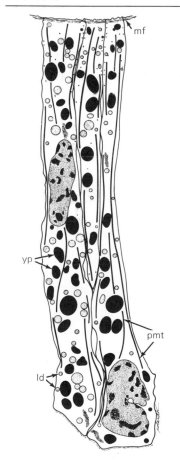

Figure 15-3. Schematic illustration of the orientation of microtubules in elongating embryonic neural plate cells. Numerous microtubules are aligned parallel to the long axis of the cell (paraxial microtubules, *pmt*). In the same plane with the apical layer of microtubules, 50 to 70 Å microfilaments *(mf)* are arranged in a circumferential bundle which encircles the cell apex in pursestring fashion. *yp*, yolk platelet; *ld*, lipid droplet. *(From B. Burnside, Dev. Biol., **26**: 434, 1971.)*

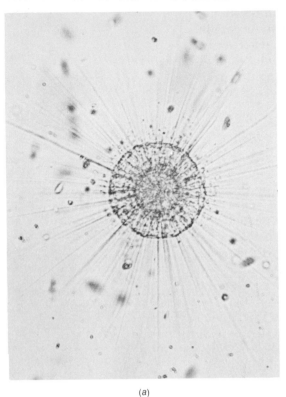

(a)

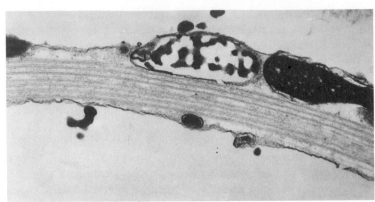

(b)

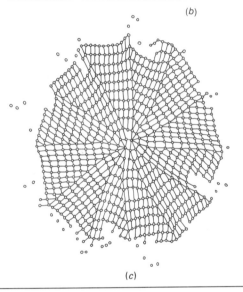

(c)

Figure 15-4. *(a)* The heliozoan *Echinosphaerium*. Slender, axopodial processes radiate from the cell body. Within each axopodium is a birefringent core, or axoneme, composed of microtubules. *(b)* Longitudinal section of an axopodium showing microtubules. *(c)* Diagram of a cross section of an axopodium showing spiraling arrangement of microtubules. *(Courtesy of L. G. Tilney.)*

A particularly striking example of the maintenance of cell shape by microtubules has been revealed by the ultrastructural analysis of the processes, termed *axopods* (Fig. 15-4), of the heliozoan *Echinosphaerium*. The numerous microtubules of these processes are arranged in two interlocking coils or spirals with individual microtubules traversing the entire length of the axopod. In other types of cells, the analysis of successive cross sections indicates that single microtubules do not generally extend the entire length of a cell but rather overlap with other microtubules. High-resolution electron microscopy often reveals the presence of material between adjacent microtubules, which is believed to help in holding the collective structure together. In their supportive capacity, microtubules appear to act in a manner analogous to the poles that maintain the shape of a tent. Since the ability of a column to act as a support is proportional to its cross-sectional area, a tubular column provides equivalent support to one that is solid.

Microtubules have also been implicated in the movement of a variety of particles within cells as well as the movement of cells themselves. In this latter capacity, i.e., cell motility, the microtubules are components of more complex structures, specifically cilia and flagella, and their activity will be considered further in the next chapter. Probably the least understood function of microtubules concerns their role in the transport of intracellular materials. It was mentioned above that microtubules are present in large numbers within the axons of nerve cells. The evidence suggests that they not only serve in a support capacity within these cells, but also as agents involved in the transport of both macromolecules and subcellular organelles (Fig. 15-5). In terms of their shape, nerve cells are highly unusual, extending in some cases for several feet. A cell whose cytoplasm is stretched out over such great lengths would be expected to face severe intracellular communication problems.

A great deal of research has focused on the means by which substances are moved around within various nerve cells. If one injects a labeled substance into the *cell body* of the neuron (the part containing the nucleus) and then follows the movement of that substance with time, the label is found to gradually move down the length of the axon. When a variety of substances were followed in this manner, it appeared that there were two types of transport phenomena in operation within nerve cells. Some substances traveled very slowly (at rates of approximately 1 to 2 mm/day) while other substances (or cell organelles) traveled much

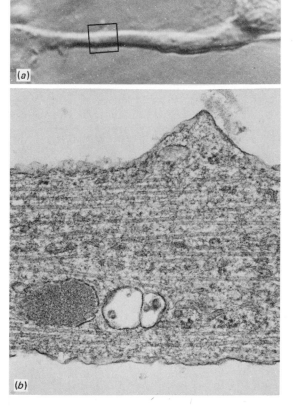

Figure 15-5. *(a)* High-resolution Nomarski micrograph of a living, isolated chick neuron. *(b)* Electron micrograph of a portion of the same cell shown in *a* (box indicates EM coverage). The particle located in this area of the Nomarski micrograph was moving until the moment of fixation. The electron microscope reveals that the particle corresponds to a lysosome adjacent to a multivesicular body. The numerous parallel microtubules (neurotubules) are believed to have an important role in these types of movements. *(From A. C. Breuer, C. N. Christian, M. Henkart, and P. G. Nelson, J. Cell Biol., 65: 568, 1965.)*

faster (at rates of about 400 mm/day). Whereas the slow transport reflects the gradual movement of the entire cytoplasmic content of the nerve cell along the axon, i.e., *bulk flow*, the fast transport of materials is selective and apparently mediated by microtubules. The evidence which implicates microtubules in the faster transport activity is based upon the two criteria mentioned above. Microtubules are often found in close proximity to cytoplasmic organelles (Fig. 15-5) and in some cases contacts between microtubules and various types of structures have been resolved. Secondly, the depolymerization of neuronal microtubules by various treatments is generally accompanied by the loss of the capacity for the rapid transport. Even if we assume that microtubules are involved in axoplasmic transport, their precise role in this process remains unanswered. On one hand, they may actually provide the motive force for the movement of the materials, while, on the other hand, they may simply represent tracks along which the materials are pushed by some other type of force-producing agent. The distinction between an "active" and a "passive" role for microtubules in the movement of particles will be explored in greater depth when we consider the movement of chromosomes by the mitotic spindle.

The role of microtubules in the movement of particles within cells is not confined to nerve cells but may be very widespread. In pigment cells, for example, microtubules are involved in the directed movement of cytoplasmic pigment granules which, in turn, are involved in the rapid changes in skin coloration. In amphibians, for example, the pigment cells of the skin have long cytoplasmic processes into which pigment granules can be directed. When the granules are concentrated in the center of the cell (Fig. 15-6a, right), the skin appears light in color while the dispersal of the granules throughout each of the many pigment cells (Fig. 15-6a, left, 15-6b) results in the darkening of the skin. The response by pigment cells is under the control of a hormone termed *melanocyte-stimulating hormone* (MSH), which acts at the cell surface to alter the intracellular concentration of cAMP (page 245). Although the manner in which cAMP brings about the movement of pigment granules is not understood, the movement itself is mediated by a collection of microtubules which radiate out from the center of the cell into the various cell processes. The movement of the granules back and forth along these tracks is responsible for the changes in coloration. In related capacities, microtubules have been implicated in the movement of granules in secretory cells and the movements (or restriction of movement) of integral proteins within the plasma membrane (page 227).

Assembly and Disassembly

All microtubules, regardless of their stability, are polymers which have arisen by the association of monomeric subunits. As is usually the case for multimeric protein complexes, the individual subunits are held together by noncovalent bonds; consequently, the assembly and disassembly of these structures can be studied. One of the earliest and best-studied microtubular structures is the mitotic spindle whose analysis has revealed

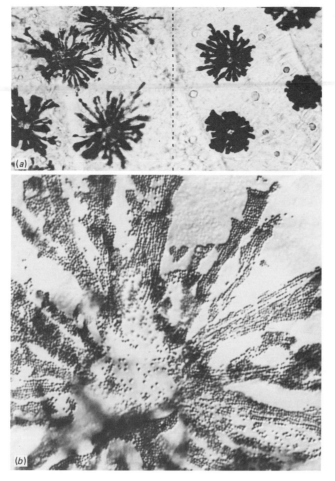

Figure 15-6. *(a)* Melanophores (pigment cells) of a fish. The pigment in the melanophores on the right is aggregated about the center of each cell (aggregated state) while that in the cells on the left is dispersed throughout the processes of each cell (dispersed state). *(b)* Detail of a melanophore in the dispersed state revealing its numerous processes and the granule-free central zone. Each process contains hundreds of granules arranged in linear files which radiate from the cell center. *(From D. B. Murphy and L. G. Tilney, J. Cell Biol.,* **61***:759, 1974.)*

a great amount of information on microtubular formation and function. The use of the word "spindle" in describing the basket of microtubules responsible for chromosome separation reflects the overall shape of the apparatus which is wide at the equator and converges at the two poles (Fig. 15-7a). Since the size of the mitotic spindle is roughly proportional to the size of the dividing cell, the large, relatively transparent eggs of marine organisms such as the sea urchin have been a favorite subject for both descriptive and experimental observations in this area. In addition to being easily observed within these cells, the mitotic spindle can be isolated from the cells in a relatively intact state for further biochemical studies. Morphological studies of the mitotic spindle have also been carried out, primarily by Shinya Inoué and coworkers, using the polarization microscope. As described in Chapter 2, the construction of an image in the polarization microscope depends upon the geometrical organization of the structure being observed. In the case of the mitotic spindle, the microtubules are aligned approximately parallel to one another and organized into bundles termed *spindle fibers*. The parallel nature of the components of the spindle fibers causes the entire mitotic (or meiotic in Fig. 15-7b) apparatus to appear strikingly birefringent against the nonstructured cytoplasm.

If one examines a sea urchin egg soon after fertilization, there is no indication of the presence of any type of birefringent cytoplasmic organelle resembling the mitotic spindle. If, however, one were to analyze the proteins present within that same fertilized egg, a large amount of tubulin

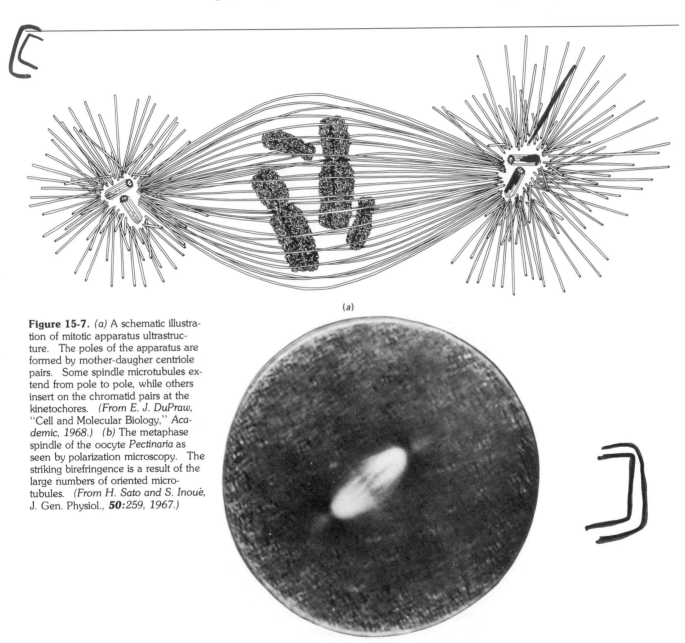

(a)

Figure 15-7. (a) A schematic illustration of mitotic apparatus ultrastructure. The poles of the apparatus are formed by mother-daugher centriole pairs. Some spindle microtubules extend from pole to pole, while others insert on the chromatid pairs at the kinetochores. *(From E. J. DuPraw, "Cell and Molecular Biology," Academic, 1968.)* (b) The metaphase spindle of the oocyte *Pectinaria* as seen by polarization microscopy. The striking birefringence is a result of the large numbers of oriented microtubules. *(From H. Sato and S. Inouė, J. Gen. Physiol., **50**:259, 1967.)*

(b)

can be shown to be present. In other words, although there is a large amount of microtubular protein present in the cell, very little of it is polymerized. If one waits an hour or so until the cell begins mitosis, there is no significant increase in the amount of tubulin present in the cell, but the polarization microscope clearly indicates the presence of an organized mitotic spindle, i.e., the subunits have become polymerized into microtubules. Similarly, after mitosis has been completed, the disassembly of the mitotic spindle occurs without the destruction of the tubulin itself. The analysis of the mitotic spindle in dividing cells has revealed one of the most striking examples of the rapid assembly and disassembly of a major cytoplasmic organelle.

How does the cell control the state of an organelle which shows such striking changes in structure? Based upon a number of lines of research, Inoué proposed that a dynamic equilibrium exists between the monomeric and polymeric form of the microtubule protein. Conditions which affected that equilibrium could shift the reaction toward the assembly of the polymer, as in the case of the formation of the mitotic apparatus, or toward the disassembly of the polymer, as occurs after mitosis (or often during cell fixation). A particularly important line of research in the field of mitotic spindle formation has been generated by the discovery that a variety of agents inhibit the formation of microtubules or disrupt the structure of ones that have already formed. These treatments, which include cold temperatures, hydrostatic pressure, colchicine, and vinblastine, are believed to upset the equilibrium between the monomeric and polymeric state of tubulin in such a way that the reaction is driven toward the depolymerized state. A brief consideration of the action of cold temperatures and colchicine will illustrate the general principles.

Cold temperature has been found to result in the dissociation of a variety of multimeric protein complexes, including various enzymes possessing quaternary structure. Complexes whose structure is sensitive to cold temperatures are generally ones whose subunits associate with one another by hydrophobic interactions. Associations of this type are endothermic (heat-requiring) and entropy driven. It was stated in Chapter 4 that a thermodynamically favored reaction was one having a $-\Delta G$. The free-energy difference between the reactants and products is dependent upon the difference in the heat content (ΔH) and their entropy (expressed in the term $T \Delta S$). In the assembly of microtubules, the increased heat content of the polymer is counterbalanced by the increased entropy of the system, a consequence of the increased disorder of the water molecules surrounding the polymer as opposed to the monomer. As the temperature of the cell drops, the relative values of ΔH and $T \Delta S$ change until a point is reached where ΔG becomes positive, i.e., the direction of the reaction shifts in favor of depolymerization, and the mitotic spindle falls apart.

Colchicine acts to dissolve microtubules in a very different way, one which illustrates the reversible nature of the reaction leading to the assembly of microtubules. Colchicine acts by specifically binding to the tubulin subunits; it cannot bind to tubulin that is present in the polymerized state (except to the terminal subunit). The interaction between colchicine and tubulin interferes with the ability of that subunit to associate with other subunits to form the polymer. Therefore, the addition of colchicine to a cell, in effect, removes those subunits complexed to colchicine from the polymerizing reaction. If the colchicine becomes attached to the subunit at the tip of the microtubule, no further subunits can be added. It would seem that, at least in the case of the labile microtubules such as those of the mitotic spindle, there is a continual association-dissociation process underway. Whether the assembly or disassembly process is favored at a given time in the cell depends upon the local conditions present at the time.

As more information became available concerning the nature of tubulin and its polymerization, it became evident that microtubule formation probably occurred via self-assembly and a number of laboratories began attempts to induce microtubule formation in vitro. The first successful approach to the in vitro assembly of microtubules was made in 1972 in the laboratory of Richard Weisenberg. Reasoning that cell homogenates should possess all of the necessary ingredients for the assembly process, Weisenberg obtained tubulin polymerization in brain homogenates by the addition of Mg^{2+}, GTP, and EGTA (a chelator which removes Ca^{2+}, an inhibitor of polymerization when present above a certain concentration). Within the past few years a large number of studies have concentrated upon the interaction of tubulin and these three substances (Mg^{2+}, Ca^{2+}, and guanine-containing nucleotides). At the present time the roles of these various factors are not fully understood, but they are believed to be important determinants of microtubule structure within the cell. The microtubule-associated proteins are also believed to be involved in microtubule assembly (as well as microtubule function).

The function of a microtubule is totally dependent upon its location and orientation within the cell. Consequently, it is important to understand why a microtubule forms in one place as opposed to another. Evidence from the studies of in vitro assembly suggest that the formation of microtubules from subunits involves two rather distinct phases, that of initiation and that of elongation. The initiation event appears to involve the formation of some type of multimeric structure upon which additional subunits can be added. The in vitro assembly of microtubules is aided greatly by the addition of pieces of microtubules or structures which contain microtubules (Fig. 15-8)

Figure 15-8. When solutions of brain tubulin are incubated with existing microtubular structures, neurotubules can form by in vitro assembly onto the intact nucleating structure. *(a)* In this negatively stained electron micrograph, the brain microtubules have assembled in continuity with the microtubules present in an axoneme from the flagellum of a protozoan *Chlamydomonas. (Courtesy of L. I. Binder and J. L. Rosenbaum.) (b)* In this case, pig brain tubulin has polymerized onto a centriole-procentriole pair from a Chinese hamster cell. The perpendicular orientation of the newly assembled neurotubules reflects the perpendicular orientation of the centriole-procentriole (see Fig. 15-16). *(From R. R. Gould and G. G. Borisy,* J. Cell Biol., **73:**610, 1977.)

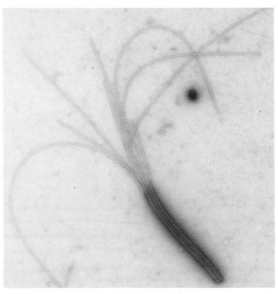

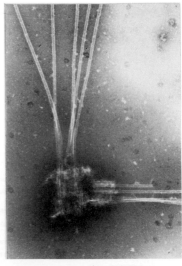

(a) (b)

which act as *seeds* upon which elongation can occur by the ordered addition of subunits. Electron microscopic examination of microtubules in different stages of in vitro assembly have consistently indicated that some type of ring structure is involved in both the initiation and elongation phases of in vitro construction. The ring is believed to be composed of a string of longitudinally associated subunits, i.e., a portion of a protofilament, that is curled up as shown in Fig. 15-9. The rings appear to be capable of attaching to the growing ends of protofilaments previously incorporated into the microtubular wall. As the individual protofilaments become extended longitudinally, the wall becomes curved and subsequently forms a cylinder by the association of the edges of the growing sheet. In the case of in vitro assembly, the microtubules generally do not contain the normal number of 13 subunits to the circumference. The microtubule shown in Fig. 15-9, for example, contains only 11 protofilaments. The steps occurring during the initiation of microtubule formation *in the cell* are totally obscure. In some cells, cytoplasmic microtubules appear to emerge from relatively unstructured cytoplasm, often from an electron-dense amor-

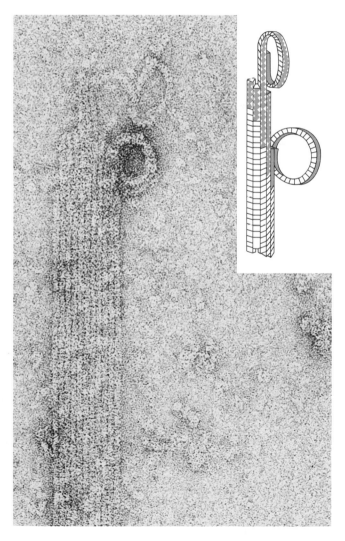

Figure 15-9. Electron micrograph of an incomplete microtubule wall (composed of 11 protofilaments) showing the attachment of rings in vitro and their continuity with protofilaments. The inset shows a schematic diagram of the type of structure seen in the electron micrograph. *(From H. P. Erickson, J. Supramol. Struct., 2:393, 1974.)*

phous material. In the case of the two most complex types of microtubular strucures (the mitotic spindle and ciliary-type organelles), the formation of microtubules generally seems to occur in the vicinity of a preexisting microtubule-containing structure.

Cilia and flagella arise from structures termed *basal bodies*, which have a similar cross-sectional appearance as that of the cilium or flagellum, itself. In this case, microtubules of the basal body appear to be elongated to form the microtubules of the cilium or flagellum. The microtubular structure associated with the formation of the mitotic spindle is termed a *centriole* (page 701), whose cross-sectional appearance is also similar to that of the basal bodies, cilia, and flagella. Even though centrioles form excellent nucleating centers in vitro (Fig. 15-7*b*) for the polymerization of microtubules, the relationship between the centriole and the mitotic spindle in the cell is unclear. Although the microtubules of the mitotic spindle do seem to originate in the region of the centriole (Fig. 15-8*a*), there is no actual physical continuity between the centriole and the spindle, nor is there any obvious geometrical relationship between the microtubules of the two structures. Most importantly, the entire spindle apparatus can form in the complete absence of a centriole. The cells of higher plants, for example, lack centrioles or similar structures, yet a mitotic spindle very similar to that found in animal cells is assembled. The structure of microtubules, their function, and their assembly will become more apparent if we consider these aspects in the context of a process in which they are intimately involved, that of mitosis.

MITOSIS

Mitosis is the process whereby eukaryotic cells bring about the separation of replicated chromosomes into two nuclei. In bacterial cells, the replicates of the single chromosome are separated by the growth of membrane inserted between the attachment sites of each chromosome (see Fig. 14-30). In eukaryotic cells a vastly more complex mechanism has evolved to ensure that the duplicates of the many chromosomes are precisely segregated. The process of mitosis is generally divided into several distinct stages (Fig. 15-10): *prophase, prometaphase, metaphase, anaphase,* and *telophase,* each of which is characterized by a particular series of events. Each of these stages represents a piece from a process which is a continuum; the division of mitosis into arbitrary sections is done only for the sake of discussion and experimentation. The process is best appreciated by watching time-lapse movies of living cells undergoing division. Figure 15-11 shows the progression of events as revealed in Nomarski interference micrographs taken by Andrew Bajer, one of the pioneers in this area.

In the nucleus of an interphase cell the tremendous lengths of chromatin fiber are spread throughout the entire space of the nuclear compartment. The nucleus of a human G_2 cell contains approximately 4 meters of DNA scattered throughout 46 duplicated chromosomes. A great deal of effort must be exerted by the cell in order for this length of chromatin to be separated into two identical packets. Therefore, preparation for mitosis is one of the major activities of the G_2 cell. The transition from G_2 to M appears to result from the presence of some type of factor or factors pro-

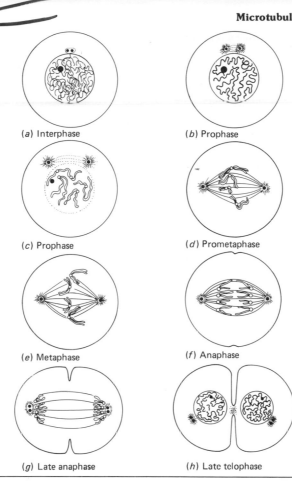

(a) Interphase

(b) Prophase

(c) Prophase

(d) Prometaphase

(e) Metaphase

(f) Anaphase

(g) Late anaphase

(h) Late telophase

Figure 15-10. Diagram illustrating the successive cellular events in mitosis, as described in the text. *(From W. M. Copenhaver, D. E. Kelly, and R. L. Wood, "Bailey's Textbook of Histology," 17th ed., William and Wilkins, 1978.)*

duced in the cell. This conclusion is based primarily on studies in which mitotic cells are fused with cells in other stages of the cell cycle. In every case, the mitotic cell induces the condensation of the chromatin in the nucleus of the nonmitotic cell. If a G_1 and an M cell are fused, the chromatin of the G_1 nucleus undergoes *premature chromosomal condensation* to form an elongated single condensed chromosome. If the nonmitotic cell is in the G_2 stage, then the condensation produces chromosomes that are visibly doubled in structure reflecting the fact that replication had previously occurred. Although both the G_1 and G_2 chromatin undergo extensive condensation, the packing ratio never reaches that which is found in the true mitotic chromosome. Fusion of a mitotic cell with an S-phase cell also produces condensation of the S-phase chromatin. However, chromatin of a replicating cell appears to be particularly sensitive and the condensation results in the formation of "pulverized" chromosomal fragments rather than intact compacted chromosomes. It would appear that, like the transition from G_1 to S (page 656), the transition from G_2 to M is also under positive control, i.e., is induced by the presence of some stimulatory agent.

The Formation and Structure of the Mitotic Chromosome

During the first stage of mitosis, that of prophase, the chromosomal duplicates are prepared for segregation and the mitotic machinery is assembled. The most prominent event occurring during prophase is the

OUTLINE MITOSIS ON PAPER

Figure 15-11. Photomicrographs from 16-mm time-lapse film taken of mitosis in the endosperm of *Haemanthus* by Nomarski optics. *(Courtesy of A. Bajer.)*

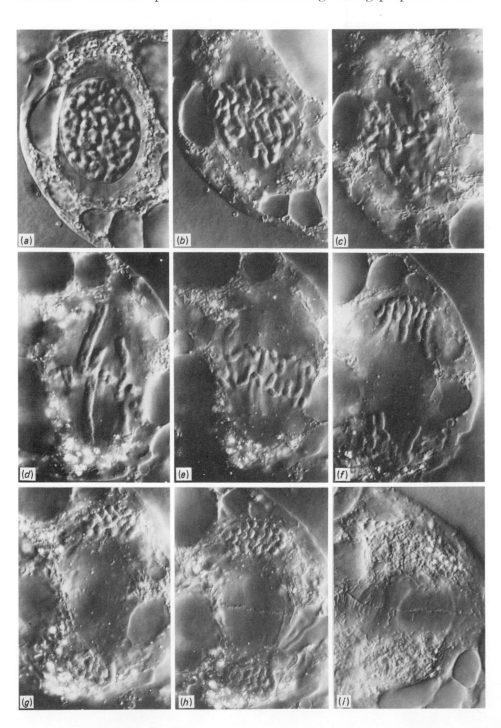

condensation of the chromatin to form the compact mitotic chromosomes. Whereas the chromatin of an interphase chromosome may extend for approximately a centimeter, that same chromatin fiber must be packed into a mitotic chromosome hundreds of times shorter in length. As described in an earlier chapter, the chromatin fibers of the interphase cell are composed of a string of nucleosomes of approximately 100 Å diameter. The development of techniques, particularly by Hans Ris and E. J. DuPraw, for the examination of chromosomes in the electron microscope has revealed that mitotic chromosomes are also fibrous in nature. The knobby fibers of the mitotic chromosome (Figs. 15-12a and 13-16) are approximately 200 to 300 Å in diameter and presumably represent some type of higher level of organization of the thinner interphase fiber. Several models (such

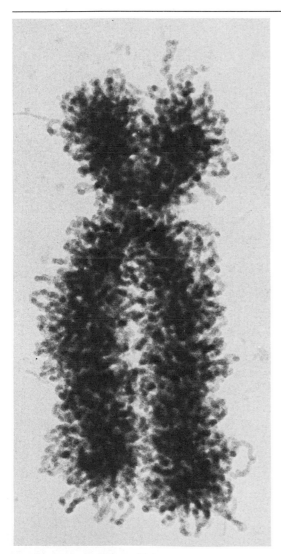

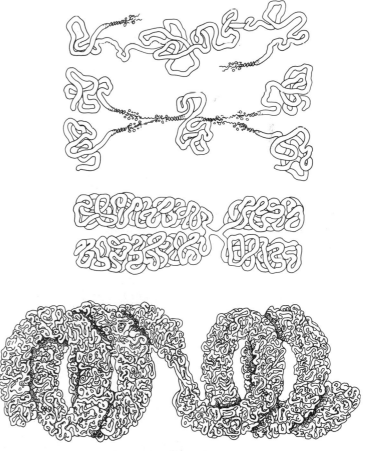

Figure 15-12. (a) Electron micrograph of a whole-mount preparation of a human mitotic chromosome. The structure is seen to be composed of a knobby fiber of 200 to 300 Å diameter. (b) A model suggested to account for the packaging of a deoxyribonucleoprotein fiber to form the thicker fibers seen in electron micrographs. (From E. J. DuPraw, "DNA and Chromosomes," Holt, Rinehart, and Winston, 1970.)

(a)

(b)

as that of Fig. 15-12*b*) have been presented which attempt to explain the ordering of interphase chromatin into the mitotic condition. In this regard, some evidence has accumulated for a role of the H1 histone in the condensation process. The phosphorylation of H1 histone molecules, which occurs prior to condensation, may be of particular importance in the overall process and a primary avenue by which cyclic AMP levels are involved in the control of cell division (page 645).

Another aspect of the structure of the mitotic chromosome has recently come to light from studies in which various components of the chromosome are selectively removed. The electron micrograph of Fig. 15-13 portrays a mitotic chromosome from which the histones as well as the majority of nonhistone proteins have been removed. It would appear from these studies that a small residual group of nonhistone proteins form a type of backbone or scaffolding which is responsible for maintaining the basic structure of the mitotic chromosome. A closer look at these protein-depleted mitotic chromosomes indicates that the DNA fibers are organized into loops of at least 10 to 30 μm in length, which are attached at their base to the nonhistone protein scaffold. Based upon these types of observations, it appears that the condensation process operates at one level to

Figure 15-13. Appearance of a mitotic chromosome after the histones and most of the nonhistone proteins have been removed. The residual proteins form a scaffold, which appears to consist of a dense network of fibers. Loops of DNA emerge from the scaffold. *(From J. R. Paulson and U. K. Laemmli,* Cell, ***12:*** *820, 1977.)*

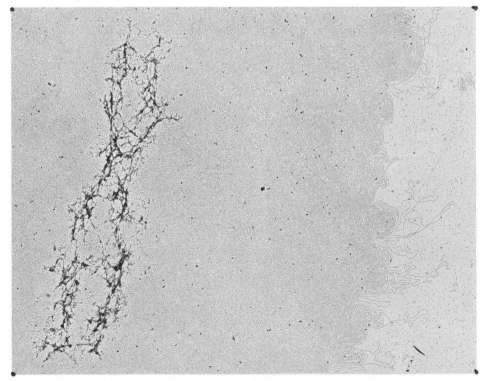

organize the chromatin fiber into some type of thicker structure and at another level to organize the strands into attached loops.

Regardless of the internal structure of the mitotic chromosome, its overall appearance has been put to use in a number of practical ways. Mitotic chromosomes, for example, have arisen in previous discussions in connection with in situ hybridization (page 466) and the localization of genes on particular human chromosomes following cell fusion (page 649). Mitotic chromosomes can be used for these types of experiments because in many cases each of the homologous pairs of chromosomes can be identified by their size, or shape, or position of their centromere. For example, a preparation of human mitotic chromosomes is shown in Fig. 15-14b. A preparation of this type is made (Fig. 15-14a) by gently lysing a mitotic cell (one generally fixed in methanol-acetic acid) on a wet slide. The cell bursts open and the mitotic chromosomes settle and attach to the surface of the slide over a very small area. If the individual chromosomes in a photograph of this type are cut out, they can be matched up into homologous pairs (23 in man) and portrayed as a karyotype as shown in Fig. 15-14b. Preparations of mitotic chromosomes are routinely made from cultures of blood cells and used to screen individuals for chromosomal abnormalities. As shown in Fig. 15-28, the occurrence of extra, missing, or grossly altered chromosomes can be detected in this manner. In recent years, new types of staining procedures have been developed which, when used on mitotic chromosomes, cause the chromosomes to have a banded appearance (Fig. 15-14c). Although the basis for the bands is not fully understood, the patterns are characteristic for a particular chromosome and provide a highly sensitive criterion for the identification of any chromosome from virtually any organism.

The Formation of the Mitotic Spindle

While the process of chromosome condensation is proceeding within the nucleus, the machinery for chromosomal separation is being formed outside the nucleus. In the cells of higher plants, which lack centrioles, the formation of the mitotic spindle begins with the formation of a clear zone around the nucleus. Microtubules rapidly appear in this zone, first in a disarranged condition and then in a more oriented state. When the nuclear envelope breaks at the end of prophase, the microtubules become organized into a mitotic spindle. The mitotic spindle of animal cells appears to be organized around the pair of centrioles present at each pole. Before discussing the formation of the mitotic spindle in animal cells, it is necessary to consider the nature of the centrioles. Centrioles (Fig. 15-15) are small cylindrical structures about 2 Å in diameter and about twice as long. Nine evenly spaced fibrils run the length of the centriole; each fibril appears in cross section as a band of three microtubules, designated the A, B, and C subfibrils. Each band of three microtubules is inclined at an angle to the surface of the centriole. At one end, identified as the proximal or old end, the central part is usually occupied by a characteristic pinwheel structure that consists of a central cylindrical hub and nine duplicate spokes extending toward the peripheral fibers. Centrioles

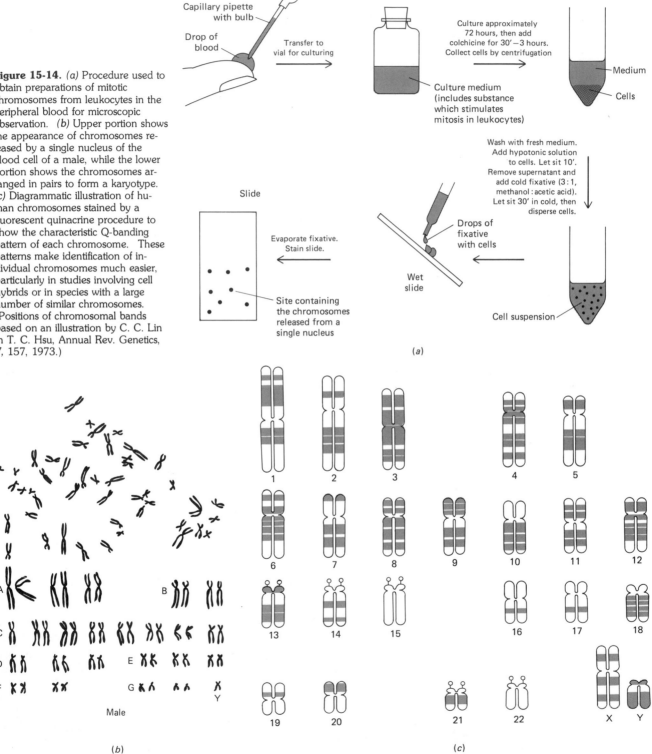

Figure 15-14. (a) Procedure used to obtain preparations of mitotic chromosomes from leukocytes in the peripheral blood for microscopic observation. (b) Upper portion shows the appearance of chromosomes released by a single nucleus of the blood cell of a male, while the lower portion shows the chromosomes arranged in pairs to form a karyotype. (c) Diagrammatic illustration of human chromosomes stained by a fluorescent quinacrine procedure to show the characteristic Q-banding pattern of each chromosome. These patterns make identification of individual chromosomes much easier, particularly in studies involving cell hybrids or in species with a large number of similar chromosomes. (Positions of chromosomal bands based on an illustration by C. C. Lin in T. C. Hsu, Annual Rev. Genetics, 7, 157, 1973.)

Capillary pipette with bulb

Drop of blood

Transfer to vial for culturing

Culture medium (includes substance which stimulates mitosis in leukocytes)

Culture approximately 72 hours, then add colchicine for 30'–3 hours. Collect cells by centrifugation

Medium

Cells

Wash with fresh medium. Add hypotonic solution to cells. Let sit 10'. Remove supernatant and add cold fixative (3:1, methanol:acetic acid). Let sit 30' in cold, then disperse cells.

Slide

Evaporate fixative. Stain slide.

Site containing the chromosomes released from a single nucleus

Drops of fixative with cells

Wet slide

Cell suspension

(a)

Male

(b)

1 2 3 4 5
6 7 8 9 10 11 12
13 14 15 16 17 18
19 20 21 22 X Y

(c)

are nearly always found in pairs with each of the members situated at right angles to one another (Fig. 15-16). During interphase the bipartite centriole "replicates" so that two *pairs* of centrioles are present in the cell close to one another outside of the nucleus. The formation of a new centriole begins with the appearance of a "procentriole" next to an existing centriole. The procentriole has the same cross-sectional appearance as a mature centriole, but is much shorter in length. The procentriole appears adjacent to the preexisting centriole and at right angles to it. Subsequent elongation of the tubules of the procentriole converts it into a mature centriole. Although the mother and daughter centrioles are oriented to one another in a highly predictable manner, there is no obvious continuity between them, nor is there any evidence that the daughter arises by any type of division process from the preexisting structure. Based on numerous studies, it appears that centrioles have the ability to arise *de novo*, i.e., from unformed materials, in the cytoplasm.

During prophase in animal cells, microtubule polymerization is initiated in the region of the centrioles. The microtubules tend to radiate out from this region forming what is termed an *aster* (Fig. 15-7*a*). The

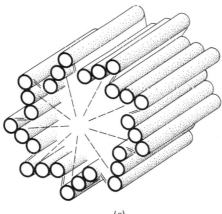

(a)

Figure 15-15. (*a*) Schematic illustration of a centriole. The primary components consist of nine peripheral microtubular triplets without any central microtubular structure. (*b*) Cross section of the centriole of an interphase cell. (*c*) Longitudinal section of a centriole of a metaphase cell. (*Photos by B. R. Brinkley, from L. Weiss, in R. O. Greep and L. Weiss, eds., "Histology," 3d ed., McGraw-Hill, 1973.*)

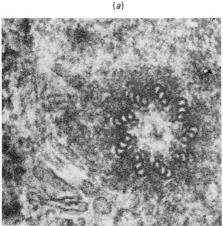

(b)

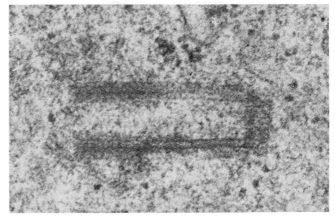

(c)

process of aster formation is followed by the separation of the pairs of centrioles from one another by migration around the nucleus. As the pairs of centrioles separate from each other (Fig. 15-10c), the microtubules stretching between them elongate due to the incorporation of additional subunits at their extremities. In addition to the microtubules, which extend between the centrioles, there are also astral microtubules which radiate out from each centriole in a sunburst arrangement. Eventually, the two centrioles reach points opposite one another and the two poles of the mitotic spindle have been determined.

The mitotic spindles of plant and animal cells are similar in structure. In both types of cells there are basically two groups of microtubules, the *chromosomal microtubules* and the *interpolar* (or *pole-to-pole) microtubules* (Fig. 15-17a). The chromosomal microtubules are organized into bundles (the chromosomal fibers which can be visualized in the polarization microscope) which connect the poles to the chromosome. The point of attachment of the fibers to the chromosome occurs at the kinetochore (Fig. 15-17b), a structure found at the primary constriction (centromere) of each mitotic chromosome. Electron microscope descriptions of the kinetochore of metaphase chromosomes have been variable. In one type of plant cell it has been described as a ball-shaped structure (5000 Å diameter) containing bands of lighter and darker material. In a rat cell it has been described as a disk-shaped structure composed of three distinct layers. In a mouse cell the kinetochore is reported to be a fibrillar structure consisting of a dense core about 200 to 300 Å diameter surrounded by a less dense zone, 200 to 600 Å wide. Regardless of the nature of the kinetochore, electron micrographs reveal that the microtubules of the chromosomal fibers penetrate deeply into the structure to which they are anchored. The interpolar microtubules form a structural basket which connects one pole to the other. Individual microtubules do not necessarily bridge the entire gap between the two poles or even between one pole and the equator. Serial sections through the mitotic apparatus indicate that many of the interpolar microtubules either begin at one pole and end at some distance, or exist free within the body of the spindle and not connected to either pole.

The end of prophase is marked by the rapid dissolution of the nuclear envelope and the subsequent appearance of the definitive mitotic spindle. As stated above, two types of microtubules (chromosomal and interpolar) can be distinguished. Presumably, the interpolar microtubules are derived directly from the microtubules stretched between the two centrioles of the prophase spindle. The origin of chromosomal fibers is less clear. There is evidence that some of them are assembled at the kinetochore of the chromosome and are elongated in a polar direction. Others may be elongated from the poles toward the chromosome, or represent sections derived from the interpolar microtubules.

Prometaphase and Metaphase

At the end of prophase, the condensed chromosomes are scattered throughout the space that was the nuclear region. As the microtubules

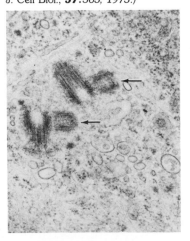

Figure 15-16. Tissue-cultured cell fixed four hours after mitosis. Two centriolar duplexes are located in an invagination on the nuclear surface. Each daughter centriole (arrow) is approximately 0.25 μm in length. *(From J. B. Rattner and S. G. Phillips, J. Cell Biol., **57:**363, 1973.)*

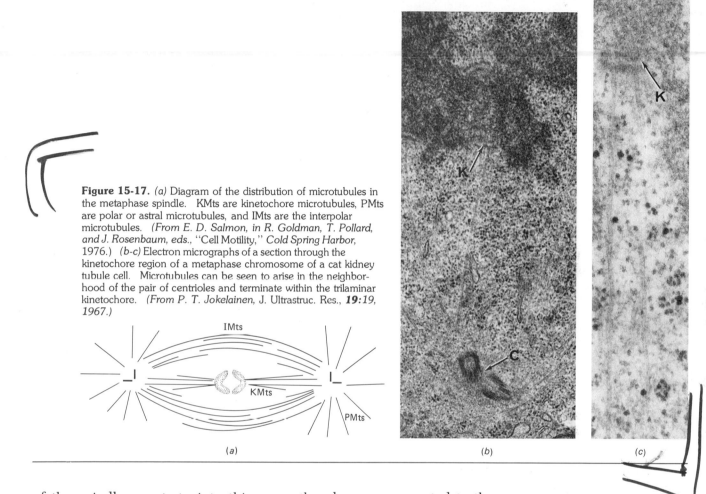

Figure 15-17. *(a)* Diagram of the distribution of microtubules in the metaphase spindle. KMts are kinetochore microtubules, PMts are polar or astral microtubules, and IMts are the interpolar microtubules. *(From E. D. Salmon, in R. Goldman, T. Pollard, and J. Rosenbaum, eds., "Cell Motility," Cold Spring Harbor, 1976.) (b-c)* Electron micrographs of a section through the kinetochore region of a metaphase chromosome of a cat kidney tubule cell. Microtubules can be seen to arise in the neighborhood of the pair of centrioles and terminate within the trilaminar kinetochore. *(From P. T. Jokelainen, J. Ultrastruc. Res.,* **19:19,** *1967.)*

IMts

KMts

PMts

(a)

(b)

(c)

of the spindle penetrate into this space they become connected to the chromosomes which are then moved toward the equator of the cell where they become aligned in a plane termed the *metaphase plate*. The directed movements of the chromosomes into position at the metaphase plate is termed *congression,* and the stage in which it occurs is termed *prometaphase*. Chromosomal movement during prometaphase is the direct responsibility of the chromosomal microtubules. If the linkage between a chromosome and its connecting fibers is broken mechanically, the chromosome cannot move until new connections with microtubules are made.

As the chromosomes undergo condensation during prophase and prometaphase, each can be seen to be made of two members, termed *chromatids,* which reflect the duplication process that had occurred during the previous interphase. The two chromatids of each chromosome are connected to each other at their centromeres. As the chromosomes are moved to their position in the metaphase plate, each is seen to be aligned so that each chromatid faces one of the poles, attached to the chromosomal fibers

705

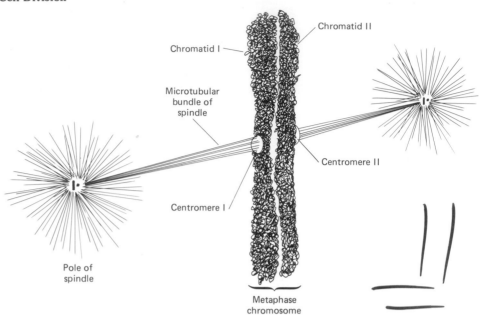

Chromatid II

Chromatid I

Microtubular
bundle of
spindle

Centromere II

Centromere I

Pole of
spindle

Metaphase
chromosome

Figure 15-18. Orientation of the centromeres at metaphase. Two centromeres are present. These are directed toward and make fiber connections to opposite poles of the spindle. *(From "Biology of the Cell" by Stephen L. Wolfe. © 1972 by Wadsworth Publishing Company, Inc., Belmont, California 94002. Reprinted by permission of the publisher.)*

at their kinetochores (Fig. 15-18). In some cases the proper alignment of a chromosome does not occur in the initial stage of orientation and subsequent activities cause the realignment of that chromosome until each of the two chromatids is in position to be drawn toward a different pole. If a chromosome is pulled out of its alignment in the metaphase plate, and its chromosomal fibers are severed, new chromosomal fibers can still form and move an inappropriately located chromosome back to its proper location. Metaphase represents the stage at which the chromosomes are aligned in a plane awaiting the separation event. The electron micrograph of Fig. 15-19 shows the components of the metaphase spindle as they appear in a cell that has been lysed under controlled conditions in the presence of detergent. The chromosomes are seen to be lined up at the metaphase plate with the microtubules converging at the opposite poles.

Anaphase and Telophase

Electron micrographs of anaphase chromosomes show that each chromatid of a replicated pair has its own distinct kinetochore, but the kinetochores of sister chromatids are firmly attached to one another. Anaphase begins with the sudden splitting apart of the kinetochores of the two sister chromatids of each chromosome. The actual splitting process does not seem to be a result of microtubule activity since it occurs even after the contact between the chromosome and its fibers are broken. All of the chromosomes of the metaphase plate are split in relative synchrony and the chromatids (now referred to as chromosomes since they are no longer attached to their duplicate) begin their poleward migration. Since each chromosome is being pulled by its attached microtubules, its centromere appears at the leading edge of the chromosome with the arms of the chro-

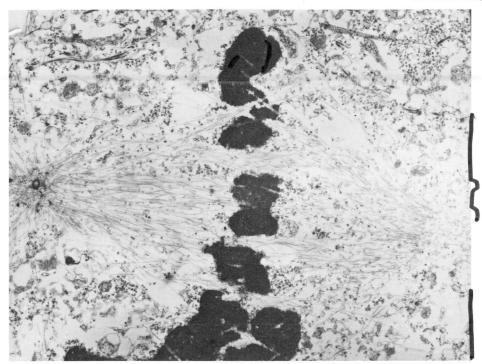

Figure 15-19. Electron micrograph of a metaphase cell that had been gently lysed in a solution of tubulin which maintains the integrity of the spindle microtubules. Preparations of this type retain the ability to undergo anaphaselike chromosome movements in the presence of added ATP. *(From J. A. Snyder and J. R. McIntosh, J. Cell Biol., **67**:746, 1975.)*

mosome trailing behind. As the chromosomes move toward their respective poles, the two poles actually move farther apart. This latter event occurs due to the lengthening of the entire mitotic spindle, presumably as a result of the lengthening of the interpolar microtubules.

The movement of the chromosomes toward opposite poles is a very slow process relative to other types of movements, proceeding at approximately 0.2 to 5.0 μm/minute and generally requiring anywhere from 2 to 60 minutes for completion. Once the chromosomes near their respective poles, events designed to return the chromosomes to the interphase condition begin. This final stage of mitosis, termed *telophase,* cannot be clearly demarcated from anaphase. Telophase is characterized by the uncoiling and dispersal of the chromatin of each chromosome and by the reformation of the nucleolus and nuclear envelope (page 631). It should be kept in mind that the preceding account of mitosis is that which occurs in the typical eukaryotic cell. As might be expected, there are many different variations, particularly among the more simple eukaryotes such as the fungi, algae, and protozoa (see the reviews by Harald Fuge, Melvin Fuller, and Jeremy Pickett-Heaps, for information on these organisms).

Forces Involved in Chromosome Movement

Although considerable information exists concerning the nature of the mitotic apparatus, the precise role of the microtubules in the overall event remains unclear. The need for microtubules in chromosome movement is well established. Most importantly, they are the only structures attached to the chromosomes as they move. If the microtubules are disrupted, the movement of the chromosomes stops. Even though it appears certain that the chromosomes are pulled toward the poles as a result of their attachment to the microtubules, there is considerable controversy as to the means by which the forces are generated.

One of the earlier theories of the role of microtubules in mitosis was formulated by Shinya Inoué based upon his observations of the process in living cells using the polarization microscope. Inoué concluded that the chromosomal fibers were shortening as the chromosomes moved poleward, and that the shortening was the result of the loss of material from the end of the fiber in the region of the poles. He concluded that the chromosomal fibers were decreasing in length as the result of the loss of subunits from the fiber at the pole, and he proposed that this disassembly was not simply a consequence of chromosome movement but the cause of it. Inoué proposed that disassembly could generate sufficient force to pull the chromosomes forward. One would not expect the dissolution of a fiber at one end to provide much pulling force; however calculations reveal that very little force is required to move an object as small as a chromosome such a short distance. It is believed that a force of only 10^{-8} dynes (equivalent to 20 to 30 molecules of ATP) is required for the average chromosome.

Experimental support for the disassembly model has come from studies on the treatment of metaphase cells with microtubular depolymerizing agents. Under controlled conditions, these treatments cause the gradual disassembly of the microtubules of the spindle at their polar end. As the disassembly occurs, the chromosomal fibers shorten and the chromosomes move to the poles (Fig. 15-20). The movement of the chromosomes occurs slowly, at approximately the same rate as in the cell during mitosis. It would appear from this experiment that microtubule disassembly can generate sufficient forces for chromosome movement. It does not, however, exclude the action of other factors operating in the cell. Even if other mechanisms are found to be responsible for generating the force involved in chromosome movement, the disassembly of the chromosomal fibers would still be required since the microtubules attached to the chromosomes are clearly shortened. Since the disassembly event would be expected to be the slowest step in the overall process, the shortening of the fibers may be rate-limiting, regardless of how the force is generated.

As will be discussed in the following chapter, the movement of cilia and flagella as well as the contraction of muscle tissue is believed to result from the sliding of certain elements across one another. This concept of sliding filaments has also been called upon to explain chromosomal movements during mitosis, primarily by J. Richard McIntosh. In this case it is proposed that the microtubules of the chromosomal fibers slide relative to the interpolar microtubules. The sliding would be accomplished by the directed movement of cross bridges connecting the two types of microtubules. These cross bridges would be composed of microtubule-associated proteins. As in the case of the disassembly theory of chromosome movement, the evidence for the sliding theory is largely indirect. Elec-

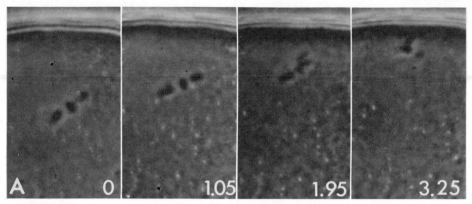

Figure 15-20. The movement of chromosomes in an oocyte as a result of the application of hydrostatic pressure. Oocytes of this stage are arrested at meiotic metaphase waiting to be fertilized, an event which triggers the completion of meiosis. In this case, the chromosomes are caused to move toward the surface of the oocyte due to the depolymerization of the attached spindle microtubules. Time in minutes is indicated on each frame. *(From E. D. Salmon, in R. Goldman, T. Pollard, and J. Rosenbaum, eds., "Cell Motility," Cold Spring Harbor, 1976.)*

tron microscopy has revealed the presence of material between microtubules which may represent the required cross bridges. If microtubules are sliding across one another during mitosis, this should be reflected by an increasing number of microtubules in the region of overlap. Counts of microtubules in serial cross sections through cells in different stages of mitosis has provided evidence for the sliding of these structures, although this type of evidence does not indicate that the sliding is actively responsible for the movement rather than simply a consequence of it.

The most recent theory of chromosomal movement has emerged from a number of observations which have revealed the presence of the contractile proteins actin and myosin within the mitotic spindles. It has been proposed that these proteins (discussed in the following chapter) are directly responsible for generating the force for chromosome movement. If this were the case, the microtubular basket of the mitotic spindle would be considered as primarily of structural importance. The development of in vitro "models" (as in Fig. 15-19) capable of undergoing chromosome movement may be of particular importance in the analysis of the underlying forces.

MICROFILAMENTS*

Microfilaments are elongated structures of very fine diameter that appear to be present in virtually every eukaryotic cell. Microfilaments have a diameter of 40 to 70 Å and are invariably composed of one particular protein, actin. Actin was identified many years ago as one of the major muscle

---

* In this chapter and throughout the book, the term *microfilament* is used to refer to a 40 to 70 Å filament composed of actin. Another type of microfilament of thicker diameter (approximately 100 Å), generally termed an *intermediate filament*, is also present in various cell types, especially neurons. The function of the intermediate filament, which is composed of a protein distinct from actin, is unknown.

proteins, and one intimately involved in muscle contractility (described fully in Chapter 16). Actin is a globular protein termed *G-actin* (molecular weight of 42,000 daltons), which is capable of assembling in vitro or in vivo into a very characteristic type of filament (Fig. 16-7), termed *F-actin*. In muscle cells these actin-containing thin filaments (approximately 60 to 70 Å diameter) become associated with another filament-forming protein, myosin, and together these two proteins cooperate to generate the forces involved in muscle contraction. The interaction between actin and myosin is highly specific and this specificity has been utilized to develop a diagnostic test for the presence of actin filaments. To carry out the test one adds myosin to cells suspected of containing actin filaments. If the filaments are made of actin, and virtually all cytoplasmic filaments of this approximate diameter (e.g., 60 Å) appear to be, then the myosin forms a complex with the filaments which can be identified in the electron microscope (see Fig. 16-14).

Observations made on a wide variety of cells have indicated that microfilaments form the major component of a cell's contractile machinery. These filaments have been implicated in such diverse processes as cell division, cell movement, nerve outgrowth, tubular gland formation, gastrulation, and neurulation. Whereas microtubules are generally implicated in maintaining a particular cell shape, the microfilaments are often responsible for causing that shape to be generated. In each case, changes in cell shape occur as the result of a contractile process within the cytoplasm. In most cases, the evidence for the involvement of microfilaments is indirect, but substantial; the filaments are found in the expected location of the cell and oriented in a manner that would be predicted to accomplish the process. Furthermore, the event is generally found to be blocked by the addition of cytochalasin B (Fig. 16-18) which causes the disappearance of the microfilaments. However, as is commonly the case for inhibitors, cytochalasin B is known to have effects on processes other than the one being studied, in this case on membrane transport. This makes it difficult to be sure the effect observed is due to the loss of microfilaments. Regardless, there is general agreement on the role of microfilaments in cell contractility. One of the best-studied processes known to involve microfilaments is cytokinesis.

Cytokinesis

The end product of cell division is the production of two cells from one, an event which requires the separation of the cytoplasm as well as the nuclear contents. The division of the cytoplasm into two compartments is brought about by a process termed *cytokinesis* which has recently been shown to be a contractile event. The splitting of one cell into two is a remarkable process, one which has intrigued cell biologists for many years. Due to the large size and predictable, rapid sequence of cell divisions, the cleavage of animal eggs has been a favored system for the study of cytokinesis. If one watches a sea urchin egg undergoing cell division, the first hint of cytokinesis is seen during late anaphase as an indentation of the cell surface around the egg. As time progresses, the indentation deepens and becomes a furrow completely encircling the egg. The plane of the furrow lies in the same plane previously occupied by the chromosomes of the metaphase plate. In other words, the plane of the furrow is perpendicular to the long axis of the mitotic spindle and therefore ensures that the two sets of chromosomes will ultimately be separated into two cells. The furrow continues to deepen until opposing surfaces make contact with one another in the center of the cell and the cell is split into two.

Various theories have been proposed to explain the advancement of the cleavage furrow, placing the forces variously with the mitotic spindle, the cytoplasm, or the plasma membrane itself. In recent years, one of the theories has continued to gain favor—the contractile ring theory (Fig. 15-21a) proposed by Douglas Marsland in the 1950s. In this theory the force generated to cleave the egg resides in a thin strip of protoplasm just beneath the plasma membrane, the *cortex*, in the region of the furrow itself. Electron microscopic examination of the cortex beneath the furrow of a cleaving egg has revealed the presence of a large number of microfilaments (Fig. 15-21b) with their long axes oriented parallel to the plane of the furrow. Although myosin filaments (page 736) have not been directly observed in the electron microscope, myosin does seem to be present

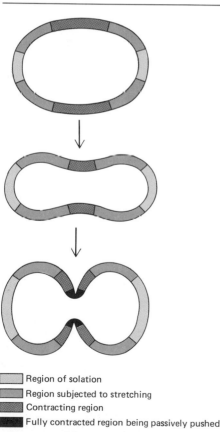

Region of solation

Region subjected to stretching

Contracting region

Fully contracted region being passively pushed

(a)

Figure 15-21. (a) The cortical gel contraction theory as originally proposed. *(From D. Marsland,* Biol. Revs., *33:109, 1958, after Marsland and Landau.)* (b) Light micrograph of cleaving sea urchin egg. (c) Electron micrograph of the furrow region cut in the same plane as that of b. The layer of "dots" beneath the plasma membrane is the contractile ring, whose microfilaments are seen in cross section when cut in this manner. *(Courtesy of T. E. Schroeder.)*

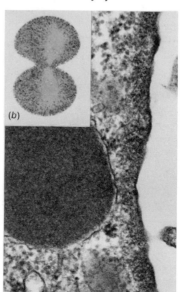

(b)

(c)

in the contractile ring since antibodies prepared against myosin are found to bind material (presumably myosin) in this part of the cell. The presence of both actin and myosin beneath the cleavage furrow provides indirect evidence that the force-generating mechanism operating during cytokinesis is similar to the one in muscle cells. The effect of the filaments of the contractile ring is to constrict the egg in two, much like a pursestring narrows the diameter of an opening.

One of the most remarkable aspects of the layer of microfilaments is the rapidity with which it is assembled prior to cytokinesis and disassembled upon its completion. It appears that some mechanism within the cell selects the appropriate ring of cortical cytoplasm to be the site of the future furrow and then causes the polymerization of actin to occur in this region, just before the initial indentation becomes visible. If cytochalasin B is added to an egg undergoing cytokinesis, the cortical microfilaments are disrupted and the furrow retracts, leaving the egg uncleaved. If the eggs are rapidly washed free of the drug, the furrow reforms and splits the egg.

Up to this point, the discussion has been restricted to cleaving eggs. It appears from recent observations on cell division in cultured cells that actin-containing microfilaments are widely responsible for cytokinesis. As in the case of cleavage, microfilaments capable of binding myosin appear in cultured cells at the site of the future furrow prior to the onset of cytokinesis and disappear once the process has been completed. Although less information is available on division in animal cells located within tissues. there is no reason to believe that the same mechanisms are not involved. As will be described in the next section, the splitting of plant cells in two occurs by a very different mechanism.

CELL PLATE FORMATION

Plant cells with their inextensible cell wall enclosure, go about separating the daughter cells of mitosis in a very different manner from that just described in animal cells. Rather than having a dividing cell split in two by a furrow which advances from the existing surface, plant cells build a cell membrane and cell wall beginning near the center of the dividing cell and growing out to the lateral walls. The construction of a new cell plate has been followed in studies that have coordinated observations on living cells with the light microscope and fixed cells in the electron microscope. The best-studied cell is that of the endosperm of the African blood lily, *Haemanthus*, which can be removed from the seed and cultured in vitro through an additional cell division. The first sign of cell plate formation can be seen in late anaphase–early telophase cells with the appearance of dense material roughly aligned in the equatorial plane of the previous metaphase plate. In the first stages this material is restricted to the central region of the cell. Examination of this region, which is termed the *phragmoplast,* with the electron microscope (Fig. 15-22*a*) shows it to consist of clusters of microtubules oriented perpendicular to the future plate together with some type of associated electron-dense

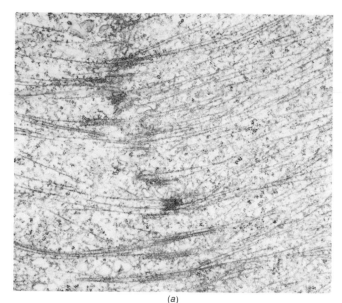

(a)

Figure 15-22. Electron microscopy of the formation of the cell plate. *(a)* Phragmoplast showing the bundles of microtubules which cross the center of the cell in an overlapping fashion. Within each bundle, a denser region can be noted which falls roughly in the plane of the future cell plate. *(From P. K. Hepler and W. T. Jackson, J. Cell Biol., **38**:442, 1968.)* *(b)* Accumulation of Golgi-derived vesicles in the equatorial region. Note the presence of Golgi cisternae and mitochondria within the region of the phragmoplast. *(c)* The arrangement of the Golgi vesicles along the equatorial plane. The vesicles are beginning to fuse in formation of the new cell plate and plasma membranes. *(d)* The middle lamella of the new cell wall within the plasma membranes. The wall will grow in thickness by the continued apposition of materials. *[(b)–(d) From A. Frey-Wyssling, J. F. López-Sáez, and K. Mühlethaler, J. Ultrastruc. Res., **10**:422, 1964, courtesy of K. Mühlethaler.]*

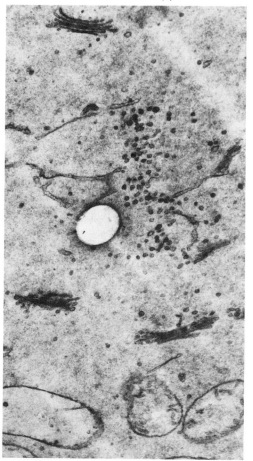

(b)

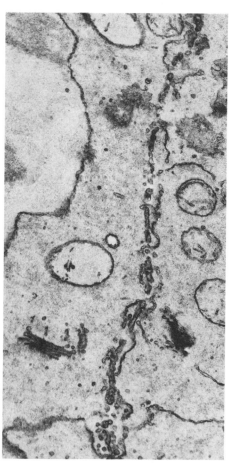

(c)

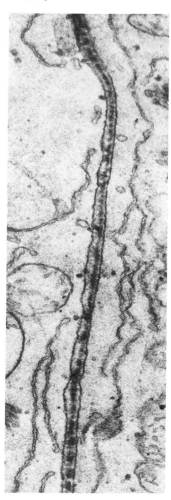

(d)

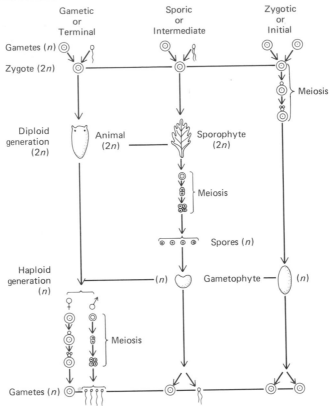

Figure 15-23. Diagram comparing the three known types of meiosis, as described in the text. (From E. B. Wilson, "The Cell in Development and Heredity," *3d ed., 1925. Reprinted with permission of Macmillan Publishing Co., Inc.)*

amorphous material. The microtubules of the clusters probably arise from remnants of the mitotic spindle as well as by *de novo* formation from tubulin subunits. With time, the dense material of the microtubule clusters becomes more perfectly aligned into a plane which foreshadows the plane of the future cell wall. After the formation of the phragmoplast, small vesicles move into the region (Fig. 15-22b), probably transported by the microtubules themselves. The vesicles, which arise from the ER-Golgi membrane system, contain material for the cell plate, i.e., the forerunner of the mature cell wall. As the vesicles move into position in the established plane of the phragmoplast, they fuse with one another to form long chains (Fig. 15-22c). The membrane of the vesicles becomes the plasma membrane of the two adjacent daughter cells, while the material within the vesicle becomes the intervening cell plate (Fig. 15-22d). The process begins near the center of the cell and then progresses laterally as clusters of microtubules and vesicles form at the lateral edges of the advancing cell plate. The cell plate rapidly becomes extended until it reaches the already established walls of the cell. Once the formation of cell plate is completed, additional materials are added in the building of the mature cell wall.

MEIOSIS

The production of offspring by sexual reproduction involves the union of two cells, each with a complete set of chromosomal information. As discussed in Chapter 10 (which should be consulted as a background to this section), the doubling of the chromosome number at fertilization must be compensated by an equivalent reduction of the number of chromosomes at some stage prior to the formation of the gametes. The process by which the chromosome number is reduced, i.e., cells are formed which contain only one member of each pair of homologous chromosomes (page 427), is termed *meiosis*. Unlike mitosis, in which the chromosomes are doubled and then divided between two daughter nuclei, the doubling of the chromosomes prior to meiosis is followed by *two* sequential divisions which distribute the chromosomes among four cells rather than two. In order to ensure that each of the daughter nuclei formed by meiosis has one member of each pair of homologs, an elaborate process of chromosome pairing occurs which has no counterpart in the mitotic division. During the period in which the homologous chromosomes are paired, genetic exchange may occur between them to produce chromosomes with new combinations of alleles. As will be apparent from the following discussion, there is very little evidence concerning the mechanisms underlying the pairing and recombination events, although there is no scarcity of speculation on the subject. Considering both the importance of the meiotic process and the challenging problems in molecular biology that it poses, it might be expected that considerable research effort will come to be focused on these events.

A survey of various eukaryotes indicates that marked differences exist with respect to the stage within the life cycle at which meiosis occurs. The following three groups (Fig. 15-23) can be identified on this basis:

1. *Gametic or terminal meiosis.* In this group, which includes all multicellular animals, many protozoa, and a few lower plants, the meiotic divisions are closely linked to the formation of the gametes themselves (Fig. 15-24). In the

Figure 15-24. Diagrams showing the formation of the sperm and egg. In each case a relatively small population of primordial germ cells proliferates to form a population of gonial cells from which the gametes will differentiate. In the male (a) meiosis occurs before differentiation while in the female (b) it occurs after differentiation. Each primary spermatocyte generally forms four viable gametes while each primary oöcyte forms only one fertilizable egg. (*From E. B. Wilson, "The Cell in Development and Heredity," 3d ed., 1925. Reprinted with permission of Macmillan Publishing Co., Inc.*)

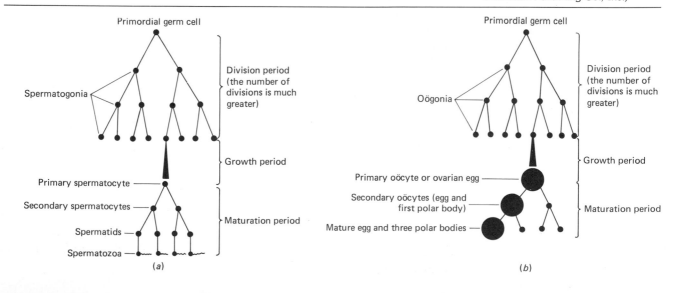

(a)

(b)

male animal (Fig. 15-24a), meiosis occurs just prior to the differentiation of spermatozoa. *Spermatogonia* become *primary spermatocytes* which then undergo their two meiotic divisions to form four relatively undifferentiated *spermatids*, each of which undergoes a complex transformation to become the highly specialized spermatozoa. In contrast, in the female animal (Fig. 15-24b), the oogonia become primary oocytes which then enter a greatly extended meiotic prophase. During this prophase, the primary oocyte grows and becomes filled with yolk and other materials (see Fig. 12-7). It is only after the differentiation of the oocyte is complete, i.e., the oocyte has reached essentially the same state as when it is fertilized, that the meiotic divisions occur. In vertebrates, for example, eggs are typically fertilized at a stage in the middle of meiosis (generally metaphase I). Meiosis is completed after fertilization while the sperm moves through the cytoplasm.

2. *Zygotic or initial meiosis.* In this group, which includes only a few algae and protozoa, the meiotic divisions occur just after fertilization. Consequently, all of the cells are haploid; the diploid stage of the life cycle is restricted to a brief period after fertilization.

3. *Sporic or intermediate meiosis.* In this group, which includes all of the higher plants and some lower ones, the meiotic divisions take place at a stage unrelated to either gamete formation or fertilization. If we begin the life cycle with the union of a male gamete (the pollen cell) and a female gamete (the ovum), the diploid zygote which is formed undergoes mitosis and develops into a diploid *sporophyte.* At some stage in the development of the sporophyte, sporogenesis (which includes meiosis) occurs, producing spores which germinate directly into a haploid *gametophyte.* The gametophyte can be either an independent stage or, as in the case of seed plants, a tiny structure which is retained within the ovules. Regardless, the gametes are produced by the haploid gametophyte by mitosis.

The Stages of Meiosis

As is the case for any cell division, the prelude to meiosis includes the replication of the DNA. The premeiotic S phase is unusual in more than one way. First of all it is typically several times the length of a premitotic S phase, and it is suspected that meiosis-specific activities (of an unknown nature) occur during this stage. In addition, it appears that a small percentage (approximately 0.3%) of the DNA of the chromosomes is not replicated during the premeiotic S phase, but rather is delayed until prophase. This surprising situation was discovered by Herbert Stern and Yasuo Hotta during their studies on meiosis in the lily and will be discussed in further detail below. Even though meiosis-specific activities may occur during the premeiotic S phase, experiments in the lily indicate that cells which have just completed DNA synthesis can still be caused to undergo a normal *mitosis* if placed under the appropriate culture conditions. However, if one waits until the beginning of prophase before culturing the cells, they will continue in their preprogrammed activities and undergo a normal meiosis. It would appear that some event(s) occurs early during premeiotic G_2 phase which commits the cell to its subsequent meiotic divisions.

Whereas the premeiotic S phase is extended somewhat, relative to its mitotic counterpart, the meiotic prophase is lengthened in extraordinary fashion. In the human female, for example, every oocyte in the ovary has

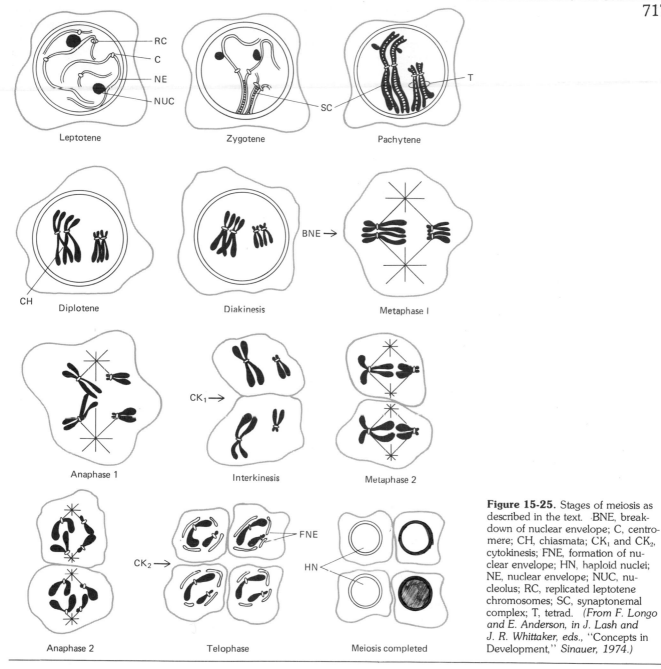

Leptotene   Zygotene   Pachytene

Diplotene   Diakinesis   Metaphase I

Anaphase 1   Interkinesis   Metaphase 2

Anaphase 2   Telophase   Meiosis completed

Figure 15-25. Stages of meiosis as described in the text. BNE, breakdown of nuclear envelope; C, centromere; CH, chiasmata; CK_1 and CK_2, cytokinesis; FNE, formation of nuclear envelope; HN, haploid nuclei; NE, nuclear envelope; NUC, nucleolus; RC, replicated leptotene chromosomes; SC, synaptonemal complex; T, tetrad. *(From F. Longo and E. Anderson, in J. Lash and J. R. Whittaker, eds., "Concepts in Development," Sinauer, 1974.)*

entered meiotic prophase at approximately the time of birth. Since no additional oocytes are produced during an individual's lifetime, many of these oocytes will remain in the same approximate stage of prophase for several decades. Since the first meiotic prophase, i.e., the prophase preceding the first meiotic division rather than the second, is so long and complex, it is customary to divide it into several stages (Fig. 15-25).

The first stage of prophase I is *leptotene,* which is characterized by the gradual appearance of the chromosomes in the light microscope. Although the chromosomes are known to have replicated at an earlier stage, in most species there is no indication that each chromosome is actually composed of a pair of identical chromatids. Condensation continues through leptotene with the chromosomes often appearing as beaded structures. The beads, termed *chromomeres,* are believed to represent regions of the chromosome that have become particularly coiled relative to the interchromeric sections. An important feature of the leptotene chromosome is its association with the nuclear membrane, particularly at the ends of the chromosomes (the telomeres). Although the precise role of the nuclear membrane in the activities of the meiotic chromosome remains uncertain, a number of proposals have been made which give it a central role in the replication and/or pairing of the chromosomes during meiosis. Another feature of importance in the leptotene chromosome is the presence of a lateral proteinaceous component which runs the length of each chromosome. This lateral fiber is believed to lie in the groove between the sister chromatids of each chromosome.

When leptotene ends, the chromosomes are partially condensed and the homologs are ready to associate. The process by which homologs become joined to one another is termed *synapsis* and occurs during *zygotene,* the second stage of prophase I. Synapsis is an intriguing event marked by important unanswered questions: By what basis do the homologs recognize one another? How do the pairs become so perfectly aligned? As mentioned above, the leptotene chromosomes are attached to the nuclear membrane at their telomeres. Considering the fluidity of cell membranes, it is not unreasonable to assign to the nuclear membrane a role in bringing the homologous chromosomes together.

The pairing of homologous chromosomes appears to occur in two stages. In the first stage, the chromosomes become roughly aligned but remain approximately 3000 Å from one another. In the second stage, the chromosomes approach one another more closely and a precise point-to-point association is brought about. In this latter state, the homologs are still separated from one another by approximately 1000 Å. Electron micrographs of synapsing chromosomes indicates that the process occurs by the formation of a new structure, the *synaptonemal complex.* The synaptonemal complex (SC) is a ladderlike structure (Fig. 15-26) composed of three parallel bars with many cross fibers connecting the central bar with the two lateral ones. The chromatin of the pairs of chromosomes is intimately associated with the lateral bars of the SC which are separated by the 1000 Å space. Observations on the assembly of the SC indicate that its lateral elements are derived directly from the lateral elements of the leptotene chromosomes. The role of the synaptonemal complex in meiosis remains uncertain. In the view of some researchers, the bars of the SC are involved simply in stabilizing the paired condition, whereas in other proposals they are assigned a role in determining the precise chromosomal alignment. Since it has been reported that a synaptonemal complex can form between genetically dissimilar chromosomes, it would appear that the SC is not directly responsible for bringing homologous DNA sequences into register.

In the view of certain investigators, the alignment process that occurs during synapsis must involve the association, at least transiently, of a

Figure 15-26. *(a)* Diagram of a synaptonemal complex (SC). The lateral elements are 1000 Å apart. *(From P. B. Moens,* Int. Rev. Cytol., **35:**130, 1973.) *(b)* Electron micrograph of a section through a bivalent in a grasshopper spermatocyte. The homologous chromosomes are held together by the synaptonemal complex whose dark-staining lateral elements and more opaque central element are clearly visible. The SC has a number of twists in it which is commonly found at this stage (midpachytene). The very dark-staining material at the periphery of the chromosome is part of the nucleolus which is attached to this particular chromosome in this species. *(Courtesy of P. B. Moens.)* *(c)* Electron micrograph of the synaptonemal complex after treatment with deoxyribonuclease to remove chromosomal fibers. The structure remaining represents the proteinaceous, ladderlike SC itself. *(From D. Comings and T. Okada,* Exp. Cell Res. **65:**104, 1971.)

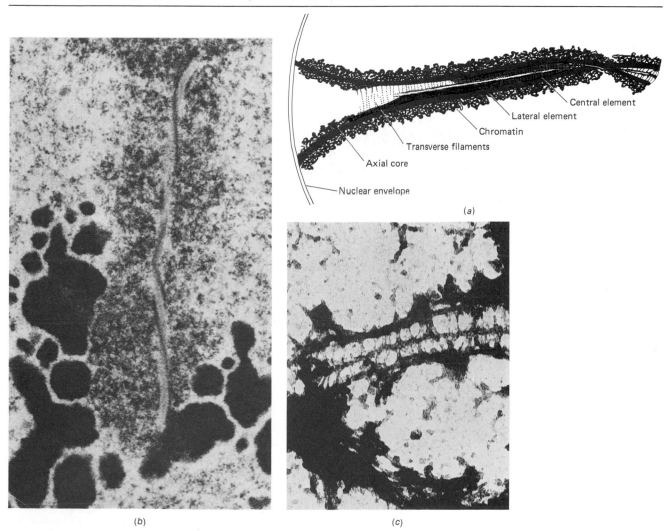

Central element

Lateral element

Chromatin

Transverse filaments

Axial core

Nuclear envelope

(a)

(b)

(c)

selected number of complementary single-stranded segments of DNA from each member of the homologous pair. It was mentioned above that a small fraction of the DNA is replicated in prophase rather than the preceding S phase. This DNA, approximately 0.3% of the genome, is replicated during zygotene and has been implicated in the synaptic process. For example, if zygotene DNA synthesis is blocked by the addition of the appropriate inhibitor, the normal pairing of homologous chromosomes does not occur. In the view of Stern and Hotta, specific DNA sequences are present at the surface of the compacted meiotic chromosomes which interact to mediate the pairing event. They go on to suggest that the DNA made during zygotene may have a role in this process.

The end of synapsis marks the end of zygotene and the beginning of the next stage of prophase I, *pachytene*. Pachytene is the first of the stages of prophase which tends to be inordinately long. Whereas leptotene and zygotene generally last for a few hours (though they can be much longer), pachytene is often extended for a period of days or weeks. The most noticeable change in the chromosomes during pachytene is their increased compactness. Unfortunately, crossing-over, which is the most important event believed to occur during pachytene, cannot be directly visualized. Rather than interrupt our discussion of meiosis at this point, we will postpone a further consideration of the mechanism of genetic recombination as it occurs during pachytene until the following section.

The beginning of *diplotene,* the next stage of meiotic prophase I, is generally recognized by the tendency of the homologous chromosomes of the bivalents (tetrads) to pull away somewhat from each other. Although the mechanism behind this event is not understood, it is generally described as a type of repulsion between the homologs. As the homologous chromosomes move apart, they are seen to remain attached to one another at specific points, termed *chiasmata* (see Fig. 10-12). The chiasmata are generally located at the sites within the chromosomes at which genetic exchange during crossing-over had previously occurred. Although a one-to-one correlation does not always exist between crossing-over and chiasmata, the presence of these points of attachment provides a striking visual portrayal of the extent to which recombination has occurred.

During the development of most oocytes, diplotene is an extremely extended phase of oogenesis, one in which the bulk of the growth of the cell occurs. In other words, the diplotene stage, occurring as it does in the middle of the meiotic prophase, is a period of intense metabolic activity. This activity can be followed in the behavior of the chromosomes as well as the growth of the mass of the cell. Whereas the pachytene chromosomes are highly compact, the diplotene chromosomes of both male cells *(primary spermatocytes)* and female cells *(primary oocytes)* are typically diffuse and difficult to observe microscopically.

In many spermatocytes and oocytes, the chromosomes become dispersed into a particular configuration which is not found at any other time during the life cycle of the organism. Chromosomes in this configuration are termed *lampbrush chromosomes* (Fig. 15-27) and are characterized by having an axial backbone from which pairs of loops extend out

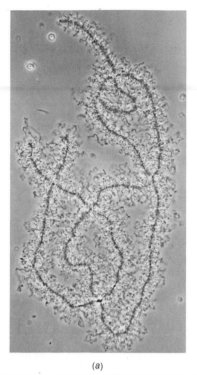

(a)

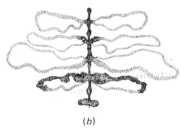

(b)

Figure 15-27. (a) Photomicrograph showing lampbrush chromosomes. Hundreds of loops can be seen projecting from the axis of the chromosome. (Courtesy of J. G. Gall.) (b) Diagram of a portion of a lampbrush chromosome.

in opposite directions. Loops arise in pairs because each member of the pair is part of one of two chromatids of each chromosome. The two homologous chromosomes still remain attached to each other at the chiasmata. The backbone of the lampbrush chromosomes contains DNA and tightly associated protein and is transcriptionally inactive. The evidence suggests that the loops are made of one double-stranded DNA molecule attached at both ends to the backbone at specific sites. At the base of the loops one can see a swelling (also containing DNA and protein), termed a *chromomere,* from which the DNA of the loop is believed to be "spun out." Associated with the DNA of the loop is a considerable amount of RNA and protein, which gives these loops a variety of appearances under the microscope. In the newt *Triturus* there are approximately 5000 loops present per haploid set of chromosomes. Each of these loops contain from 50 to 100 μm of DNA, an amount sufficient to code for many proteins. Autoradiographs prepared after incubating the lampbrush chromosomes in ^{3}H-uridine indicate that the loops are sites of intense RNA synthesis. The discovery of lampbrush chromosomes and the visualization of their template activity, provided (along with the puffs of the giant salivary chromosomes) an excellent system for the early studies on transcription at selected sites of eukaryotic chromosomes. In the case of the oocyte, the RNA made by diplotene lampbrush chromosomes is utilized for the production of proteins both during oogenesis and during embryonic development following fertilization.

The final stage of meiotic prophase I, termed *diakinesis,* prepares the

chromosomes for the attachment of the meiotic spindle fibers. In those species where the chromosomes become highly dispersed during diplotene, they become recondensed during diakinesis. In many cases, the chiasmata disappear by sliding down the length of the chromosome during a process referred to as *terminalization*. Diakinesis ends with the disappearance of the nucleolus, the breakdown of the nuclear envelope, and the movement of the tetrads to the metaphase plate. Metaphase I finds each pair of homologous chromosomes oriented in such a way that both chromatids of one chromosome face the same pole. In other words, the joined kinetochores of sister chromatids lie side-by-side to one another relative to the long axis of the spindle. The spindle fibers from a given pole are therefore connected to both chromatids of a single chromosome. During anaphase I, the homologous chromosomes separate from one another. Since there is no interaction between one tetrad and another, the chromosomes of each tetrad segregate into the two daughter cells *(secondary spermatocytes* or *secondary oocytes)* independent of the other chromosomes. Anaphase I is the cytological event which corresponds to Mendel's law of independent assortment (page 422). During the anaphase movement, each chromosome (now termed a *dyad*) can be seen to be made of two associated chromatids, still connected to one another by their kinetochores.

On occasion the homologous chromosomes of a given tetrad fail to separate from one another and consequently, the entire tetrad moves to one pole. This event, termed *primary nondisjunction,* results in the production of gametes with an abnormal chromosome number; two of the four meiotic products will have an extra chromosome and the other two will be deficient for that chromosome. A variety of human congenital disorders are the direct result of meiotic nondisjunction. In almost every case (the exception being the Y chromosome), if a zygote is formed from a gamete that is deficient in a chromosome, the embryo that forms from that zygote undergoes an early death. In contrast, if a zygote is formed from a gamete having an extra chromosome, the embryo often survives (depending upon which chromosome is involved) and develops into an individual whose cells contain an additional chromosome. Down's syndrome (mongolism), for example, results from the presence of an extra chromosome (number 21) in each cell (Fig. 15-28). The severity of the abnormalities of this syndrome (as well as those from other chromosome *trisomies)* illustrates how sensitive cellular processes are to chromosomal imbalance. The presence of extra sex chromosomes (producing such conditions as XXY, XXXY, XYY, etc.) have less widespread effects upon individuals, yet do produce severe abnormalities particularly with regard to the proper development of reproductive tissues. An oral epithelial cell from an XXXX individual is shown in Fig. 13-24.

Telophase I of meiosis produces less dramatic changes than telophase of mitosis since the chromosomes generally do not revert to a true interphase condition. Although in many cases the chromosomes do undergo some dispersion, they do not reach the extremely extended state of the

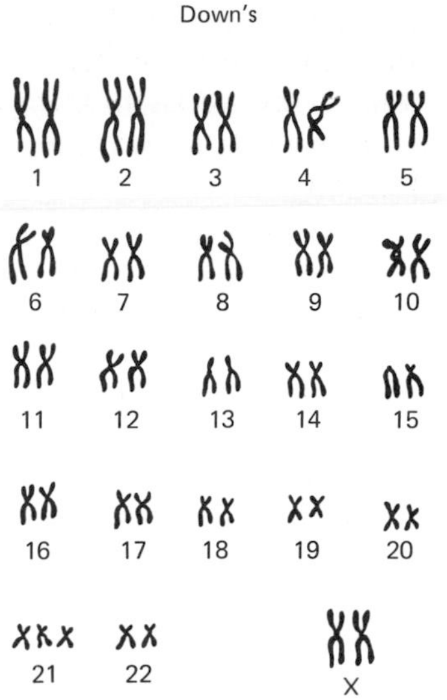

Down's

1	2	3	4	5
6	7	8	9	10
11	12	13	14	15
16	17	18	19	20
21	22		X	

Figure 15-28. The karyotype of Down's syndrome is characterized by an extra chromosome (trisomy) number 21.

interphase nucleus. Similarly, the nuclear envelope may or may not reform during telophase I. The stage between the meiotic divisions is termed *interkinesis* and is generally short-lived. In an animal, cells in this stage are referred to as secondary spermatocytes or secondary oocytes. They are characterized as having a diploid amount of nuclear DNA but a haploid number of chromosomes, since each chromosome is still represented by a pair of attached chromatids.

Interkinesis is followed by prophase II, a much simpler prophase than its predecessor. The chromosomes merely recondense and line up at the equatorial plane. In contrast to their orientation in metaphase I, the kinetochores of sister chromatids of metaphase II face opposite poles and become attached to opposing sets of spindle fibers. The chromatids become separated at anaphase II as each of the members of a pair of sister kinetochores become split from one another. The failure of a given pair of chromatids to separate from one another at anaphase II is termed *secondary nondisjunction*. If secondary nondisjunction occurs, *one* of the four haploid cells produced will have an extra chromosome, i.e., two copies of one type, another will be missing that chromosome, and two of the products will have a normal chromosomal complement. The products of normal meiosis are haploid cells, haploid in both nuclear DNA level and chromosome number.

Genetic Recombination

The topic of recombination has arisen several times in previous chapters in connection with crossing-over of meiotic chromosomes (page 434), bacteriophage infection (page 479), bacterial conjugation (page 477), and bacterial transformation (page 442). In all of these cases, it appears that some type of movement of genetic markers occurs between two distinct chromosomes. Although these various types of genetic recombination do not occur by precisely the same mechanism, it is generally believed that important similarities do exist among these diverse phenomena. If we continue on the assumption that the molecular basis of recombination is basically similar in all forms, then studies performed on viral and bacterial chromosomes, which consist essentially of naked DNA molecules, might be expected to provide basic information on the underlying mechanism. Before considering the molecular events which may occur during recombination, it is worthwhile to reexamine some of the earlier experiments which have shaped our present concept of the phenomenon.

In the early view of T. H. Morgan and other classical geneticists, the tension generated by the twisting of the chromosomes around each other during meiotic prophase was believed to cause the chromosomes to become fractured in such a way that pieces of chromatids were exchanged between homologous chromosomes. In this view, crossing-over involves an actual physical *breakage and reunion* of chromosomal material. Examination of the effects of irradiation upon chromosomes indicated that chromosomes could actually be broken and that isolated pieces were capable of attaching themselves to other chromosomal elements. The difficulty with the breakage and reunion theory as conceived by the early geneticists resides in the problems associated with having two large pieces of chromosome break at precisely corresponding regions and exchange homologous segments of genetic material.

During the 1930s a new concept of genetic recombination, termed *copy-choice*, was proposed by John Belling as an alternative to the breakage-reunion scheme. In the copy-choice model recombination occurred during replication. Whatever the mechanism responsible for duplicating the chromosomes, Belling suggested that the newly formed chromosomes could contain a mixture of information from the two parental chromosomes. Rather than discuss the model as it was presented in 1931, we can briefly examine the concept in terms of the semiconservative replication of a pair of DNA helices of two homologous chromosomes (Fig. 15-29a). If copy-choice is actually responsible for recombination, we would expect that during replication the enzymes responsible for DNA synthesis would switch templates at some point from one homologous duplex to the other. During the 1950s, Seymour Benzer showed (page 479) that recombination could occur between two viral genomes, within a given gene, without the loss or addition of a single nucleotide. With the demonstration of the incredible precision with which the joining of genetic markers from two separate sources could occur, the breakage-reunion model of recombination (Fig. 15-29b) fell into general disfavor. It therefore came as a surprise when experiments by Matthew Meselson and Jean Weigle in 1961 clearly demonstrated that recombination in viral systems occurred by a process of breakage and reunion rather than copy-choice. These experiments provided the first important insight into the molecular activities of recombination.

The early studies in bacteriophage genetics had found that if bacterial

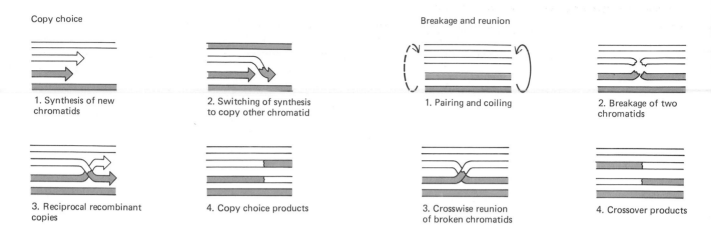

Copy choice

1. Synthesis of new chromatids

2. Switching of synthesis to copy other chromatid

3. Reciprocal recombinant copies

4. Copy choice products

Breakage and reunion

1. Pairing and coiling

2. Breakage of two chromatids

3. Crosswise reunion of broken chromatids

4. Crossover products

Figure 15-29. Diagrammatic representation of two possible mechanisms of crossing-over. *(From J. D. Watson, "The Molecular Biology of the Gene," W. A. Benjamin, 3d ed., 1976.)*

cells were simultaneously infected by two different strains of viruses (for example, +,+ and a,b) then recombinants +,b and a,+ could be found among the progeny. In order to prove that these recombinant viral genomes arose by the actual fusion of parts of two different DNA molecules (as opposed to arising during replication by being copied successively from two different template strands), Meselson and Weigle conducted an experiment in which they could distinguish between the two parental strains of virus on the basis of physical as well as genetic differences. In order to distinguish between the DNAs of the two viral strains, one contained the heavy ^{13}C and ^{15}N isotopes, whereas the DNA of the other strain contained only the light isotopes of these atoms. The two types of viruses were allowed to jointly infect a culture of bacterial cells, the infection occurring in a medium containing only light isotopes of carbon and nitrogen. The phage progeny produced by the infection were collected and the density of their DNA determined by equilibrium sedimentation in cesium chloride. Since the density of the DNA is a direct measure of the percentage of heavy isotopes it contains, the position of a particular DNA molecule in the gradient provides a measure of how much of the original parental heavy DNA it contains. If recombination occurred by copy-choice, then recombinant DNA strands should have arisen during replication and, therefore, should be composed solely of light isotopes. In contrast, if the recombination occurs via breakage and reunion, then the original recombinant strands (which will be present among the progeny molecules) should contain sections of heavy as well as light DNA. The results of the experiment indicated that some of the recombinant DNA molecules did indeed contain strands of DNA composed of both heavy and light DNA. Furthermore, the amount of heavy DNA contained in these recombinant molecules correlated with that which would be expected as a result of a break occurring between the two genes being used as markers.

At the present time there are a number of different models which have been proposed to account for the recombination event. In addition, there

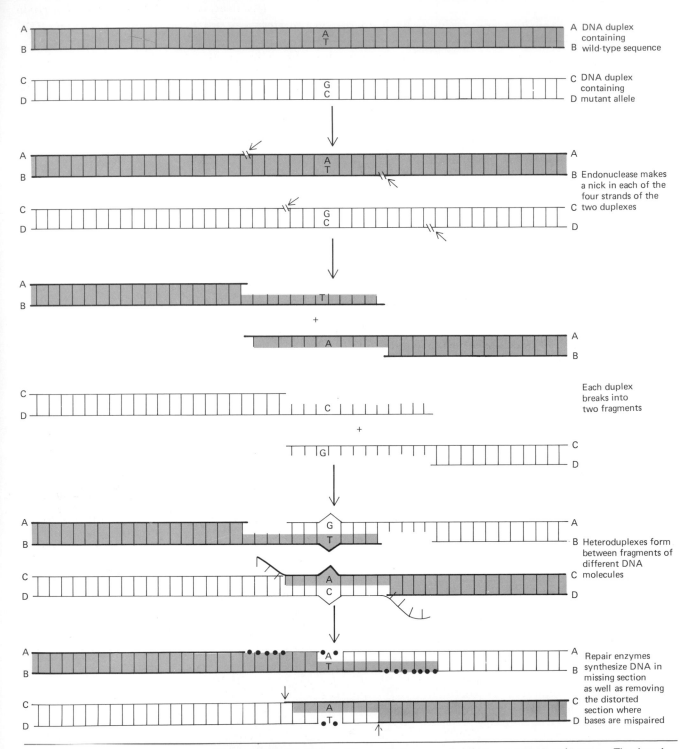

Figure 15-30. Postulated sequence of steps occurring during genetic recombination. The dotted sections in the lower part indicate the sites of newly synthesized single DNA strands and the arrows indicate the sites of nuclease action. In the particular case shown here, the mutant DNA strands happen to be the ones selected for DNA repair and, therefore, the wild-type condition results.

is good evidence that the steps which occur during recombination may be quite different in different systems (meiotic crossing-over, transformation, transduction, conjugation, etc.). The subject rapidly becomes very complex and highly speculative. Rather than attempting to discuss these various models, none of which may be correct, we will simply point out a few salient features which tend to be included in a number of them. Figure 15-30 illustrates one possible series of steps which might occur during recombination. Regardless of the correctness of this hypothetical diagram, it serves to indicate the types of events one might expect to occur during the process.

First of all, it is generally believed that the initial specific interactions between two double helices leading to recombination are mediated by single-stranded elements. Recombination presumably begins by having an endonuclease make a break in one strand of one of the two adjoining duplexes. In some manner, a section of single-stranded DNA becomes displaced from that helix and becomes capable of interacting with a complementary strand of the other duplex. In order for this interaction to occur, a second nuclease-mediated break would presumably have to occur in this neighboring molecule, although the two breaks do not have to occur in precisely the same site (Fig. 15-30). Hydrogen bond formation between the strands of two different DNA molecules would provide the specificity that is known to occur during recombination. The existence of joined DNA molecules (Fig. 15-31a) has been directly visualized with the electron microscope (Fig. 15-31b). In some cases the two homologous chromosomes would be expected to contain genetic differences in the site

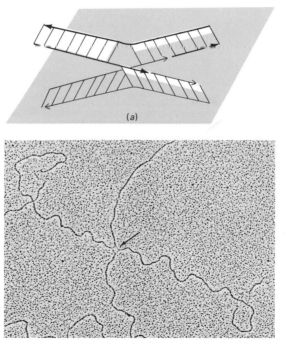

(a)

(b)

Figure 15-31. (a) Postulated structure of a pair of bridged double helices formed as an intermediate during recombination. (From L. Hood, J. H. Wilson, W. B. Wood, "Molecular Biology of Eukaryotic Cells," copyright © 1975 by the Benjamin/Cummings Publishing Company, Inc., Menlo Park, California.) (b) Electron micrograph of two lambda phage molecules bridged at a site of crossing-over. [From M. Valenzuela and R. B. Inman, Proc. Natl. Acad. Sci. (U.S.), **72:3024,** 1975.]

of hydrogen bond formation, resulting in the formation of a heteroduplex. Once the bridge between two DNA molecules has formed it is believed that the branch point between them is capable of migration linearly down one or the other molecules. The exchange of more and more bases between the two duplexes by branch migration would serve to greatly expand the segment involved in recombination.

In the diagram of Fig. 15-30, the breakage of the two helices at different sites requires that a section of one of the strands be trimmed off by nuclease action, and that corresponding segments be resynthesized by a repair-type mechanism. Although there are numerous ways in which one could visualize recombination occurring, the synthesis of pieces of DNA is assumed to occur in a number of the models. Among other roles, the involvement of DNA synthesis in the recombination process allows the position of the breaks in the two duplexes to be less precise. As long as the breaks are in the same general region of the two molecules, the nuclease-polymerase activities of various enzymes could allow the process to be completed without the loss or addition of genetic information. Direct evidence of both a biochemical and genetic nature has accumulated for the occurrence of DNA synthesis during recombination. An important consequence of the synthesis of a limited amount of DNA by a repair-type mechanism during recombination is the potential it provides for a change in nucleotide sequence of the strands. For example, in the recombination event depicted in Fig. 15-30 the formation of heteroduplexes containing a mispaired section activates repair enzymes to replace the distorted section. One of the strands of each heteroduplex is *randomly* selected to undergo excision and repair in the distorted region. In the particular case shown in Fig. 15-30, the nucleotide of the mutant strand happened to be the one replaced in both molecules thereby causing the molecules to assume the wild-type state.

In the case shown in Fig. 15-30, it is apparent that the synthesis of new DNA accompanying recombination has resulted in the "conversion" of one allele into another. The occurrence of gene conversion during recombination had been discovered in the 1930s during genetic studies on eukaryotic microorganisms such as *Neurospora* and yeast and provided a rather baffling element that had to be accounted for in any general model of the mechanism of recombination. As is apparent from the following discussion, organisms such as *Neurospora* offer special opportunities for genetic analysis which are not available in most types.

Based on the information presented in Chapter 11, one would expect that, if crossing-over were to occur between two gene loci, recombination would always be reciprocal. For example, if one of the homologous chromosomes of a tetrad was wild type (+,+) and the other was doubly mutant (a,b), then a cross-over occurring between two chromatids of the tetrad between these genes should produce one gamete with the a,+ genotype and one with the +,b genotype. We can make the example even simpler and consider a single gene locus, x. Throughout our previous discussion of Mendelian genetics, it was assumed that a heterozygote x/+ produced an equal number of x-bearing and +-bearing gametes. Since it is generally impossible to directly observe the genotype of a gamete, we have to examine the genetic nature of the offspring and, on this basis, deduce the genotypic ratios of the gametes. When this is done one generally concludes that the ratio of the two types of gametes was 1 : 1, suggesting that in each case meiosis produced 4 cells, 2 of each type. It should be noted, however, that these conclusions are based strictly on statistical considerations since most gametes of an individual do not participate in the production of offspring.

Fortunately for genetic investigation, organisms do exist where the genetic constitution of the products of a *particular* meiosis can be determined. In *Neurospora*, for example, the products of meiosis are haploid spores (termed *ascospores*) which remain together in one sac (termed an *ascus*). When the segregation of particular alleles is followed during individual meiotic events in *Neurospora*, one finds that ratios of 3 : 1 rather than 2 : 2 will appear on occasion. One of the alleles has apparently been converted to the other one as described previously.

Just as there is evidence that recombination in microorganisms involves DNA synthesis, some evidence also exists for this type of synthesis during recombination in higher organisms. It was mentioned previously that a small amount of semiconservative DNA synthesis occurs during the zygotene stage of meiosis in the lily. Further analysis of this system has revealed that a limited amount of DNA synthesis also occurs during pachytene, the time when crossing-over would be expected to occur. Unlike the case of zygotene DNA synthesis, that which occurs during pachytene is of the repair-type and results in the synthesis of DNA sequences that had already been replicated during the previous premeiotic S phase. These properties are precisely those which would be expected if recombination involved DNA synthesis as shown in Fig. 15-30.

REFERENCES

Bajer, A., Chromosoma **24**, 383–417, 1968. "Fine Structure Studies on Phragmoplast and Cell Plate Formation."

——— and Molè-Bajer, J. Adv. Cell Mol. Biol. **1**, 213–266, 1971. "Architecture and Function of the Mitotic Spindle."

——— and ———, Int. Rev. Cytol. Supp. **3**, 1972. "Spindle Dynamics and Chromosome Movement."

Bearn, A. G., and German, J. L., III, Sci. Am. **205**, 66–76, Nov. 1961. "Chromosomes and Disease."

Brinkley, B. R., and Stubblefield, E., Chromosoma **19**, 28–43, 1966. "The Fine Structure of the Kinetochore of a Mammalian Cell in Vitro."

——— and ———, Adv. Cell Mol. Biol. **1**, 119, 1970. "Ultrastructure and Interaction of the Kinetochore and Centriole in Mitosis and Meiosis."

Comings, D. E., and Okada, T. A., Adv. Cell. Mol. Biol. **2**, 309–384, 1972. "Architecture of Meiotic Cells and Mechanisms of Chromosome Pairing."

Erickson, R. O., Science **181**, 705–716, 1973. "Tubular Packing of Spheres in Biological Fine Structures."

Frey-Wyssling, A., Lopez-Saez, J. F., and Muhlethaler, K., J. Ultrastruc. Res. **10**, 422–432, 1964. "Formation and Development of the Cell Plate."

Fuge, H., Protoplasma **82,** 289–320, 1974. "Ultrastructure and Function of the Spindle Apparatus and Chromosomes during Nuclear Division."

Fuller, M. S., Int. Rev. Cytol. **45,** 113–153, 1976. "Mitosis in Fungi."

Gillies, C. B., Ann. Rev. Genetics **7,** 91–110, 1975. "Synaptonemal Complex and Chromosome Structure."

Goldman, R., Pollard, T., and Rosenbaum, J., eds., *Cell Motility.* Cold Spring Harbor, 1976.

Hepler, P. K., and Palevitz, B. A., Ann. Rev. Plant Phys. **25,** 309, 1974. "Microtubules and Microfilaments."

Huberman, J. A., Ann. Rev. Biochem. **42,** 355–378, 1973. "Structure of Chromosome Fibers and Chromosomes."

Inoué, S., and Stephens, R. E., eds., *Molecules and Cell Movement.* Raven, 1975.

Johnson, R. T., and Rao, P. N., Biol. Revs. **46,** 97–156, 1971. "Nucleo-Cytoplasmic Interactions in the Achievement of Nuclear Synchrony in DNA Synthesis and Mitosis in Multinucleate Cells."

Ledbetter, M. C., and Porter, K. R., J. Cell Biol. **19,** 239–250, 1963. "A 'Microtubule' in Plant Cell Fine Structure."

Luykx, P., Int. Rev. Cytol. Supp. **2,** 1970. "Cellular Mechanisms of Chromosome Distribution."

Mazia, D., Sci. Am. **205,** 100–120, Sept. 1961. "How Cells Divide."

Meselson, M., and Weigle, J. J., Proc. Natl. Acad. Sci. (U.S.) **47,** 857–868, 1961. "Chromosome Breakage Accompanying Genetic Recombination in Bacteriophage."

Moens, P. B., Int. Rev. Cytol. **35,** 117–134, 1973. "Mechanisms of Chromosome Synapsis at Meiotic Prophase."

Moses, M. J., Ann. Rev. Genetics **2,** 363–412, 1968. "Synaptonemal Complex."

Nicklas, R. B., Adv. Cell Biol. **2,** 225–297, 1971. "Mitosis."

Olmstead, J. B., and Borisy, G. G., Ann. Rev. Biochem. **42,** 507–540, 1973. "Microtubules."

Pickett-Heaps, J., Bioscience **26,** 445–450, 1976. "Cell Division in Eukaryotic Algae."

Rao, P. N., and Johnson, R. T., Adv. Cell Mol. Biol. **3,** 136–189, 1974. "Induction of Chromosome Condensation in Interphase Cells."

Rappaport, R., Int. Rev. Cytol. **31,** 169–213, 1971. "Cytokinesis in Animal Cells."

Rhoades, M. M., "Meiosis," in *Cell,* vol. III, J. Brachet and A. Mirsky, eds., pages 1–75, Academic, 1960.

Roberts, K., Prog. Biophys. Mol. Biol. **28,** 371–420, 1974. "Cytoplasmic Microtubules and Their Functions."

Snyder, J. A., and McIntosh, J. R., Ann. Rev. Biochem. **45,** 699–720, 1976. "Biochemistry and Physiology of Microtubules."

Soifer, D., ed., Ann. N.Y. Acad. Sci. **253,** symposium on "The Biology of Cytoplasmic Microtubules."

Spooner, B. S., Bioscience **25,** 440–451, 1975. "Microfilaments, Microtubules, and Extracellular Materials in Morphogenesis."

Stern, H., and Hotta, Y., Ann. Rev. Genetics **9**, 37–66, 1972. "Biochemical Controls of Meiosis."

Stubblefield, E., Int. Rev. Cytol. **35**, 1–60, 1973. "The Structure of Mammalian Chromosomes."

Wessels, N. K., Sci. Am. **225**, 76–85, Oct. 1971. "How Cells Change Shape."

Westergaard, M., and von Wettstein, D., Ann. Rev. Genetics **6**, 71–110, 1972. "The Synaptonemal Complex."

Wettstein, R., and Sotelo, J., Adv. Cell Mol. Biol. **1**, 109–152, 1971. "The Molecular Architecture of Synaptonemal Complexes."

Wilson, L., and Bryan, J., Adv. Cell. Mol. Biol. **3**, 21–72, 1973. "Biochemical and Pharmacological Properties of Microtubules."

Yeoman, M. M., ed., *Cell Division in Higher Plants.* Academic, 1976.

CHAPTER SIXTEEN
Contractility and Cell Movement

Movement among animals is highly varied. Single-celled organisms loco-mote with the aid of organelles such as cilia and flagella, or move about by a process of cytoplasmic streaming as seen in an amoeba. Multicellular organisms walk, fly, swim, and wriggle in myriad different ways. Motility is equally important within an animal as evidenced by the pumping of the heart, the peristaltic contractions of the gut, or the movement of mucus by the ciliary lining of the respiratory tract. More subtle types of move-ments occur within cells, such as the uptake of materials during phago-cytosis, or the splitting of one cell into two during cytokinesis, or the movement of the microvilli at the surface of the intestine. Analysis of the mechanisms underlying a wide variety of different types of movements has revealed a remarkable similarity among them at the molecular level. Most importantly, they are all essentially contractile events. The best-understood contractile process is that which occurs in vertebrate skeletal muscle cells. Although it has been the general rule in this book to avoid a detailed consideration of differentiated cells with specialized function, it is necessary in this case to describe certain studies of vertebrate skeletal muscle cells which have led to a better understanding of the mechanisms underlying cell motility in general.

MUSCLE CONTRACTILITY

Vertebrate muscle cells are generally grouped into two categories, *striated* or *smooth.* Striated muscle, which is characterized by the presence of distinct bands, makes up the tissue of the heart and skeletal muscles. In the following discussion, we will concentrate on the skeletal muscle cell whose highly ordered structure has received the most attention. The precise organization of the components of the skeletal muscle cell has provided an unparalleled opportunity for investigators to correlate the structure of a cell with its function and has led to the formulation of a model for muscular contraction which appears to be valid not only for the other types of muscle cells but for many nonmuscle systems as well. The first step in the discussion of this model is an analysis of the structure of the cells themselves.

The Structure of Muscle Cells

If one considers a cell to be that unit of protoplasm enclosed within a continuous plasma membrane, then the cells of the typical vertebrate skeletal muscle are highly unorthodox (Fig. 16-1). These cells are very large (10 to 100 μm thick and up to a few centimeters in length), they contain many nuclei, and they have a marked *striated,* i.e., striped, appearance. If one traces these giant cells back to their embryonic origins, they are found to arise as the result of the fusion of normal-sized, mononucleated cells termed *myoblasts.* Once the embryonic myoblasts have fused to form the multinucleate skeletal muscle cells, i.e., skeletal *muscle fibers,* the cytoplasm becomes filled with a variety of muscle proteins which give the living fiber its characteristic striated appearance when viewed in the polarization microscope. As will become apparent in the following discussion, the stripes reflect an organization of the cytoplasm in which the elements are lined up in precise register across the entire thickness of the fiber. The situation is analogous to that of the giant salivary gland chromosomes whose banding pattern reflects a precise organization of DNA and protein in each of the fibers which makes up the chromosome. In both cases, those regions with more stainable macromolecular material appear darker.

Examination of a muscle fiber under the microscope (Figs. 16-1, 16-2) reveals an organized substructure of basic importance to muscle function. The fiber is seen to be a type of cable made up of thousands of thinner strands, termed *myofibrils,* each of which runs the entire length of the fiber and is surrounded by cytoplasm separating it from adjacent myofibrils. The cytoplasm of the muscle cell is termed *sarcoplasm* and is seen to contain an elaborate system of intracellular membrane as well as mitochondria, lipid droplets, and glycogen granules. If one examines the transverse bands of a stained muscle fiber from one end to the other, one sees a striking pattern which repeats itself many times along the length

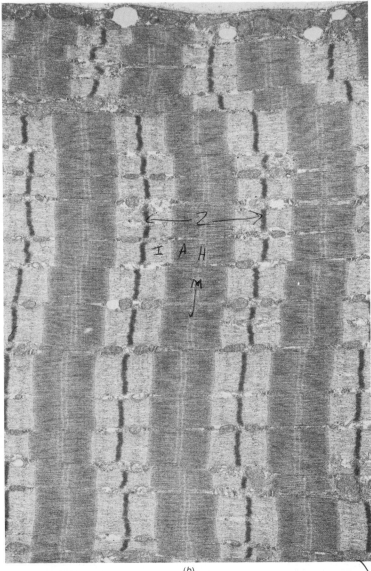

Figure 16-1. (a) Light micrograph of a longitudinal section of several skeletal muscle fibers of a cat tongue. The fibers have a characteristic striated appearance and contain numerous nuclei located toward the periphery of each fiber. *(From G. F. Gauthier, in R. O. Greep and L. Weiss, eds., "Histology," McGraw-Hill, 3d ed., 1973.)* (b) Low-power electron micrograph through a part of a soleus muscle of a guinea pig. *(From B. R. Eisenberg, A. M. Kuda, and J. B. Peter, J. Cell Biol., 60:743, 1974.)* (c) Higher-power electron micrograph of a single sarcomere with the bands lettered. *(From G. F. Gauthier, in R. O. Greep and L. Weiss, "Histology," 3d ed., McGraw-Hill, 1973.)*

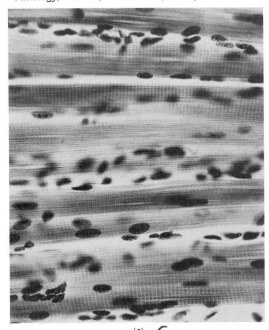

(a)

(b)

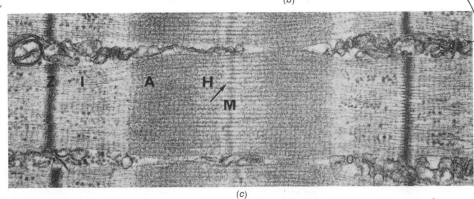

(c)

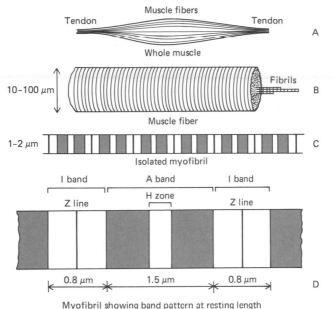

Figure 16-2. Levels of organization of skeletal muscle. *(From H. E. Huxley, in J. Brachet and A. E. Mirsky, eds., "The Cell," vol. 4, Academic, 1960.)*

of the fiber (Fig. 16-1a). Each repeating unit is approximately 2 5 μm in length and is composed of several bands which have been named with letters of the alphabet (Fig. 16-1b, c).

For reasons which will become apparent in the following discussion, the functional unit of the myofibril extends from one Z line to the next Z line and is termed a *sarcomere*. Within each sarcomere adjacent to each Z line is an I band and between the two I bands of a sarcomere lies the A band. The terms A and I stand for *anisotropic* and *isotropic* which reflects their appearance in the polarization microscope. The A band, which is positively birefringent, appears bright under this type of illumination while the I bands appear dark (Fig. 16-3). In the center of the A band, as viewed under bright-field illumination, a less densely stained region is apparent, which is termed the H band (Fig. 16-1c). At the center of the H band, a slightly more densely stained band, the M line is found. Examination of muscle tissue with the electron microscope during the 1950s (particularly by Hugh Huxley and coworkers) showed the banding

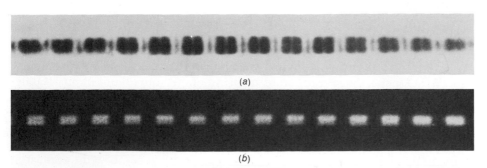

Figure 16-3. The appearance of an isolated glycerinated myofibril observed with phase contrast optics *(a)* and in polarized light *(b)*. In the polarization micrograph of *b*, the A band is bright and the I band dark. *(From R. H. Colby, J. Cell Biol., **51**: 765, 1971.)*

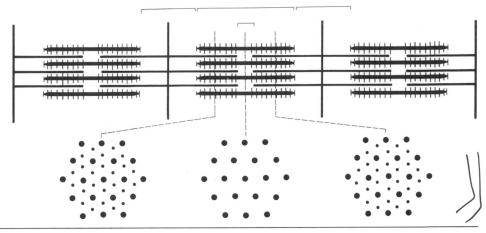

Figure 16-4. *(a)* Diagrammatic representation of part of a myofibril showing the overlapping array of thin actin- and thick myosin-containing filaments. The small transverse projections on the myosin fiber represent the cross bridges. The relationship between the regions of the sarcomere in this diagram and the various bands can be determined from Fig. 16.2. The hexagonal arrangement of the thin filaments around each thin filament in the lateral regions of the A band is apparent in the cross sections. *(After H. E. Huxley,* Science, ***164:*** *1357, 1969. Copyright by the American Association for the Advancement of Science.)*

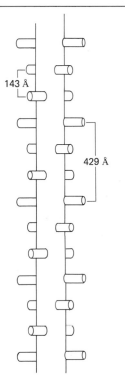

Figure 16-5. The disposition of the myosin cross bridges about the thick filament. *(From H. E. Huxley and W. Brown,* J. Mol. Biol., ***30:***383, *1967.)*

143 Å

429 Å

pattern to be the result of the partial overlap of two distinct types of filaments (Fig. 16-4). Both types of filaments are arranged parallel to the long axis of the myofibril and in register across its entire thickness. Similarly, the filaments of one myofibril are in register with all those of the laterally adjoining myofibrils across the entire fiber.

One type of muscle filament, the thin filament (approximately 50 to 60 Å diameter) was found to extend from the Z line on each sarcomere toward its center. The H band was found to represent that portion in the center of the sarcomere of a relaxed fiber which is not penetrated by the thin filaments (Fig. 16-4). The other type of filament, the thick filament (approximately 100 Å diameter), was found within the center of each sarcomere in a position equivalent to that of the A band. Cross sections through various regions of the sarcomere confirm this interpretation (Fig. 16-4); the I bands contain only thin filaments, the H band only thick filaments, and that part of the A band on either side of the H zone represents the region of overlap and contains both types of filaments. Cross sections through the region of overlap show that the thin filaments are organized in a hexagonal array around each thick filament while longitudinal sections show the presence of projections from the thick filaments at regularly spaced intervals. The projections represent cross bridges capable of forming attachments with neighboring thin filaments. The projections arise in pairs which extend from the thick filament in opposite directions, i.e., 180° from one another. The pairs of projections occur every 143 Å along the filament, with each successive pair being displaced 120° around the filament from the previous one (Fig. 16-5).

The Mechanism of Muscle Contraction

All skeletal muscles operate to bring about movement of an attached load by shortening; there is no other way in which they can perform work. The shortening of an entire muscle reflects the shortening of its component fibers, which in turn reflects the shortening of its component myofibrils, which in turn reflects the shortening of its component sarcomeres. The units of shortening are the sarcomeres, whose combined decrease in length accounts for the decrease in length of the entire muscle. The most important clue as to the mechanism underlying muscle contraction came from observations of the banding pattern at different stages in the contractile process. When muscle fibers were examined in the process of shortening, the A band was found to remain essentially constant in length while the width of the H band and I bands of each sarcomere decreased in proportion to the degree of shortening (Fig. 16-6). If the contraction were sufficiently strong, these latter two bands would disappear altogether. As the individual sarcomeres shortened and the I bands decreased in width, the Z lines came closer and closer toward the outer edge of the A band until contact was made.

Based upon these types of observations, two groups of investigators, namely Andrew Huxley and R. Niedergerke and Hugh Huxley and Jean Hanson, working on different types of contractile systems, proposed a far-reaching model to account for muscle contractility. They proposed that the shortening of individual sarcomeres did not result from the shortening of the filaments, but rather their sliding over one another. The sliding of the thin filaments toward the center of the sarcomere would result in the observed increase in overlap between the filaments and the decreased width of the I and H bands.

The sliding-filament model of muscle contraction has met with widespread approval and a large body of evidence has accumulated in its favor. The most important question raised by the model was the means by which the filaments were caused to slide across one another. The cross bridges that projected from the thick filaments were in a perfect position to serve in this capacity, and it was proposed that their attachment to the thin filaments could provide the vehicle whereby the hydrolysis of ATP could be utilized in the movement of the thin filaments. Several lines of evidence supported this contention. It was found, for example, that the tension a muscle fiber was capable of generating was related to the number of cross bridges that could form between the filaments. If a muscle fiber were stretched to the point that the thick and thin filaments no longer overlapped, no cross bridges could form and the fiber was incapable of

Figure 16-6. A schematic illustration of the successive steps in the shortening of a sarcomere according to the sliding-filament model. (a) Provides a concept of the entire sarcomere, (b) indicates the changes that would occur in the banding pattern as the I bands and H bands disappear. (From "The Mechanism of Muscular Contraction" by H. E. Huxley. Copyright © Dec. 1965 by Scientific American, Inc. All rights reserved.)

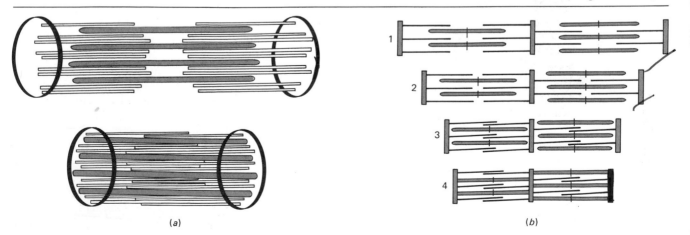

(a)   (b)

developing any tension whatsoever. If, however, the muscle fiber were stretched only to the point where a small amount of overlap remained, the fiber was still able to contract upon stimulation. The manner in which the cross bridges are able to bring about the relative movement of the filaments will be considered in detail below.

The explanation of muscle contraction in terms of the sliding-filament theory provides an elegant demonstration of the relationship between structure and function. It is the geometrical organization of the muscle filaments which facilitates muscle shortening. It can be demonstrated, in fact, that the soluble components, i.e., the invisible ones, of the muscle fiber are not even required for shortening under the appropriate conditions. As in the case of most other biological processes, contractility can also be demonstrated in a nonliving in vitro state. In the 1940s, one of the pioneers in the study of muscle biochemistry, Albert Szent-Györgyi, found that muscle fibers could be extracted with glycerol in such a way as to destroy the integrity of the plasma membrane and extract all of the soluble components, leaving the filamentous lattice of the sarcomeres essentially intact. When ATP was added to these glycerinated preparations (which were termed *models*) in the presence of the necessary ionic environment, these nonliving models would respond by shortening. All of the information for contraction is present in the visible structure.

The Biochemistry of Contractility

During the 1930s and 1940s a large effort was made toward understanding the basic biochemistry of the proteins involved in muscle contraction. Two proteins, actin and myosin, were found to be the primary proteins of muscle cells and each was shown to be present in a polymerized state. It was found that solutions which extracted myosin from the fibers resulted in the disappearance of the thick filaments. Subsequent extraction of actin from the fibers was accompanied by the disappearance of the thin filaments as well. Most importantly, actin and myosin were found to interact with one another in vitro to form an actomyosin complex, which could be prepared as an artificial fiber that would shorten in the presence of ATP. This latter finding demonstrated that the basis of muscle contractility resided in the interaction between these two proteins. Myosin was found to be an ATPase, and it was presumed that this enzymatic activity was responsible for releasing the free energy stored in ATP to be used in mechanical work.

As implied above, the thin filaments of the muscle cell are composed primarily of actin. In its monomeric form, actin is a globular protein of about 42,000 molecular weight and is termed *G-actin* (for globular actin). Monomers of G-actin polymerize into thin filaments, termed *F-actin* (for fibrous actin). The filaments, which measure about 60 Å, consist of two chains of actin monomers wound around each other in a double helix (Fig. 16-7a). As we will see later, an important feature of an actin filament is its polarity. Each of the monomers has an asymmetric structure and, since all of the monomers are lined up in the same direction, this polarity is superimposed on the entire filament. In the intact muscle cell,

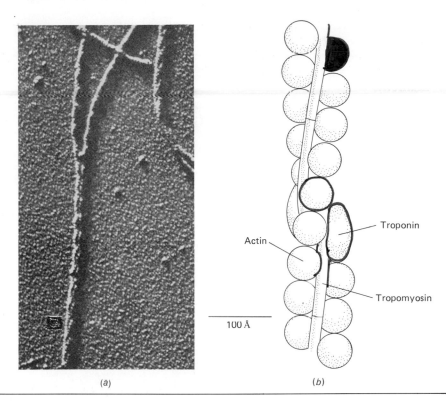

Actin

Troponin

Tropomyosin

100 Å

Figure 16-7. *(a)* Electron micrograph of a replica of fibrous actin showing the double helical nature of the filaments. *(From R. H. Depue and R. V. Rice,* J. Mol. Biol., **12:**302, 1965.) *(b)* Organization of a thin filament. *(After S. Ebashi, M. Endo, and I. Ohtsuki,* Quart. Rev. Biophys., **2:**164, 1969.)

(a)　　　　　*(b)*

the thin filaments contain at least two additional proteins, tropomyosin and troponin (Fig. 16-7*b*). Tropomyosin is an elongated molecule (approximately 400 Å in length) which fits securely into the groove between the two actin chains of the thin filament. Each rod-shaped tropomyosin molecule is associated with seven actin monomers linearly disposed along the F-actin chain (Fig. 16-7*b*). Tropomyosin molecules are situated end-to-end along the entire thin filament. Troponin is a globular protein composed of three subunits, each having an important and distinct role in the overall function of the molecule. Troponin molecules are spaced approximately 400 Å apart along the thin filament and situated in contact with both the actin and tropomyosin components of the filament. Tropomyosin (along with another protein α-actinin) is also present in the Z line where it is likely involved in holding the actin filaments of adjacent sarcomeres together.

The thick filaments are composed primarily of myosin, although at least two other proteins (the C protein and the M line protein) are present in small amounts. Myosin, like actin, is capable of assembling in vivo or in vitro into filaments, but the appearance of a myosin-containing filament is very different from one made of actin. In order to understand the structure of the myosin filament, we have to consider the myosin monomer from which it is made. Each myosin molecule is composed of six polypeptide chains organized in such a way as to produce a highly asymmetric protein (Fig. 16-8*a*). There is a very long tail portion with two heads

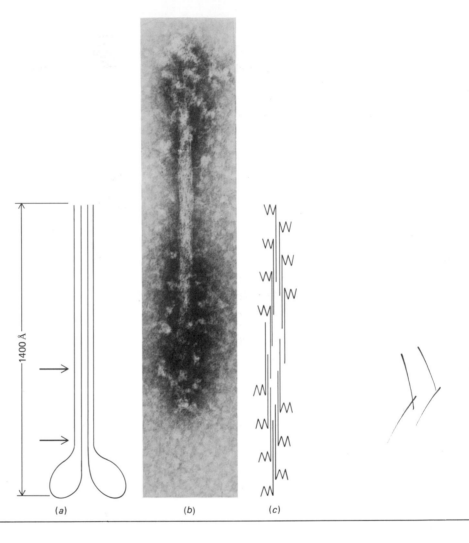

Figure 16-8. Structure of the myosin molecule. *(a)* The two major poly-peptide chains of a myosin molecule each consist of a rod-shaped portion and a globular head, the latter representing the cross bridge. The arrows indicate the relative sites along the rod which are sensitive to protease action. These two susceptible regions of the molecule represent non-helical sections of the polypeptide chain and are believed to provide the molecule with flexibility and act as a type of "hinge." *(From L. D. Peachey, in "Biocore XVIII," McGraw-Hill, 1974.) (b)* Examples of the formation of filaments in vitro by the aggregation of myosin molecules. The filaments formed are bipolar having the cross bridges located toward the ends and a cross bridge-free section in the filament's center. *(Courtesy of H. E. Huxley). (c)* Diagrammatic illustration of the type of myosin filament shown in *b. (From H. E. Huxley,* Science, **164:**1359, 1969. Copyright by the American Association for the Advancement of Science.)

1400 Å

(a)   *(b)*   *(c)*

projecting out from the molecule at one end. When myosin molecules aggregate to form filaments they do so in a very characteristic way. All of the monomers aggregate so that their tails point toward the center of the filament and their heads point away from the center (Fig. 16-8*b, c*). The thick filament is said to be *bipolar,* indicating there is a reversal of polarity at the filament's center, i.e., the center of the A band. Since the exact center of the filament is composed of the opposing tail regions of abutting molecules, this small region of the filament is devoid of heads. The remainder of the filament has heads projecting out sideways from the filament forming the cross bridges alluded to earlier.

Treatment of preparations of myosin with proteolytic enzymes has helped elucidate the role of various parts of the molecule. Brief treatment of myosin with trypsin splits the molecule into two fragments referred to as *heavy meromyosin (HMM)* and *light meromyosin (LMM).* The HMM fragment is about 500 Å long and the LMM fragment about 900 Å long.

The site at which the molecule is cleaved is shown in Fig. 16-8*a*. The HMM fragment contains the heads of the myosin molecule, i.e., that part which carries out the interaction with an actin molecule of an adjacent thin filament. If a preparation of HMM fragments is treated with another proteolytic enzyme, papain, the fragment is split into two parts. One of these parts, termed the S_1 fragment, represents the head of the protein and is found to contain all of the ATPase activity of the original myosin molecule.

The two sites on the rod-shaped portion of the myosin molecule which are sensitive to proteolytic cleavage are believed to have an important role in the functioning of the molecule. The rod-shaped portion of a myosin molecule consists almost exclusively of alpha helix, giving that section of the molecule its shape and making it inflexible. The two protease-sensitive sites are believed to represent places where the polypeptide chain is flexible and capable of providing a hinge for the lateral extension of the head of the molecule (Fig. 16-9). A comparison of the position of the S_1 heads at different stages in the contractile process confirms the expectation of a conformational change in these "hinge-regions" of the molecule. In the relaxed state, the heads do not make contact with the actin filaments. During contraction, the heads are extended laterally to make contact with the thin filament. Once the heads attach to the actin filament, a conformational change appears to occur in the myosin head so that the angle of the cross bridge relative to the actin molecule is changed (Fig. 16-10*a*, *b*). This tilting movement of the myosin head moves the

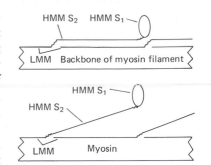

Figure 16-9. Postulated mechanism by which the flexible regions of the myosin molecule allow the S_1 head (cross bridge) to move toward the thin filament. Intermittent contacts between the myosin heads and the actin filaments provide the basis by which the sliding occurs. *(After H. E. Huxley, Science, **164**:1363, 1969. Copyright by the American Association for the Advancement of Science.)*

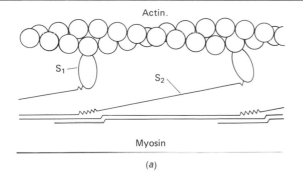

(a)

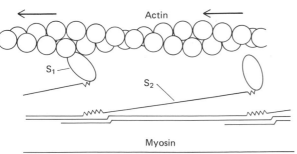

(b)

Figure 16-10. Postulated mechanism by which an active, i.e., energy-requiring, change in the angle of attachment of cross bridges to actin filaments could produce the relative sliding movement between filaments maintained at constant lateral separation. *(a)* Left-hand bridge has just attached; other bridge is already partly tilted. *(b)* Left-hand bridge has just come to the end of its working stroke; other bridge has just detached and will probably not be able to attach to this actin filament until further sliding brings helically arranged sites on actin into favorable orientation. *(From H. E. Huxley, Nature, **243**:446, 1973.)* *(c)* Diagram illustrating the reversal of polarity of the thin and thick filaments on each side of the center line in each sarcomere. *(From H. E. Huxley, in R. Goldman, T. Pollard, and J. Rosenbaum, eds, "Cell Motility," Cold Spring Harbor, 1976.)*

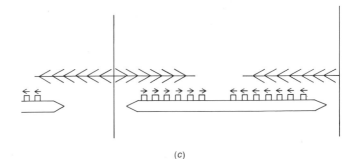

(c)

actin filament a perceptible distance. Before continuing further with the details of the actin-myosin interaction, it is important to reconsider the polarity of the two filaments as it relates to the shortening of the length of the sarcomere.

The shortening of each sarcomere results from the movement of the thin filaments toward its center (Fig. 16-10c). In order for this to occur it is essential that the polarity of both types of filaments be reversed on opposite sides of the center line. The reversal of polarity of the thick filament is built directly into its bipolar nature. Since actin filaments are polarized in one direction, the actin filaments of each sarcomere are directed toward the center from the two Z lines. In other words, actin filaments on either side of a Z line point away from each other toward the center of their respective sarcomeres.

During a typical contraction, a muscle fiber will shorten to about 65% of its relaxed length. Since each sarcomere is approximately 2.5 μm, the thin filaments would be expected to slide in the order of a micron during the contraction. Since a single conformational change in a myosin head could only be expected to move an actin filament approximately 50 to 100 Å, it becomes apparent that a single contraction must require a repetitive series of interactions between the cross bridges and the thin filaments. Each cycle would involve the attachment of the cross bridge, its tilting motion to advance the thin filament, the detachment of the cross bridge and its return to its original position where it can reattach to the actin filament closer to its terminus at the Z line. In some respects the situation is analogous to the pulling of a rope by a hand-over-hand type of operation since the entire distance moved is accomplished by a series of smaller displacements. It has been estimated that a given cross bridge is capable of undergoing 50 to 100 cycles of movement per second.

The above account of the mechanism by which muscle filaments are caused to slide across one another raises a number of basic questions, foremost among these being the manner in which the chemical energy stored in ATP is converted to mechanical energy involved in filament movement and the means by which the process is controlled. Although the precise manner in which energy transduction in muscle occurs remains unknown, a considerable amount of information is available about the process. It is generally agreed that the actual movement of the thin filaments toward the center of the sarcomere is accomplished by the conformational change in the myosin head or cross bridge previously described. Similarly, it is generally believed that the energy required to drive the conformational change results from the hydrolysis of ATP by the ATPase activity present within the cross bridge itself. Being both a structural protein and an enzyme, myosin is perfectly suited for the job at hand. The rod-shaped section of the molecule allows it to form filaments which are required to maintain the cross bridges in a rigid spatial position at the proper intervals. The myosin heads allow the molecule to tap the energy stored in ATP as well as to physically move the actin filaments. The most unresolved questions concern the manner in which ATP hydrolysis brings about filament movement.

Early studies on isolated myosin filaments indicated they possessed detectable ATPase activity, and this activity was greatly stimulated by the addition of actin under conditions which promoted the formation of actomyosin complexes. Further studies suggested that the actin was not stimulating the *hydrolysis* of ATP molecules, but rather promoting the *release* of the bound products, ADP and P_i. In other words, it would appear that myosin by itself will readily hydrolyze ATP, but the products remain bound to its active site until the interaction with actin stimulates their release. If we attempt to relate the evidence from in vitro studies of the properties of the myosin ATPase system to events occurring within the cell, the following picture emerges. In the repeating cycle of attachment and detachment of the cross bridges with the actin filaments, the detached stage would be represented by having the cross bridge hydrolyze the ATP without the release of the bound products. At this point, the cross bridge is thought to be in an energized state, one analogous to a stretched spring capable of spontaneous movement. This energized myosin, which can be written $M^\circ \cdot ADP \cdot P_i$, is believed to attach to the actin molecule and use its stored free energy to change its conformation and shift the actin filament in the appropriate direction. This step would represent the power stroke. The deenergized myosin would then be able to release the bound products of ATP hydrolysis which would then be replaced by new ATP molecules. The attachment of ATP to the myosin head results in the dissociation of the cross bridge from the actin filament facilitating its return to the initial position so that a new cycle can begin.

The role of ATP in separating the myosin head from the actin filament is best demonstrated using nonliving glycerinated muscle fibers. As stated above, these glycerol-extracted muscle fibers retain the capacity to shorten in the presence of ATP. If, however, the ATP is suddenly removed from solution, the muscle fiber becomes rigid and inflexible. In this state, which is termed *rigor*, all of the myosin heads in a position to form cross bridges with an actin filament become locked in the bridged condition. If ATP is added back to the preparation, the cross bridges can detach from the thin filaments and continue in their cyclic activity leading to muscle shortening. The inability of myosin cross bridges to detach in the absence of ATP is the basis for the condition of *rigor mortis* which ensues following death.

The Regulation of Muscle Contraction

If one considers the precision with which muscular activity occurs within an animal, it is apparent that a very finely tuned control system must exist to ensure that a particular muscle cell is stimulated to contract at a particular time. Since vertebrate muscular activity is under the control of the nervous system, a discussion of the regulation of contractility has to consider various steps (Fig. 16-11) leading from the activation of the muscle by a nerve to the attachment of a particular myosin cross bridge to its adjacent thin filament. These steps will be described below without consideration of the experimental evidence on which they are based.

Muscle fibers are grouped into units termed *motor units* which are jointly innervated by single motor neurons. As the neuron approaches the muscle, it sends out branches such that each muscle fiber becomes associated with a single branch. All of those fibers constituting a motor unit undergo contraction simultaneously upon stimulation by an impulse transmitted along that neuron. The point of contact of a terminus of an

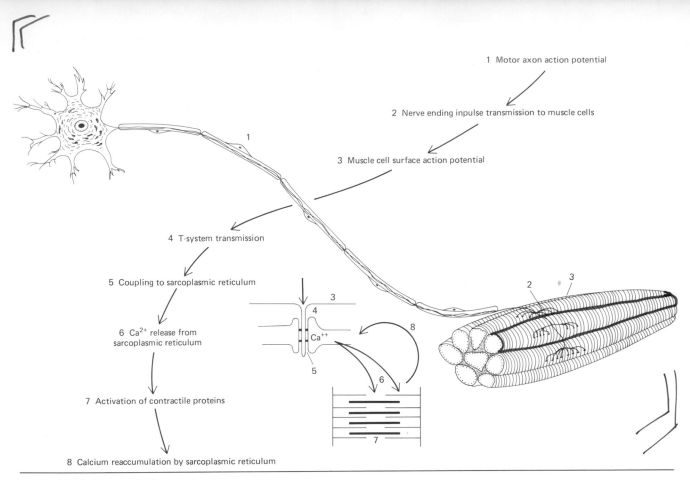

1 Motor axon action potential

2 Nerve ending inpulse transmission to muscle cells

3 Muscle cell surface action potential

4 T-system transmission

5 Coupling to sarcoplasmic reticulum

6 Ca^{2+} release from sarcoplasmic reticulum

7 Activation of contractile proteins

8 Calcium reaccumulation by sarcoplasmic reticulum

Figure 16-11. Summary of the events in the contraction of a skeletal muscle motor unit. *(From L. D. Peachey, in "Biocore XVIII," McGraw-Hill, 1974.)*

axon with a muscle fiber is termed a *neuromuscular junction.* The neuromuscular junction is a site of transmission of the nerve impulse from the axon across a synaptic cleft to the muscle fiber whose membrane is also excitable and capable of conducting an action potential (a nerve impulse).

The next step in the sequence of events leading to contraction concerns the means by which the electrical activity of the fiber's surface is transmitted to the thousands of myofibrils which make up the body of the fiber. Electron microscopic and physiological studies have revealed the presence of folds of the plasma membrane which penetrate into the depths of the muscle fibers. These folds, which constitute what is termed the *T system* (Fig. 16-12), are directed transversely into the fiber and are capable of transmitting the action potential from the fiber's surface to the myofibrils. Each myofibril is surrounded by a membranous system termed the *sarcoplasmic reticulum* (SR) which is derived from endoplasmic reticulum of the embryonic myoblast. The sarcoplasmic reticulum consists of tubules which run parallel to the long axis of the myofibril. In the frog, the tubules of the SR come together at each edge of the sarcomere and fuse to form a terminal *cisterna* which runs parallel to the adjacent Z line (Fig. 16-12). The two terminal cisternae from adjoining sarcomeres are associated at the Z line with one of the transverse tubules of the T system. These sites of association of the sarcoplasmic reticulum with the T tubules

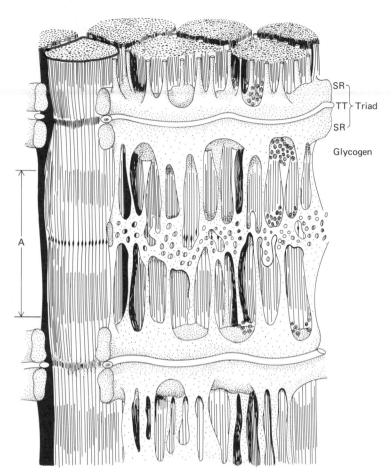

SR
TT $\}$ Triad
SR

Glycogen

A

Figure 16-12. Diagrammatic representation of the membrane surrounding a myofibril in a frog. The sarcoplasmic reticulum is present through most of the sarcomere as tubules which fuse near the Z line to form a membranous complex with the transverse tubules. Together these membranes form a conducting system for the impulse of the neuron to be transmitted to the sarcoplasmic reticulum surrounding the myofibril. The release of calcium by the SR in response to stimulation leads to the contractile event. *(From L. D. Peachey, J. Cell Biol., **25**:222, 1965.)*

are termed *triads* and provide the means by which the stimulation from the sarcolemma (plasma membrane) is transmitted to the myofibril.

In order to understand the significance of electrical events occurring at the triads, we need to consider the major activity of the SR itself, that of concentrating calcium ions. The most important protein of the SR membranes is a calcium-transporting ATPase. This integral membrane protein has the responsibility of actively transporting calcium ions from the cytoplasm outside of the SR to the internal space within the SR itself. The capability of the SR transport system is best appreciated in preparations of isolated SR vesicles prepared from broken muscle cells. When these closed, membranous vesicles are incubated in the presence of Ca^{2+}, there is a dramatic uptake of the ions from the medium into the vesicles. The SR membranes are capable of reducing the external calcium levels to the micromolar ($10^{-6}\ M$) range, several orders of magnitude below that within the vesicles. In the muscle fiber at least two calcium-binding proteins are present on the inner side of the SR membranes which are capable of removing the ions from the free state. One of the proteins, calsequestrin, is capable of binding 43 calcium ions per protein molecule.

The importance of calcium in muscle contraction was first shown by L. V. Heilbrunn and Floyd Wiercinsky in 1947 when they injected calcium into a muscle fiber and found it caused the fiber to contract. Along with Mg^{2+} and ATP, calcium is one of the small molecular weight components required for contraction and is *the* component involved in its regulation. In the relaxed state, the Ca^{2+} levels within the cytoplasm (sarcoplasm) are very low (approximately 10^{-7} M), below the threshold concentration required for contraction. The calcium content of the cytoplasm is maintained at this very low level by the activity of the active transport system of the SR membranes. Since ATP is needed for Ca^{2+} uptake by the SR, it is apparent that ATP hydrolysis occurs during muscle relaxation as well as during muscle contraction. With the arrival of electrical activity at the triad, the permeability of the SR is somehow altered so that calcium ions are free to diffuse out of the SR space and over the short distance to the myofibrils. The Ca^{2+} levels in the sarcoplasm are estimated to rise from about 0.3 μM to about 50 μM at this time. As will be described in the next section, calcium plays the role of a second messenger (page 245) in carrying information for contraction to the myofibrils.

In order to understand the nature of the repression to which muscle fibers are subjected in the relaxed state, we need to consider, once again, the protein makeup of the thin filaments. The tropomyosin molecules, which are located in the groove of the actin double helix (see Fig. 16-7b), are believed to be situated such that they block the access of the actin molecules to the myosin cross bridges in the relaxed state. The position of the tropomyosin within the groove appears to be under the control of the attached troponin molecule. An important property of one of the subunits of troponin is its ability to bind Ca^{2+}, an event which causes a conformational change in another subunit of the troponin molecule which, in turn, is transmitted to the adjacent tropomyosin. When the Ca^{2+} levels of the sarcoplasm rise, the calcium-troponin association is favored, and the tropomyosin molecules are caused to move approximately 15 Å closer toward the center of the filament's groove. This shift in the position of the tropomyosin appears to open the myosin-binding site of the adjacent actin molecules in preparation for the sliding event. Each troponin molecule controls the position of one tropomyosin molecule which, in turn, controls the binding capacity of seven actin monomers linked together in the thin filament.

Once the electrical stimulation from the innervating nerve fiber stops, the transport system of the SR once again lowers the sarcoplasmic Ca^{2+} level below the values needed for its association with troponin, and the system moves back to the repressed condition typical of the resting fiber. The process of relaxation can be thought of as a battle for calcium between the transport protein and troponin; the transport protein has a greater affinity for the ion so it preferentially removes it from the sarcoplasm, leaving the troponin in the unbound state. Although the troponin-tropomyosin system is found throughout the animal kingdom, a number of invertebrate muscles have been shown to be regulated by another system, one which acts (via Ca^{2+}) on the thick filaments rather than the thin fila-

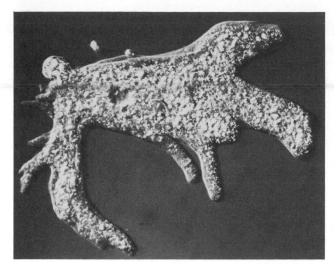

(a)

Figure 16-13. (a) Photomicrograph of a living amoeba moving along a substrate. (Courtesy of the Carolina Biological Supply Co.) (b) Schematic diagram of postulated processes in amoeboid locomotion. In the theory suggested by this diagram, the force for locomotion occurs during a frontal contraction which pulls on a central column of stabilized cytoplasm (SS). As the column of cytoplasm reaches the front end, it is converted to a contracted state (CS) and becomes part of the plasmagel (PGS) which is moved posteriorly as gelated ectoplasm (GE) until, at the rear, it is converted to a relaxed state (RS). The velocity (V) profiles show the patterns of plasmalemma (PL) displacement and cytoplasmic streaming in the front, middle, and rear regions. (From D. L. Taylor, J. S. Condeelis, P. L. Moore, and R. D. Allen, J. Cell Biol., **59**:391, 1973.)

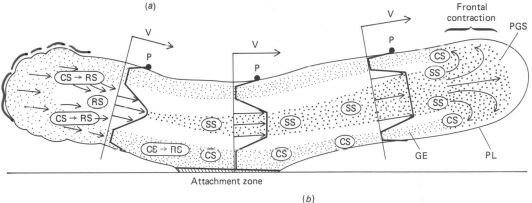

(b)

ments as described above. The mechanism by which the thick filament control is exerted is not yet understood.

NONMUSCLE MOTILITY

Amoeboid Movement

Anyone who has watched an amoeba crawl along the surface of a microscope slide can appreciate the ability of cytoplasm, at least under certain circumstances, to move within a cell. As the organism moves, extensions of the surface, termed _pseudopods_, are seen to be formed by the internal flowing cytoplasm (Fig. 16-13). As the cytoplasm streams into the advancing pseudopods, the amoeba slowly moves in one direction or another. This type of flowing cytoplasmic movement, generally referred to as _amoeboid movement_, is not restricted to single-celled organisms, but similar types of movement are observed in single cells throughout the animal kingdom. White blood cells, for example, crawl through our own bodies by a process which, though quite different from the movement of an amoeba, involves cytoplasmic flow.

Examination of a moving amoeba under the light microscope suggests to the observer that there are two types of cytoplasmic regions in the cell. The outer crust (termed *ectoplasm*) of cytoplasm, which is particularly evident at the sides of the pseudopod, is relatively devoid of granules or vacuoles. In contrast, these components are rather densely distributed through the bulk of the cytoplasm (termed *endoplasm*). A number of early observations suggested that the ectoplasm was more gelated, i.e., had a greater consistency or viscosity, than the endoplasm. The *gel*like nature of the ectoplasm was such that it resisted the penetration of cytoplasmic particles which therefore became restricted to the more *sol*like, i.e., more fluid, endoplasm. Observations of the movement of giant amoebae suggested that a cyclic activity was occurring. Endoplasm was flowing forward into advancing pseudopods toward the tip of the extension, then moving laterally where it became converted to stationary ectoplasm near the amoeba's surface. The picture is analogous to the movement of water emerging from a fountain; it moves upward until it reaches its upper height and then flows outward in all directions before it falls back to the ground. In the amoeba, the endoplasm-to-ectoplasm conversion at the tips of the pseudopods is matched by the reverse transformation in the rear of the cell (a region termed the *uroid*). The endoplasm formed in the uroid then moves forward in a stream to round out the cycle.

It has long been felt that the key to the mechanism underlying ameboid movement resided in the sol-gel transformations of the cytoplasm. It was found, for example, that if the gelated cytoplasm were caused to lose its viscosity, a change which can be brought about by subjecting an amoeba to cold temperature or high hydrostatic pressure, its movement would stop and the cell would round up. Experiments of this type gave support to the concept that the more viscous ectoplasm was involved in generating some type of motile force. However, a great deal of disagreement has existed as to the nature and location of the force-generating reaction. Two broad theories have been formulated which have divided the support of investigators in this field. These two theories are characterized by diametrically opposed suggestions, although a large body of evidence has been put forward in defense of one or the other.

In the earlier proposal, that of the *ectoplasmic tube contraction theory* formulated by S. O. Mast in the 1920s, the contractile force is generated in the rear of the amoeba by the ectoplasm against the endoplasm, which is pushed forward. The theory suggests that amoeboid movement results from hydraulic pressures, the rear of the cell becomes a site of increased pressure pushing the endoplasm forward toward a region of lesser pressure. In the other major proposal, the *front zone contraction theory* of Robert Allen, shown in Fig. 16-13b, the force for movement is generated by protoplasmic contraction occurring near the tip of the advancing pseudopods and the remainder of the amoeba is pulled forward by the contraction. In this theory it is assumed that the endoplasm has sufficient consistency (tensile strength) so that a pull (tension) exerted on its forward edge can be transmitted through it and drag the remainder of the endoplasm forward.

Once the analysis of the molecules involved in muscle contraction had begun, a few investigators began to apply some of the same techniques to questions of nonmuscle motility. In 1952, Ariel Loewy reported on a study of motility in a giant multinucleated slime mold (a plasmodial stage) which moves about by a process of cytoplasmic streaming. A crude extract from these organisms was found to increase in viscosity upon the hydrolysis of added ATP. This is the same response that had been obtained earlier with a similar extract from muscle cells and Loewy proposed that actomyosinlike substances were also present in the cytoplasm of this organism. A short time later, Hartmut Hoffman-Berling made glycerinated preparations (page 738) of amoebae and mammalian fibroblasts and found these glycerinated cells (termed *models*) would actively contract upon the addition of Mg^{2+} and ATP; another observation paralleling the findings in muscle cells. A major stumbling block in the analysis of contractility in nonmuscle cells was the absence of an elaborate assembly of filaments as was so readily seen in the examination of a muscle cell. Although some electron microscopic evidence became available suggesting that the gelated regions of amoebae and slime mold plasmodia represented regions containing large numbers of threadlike structures, the cytological side of the story received relatively little attention until recent years. We will return to the ultrastructural appearance of contractile cytoplasm later in the discussion after first considering the nature of the molecules involved.

Vertebrate Nonmuscle Cells

Since the early observations implicating the proteins actin and myosin in nonmuscle movement, a large number of studies have proven the existence of these molecules in nonmuscle cells. Actin* generally accounts for about 5 to 10% of a nonmuscle cell's protein, a value which by itself suggests a fundamental widespread involvement in the cell's activities. Actins have been found to be remarkably similar in amino acid sequence among widely divergent organisms as well as between muscle and nonmuscle cells in the same organism. Actin has been known for a long time to be relatively nonimmunogenic. In other words, actin extracted from one species, such as a frog, and injected into another, such as a rabbit, generally does not stimulate the production of antibodies against the protein. The basis for this relative lack of antigenicity is the similarity between the frog actin and the rabbit's own actin, so that the rabbit does not recognize the injected protein as being foreign. Actins from protozoa to mammals differ by only a few relatively unimportant amino acid residues. All actins are capable of polymerizing to form the F-actin double-stranded helical filament, and all of them are capable of binding to vertebrate muscle myosin and activating its ATPase activity.

The best criterion as to whether or not a particular type of cytoplasmic filament is composed of actin is to take advantage of the fact that actin

---

* In the following discussion, the structure and function of cytoplasmic actin is considered. It should be noted that actin has also been found within cell nuclei where its role is obscure. It should also be noted that cytoplasmic actin, of which there appears to be more than one species, differs somewhat in amino acid sequence from muscle actin.

filaments, regardless of their source, are capable of interacting with myosin in a highly specific manner. In order to carry out the reaction, heavy meromyosin fragments (page 740) are prepared from muscle myosin and the HMM is then incubated with a tissue section or a glycerinated (thus permeable) preparation of the cells in question. If actin filaments are present in the cell, the HMM will attach to them to form a highly characteristic complex which is identifiable in the electron microscope (Fig. 16-14). Since both actin and the HMM fragment are asymmetrical molecules, the actin-HMM complex has a definite polarity which reveals the underlying polarity of the actin filament. Actin localization can also be made with the light microscope using fluorescently labeled HMM molecules.

Nonmuscle myosins are more diverse in their structure though nearly all of them share certain basic characteristics. All of the myosins that have been studied (except one from the protozoan *Acanthamoeba*) have a rod-shaped portion and a globular head which has ATPase activity. These various myosins are capable of aggregating to form bipolar filaments as in the case of muscle myosin, although the filaments are shorter and thinner. Most importantly, all of the myosins are capable of interacting with actin filaments. While actin is present in large amounts in nonmuscle cells, myosin is generally found to consist of less than 0.5% of the cell's protein. The reason for actin being present at about one hundred fold excess over myosin is not understood. It has been suggested that a large percentage of the actin is involved in activities that do not require myosin, i.e., noncontractile activities, and that only a few myosin molecules are needed to generate the small forces involved in nonmuscle contractility. Regardless of the reason, the thicker myosin filaments are generally not observed in the electron microscope, even though actin filaments may be abundant. These points will be considered further in the following discussion.

The finding that actin and myosin are present in nonmuscle cells does not prove that they are involved in cell motility. In fact, this point has still not been settled since the evidence is largely indirect, consisting primarily of correlations between movements and the presence of these

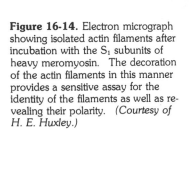

Figure 16-14. Electron micrograph showing isolated actin filaments after incubation with the S_1 subunits of heavy meromyosin. The decoration of the actin filaments in this manner provides a sensitive assay for the identity of the filaments as well as revealing their polarity. *(Courtesy of H. E. Huxley.)*

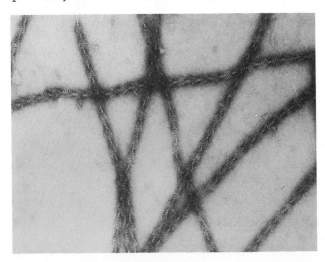

proteins. One of the strongest correlations is the location of actin and myosin at the sites of expected contractile activity. This is best demonstrated in studies on cultured cells examined under different conditions.

When tissue-cultured cells are lifted from the surface of the culture dish, they are swept into suspension as rounded cells. When these rounded cells are examined under the electron microscope, thin (approximately 60 Å) filaments are seen beneath the surface of the cells, but they are not organized in any obvious manner. These filaments, which can be shown to be composed of actin on the basis of their myosin-binding activity, form a loose randomly interconnected meshwork. If these cells are allowed to settle back on the bottom of the dish, they start to flatten and spread themselves out in order to reattach to the substrate. If cells are fixed and examined in the electron microscope about 30 minutes after attachment, a very different pattern of thin filaments is observed. Rather than being randomly oriented, they are present in bundles which are termed *stress fibers* (Fig. 16-15). The stress fibers, which are large enough to be visible in the light microscope, are oriented parallel to the membrane and are found localized in the region where the cell is in contact with the substrate. The function of the stress fibers is not known for certain; they may be involved in cell shape changes, and/or adhesion, and/or migration. Regardless, these fibers illustrate the dramatic change in organization to which microfilaments are subject. When these cells enter mitosis, they round up (Fig. 16-21) and loosen their attachment to the substrate. When mitotic cells are examined, the stress fibers have disappeared and the actin filaments appear within the mitotic spindle and then in the contractile ring during cytokinesis. The role of the filaments in mitosis is controversial (page 709), but their role in cytokinesis seems firmly established (page 711).

If actin filaments are involved in contractile activities in these various cellular locations, then we would expect that myosin should also be localized in the same places. Using fluorescent antibodies against myosin, various investigators have recently shown that myosin is concentrated in each of the locations in which actin is found (Fig. 16-16): in the cortical network of the suspended cells, the stress fibers of the attached cells, and the mitotic spindle and contractile ring of the mitotic cells. Even though the fluorescent antibodies (which are highly specific and sensitive) indicate the presence of myosin in these locations, thick myosin filaments have not been demonstrated. Whether this lack of electron microscopic visibility reflects their rarity among all of the thin filaments, their destruction during fixation, or their true absence in the cell, remains uncertain. The localization of actin and myosin together in these various locations argues for their having a contractile role; there is no other function that actomyosin is known to have. Once again, the explanation for the great excess of actin is not obvious. It has been suggested by Thomas Pollard that most of the actin filaments in the cell may be involved in some type of structural role, i.e., as a cytoskeleton.

Although actin and myosin of muscle and nonmuscle cells are very similar, the types of assemblies formed are quite different. Whereas muscle cells possess a highly stable, precisely organized system of filaments,

Figure 16-15. *(a)* Electron micrograph of a portion of a cultured cell flattened against the substrate. Bundles of microfilaments (mfb), termed stress fibers, are seen to terminate near the plasma membrane. Regions that are flattened and smooth are indicated by the arrows. Some of these regions are associated with the end of microfilament bundles, whereas others do not seem to be clearly associated with them. These areas may represent attachment points of the cell to the substrate. Openings of pinocytotic vesicles (pin) are indicated. *(From J. P. Revel and K. Wolken, Exp. Cell Res., **78**:8, 1973.)* *(b)* Electron micrograph of a thin section through a spread cultured cell taken just below the plasma membrane at the level of cell-substrate contact and through the microfilament bundles. *(From R. D. Goldman, J. A. Schloss, and J. M. Starger, in R. Goldman, T. Pollard, and J. Rosenbaum, eds., "Cell Motility," Cold Spring Harbor, 1976.)* *(c)* The distribution of actin filaments in a human skin fibroblast as revealed by localization of fluorescent antiactin antibodies. *(From E. Lazarides, J. Cell Biol., **65**:553, 1975.)*

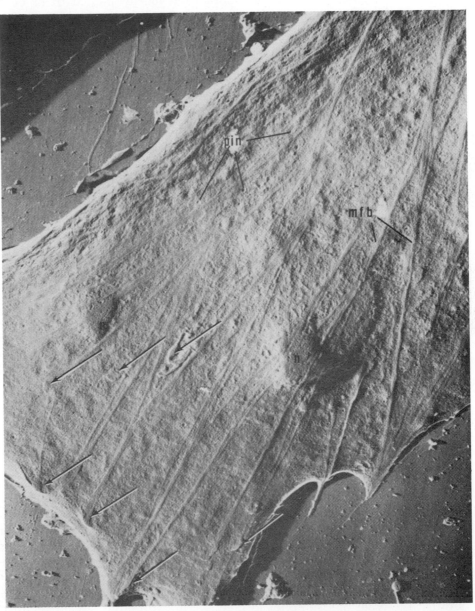

(a)

the filaments of nonmuscle cells are transient, i.e., they appear and disappear, and are much less precisely organized. In addition, their organizational state can vary from one situation to the next as exemplified by the switch from the ill-defined loose network to the bundles of stress fibers. The appearance and disappearance and change in organizational state of cytoplasmic microfilaments indicates that some type of control system must exist to coordinate the necessary events. An important difference between muscle and nonmuscle contractile systems is the presence of unpolymerized actin in the nonmuscle cells. Extractions of nonmuscle cells typically indicate that 50% or more of the actin is in the monomeric G form. It would seem that the cell maintains a store (or pool) of actin monomers which it can utilize when needed. Very little is known about the manner in which microfilament assembly is controlled; however, some evidence suggests that the plasma membrane may be involved. Although the means by which nonmuscle contractility is regulated remains a mysterious topic,

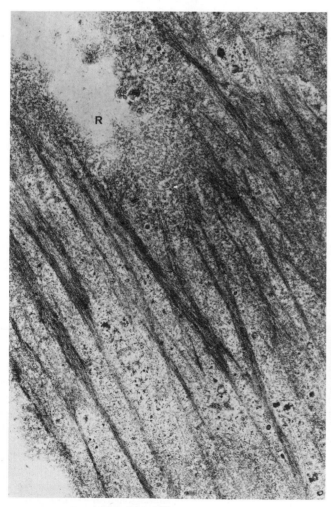

(b)

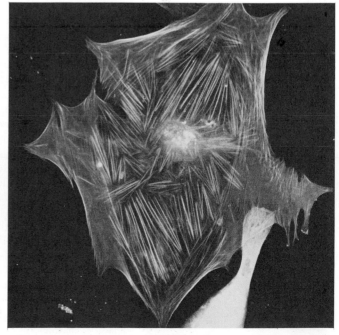

(c)

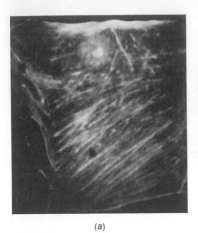

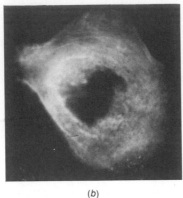

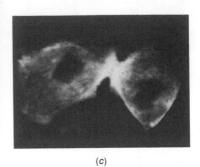

(a)   (b)   (c)

Figure 16-16. The localization of myosin by fluorescent antibodies in different regions of the cell at different stages of the cell cycle, and its correlation with the presence of actin. Myosin localization in the stress fibers of the interphase cell (a), in the mitotic spindle of the metaphase cell (b), and in the cleavage furrow of a cell undergoing cytokinesis (c). (From J. W. Sanger and J. M. Sanger, in R. Goldman, T. Pollard, and J. Rosenbaum, eds., "Cell Motility," Cold Spring Harbor, 1976.)

considerable evidence has accumulated implicating calcium ions in the process. The involvement of calcium is best illustrated by considering experiments on in vitro systems.

In Vitro Systems

As in the case of many other processes in cell biology, the development of in vitro reconstituted systems has been of particular importance in the analysis of nonmuscle motility. In 1960, Robert Allen and his colleagues found that streaming could still continue in cytoplasm that was no longer enclosed within a membrane. In this case, giant amoebae were trapped inside quartz capillary tubes, the tubes were then sealed with inert paraffin oil to keep the cytoplasm inside the tube, and the amoebae were broken open to release their contents. Observations of the naked cytoplasm within the tube showed that streaming could continue for upwards of an hour. In more recent experiments of a related nature, investigators have mechanically ruptured the membrane of single amoebae after placing the organisms into different types of physiological solutions. When the amoebae are broken open in a medium lacking Ca^{2+} and ATP, the cytoplasm goes into rigor, i.e., a stabilized inflexible state. When the amoebae are ruptured in a medium containing ATP but lacking calcium, the cytoplasm remains in a relaxed state. If the Ca^{2+} concentration exceeds about $7 \times 10^{-7} M$, the cytoplasm is capable of contraction and orderly streaming (Fig. 16-17). These are essentially the same results one obtains for glycerinated muscle preparations.

Numerous other types of in vitro studies have also suggested that Ca^{2+} is a key component in the nonmuscle contractile process and must be present at approximately $10^{-6} M$ in order for contraction to occur. As previously described, calcium is required in vertebrate skeletal muscle contraction for its ability to relieve the repression exerted by the troponin-tropomyosin system. There is no compelling reason to believe this type of repression is widespread in nonmuscle cells, so the calcium is likely exerting its effect in some other way. As in the case of muscle cells, it is believed that the free calcium level in nonmuscle cytoplasm is regulated by membranes. Membranous vesicles

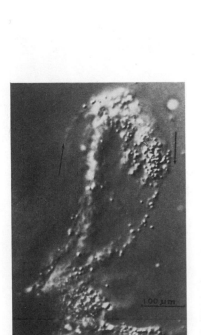

Figure 16-17. When placed in the proper solution (discussed in the text), isolated cytoplasm is capable of continuing its streaming activity as shown in this photomicrograph. (From D. L. Taylor, J. S. Condeelis, P. L. Moore, and R. D. Allen, J. Cell Biol., **59**:387, 1973.)

can be prepared from a variety of cells that will actively concentrate calcium ions much like the sarcoplasmic reticulum of muscle cells. In addition to these cytoplasmic vesicles, the mitochondria, which are also accumulators of free Ca^{2+} (page 374), may be involved in the regulation of contractility.

The presence of some type of calcium concentration mechanism can be demonstrated in vitro by injecting small amounts of calcium into large cells and showing that it remains in the same area of the cell in which it was injected rather than diffusing throughout its volume as would be expected if it were free to move about. The presence of free Ca^{2+} in this last experiment (or numerous others) can be revealed by the use of an ingenious new technique. It has been found that a protein prepared from jelly fish, termed *aequorin,* is capable of binding free Ca^{2+} and undergoing luminescence in the calcium-complexed state. If aequorin is injected into a cell, its level of luminescence indicates the level of free calcium. In the experiment just described, the lack of diffusion of the injected calcium is revealed by the lack of widespread luminesence. The use of aequorin has also provided some evidence on the importance of Ca^{2+} in cell motility in vivo. If aequorin is injected into moving amoebae, luminiscence is found to parallel cellular motility, suggesting that motility is accompanied by an increase in the local cytoplasmic calcium concentration.

In the last few years a different type of in vitro study has led to a new insight into the molecular basis of cytoplasmic viscosity and contractility. It has been found that, when isolated cytoplasm from a wide variety of cells (sea urchin eggs, amoebae, macrophages, HeLa cells, etc.) is slowly warmed to room temperature in the presence of Mg^{2+} and ATP (without Ca^{2+}), it becomes markedly gelated. The gel is of sufficient viscosity that, if formed in a small test tube (Fig. 16-18a), the tube can be turned upside down without the con-

Figure 16-18. (a) Test tubes containing an extract from macrophage cells that had been warmed from 0°C to 25°C in the presence of ATP. The extract on the right contains cytochalasin B which inhibits gelation. (From J. H. Hartwig and T. P. Stossel, J. Cell Biol., **71:**296, 1976.) (b) Nomarski interference micrograph of a protoplasmic gel formed in vitro. (From R. E. Kane, J. Cell Biol., **71:**709, 1976.)

(a)   (b)

tents flowing out. Analysis of the gelated extract in the electron microscope indicates it is made up of a huge network of interconnected (cross-linked) thin filaments (Fig. 16-18*b*) shown to consist of actin. Protein analysis of the gelated cytoplasm generally indicates that the actin is not involved by itself in the formation of the gel, but is present in association with one or more high molecular weight proteins. These other types of proteins have been generally termed *actin-binding proteins* and they are believed to be an important component in cross-linking the actin filaments together. Addition of Ca^{2+} to these actin-containing gels leads to their contraction, a response which illustrates the inherent contractility of nonmuscle cytoplasm.

The types of experiments described above have led to the proposal that the cytoplasm of most cells may exist in the filamentous state in which actin is the major structural component. The existence of structured cytoplasm, or "ground plasm," would account for the large amounts of actin found in nonmuscle cells. According to this line of reasoning the bulk of the actin in the cell would be more involved in forming this structural cytoskeleton than directly in motility. If this theory does prove to be substantiated, then actin filaments will join microtubules as the major structural elements within cells. This does not mean that the actin filaments would not be involved in contractility, but rather they would have two related but distinct roles. The contractility of the actin filaments would involve at least two other components, calcium ions and myosin. Where these two were present locally in the cell, the thin filaments could be contractile in function.

One of the major reasons for working with in vitro extracts was the difficulty in observing any visible structure in the cytoplasm of most cells. Recent analysis of whole cells prepared by critical point drying (page 66) and examined with the high-voltage electron microscope (page 67) has provided direct morphological evidence supporting the concept of a highly structured matrix within the cytoplasm. In these studies, carried out by Keith Porter and his colleagues, the cytoplasm is seen to be filled with a filamentous branched network, the *microtrabecular network* (Fig. 16-19). It is Porter's belief that this network or lattice serves to interconnect all of the cytoplasmic organelles, with the exception of the mitochondria, which do not appear to be contained within it. Furthermore, it appears that the lattice exists in a dynamic state and is capable of undergoing movements generated by actomyosin components of the structure, as well as being subject to disassembly and reassembly so as to give the cell control over its very fabric. The dynamic nature of the microtrabecular network has been particularly well revealed in studies on the movement of pigment granules within pigment cells (see Fig. 15-6). In at least one case (H. R. Byers and K. R. Porter, J. Cell Biol. **75**, 541, 1977), the assembly, disassembly, and contractility of these cytoplasmic filaments (in conjunction with microtubules) appears responsible for the aggregation and dispersion of the pigment granules.

Locomotion by Lamellipodia and Filopodia

The locomotion of single cells of higher organisms is not generally accomplished by obvious cytoplasmic flow. A quick look at a mammalian fibroblast moving along the surface of a culture dish will serve to ex-

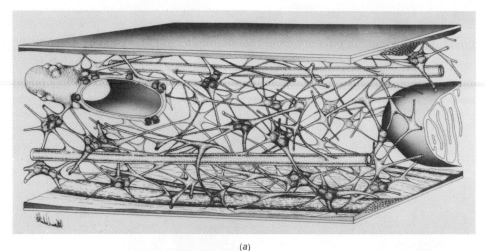

(a)

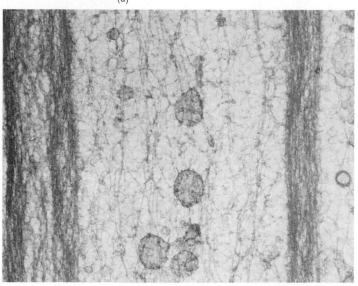

(b)

Figure 16-19. *(a)* A model of the cytoplasmic ground substance showing lattice and contained microtubules. The actin filament bundles are depicted here as part of the cytoplasmic cortex, i.e., the region just below the plasma membrane. *(b)* High-voltage electron micrograph of a thin portion of a human fibroblast. Two bundles of actin filaments run vertically in the image and represent stress fibers. Vesicles of the ER as well as a few microtubules populate the space between them. The rest of the image comprises a lattice of microtrabeculae which attach to the surfaces of all the other elements in the image. *(From K. R. Porter, in R. Goldman, T. Pollard, and J. Rosenbaum, eds., "Cell Motility," Cold Spring Harbor, 1976.)*

emplify a different type of locomotion. As it moves, the fibroblast is greatly elongated (parallel to the direction of movement) and flattened out close to the substrate. Its movement is erratic and jerky, sometimes advancing and other times withdrawing. The key to the fibroblast's ability to locomote is seen in an examination of its leading edge, which is extended out from the cell as a broad, flattened projection termed a *lamellipodium*. In many cases, the lamellipodia (Fig. 16-20) are ruffled in appearance and the edge can be seen to carry out an undulating motion. As the lamellipodia are extended from the cell, they appear to adhere to the substrate at specific points. Electromicroscopic examination show the lamellipodia to be filled with microfilaments, particularly near those points involved in adhesion. In addition to the microfilaments, a dense plaque of amorphous material is concentrated just within the membrane at the sites of contact with the substrate.

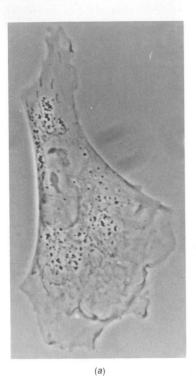

(a)

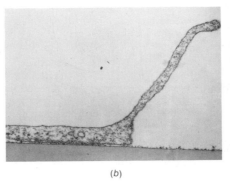

(b)

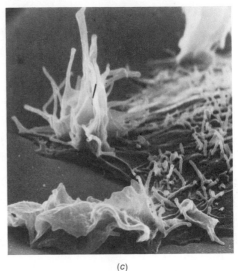

(c)

Figure 16-20. *(a)* Embryonic chick fibroblast cultured on a glass surface. The broad leading lamella in front of the nucleus is bordered anteriorly by a number of linear thickenings which appear as dark lines in phase contrast. These are the ruffles. *(Courtesy of M. Abercrombie.)* *(b)* Vertical longitudinal section of lamellipodium parallel to the substrate. The cell is moving from left to right. *(From M. Abercrombie, J. E. M. Heaysman, and S. M. Pegrum, Exp. Cell Res., 67:362, 1971.)* *(c)* A scanning electron microscope of the leading edge of a cultured cell. Two ruffled lamellipodia are seen as veillike extensions of the peripheral region of the cell. In this type of preparation, one can look underneath the ruffles and see attachments of the cell membrane to the substratum. *(From J. P. Revel, Symp. Soc. Exp. Biol., 28:447, 1974.)*

As in the case of amoeboid movement, the fibroblast too seems to depend upon its microfilaments to provide the motile force. If the microfilaments are disrupted by the addition of cytochalasin B, movement is brought to a stop. As a lamellipodium stretches forward and adheres to the substrate, the elongated cell can be seen to be stretched and under tension. After a time, the trailing edge of the cell is ripped loose from its attachment, and it springs forward. In many cases the detachment of the rear end of the cell reveals the presence of thin protoplasmic strands, termed *retraction fibers* (Fig. 16-21), connecting it to the substrate. These retraction fibers illustrate the manner in which a cell adheres to its substrate at discrete points. Once the rear of the cell has moved forward, a new lamellipodium can be formed and the cell can move off in the same or some different direction.

An important feature of the locomotory properties of cells can be seen when one of these cultured fibroblasts makes contact with another cell. As first observed by Michael Abercrombie in the early 1950s, contact between cells is followed by a dramatic cessation of their locomotor activi-

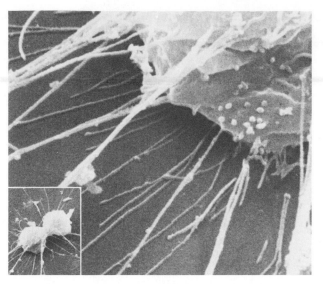

Figure 16-21. Scanning electron micrograph of a telophase cell. Cultured cells become rounded in mitosis and lose much of their adherance to the substrate. As the cells round up, they can be seen to remain attached to the substrate by retraction fibers. *(From S. Stenman, J. Wartiovaara, and A. Vaheri, J. Cell Biol.,* **74:**465, *1977.)*

ties. If one observes the ruffled lamellipodia of the two cells, one sees an immediate inhibition of its undulatory behavior at those points at which contact is made. This behavior of cells is termed *contact inhibition of movement*. The initial contact between cells is generally followed by a more stable cohesion which must be broken if the cell is to move off again. As a result of contact inhibition, cells are prevented from "walking" on top of one another and will, therefore, tend to move away from one another and spread out over an available surface. For example, when a small piece of tissue is placed in culture, migratory cells generally move out from the body of the mass and spread themselves out in an ever-increasing circle. Rather than piling up on one another, the cells form a monolayer and cover the dish (page 641).

One of the important questions in the study of movement of cultured cells is the degree to which their characteristics reflect their behavior within the organism. In most cases, cell locomotion is impossible to observe in vivo since it occurs within opaque tissues by cells which cannot be distinguished from their cellular surroundings. Since the nature of the substrates available within the body are very different from the glass or plastic dishes (regardless of their coating) used in studies on cultured cells, as is their chemical environment, it becomes very difficult to be sure of the relevance of a particular feature of in vitro locomotion. Fortunately, a few systems do exist where cell migration within an organism can be followed. In nearly every case the best-studied cell migrations occur within embryos. These include the migration of fibroblasts along the corneal stroma (page 270), the movement of the deep cells during fish development (see the paper by John Trinkaus, 1973), and the movement of primary mesenchyme cells of the sea urchin embryo (see below).

In the sea urchin, the primary mesenchyme cells originate from within the wall of the blastula at one end of the embryo and are released into the hollow cavity of the embryo, the *blastocoel*. This in itself is an interesting phenomenon since it illustrates that cells can change from a condition where they are adhering strongly to other cells (in this case other cells of the wall), to one where they are nonadhesive, causing them to be forced out of the wall by the other cells. In other words, loss of adhesion can be as important as its attainment. From their initial site at one end within the blastocoel, the primary mesenchyme cells migrate along the inner side of the embryo wall and move to form clusters at predictable locations. Since the single-celled wall of the sea urchin embryo is transparent, the migratory cells can be easily followed and time-lapse movies can be taken of their progress. Analysis of these movies has revealed a fascinating course of activities. The primary mesenchyme cells (Fig. 16-22a) move by the extension of fine elongate pseudopods, or *filopodia*, which make contact with the inner surfaces of the cells of the wall. One gains the impression that these cell extensions "feel" their way along the wall, making intermittent contacts with the cell surfaces of the wall, breaking these contacts, and making new ones. As might be expected, filopodia contain large numbers of both microfilaments and microtubules (Fig. 16-22b). The tips of the filopodia appear to move over the inner surface of the cells of the embryo wall, making attachments at selected sites. Once firm attachments are made,

Figure 16-22. *(a)* Light micrograph of the primary mesenchyme cells within the blastocoel of a living sea urchin embryo. These cells move by the extension of long processes which make attachments to each other and to the inner surface of the wall. In this micrograph, a cablelike complex of these filopodia is very apparent. *(From T. Gustafson and L. Wolpert, Exp. Cell Res., **24**:71, 1961.) (b)* Electron micrograph of a primary mesenchyme cell fixed during its exploratory movements in the blastocoel. Extending from this cell is a long pseudopodium which has been implicated in the movements of these cells. The inset shows a portion of the pseudopodium which contains numerous ribosomes, small vesicles, and, most prominently, a large number of microtubules lying parallel to the long axis of the process. *(From J. R. Gibbins, L. G. Tilney, and K. R. Porter, J. Cell Biol., **41**:214, 1969.)*

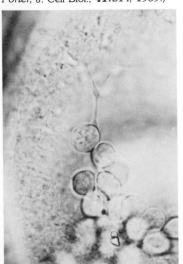

(a)

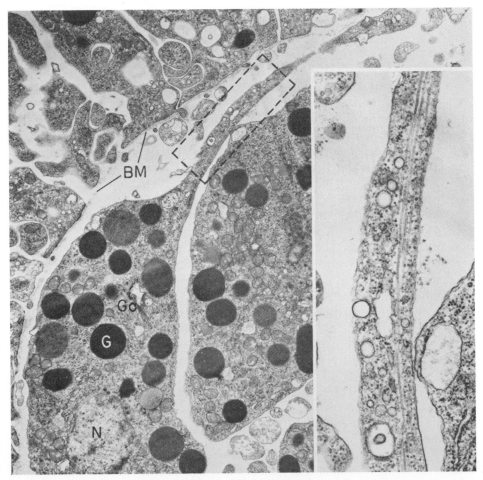

(b)

contractile processes within the filopodia are able to pull the more rounded cell body forward. In this manner, the mesenchyme cells move through the fluid-filled embryonic space and ultimately take up characteristic predetermined positions from which other types of activities can be initiated.

In this discussion of nonmuscle motility, we have concentrated upon the cytoplasmic elements that appear responsible for generating the motile force. The movement of a cell along a substrate involves contact between the substrate and the cell surface, and numerous studies have indicated that the cell coat and associated extracellular materials (page 267) can play an important role in cell motility. For example, the movement of the primary mesenchyme cells of the sea urchin embryo depends upon the synthesis of acid mucopolysaccharide destined for the cell surface. If the production of these materials is inhibited, the primary mesenchyme cells appear in the blastocoel but do not migrate along the inner surface of the wall (Fig. 16-23a). In the case depicted in Fig. 16-23, inhibition of synthesis is brought about by raising the embryos in sea water lacking sulfate ions, sulfate being an integral component of these polyanionic macromolecules. Similar results are obtained by raising sea urchin embryos in sea water containing tunicamycin, a drug which inhibits the synthesis of many cell surface macromolecules by blocking the addition of sugar groups to the protein portions of the molecule. Scanning electron micro-

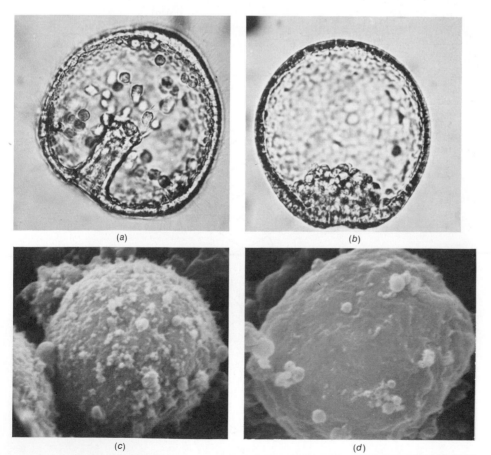

(a)

(b)

(c)

(d)

Figure 16-23. These micrographs illustrate the effect of inhibition of acid mucopolysaccharide synthesis on the migratory activity of primary mesenchyme cells. (a) A normal sea urchin gastrula raised in complete sea water. (b) An embryo of the same age as that of a that had developed in sulfate-free sea water. Primary mesenchyme cells collect in the blastocoel without migrating. (c)–(d) Scanning electron micrograph of primary mesenchyme cells from an embryo raised in normal (c) and sulfate-free (d) sea water. (From G. C. Karp and M. S. Solursh, Dev. Biol., **41:**110, 1974.)

scopic examination (Fig. 16-23b) of the surface of the primary mesenchyme cells of embryos raised in sulfate-free sea water reveals a relative deficiency of extracellular materials as compared to the mesenchyme cells of control embryos. It is presumed that the immobility of the mesenchyme cells is a consequence of their altered surface morphology which, in turn, reflects the relative absence of newly synthesized acid mucopolysaccharide.

Before leaving the subject of cytoplasmic-based mobility, it should be pointed out that other types of systems have also been studied and have provided valuable information. Since space is limited, certain of these systems have escaped coverage but should not be ignored. Foremost among these are the cytoplasmic streaming activities (termed *cyclosis*) which occur in certain giant algal cells, such as *Nitella,* and the movements of the microvilli of intestinal cells. Information on cyclosis can be obtained in the articles by Nina Allen and by Noburo Kamiya; microvilli by Mark Mooseker and Lewis Tilney.

CILIA AND FLAGELLA

In the previous section we considered a type of single cell movement in which the locomotory organelles are transient and relatively unstructured and the rates of locomotion are low. A very different type of locomotion is found among cells containing cilia and flagella. The terms *cilia* and *flagella* refer to a group of organelles which project out from a cell as an elongated process of small diameter which is capable of exerting a force upon the surrounding fluid medium. Although the term flagella is used in

Figure 16-24. The various stages in the beat of a cilium (a) and a flagellum (b). [(a) From E. Aiello and M. A. Sleigh, J. Cell Biol., **54**:498, 1972; (b) from T. L. Jahn, M. D. Landman, and J. R. Fonseca, J. Protozool., **11**:293, 1964.]

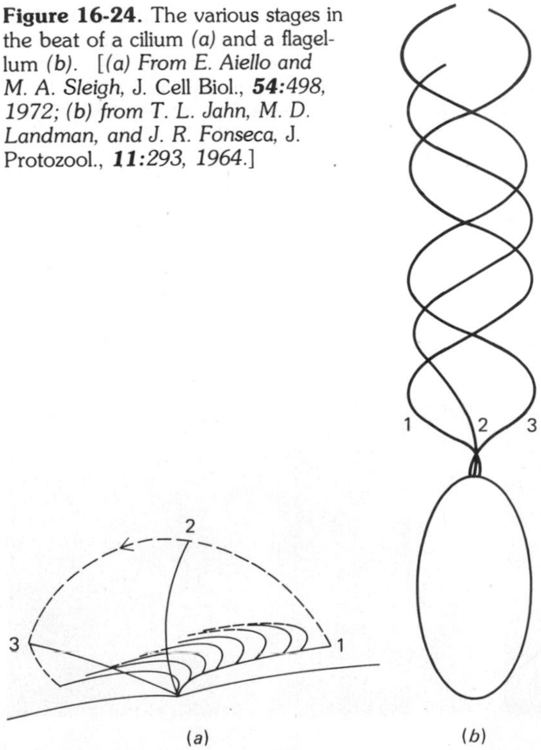

connection with locomotor organelles of both prokaryotic and eukaryotic cells, there is no structural (or evolutionary) relationship whatsoever between these two types of flagella. In contrast, all flagella and cilia of eukaryotic cells are remarkably similar. The basic structure of these organelles has been retained with very little modification during evolution all the way from the most primitive flagellated protozoa to the most complex multicellar organisms.

The distinction between cilia and flagella is not always obvious. At the far ends of the spectrum, cilia and flagella are best distinguished by their motion, i.e., the nature of their beat, which has direct consequences upon the type of movement they produce. A cilium can generally be likened to an oar in that it has an effective power stroke perpendicular to the cilium itself. The medium is consequently moved parallel to the cell surface. In its power stroke the cilium (Fig. 16-24a) is maintained in a rigid state using the surrounding medium to resist the movement of the cilium, thereby generating a force in the opposite direction. In its recovery stroke, the cilium is maintained in a flexible state offering relatively little resistance to the medium as it returns to its initial position. Cilia tend to occur in large numbers on a cell's surface and their beating activity is usually coordinated (Fig. 16-25a). In multicellular organisms, cilia are no longer used for locomotion of cells but rather for the movement of fluids past the cell's surface (Fig. 16-25b). The principles of cell movement and fluid movement are the same; in both cases force is being exerted upon the medium in a direction parallel with the cell surface. In humans, cilia occur on the linings of the respiratory tract, the paranasal sinuses, the ependymal lining of the ventricles of the brain and the spinal cord, and in the oviduct.

Unlike a cilium, the beat of a flagellum (Fig. 16-24b) is typically symmetrical and undulatory in nature with more than one wave present in a

Figure 16-25. *(a)* Metachronal waves of cilia on the surface of the ciliate *Opalina*. RS, cilia in recovery stroke; ES, cilia in effective stroke. *(From G. A. Horridge and S. L. Tamm, Science, **163**:818, 1969. Copyright by the American Association for the Advancement of Science.)* *(b)* Cilia on the surface of the fimbria of a mouse oviduct. *(From E. R. Dirksen, J. Cell Biol., **62**:901, 1974.)*

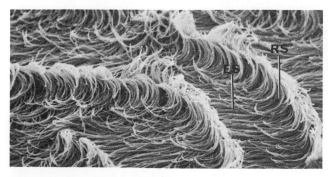

(a)

(b)

flagellum at one time (see Fig. 16-31). The effect of flagellar activity is to produce a force which is parallel to the flagellum, thereby directing the organism in approximately the same direction as the long axis of the flagellum itself. Flagella are usually present in small numbers on a cell's surface. They are utilized by higher organisms primarily as a means to propel the single-celled sperm through the medium toward the egg. Flagella are found to vary greatly in length, from approximately one micrometer to several millimeters, though they are invariably of similar diameter (approximately 0.2 μm). The constancy of their diameter reflects the invariant nature of their cross section, which will be discussed at length below. Ciliary organelles are also found in a modified form as a basic part of various sensory organs, where they do not function in a locomotor capacity.

Bacterial Flagella

Bacterial flagella (Fig. 16-26) are composed of two main parts, the elongated filament which extends out into the medium, and the basal part which anchors the flagellum to the cell and generates the forces by which it moves. The extended part of the flagellar apparatus is composed of a polymer of the protein flagellin. Purified preparations of flagellin can polymerize in vitro to form a flagellar filament, thereby providing an important system for the study of self-assembly (page 99). The base of the flagellum is much more complex, being composed of several components and numerous proteins. The inner ring of the base is connected to the cell membrane, which in turn is responsible for providing the energy utilized in the movement. As described in Chapter 8, the plasma membrane of the bacterium is in a high-energy state as the result of the maintenance of a proton gradient (page 374). This high-energy state is directly responsible for driving the flagellum without the need for an ATP intermediate. The typical bacteria flagellum is approximately 150 Å in diameter and 10 to 15 μm in length and is capable of rotating in either a clockwise or a counterclockwise direction. The nature of the rotation has been best observed in experiments where the flagella are immobilized by

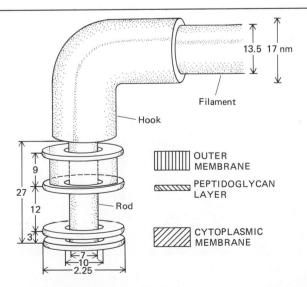

Figure 16-26. Model of the flagellar base of *E. coli*. Dimensions are in nanometers. *(From J. Adler, in R. Goldman, T. Pollard, J. Rosenbaum, eds., "Cell Motility," Cold Spring Harbor, 1976.)*

attachment to a glass slide via an antibody linkage. Under these conditions, the activity of the flagellar "motor" causes the bacterial cell to rotate which, unlike the movement of the very small flagella, can be followed in the light microscope.

The direction of rotation of the flagella has been shown to have important consequences for the movement of the cell. Consider the position of a bacterium, a small isolated cell without any type of advanced sensory apparatus to guide its way. If the ability to locomote is to be of any value to the organism, it needs some type of guidance system to direct it toward more hospitable environments and away from deleterious ones. The behavior of simple organisms is usually described in terms of *taxes,* i.e., responses either toward or away from a stimulus. On the basis of their movement toward positive stimuli and away from negative stimuli, these organisms will ultimately find themselves in the most appropriate region of their environment.

The basis for the locomotor behavior of *E. coli,* which has a few flagella at one pole of the cell, appears to be determined by its ability to reverse the direction of flagellar rotation. Observations of bacteria indicate that they generally move in a straight line for brief periods of time and then undergo a type of random tumbling motion which puts them in a new position to move in a different direction. The movement of a bacterium in a straight line occurs from a counterclockwise rotation of its flagella, while the tumbling motion results from a clockwise rotation. When the flagella rotate in the counterclockwise direction, they remain associated with one another in a bundle (Fig. 16-27a) and thereby act as if they were a single multistranded organelle. However, when rotating in the clockwise direction, each flagellum acts inde-

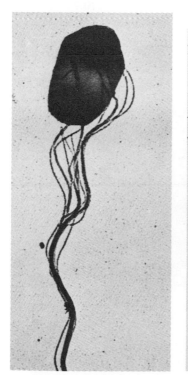 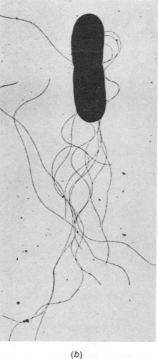

(a)   (b)

Figure 16.27. Electron micrographs of *Salmonella* cells with intact flagella. *(a)* Flagella directed to one end of a cell have coalesced to form a bundle with a characteristic wave form. *(b)* Flagella radiating from periphery in an independent manner. *(From B. R. Gerber, L. M. Routledge, and S. Takashima, J. Mol. Biol., 71:322, 1972.)*

pendently (Fig. 16-27*b*), and together they produce the tumbling behavior. If a bacterium is exposed to a chemical which acts as an attractant, i.e., one which it will move toward, the rotation of the flagella will be maintained in a counterclockwise manner. In contrast, if a chemical known to repel bacteria is added, the clockwise rotation will be initiated, causing the bacterium to tumble and move off in a new direction.

The ability of an organism to move toward higher concentrations of a particular chemical (Fig. 16-28) is termed *chemotaxis* and is believed to be of widespread importance in directing cells toward particular locations. For example, it has been shown that the sperm of certain invertebrates are guided toward the surface of an egg by a chemotactic response. Studies on bacteria have recently moved toward an understanding of the mechanism of chemotaxis in these simpler organisms. (The reader can consult papers by Julius Adler, Howerd Berg, and Daniel Koshland.) It appears that specific chemical changes in the protein of the cell membrane are involved in mediating the response by the flagellum to the presence of chemicals in the environment.

The Structure of Eukaryotic Cilia and Flagella

An electron micrograph of a cross section of a cilium or flagellum is one of the most characteristic images in cell biology. A cilium is composed primarily of a set of continuous fibers which run longitudinally through the entire cilium. In nearly every case, there are nine peripheral fibers, each made of two subfibers joined together to form a doublet, and two central fibers which occur as singlets. As described in Chapter 15, the fibers are composed of microtubules made of polymerized tubulin. The peripheral doublets are composed of one complete microtubule, the A subfiber, and one incomplete microtubule, the B subfiber, which contains 10 (sometimes 11) subunits, rather than the usual 13. The early

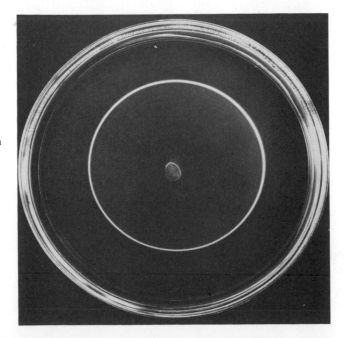

Figure 16-28. When bacterial cells are plated at the center of an agar plate containing a chemical such as galactose to which the cells are attracted, they form a ring as shown in the photograph. Beginning in the center of the plate, the cells consume the substance and are attracted toward increasing concentrations in the direction away from the center. As they consume the substrate, they form an expanding ring on the dish. *(Courtesy of J. Adler.)*

examination of whole mounts of flagella and cilia in both the light and electron microscopes revealed the fibrous nature of the organelle, but it was not until a 1954 study by Don Fawcett and Keith Porter that their cross-sectional organization was shown in detail. As the resolution in electron microscope work improved, certain of the less obvious components became visible (Fig. 16-29). A pair of "arms" was found to project from the A subfiber in a clockwise direction (when viewing the cilium from the base toward the tip). Projections from the central tubules form what is termed the *central sheath* and *radial spokes* can be seen connecting the central sheath with the A subfiber of each of the peripheral doublets. The doublets are connected to each other by a bridge composed of a protein, nexin.

If one disregards the arms of the A fiber, the cilium or flagellum can be bisected into two approximate mirror halves by a plane passing through the length of the structure, perpendicular to a line drawn between the two central fibers. This plane of symmetry would cut one of the doublets in half (the number 1 doublet) and slice between two doublets (the 5 and 6 doublets) on the other side. As will be described below, there is a constant relationship between the locations of specific doublets and the nature of the organelle's beat. In some species other minor components are also visible. The entire ciliary projection is covered by a membrane which is continuous with the plasma membrane of the cell. The part of the cilium

Figure 16-29. *(a)* Cross section of the microtubules of a sperm flagellar axoneme. The subunits are apparent in most microtubules. *(Courtesy of L. G. Tilney and K. Fujiwara.)* *(b)* Diagram of a portion of the 9 + 2 axoneme of a lamellibranch gill cilium as viewed from base to tip. *(From F. D. Warner, in R. Goldman, T. Pollard, J. Rosenbaum, eds., "Cell Motility," Cold Spring Harbor, 1976, after Satir and Warner.)*

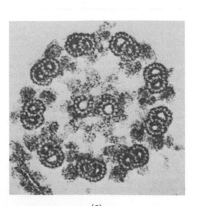

(a)

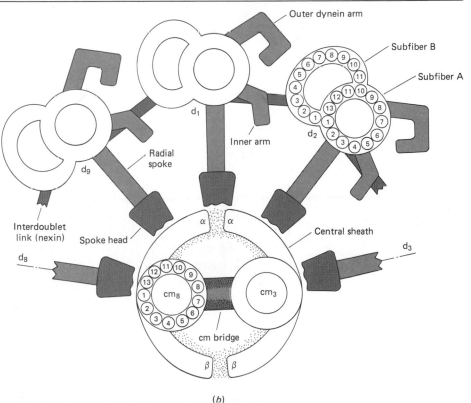

(b)

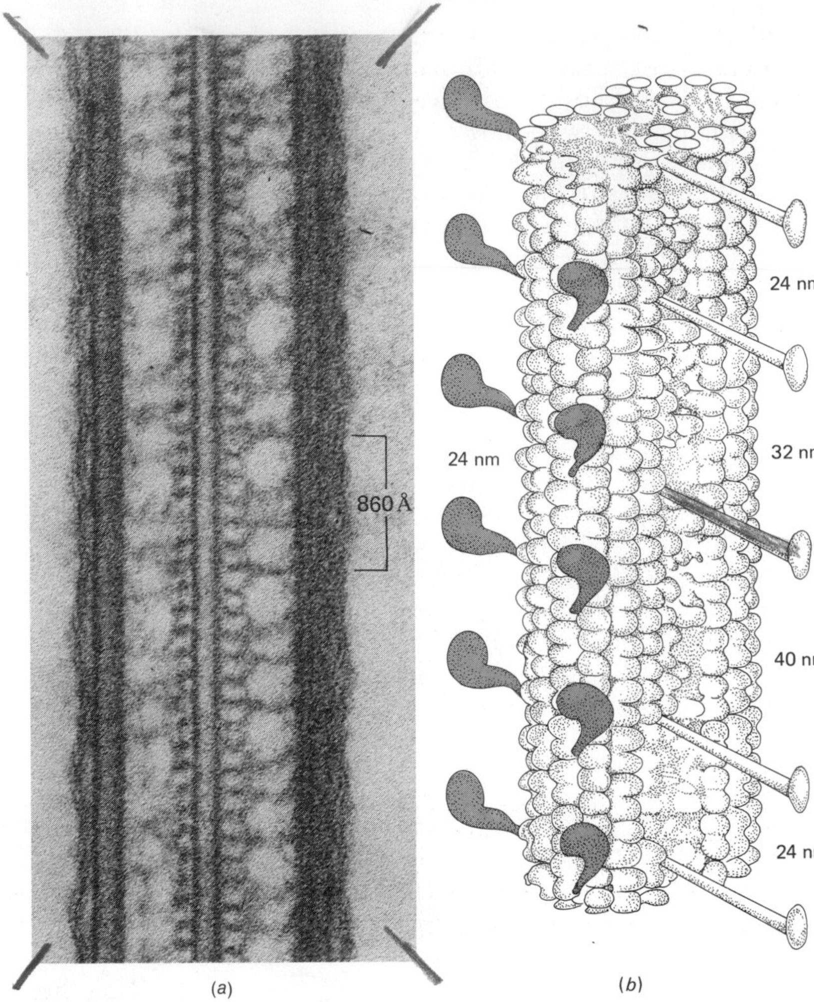

Figure 16-30. *(a)* Median longitudinal section of a straight region of a bent cilium. The radial spokes are seen joining the A subfiber of the doublet on the right. The spokes lie in groups of three with spacings of 285, 205, and 370 Å, thereby forming a major repeat of 860 Å. *(From F. D. Warner and P. Satir,* J. Cell Biol., **63**:41, 1974.) *(b)* Model of a section of a flagellar doublet tubule, tip end uppermost. Radial spokes emerging to the right occur in groups of three, which repeat (at 960 Å in this case) along the length of the microtubule. Dynein arms emerging to the left repeat every 240 Å. *(From F. D. Warner, in R. Goldman, T. Pollard, and J. Rosenbaum, eds.,* "Cell Motility," *Cold Spring Harbor, 1976.)*

within the membrane, i.e., the tubular component, is termed the *axoneme*. In a few cases the axoneme has a combination of tubules other than the $9 + 2$ pattern just described. For example, a $9 + 1$ pattern is found in flatworms and a $9 + 0$ pattern in the mayfly. In some cases, flagella lacking the central tubules have a fibrous structure in its place. Those lacking any type of central element are nonmotile.

Although the cross-sectional appearance of the axoneme reveals the major structures, it fails to indicate certain aspects of their organization which are best seen in a longitudinal section, i.e., one cut through the axoneme parallel to its long axis (Fig. 16-30). Longitudinal sections indicate the continuous nature of the microtubular fibers and the discontinuous nature of the other elements. The arms, for example, appear as repeating projections on the A subfiber with a period of about 230 Å. All of the radial spokes are seen to be organized in groups of three with different spaces between each member of the three spokes. The entire group repeats with a period of 960 Å. The base of the cilium or flagellum is composed of a structure whose cross-sectional appearance is identical with that of a centriole. This structure, which is usually called a *basal body*, contains nine peripheral fibers, each made of three microtubules and no central tubules. A terminal plate separates the basal body from the rest of the axoneme. The basal body usually has processes, termed *rootlets*, which connect it to deeper layers of cytoplasm.

The Mechanism of Ciliary and Flagellar Function

With a description of the structure of cilia and flagella behind us, we can consider the functions of certain of the components and the manner in which they are believed to interact to bring about the locomotor activity. Various studies have clearly indicated that the machinery involved in ciliary locomotion was contained within the cilium itself. Unlike a bacterial flagellum, the force-generating structure is not in the base of the organelle but rather in the axoneme itself. In one study, for example, cilia were severed from the cell by use of a laser microbeam, and the cilia were found to continue their normal beating activity in isolation. In addition, the membrane can be removed from isolated cilia or flagella and the bare axoneme is capable of sustained beating activity (Fig. 16-31) in the presence of added ATP. The greater the ATP concentration, the greater the beat frequency. The protein responsible for the conversion of the chemical energy of ATP into the mechanical energy of ciliary locomotion was isolated in the 1960s by Ian Gibbons and coworkers. The experiments by Gibbons provide an elegant example of the relationship between structure and function in biological systems and the means by which the relationship can be revealed by experimental analysis.

By the use of various solutions capable of solubilizing different components, Gibbons was able to carry out a chemical dissection of the cilia from the protozoan *Tetrahymena* (Fig. 16-32). The dissection of the axoneme from the enclosing membrane was accomplished by dissolving the membrane in the detergent digitonin. The isolated axonemes were further dissected by immersing them in a solution containing EDTA, a chelator of divalent ions. When EDTA-treated axonemes were examined in the electron microscope, the central tubules were missing, as were the arms projecting from the A tubules. Simultaneous with the loss of these structures, the axonemes lost their ability to hydrolyze ATP, while the supernatant gained it. The ATPase which had been solubilized from the axonemes was a protein termed *dynein*. When the insoluble axoneme was mixed with the soluble protein in the presence of Mg^{2+}, the ATPase activity once again became associated with the insoluble material in the mixture. Examination of the axonemes after mixture with dynein and Mg^{2+} showed that the arms had once again reappeared on the A fiber. Gibbons concluded that the arms, which could be visible in ciliary cross sections, represented the ATPase involved in energy utilization during locomotion. It has recently been noted that persons suffering from certain types of sterility produce sperm devoid of these arms. Further examination indi-

Figure 16-31. A sea urchin spermatozoon reactivated with 0.2 mM ATP following demembranation with the detergent Triton X-100. This multiple-exposure photomicrograph was obtained with 5 flashes at 58.0 Hz; the beat frequency of the flagellum is 14.5 Hz. *(From C. J. Brokaw and T. F. Simonick, J. Cell Biol., **75:**650, 1977.)*

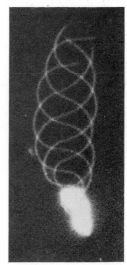

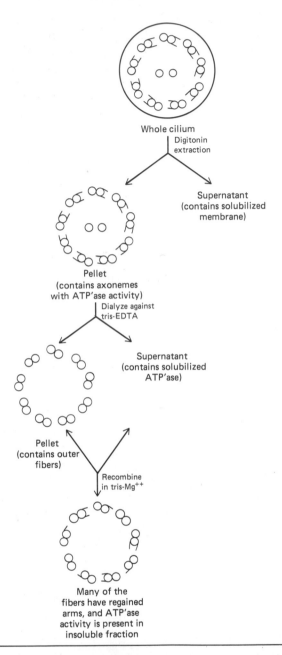

Figure 16-32. Steps in the chemical dissection of cilia from the protozoan *Tetrahymena*.

cated that these persons generally suffered from respiratory ailments as a result of the absence of dynein arms on respiratory tract cilia. One can conclude that cilia and flagella lacking the arms are nonfunctional.

The most important question to consider from a mechanistic standpoint is the means by which the components of the axoneme bring about locomotor activity. The movement generated by a cilium or flagellum is dependent upon the organelle's ability to bend in specific regions, i.e., locally, along its length. Given a structure composed of elongated continuous fibers, there are two mechanisms which can be considered as logical explanations for the bending activity. Bending might be brought about by a lengthening of the fibers on one side and a shortening of the fibers on the other side. In the case of the cilium, we are dealing with microtubular fibers which would have to shorten and lengthen. In order to account for changes in length of a microtubule one would have to assume that the subunits were capable of conformational changes such that their dimensions along the longitudinal axis of the microtubule would increase or decrease. An alternative possibility to account for ciliary bending is to assume that there is a sliding of the microtubular fibers relative to one another (Fig. 16-33a). If the fibers on one side of the cilium slid toward the tip and the fibers on the other side slide toward the base, the cilium would bend toward the latter side.

Research on ciliary locomotion over the past fifteen years has steadily accumulated evidence in favor of the sliding-microtubule theory which was first proposed by Bjorn Afzelius in 1959. Most of the evidence has been accumulated by Ian Gibbons, Peter Satir, and Fred Warner and their colleagues. If ciliary bending results from sliding rather than shortening, certain requirements should be met. The major assumptions upon which the theory is based and some of the evidence which supports them are described below.

1. *The overall length of the fibers should remain constant.* One way in which this assumption can be verified is to utilize structures such as the radial spokes as markers of specific sites along the longitudinal axis of the fibers. If shortening or lengthening of the fibers occurs during ciliary bending, then the periodicity of the attachment points of the spokes to the fibers should be altered. Since the periodicity remains unchanged along the length of the fiber, one can assume the microtubules are not shortening or lengthening.

2. *The relative positions of the doublets should change relative to one another during beating activity.* Extensive electron microscopic examination of the tips of cilia lining the gills of a clam have revealed that the pattern of fibers seen in a cross section through the tip varies with different positions in the beat of the cilium, but is constant for cilia in the same stage of their beat. In other words, as the cilium moves through its stroke, the relative positions of the nine outer fibers change as expected if the sliding tubule hypothesis is correct. The plane of symmetry of the cilium which passes through fiber 1 and between fibers 5 and 6 is also the plane of symmetry of the beat. In the cilia of the clam gills, the 5-6 doublets are always at the leading edge of the cilium during its effective stroke, while the number 1 doublet is at the leading edge in the recovery stroke. Examination of the tips of the cilia has shown that when each cilium is bent so that the 5–6 doublets are at the leading edge, they extend well past the number 1–9 doublets. When the cilium is bent in the opposite direction, the number 1–9 doublets extend past the pair of 5–6 doublets (Fig. 16-33b).

3. *Some type of movable bridge structure must exist to connect adjacent doublets.* With the discovery of the arms of the A subfiber, it was

Figure 16-33. (a) Sliding-filament hypothesis of ciliary motility. The distal ends of two of the doublets (numbers 1 and 6) are shown at three different positions of the cilium. The B subfibers are drawn as solid tubules, the A subfibers are unshaded. The cross sections shown reveal which subfibers of the various doublets extend to the tip of the cilium at its different positions. When the cilium is in a straight-up position, all of the B subfibers end at the same point while the A subfibers continue as naked singlets of variable length. However, when the cilium is bent in its effective stroke (right), doublet 6 is closer to the tip than doublet 1 as revealed by the lack of the B subfiber of doublet 1 in the cross section. In contrast, when the cilium is bent in its recovery stroke, doublet 1 is extended relative to doublet 6 as revealed by the lack of the B subfiber of doublet 6 in the cross section near the tip. The displacement, Δl_{η}, can be determined by the formula indicated. *(From P. Satir, J. Cell Biol., **39**:79, 1968.)* (b) A representative ciliary tip from organized, splayed axonemes. The axoneme attaches and opens at random in the process of splaying out. In this micrograph, doublet 9 appears to be longer than doublet 5 by about 4 spoke groups or 0.36 μm. This amount of tip displacement would correspond to a bend of just over 100°. The microtubule displacement observed in these types of preparations fits that predicted from the geometry of the sliding-microtubule model. *(From W. S. Sale and P. Satir, J. Cell Biol., **71**:598, 1976.)*

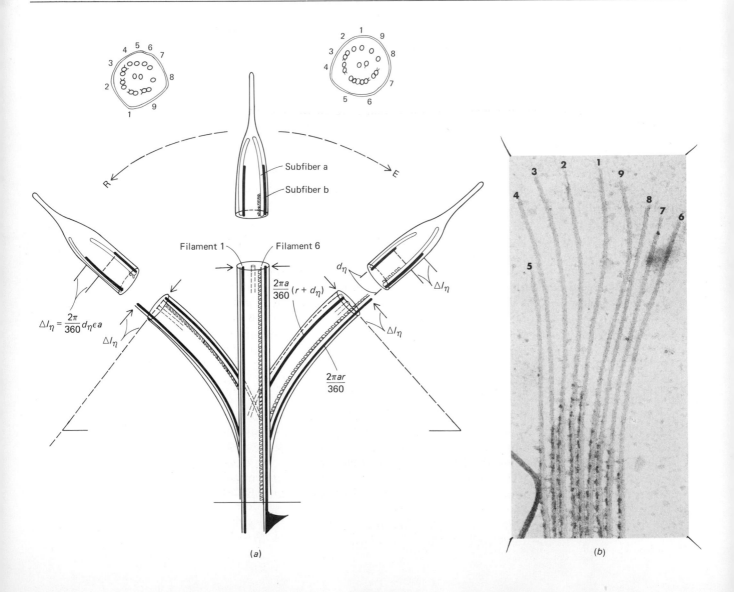

(a)

(b)

proposed that these might provide the bridges that connect adjacent fibers. If so, they might function in a manner analogous to that of the myosin cross bridges in muscle contractility. With the discovery of their ATPase activity, the relationship of the arms to the myosin heads was greatly strengthened. More recent electron micrographs have shown the attachment between the A subfiber of one doublet and the B subfiber of another doublet does occur via the dynein arms. This is most readily seen under a particular set of in vitro conditions. As stated above, cilia can be isolated and stripped of their membrane and the naked axonemal core is capable of undergoing rhythmic beating activity in the presence of ATP. If such preparations are suddenly deprived of their ATP content by dilution, a condition persists which is very similar to that occurring in isolated muscle fibrils — the cilia go into rigor. In the rigor state, the dynein arms remain permanently linked to the adjacent fiber. In the living state it is presumed that the attachment and breakage of the arms and fibers occurs in a cyclic way in a manner similar to its occurrence in muscle. By virtue of this repeating activity, the arms of one fiber, in essence, "walk" along the neighboring fiber resulting in the relative displacement of the two doublets.

4. *In order to convert a sliding activity (a shearing force) into a bending motion, there has to be some element of the cilium to resist the sliding, but at the same time be capable of some deformation to allow the bending to occur.* In other words, some type of elastic resistance must be present in those regions of the organelle that happen to be bent at any given time. If there were no resistance, the fibers would simply slide without causing a bend. Since bending is localized in the cilium or flagellum, the resistance must also be localized. Similarly, since the bending is propagated along the entire length of the organelle, it would appear that the resistance to sliding is also propagated in a coordinated manner. Longitudinal sections through bent cilia show characteristic differences between the bent and straight regions of a cilium. The radial spokes are oriented perpendicularly to the A subfiber in the straight regions and oriented at a distinct angle (maximum spoke tilt of 26° to 33°) in bent regions. Based on these types of observations, it has been concluded that the spokes of the A subfibers are not attached to the central sheath in the straight regions so that no resistance to sliding occurs in this part of the organelle. The angular orientation of the spoke in the bent region indicates its attachment to the central sheath as well as its deformation as a result of the sliding process. Since the sliding that occurs is much greater than can be accounted for by the displacement of the radial spokes seen in electron micrographs of the bent region, it is believed that the spokes in the bent region are capable of detachment upon sufficient shearing tension. In other words, it is believed that as the fibers slide, they generate tension on the attached spokes, i.e., the ones in an area where bending is supposed to occur. When a maximum angular displacement of a spoke is reached, that spoke detaches from the central sheath, thereby allowing the fibers to slide past each other. After the spokes have detached, they would reattach and once again promote bending from the sliding fibers. As the bend propagates along the fiber, the radial spokes at the leading edge of the bend would become attached so as to cause resistance. Similarly, spokes would become detached from the central sheath at the trailing edge of the bend so as to allow the organelle to straighten in that region.

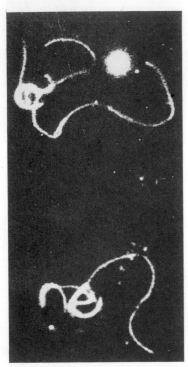

Figure 16-34. Dark-field photomicrographs of axonemes that had disintegrated following treatment with trypsin in the presence of ATP. Two axonemes are shown that elongated several times their length due to sliding of outer doublets past one another. *(From G. B. Whitman, J. Plummer, and G. Sander, J. Cell Biol., 76:743, 1978.)*

A striking confirmation of the existence of some structure of the axoneme which resists sliding has been obtained from in vitro experiments. When isolated axenomes are incubated with ATP, normal movement occurs. If these axonemes are first treated under mild conditions with the proteolytic enzyme trypsin so as to destroy the integrity of the radial spokes and the nexin fibers (see Fig. 16-29b), the addition of ATP is followed by the disintegration of the structure of the axoneme. When this process is observed in the light microscope under special illumination, the subfibers can actually be seen to be moving away from each other (Fig. 16-34). In the absence of the resistance afforded by the spokes (and possibly the nexin fibers), the sliding activity is no longer held in check and the subfibers simply slide past one another until they are no longer associated.

If one considers that cilia beat from 10 to 40 times per second, that each stroke has a precise form, and that their motion is usually coordinated so that thousands of cilia act together to produce the desired fluid movement, it becomes clear that this type of locomotor activity must be much more tightly regulated than that involving protoplasmic flow. Remarkably, it appears that much of the coordination is handled within the organelle itself since isolated cilia and flagella have essentially normal beating form. Since the propagation of a bend along the structure is a relatively complex process involving attachments and detachments of the arms and probably the radial spokes, this is a considerable feat. Unlike the situation that exists in the regulation of contractility involving actin and myosin filaments, calcium ions do not seem to be required for locomotor activity of cilia and flagella. However, calcium ions have been implicated in certain aspects of their function, such as the reversal of the direction in which cilia beat or the form of the undulating wave in sperm flagella.

REFERENCES

Adler, J., Sci. Am. **234,** 40–47, April 1976. "The Sensing of Chemicals by Bacteria."

Afzelius, B. A., "Ultrastructure of Cilia and Flagella," in *Handbook of Molecular Cytology,* A. Lima-de-Faria, ed., Elsevier, 1969.

Albrecht-Buehler, G., Sci. Am. **238,** 68–76, 1978. "The Tracks of Moving Cells."

Allen, N. S., J. Cell Biol. **63,** 270–287, 1974. "Endoplasmic Filaments Generate the Motive Force for Rotational Streaming in *Nitella.*"

Allen, R. D., Sci. Am. **206,** 112–122, Feb. 1962. "Ameboid Movement."

———, Cooledge, J. W., and Hall, P. J., Nature **187,** 896–899, 1960. "Streaming in Cytoplasm Dissociated from the Giant Amoeba, *Chaos chaos.*"

——— and Kamiya, N., eds., *Primitive Motile Systems in Cell Biology,* Academic, 1966.

Berg, H. C., Sci. Am. **233,** 36–44, Aug. 1975. "How Bacteria Swim."

Blake, J. R., and Sleigh, M. A., Biol. Revs. **49,** 85–126, 1974. "Mechanics of Ciliary Locomotion."

Carlson, F. D., and Wilkie, D. R., *Muscle Physiology.* Prentice-Hall, 1973.

Clarke, M., and Spudich, J. A., Ann. Rev. Biochem. **46,** 797, 822, 1977. "Nonmuscle Contractile Proteins."

Constantin, L. L., Prog. Biophys. Mol. Biol. **29,** 197–224, 1975. "Contractile Activation in Skeletal Muscle."

Davson, H., *A Textbook of General Physiology*, 3d ed. Churchill, 1970.

Ebashi, S., Ann. Rev. Physiol. **38,** 293–313, 1976. "Excitation-Contraction Coupling."

Endo, M., Physiol. Revs. **57,** 71–108, 1977. "Calcium Release from the Sarcoplasmic Reticulum."

Goldman, R., Pollard, T., and Rosenbaum, J., eds., *Cell Motility.* Cold Spring Harbor, 1976.

Hitchcock, S. E., J. Cell Biol. **74,** 1–15, 1977. "Regulation of Cell Motility in Nonmuscle Cells."

Hoyle, G., Sci. Am. **222,** 84–93, April 1970. "How Is Muscle Turned On and Off?"

Huxley, H. E., Sci. Am. **213,** 18–27, Dec. 1965. "The Mechanism of Muscular Contraction."

————, Science **164,** 1356–1366, 1969. "The Mechanism of Muscular Contraction."

Inoué, S., and Stephens, R. E., eds., *Molecules and Cell Movement.* Raven, 1975.

Jeon, K., ed., *Biology of Amoeba.* Academic, 1973.

Kamiya, N., Protoplasmatologia **8,** 1959. "Protoplasmic Streaming."

Komnick, H., Stockem, W., and Wohlfarth-Botterman, K. E., Int. Rev. Cytol. **34,** 169–249, 1973. "Cell Motility: Mechanisms in Protoplasmic Streaming and Ameboid Movement."

Korn, E. D., Proc. Natl. Acad. Sci. (U.S.) **75,** 588–599, 1978. "Biochemistry of Actomyosin-Dependent Cell Motility" (a review).

Loewy, A. G., J. Cell. Comp. Phys. **40,** 127–156, 1952. "An Actomyosin-Like Substance from the Plasmodium of a Myxomycete."

MacLennan, D. H., and Holland, P. C., Ann. Rev. Biophys. and Bioeng. **4,** 377–404, 1975. "Calcium Transport in Sarcoplasmic Reticulum."

Mannherz, H. G., and Goody, R. S., Ann. Rev. Biochem. **45,** 427, 1976. "Proteins of Contractile Systems."

Margaria, R., Sci. Am. **226,** 84–91, March 1972. "The Sources of Muscular Energy."

"The Mechanism of Muscle Contraction," Cold Spring Harbor Symp. Quant. Biol. **37,** 1973.

Mooseker, M., and Tilney, L. G., J. Cell Biol. **67,** 725–743, 1975. "The Organization of an Actin Filament-Membrane Complex."

Murray, J. M., and Weber, A., Sci. Am. **230,** 58–71, Feb. 1974. "The Cooperative Action of Muscle Proteins."

Peachey, L. D., "Muscle and Motility" in *Biocore,* chapter XVIII, McGraw-Hill, 1974.

Perry, S., Margreth, A., and Adelstein, R., eds., *Contractile Systems in Nonmuscle Tissues.* Elsevier, 1977.

Pollard, T. D., J. Supramol. Struc. **5**, 317–334, 1976. "Cytoskeletal Functions of Cytoplasmic Contractile Proteins."

Porter, K. R., and Franzini-Armstrong, C., Sci. Am. **212**, 72–81, March 1965. "The Sarcoplasmic Reticulum."

Satir, P., Sci. Am. **231**, 44–52, Oct. 1974. "How Cilia Move."

Squire, J. M., Ann. Rev. of Biophys. and Bioeng. **4**, 136–164, 1975. "Muscle Filament Structure and Muscle Contraction."

Symposia on "Microtubules and Motility" and "Molecular Basis of Motility." *International Cell Biology 1976–1977*, pages 343–406, Rockefeller University Press, 1977.

Taylor, D. L., et al., J. Cell Biol. **59**, 378, 1973. "The Contractile Basis of Amoeboid Movement."

Trinkaus, J. P., Dev. Biol. **30**, 68–103, 1973. "Surface Activity and Locomotion of *Fundulus* Deep Cells during Blastula and Gastrula Stages."

Weber, A., and Murray, J. M., Phys. Revs. **53**, 612, 1973. "Molecular Control Mechanisms in Muscle Contraction."

Wiederhold, M. L., Ann. Rev. Biophys. Bioeng. **5**, 39–62, 1976. "Mechanosensory Transduction in 'Sensory' and 'Motile' Cilia."

CHAPTER SEVENTEEN
Cytoplasmic Genes and Their Expression

Mitochondrial DNA
Transcription
Translation
Origin of Mitochondria
    The Endosymbiont Theory
    The Theory for Direct Evolution

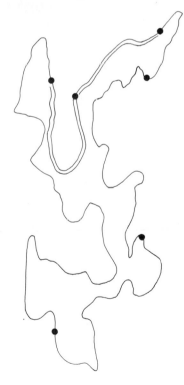

Within a decade after the rediscovery and general acceptance of the work of Mendel, a pair of genetic studies were carried out in plants that yielded results incompatible with Mendelian predictions. In both cases the phenotype being investigated was a type of leaf variegation in which mutant leaves appeared striped due to alternating sections containing pigmented and nonpigmented chloroplasts. In one case, for example, it was found that a cross between a wild-type female parent and a variegated male parent produced progeny that were totally wild type. Conversely, if the female parent were variegated and the male were wild type, all of the progeny were variegated. The trait was being inherited in a strictly maternal non-Mendelian fashion. Over the next several decades a number of studies on plants, particularly those on corn by Marcus Rhoades, confirmed the existence of genetic traits that were not transmitted in the same manner as were the nuclear chromosomal traits. However, it was not until research began on various microorganisms (particularly yeast, *Neurospora, Chlamydomonas*) that convincing information began to accumulate concerning the cellular and molecular nature of these extrachromosomal genes. We will begin by briefly considering an early study on yeast.

In 1949 Boris Ephrussi and his colleagues found that a typical population of baker's yeast contained a small percentage (approximately 1%) of cells which gave rise to particularly small colonies. Cells from these colonies could be isolated and each would develop into dwarf colonies; the change was the result of a stable, inheritable, irreversible mutation. Further analysis of these *petite* mutants, as they were called, revealed that the cells were unable to carry out aerobic respiration; they were deficient in a number of proteins (such as cytochrome oxidase) that were required for oxidative phosphorylation. As might be expected, the ability of yeast cells to grow under totally anaerobic conditions (given a fermentable substrate such as glucose) makes them an excellent system for the study of mitochondrially deficient mutations.

In addition to being able to grow anaerobically, yeast cells offer the genetic advantage of growing in both a haploid or diploid state. Haploid yeast cells exist in either of two different mating types. When cells of opposite mating type are mixed, they fuse to form a diploid zygote which grows vegetatively, i.e., by mitosis, to form a colony of diploid cells. Under the appropriate conditions, each diploid cell undergoes meiosis to produce four haploid ascospores which germinate into haploid cells that will proliferate to form colonies. Ephrussi and coworkers found that it made little difference whether the cells were in a haploid or a diploid stage in their life cycle, a similar percentage exhibited the *petite* phenotype when grown into separate colonies. Since it did not seem to matter how many copies of the chromosomes were present in the nucleus, it was suggested that the trait was housed in a cytoplasmic genetic element. In subsequent experiments it was shown that the *petite* mutation did actually segregate independently from the nucleus (and therefore all of the nuclear chromosomes), and the cytoplasmic nature of the gene was confirmed. It was finally shown that genes were localized within the mitochondria by demonstrating that genetic markers could be transmitted by the injection of mitochondria from one strain of *Neurospora* or *Paramecium* into the cells of another strain.

MITOCHONDRIAL DNA

By 1960, the concept of DNA as the genetic material was firmly entrenched and it was widely believed that cytoplasmic inheritance would also depend upon information in DNA. In 1963–1964 the first convincing evidence was obtained for the presence of DNA in chloroplasts and mitochondria. The electron microscope revealed the presence in these organelles of DNase-sensitive fibrils and biochemical studies were successful in isolating the DNA and characterizing it with respect to base composition, buoyant density, etc. During the decade that followed, a large number of genetic, morphological, and biochemical studies provided a relatively complete picture of the genetic activities of mitochondrial and chloroplast DNAs in a number of organisms including yeast, *Neurospora, Chlamydomonas,* an amphibian *(Xenopus),* and the human HeLa cell. The basic findings appear to be very similar in all of these diverse species.

Mitochondria and chloroplasts contain a small amount of genetic material which is replicated, transcribed, and translated within the organelle itself. The products of this genetic system are utilized as part of a small number of components present in the inner membrane of the mitochondrion or the thylakoid membrane of the chloroplast. In both types of organelles, the bulk of the membrane components, as well as the numerous soluble proteins, are products of the cytoplasmic protein-synthesizing system and coded for by the nuclear chromosomal DNA. It becomes apparent that the construction of these cytoplasmic organelles results from the cooperation of two essentially independent protein-synthesizing systems. In the following discussion we will concentrate on results obtained in studies on mitochondria.

The DNAs of mitochondria are usually (though not always) circular duplexes (see Fig. 8-6). The DNA of animal cell mitochondria are generally about 5 μm (approximately 15,000 base pairs) in circumference, much smaller than the mitochondrial DNA of protists (typically about 25 μm) or that of plants (50 μm or greater). Although there is considerable difference in the amount of DNA present in a mitochondrion of diverse organisms, there appears to be relatively little difference in actual genetic information. Those species with larger mitochondrial genomes contain a large percentage of spacer DNA without genetic function. Essentially all of the DNA of animal cell mitochondria can be accounted for in terms of specific known function. Molecular hybridization studies between nuclear and mitochondrial DNA have established the unique identity of each of these genetic systems; there is no significant overlap between the two sets of DNA sequences. In addition, there is little evidence for heterogeneity among the mitochondrial DNA molecules of a given individual organism, i.e., all of them appear to be equivalent.

TRANSCRIPTION

As in the case of nuclear DNA, the DNA of the mitochondrion also codes for ribosomal, transfer, and messenger RNA. The synthesis and processing of mitochondrial transcripts is best understood in animal cells whose mitochondrial DNA is small and offers several advantages. When mitochondrial DNA of animal cells is isolated and thermally denatured, the two different single-stranded components of each duplex can be separated from one another on the basis of their buoyant density (Fig. 17-1a). Preparations of one or the other of these two types of strands, termed the heavy (H) and light (L) strand, can then be used in hybridization experiments designed to identify the sites at which particular mitochondrial RNAs are transcribed (Fig. 17-1b). When cells are labeled for very short periods (e.g. 1 to 5 minutes) in the presence of a radioactive RNA precursor, the labeled mitochondrial RNA produced is found to hybridize to essentially all sites in both the H and L DNA strands. In other words, it appears that RNAs are synthesized using all of both mitochondrial DNA strands as the template. It would appear that the mitochondrial RNA polymerase is not selective in its choice of DNA template strand, but rather transcribes both symmetrically.

It was pointed out in Chapter 13, that the control of information transfer in eukaryotic cells operated at several distinct levels. In the nuclear-

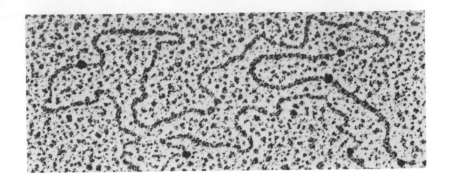

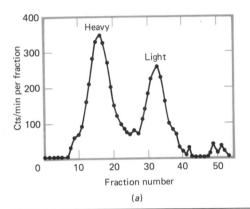

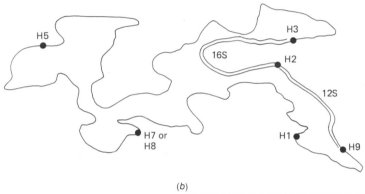

(a)   (b)

Figure 17-1. (a) Separation of the complementary strands of ¹⁴C-labeled HeLa cell mitochondrial DNA in an alkaline CsCl density gradient. The basic pH (12.4) of the gradient causes strand separation. *(From Y. Aloni and G. Attardi, J. Mol. Biol., **55:285**, 1971.)* (b) The results of a hybridization experiment with isolated H strand DNA and purified mitochondrial RNA species. The location of DNA sequences complementary to several ferritin-conjugated mitochondrial tRNAs (indicated by the position of the ferritin) and the two ribosomal RNAs (indicated by the double-stranded region) is revealed. *(From M. Wu, N. Davidson, G. Attardi, and Y. Aloni, J. Mol. Biol., **71:81**, 1972.)*

cytoplasmic genetic system, the primary control over gene expression operates at the transcriptional level in that only a small percentage of the DNA sequences are transcribed in most cells at a given time. It appears that control over mitochondrial gene activities operates at a posttranscriptional level; the entire mitochondrial genome is transcribed into giant primary RNA transcripts which are then processed in some unknown manner to produce a number of final products which include tRNAs, rRNAs, and mRNAs. The extensive processing of mitochondrial RNAs is seen by comparing the hybridization capacity of mitochondrial RNA labeled for a very brief pulse as opposed to a much longer period (e.g. an hour or more). As was stated above, short pulse-labeled mitochondrial RNA hybridizes to virtually the entire mitochondrial genome. In contrast, radioactive RNA extracted after a lengthy labeling period no longer hybridizes in this extensive manner. Some of the labeled RNA sequences complementary to the H strand and most of the RNA sequences complementary to the L strand are no longer present in the mitochondria. Those sequences no longer present have either been selectively degraded or, less likely, transported out of the mitochondria. Unlike the RNA polymerases of prokaryotes or eukaryotic nuclei, which have numerous subunits, the enzyme responsible for the transcription of mitochondrial DNA appears to be composed of a single polypeptide chain. This is a surprising finding since enzymes having such a similar and complex function in different types of organisms are usually rather similar in construction. Since the RNAs made in the mitochondria are involved in the synthesis of protein in that organelle, it is necessary to consider the mitochondrial translation machinery before continuing further with the topic of mitochondrial transcription.

TRANSLATION

Although the idea of protein synthesis occurring in the mitochondrion and chloroplast was proposed as early as 1958, little attention was paid to the concept until after the demonstration of organelle DNA. Even after the finding of DNA in the mitochondria and chloroplasts, the levels of protein synthesis observed in isolated organelle preparations were so low that they were widely believed to be the result of protein synthesis by contaminating bacteria. The skepticism seems particularly reasonable when one considers that the protein synthetic activity of these isolated mitochondrial preparations is sensitive to chloramphenicol, a powerful inhibitor of protein synthesis in prokaryotic cells, but one without such effects on eukaryotic cytoplasmic protein synthesis. However, many studies over a period of several years established beyond a doubt that protein synthesis was occurring within these organelles. The fact that mitochondrial and chloroplast protein-synthesizing machinery is sensitive to inhibition by chloramphenicol has become one of the important points in the argument that these organelles are the descendants of a prokaryotic symbiont (discussed later in this chapter). The demonstration of a mitochondrial protein-synthesizing system raises a number of questions concerning the nature of the RNA templates, the nature of the polypeptides formed, and the relationship between the nuclear-cytoplasmic and mitochondrial systems.

As might be expected, the process of protein synthesis within the mitochondrion and chloroplast is quite similar to that of the cytoplasm; however, the specific components of the two systems are quite distinct. The mitochondrial ribosomes (or *mitoribosomes*) from a variety of species have been purified and their composition analyzed. The task of analyzing mitoribosomes is made difficult by the relative scarcity of these structures within a cell, thereby requiring a large quantity of tissue (or cultured cells) to begin the isolation. As in the case of cytoplasmic and prokaryotic ribosomes, those of mitochondria (and chloroplasts) contain two subunits each made of RNA and protein. However, further analysis indicates that the mitoribosomes are quite different from both of these other types. The mitochondrial ribosomes of animal cells have a sedimentation velocity of approximately 55S, considerably less than either prokaryotic or cytoplasmic ribosomes. As it turns out, the sedimentation velocity is somewhat misleading since the electron microscope indicates that the mitoribosomes are of equivalent dimensions to that of the bacterial cell; both have a molecular weight of approximately 2×10^6 daltons. The sedimentation velocity of the mitoribosome is decreased relative to other types of ribosomes as a result of its unusually high protein/RNA ratio. Whereas 65% of the weight of a bacterial ribosome can be accounted for by its RNA content, only about 30% of the animal cell mitoribosome consists of RNA. The 55S animal mitoribosome is composed of a 39S and 28S subunit containing a 16S and 12S ribosomal RNA molecule, respectively. These represent the smallest ribosomal RNA species known. Some evidence indicates that the large subunit contains a second rRNA species (presumably analogous to the 5S rRNA of bacterial and cytoplasmic ribosomes), although this point remains controversial. Not only are the mitoribosomes different from both prokaryotic and cytoplasmic ribosomes, they are a very heterogeneous group themselves. If one compares the mitoribosomes from higher animals, fungi, protozoa, and plants, they are found to range in sedimentation velocity from 55S to 80S, in buoyant density from 1.40 to 1.61 g/cc, and in RNA base composition from 19 to 43% G + C. Similarly, when the proteins of the mitoribosomes and cytoplasmic ribosomes from the same animal cell are fractionated (most readily by two-dimensional electrophoresis), one sees virtually no similarity between the protein patterns of the two types of ribosome. There is no evidence that any of the proteins of the two types of ribosomes of a eukaryotic cell are shared. Analysis of the transfer RNAs and aminoacyl-tRNA synthetases of the mitochondrion indicate that each and every one of these molecules is also distinct from its counterpart in the cytoplasm. These results indicate that all of the various components of the protein-synthesizing machinery of the mitochondrion are different from their counterparts in the cytoplasm. Where is the genetic information for these various components stored in the nucleus or the mitochondrion?

Hybridization studies suggest that most, if not all, of the tRNAs present in the mitochondrion are transcribed from mitochondrial DNA. The location of the mitochondrial tRNA genes on the H or L strand of the mitochondrial genome has been determined by hybridizing specific ferritin-conjugated tRNAs to the DNA. The location of a particular tRNA gene on the circular DNA thread is seen in the electron microscope by the position of the electron-dense ferritin molecule (Fig. 17-1b). In other experiments, tRNAs charged with radioactively labeled amino acids have been used to hybridize with mitochondrial DNA. Those mitochondrial tRNAs that bind to the DNA can be determined by the binding of the associated radioactivity. By the use of these types of techniques it has been shown that (at the time of this writing) at least 19 different tRNA genes were present in the HeLa cell mitochondrial genome and at least 21 different tRNAs in the corresponding yeast genome. The 19 genes in the HeLa cell accounted for the tRNAs of 16 different amino acids. In yeast, the tRNAs for 19 of the 20 coded amino acids have been accounted for. Suggestive evidence exists for the presence of additional tRNA species and, at the present time, there is no reason to believe that any of the mitochondrial tRNAs are imported from the cytoplasm. One of the tRNAs found in yeast is specific for the initiation codon AUG and carries with it the amino acid N-formylmethionine (page 513).

A complete protein-synthesizing system requires the participation of well over 100 different protein molecules (ribosomal proteins, aminoacyl-tRNA synthetases, and soluble protein factors). Since none of these proteins appear to be shared between the cytoplasmic and mitochondrial systems, it is apparent that the production of this second synthetic system represents a costly output for the cell in terms of required information and regulation. Are these various proteins synthesized within the mitochondrion or are they imported from the cytoplasm? Similarly, are the mRNA templates for these proteins coded for by the mitochondrial genome or within the nuclear chromosomes? This latter question can be answered without performing any experiments. The 5 μm of mitochondrial DNA in animal cells is not nearly sufficient to code for the proteins just described. Furthermore, if one examines a map of the HeLa mitochondrial genome after the locations of the tRNA and rRNA genes have been made, there is only room enough on the DNA to code for 13 average-sized polypeptides (Fig. 17-2). The evidence suggests that the mRNAs for all of the proteins needed for the synthesis of mitochondrial DNA, RNA, and pro-

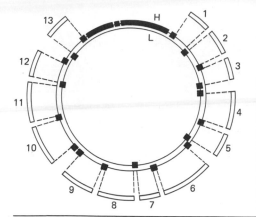

Figure 17-2. Genetic map of HeLa mitochondrial DNA. Solid bars indicate the position and the extent of genes for the rRNAs (long bars) and tRNAs (short bars) on both the heavy (H) and light (L) strands of the circular DNA molecule. Intergene gap regions of significant size have been extended to indicate the location and maximum span of potential mRNA genes (open bars). (From T. W. O'Brien, Int. Symp. Cell Biol., B. R. Brinkley and K. R. Porter, eds., 1977.)

tein are produced in the nucleus on chromosomal DNA sequences. The site within the cell where these proteins are synthesized can be determined in a variety of experiments. For example, if one or all of the ribosomal proteins are synthesized in the cytoplasm one would expect that:

1. The synthesis of these proteins would not occur in isolated mitochondria

2. The synthesis of these proteins would not occur in cells treated with an inhibitor such as cycloheximide or anisomycin which selectively block the synthesis of proteins on cytoplasmic ribosomes

3. The synthesis of these proteins would still occur in cells treated with inhibitors such as chloramphenicol or erythromycin which selectively block mitochondrial protein synthesis

When these criteria are applied, it is found that all polymerases, synthetases, soluble factors, and ribosomal proteins (with the possible exception of one ribosomal protein in *Neurospora*) are produced on cytoplasmic ribosomes and, therefore, imported by the mitochondrion.

Not only is the synthesizing machinery of the mitochondrion resistant to agents which block mitochondrial RNA and protein synthesis, so too is the gross formation of the organelle. For example, HeLa cells can be grown in the presence of chloramphenicol and they will continue to grow and divide (although at a decreasing rate) for approximately four generations. During this time, mitochondria, which appear essentially normal in the electron microscope (Fig. 17-3), continue to increase in number as the cell number increases. The same type of result is obtained in yeast which have completely lost their mitochondrial DNA (*rho°* mutants); they continue to form new, nearly normal mitochondria in the absence of the mitochondrial genome.

Up to this point we have indicated some of those aspects of mitochondrial physiology which are not dependent upon a mitochondrial genome or a mitochondrial protein synthesizing system without mention-

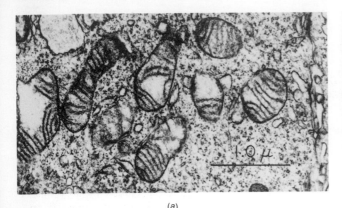

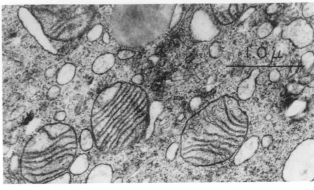

(a)   (b)

Figure 17-3. Morphological appearance of mitochondria in HeLa cells grown in the absence *(a)* or presence *(b)* of 200 μg/ml chloramphenicol for four days. *(From B. Storrie and G. Attardi, J. Cell Biol., **56**:825, 1973.)*

ing those aspects which do require the genome and its expression. The mitochondria of HeLa cells grown in chloramphenicol or those of the *rho°* yeast are unable to carry out normal aerobic respiration, a consequence which slows the growth rate of the yeast cells and stops the growth of the HeLa cells. Eight different polypeptides have been conclusively shown to be synthesized in the mitochondria, and all of them are believed to be coded by the mitochondrial genome. These polypeptides, which account for about 10 to 15% of the inner mitochondrial membrane, include 3 of the subunits of cytochrome oxidase (complex IV, Fig. 8-12), one of the subunits of the coenzyme QH_2-cytochrome *c* reductase (complex II, Fig. 8-12), and four of the subunits of the ATP-synthesizing complex. In each of these three multiprotein complexes, the polypeptides produced within the mitochondrion represent the hydrophobic portion of the complex, i.e., that portion which is located firmly within the membrane (Fig. 17-4). For example, when the structure of ATP synthetase was discussed in Chapter 8, it was indicated that the entire assembly contained three parts (Fig. 8-17); that portion in the membrane is synthesized in the mitochondria. In the case of cytochrome oxidase, the three subunits of the enzyme represent integral proteins of the membrane while the remaining three or four subunits (which are synthesized in the cytoplasm) are relatively hydrophilic in nature and contain the catalytic activity of the complex. It is believed that the hydrophilic polypeptides are synthesized on ribosomes that are directly attached to the outer surface of the outer membrane. The nascent chains apparently penetrate the membrane as they are synthesized in a manner analogous to the proteins synthesized by membrane-bound ribosomes of the rough ER (Fig. 7-9).

As suggested above, the mRNAs which serve as the templates for the proteins just discussed are believed to be synthesized from the mitochondrial DNA. If isolated mitochondria are incubated in a radioactive RNA precursor, a number of labeled RNA species are found to be associated with poly A, though the poly A fragment is smaller than that associated with nuclear transcripts. When the poly A-mRNAs of HeLa cell mitochondria are subjected to acrylamide gel electrophoresis, eight distinct species, ranging from 9×10^4 to 5.3×10^5 daltons, can be distinguished. The smallest of the eight mRNAs is complementary to the L strand while

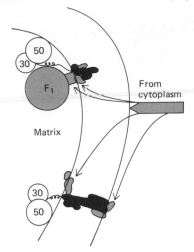

Figure 17-4. Hypothetical scheme concerning the cooperation between mitochondrial and cytoplasmic protein synthesis in the assembly of cytochrome oxidase and ATP synthetase. Mitochondrially synthesized subunits are shown in solid black (in the inner membrane) and cytoplasmically synthesized subunits are shaded. *(From E. Ebner, T. L. Mason, and G. Schatz,* J. Biol. Chem., **248**:*5378, 1973.)*

the other seven are complementary to the H strand. Although there happen to be eight separable mRNAs and eight known mitochondrially synthesized proteins, it should not be assumed that each of the mRNAs codes for one of the proteins, since some evidence has been obtained for the existence of one or two other polypeptides. Regardless of whether or not there is a one-to-one correlation between the known mRNAs and the known polypeptides of the mitochondria, it is apparent that the major synthetic activities of the mitochondria can be accounted for.

ORIGIN OF MITOCHONDRIA

It has been noted in several chapters that speculation exists as to the origin of the mitochondrion (and chloroplast). Since it will never be possible to directly test one or another hypothesis, it is safe to assume that the definitive answer will never be known and, therefore, that this topic will continue to invite new speculations. There are basically two opposing general theories with regard to the origin of the mitochondrion of eukaryotic cells. In one theory, the organelle arose as a result of the association of two previously independent organisms, one of which evolved into the organelle itself. In the other theory, the mitochondrion evolved gradually from preexisting cellular components without the intervention of another organism. We will consider briefly these two opposing concepts.

The Endosymbiont Theory

In the endosymbiont theory, the ancestor of the eukaryotic cell (we will call this organism a *protoeukaryote*) was a large anaerobic, heterotrophic, prokaryote that obtained its energy by a glycolytic pathway. Unlike present-day bacteria, this organism is proposed to have had the ability to take up particulate material. At some point in time (possibly as early as 2.5 billion years ago), photosynthetic prokaryotes emerged on the scene and began to use water as a source of reducing electrons, an event which was accompanied by the release of oxygen. This type of prokaryote is believed to have resembled the blue-green algae of the present time. With the availability of oxygen in the atmosphere, a new and more energetic type of metabolism became possible and prokaryotes evolved aerobic pathways to take advantage of this new type of electron transport "sink." The endosymbiont theory postulates that a condition arose in which a large, particularly complex, *anaerobic* prokaryote took up a small aerobic prokaryote into its cytoplasm and retained it in a permanent state. Since numerous examples are known of existing symbiotic relationships between two types of cells, for example, the coexistence of the alga *Chlorella* within the cytoplasm of *Paramecium,* it is not unreasonable to presume that such an event could have taken place. In the case of the symbiotic proto-eukaryote we have been discussing, given a long period of evolutionary coexistence, certain of the genetic functions of the aerobic symbiont were transferred to the nuclear genome of the host cell, leaving the mitochondrial genetic information and protein-synthesizing machinery for the production of a few hydrophobic proteins within the organelle itself. Some of the evidence used in favor of the endosymbiont theory includes:

1.   The existence of intracellular endosymbionts among current organisms points to the feasibility of the theory.

2.   The presence of a genome and a utilization apparatus enclosed within a membrane is reminiscent of the basic organization of an intact organism. Proponents of this theory would argue that this arrangement could not have evolved in other ways.

3.   The properties of the protein-synthesizing machinery of the mitochondrion closely ally it with the prokaryotic apparatus. For example, the inhibitors which block protein synthesis in prokaryotes are effective on mitochondria while those which block protein synthesis by cytoplasmic ribosomes do not inhibit mitochondrial activity. Both mitochondrial and bacterial systems use N-formylmethionine as the initiating amino acid. The soluble initiation and elongation factors from bacterial systems can substitute for those from the mitochondria, while the comparable factors from eukaryotic cytoplasmic systems cannot be used in mitochondrial protein synthesis.

4.   The striking similarity that exists in the structure and function of the inner mitochrondrial membrane and the plasma membrane of bacteria.

The Theory for Direct Evolution

In this second hypothesis it is proposed that mitochondria evolved gradually, rather than being obtained in some "quantum" step of swallowing another organism. In this theory, the forerunner of the eukaryotic cell was a large, complex, *aerobic* prokaryote. This prokaryote would be expected to have all of the cytochromes and other machinery necessary for aerobic metabolism, with the bulk of it located within the plasma membrane where it is found in existing prokaryotes. In this theory, it is the plasma membrane of the protoeukaryote that would become the inner membrane of the mitochondrion. The transformation from plasma membrane to mitochondrial membrane would occur gradually over time. The selective pressure for the change would result from the need for a larger and larger respiratory surface to accommodate the needs of an increasingly larger and more complex organism. It is presumed that the development of a membrane-bound mitochondrion arose as a result of infoldings of the plasma membrane as occurs in existing bacteria in the form of mesosomes. It is generally assumed in this theory that the ribosomes required for mitochondrial protein synthesis were those that were originally enclosed within the infolded plasma membrane. In other words, the machinery for protein synthesis was derived from that present in the cytoplasm of the prokaryote. After a period of evolutionary time, the two types of protein-synthesizing systems, i.e., that of the mitochondria and the cytoplasm, diverged from one another. It is postulated that the system that remained within the mitochondrion retained more of the characteristics of the bacterial system, a point which would explain its chloramphenicol sensitivity, etc.

The last major point in this theory that needs to be considered is the origin of the mitochondrial genome. It has been proposed that the DNA of the mitochondrion arose by the introduction of a plasmid into the cytoplasmic organelle. The existence of semi-independent circular DNA molecules in both bacteria (as plasmids) and eukaryotic cells (as in the amplified nucleoli of oocytes) is well established; therefore, the proposition that such a molecule gave rise to the mitochondrial genome is certainly feasible. Once the DNA was present in the organelle, along with the ribosomes and other components, the basic organization of the mitochondrion as we presently know it had evolved.

Proponents of the second theory suggest that the resemblance between mitochondrial and bacterial protein synthesis is superficial. They argue that, since both the mitochondrial and cytoplasmic systems are derived from prokaryotic ancestors, it is not surprising that similarities should exist. Those features, such as chloramphenicol sensitivity, that are found in mitochondrial systems are simply ones which were subsequently lost in the cytoplasmic system. Furthermore, they argue on the basis of mito-ribosome structure that the mitochondrial system is, in many ways, no more similar to the bacterial system than to that found in the cytoplasm. Furthermore, proponents of the direct evolution theory believe that the protoeukaryote could not have been a "primitive" anaerobic cell but rather must have been a relatively advanced aerobic form. In addition, these theorists believe that the transfer of genetic information from a cytoplasmic endosymbiont to the nucleus is an unlikely occurrence, at least more unlikely than the transfer in the other direction of a semi-independent nuclear genome, i.e., a plasmid, to the organelle.

REFERENCES

Bandlow, W., et al., eds., *Genetics, Biogenetics, and Bioenergetics of Mitochondria.* Walter de Gruyter, 1976.

Birky, C. W., Jr., Bioscience **26,** 26–33, 1976. "The Inheritance of Genes in Mitochondria and Chloroplasts."

———, Perlman, P. S., and Byers, T. J., eds., *Genetics, and Biogenesis of Mitochondria and Chloroplasts.* Ohio State University Press, 1975.

Bogorad, L., Science **188,** 891–898, 1975. "Evolution of Organelles and Eukaryotic Genomes."

Brinkley, B. R., and Porter, K. R., eds., symposium on "Biogenesis of Mitochondria," *International Cell Biology 1976–1977,* pages 235–263, Rockefeller University Press, 1977.

Ellis, R. J., Biochim. Biophys. Acta **463,** 185–215, 1977. "Protein Synthesis by Isolated Chloroplasts."

Gillham, N. W., Ann. Rev. Genet. **8,** 347–392, 1974. "Genetic Analysis of the Chloroplast and Mitochondrial Genomes."

Goodenough, U. W., and Levine, R. P., Sci. Am. **223,** 22–29, Nov. 1970. "The Genetic Activity of Mitochondria and Chloroplasts."

Kroon, A. M., and Saccone, C., eds., *The Biogenesis of Mitochondria.* Academic, 1974.

Lloyd, D., *The Mitochondria of Microorganisms.* Academic, 1974.

Margulis, L., Sci. Am. **225,** 48–57, Aug. 1971. "Symbiosis and Evolution."

———, *Origin of Eukaryotic Cells.* Yale University Press, 1972.

Packer, L., and Gomez-Puyou, A., eds., *Mitochondria: Bioenergetics, Biogenesis and Membranes.* Academic, 1976.

Raff, R. A., and Mahler, H. R., Science **177,** 575–582, 1972. "The Nonsymbiotic Origin of Mitochondria."

Sager, R., Sci. Am. **212,** 70–79, Jan. 1965. "Genes Outside the Chromosome."

———, *Cytoplasmic Genes and Organelles.* Academic, 1972.

Schwartz, R. M., and Dayhoff, M. O., Science **199,** 396–403, 1978. "Origins of Prokaryotes, Eukaryotes, Mitochondria, and Chloroplasts."

Tedeschi, H., *Mitochondria: Structure, Biogenesis and Transducing Functions.* Springer-Verlag, 1976.

Tzagoloff, A., Bioscience **27,** 18–23, 1977. "Genetic and Translational Capabilities of the Mitochondrion."

CHAPTER EIGHTEEN
The Immune System

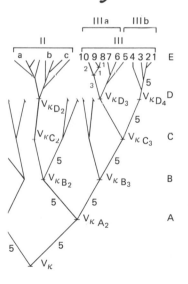

The immune system is an extremely complex subject. An introductory discussion of this nature can only serve to cover the broadest points and illustrate several important principles developed in earlier chapters. The immune system functions primarily as a system of defense at the molecular level. The targets are molecules recognized as foreign by the individual; the weapons are proteins called *antibodies* (or immunoglobulins). The cells responsible are those of the lymphoid tissue. In mammals this includes the thymus, bone marrow, lymph nodes, spleen, certain tissues of the gut (e.g. Peyer's patches), and the fetal liver. Lymphocytes armed with antibody are released from these sites to circulate through the body in the lymph and blood.

ANTIBODY STRUCTURE AND FUNCTION

Each animal has within its body, macromolecules which distinguish it from other members of the same species or a different species. The identification of one's own materials as "self" and those of an outside source as "nonself" forms the basis for the immune system's ability to react

789

against foreign substances. For example, if proteins extracted from one animal, such as a mouse, are injected into another animal, such as a rabbit, the latter will recognize these proteins as being foreign, i.e., distinct from its own proteins. It will respond to them by the production of antibodies. Any substance (whether protein, carbohydrate, nucleic acid, or other) that evokes the production of antibodies is termed an *antigen*. A closer look at a large molecular weight antigen indicates that the molecule is composed of a number of distinctive subregions, each having a defined spatial and electronic configuration. These smaller sections of a large molecule are termed *antigenic determinants* and provide the actual stimulus for the eventual production of a particular antibody. In other words, the combining sites of antibody molecules are directed against small sections of a macromolecule rather than the entire molecule itself.

Once produced, the antibody is capable of reacting chemically with the antigen responsible for its production or any other molecule having the same determinant. One of the most remarkable features of the immune system is its specificity. If a particular antigen, such as measles virus, is injected into an animal, the antibodies formed in response to that injection are highly specific: they will combine only with measles virus protein and not with protein obtained from any other source. If one tests the combining powers of antibodies present in an immune serum, one often finds that two proteins having only a single amino acid difference can be distinguished. Several very important questions are raised by these observations. How can certain antibody molecules combine with one antigen and not with another? That is, what is the basis of antibody specificity? This question can be answered by considering antibody structure and variability as discussed below. How does a given antigen evoke the production of an antibody that can specifically combine with it? This latter question will be considered in a following section.

Antibodies are proteins (termed *immunoglobulins*) built of two types of polypeptide chains. These two chains, referred to as heavy (H) chains (molecular weight of 50,000 to 70,000) and light chains (molecular weight of 23,000) occur in pairs linked to one another by disulfide bonds. The most widely studied class of immunoglobulin, IgG, is composed of two light chains and two heavy chains arranged in the manner shown in Fig. 18-1. The light chains of IgG (or any class of Ig) are of two distinct types, kappa (κ) and lambda (λ). In contrast, there is only one type of H chain found in all IgG molecules.

Five different classes of immunoglobulin (IgA, IgD, IgE, IgG, and IgM have been identified), each having a somewhat different structure (Table 18-1) and, presumably, a different set of functions. Each of these types of immunoglobulin is built from pairs of heavy and light chains. All classes utilize the same two types of light chains, though each class has its own distinct heavy chain. The three classes representing the bulk of the antibody content of serum are IgM, IgG, and IgA. IgM is a very large complex containing five pairs of L-H subunits held together by a totally different polypeptide termed the *J chain*. IgM molecules appear in the serum most rapidly after contact with a given antigen. With the

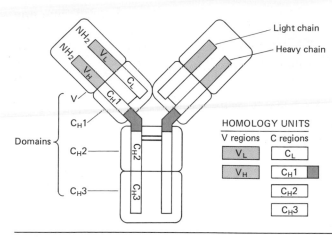

Light chain

Heavy chain

HOMOLOGY UNITS

V regions	C regions
V_L	C_L
V_H	C_H1
	C_H2
	C_H3

Figure 18-1. A model depicting the basic structure of the immunoglobulin molecule, IgG. The polypeptide chains are indicated by the bars in the diagram, while the enclosing boxes represent the domains of the molecule. Each domain (composed of a segment of two chains) folds independently of other regions of the molecule and has specific functions to carry out. It is the V domain which provides the molecule with its combining site specificity. *(Reproduced, with permission, from L. Hood, J. H. Campbell, and S. C. P. Elgin,* Annual Reviews of Genetics, *Volume 9. © 1975 by Annual Reviews Inc.)*

passage of time, the IgM molecules are replaced by IgG species having the same antigen-combining sites. IgA molecules which also have the basic subunit structure shown in Fig. 18-1 (with the gamma H chain replaced by an alpha H chain) can exist as monomers, dimers, trimers, or even higher-order complexes. Although IgA is also found in the serum, it is present as the primary immunoglobulin in the secretions of various exocrine gland cells. These secretions include nasal mucus, tears, and milk.

The basis of antibody specificity is revealed upon consideration of the amino acid sequences of the component polypeptide chains. The first step in amino acid sequence analysis (page 82) is to obtain a preparation of purified protein. However, under normal conditions it is impossible to obtain a purified preparation of a given antibody since each individual produces many different antibody molecules, all of which are similar enough in structure to be very difficult to separate. Even after the injection of one antigen (or a molecule containing only one antigenic determinant), antibodies with many different combining sites are produced, all of which are capable of combining with the antigen. A way out of this dilemma has been found by taking advantage of a particular disease condition.

TABLE 18-1
Properties of Immunoglobulin Classes

Class	Serum concentration (mg/ml)	Molecular weight	Sedimentation coefficient(s)	Light chains	Heavy chains	Chain structure
IgG	12	150,000	7	κ or λ	γ	$\kappa_2\gamma_2$ or $\lambda_2\gamma_2$
IgA	3	180,000	7, 10, 13	κ or λ	α	$(\kappa_2\alpha_2)_n$ or $(\lambda_2\alpha_2)_n$
		500,000				
IgM	1	950,000	18–20	κ or λ	μ	$(\kappa_2\mu_2)_5$ or $(\lambda_2\mu_2)_5$
IgD	0.1	175,000	7	κ or λ	δ	$\kappa_2\delta_2$ or $\lambda_2\delta_2$
IgE	0.001	200,000	8	κ or λ	ϵ	$\kappa_2\epsilon_2$ or $\lambda_2\epsilon_2$

NOTE: $n = 1$, 2, or 3.
SOURCE: BIOCHEMISTRY by Lubert Stryer. W. H. Freeman and Company. Copyright © 1975.

Amino acid sequence data have been obtained from immunoglobulins of patients with multiple myelomas, tumors of lymphoid cells. In patients with these tumors, large amounts of a single class and molecular species of antibody are produced and secreted. The particular species of immunoglobulin produced depends upon the particular cell that becomes malignant (as described below, a given lymphocyte is destined to make only one species of antibody). Different patients produce distinct immunoglobulin molecules. When the amino acid sequences of several light and heavy chains of IgG from different myeloma patients were compared, an important pattern was revealed (Fig. 18-2). In the group of kappa light chains that were studied, it was found that one-half of each chain was constant in amino acid sequence among all the kappa chains while the other half was variable from patient to patient. Similarly, comparison of the amino acid sequences of several lambda chains from different patients revealed that they too consisted of a section of constant sequence and a section whose sequence varied from immunoglobulin to immunoglobulin. Further analysis indicated that the heavy chains of these antibodies also contained a variable (V) and a constant (C) portion. In the case of the heavy chain, approximately one-quarter (110 amino acids at the amino end) has a variable amino acid sequence; the remaining three-quarters being constant for all species of IgG. Regardless of the Ig class, the constant portion of the heavy chain for that class can be divided into sections of approximately equivalent length which are clearly homologous to one another. There are three such *homology units* in the H chain for an IgG molecule (Figs. 18-1, 18-2) designated C_H1, C_H2, and C_H3. It would appear that the component sections of the C part of the heavy chains arose during evolution by the duplication of an ancestral gene. Similarly, the amino acid sequence of the constant portion of light chains bears sufficient similarity to the sequences of the homology units of the heavy chain to indicate an evolutionary relationship as well. A closer look (in both humans and mice) has revealed that the variable portions of both the heavy and light chains contain subregions which are highly variable, i.e., *hypervariable*, from Ig to Ig. The hypervariable portions of the chains (three in the L and four in the H) contain deletions and insertions of amino acids as well as substitutions.

As shown in Fig. 18-1, each IgG molecule contains two *pairs* of polypeptide chains, each consisting of one H and one L. Various types of analyses have indicated that in the complete immunoglobulin molecule the V part of each L chain is associated with the V part of one of the H chains. The specificity of a particular antibody molecule is determined by the amino acids present in the antigen-combining sites (two identical sites

Figure 18-2. The general structure of the heavy and light chains of an IgG molecule. Each half of the light chains contains approximately 110 amino acid residues. The heavy chain has approximately 450 amino acid residues, about one-quarter of which compose the variable region. The constant portion of the heavy chain in IgG molecules is composed of three homology units. The location of the hypervariable regions in this diagram (indicated by black bands in V regions) has no relationship to their positions in the actual amino acid sequence.

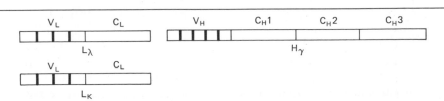

per IgG molecule). Each combining site is made of the variable portion of both a heavy and light chain. Given the availability of large amounts of unique myeloma immunoglobulin molecules, the question was raised as to whether or not the immunoglobulins secreted by these tumors were capable of combining with any known substance, i.e., were they indeed antibody molecules? If such an antigenic determinant could be found, then the opportunity for studying the interaction between a given antigen and antibody would be greatly improved. The search for "antigens" complementary to purified immunoglobulins has been successful in a few cases. For example, one human myeloma protein was found to bind vitamin K in a highly specific manner. The discovery of suitable "antigens" has recently led to crystallographic studies which have defined the precise, three-dimensional organization of the combining sites of antibody molecules. As expected, it is the hypervariable regions of each chain which make up the walls of the combining site and are therefore most intimately involved in antigen binding.

THE BASIS OF ANTIBODY DIVERSITY

An important question in the study of the immune system concerns the number of antibodies with different combining sites that an individual can make. Although there is no current answer to this question, it is generally estimated that a person might be able to synthesize over 1000 different light or heavy chains. If we assume that any H chain can combine with any L chain, we would conclude that a person might be able to synthesize in the neighborhood of 10^6 to 10^7 different species of immunoglobulin. Whatever the actual number might be, it is clear that the ability of the immune system to produce a tremendous variety of different antibodies is one of the most remarkable biochemical feats of living systems, and many important questions about mechanisms arise.

In the previous discussion, combining-site specificity was attributed to the V portions of the polypeptide chains that make up each site, while the C portions contribute to that part of the immunoglobulin molecule that is similar among antibodies. If the capacity to produce thousands of different antibodies exists within each individual as a result of the large numbers of different variable portions of each chain which can be synthesized, then an equivalent number of genes that code for these polypeptides must also be present.

One question of basic importance is whether the large number of genes required for immune function are present at the beginning of development in the DNA of the sperm and the egg, or whether this great diversity arises during the development of each animal. Since the amino acid sequences of the V regions of both the L and H chains are similar to one another, it is believed that the genes are related and that the DNA sequences which code for them arose as a result of duplication, mutation, and recombination. The question centers on the events surrounding the generation of the large numbers of variable sequences. In the *germ-line theory*, the variable gene diversity has arisen during evolution causing the entire spectrum of sequences to be present in the DNA of the gametes. In the *somatic-recombination theory*, the diversity arises by some special process in the immune cell line during development and differentiation. At the present time a considerable body of evidence has been gathered, but the definitive conclusions can not yet be made.

Attempts to answer the question of the origin of antibody diversity have been largely based on two experimental approaches, the analysis of amino acid sequence differences and the analysis of nucleic acid hybridization kinetics. In the former type of study, the polypeptides are organized into genealogical trees (Fig. 18-3) on the basis of amino acid sequence differences. Attempts are then made to determine how much of this diversity could be expected to occur within a single individual and how much during evolution. In the latter type of analysis, attempts are made to determine the number of copies of a particular DNA sequence present in the germ-line DNA that are complementary to a particular mRNA (or cDNA) probe. Both types of analyses make assumptions which limit the reliability of the conclusions. Although the topic remains controversial, it appears that the correct assessment of the origin of antibody diversity might lie between the extremes of a strict germ-line or strict somatic-recombination viewpoint. On the basis of the bulk of the evidence, it appears that a relatively small number of variable genes are inherited through the gametes (rather than one or a large number), and that this small collection of genes is then amplified by some type of somatic process of duplication and mutation to form a much larger number in the adult. For example, it may be found that the hypervariable stretches of the V portion of the H and L chains arise by somatic generation while the differences in the remaining parts of the variable regions (forming subgroups) reflect an actual diversity in the germ-line DNA. Whatever the precise situation may be (and it may vary from one type of L or H chain to another), it is likely that some new types of molecular mechanisms will have to be uncovered to explain the events.

Figure 18-3. A hypothetical genealogic tree for human V regions. The tree is constructed from a hypothetical series of amino acid sequences of V regions. The genetic events responsible for generating this genealogic pattern could occur, in part, during somatic differentiation (somatic theory) or entirely during the evolution of the species (germ-line theory). The fine details of one region of the tree are represented by V regions 1–10. *(From L. Hood et al.,* Cold Spring Harbor Symp. Quant. Biol., **41:**819, *1976.)*

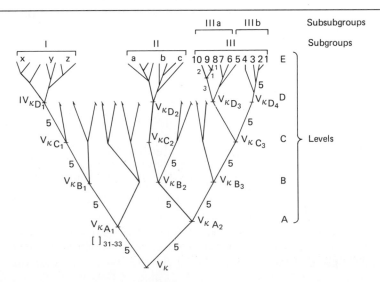

TWO GENES, ONE POLYPEPTIDE

Another important question in immunology concerns the presence in a given H or L polypeptide chain of two regions (a V and a C) which appear to have separate genetic origins. Analysis of mRNAs for the L and H chains of immunoglobulins indicate that a single mRNA molecule contains the codes for both the V and C regions. In other words, the L or H polypeptide chain is *not* formed by the joining of completed V and C polypeptides. Therefore, if all of the light chains (of a given type, κ or λ) produced by an individual have identical C portions but diverse V portions, then one or the other of the following conditions must exist.

1. There are many copies of each type of C gene so that each variable gene is present next to one of these copies. In this way an mRNA containing both sequences in tandem can be synthesized.

2. There is one (or a few) copy of each type of C gene and it can become associated with different V genes in different cells. In other words, some type of DNA-joining mechanism exists whereby two separate DNA sequences are brought together in such a way that a single mRNA containing both their information can be transcribed.

The evidence obtained so far on this question strongly suggests that the second alternative is the correct one. Hybridization studies, for example, indicate that the number of C genes is very limited and, therefore, the same segment of DNA must be associated with different V genes in different cells. Evidence of a more direct nature has also been obtained. In a recent hybridization study it has been found that the V and C genes complementary to a particular kappa chain mRNA appear to be separated from one another in the DNA of the embryo while they are much closer in the DNA of the lymphoid tumor cells responsible for secreting this polypeptide.* Several possible models by which the movement of a C gene next to a V gene could occur are shown in Fig. 18-4. Regardless of the mechanism, it would appear that the immunoglobulin is a highly unusual protein in having its polypeptide chains coded for by two genes.

THE CLONAL SELECTION THEORY

Regardless of the means by which cells of the immune tissue acquire a great number of different genes that code for immunoglobulins, some mechanisms must exist whereby appropriate antibodies are produced in response to specific antigens. The theory presented in detail by Sir Macfarlane Burnet in 1957 to explain the general basis of antibody production, the *clonal selection theory,* has gained virtually complete acceptance. The theory is based on several premises.

---

* Recent evidence suggests that, even though the V and C DNA sequences move near one another, a portion of untranslated (but presumably transcribed) DNA remains between them constituting a gene insert (page 555) of approximately 1250 bases.

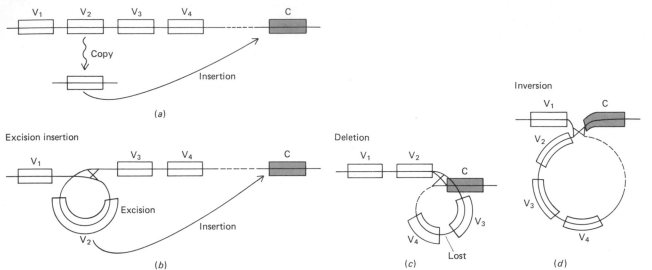

Copy insertion

(a)

Excision insertion

(b)

Deletion

(c)

Inversion

(d)

Figure 18-4. Several possible models for V-C joining at the DNA level. *(From S. Tonegawa, N. Hozumi, G. Matthyssens, and R. Schuller,* Cold Spring Harbor Symp. Quant. Biol., *41:881, 1976.)*

1.   As the cells of the immune system differentiate, they become capable of producing only one species of antibody molecule, i.e., Igs with a *particular* V region in both their H and L chains.

2.   The entire spectrum of possible antibody-producing cells is present within the lymphoid tissues *prior to stimulation by antigen* and, therefore, independent of the presence of foreign materials (Fig. 18-5).   In other words, the preliminary step in the differentiation of specific lymphocytes, i.e., that step in which they become specified to produce only one type of antibody molecule, occurs in the absence of a potential antigen for that antibody.

3.   The ability of an antigen to elicit the production of a complementary antibody molecule results from the recognition, i.e., the *selection* by the antigen of the appropriate antibody-producing cell.   Those cells selected by antigen respond by proliferation (Fig. 18-5) to produce a clone of cells capable of producing the suitable antibody.   The basis for the selection process resides in the cell membranes of the lymphocytes.   Receptor molecules (discussed below) specific for a particular antigenic determinant are embedded in the membrane and are capable of interaction with antigen.

All aspects of the clonal selection theory have been confirmed.   Rather than discuss the types of evidence that led up to the proposal and acceptance of this very important concept in cell biology, one recent experiment by G. J. V. Nossal and coworkers will be described which illustrates the theory's major tenets.   It is estimated that, among a normal population of mouse spleen lymphocytes, approximately 1 in 20,000 cells can be triggered to form antibody to any given antigenic determinant (termed a *hapten* when present as an isolated molecular group attached to an "inert" carrier).   In the first step of the experiment, a particular population of lymphocytes was selected on the basis of its interaction with a particular hapten, in this case NIP. This step should mimic the selection of the appropriate lymphocyte by antigen during the normal course of an immune response in vivo. In order to select a population capable of interacting specifically with NIP, this hapten was complexed with gelatin and prepared as a thin layer on the bottom of a petri dish. Approximately 10^8 spleen cells were added to the dish at 4°C and the dish was rocked to allow the cells and the immobilized hapten to come into contact with one another. After about 15 minutes it was found that approximately 1 in 2,000 cells were firmly bound to the haptenated gelatin layer, while all of the other cells could be removed by washing. The cells attached to the NIP-gelatin were then removed from the dish simply by melting the gelatin. When this very small subpopulation of cells was tested for their ability to produce antibodies, it was found that approximately 1 in 3 cells began to synthesize anti-NIP antibody. A tremendous enrichment with respect to the formation of anti-NIP antibody (from 1 in 20,000 to 1 in 3) had been accomplished solely on the basis of an affinity of the cell surface receptor for the chosen antigen.

T AND B LYMPHOCYTES

The results of a large variety of studies have indicated that there are two major groups of lymphocytes, T lymphocytes and B lymphocytes. Although these two types of cells may appear morphologically identical, they are readily distinguished by the presence or absence of several cell surface proteins. As will be apparent in the following discussion, these two types of cells carry out very different immune functions. The letters T and B used in referring to these cells denote the sites within the body in which

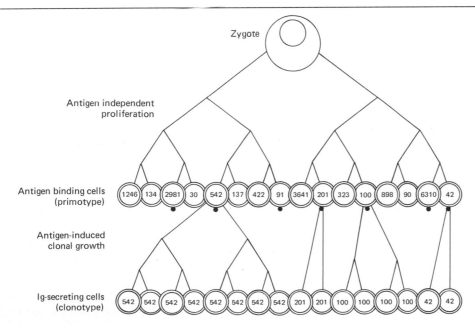

Figure 18-5. A model of the clonal selection theory. In the first phase of differentiation, a large variety of lymphocytes are generated, each capable of synthesizing a single antibody type (indicated by the particular number). The diversity of DNA sequences could arise by a somatic process (which this diagram was initially intended to illustrate) or by an evolutionary one. The second phase of differentiation involves the selection of specific lymphocytes (B cells in this case) which are induced to proliferate and form antibody-secreting clones. *(From G. M. Edelman,* Cold Spring Harbor Symp. Quant. Biol., **41:892,** *1976.)*

these lymphocytes undergo their primary antigen-independent differentiation, thereby becoming restricted to the production of a single species of antibody. T lymphocytes are those lymphoid cells whose progenitors ("ancestors") had migrated into the thymus at an early stage wherein they underwent a period of proliferation and then differentiation into immunocompetent cells. From the thymus, these T cells migrate into the circulation to populate the peripheral lymphoid tissues such as the spleen and lymph nodes. If the thymus is removed from a bird or mammal at an early stage, T lymphocytes are absent. In contrast, B lymphocytes undergo their differentiation into immunocompetent cells at some site other than the thymus. In birds, B lymphocytes differentiate in a structure known as the bursa of fabricius, whose early removal results in the absence of B lymphocytes. The counterpart of the bursa in mammals has not yet been identified with certainty, although some evidence indicates that the fetal liver may perform this function.

Humoral Immunity

Just as there are two broad classes of lymphocyte, there are two broad categories of immunity, *cell-bound* and *humoral*. In cell-bound immunity, the antibodies remain at the surface of a T lymphocyte and the destruction of the target antigen involves the direct participation of that lymphocyte. Humoral immunity, which is carried out by B cells, refers to the secretion of immunoglobulin into the blood plasma or lymph as soluble antibody. These two groups of immune response can be dissociated to a large extent. In humans, for instance, there is a disease (congenital agammaglobulinemia) in which humoral antibody is deficient and cell-bound immunity is normal. In contrast, congenital thymus deficiencies produce individuals with greatly impaired cell-bound immunity but relatively high serum antibody levels. The two types of immunity can be correlated with the two type of lymphocytes.

The ability of an individual to secrete specific antibodies into the blood stream is one of the bases of our defense against invading pathogenic organisms, whether as a response to an infection itself or an immunization. The cells responsible for the production of circulating immunoglobulins are the plasma cells, descendants of the B cells. The plasma cells are the end products of the B cell line, cells which have differentiated (in response to the presence of antigen) in order to secrete large quantities of antibody. Although the precise nature of the steps involved is not known, the combination of antigen with receptors on the surface of the B lymphocyte results in the proliferation of the B lymphocyte to form a clone of cells (called *blast* cells), all of which are specified to make the same antibody. Some of these blast cells will then differentiate into plasma cells and begin to secrete antibody, while others will remain in the lymphoid tissues as "memory" cells to respond rapidly at some later date if that antigen becomes reintroduced. It is the memory aspect of the immune response to which booster immunizations are geared; the reintroduction of an antigen can cause a much more rapid production of antibody than occurred after the initial injection.

The nature of the receptor on the surface of the B lymphocyte (that molecule which was capable of interacting with the NIP-gelatin in the experiment described above) has been clearly identified. As might be expected, the receptors on the surface of the B lymphocyte are immuno-globulin molecules having the identical combining site as the antibody which the descendent plasma cells will eventually secrete. It is estimated that each B lymphocyte has between 30,000 and 100,000 Ig receptors. Once the B cell is converted into an active plasma cell, it is estimated that approximately 10^7 antibody molecules are produced and secreted per hour.

Cell-mediated Immunity

In contrast to the B lymphocytes, T lymphocytes are involved in a wide variety of immune functions, all of which involve the participation of the T lymphocyte itself. In all of its various roles, the T lymphocyte carries out its activity by direct interaction with some other type of cell. The interacting cell(s) can be another T lymphocyte, a B lymphocyte, a macro-phage, or a target cell. In the latter case, the antigen to be recognized is generally found at the surface of a target cell. Fortunately, many of these types of intercellular interactions can be made to occur in an in vitro cul-ture system and are therefore more readily investigated. The existence of cell-mediated immunity can be clearly demonstrated in connection with the rejection of foreign cells. The ability of animals to reject grafts ob-tained from another member of the species has been known for a long time. Many of the glycoproteins located on cell surfaces are highly specific an-tigenic markers to which the immune system can respond. One group of cell surface glycoproteins, the *histocompatibility antigens,* is particularly important in the rejection of foreign cells. Histocompatibility antigens, which are widespread among tissues of the body, are coded for by a num-ber of different genes for which many different alleles of each are present in the population. In man the two most important histocompatibility gene loci are the HLA-A and HLA-B sites (the comparable sites in the mouse are the H-2K and H-2D loci). Since there are so many possible alleles which can occur at these loci, the probability that two individuals will have the same alleles is very unlikely, even among nonidentical siblings. Graft rejection is therefore highly likely after tissue transplantation in all but identical twins, or individuals whose tissue types have been sero-logically matched, or highly inbred strains of laboratory animals.

The basis of graft rejection is an attack by T lymphocytes which have proliferated in response to the presence of foreign tissues. In mice that have been deprived of thymus tissue from birth, graft rejection does not occur and such grafts will remain in place. The ability of T lymphocytes to destroy cells containing foreign surface molecules is believed by many to play an important role in the detection and destruction of potential tumor-forming cells. The conversion of a normal cell to the malignant state is accompanied by changes in the nature of the cell surface (page 808), changes that should make that cell vulnerable to attack by immunocom-petent T lymphocytes. Although the topic remains controversial, T lym-phocytes are believed by many investigators to carry out a process termed *immunological surveillance,* whereby the surfaces of cells are continually examined for the appearance of antigenic molecules. Cells bearing inappropriate antigens would be destroyed. Inherent in this theory is the concept that tumor cells are being formed and destroyed throughout life and only those malignant cells that escape immune destruction can lead to the formation of tumors. Similarly, the infection of cells with various types of viruses also leads to alterations of the cell surface which can be recognized by immunocompetent T lymphocytes. It is the concept of immunological surveillance that provides investigators with the primary reason for the existence of such a complex cell-mediated immune function.

Not all T lymphocytes are cytotoxic, i.e., involved in killing other cells. Other populations of T lymphocytes have been identified as having very different roles in immune function. For example, a distinct subpopulation of T lymphocytes is involved in "helping" B lymphocytes mount an effective humoral response. Whereas some antigens (termed *thymus-independent antigens*) seem to be able to evoke the proliferation and differentiation of B lymphocytes without the involvement of T lymphocytes, the response to other antigens *(thymus-dependent antigens)* by B lymphocytes is dependent upon helper T lymphocytes. In other cases, a different subpopulation of T lymphocytes becomes active and inhibits the response by the B lymphocyte. In this case the T lymphocyte is termed a *suppressor* cell. In these various interactions with other cells, there is evidence for the actual contact between cells as well as evidence for the secretion of active substances by the T lymphocyte, substances which can lyse another cell, or cause it to divide, or produce some other response. In addition to reacting with target cells or other lymphocytes, T lymphocytes also interact with macrophages, cells which are involved in phagocytizing debris. Macrophages appear to take up antigenic material and "present" the antigens to lymphocytes in such a manner as to make it extremely immunogenic to the lymphocyte. In the absence of the macrophage, the antigen may be much less efficient in stimulating an immune response. As in the case of the T lymphocyte, macrophages also secrete a variety of factors that affect lymphocyte function.

The interaction between a T lymphocyte and various other cells has generated a great deal of research during the past few years and the subject has become increasingly complex. Part of the attention centered on T cells has concerned the nature of their antigen-specific receptors. Whereas the immunoglobulin nature of the antigen receptor of the B lymphocyte is well established, its counterpart on the T cell has been resistant to analysis. Recent results suggest that the T cell receptor is a type of heavy chain containing a variable region which provides the antigen-combining site and a constant portion unlike those previously described. The T cell receptor does not appear to contain a light chain and may exist in the form of a dimer at the cell surface.

Another aspect of current research on immune cell biology concerns a collection of genes which appear to play a major determining role in T lymphocyte function. These genes, termed *Ir* (Immune response) *genes,* are located within the major histocompatibility complex of chromosome number 6 in man (chromosome 17 in the mouse) where they control the ability of T lymphocytes to respond to particular thymus-dependent antigens. If a mouse, for example, carries a mutation in a particular Ir gene, it may find itself unable to respond to the presence of a particular antigen. The inability to respond pertains to all aspects of T cell action (cytotoxic, helper, suppressor) in which that particular antigen is involved. Those antigens which are thymus-independent, i.e., ones to which B lymphocytes can respond without aid, do not require the activity of any of the Ir genes. Although over 30 different Ir loci have been detected, the means by which their products exert control over certain populations of T lymphocytes remains unclear.

In the brief discussion of graft rejection above, it was pointed out that T lymphocytes were responding to foreign histocompatibility antigen on the target cells. More recent experiments have revealed that cell surface antigens coded within the major histocompatibility complex of the genome are involved in other T cell interactions beside graft rejection (or immune surveillance). It has been found that products of both the I region and the H-2K and H2-D regions of the mouse serve as important recognition markers for T cells in their interactions with B cells, macrophages, target cells, or other T cells. For example, in order for a T lymphocyte to destroy a cell infected with a particular sensitizing virus, that infected cell must share identical H-2K or H2-D antigens. Similarly, certain types of interactions between B and T lymphocytes require that the two cells share identical I region antigens. In either case, it appears that the T cell must recognize two distinct types of molecules on the surface of the cell with which it interacts (Fig. 18-6). One of these molecules is the sensitizing antigen and the other is the cell surface histocompatibility protein which it must match. These points serve to illustrate the complex nature of specific interactions involving immunocompetent cells.

It is apparent from this discussion that the histocompatibility gene products play an important, but poorly understood, role in immune function. Interestingly, it has been found that certain human disease states (including multiple sclerosis) can be correlated with the presence of specific alleles of the HLA-A and HLA-B gene loci. The most striking correlation has been made on studies of persons suffering from a crippling disease termed ankylosing spondylitis. It has been found that virtually all of these individuals carry the HLA-B27 allele (which codes for the B27 antigen). It appears that these individuals have some type of immune deficiency which causes them to produce antibodies capable of interacting with their own tissue. This type of condition, termed *autoimmunity,* occurs in numerous other diseases including rheumatoid arthritis, rheumatic fever, and thyroiditis.

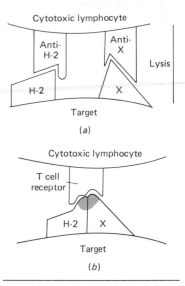

Figure 18-6. Two models for the recognition of H-2 and viral antigens. In both models X represents a viral antigen. In *a* the T lymphocyte has two independent receptors, one for protein of the histocompatibility gene and the other for the antigen. In *b* a single T cell receptor recognizes an altered histocompatibility protein on the target cell surface. *(From J. W. Schrader, R. Henning, R. J. Milner, and G. M. Edelman,* Cold Spring Harbor Symp. Quant. Biol., **41:554, 1976.)**

REFERENCES

Beale, D., and Feinstein, A., Quart. Rev. Biophys. **9,** 135–180, 1976. "Structure and Function of the Constant Regions of Immunoglobulins."

Benacerraf, B., ed., *Immunogenetics and Immunodeficiency.* University Park Press, 1975.

————, Hospital Practice **13,** 65–75, 1978. "Suppressor T Cells and Suppressor Factor."

Boyse, E., and Cantor, H., Hospital Practice **12,** 81–90, April 1977. "Surface Characteristics of T-Lymphocyte Subpopulations."

Brinkley, B. R., and Porter, K. R., eds., symposium on "Cell Surface Immuno-receptors, *International Cell Biology 1976–1977,* pages 103–127. Rockefeller University Press, 1977.

Burnet, M., Sci. Am. **204,** 58–67, Jan. 1961. "The Mechanism of Immunity."

Capra, J. D., and Edmundson, A. B., Sci. Am. **236,** 50–59, Jan. 1977. "The Antibody Combining Site."

Cunningham, B. A., Sci. Am. **237,** 96–107, Oct. 1977. "The Structure and Function of Histocompatibility Antigens."

Davies, D. R., Padlan, E. A., and Segal, D. M., Ann. Rev. Biochem. **44,** 639–668, 1975. "Three-Dimensional Structure of Immunoglobulins."

D'Eustachio, P., Rutishauser, U. S., and Edelman, G. M., Int. Rev. Cytol. Supp. **5,** 1–60, 1977. "Clonal Selection and the Ontogeny of the Immune Response."

Edelman, G. M., Sci. Am. **223,** 34–42, Aug. 1970. "The Structure and Function of Antibodies."

———, Science **180,** 830–840, 1973. "Antibody Structure and Molecular Immunology."

Feldman, M., ed., *Immune Reactivity of Lymphocytes.* Plenum, 1976.

Friedman, H., ed., Ann. N.Y. Acad. Sci. **249,** 1975. Symposium on "Thymus Factors in Immunity."

Golub, E. S., *The Cellular Basis of the Immune Response: An Approach to Immunobiology.* Sinauer, 1977.

Hood, L., Campbell, J. H., and Elgin, S. C. R., Ann. Rev. Genet. **9,** 305–354, 1975. "The Organization, Expression, and Evolution of Antibody Genes."

Hozumi, N., and Tonegawa, S., Proc. Natl. Acad. Sci. (U.S.) **73,** 3628–3632, 1976. "Evidence for Somatic Rearrangement of Immunoglobulin Genes Coding for Variable and Constant Regions."

Jerne, N. K., Harvey Lecture **70,** 93–110, 1976. "The Immune System: A Web of V-Domains."

Kunkel, H. G., and Dixon, F. J., eds., *Advances in Immunology,* vol. 1, beginning 1961, Academic.

Lerner, R. A., and Dixon, E. J., Sci. Am. **228,** 82–91, June 1973. "The Human Lymphocyte as an Experimental Animal."

Marchalonis, J. J., ed., *The Lymphocyte.* Dekker, 1977.

———, Science **190,** 20–29, 1975. "Lymphocyte Surface Immunoglobulins."

Marx, J. L., Science **188,** 245, 1975. Research News. "Suppressor T Cells: Role in Immune Regulation."

———, Science **189,** 1075, 1975. Research News. "Antibody Structure: Now in Three Dimensions."

———, Science **191,** 277, 1976. Research News. "Immunology: Role of Immune Response Genes."

Mayer, M. M., Sci. Am. **229,** 54–66, Nov. 1973. "The Complement System."

Moller, G., ed., Transplantation Reviews (Immunological Reviews), Munksgaard, Copenhagen.

Munro, A., and Bright, S., Nature **264,** 145–152, 1976. "Products of the Major Histocompatibility Complex and their Relationship to the Immune Response."

"Origins of Lymphocyte Diversity," Cold Spring Harbor Symp. Quant. Biol. **41,** 1977.

Paul, W. F., chairman, Federation Proc. **35,** 2044–2072, 1976. Symposium on "Cellular and Soluble Factors in the Regulation of Lymphocyte Activation."

———— and Benacerraf, B., Science **195,** 1293–1300, 1977. "Functional Specificity of Thymus-Dependent Lymphocytes."

Poljak, R. J., Amzel, L. M., and Phizackerley, R. P., Prog. Biophys. Mol. Biol. **31,** 67–93, 1976. "Studies on the Three-Dimensional Structure of Immunoglobulins."

Richards, F. F., Konigsberg, W. H., Rosenstein, R. W., and Varga, J. M., Science **187,** 130–137, 1975. "On the Specificity of Antibodies."

Rowley, D. A., et al., Science **181,** 1133–1141, 1973. "Specific Suppression of Immune Responses."

Sasazuki, T., McDevitt, H. O., and Grumet, F. C., Ann. Rev. Med. **28,** 425–452, 1977. "The Association between Genes in the Major Histocompatibility Complex and Disease Susceptibility."

Unane, E. R., Federation Proc. **34,** 1723–1748, 1975. Symposium on "Function of Macrophages."

Vitetta, E. S., and Uhr, J. W., Science **189,** 964–969, 1975. "Immunoglobulin-Receptors Revisited."

Williamson, A. R., Ann. Rev. Biochem. **45,** 467–500, 1976. "The Biological Origin of Antibody Diversity."

CHAPTER NINETEEN
Cancer and Aging

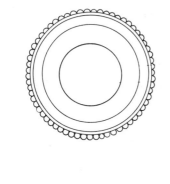

In the preceding chapters we have considered a wide range of activities that occur within cells. Given the complexity of most cellular processes, it is not surprising that cells undergo changes which lead to various physiological disturbances. Two of the most serious physiological problems that we all face are the possibility of cancer and the inevitability of aging. Both of these conditions are believed to have an underlying cellular basis, one which remains obscure in many ways. In this chapter we will examine the properties of the malignant and the aging cell, attempting to bring some of the information previously discussed to bear on these matters.

CANCER

Cancer can be defined as a disease involving heritable defects in cellular control mechanisms, resulting in the formation of malignant and usually invasive tumors. In almost every case that has been examined, the tumor appears to be *monoclonal*, i.e., derived from a single cell. The clinical picture will not concern us here; rather, we will focus our attention on that single aberrant cell and attempt to examine some of the factors that might have caused its conversion from the normal to the malignant state, some of its newly found properties as a malignant cell, and some of the reasons why these altered properties might cause it to develop into a tumor. It is important to keep in mind that we will concentrate in our discussion on cells growing in culture as opposed to cells growing within a tumor in vivo. Although the properties of malignant cells are much more readily studied in cell culture, the system is highly artificial and one must be very careful when attempting to correlate cell behavior under these very different conditions.

The most prominent characteristic of the cancer cell, whether growing in vitro or in vivo, is its loss of growth control. The *capacity* for growth, division, and movement is much the same for normal and malignant cells, at least in many cases. In tissue culture, normal cells released from inhibitory factors will grow and divide at a rate similar to that of malignant cells, as long as certain conditions are met. The difference is that, in the body, normal cells grow at a maximal rate only under controlled conditions within tissues having a high rate of turnover. Malignant cells grow indefinitely, as though the body were only a culture medium serving as a nutritive support. The difference between the rapidly growing normal and malignant cell can be illustrated using liver cells. If the greater part of a rat's liver is removed, the remaining portion regenerates rapidly, growing at the rate of about a billion cells a day during the first four days. The rate is faster than that of virtually any malignant growth. However, growth and proliferation soon slow down until by the seventh, most of the amputated liver has been regenerated. In hepatomas, i.e., primary liver tumors, growth is slower but continues indefinitely. The uncontrolled growth is one aspect of the lethality of the cancerous state, another is the tendency for clusters of malignant cells to break away from the primary growth and enter the circulation whereby they become lodged in other parts of the body. This process of cell release, termed *metastasis*, results in secondary tumors at distant locations in the body, thereby making it impossible to stop the progress of the disease by surgery. Many investigators believe that both of the above-mentioned properties of malignant cells, i.e., their loss of growth control and their detachment from other cells of the tumor, reflect certain basic changes in the cell membrane. We will return to this point below, but first we will consider the types of stimuli which can cause a cell to become malignant.

Carcinogenic Stimuli

Various types of agents—chemical, physical, and biological—have been shown to be capable of converting a normal cell in culture into a malignant one, a process termed *transformation*. Since the same agents can be used to induce tumors in the organism, it would appear that the study of the transformation of cells in vitro by these various agents has relevancy for the development of tumors in the body. Similarly, in most cases, cells transformed in culture will cause tumors when injected into an appropriate host. In most cases, the test recipient is a *nude* mouse, (one carrying a genetic defect which prevents the development of a thymus gland), which does not reject the injected cells. The problem in the study of "spontaneous" human malignancy is the unknown nature of the carcinogenic stimulus. It is generally believed that various chemicals present in our complex environment are responsible for a large percentage of human cancers. The tremendous diversity in molecular structure of the chemicals known to be carcinogenic has led to a search for some unifying principle. The common denominator which has emerged in recent years is the finding that, in order to be carcinogenic, a chemical must be either directly or indirectly *mutagenic*. Direct mutagenesis is easy to understand; compounds such as alkylating agents or acridine described in Chapter 11 result in alterations of the nucleotide sequence of the DNA. However, it was found in the early studies that many of the most potent carcinogens, such as the polycyclic aromatic hydrocarbons, were not capable of inducing mutations in susceptible bacterial cells. After considerable study, it was found that these chemicals become carcinogenic only upon activation within the body. When the subject of the endoplasmic reticulum was discussed in Chapter 7, it was pointed out that this organelle contains various enzymes which serve to detoxify drugs and other compounds. Ironically, in the process of modifying chemicals, various cellular oxidases actually convert them into carcinogenic species. The actual carcinogens are believed to be electrophilic epoxide stages which interact covalently with DNA. It is difficult to consider any target in the cell other than DNA that would allow the effect of chemical exposure, to cause cancer to appear many years later, or to allow the malignant phenotype to be retained through each successive cell cycle during the development of the tumor. One of the advances in the area of chemical carcinogenesis in recent years has been the development of an in vitro system to screen large numbers of compounds for potential carcinogenicity. The assay involves subjecting sensitive bacterial cells to chemicals in the presence of rat liver microsomal enzymes. If a given chemical proves to be mutagenic under these circumstances, it has a very high likelihood (e.g., 90%) to be carcinogenic in further tests.

Other agents capable of converting a normal cell to a malignant one include radiation and infection by tumor viruses. Radiation presumably exerts its effect via direct mutagenesis. The role of tumor viruses in the etiology of cancer is a very controversial one. As will be discussed below in some detail, a number of different types of viruses are known which can transform cells.

What is the relationship among these various cancer-causing agents? Does a cell transformed by different chemicals have the same defect, and is this defect the same as the one induced by radiation or tumor viruses? Or rather, are there various genetic sites which can be altered in different malignant cells, all of which will produce a cell with a similar loss of growth control and the ability to form tumors? These are very important questions which cannot be answered at present with certainty. One possibility that has been raised is that all forms of cancer involve the presence of a tumor virus and that the role of the chemical or physical carcinogen is simply to activate the viral agent.

The Phenotype of the Transformed Cell

It is generally assumed that, in order to understand the cancer cell, its properties relative to the normal cell must be determined. Toward this end a large number of differences, both structural and biochemical, have been catalogued. In fact, it is the diversity of the observations which poses one of the major problems. Which of the changes that occur when a normal cell becomes malignant is a primary change and which is a secondary alteration? Similarly, are any of the differences that have been detected directly responsible for the altered behavior of the cells themselves? One point should be kept in mind in the following discussion: nearly all of the observations are made upon cells transformed by oncogenic (tumor-causing) viruses, and there are questions as to the general nature of the properties of virally transformed cells. Viruses are used in these studies for the ease with which they transform a wide variety of cells growing in culture, for the predictable nature of the events that follow addition of the virus at well-defined times, and for the availability of mutants, particularly temperature-sensitive ones (page 674). By the use of temperature-sensitive viruses, transformation and reversion can be studied by simple temperature shifts in one direction or the other. In addition, transformation by viruses occurs in response to the integration of a small, well-studied genetic element. In at least a few cases, it has been shown that the transformation process, with all of its attendant changes, is the result of the activity of a *single* viral gene. Scattered evidence from spontaneous or chemically transormed cells has corroborated much of the observations on the altered properties of virally transformed cells. Another very important point to consider is the extreme variation that exists with respect to the properties of one type of malignant cell to another. Although one would hope that all malignant cells might share some basic property which reflected their altered growth control, no such universal property has yet been discovered. At best, there appear to be tendencies which are revealed by a greater or lesser variety of tumor cells.

Within the cell nucleus, the most striking and widespread alterations occur to the chromosomes. Normal cells are generally fastidious in maintaining their normal diploid chromosomal complement as they grow and divide, whether in vivo or in vitro. In contrast, transformed cells often have highly aberrant chromosome complements, a condition termed *aneuploidy*. These very abnormal karyotypes are presumably a *result* of the type of growth of the transformed cells rather than a cause of it. Regardless, it is evident that transformed cell growth is much less dependent upon a normal chromosome content than is normal cell growth. In some cases, rather characteristic chromosome changes are seen in malignant cells as evidenced by the presence of an extra band in one of the chromosomes of most cells from a Burkitt's lymphoma. The most widespread alterations involving the cytoplasm of the transformed cell concern its enzymatic properties, particularly those related to glycolysis. Malignant cells generally have a much greater level of glycolytic activity (regardless of the oxygen levels) than do normal cells. The importance of this observation is not understood. The most striking morphological changes that

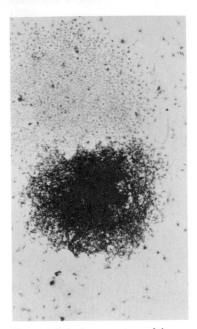

Figure 19-1. A comparison of the morphologies of two colonies, one of polyoma-transformed hamster embryo cells (bottom) and one of normal hamster embryo cells (top), at seven days after plating. *(From L. Sachs, in M. Sela, ed., "New Perspectives in Biology," Elsevier, 1964.)*

occur in the cytoplasm following transformation center around the cytoskeletal elements. The existence of an extensive interconnected network of actin-, myosin-, and tubulin-containing proteins has been referred to in several places in previous chapters. Although the exact nature and role of these structural and contractile proteins is not understood, their organization seems to be greatly disrupted following transformation. Whereas the normal cell generally contains a well-organized and oriented network of these microfilaments and microtubules, the transformed cell tends to have a reduced and/or disorganized cytoskeleton. This point will be brought up again below with regard to the possible role of this change in the loss of growth control. One of the consequences of the alterations in the cytoskeleton is a change in the overall morphology of the transformed cell and the colonies it forms (Fig. 19-1). The most extensive alterations that have been found to occur upon transformation concern the cell surface, an observation not altogether surprising considering the role of the cell surface in growth control and cell-cell interaction.

At the biochemical level, the changes at the cell surface are marked by the appearance (or increase) and disappearance (or decrease) of particular components. Although there is little evidence of widespread changes in the phospholipid nature of the bilayer, there are reports of various changes in the glycolipids, and glycoproteins of the membrane, as well as the glycosaminoglycans of the fuzzy layer (page 268). Some of the surface changes may result from the increased activity of proteolytic enzymes in the medium. One of the widespread reports on a variety of transformed cells is the disappearance (or marked reduction) from the cell surface of a high molecular weight glycoprotein. This molecule is generally referred to as LETS (Large molecular weight-External Transformation-Sensitive protein). In addition to the loss of this type of protein, transformed cells generally possess new cell surface proteins, proteins that are generally referred to as antigens because they can induce the formation of antibodies directed against the cell. In some cases these antigens are coded by the transforming virus, while in other cases they appear to be host cell derived. In some cases, proteins typical of embryonic cells appear on the surface of transformed adult cells.

In addition to the biochemical changes, numerous physiological or behavioral differences involving the cell surface have been reported. One of the most characteristic features of transformed cells in culture is their lack of contact inhibition of movement. As pointed out in Chapter 16, the membrane of a normal cell, when contacted by another cell, ceases its activity and movement in the original direction. When normal cells become surrounded on all sides by other cells, their motility ceases; they are not capable of moving over one another and they form a monolayer on the bottom of the dish. Another characteristic which generally accompanies contact inhibition of movement is density-dependent inhibition of growth (page 643). When normal cells reach a particular density in culture, their growth potential drops to a low value, and the number of cells on the dish remains relatively constant. In contrast, the movement of transformed cells is generally not obstructed or diverted when contacted by another cell, nor is its division potential reduced nearly as much as the cell density increases. In addition, malignant cells are much less dependent upon serum in the medium. As a result of these properties, cultures of malignant cells form clumps rather than monolayers and are capable of proliferating to many times the cell density of normal cultures. There are other important differences in the growth characteristics between normal and transformed cells. Whereas normal cells seem to have a limited capacity for cell division (page 819), cancer cells are seemingly immortal in the sense that they continue to divide indefinitely. In addition, cancer cells, unlike normal cells, can generally grow in a state of suspension in a medium made of soft agar or methyl cellulose. In this regard, cancer cells are said to have lost their anchorage dependence upon which the growth of normal cells depends. Other relatively widespread properties of transformed cells relative to their normal counterparts are an increased agglutinibility by lectins, increased mobility of cell surface receptors, increased transport of various compounds, decreased adhesiveness to the substrate, and lack of gap junctions between cells.

The Underlying Basis

In previous sections we have mentioned an extensive, though by no means exhaustive, number of alterations (Fig. 19-2) that occur upon transformation. The most important question concerns the relationship between these properties and the underlying causes of tumor formation. The changes listed above are simply correlations without any mention of cause

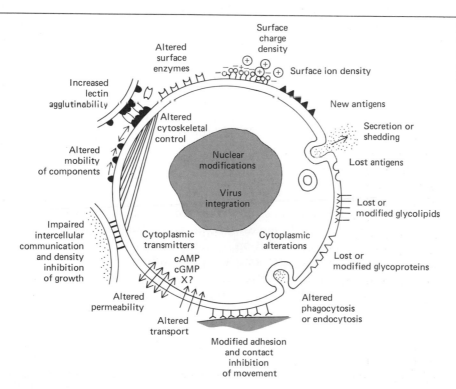

Figure 19-2. Diagrammatic illustration of various cell surface alterations found after neoplastic transformation. *(From G. L. Nicolson, Biochim. Biophys. Acta, **458**:16, 1976.)*

and effect. A particular property of transformed cells is discovered, often by chance, and various investigators begin to study it without knowing the underlying genetic basis for the change. In a sense we are dealing with several levels at which the malignant phenotype is expressed. At one level, we speak of the tumor cell's loss of growth control, or its invasiveness, or its ability to be attacked by the immune system, or the ability of tumors to metastasize. At another level, we speak of the behavior of transformed cells in culture, for example, their immortality, their anchorage independence, or their loss of contact inhibition. At the biochemical level, specific alterations in cytoplasmic elements or membrane proteins have been reported. Although these observations are generally believed to be related, the greatest challenge in this field is attempting to assign causes and effects among these various properties.

Even though we may not yet understand the basis for all of the alterations that occur after transformation, it is still important to search for the basis of the loss of growth control in the malignant cell. It was suggested in Chapters 5 and 14 that the cell surface was an important mediator of the growth potential of the cell. Two general hypotheses were presented in an attempt to explain how events at the edge of the cell might dictate policies of growth and division within the cell itself. In one theory it is suggested that the levels of the cyclic nucleotides cAMP and cGMP were of the utmost importance in determining the cell's growth potential, while in the other theory the mobility of cell surface receptors was of primary importance. In this latter theory, the information from the cell surface is carried inward by a cytoplasmic network of cytoskeletal and/or contractile proteins. Since cyclic nucleotide levels may be an important determinant of the state of the cytoskeletal machinery, the two theories may be closely interrelated.

Evidence has been obtained which supprts the concept that alterations in the levels of cyclic nucleotides and/or the cytoskeletal-contractile elements of the cell are involved in the transformation process. For example, cyclic AMP levels tend to be low in the transformed cell, as they are in the mitotic cell (page 645). Furthermore, the treatment of transformed cells with cyclic AMP causes them to temporarily develop a more normal phenotype: they become more adhesive to their substrate, they lose their increased agglutinibility, they regain their normal morphological appearance, and in some cases they regain their density-dependent inhibition of growth. The dramatic effect of cyclic AMP on cell behavior is shown in Fig. 19-3. Under certain conditions, *normal* Chinese hamster ovary cells do not exhibit contact inhibition, and they grow in multilayered clumps. Cells in these clumps are compact and randomly oriented (Fig. 19-3a). The addition of cyclic AMP to these cells converts them to elongated, fibroblastlike cells, which are contact inhibited and grow as a monolayer (Fig. 19-3b). This morphological change presumably reflects an underlying change in the distribution of cytoplasmic skeletal elements. With regard to the second theory, it has already been noted that transformed cells generally have an increased surface receptor mobility and a disorganized cytoskeletal-contractile apparatus. These are conditions one would expect to lead to a loss of growth control, on the basis of this theory.

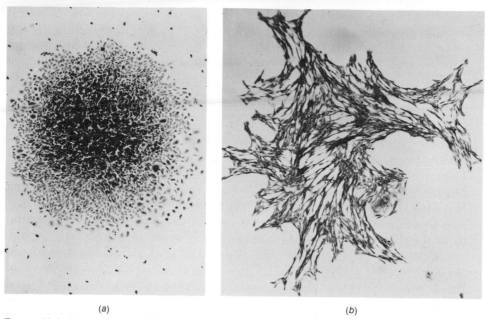

(a) (b)

Figure 19-3. Effect of cyclic AMP on the morphology of a colony of Chinese hamster ovary cells. *(a)* Grown on standard medium where contact inhibition does not occur. *(b)* Cells grown in the presence of dibutyryl AMP are now contact-inhibited. [*From A. W. Hsie and T. T. Puck, Proc. Natl. Acad. Sci. (U.S.), **68:**359, 1971.*]

Oncogenic Viruses

Viruses were first reported in tumors as early as 1908, when cell extracts were found to transmit leukemia through successive passages from fowl to fowl. In 1910 Peyton Rous began his pioneering work on the propagation of virus-induced sarcomas. Although the initial studies were met with great skepticism, the role of viruses in animal cancers was gradually established and, more recently, attention has turned to their probable involvement in some human tumors. Oncogenic viruses can be broadly divided into two heterogeneous groups based on the presence of DNA or RNA as their genetic material, i.e., the nucleic acid in the virus particle released from one cell and capable of infecting another. The transformation of a cell by an oncogenic virus is analogous in many ways to the temperate infections of bacteria (see Fig. 2-2) by bacteriophage. The viral genome becomes hidden within the host cell DNA, and the genomes undergo replication together as the virus is passed from each cell to its progeny at cell division. Whereas the typical lysogenic bacteriophage remains transcriptionally inactive (other than producing its own repressor), the oncogenic virus produces RNAs which become translated into proteins that profoundly disturb the host cell's metabolic machinery. The presence of the oncogenic virus is manifested as the transformation of a normal cell to the malignant state.

DNA Tumor Viruses

Most of the DNA tumor viruses that have been studied contain very small genomes and are not believed to be a natural factor in tumor formation. These viruses are highly infectious; cells growing in culture as well as cells growing within the body are susceptible to transformation by the addition of virus particles. The two best-studied viruses are polyoma and simian virus 40 (SV40). These two viruses, which are very similar in nature, are icosahedral types of approximately 450 Å diameter, having a circular DNA molecule (3.5×10^6 daltons) containing as few as three genetic functions, associated with a few histone molecules derived from the host cell. Two types of host cell are studied in tissue culture in connection with each virus. In the case of the permissive host cell, the virus multiplies unchecked until the cell is killed. Certain types of monkey kidney cells are permissive for SV40, while various mouse cells are permissive for polyoma. In the other case, that of the nonpermissive cell, the viral DNA becomes integrated into the host DNA and does not undergo independent replication. Rather than producing a productive infection as in the permissive cell, the virus converts the cell to the malignant state.

Of the three genes present in the SV40 genome, two code for viral capsid proteins, while the third, the A gene, codes for a protein termed the T (tumor) antigen which accumulates in the cell nucleus (Fig. 19-4a) and appears to be responsible for the transformation process. In fact, the only gene expressed in the transformed cell is that which codes for the T antigen; its presence is both necessary and sufficient for the change. It would appear that this one protein is responsible for all of the behavioral changes catalogued in a previous section. Since it is unlikely that a single protein could directly interfere with so many functions, we can conclude

Figure 19-4. *(a)* The presence of the T antigen within the nuclei of SV40-infected cells as demonstrated by the binding of fluorescent antibody. *(From B. Steinberg, R. Pollack, W. Topp, and M. Botchan, Cell, 13:21, 1978.)* *(b)–(c)* Results of an experiment which demonstrates the importance of the viral T antigen in the activation of DNA synthesis. Mouse kidney cells in a nondividing, confluent culture were injected with very small volumes of purified T antigen. The cultured cells were then incubated in ³H-thymidine for 18 to 20 hours before fixation. Those cells which incorporate ³H-thymidine *(c)* are found to be the ones into which the T antigen had been injected, as demonstrated by immunofluorescence *(b)*. [*From R. J. Tjian, G. Fey, and A. Graessmann, Proc. Natl. Acad. Sci. (U.S.), 75:1281, 1978.*]

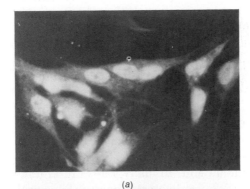

(a)

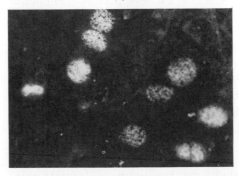

(b)

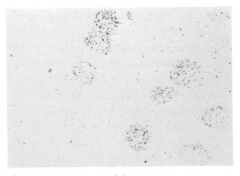

(c)

that most of the phenotypic changes that occur during transformation are secondary responses to some more basic change(s), possibly in some aspect of the cell's regulatory machinery.

The primary response to infection by polyoma or SV40 is the initiation of DNA synthesis. The role of the T antigen in this response is clearly shown by the injection of purified T protein into the cytoplasm of *normal* nondividing cells. The protein rapidly accumulates within the nucleus of the cell and soon activates DNA synthesis as revealed by its incorporation of ^{3}H-thymidine (Fig. 19-4b). Activation of DNA synthesis in a previously nondividing cell may involve the production of a variety of required enzymes (such as thymidine kinase) which precede the actual synthesis of DNA. If the cell is a permissive one, the activation of the A gene is followed by a later activation of the remainder of the genome and the subsequent production of capsid proteins. The synthesis of the components of the virus is followed by their assembly and the lysis of the infected cell. In contrast, if the cell is a nonpermissive one, the viral DNA becomes integrated into the chromosomes of the host cell, only the A gene is expressed, and the cell becomes transformed.

Another type of oncogenic DNA virus that has been widely studied is the adenovirus. These agents which cause influenzalike infections in humans are capable of forming tumors in a variety of animals. Although the genome of the adenovirus is approximately seven times that of polyoma or SV40, it has been found that only a very small portion of that genome (approximately 2400) nucleotides is actually required for transformation itself. Apart from its ability to induce tumors, adenovirus has been most valuable as a research system in providing information on the molecular biology of DNA expression in mammalian cells.

The other major group of DNA-containing viruses implicated in tumor formation, including a number of human malignancies, are the herpes viruses. The first indication that a herpes-type virus was involved in cancer came in 1964. Tumor cells from patients with Burkitt's lymphoma, a rare form of cancer endemic to Africa, were found to contain a virus, since called the Epstein-Barr virus (EBV) after its discoverers. As with other herpes viruses, a large percentage of the population has had contact with them or even harbor them, evidenced in this case by the presence of serum antibodies against this virus in most healthy individuals. It would appear that some factor in addition to the presence of the virus is responsible for the development of the cancer.

RNA Tumor Viruses

Although there are several different groups of RNA tumor viruses, all of them are basically similar and believed to be evolutionarily related. RNA tumor viruses are constructed as shown in Fig. 19-5. They consist of an inner core in which the viral RNA resides, together with a few internal proteins surrounded by an outer shell which is, in turn, surrounded by an outer envelope. The outer envelope, which contains certain virally coded glycoproteins, is involved in the attachment of the particle to a new host cell. The envelope forms from the plasma membrane of the host cell

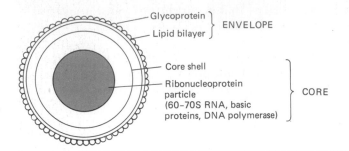

Figure 19-5. Diagram of a cross section of a typical RNA tumor virus. The diameter of the virion is about 100 to 150 nm. The diameter of the ribonucleoprotein particle is about 50 nm. *Reproduced with permission, from H. M. Temin,* Annual Reviews of Genetics, *Volume 8.* © *1974 by Annual Reviews Inc.)*

Glycoprotein ⎱
Lipid bilayer ⎰ ENVELOPE

Core shell
Ribonucleoprotein particle
(60–70S RNA, basic proteins, DNA polymerase) ⎱ CORE

by a budding process (Fig. 19-6), after that part of the membrane is occupied by viral proteins and essentially cleared of host proteins. Although it was initally believed that the RNA genome consisted of a single 70S RNA molecule, it is now known that the core contains several smaller RNAs present together in a tightly held 70S aggregate. The individual molecules are believed to be identical 35S copies which contain all of the information necessary for either transformation or productive cell infection. The best-studied core protein is the RNA-dependent DNA polymerase, commonly known as the reverse transcriptase. After the infectious particle has entered a host cell, this enzyme functions to make a complementary copy of the RNA genome in the form of DNA, which is then inserted into the DNA. Although the details have not been completely worked out, it appears that this DNA polymerase uses a tRNA molecule present in the core as a primer and produces a single-stranded DNA complement. The tRNA is then digested, a complementary DNA

Figure 19-6. Budding of a murine Friend leukemia virus from the surface of a cultured leukemic cell. *(Courtesy of E. de Harven.)*

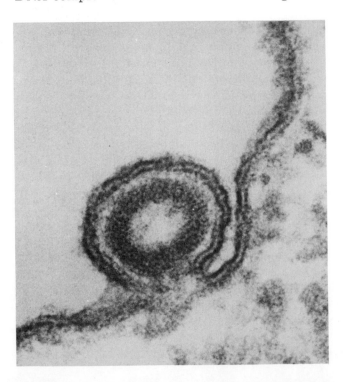

strand is synthesized to form a double-stranded DNA form of the viral genome which becomes circularized and then integrated into the host DNA. Although the genetics of RNA viruses are not as well understood as those of polyoma or SV40, it is apparent that the number of genetic functions are very limited. There are four main genetic elements known to exist in the 35S (3×10^6 daltons) RNA.

1. The gene *(env)* which codes for the viral-specific glycoprotein of the envelope.

2. The gene *(pol)* which codes for the reverse transcriptase.

3. The gene *(gag)* which codes for a precursor polypeptide that is subsequently cleaved to form several proteins of the inner core.

4. The gene *(src or onc)* which codes for a protein responsible for the transformation process. As in the case of the small DNA tumor viruses, a single genetic function appears responsible for the entire spectrum of changes in the transformed cell. Unlike the A gene of SV40, the *src* product is not required for replication of the viral genome, it seems only required for transformation itself.

Role of RNA Viruses in Cancer

The role of RNA-containing viruses in the etiology of cancer has become an extremely controversial topic. The involvement of these viruses as *natural* agents of certain types of tumors in animals, particularly rodents, is well-established. One of the most interesting aspects of the topic concerns the manner in which the virus is transmitted from one individual animal to another. Viral infections are normally thought of as being caused by the release of particles from an infected cell and their subsequent infection of other susceptible cells. Transmission of this sort, termed *horizontal* transmission, does appear to be responsible for the transmission of certain types of animal cancers, such as feline leukemia. However, there is a very different type of transmission which may be more important. The early studies on mice indicated that certain tumor viruses could be passed from mother to offspring in the milk. In these cases, the virus particles could be detected in the milk and the offspring become susceptible to development of the tumor at a later age. In other cases, it has been clearly shown that the virus can be passed from one generation to another in the DNA of the sperm and/or the egg, just as the bacteriophage can be passed from one bacterium to its progeny. The inheritance of a genetic element containing information for virus formation adds a whole new perspective to the concept of viral infection and has stimulated a great deal of research and speculation. Some investigators believe these viral genomes may actually have evolved from cellular DNA, while others believe they represent the descendants of viruses which had infected the species at some point in the evolutionary past. Regardless of their origin, the consequences to a cell harboring the information for one or more of these viruses can be quite variable. In some cases, the virus appears to result in the transformation of the cell in which it exists and the subsequent death of the individual. Inbred strains of rodents have been bred which inherit various viruses, some of which can lead to production of tumors in the individuals. In other cases, however, the virus does not appear to affect the metabolic activities of the cell. In some cells, there is no evidence of the virus other than the presence of viral DNA sequences; in other cells the presence of viral antigens can be demonstrated. In many cases mature viral particles are also found. All of these conditions, including the production of a complete virus, can occur in a "normal" cell, one that is neither malignant nor in the process of being destroyed. In some cases, embryonic cells are characterized by the production of virus particles which then stops later in life without any sign of malignancy. Generally, the viral particles released by the nonmalignant cell are not infectious to the cells of the strain from which they are derived. Whether these viruses have a role in the cell or simply represent the remnants of some evolutionary infection which is no longer harmful, remains unclear.

The point of most obvious importance in the topic of tumor viruses is their role in the etiology of human cancer. Although the subject remains controversial, RNA-containing viruses have been directly implicated in specific types of human malignancies. The first indication in this regard came when virus particles were identified in certain samples of human milk. It was soon established that human breast *tumor* cells can contain RNA in their polyribosomes that is complementary to labeled DNA probes prepared from mouse mammary tumor virus. Since normal cells did not seem to contain evidence of viral RNA, it was strongly suggested that the presence of a human virus very similar to that causing tumors in mice was responsible for human breast cancer. Subsequent experiments on human leukemias and related malignancies have also suggested that the affected cells contain viral information, both in the genome (as integrated DNA) and in the cytoplasm (as viral mRNA). Once again, the results from hybridization studies suggested that the viral nucleotide sequences were similar to those known to be responsible for causing leukemia in other animals. Most importantly, it has been found that normal cells do not appear to contain complete copies of these particular viral sequences and, therefore, that the agents responsible do not seem to be passed from parents to offspring in the germ line. The extent to which tumor viruses, both DNA and RNA, are responsible for human cancer remains one of the top-most priorities of research in this field.

Genetic Aspects of Malignancy

If one considers the types of agents capable of transforming cells in culture, namely chemicals, radiation, and viruses, it becomes apparent that the primary alterations induced by the agents are genetic. Although it is possible that radiation and chemical carcinogens cause the transformation by activating an endogenous tumor virus, it is more likely that they act by damaging, i.e., mutating, the DNA. If this is the case, then we could conclude that the transformed cell is one that is deficient in a particular genetic function which secondarily manifests itself in the altered behavior of the malignant cell. In contrast, treatment of the culture with the tumor virus would appear to *add* a genetic factor, one whose products somehow interfere with the cell's activities thereby causing it to become malignant. To some degree, these are alternative concepts: one suggesting that cancer results from a negative genetic basis, the other suggests a positive genetic basis. One possibility that can be eliminated is that tumor viruses act to produce cellular mutations which, secondarily, cause the transformation. Results with temperature-sensitive tumor viruses clearly show that the viral genome must continue to produce active protein if the transformed state is to be maintained.

It has been shown that spontaneous tumor formation can be associated with the presence of a mutant genetic condition. For example, persons with the genetic disease Fanconi's anemia are much more susceptible to the development of tumors. This condition is not simply due to some systemic problem, such as a defective immune system, since the cells from persons with this or certain other genetic diseases are much more susceptible to transformation by various agents in vitro. A similar condition was mentioned in the case of xeroderma pigmentosum in Chapter 14. Cells from these persons are known to have deficient genetic repair mechanisms, a condition which causes them to be much more susceptible to agents such as ultraviolet light. The correlation between a tendency toward cancer and the inability to repair DNA damage, can be considered independent evidence for the concept of the cancer cell as genetically deficient.

Another approach to determining whether cancer is a result of positive or negative genetic alterations is to fuse a normal cell with a cancer cell and observe the phenotype of the hybrid cells. Superficially, at least, one might expect that, if the malignant state were due to a genetic defect, the presence of the normal genes should correct the deficiency and restore the normal condition. Experiments of this type have generally tended to support the concept that the tumor cell is one suffering from the loss of genetic function. The hybrid formed between the normal and malignant cell is generally found to have greatly reduced malignant properties in culture and to be much less tumorigenic if injected into a suitable host. Regardless of these experiments, we know that the introduction of a viral genome can transform the host cell, so we are still left with the same basic question. The answer to the question may be that either type of genetic alteration can cause the transformation; the underlying similarity in all cases is the loss by the cell of growth control.

AGING

As an animal ages, the outward signs are obvious; yet the underlying basis for the deteriorative processes that occur in all animals are very poorly understood. Physiological measurements of organ function, nerve conduction velocity, muscle power, etc., indicate a decreasing capacity with age. It is generally believed that these more easily observed alterations reflect a progressive deterioration in the elements of which the tissues are composed. In this brief discussion we will consider two sites, the cellular and the extracellular, in which age-related events occur. There is no doubt that progressive changes take place both within cells and in their surrounding environment. The principal controversy in this regard concerns which of these two sites within the tissues is the primary one from which the aging of the whole animal results. This problem is similar to that previously discussed on the underlying basis of cancer. Not only is there controversy over the basic causes of aging, there is considerable disagreement over many of the observations themselves. Various laboratories working with different types of aging systems have not always agreed on even the more basic findings. In the following discussion we will simply present some of the basic observations that have been made and a little of the speculation about their significance.

Cellular Aging

Theories that attempt to explain the aging process as a result of defective intracellular processes are based upon several assumptions. The underlying assumption is that there is a finite frequency of error in the biochemical operations of every cell. Over a period of time, these errors might accumulate to produce a cell with a defective function or they might result in the death of the cell. Biochemical mistakes or damage can occur at numerous sites and manifest themselves in numerous ways. For example, in one theory of aging, damage is believed to result from the formation of highly reactive free radicals, particularly those of oxygen. In aerobic cells, certain enzymatic reactions are responsible for generating a superoxide radical (O_2^-) as a free intermediate. In the presence of water, this species is converted into the highly reactive hydroperoxyl radical $(HO_2 \cdot)$. Although there are numerous macromolecules with which this radical might react, one of its prime targets would be the unsaturated fatty acids of the phospholipids of celluar membranes. The interaction of fatty acids in the membrane with a free radical can lead to a chain reaction in which considerable lipid autoxidation can occur. In order to cope with the prospect of this type of damage, aerobic cells possess an enzyme termed *superoxide dismutase* which is responsible for the destruction of the superoxide radical in the following reaction

$$2O_2^- + 2H^+ \rightarrow H_2O_2 + O_2$$

Although lipid autoxidation may be an important consequence of aerobic metabolism, it is generally believed that the most sensitive site in the cell for the accumulation of age-related damage is the genetic material itself. Any change in the linear base sequence in the DNA will result in its altered regulatory or template properties and subsequent effect upon all processes in which that gene is involved. Alteration of the DNA template can result from damage (known to occur after most types of radiation) or from a mistake during replication by insertion of an incorrect nucleotide. Alterations in the genome of cells outside the germ line are called *somatic mutations*. Such mutations would be expected to be random, and each cell would have its information content affected in a different way.

The best evidence that individual cells undergo deterioration has been obtained with cells growing in culture. In 1965 it was reported by Leonard

Hayflick that, when human lung fibroblasts are followed in tissue culture, the number of divisions these cells can undergo is limited. To perform this experiment, a small piece of tissue is removed, the cells are dissociated, and a certain number of these single cells are transferred to a culture dish and allowed to attach to the surface and divide. After a number of divisions the cells are removed from the dish and placed into suspension; a small percentage are used to inoculate new culture dishes, just as the first had been done. These cells will continue to divide and cover the new dish, after which the process can be repeated. Each time the cells are removed from the dish and plated on a new dish at a lower concentration, they are said to be *subcultured*. If cells are subcultured numerous times, a point is reached where they stop dividing and the cell strain dies out. This has been found for many types of cells. It is generally reported that the number of divisions the strain goes through determines its life span, rather than the time that has passed during the experiment. This is clearly an example of age-related cellular death.

Tumor cells provide a significant exception to the rule that cells undergo a limited number of doublings in culture. Malignant cells have no such restriction; they continue to divide in essentially an immortal way. Tumor cells are often characterized by an abnormal chromosome number, which may be related to their immortality in vitro. Interestingly, occasional cells derived from a normal donor will continue to divide rather than become senescent, as in most cases. When the chromosomes of these peculiar cells are examined, they are generally found to be aneuploid, like the malignant cells, and they often behave in other respects as if they were tumor cells. Cells that undergo this change are said to be a *cell line* as opposed to a *cell strain,* as in cells that have a restricted division potential. The frequency with which cells become a cell line is related to the species from which they were derived. Mouse cells will frequently become altered in this way, human cells only rarely.

If aging and ultimate age-related death of an animal result from intracellular damage, it would be expected that cells from a short-lived species would have less potential for cell division than those from a long-lived animal. Similarly, cells from a young animal should be more capable of extended division than cells of an older individual. This has become a controversial topic. Based on the work of Hayflick and others, it appears that human fetal lung fibroblasts are capable of approximately 50 divisions before they lose this ability; those from an adult at maturity are capable of approximately 30; and those from more aged individuals are capable of less. These numbers, however, are averages, and in one case an 87-year-old man was reported to have cells capable of an average of 29 further divisions. In other words, there is convincing evidence that cells age, but it is equally clear that we do not run out of all types of cells and then die as a result. If we take the value of 50 to 100 divisions for all cells of the fetus, sufficient cells for many lifetimes could be produced. If the exhaustion of the capacity of cells to divide is responsible for age-related processes, a search must be made for a tissue that is particularly sensitive, which could secondarily affect the other tissues of the body in a destructive way. A candidate would be one or more of the endocrine glands, whose effects on the function of the whole body are well known. Another is the lymphoid system, which has received a great deal of attention as a potential aging pacemaker.

Up to this point we have considered the accumulation of simple, randomly occurring mutations. To understand the potential for cellular damage inherent in the occurrence of somatic mutation, we must consider the "error catastrophe" concept proposed by Leslie Orgel. In this theory it is pointed out that certain types of errors are likely to produce a great number of subsequent errors. Consider, for example, a mistake occurring in a DNA polymerase gene that results in an enzyme that will make further mistakes during replication. Such a mistake would be considered an error catastrophe, and examples of altered enzymes that cause an increased number of mistakes are known from studies of bacterial viruses. Replicational enzymes are only one place where an error catastrophe could occur. Proteins are needed for transcription and translation, and altered proteins that serve these functions could similarly be expected to rapidly fill the cell with an entire spectrum of defective proteins. At first glance, the somatic mutation theory of aging seems incompatible with the observed differences in life span among organisms. If senescence were due to the accumulation of random mutations, all cells and organisms might be expected to show a similar aging time course. However, our knowledge of mutation rates and genetic repair mechanisms among organisms is far too meager to justify this conclusion. The degree to which the accuracy of replication, or transcription, or DNA repair can vary from species to species remains to be determined.

Is there evidence that, as cells of an organism (or a culture) become older, they have an increasing content of deficient proteins? Over the past decade, hundreds of observations have been made in regard to this question, though no definitive answer has been forthcoming. In many cases it has been shown that, during the senescence of an organism (or a cell culture), certain of its enzymes undergo a decrease in specific activity. Similarly, it has been shown that aging is accompanied by an increase in the presence of inactive or unstable enzymes. More recently, specific enzymes have been purified from young and old animals and differences in activity have been noted. These types of reports provide evidence for a theory of aging based on somatic mutation, although one cannot rule out the possibility that the deterioration of these proteins results from posttranslational modification. Even though there are many reports of changes in enzyme activity with age, there are many studies in which no such changes have been found. One might conclude that different enzymes undergo deteriorative changes at different rates, a conclusion which is difficult to reconcile with a proposal based on random genetic damage. Another finding which argues strongly against the somatic mutation theory (and particularly the error catastrophe theory) comes from studies of viral production in young versus old cells. If the protein-synthesizing machinery of a cell deteriorates with age, older cells should be less able to support a viral infection. This does not seem to be the case.

If the aging process does result from the effects of somatic mutation, then it should be possible to accelerate aging by increasing the mutation frequency. Once again, results in this area are controversial, although evidence in behalf of this prediction has been reported. For example, it has been shown that *Drosophila* larvae that have been fed compounds which increase the content of defective proteins experience a markedly shortened life span. Similar findings have been reported for cells growing in culture—compounds which increase translational errors can reduce the doubling capacity of these cells.

One of the main lines of evidence cited in support of theories of cellular aging has been obtained by irradiation. It is a well-established fact that irradiation, in proportion to the dose, has life-shortening effects. Remarkably, animals whose life expectancy has been shortened as a result of such exposure undergo the signs of aging prematurely. Figure 19-7 show two groups of litter mates of mice. The group exposed to X-radiation shows the debilitating signs of age at a chronological age well in advance of the normal time for these animals. The means by which the radiation is administered and whether the effects are truly mimicking the natural aging process are controversial and beyond the scope of this book. If we assume that radiation is accelerating the natural aging events, the most likely explanation for its action is via somatic mutation, i.e., changes in the DNA.

The aging of cells as they grow in vitro has been described above. By *serial transplantation,* whereby cells from one animal are repeatedly transplanted from one host to another, the longevity of cells in vivo can be estimated. With this technique, the descendants of the cells of the original transplant can be maintained in the bodies of animals of any age rather than be exposed to a continually aging environment. In other words, the technique of serial transplantation allows one to dissociate the cell from its normal environment, yet maintain it within an appropriate physiological container. Several types of tissues, including skin, mammary gland, and spleen cells, have been treated in this way. In all these cases the proliferation of cells is limited, just as in the in vitro experiments. However, these transplanted tissues will outlive, to a considerable extent, the animal that they were originally taken from. The results of these experiments, therefore, are inconclusive. They confirm that cells undergo aging in vivo but do not establish if this phenomenon is responsible for the age-related death of the animal.

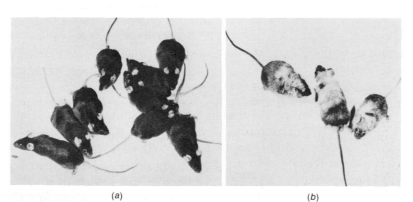

(a)　　　　　　　　(b)

Figure 19-7. Two groups of mice that were originally identical. Mice on the left are control animals; mice on the right received a large but non-lethal dose of radiation. Only three mice of the irradiated group remain alive; these are "old" and senile. *(Courtesy of H. J. Curtis.)*

Before questions of cellular aging within the mammalian body can be considered, a distinction must be made as to the nature of the cells involved (page 653). A certain percentage of the cells of an animal, including neurons and muscle cells, are formed during the developmental period of the life cycle and are not replaced during the entire life of the organism. The study of a postmitotic cell for signs of aging allows one to separate aging processes from mistakes made during division. Neurons or muscle cells, once formed, might be expected to accumulate errors during their long life. These errors might be expected to reduce the capabilities of each cell and to cause deteriorating changes in the physiological properties of the tissue. Both nervous and muscle tissue undergo age-related physiological change.

Another class of cells produced continually throughout the life of the animal includes circulating blood cells, lymphoid and bone marrow cells, skin cells, mucous membrane epithelial cells, etc. These cells are produced with a finite lifetime, from days to months, and are destroyed. Use of radioactively labeled cells indicates that in many cases these cells are not removed on a random basis; rather, as a cell's lifetime in the body increases, its chance of becoming destroyed also increases. This is an important observation because it indicates there are mechanisms by which aging cells can be marked and mechanisms whereby they can be destroyed. These short-lived cells can be shown to age in very short periods, while longer-lived cells show no such changes in these short time periods. Different cells seem to undergo different rates of age-related deterioration.

Another class of cells can be distinguished that either continue to divide or retain the capacity, to some degree, throughout the life of the organism. Examples are hepatocytes, osteoblasts, chondrocytes. In this group are the many types of stem cells that are partially differentiated, at least to the extent that they can give rise to only one or a limited variety of differentiated cells. Aging to this group of cells would be reflected in their inability to renew the tissues they are responsible for as well as in their decreased ability to function physiologically. Each type of cell must be considered individually: each may have its own error frequency, its own capacity for DNA repair, its own molecular turnover rate, etc., and each may contribute in a different way to the overall aging process of the animal.

As stated above, theories of cellular aging must explain the entire age-related spectrum of deteriorative changes. Can an increased likelihood of malignancy, increased heart disease, etc., be accounted for by underlying cellular damage? At the present time this question cannot be answered.

Extracellular Aging

A significant percentage of the dry weight of an animal resides outside the cell in the extracellular space. The extent to which a tissue is composed of extracellular material varies greatly, reaching a maximum in the supportive tissues of the body. The primary components of the extracellular space are mucopolysaccharides and fibrous proteins, particularly collagen and elastin, which are secreted out of the connective tissue cells in which they are synthesized. Collagen is estimated to account for up to 40% of the body protein, is present in extracellular spaces of virtually all tissues, and has been suggested as the primary site for age-related changes.

Several properties of collagen suggest it may be responsible for the aging process. That collagen undergoes molecular modification with age is primary to the theory. (The molecular organization and assembly processes were described in Chapter 3.) The fibrous protein molecules are polymers of collagen monomers, each composed of three polypeptide chains. In the newly polymerized molecule the collagen monomers are held together by noncovalent bonds. As collagen is maintained under physiological conditions, covalent cross-linking takes place both within the collagen monomer (among the three polypeptides) and between the monomers. These cross-linked reactions have profound effects on both physicochemical and biological properties of collagen. For example, the ease with which collagen can be extracted and dissociated drops markedly with age until it is essentially insoluble by the time of maturity in mammals. From that point through old age, cross-linking continues and its effect can be measured. One study of its changing properties has utilized collagen fibers from rat tail and their resistance to shrinkage. If collagen fibers are heated to 65°C, they shrink in length unless maintained at their original length with attached weights. The amount of weight required to prevent shrinkage is a measure of the extent of cross-linking and, therefore, of aging of the collagen molecules (Fig. 19-8). Frederic Verzár has found that 2-month-old rat-tail collagen can be maintained by 1.5 g, 5-month by 3 g, and 30-month by approximately 10 g. However, comparative studies suggest that cross-link density, by itself, cannot be used as a measure of the degree to which an animal has aged. While a 10-g weight prevents the shrinkage of collagen fibers of a 30-month rat, it takes approximately 50 g to accomplish the same task when the collagen being tested is obtained from a cat of comparable age (e.g., 15 years). It

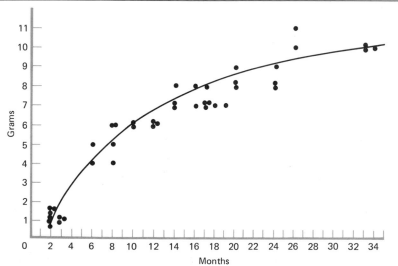

Figure 19-9. Age change in isotonic thermal contraction. The plot shows the amount of weight that is required to prevent shortening by rat tail tendon fibers (collagen fibers) when the fibers are taken from various aged animals. *(From F. Verzar, Int. Rev. Connect. Tissue Res., **2**:257, 1964.)*

would appear that the accelerated aging experienced by rats as compared to cats (or humans) is not a direct result of the accelerated aging of those collagen molecules that can be readily tested, i.e., those of tails, tendons, etc. Much less is known about the more physiologically important collagen of the extracellular matrices of the various internal organs.

The theory of aging based on collagen is analogous to the theory based on intracellular changes in that the evidence of age-related modification is indisputable. The question is whether or not all the other age-related events are explainable as a consequence. The primary chemical modifications are believed to directly affect a wide variety of physiological activities, as a result of the widespread distribution of collagen. For a cell to maintain a healthy intracellular composition, it must receive oxygen, nutrition, ions, hormones, etc., from its environment and release carbon dioxide and other waste products. The occurrence of collagen in spaces between cells suggests that it could play a critical role in the exchange activities between a cell and its environment. The presence of large amounts of collagen in the lining of all blood vessels would similarly affect movement from the blood into the extracellular space as well as affect the distensibility of the arterial wall.

REFERENCES

Adler, W. H., Bioscience **25**, 652–657, 1975. "Aging and Immune Function."

Bailey, A. J., Robbins, S. P., and Balian, G., Nature **251**, 105–109, 1974. "Biological Significance of the Intermolecular Crosslinks of Collagen."

Baltimore, D., Science **192**, 632–636, 1976. "Viruses, Polymerases and Cancer."

———, Hospital Practice **13**, 49–57, 1978. "Retroviruses and Cancer."

Brinkley, B. R., and Porter, K. R., eds., symposia on "Cell Surface and Neoplasia" and "Viral Function in Cell Transformation." *International Cell Biology, 1976–1977*, pages 131–158, 531–557. Rockefeller University Press, 1977.

Burch, P., *The Biology of Cancer.* University Park Press, 1976.

Burnet, F. M., *Immunology, Aging and Cancer.* W. H. Freeman, 1976.

Cairns, J., Sci. Am. **233**, 64–78, Nov. 1975. "The Cancer Problem."

Cristofalo, V. I., et al., eds., *Exploration in Aging.* Plenum, 1975.

Croce, C. M., and Koprowski, H., Sci. Am. **238**, 117–125, 1978. "The Genetics of Human Cancer."

Dulbecco, R., Science **192**, 437–441, 1976. "From the Molecular Biology of Oncogenic Viruses to Cancer."

Eckhart, W., Ann. Rev. Genet. **8**, 301–318, 1974. "Genetics of DNA Tumor Viruses."

Finch, C. E., Bioscience **25**, 645–650, 1975. "Neuroendocrinology of Aging: A View of an Emerging Area."

———, Quart. Rev. Biol. **51**, 49–83, 1976. "The Regulation of Physiological Changes during Mammalian Aging."

Friedman, H., and Southam, C. M., eds., Ann. N.Y. Acad. Sci. **276**, 1976. Symposium on "Immunobiology of Cancer."

Hayflick, L., Sci. Am. **218**, 32–37, March 1968. "Human Cells and Aging."

————, Bioscience **25**, 629–637, 1975. "Cell Biology of Aging."

————, New England J. Med. **295**, 1302–1308, 1976. "The Cell Biology of Human Aging."

Heidelberger, C., Ann. Rev. Biochem. **44**, 79–121, 1975. "Chemical Carcinogenesis."

Hynes, R. O., Biochim. Biophys. Acta **458**, 73–107, 1976. "Cell Surface Proteins and Malignant Transformation."

Klein, G., Federation Proc. **35**, 2202–2204, 1976. "Analysis of Malignancy and Antigen Expression by Cell Fusion."

————, and Weinhouse, A., eds., *Advances in Cancer Research*. Academic, vol. 1, beginning 1953.

Kolata, G. B., Science **190**, 39, 1975. Research News. "Cell Surface Protein: No Simple Cancer Mechanism."

Lamb, M., *Biology of Ageing*. Halsted. 1977.

Littlefield, J. W., *Variation, Senescence, and Neoplasia in Cultured Somatic Cells*. Harvard University Press, 1976.

Nowell, P. C., Science **194**, 23–28, 1976. "The Clonal Evolution of Tumor Cell Population."

Old, L. J., Sci. Am. **236**, 62–76, May 1977. "Cancer Immunology."

Orgel, L. E., Proc. Natl. Acad. Sci. (U.S.) **49**, 517–521, 1963. "The Maintenance of the Accuracy of Protein Synthesis and Its Relevance to Aging."

Pastan, I., Adv. Metab. Disorders **8**, 377–383, 1977. "Cyclic AMP and the Malignant Transformation of Cells."

Rafferty, K. A., Jr., Sci. Am. **229**, 26–33, Oct. 1973. "Herpes Virus and Cancer."

Reich, E., Rifkin, D. B., and Shaw, E., eds., *Proteases and Biological Control*. Cold Spring Harbor, 1975.

Reviews on Cancer, Biochim. Biophys. Acta **417**, 1975, **458**, 1976, and **473**, 1977.

Shildrake, A. R., Nature **250**, 381–385, 1974. "The Aging, Growth and Death of Cells."

Southam, C. M., and Friedman, H., eds., Ann. N.Y. Acad. Sci. **277**, 1976. International Conference on "Immunotherapy of Cancer."

Temin, H. M., Sci. Am. **226**, 24–33, Jan. 1972. "RNA-Directed DNA Synthesis."

————, Ann. Rev. Genet. **8**, 155–178, 1974. "On the Origin of RNA Tumor Viruses."

————, Bioscience **27**, 170–176, 1977. "The Relationship of Tumor Virology to an Understanding of Nonviral Cancers."

Tooze, J., and Sambrook, J., eds., *Selected Papers in Tumor Virology*. Cold Spring Harbor, 1974.

"Tumor Viruses," Cold Spring Harbor Symp. Quant. Biol. **39**, 1975.

Verzar, F., Sci. Am. **208**, 104–114, April 1963. "The Aging of Collagen."

Vogt, P. K., and Hu, S. F., Ann. Rev. Genet. **11**, 203–238, 1977. "The Genetic Structure of RNA Tumor Viruses."

Weinberg, R. A., Cell **11**, 243–246, 1977. "How Does T Antigen Transform Cells?"

INDEX

INDEX

Page numbers in **boldface** indicate illustrations.